Intelligence in Chip: Integrated Sensors and Memristive Computing

Editors

Alex James
Digital University Kerala

Alon Ascoli
Politecnico di Torino

Bhaskar Choubey
Siegen University

Tutorials in Circuits and Systems

For a list of other books in this series, visit www.riverpublishers.com

Series Editors

Myung Hoon Sunwoo

IEEE CASS President

Fakhrul Zaman Rokhani

Vice President - Education and Communications, IEE CASS

Hanho Lee

Vice President - Technical Activities, IEEE CASS

Published 2024 by River Publishers
Alsbjergvej 10
9260 Gistrup Denmark

www.riverpublishers.com

Distributed by Routledge
4 Park Square, Milton Park, Abingdon, Oxon OX14 4RN
605 Third Avenue, New York, NY 10017, USA

ISBN: 978-87-7022-834-3 (hardback)
ISBN: 978-87-7004-254-3 (paperback)
ISBN: 978-87-7004-724-1 (online)
ISBN: 978-8-770-04253-6 (master ebook)

Library of Congress Cataloging-in-Publication Data: May 2024
Editors: Alex James, Alon Ascoli and Bhaskar Choubey
Title: *Intelligence in Chip: Integrated Sensors and Memristive Computing*

Table of contents

Introduction 6

SECTION 1: MEMRISTIVE DEVICES

Chapter 1

Overview of redox-based resistive RAM (ReRAM) – Physico-chemical background and history 9

by Rainer Waser

Chapter 2

Deep learning inference with in-memory computing 63

by Abu Sebastian

Chapter 3

Modeling of Memristive Devices 78

by Stephan Menzel

SECTION 2: NONLINEAR MEMRISTIVE CIRCUITS

Chapter 4

Analog Neural Computing Under Variability Intense Memristor Arrays 109

by Alex James

Chapter 5

Edge of Chaos Clarifies the Origin for Yet-Unexplained Nonlinear Phenomena in Biological Systems 128

by Alon Ascoli

Chapter 6

Memristor Computing Systems: at the crossroad between Circuit Theory and Artificial Intelligence 176

by Fernando Corinto

SECTION 3: MEMRISTIVE CIRCUIT APPLICATIONS

Chapter 7

Aspects of Digital Circuit Design using Memristor Ratioed Logic (MRL) 196

by Spiros Nikolaidis

Chapter 8

Neuromorphic Computing with Memristors: From Circuits to Algorithms 216

by Cory Merkel

Chapter 9

In-memory computing meets Neuromorphic Event-Driven Vision 259

by Arindam Basu

SECTION 4: MEMRISTIVE SYSTEMS

Chapter 10

Open-Source RRAM Neuromorphic Accelerators 283

by Jason Eshraghian

Chapter 11

Mixed-Signal Neuromorphic Computing: Devices, Algorithms, and Circuits 295

by Vishal Saxena

About the Editors 326

About the Authors 330

Introduction

Welcome to the compilation of presentations from the IEEE CASS Seasonal School on Intelligence in Chip: Integrated Sensors and Memristive Computing 2022. This book showcases a diverse array of talks delivered by experts at the forefront of research and development in the fields of neuromorphic computing, in-memory computing, and memristor-based systems.

Throughout these presentations, leading scholars and practitioners explore cutting-edge topics such as deep learning inference with in-memory computing, analog neural computing under variability in-tense memristor arrays, and the intersection of circuit theory and artificial intelligence in memristor computing systems.

Speakers delve into the intricacies of edge computing, elucidate the role of memristors in neuromorphic event-driven vision, and discuss the journey from memristor circuits to algorithms in neuromorphic computing. Additionally, they provide insights into the physico-chemical background and history of redox-based resistive RAM (ReRAM) and delve into aspects of digital circuit design using memristor ratioed logic (MRL).

From modeling memristive devices to open-source RRAM neuromorphic accelerators, this collection covers a broad spectrum of topics, offering valuable perspectives on the latest advancements and challenges in the field.

We extend our gratitude to all the speakers for their invaluable contributions, and we hope that this compilation serves as a comprehensive resource for researchers, engineers, and enthusiasts alike, seeking to explore the forefront of intelligence in chip technology.

Alex James
Bhaskar Choubey
Alon Ascoli

SECTION 1:
MEMRISTIVE DEVICES

CHAPTER 01

Overview of redox-based resistive RAM (ReRAM) – Physico-chemical background and history

Rainer Waser

RWTH Aachen University

The presentation on redox-based resistive RAM (ReRAM) by Rainer Wasser from Jülich-Aachen Research Alliance JARA delves into the physico-chemical background and history of ReRAM technology, encompassing its motivations, challenges, and advancements. It begins by contextualizing the significance of ReRAM in addressing global issues such as energy consumption and the end of Moore's law. The discussion navigates through the history and evolution of ReRAM technology, highlighting seminal works and breakthroughs in understanding the underlying mechanisms, particularly focusing on the valence change mechanism (VCM). Through detailed slides, the presentation elucidates the intricate processes involved in forming memristive devices, exploring the nanoscale phenomena driving resistive switching and memristive behaviors. Furthermore, it delves into the challenges and mysteries surrounding ReRAM technology, including non-linear switching kinetics and ferroelectric tunnel junctions, offering insights into ongoing research endeavors and future directions. The abstract encapsulates a comprehensive overview of the presentation's breadth, spanning from fundamental concepts to cutting-edge research, underscoring the pivotal role of ReRAM in shaping the future of information technology.

1 Outline

1 Motivation
2 Introduction and classification
3 Redox-based resistive switching memories – the valence change mechanisms (VCM): the forming process
4 VCM-Systems – the filamentary C8W switching process
5 VCM-Systems – other variants
6 Nonlinear Switching Kinetics
7 Reliability Aspects – the Variability
8 Neuromorphic Computing for Artificial Intelligence
9 Conclusions

2 Acknowledgement for funding

DFG
SFB 917 Nanoswitches
Bundesministerium für Bildung und Forschung
RWTH AACHEN UNIVERSITY
JÜLICH FORSCHUNGSZENTRUM
HELMHOLTZ ASSOCIATION
intel
Western Digital
IBM
hp
infineon
ST
ASML
SAMSUNG

Acknowledgements for funding.

1 MOTIVATION

3 History

History

In 1896, Arrhenius (Nobel Prize 1903) calculated for the first time that a doubling of the **CO_2 content** of the atmosphere leads to a temperature increase of **4** to 6°C.

1957/58 (IGY): **first evidence** that the CO_2 concentration - man-made - in the atmosphere is actually increasing (by isotope analyses);

1970s: **National Academy of Sciences of the USA** warned for the first time about global warming

1971: **German Physical Society (DPG)** - warning of the serious consequences of man-made climate change. DPG estimate: from 280ppm CO2 in 1900 to 380ppm in 2000 (actual: 370ppm)

Climate conferences: 1979, 1988, 1992, 1997,…., 2015, 2019, …. 2022 (Egypt)

Finally: Decision of the Bundestag and the Bundesrat – **phase-out by 2038** (now possibly 2030 ... 2035)

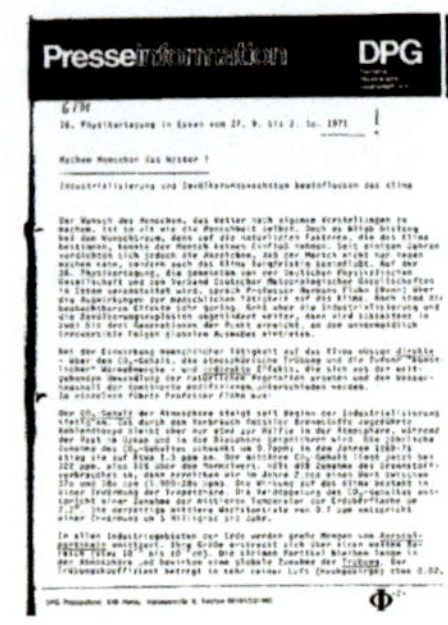

Press release DPG 1971

Autumn 2019: Structural Strengthening Act Coal Regions

This slide focuses on the history of CO2 concentration and global warming, and the decisions to overcome it.

4 Portion of ICT of electrical energy

Global consumption of electr. energy by ICT (approx. expectation)

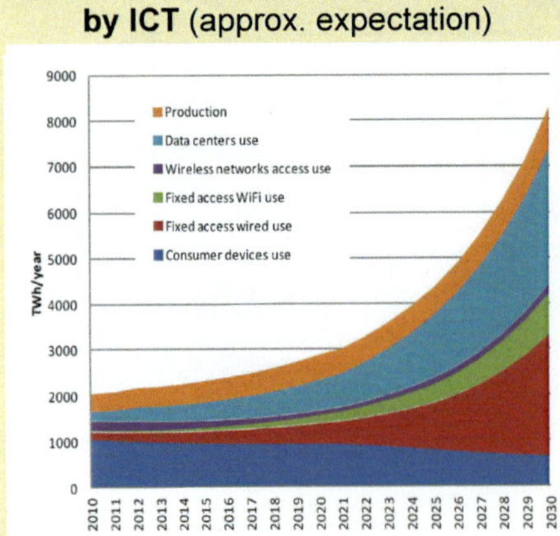

A. Andrae, Challenges, 2015

Trends

- **Video streaming & gaming**
- **5G grid**
- **autonomous cars**
- **Internet of things (IoT)**
- **Training of artificial intelligence (AI)**
- **Digitalization, digital Transformation**

⇨ **a Data Tsunami approaches us!** (Guardian, 11.12.2017)

This slide summarizes the global consumption of electrical energy by ICT and the trends in ICT.

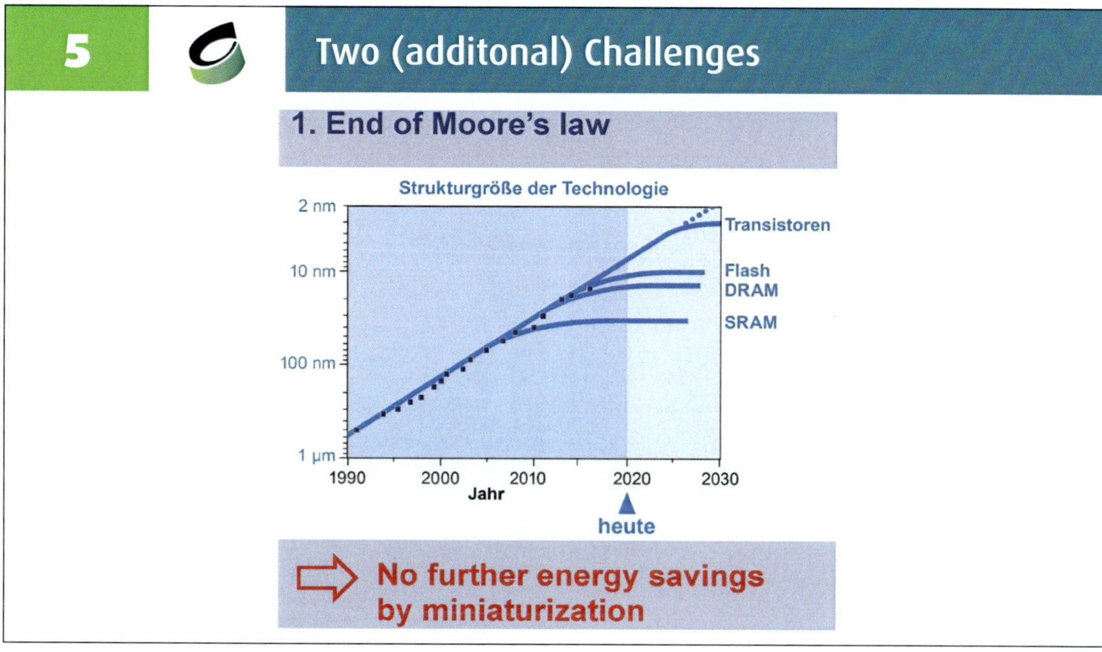

This slide presents one of the two additional challenges, the end of Moore's law. Miniaturisaton can no longer support energy savings.

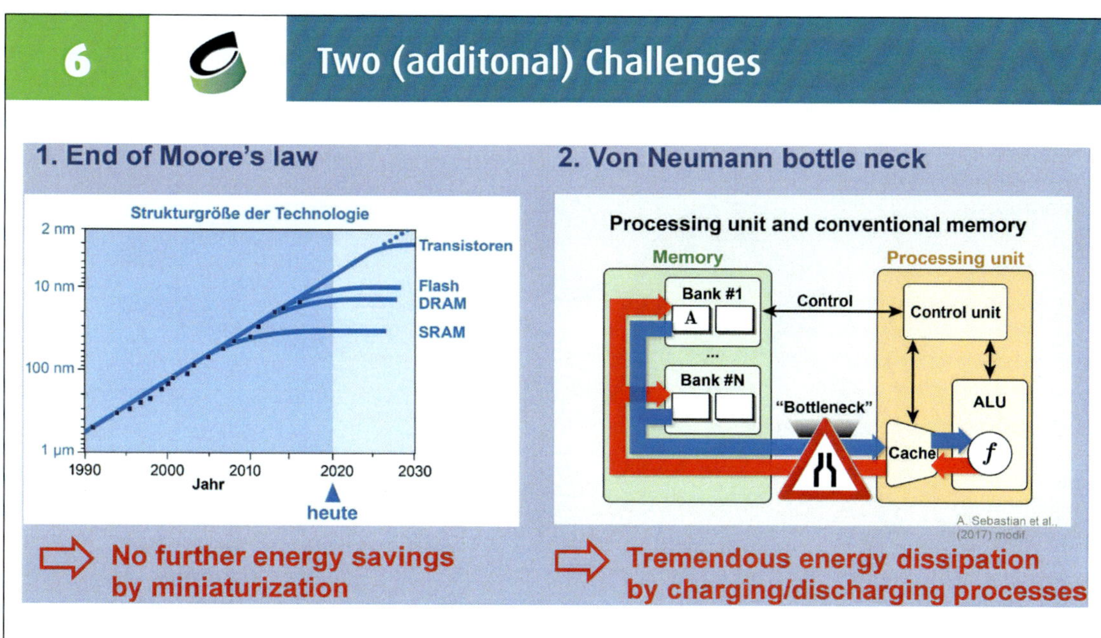

This slide presents another challenge, the Von Neumann bottleneck. Charging and discharging process cause severe energy dissipation.

7 JUQUEEN BRAIN

458.752 „Neurons"	10 bio. Neurons in the cortex
1 Mio. fixed connections	100.000 Bio. (flexible) connections
2.301.000 Watt	25 Watt

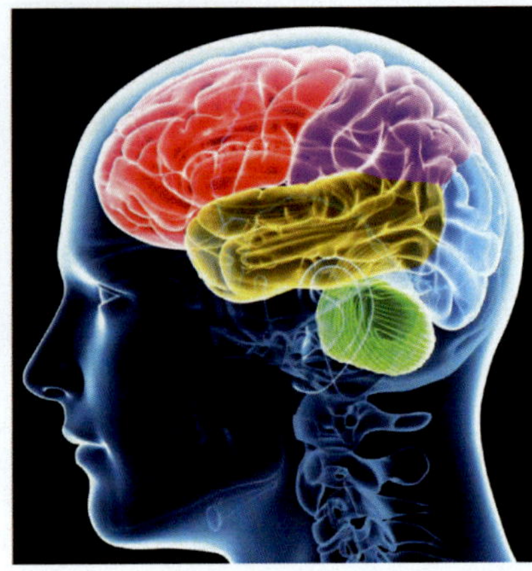

This slide highlights the difference between the supercomputer JUQUEEN and the human brain in terms of no of neurons, connections and the power consumption.

8 NOVEL DEVICES FOR NEUROMORPHIC COMPUTING

Artificial synapses for neuromorphic computers

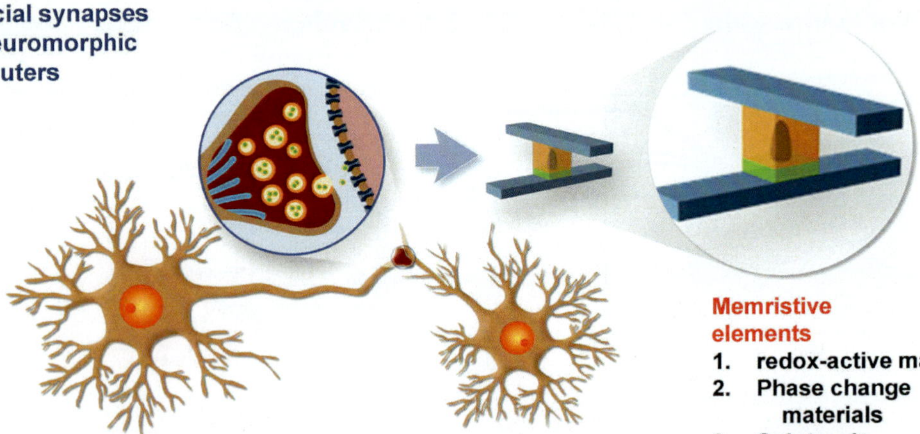

Memristive elements

1. **redox-active materials**
2. **Phase change materials**
3. **Spintronic materials**

This slide presents the novel devices for neuromorphic computing. It highlights artificial synapses and the memristive elements.

2 INDRODUCTION AND CLASSIFICATIONS

9 Nanoionic memristive phenomena in metal oxides:the valence change mechanism

This slide introduces the paper Nanoionic memristive phenomena in metal oxides: the valence change mechanism by Regina Dittmann, Stephan Menzel & Rainer Waser inAdvances in Physics, p. 1-195.

10 Memristive devices

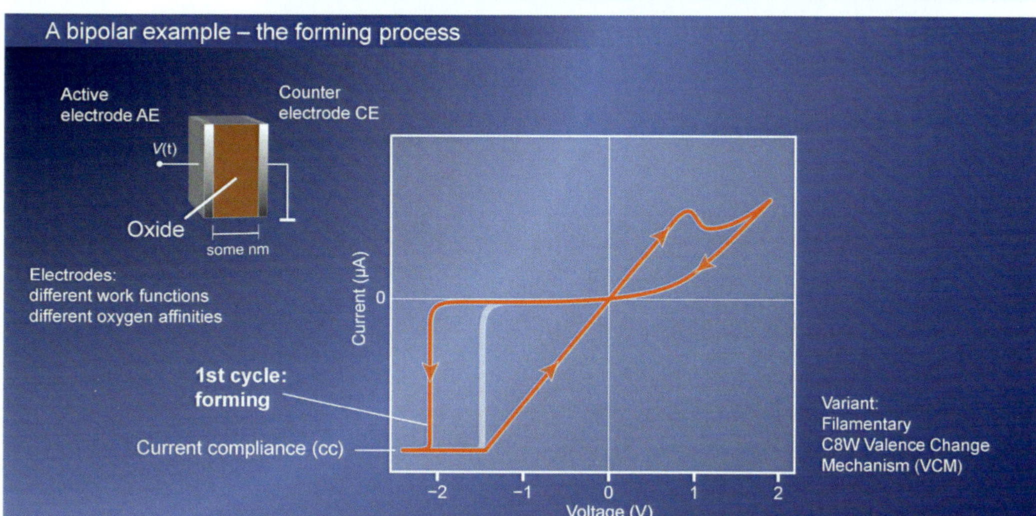

This slide discusses the forming process of memristive devices through the IV characteristics.

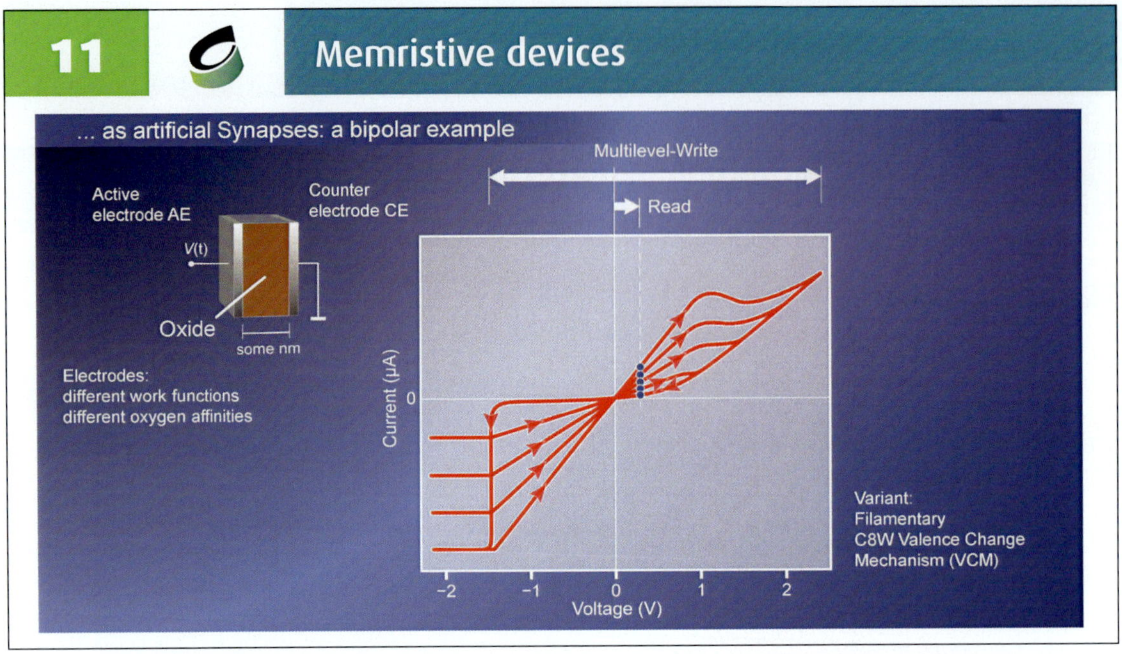

This slide discusses the memristive devices as artificial synapses with help of IV characteristics of the device.

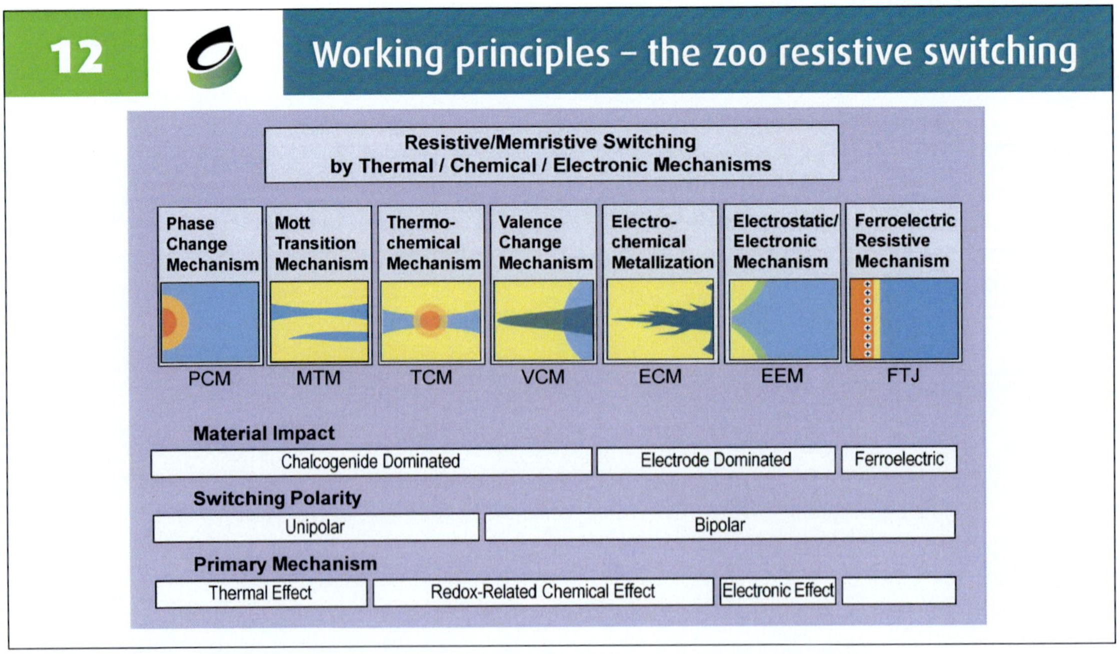

This slide discusses the various resistive switching mechanisms.

13 Working principles – new members of the zoo

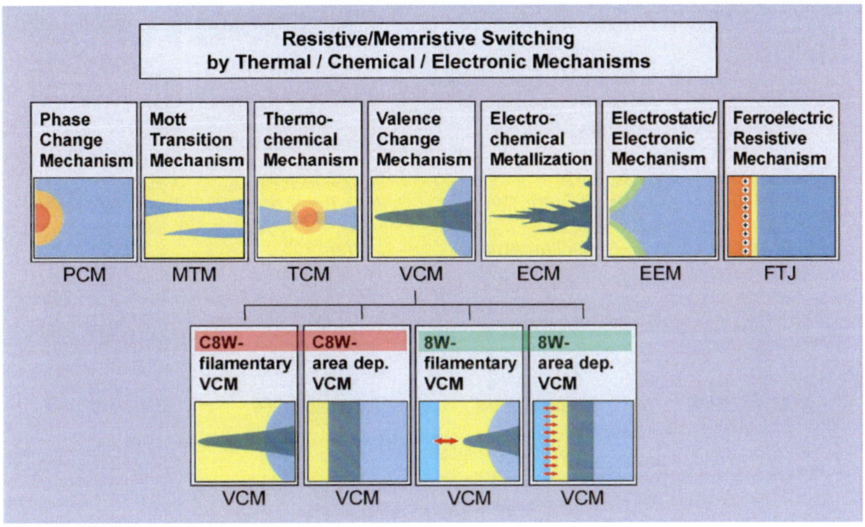

This slide introduces the new members under Valence Change Mechanism viz. C8W-filamentary VCM, C8W- area dep. VCM, 8W- filamentary VCM and 8W- area dep. VCM.

14 Observation of two VCM switching polarities

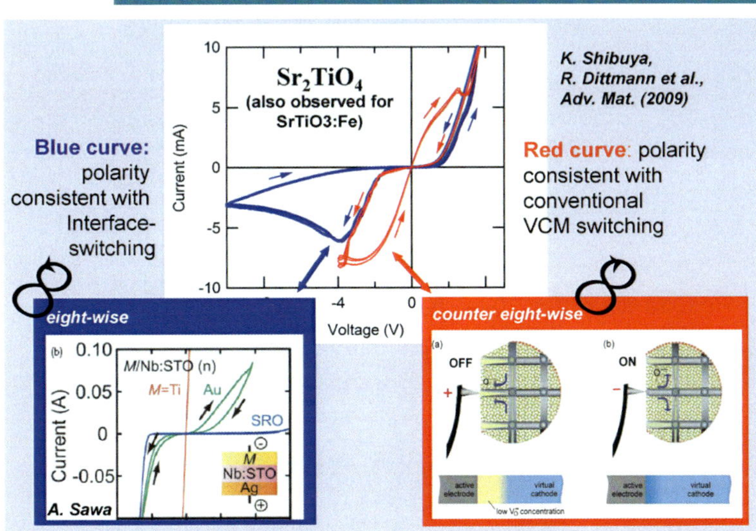

This slide presents the observation of two VCM switching polarities for Sr2TiO4.

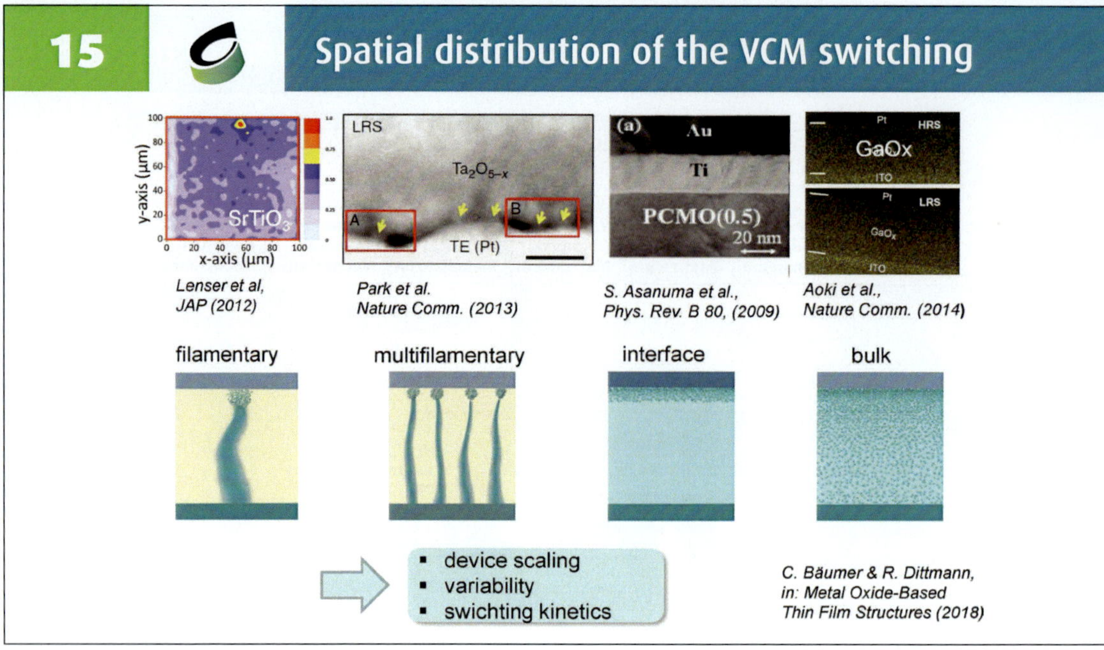

This slide presents the spatial distribution of the VCM switching.

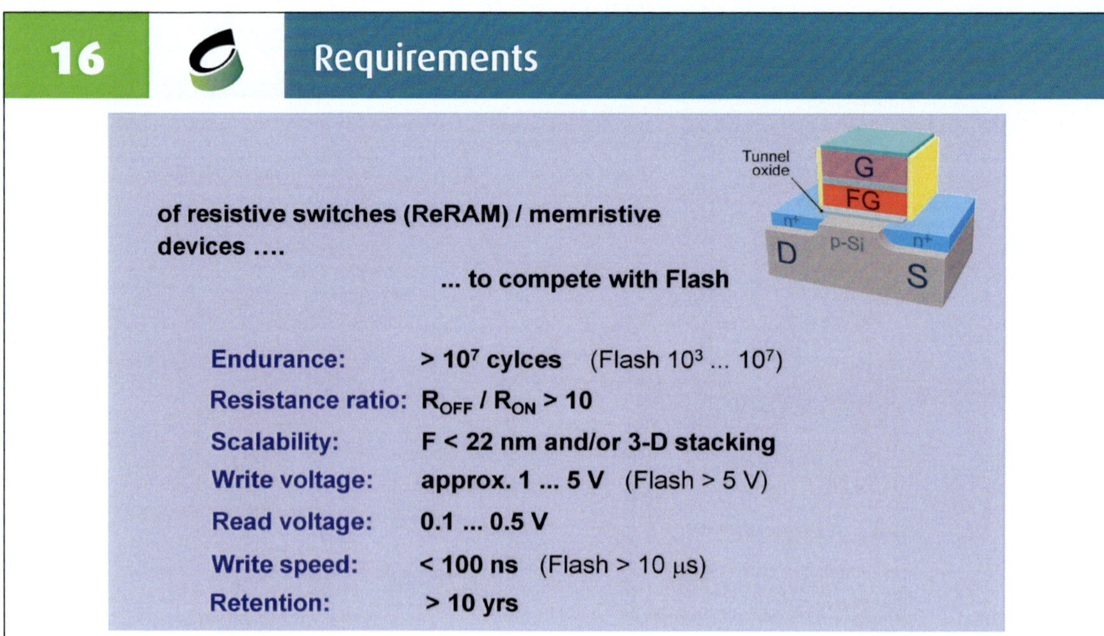

This slide looks at the requirements of resistive switches (ReRAM)/ memristive devices to compete with flash.

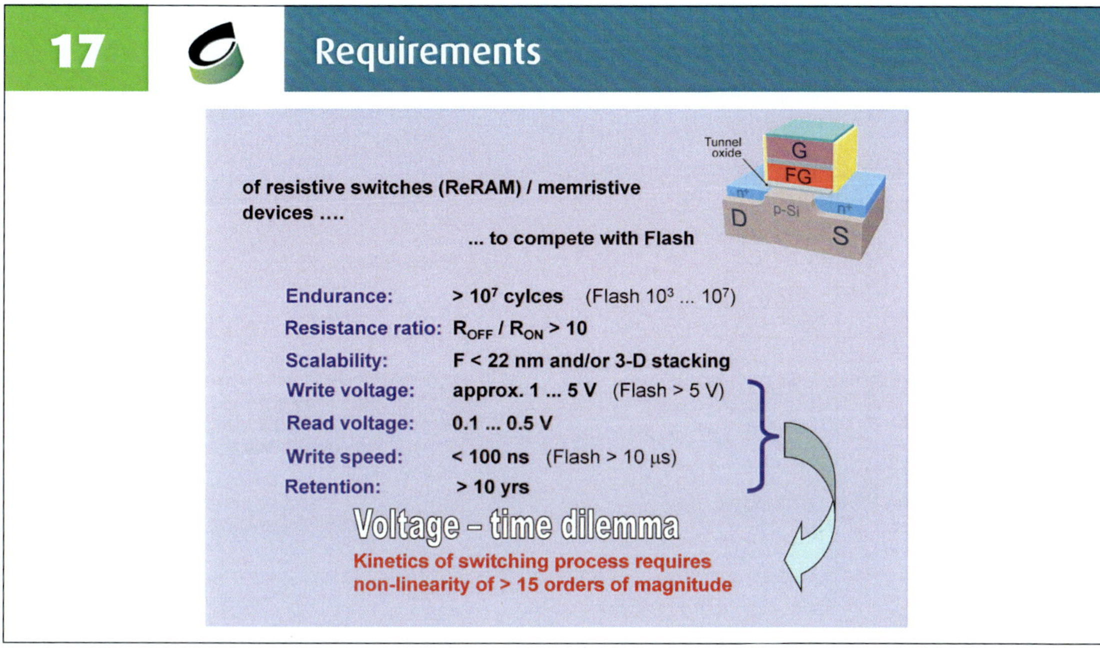

This slide tries to understand the voltage- time dilemma in implementing successful ReRAM/ memristive devices. The process demands a non linearity of >15 orders of magnitude.

3 REDOX-BASED MEMRISTIVE DEVICES - THE VALENCE CHANGE MECHANISMS (VCM): THE FORMING PROCESS

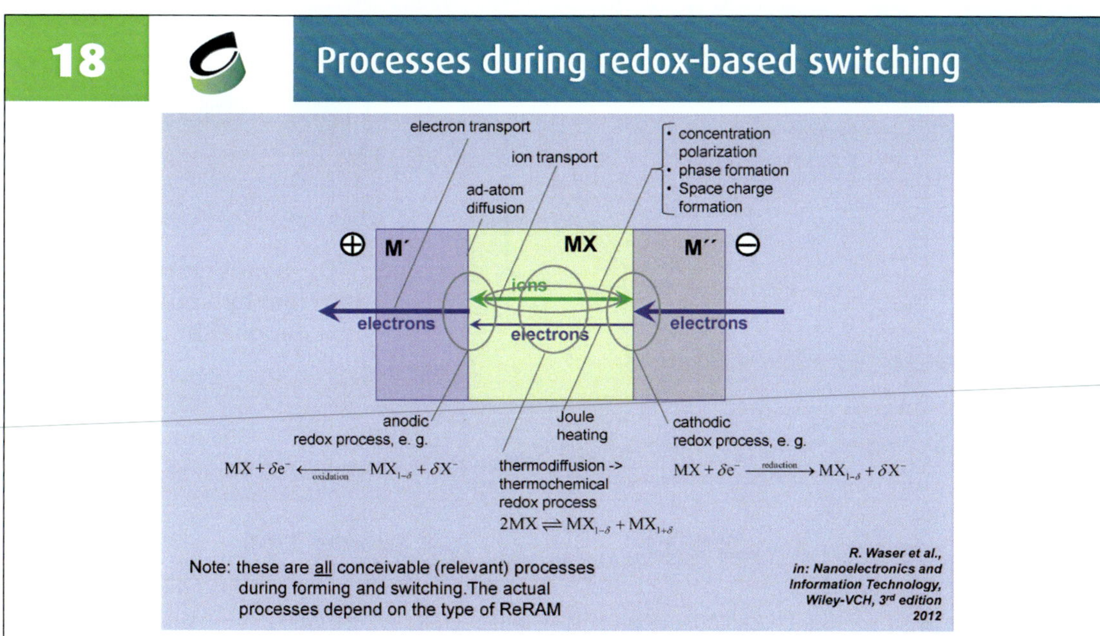

This slide explains the detailed processes during redox- based switching.

19 Thermochemical behaviourof transition metal oxides

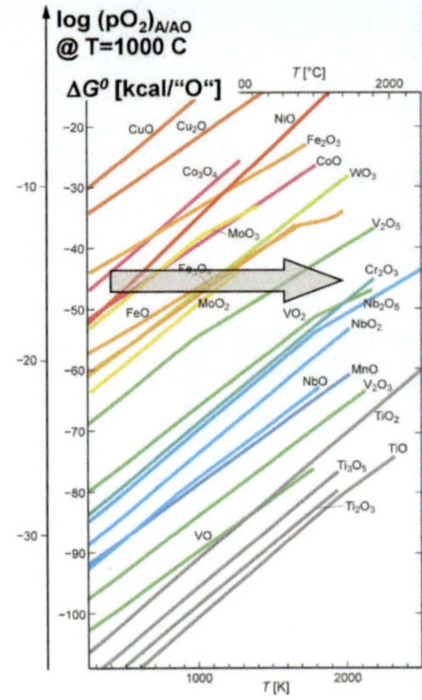

Temperature dependence of the free formation energy ΔG^0 of metal oxides:

$$\Delta G^0 = \Delta W^0 - T\,\Delta S^0$$

⇨ redox characteristics: lower valent states more stable at higher T because of the increasing dominance of the entropy term.

This slide explains the Thermochemical behaviour of transition metal oxides.

20 VCM – Valence Change Memories

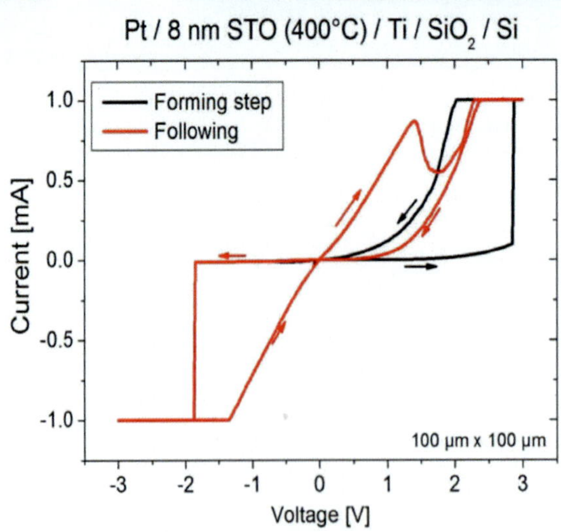

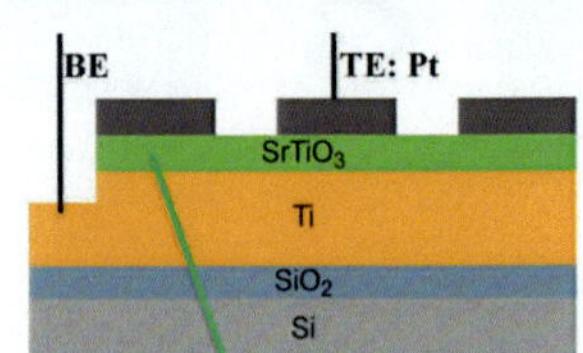

$SrTiO_3$: nanocrystalline thin film by sputter deposition

Forming into the HRS

S. Schmelzer (2013)

This slide demonstrates the characteristics of VCM- valence change memories.

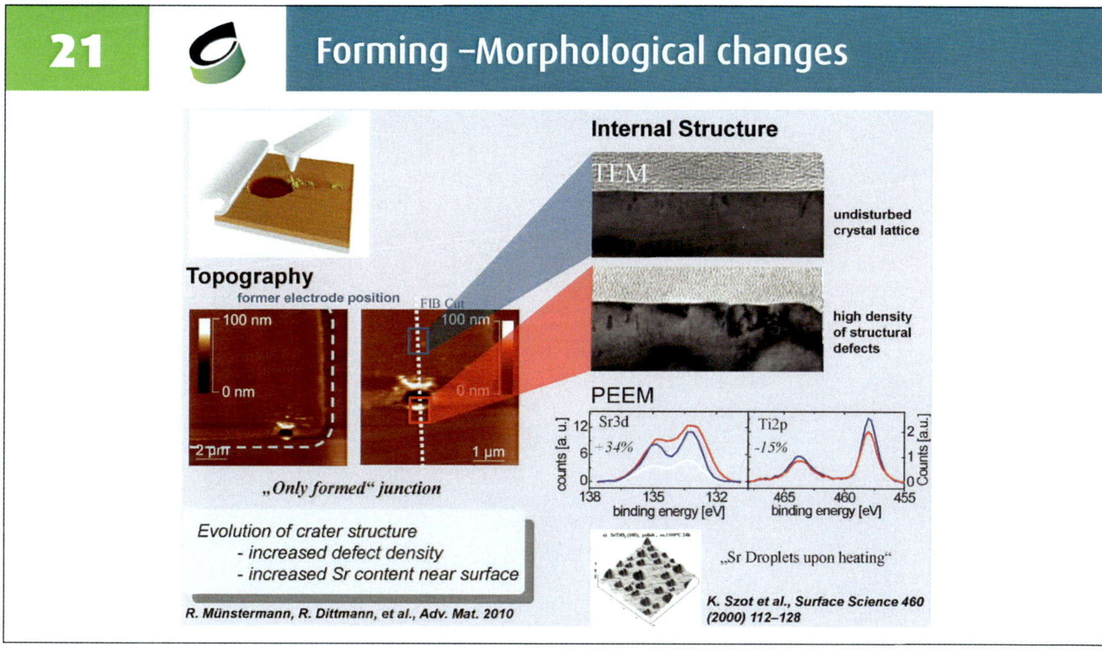

This slide focuses on the forming process, focusing on the morphological changes during the process. The topography and internal structure is shown.

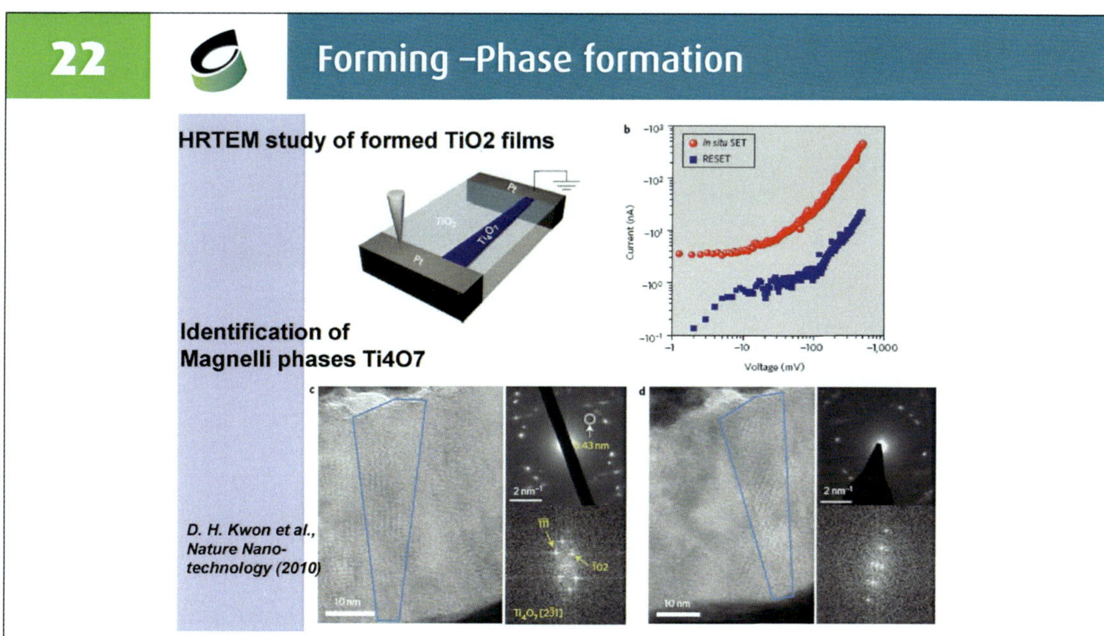

This slide focuses on the phase formation in forming process. It analyses the HRTEM study of formed TiO2 films and identification of Magnelli phases in Ti4O7.

23 Processes during formation into the OFF state

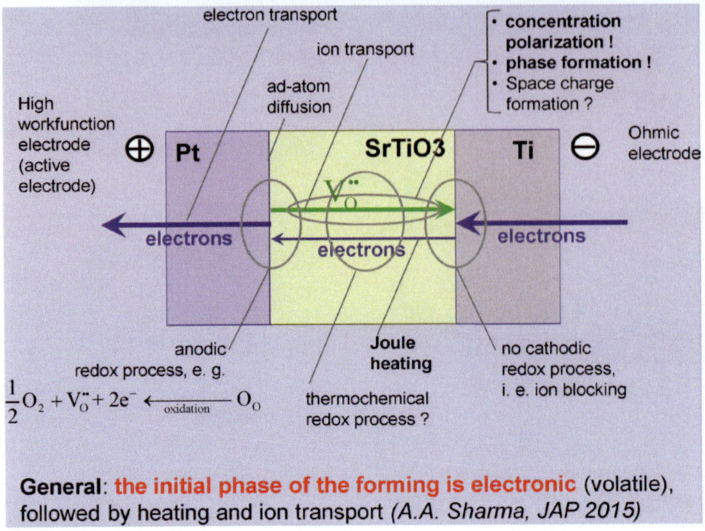

This slide explains the detailed process during the formation into the OFF state.

24 Numerical simulation

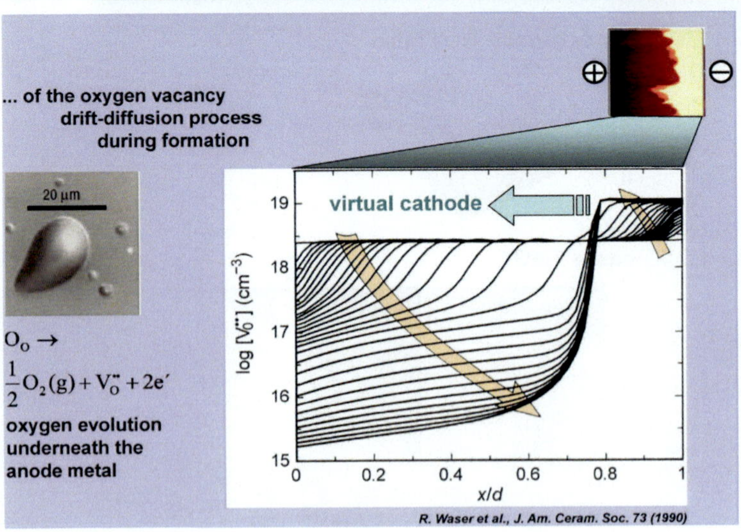

This slide shows the numerical simulation of the oxygen vacancy drift-diffusion process during formation. It shows a plot of log[V0″] vs x/d.

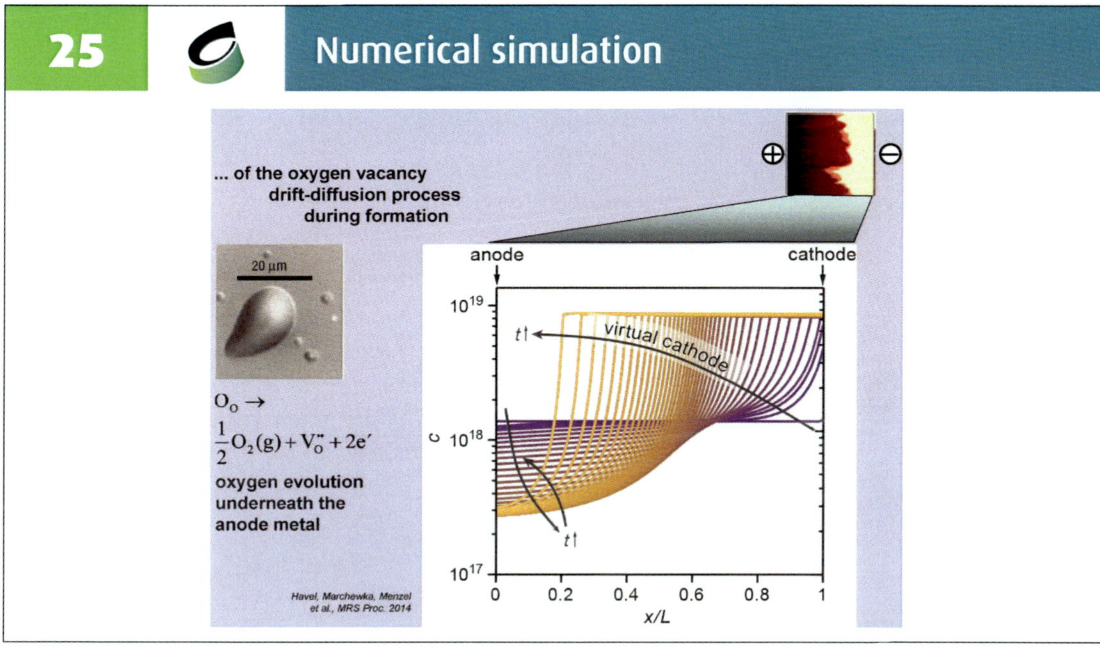

This slide shows the numerical simulation with a plot of c vs x/L.

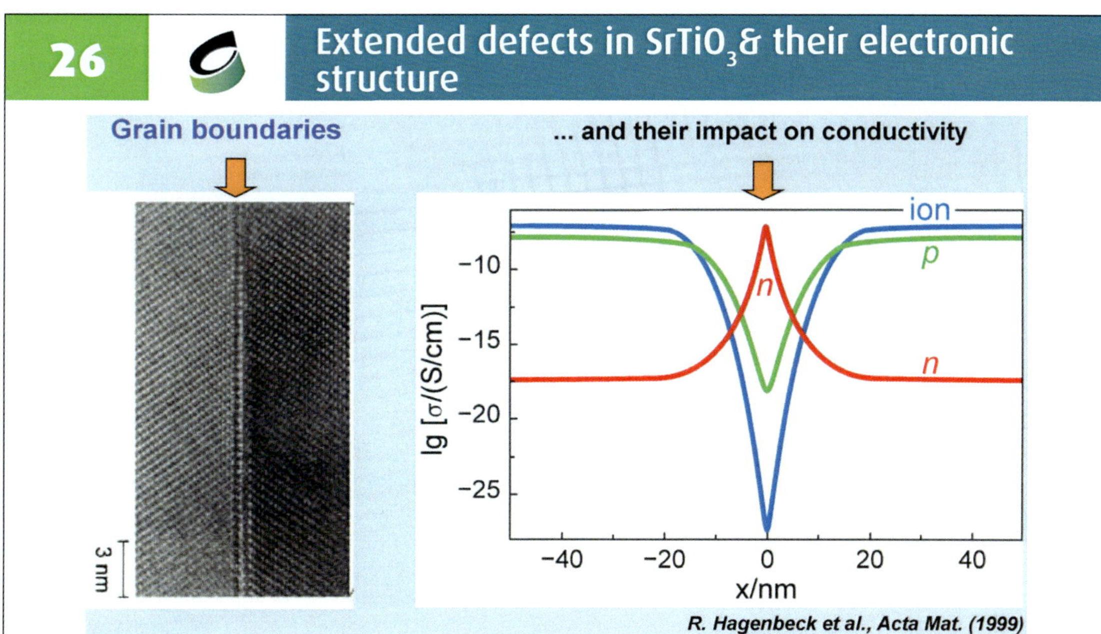

This slide discusses how grain boundaries in a material can affect how well it conducts electricity.

27 Thermal preformation by reduction annealing:conductive Tip AFM Mapping – types of I-V Characteristics

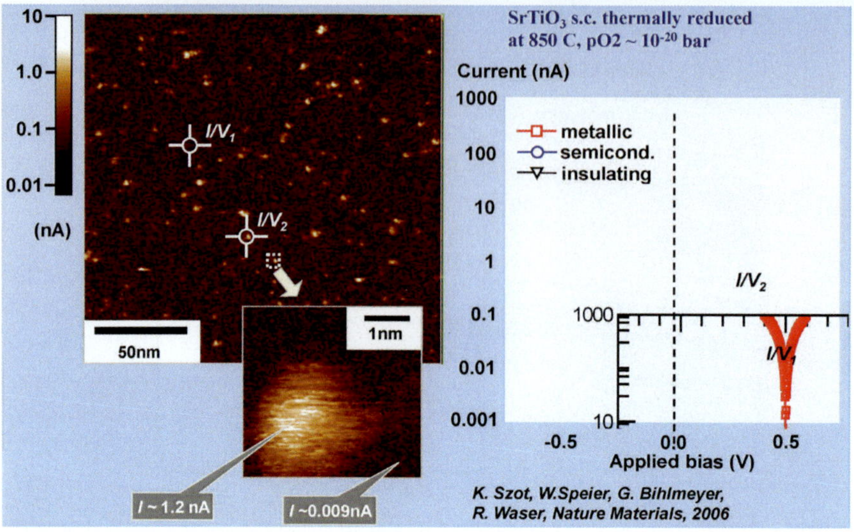

The slide shows electrical properties of a material after thermal annealing measured using conductive AFM, thermal performance by reduction annealing.

28

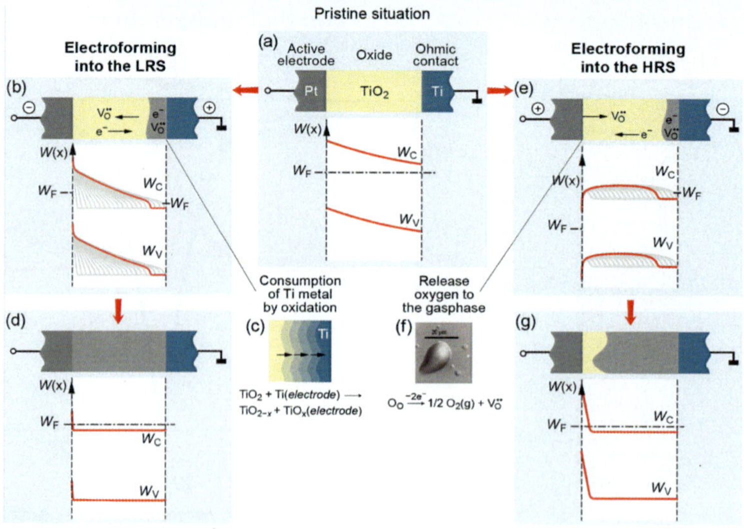

This slide illustrates how electroforming reduces resistance in a material by applying a voltage.

4 VCM-SYSTEMS – THE FILAMENTARY C8W SWITCHING PROCESS

29 Illustration of the resistive switching (filamentary C8W-VCM)

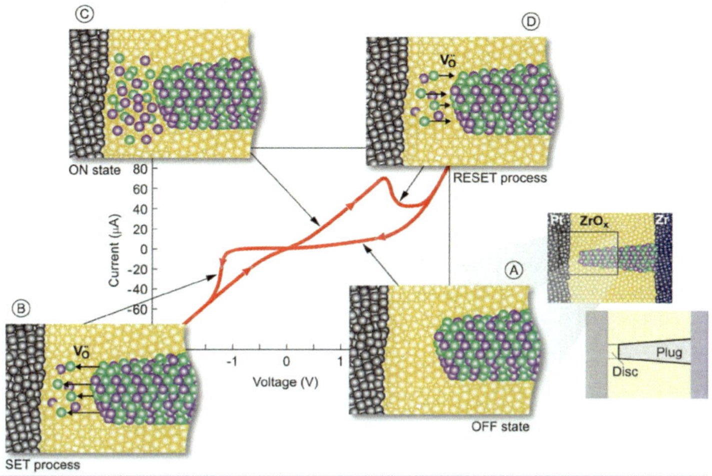

R. Waser et al. in: Nanoelectronics and Information Technology (3rd ed.), Wiley-VCH 2012

Illustrates a resistive switching process for C8W-VCM, likely referring to a memory device, showing transitions between high and low resistance states.

30 VCM - Processes during filamentary C8W switching

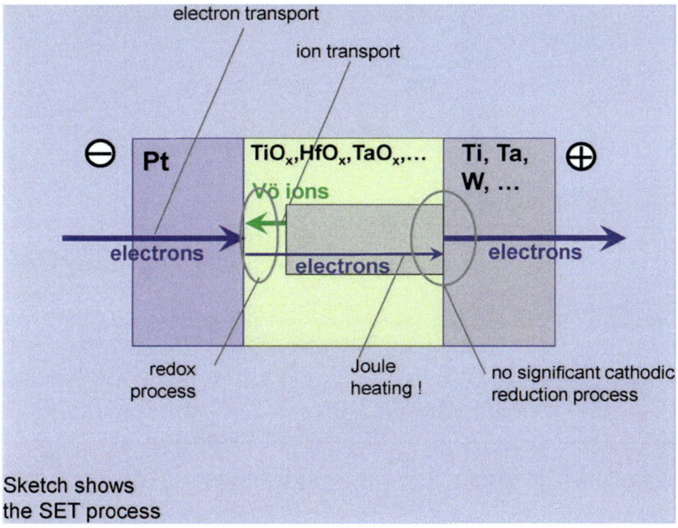

This slide depicts VCM, a type of memory, and details the SET and RESET processes involved in C8W resistive switching, which likely controls resistance through filament formation and rupture.

31 Redox-induced Schottky barrier variations

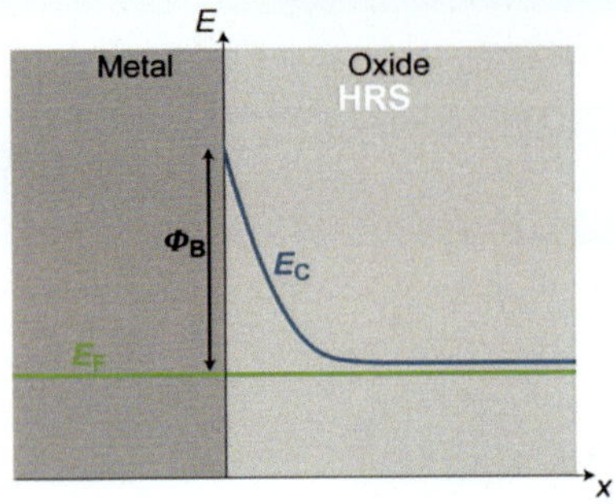

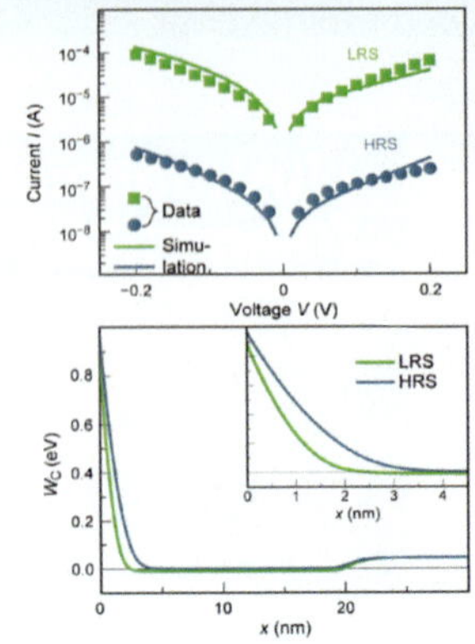

C. Bäumer et al., Nature Comm. (2016)

Transport model from A. Marchewka, SISPAD 2015

- AE interface: Vacancy concentration variation **by factor two**
- Resistance changes **by two orders of magnitude**
- Agreement between data and simulation for both states
- Schottky barrier modification responsible for switching

This slide illustrates how redox reactions can modify a Schottky barrier, potentially leading to a switch between high and low resistance states in a device.

5 VCM-SYSTEMS – OTHER VARIANTS

32 Motivation: Impact on variability

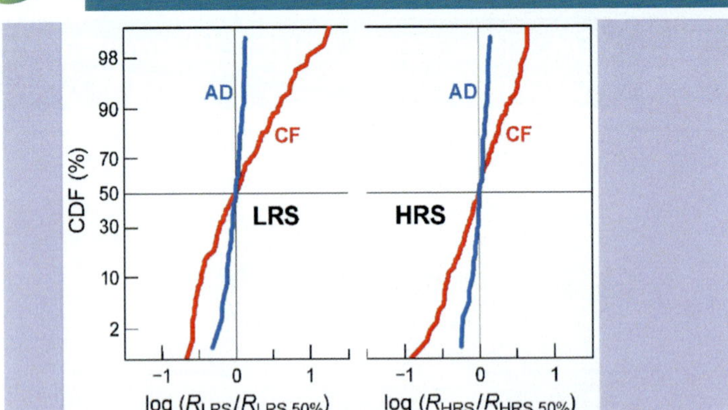

Comparison of LRS and HRS cumulative resistance distributions (CDF) for area-dependent switching cells (AD) and conducting-filament switching cells (CF)

CF: 20x20nm^2 TiN/HfO2/Hf/TiN (Fantini et al., 2013)
AD: 40x40nm^2 TiN/a-Si/TiOx/TiN (a-VMCO) cell (B. Govoreanu, 2015)

This slide compares resistance distributions in two resistive memory cell types, showing lower resistance in AD cells and lower overall resistance in LRS.

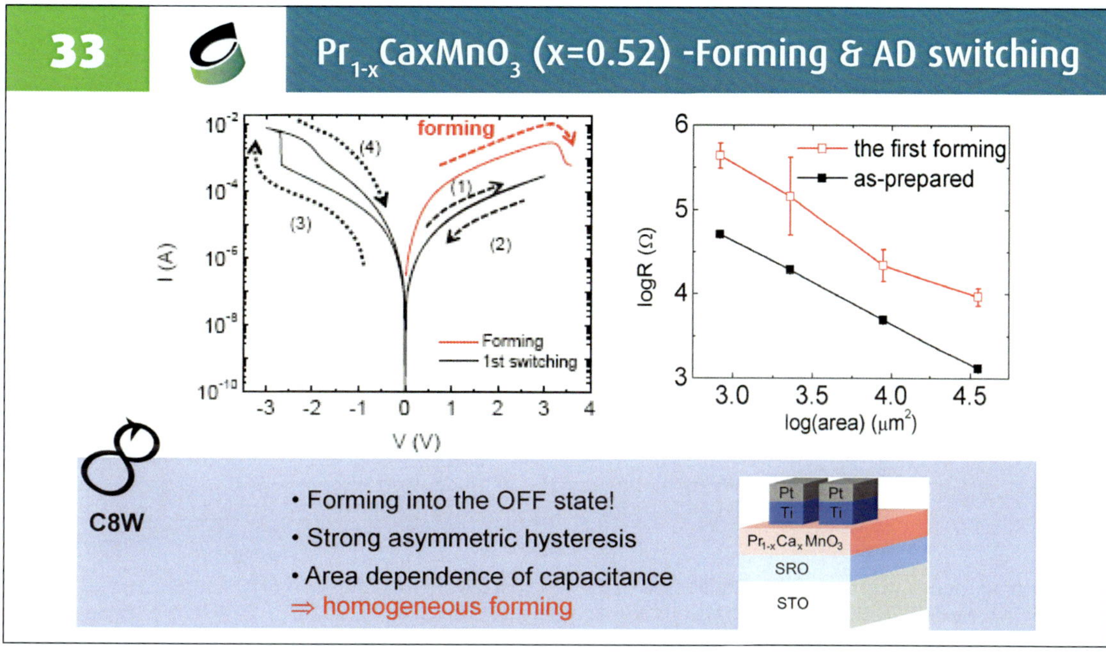

This slide shows a graph of the formation and ad switching of a compound $Pr_{1-x}Ca_xMnO_3$, with lower resistance observed during the switching process.

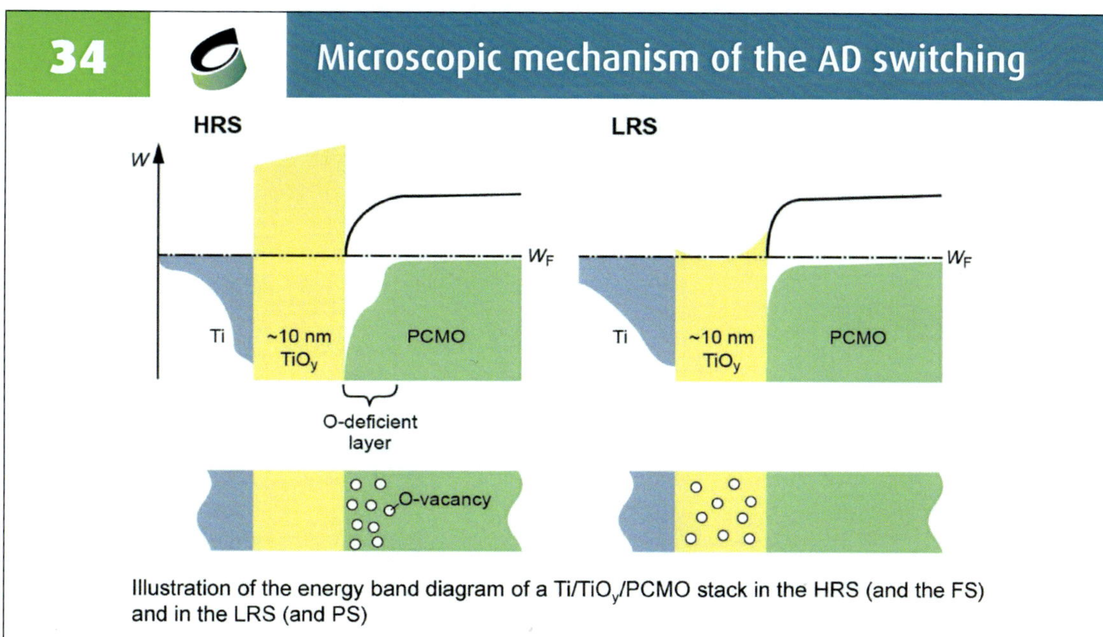

This slide illustrates a microscopic mechanism for AD switching, showing transitions between high and low resistance states using a microscope image and energy band diagrams.

35 8W switching-Coexistence of two bipolar switching polarities

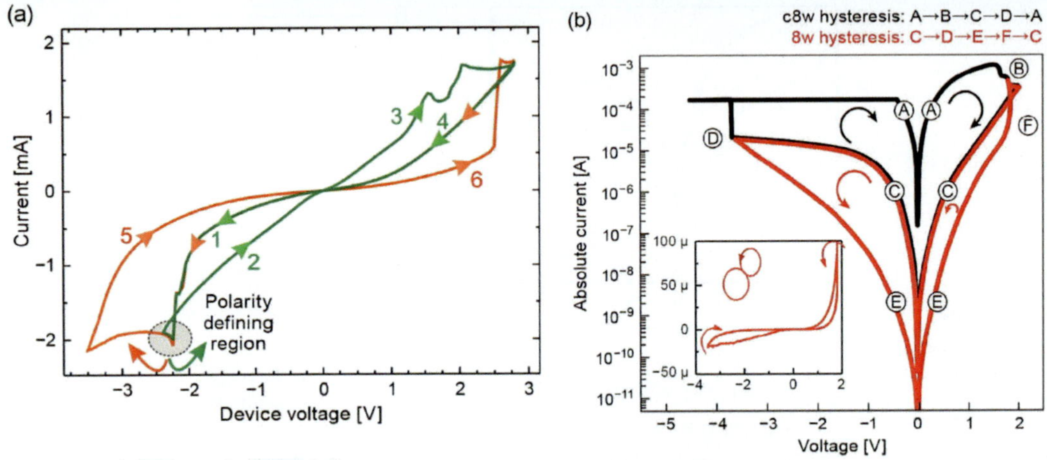

Switching modes with opposite polarities in $Pt/SrTiO_3:Fe/SrTiO_3:Nb$ cells (R. Münstermann et al.)

Pt (BE)/6 nm TiO_2/Ti/Pt (TE) (H. Zhang et al., 2018)

This slide shows two bipolar resistive switching polarities (8W and c8w) with graphs and highlights that 8W switching is a low-energy process.

36 8W switching

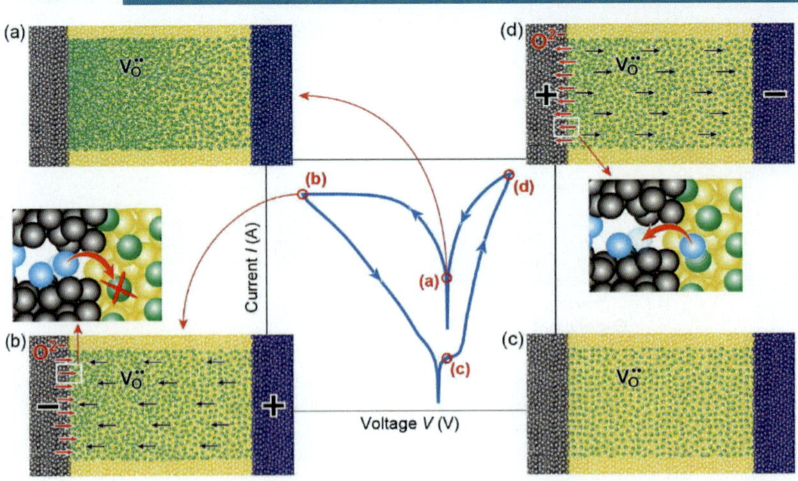

Proposed mechanism for the 8W switching through oxygen evolution and reincorporation of oxygen at the AE of $Pt/SrTiO_3:Fe/SrTiO_3:Nb$ cells (D. Cooper et al., Adv. Mater., 2017)

This slide depicts a mechanism for 8W resistive switching in a Pt/SrTiO3 device, focusing on oxygen evolution and reincorporation at the anode electrode.

37 8W switching

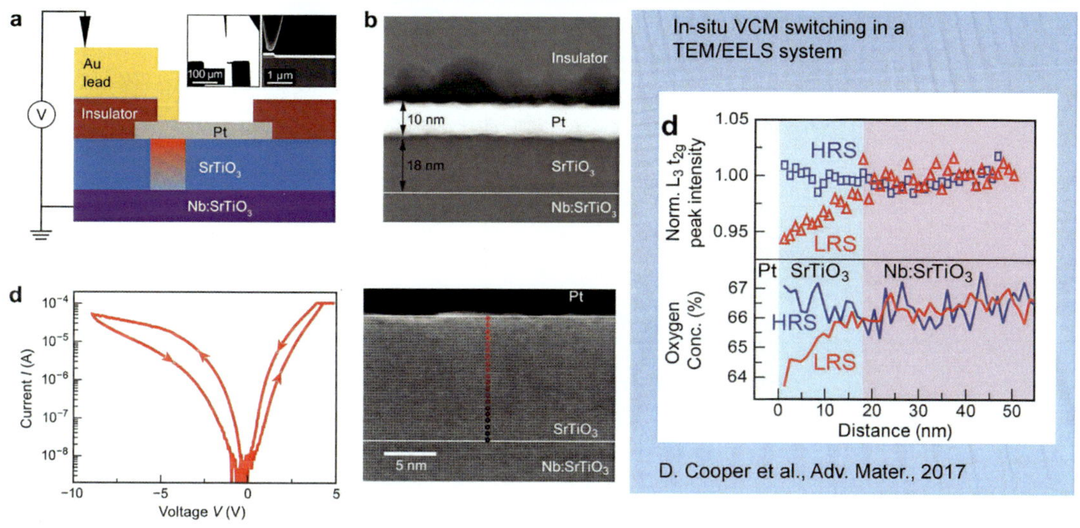

This slide shows VCM switching observed with TEM/EELS, a technique for high-resolution material investigation. The graphs likely detail electrical properties during switching.

6 NONLINEAR SWITCHTING KINETICS

38 How to solve the voltage-time dilemma of RRAM?

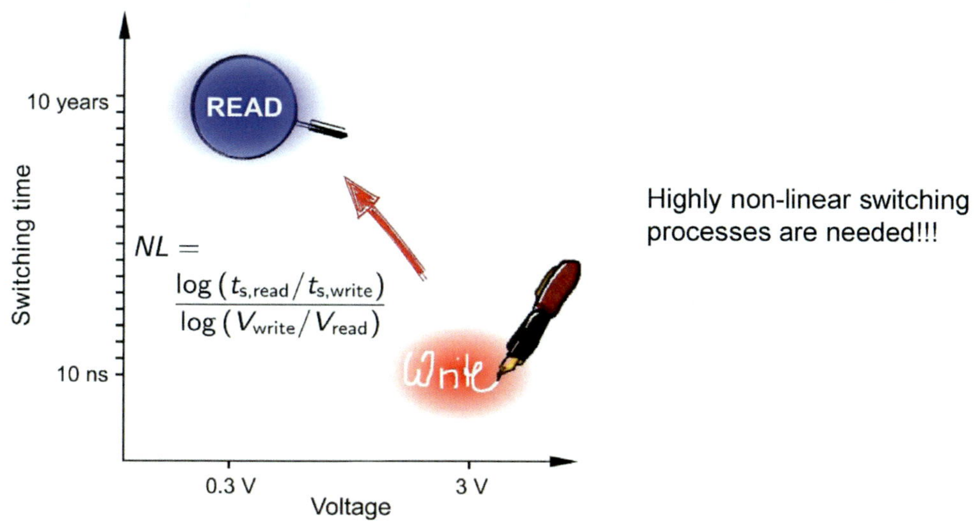

This slide discusses a challenge in RRAM technology and proposes highly non-linear switching processes as a solution.

39 What determines the switching speed?

Electric field

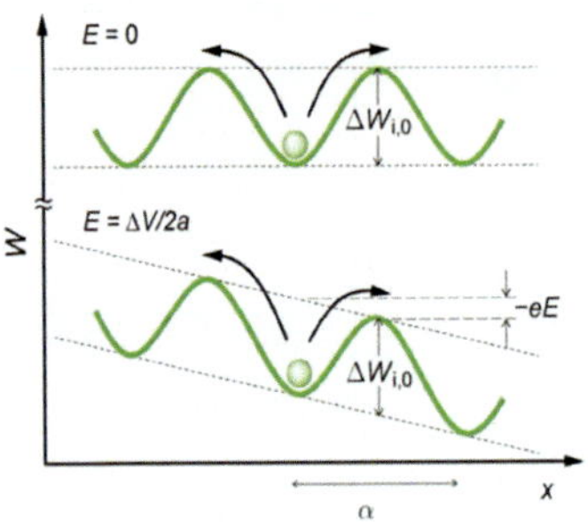

Temperature

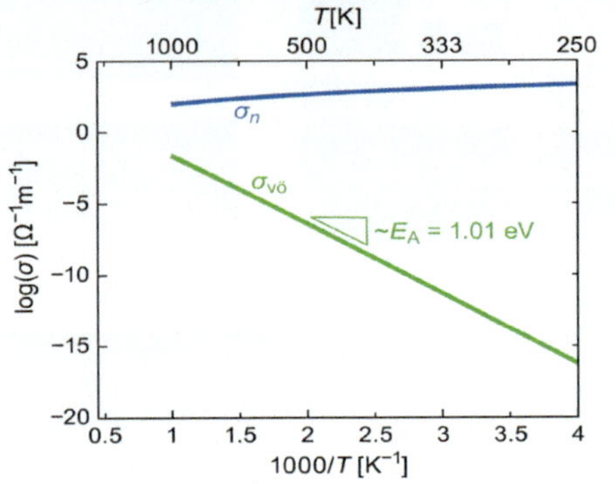

Drift velocty of oxygen ions:

$$v_D = a\omega_0 \exp\left(-\frac{\Delta W_{i,0}}{k_B T}\right)\sinh\left(\frac{a z_i e}{2 k_B T}E\right)$$

S. Menzel, R. Dittmann, R. Waser et al. (Adv. Funct. Mat. 2011)

This slide explores factors affecting switching speed in RRAM, including electric field, temperature, and oxygen vacancy properties.

40 Field enhanced ionic conduction

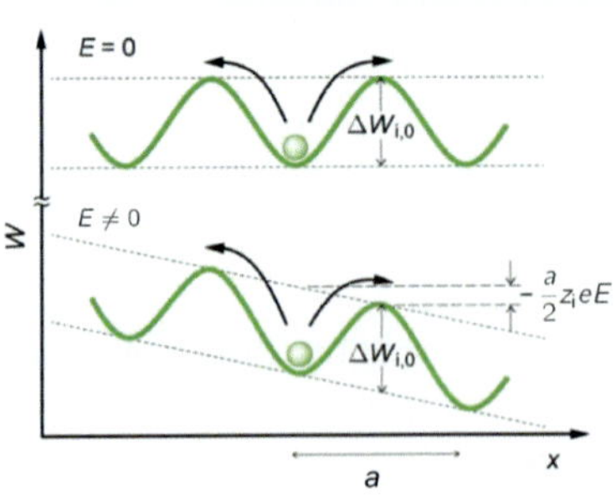

$E = 0$

$$v_{D,0} = a\omega_0 \exp\left(-\frac{\Delta W_{i,0}}{k_B T}\right)$$

$$J_\pm = J_0 = z_i e N_i v_{D,0}$$

$$J_g = J_+ - J_- = 0$$

$E \neq 0$

$$\Delta W_{i,0} \longrightarrow \Delta W_{i,0} \pm z_i e \frac{a}{2} E$$

$$J_\pm = J_0 \exp\left(\pm \frac{a}{2}\frac{z_i e}{k_B T} E\right)$$

$$J_g = 2\, J_0 \sinh\left(\frac{a}{2}\frac{z_i e}{k_B T} E\right)$$

$E \gg 2\, k_B T / a z_i e$

$$J_g \propto \exp\left(\frac{-2\,\Delta W_{i,0} + a z_i e E}{2\, k_B T}\right)$$

Temperature and/or field enhanced ionic transport !

This slide shows how an electric field can enhance ionic conduction by affecting the energy barrier for ion diffusion.

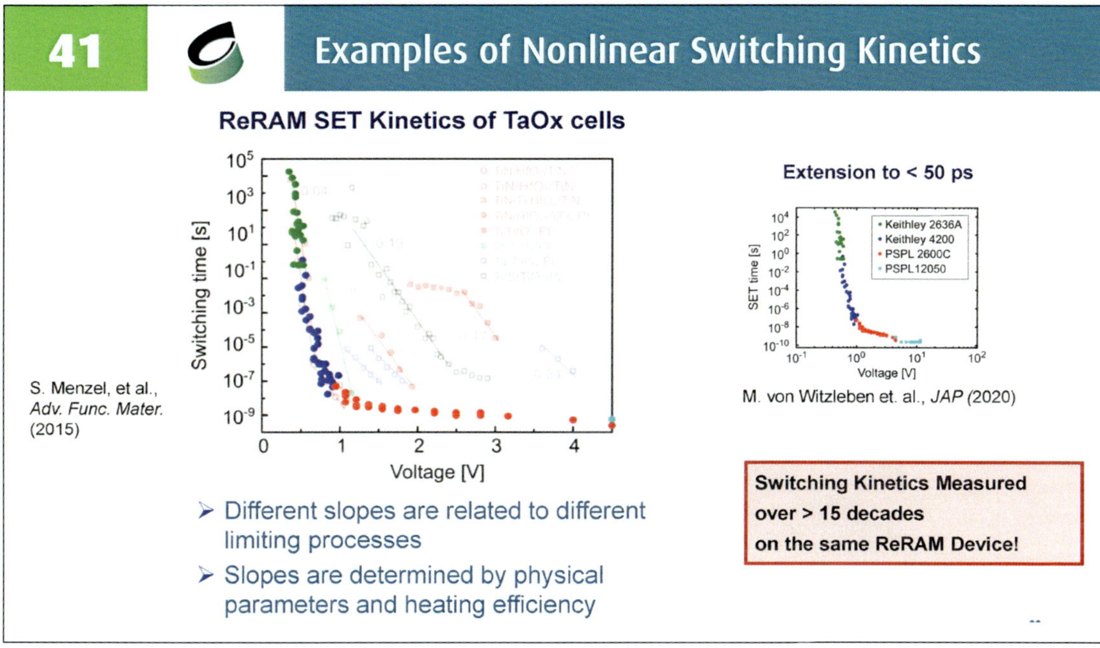

This slide shows examples of nonlinear switching kinetics in RRAM devices. The slopes in the plot correspond to different processes that limit switching speed.

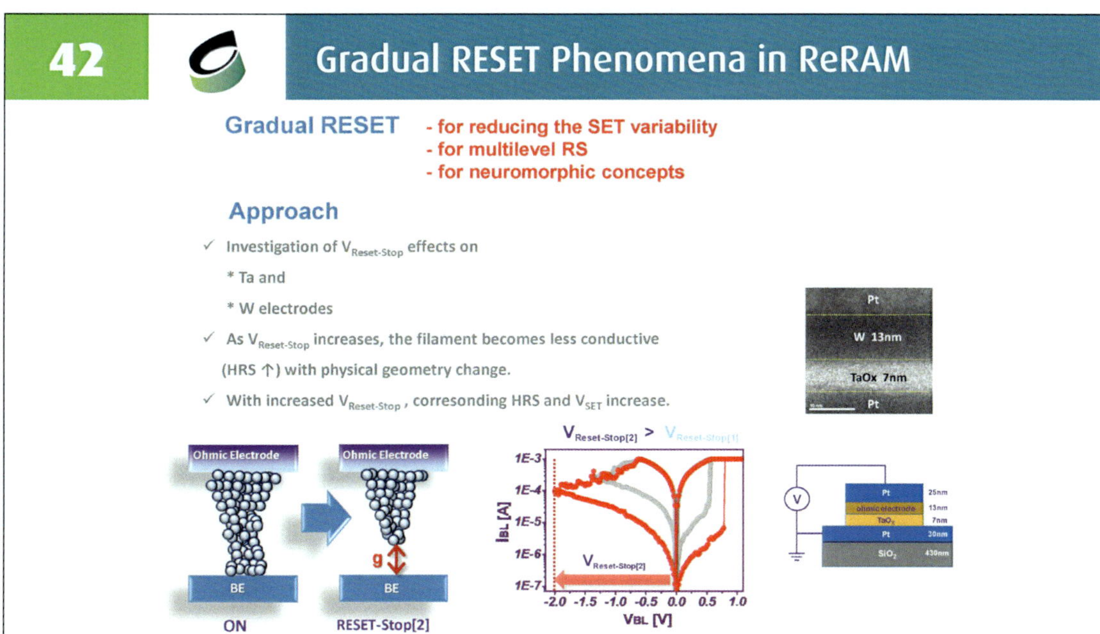

This slide describes the gradual reset phenomenon in ReRAM devices, influenced by the presence, number, and type of transistors.

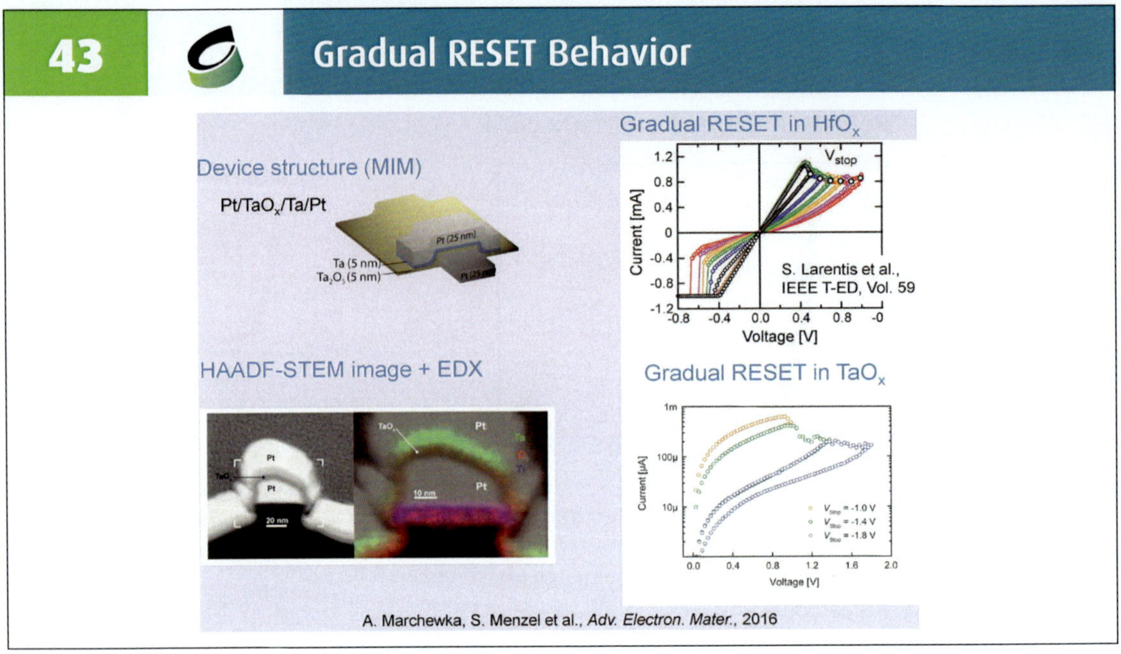

This slide shows a gradual decrease in current with applied voltage in HfO_2 resistive memory devices, representing the reset process.

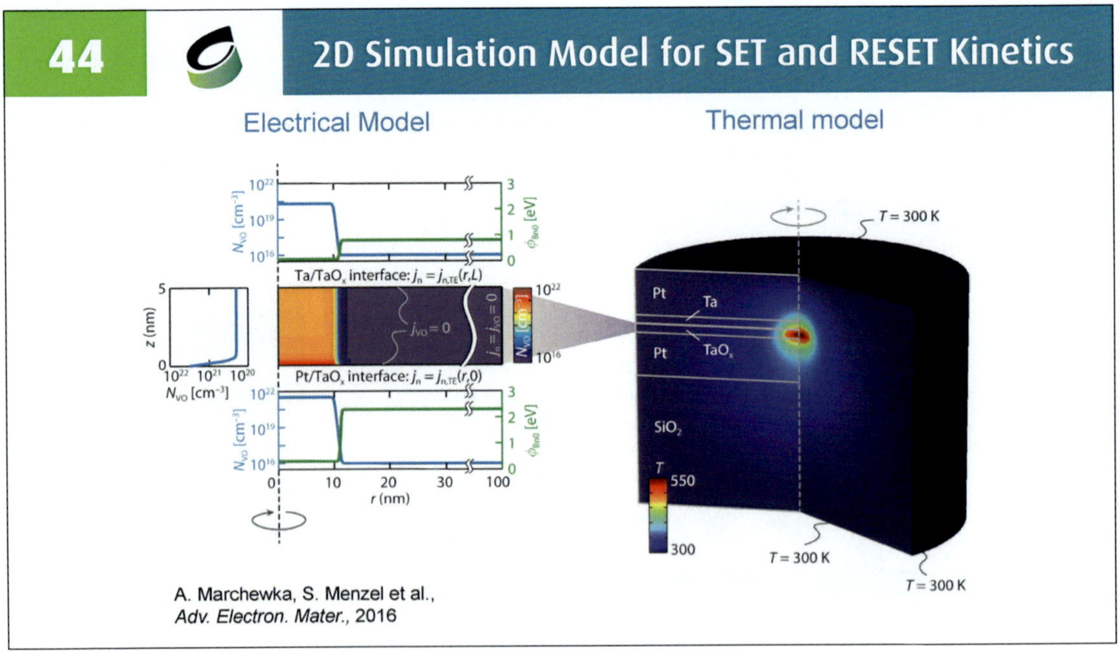

This slide presents a simulation model for understanding the electrical and thermal properties involved in setting and resetting resistive memory devices.

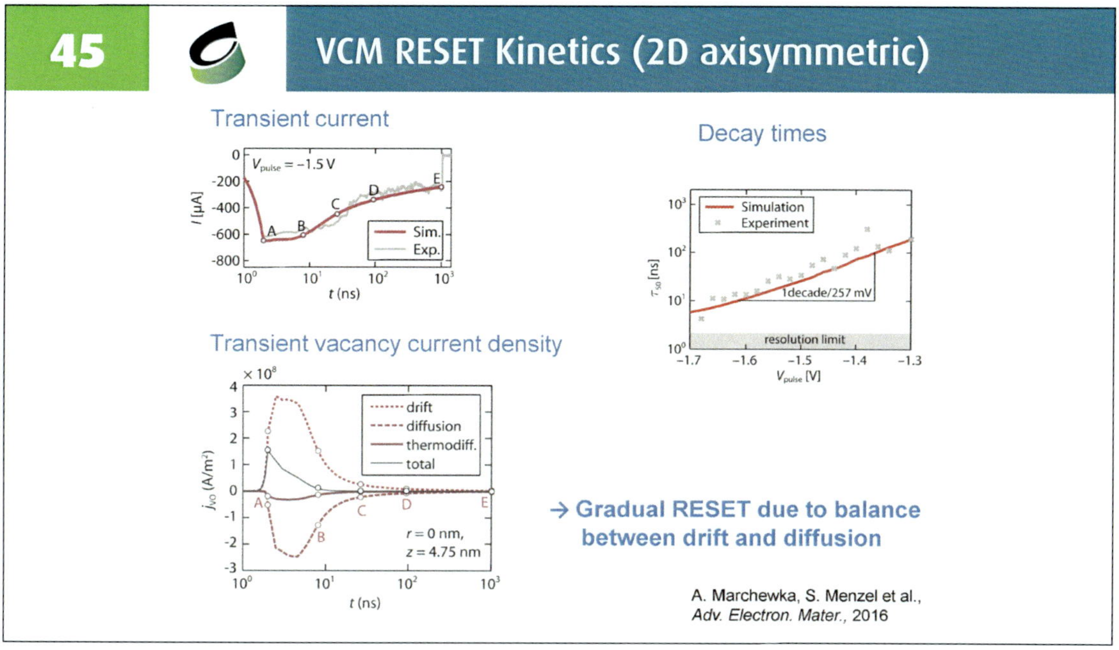

This slide shows a graph of VCM reset kinetics with transient current, decay times, and vacancy current density, It also shows a separate plot for vacancy current density.

7 RELIABILITY ASPECTS - THE VARIABILITY

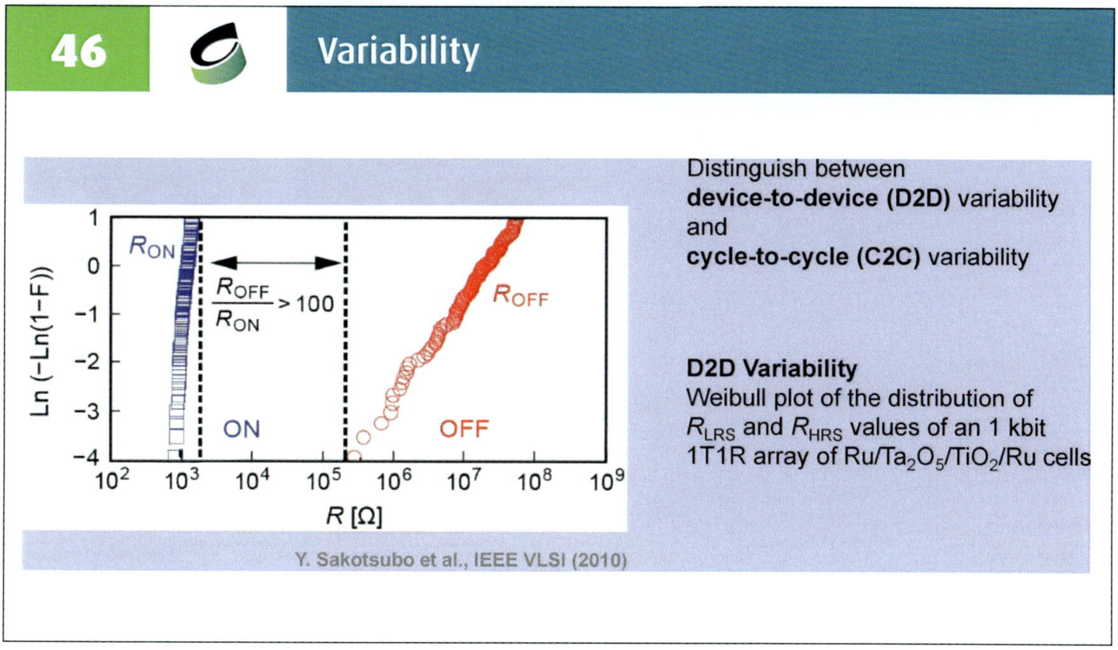

This slide uses a Weibull plot to show variations in reset voltage required to reset HfO_x-based resistive memory devices.

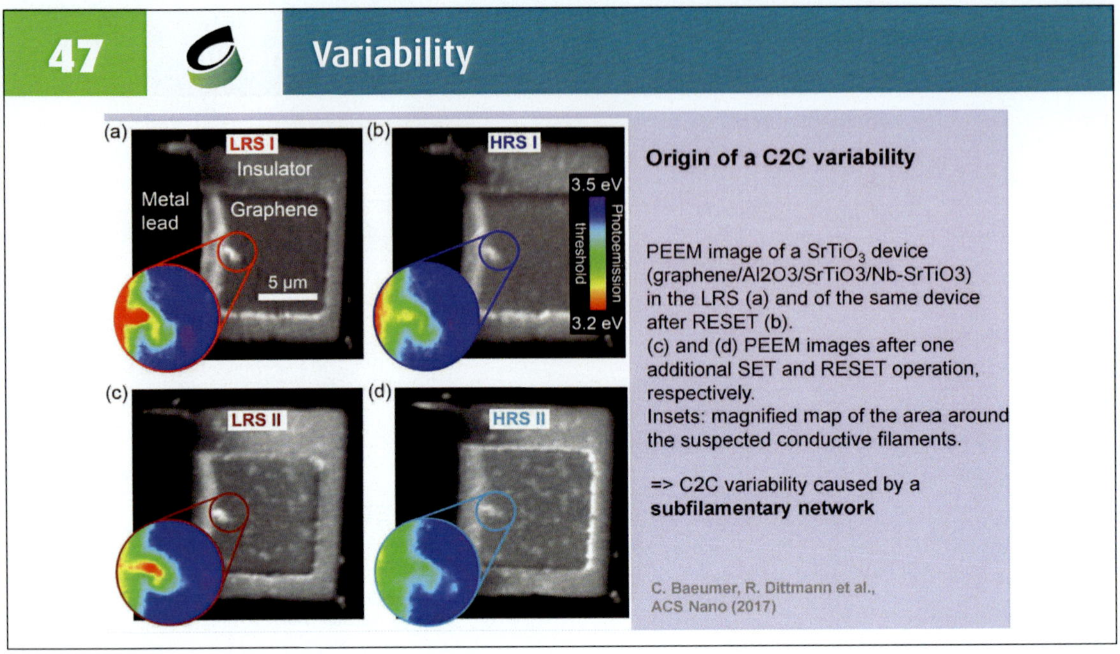

This slide explores the source of variations in conductivity observed in a graphene conductor. It compares a device before and after reset and highlights suspected conductive filaments.

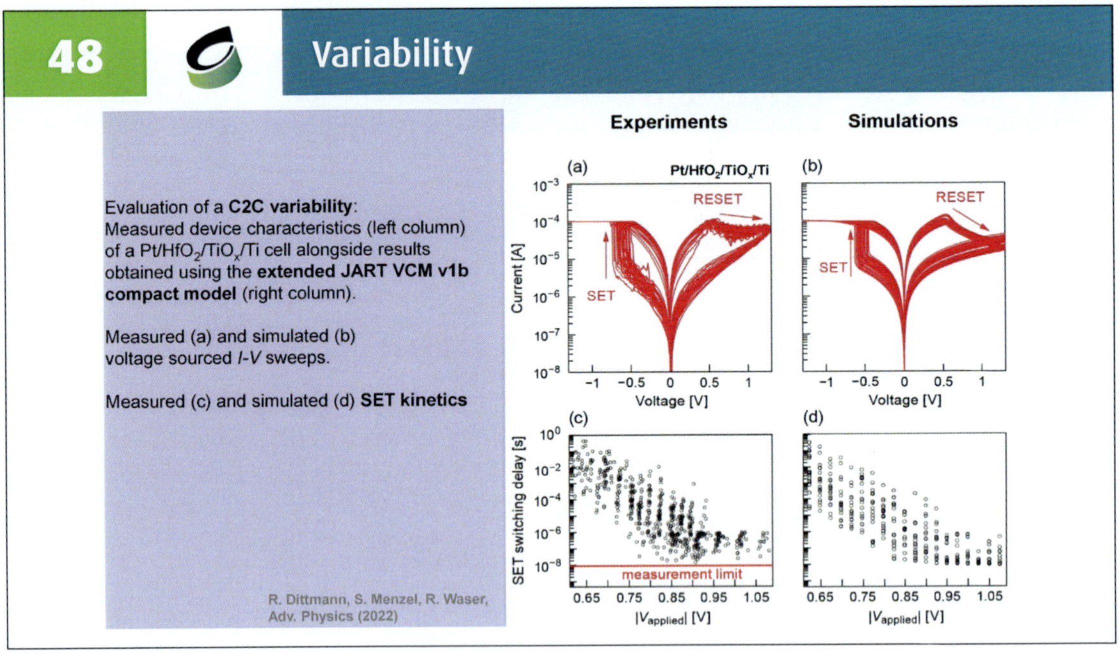

This slide compares measured and simulated electrical characteristics of a Pt/HfO2/TIO/Ti device. The plot shows current versus voltage.

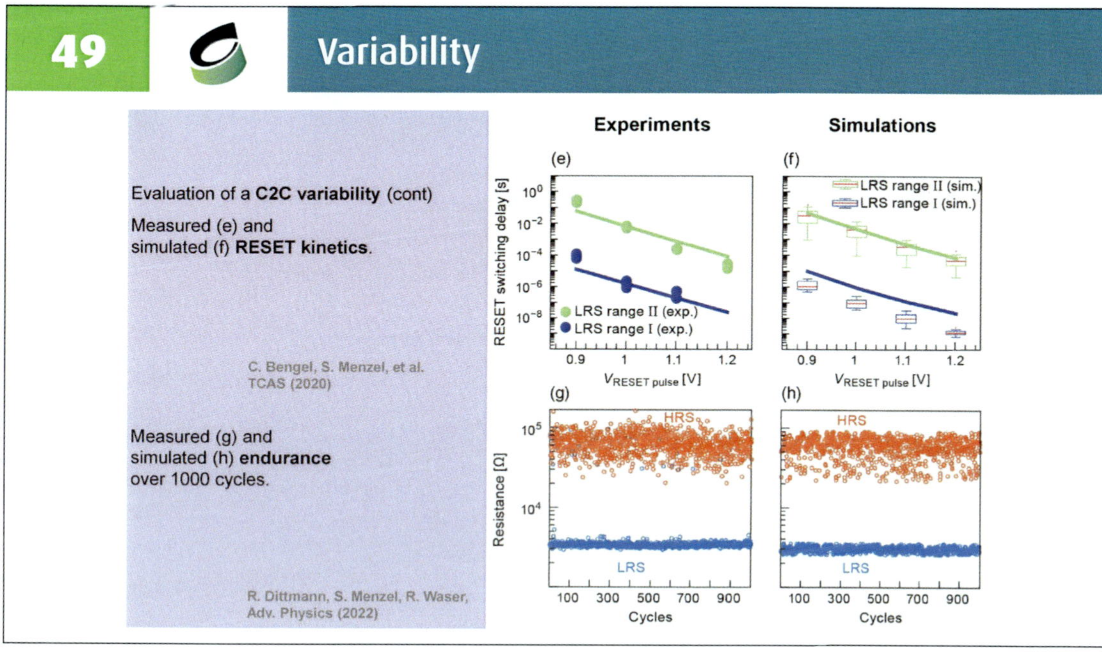

This slide illustrates transitions between low and high resistance states in an Ag_2Se memory device through filament formation and rupture.

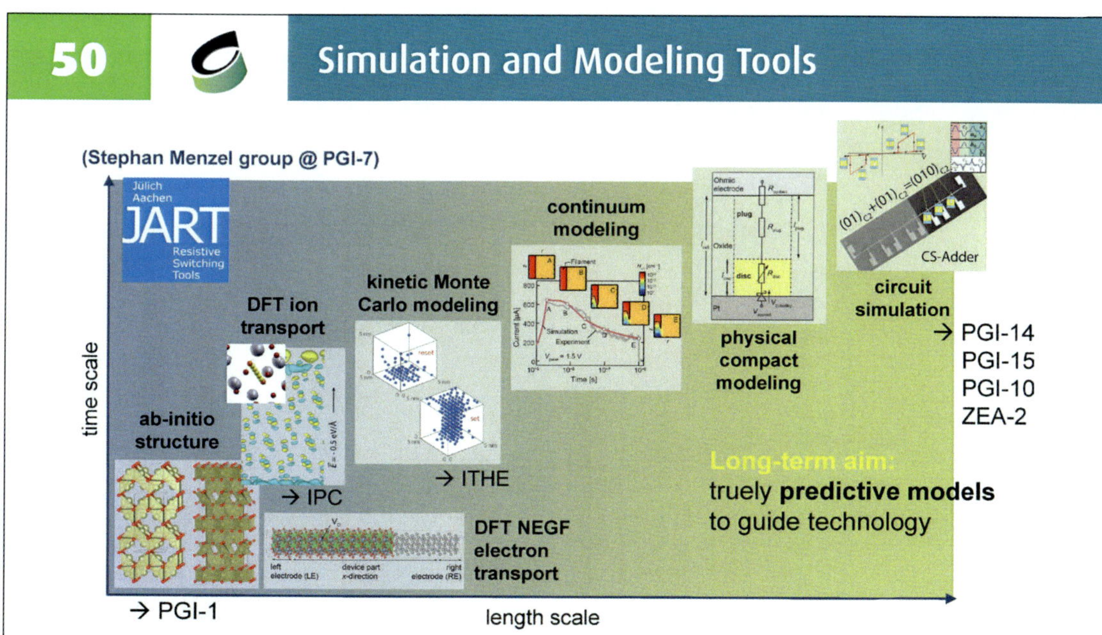

This slide depicts the ECM mechanism in a resistive memory device, where metal ion migration and deposition create a conductive filament.

51 Compact Model: JART VCM v1b var

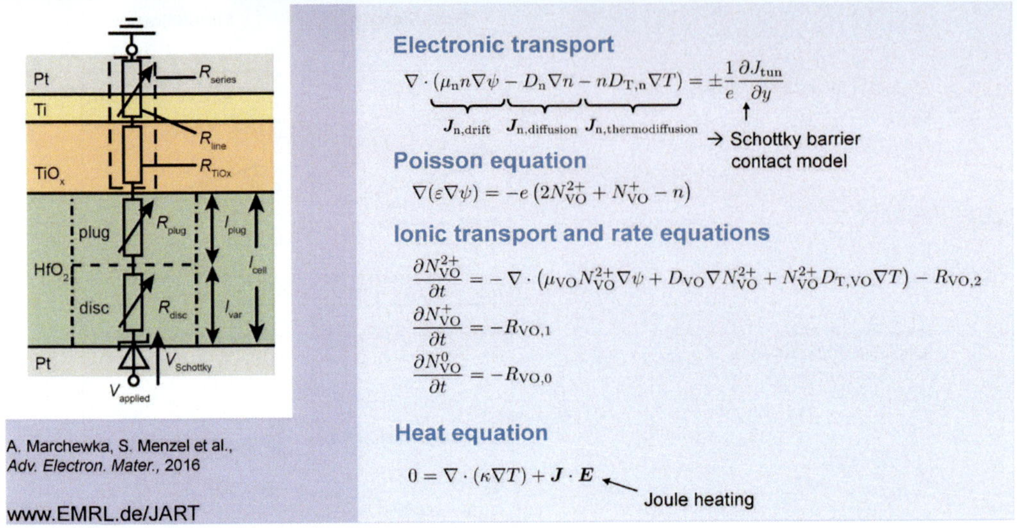

Compact model of JART VCM v1b by A. Marchewka, S. Menzel et al., Adv. Electron. Mater., 2016 where electronic transport, poisson equation, ionic transport and rate equations and heat equation is explained.

8 NEUROMORPHIC COMPUTING FOR ARTIFICIAL INTELLIGENCE

52 Modeling of biological neurons

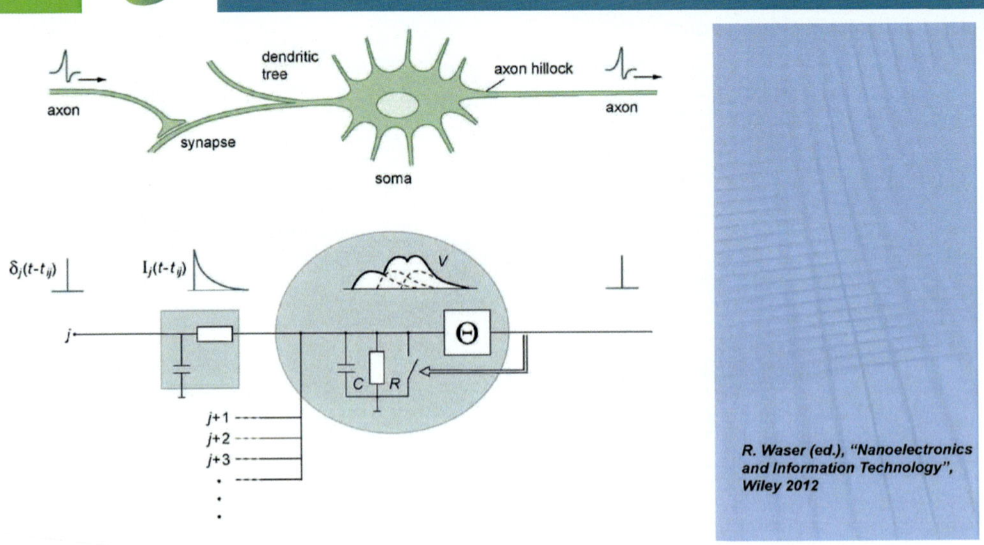

Comparison between biological and artificial neuron is explained, Figure by R. Waser (ed.), "Nanoelectronics and Information Technology", Wiley 2012.

53 From Multibit memory to artificial neurons

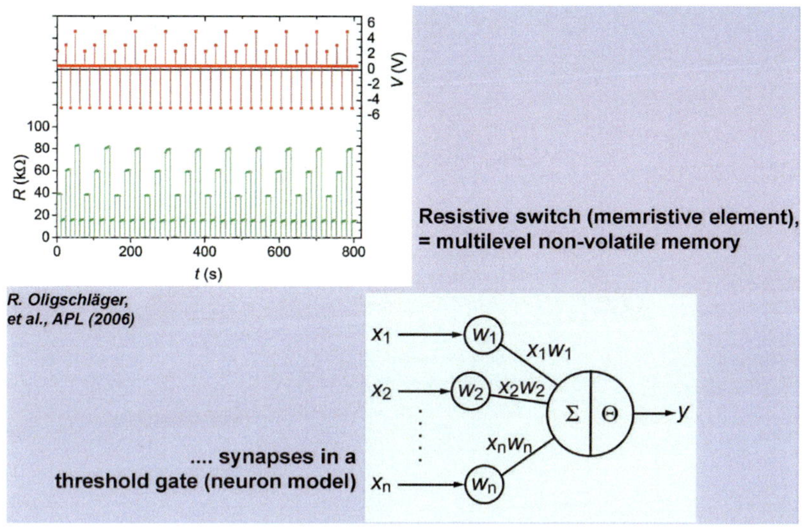

Comparison between Multibit memory to artificial neurons where Resistive switch and synapses in neuron model is explained.

54 Neuromorphe Beyond von-Neumann-Rechenkonzepte mit memristiven Bauelementen

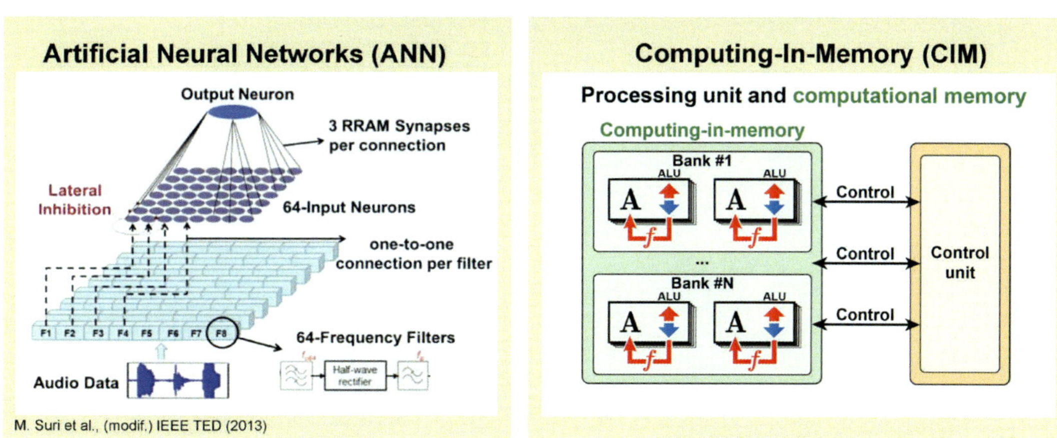

Neuromorphic Beyond von Neumann computational concepts with memristive components. Artificial neural networks and computing in memory is compared.

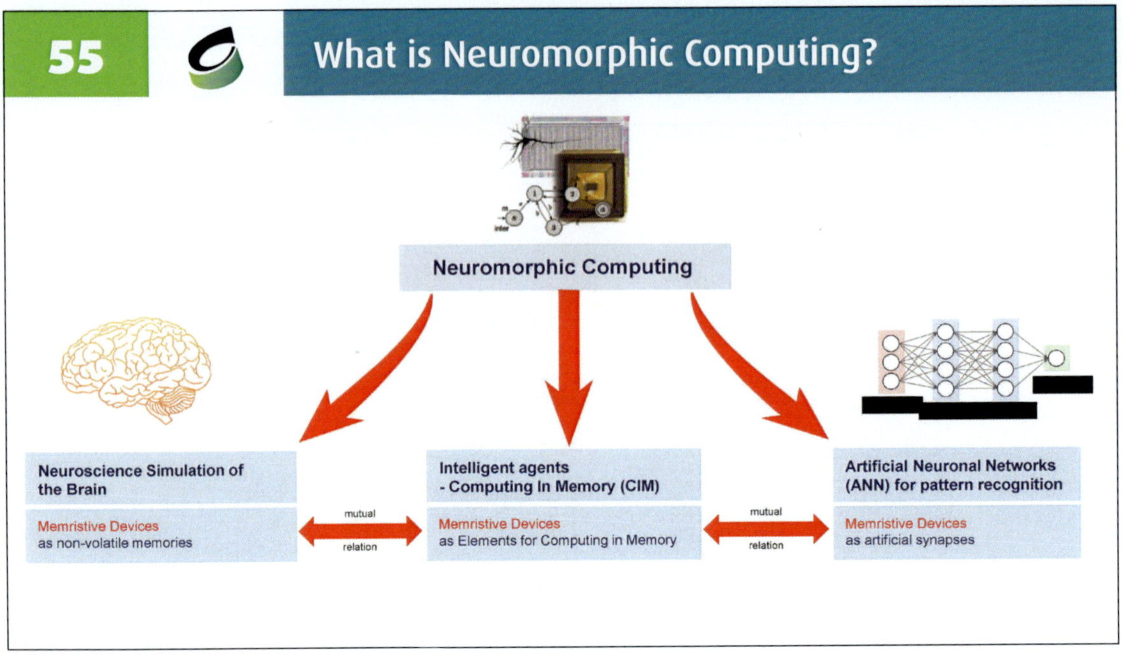

Flow chart of Neuromorphic computing which breaksdown into Neuroscience Simulation of the Brain, Intelligent agents- Computing In Memory (CIM), Artificial Neuronal Networks (ANN) for pattern recognition.

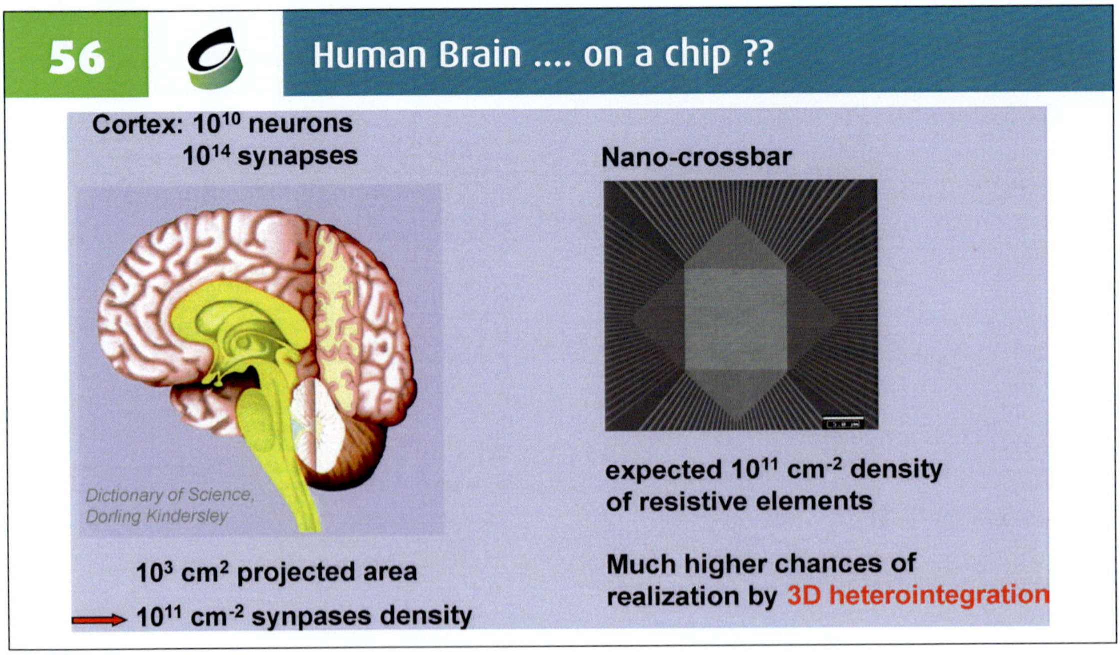

Information about containing entire human brain memory on a chip where cortex of brain has 10^10 neurons and 10^14 synapses, Nano-crossbar has expected 10^11/cm^2 of resistive elements.

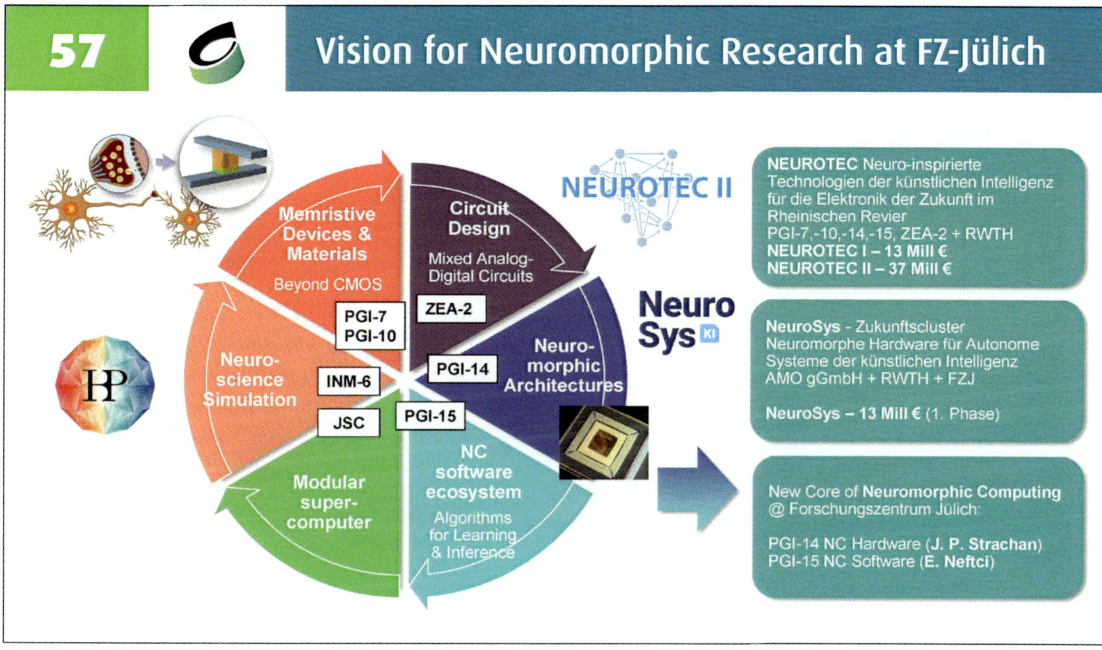

Vision for Neuromorphic Research at FZ-Jülich is explained using a pie chart which contains Memristive Devices & Materials, Neuro-science Simulation, Circuit Design, Modular ,super-Computer, NC Software ecosystem.

Headline: NEURO-INSPIRED AI TECHNOLOGIES FOR THE ELECTRONICS OF THE FUTURE.

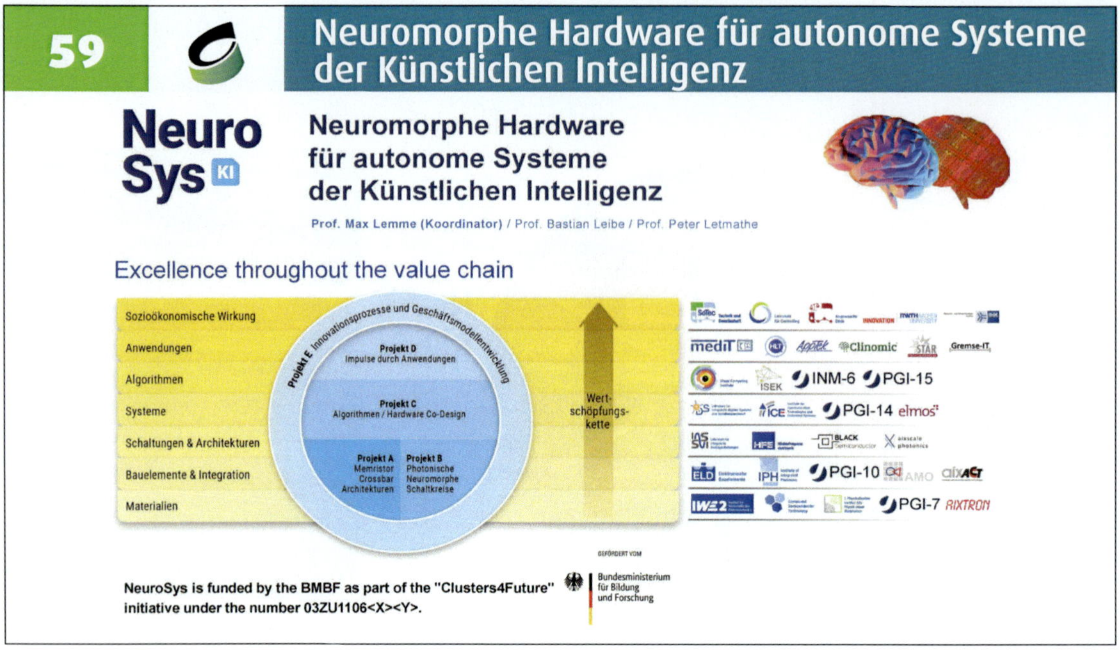

Neuromorphic hardware, designed for autonomous systems in artificial intelligence, emulates the brain's neural networks, facilitating advanced cognitive abilities and adaptive learning processes in machines. Its architecture enables energy-efficient computation and real-time processing, enhancing the performance and scalability of AI applications.

9 CONCLUSIONS

Highlights the upcoming events : International Conference on Memristive Phenomena in Chalcogenides and Beyond, International Conference on Neuromorphic Computing.

61 2022 Roadmap on Neuromorphic Computing and Engineering[1]

Section 1 – Materials and devices
Section 2 – Neuromorphic circuits
Section 3 – Neuromorphic Algorithms
Section 4 – Applications
Section 5 – Ethics

Editor-in-Chief: **Giacomo Indiveri,** University of Zurich, Switzerland

PGI affiliated authorship: Dittmann, Menzel, Valov, Strachan, Nefci

Regina Dittmann is also member of the journal's editorial board

[1] D. V. Christensen *et al.*,
"2022 roadmap on neuromorphic computing and engineering",
Neuromorphic computing and engineering.
IOP Publishing Ltd., Bristol, 2022. [Online].

Explains the 2022 Roadmap on Neuromorphic Computing and Engineering which hasSection 1 – Materials and devices, Section 2 – Neuromorphic circuits, Section 3 – Neuromorphic Algorithms, Section 4 – Applications, Section 5 – Ethics. Also gives a description of Editor-in-Chief Giacomo Indiveri.

62 Today: Computer are ubiquitous

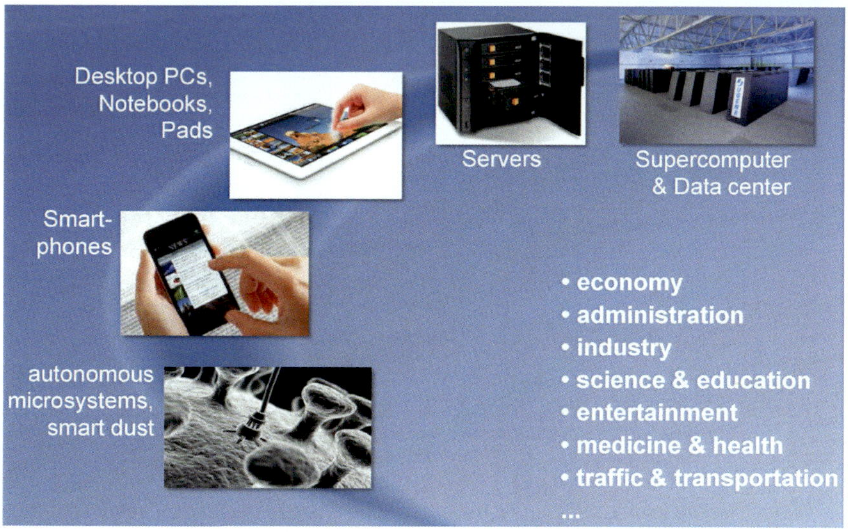

Highlights the fact that Devices are found everywhere in economy, administration Industry, science & education, entertainment, medicine & health, traffic & transportation.

63 Tomorrow: Neuromorphic Computer save energy

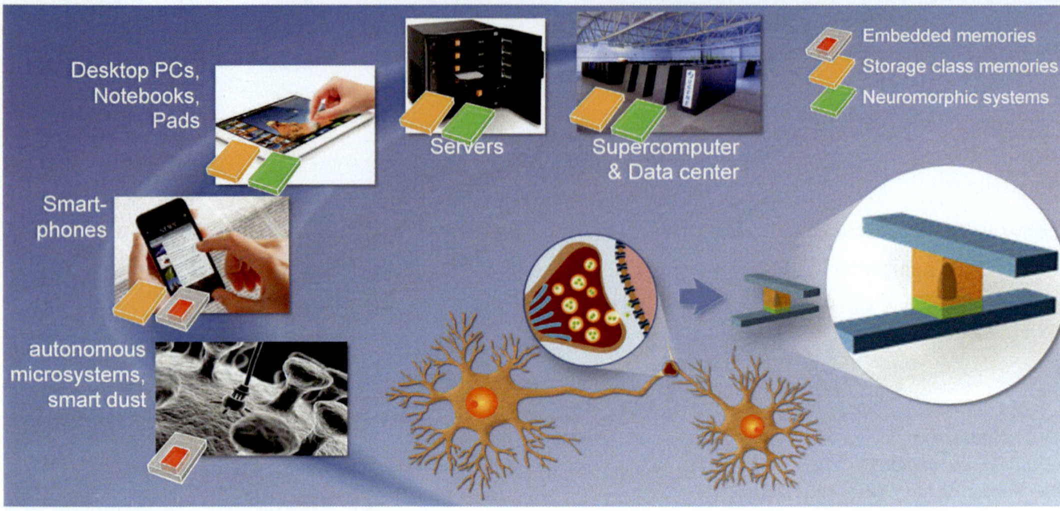

By using neuromorphic computers we can save energy. Using Embedded memories, Storage class memories, Neuromorphic systems.

64

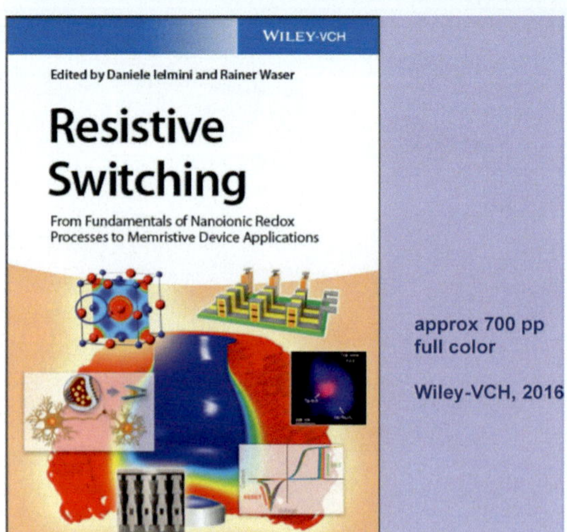

Highlights the book Resistive Switching written by Daniele Ielmini and Rainer Waser.

Wraps the session with gratitude towards the audience.

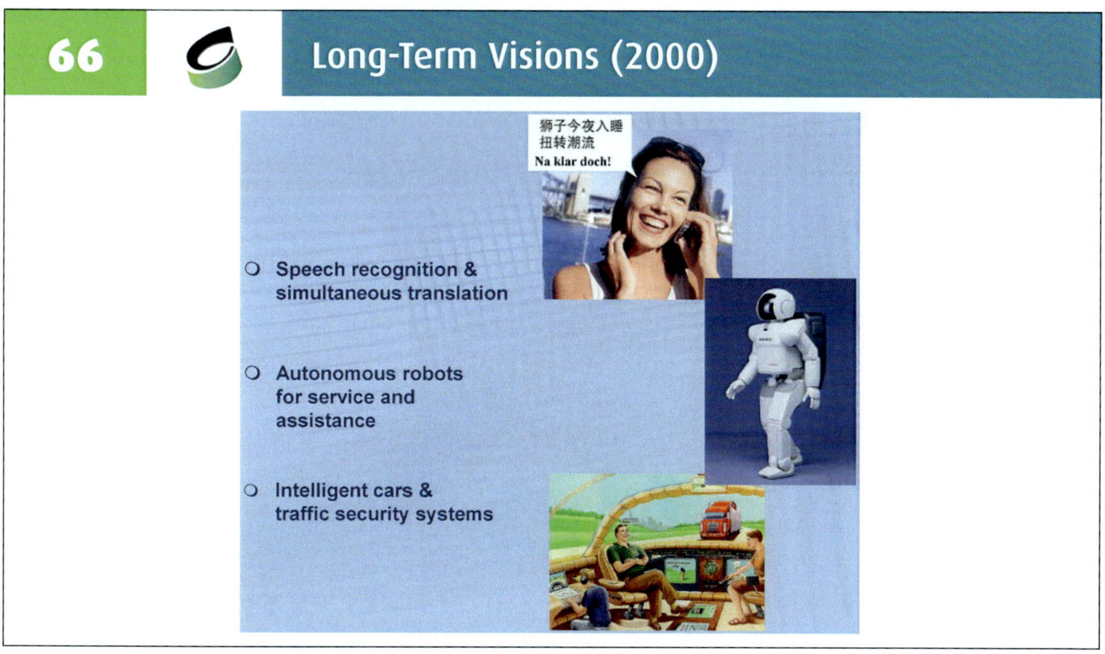

Explains the Long-term visions some of them are: Speech recognition and simultaneous translation, Autonomous robots for service and assistance, Intelligent cars & traffic security systems.

3 HISTORY OF RESISTIVE SWITCHING AND MEMRISTIVE DEVICES

67 Resistive („memristive") switches

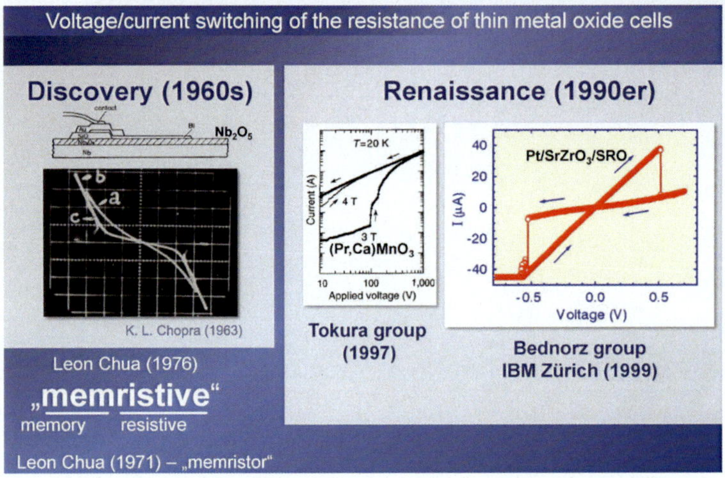

Resistive switches, also known as "memristive" devices, alter their resistance via voltage/current changes in thin metal oxide cells. Their discovery dates back to the 1960s, with Leon Chua coining the term "memristor" in 1971, leading to a renaissance in the 1990s and subsequent advancements by research groups like IBM Zurich and pioneers such as the Tokura and Bednorz groups.

68 Brief history of ReRAM(VCM type)

1960/70s	**many metal oxides**	**Many phenomenological studies, no clear mechanism**
1997	**(Pr,Ca)MnOx** **A. Asamitsu et al.**	**Field-induced charge-ordered to metallic ferromagnetic transition**
2000 ...	**$SrZrO_3$, $SrTiO_3$** **A. Beck et al., etc.**	**Electronic charge-transfer or charge injection process**
2004	**Ti/(Pr,Ca)MnOx** **A. Sawa et al.**	**Charging/discharging of interface states to modify Schottky barrier**
2006	**$SrTiO_3$** **K. Szot et al.**	**Oxygen ion drift and coupled redox process (at extended defects)**
2007	**TiO_2** **D. Jeong et al.**	**Coexistance of unipolar (TCM) and bipolar (VCM) resistive switching**
2008	**TiO_2** **D. Strukov et al.**	**Description of RS by the EE concept of memristors / memristive devices**
20xx-	**... many oxides** **(few thousands papers)**	

ReRAM (Resistive Random Access Memory) of the VCM (Voltage-Controlled Memory) type has evolved significantly since its initial discovery in the 1960s and 70s with various metal oxides. Notable milestones include the transition to (Pr,Ca)MnOx in 1997, advancements in charge transfer processes, interface state modification, and elucidation of resistive switching mechanisms, paving the way for a multitude of oxide materials and extensive research in subsequent years.

69 Link between device theory and physics

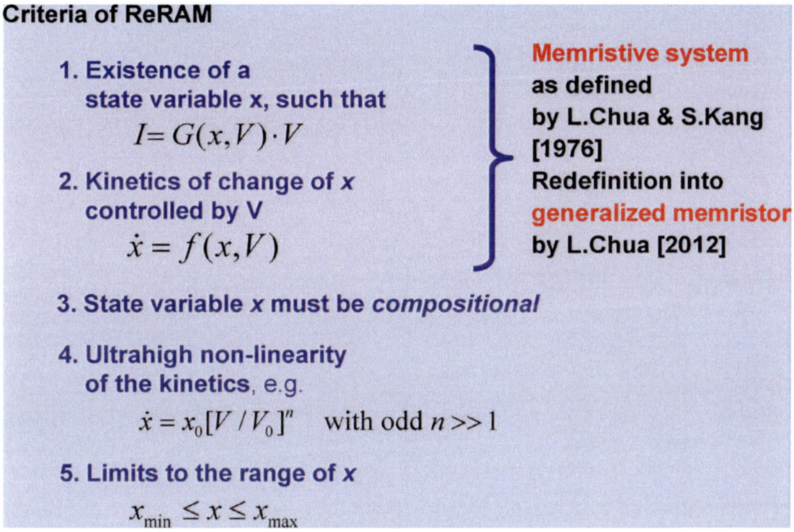

The link between device theory and physics is established through the definition of a state variable x, where the current-voltage relationship is governed by its kinetics controlled by the applied voltage. This framework, pioneered by Chua and Kang in 1976 and further refined into the generalized memristor concept by Chua in 2012, elucidates the ultrahigh non-linearity of kinetics and sets boundaries on the range of x, facilitating a deeper understanding of memristive systems at the intersection of theory and physics.

5 PITFALLS AND WISHFUL THINKING

70 Mystery #1: Charge-induced Insulator Metal Transitions ?

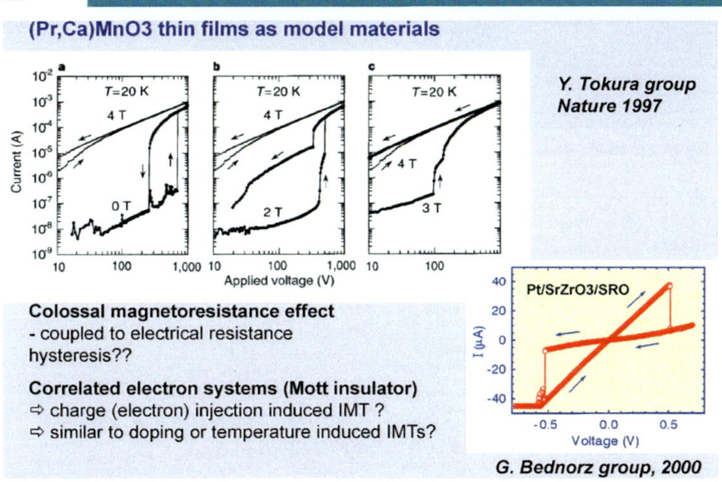

It explains the mystery of charge-induced insulator-metal transitions (IMTs) in (Pr,Ca)MnO3 thin films, characterized by their colossal magnetoresistance effect and correlated electron systems resembling Mott insulators, prompts investigation into whether electron injection triggers IMTs akin to doping or temperature-induced transitions. Understanding this phenomenon could offer insights into manipulating material properties for advanced electronic applications, shedding light on the interplay between charge dynamics and phase transitions in complex materials.

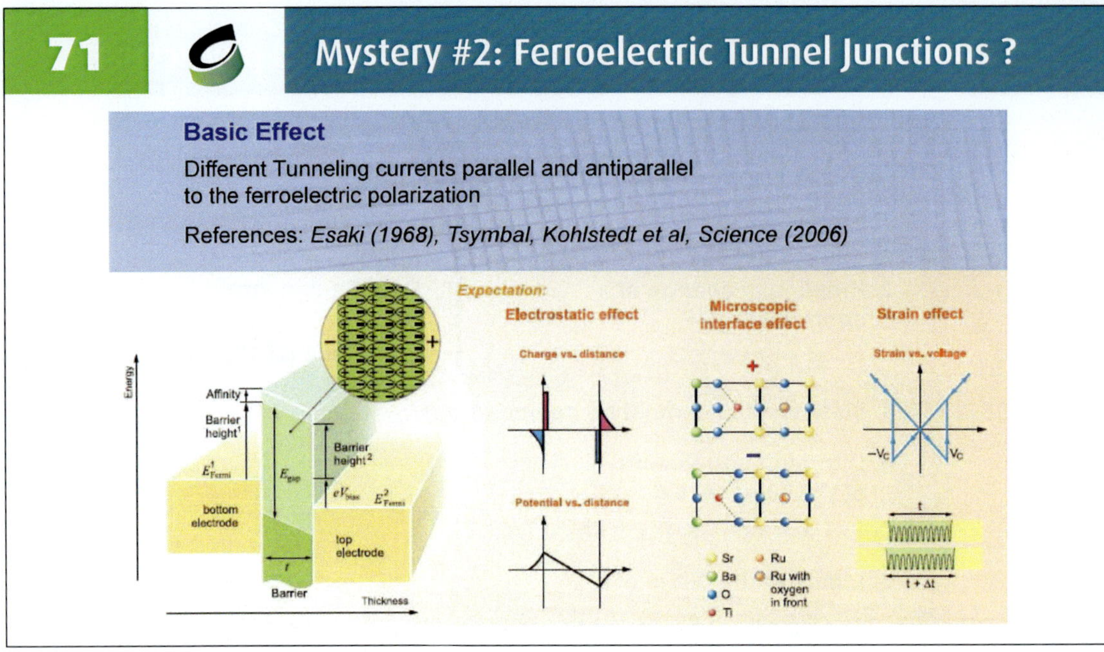

The mystery of ferroelectric tunnel junctions centers on the observation of distinct tunneling currents aligned parallel and antiparallel to the ferroelectric polarization. This phenomenon, first noted by Esaki in 1968 and extensively explored by researchers like Tsymbal, Kohlstedt et al., in 2006, raises questions about the underlying mechanisms governing electron transport in these unique nanostructures, holding promise for novel applications in data storage and computing.

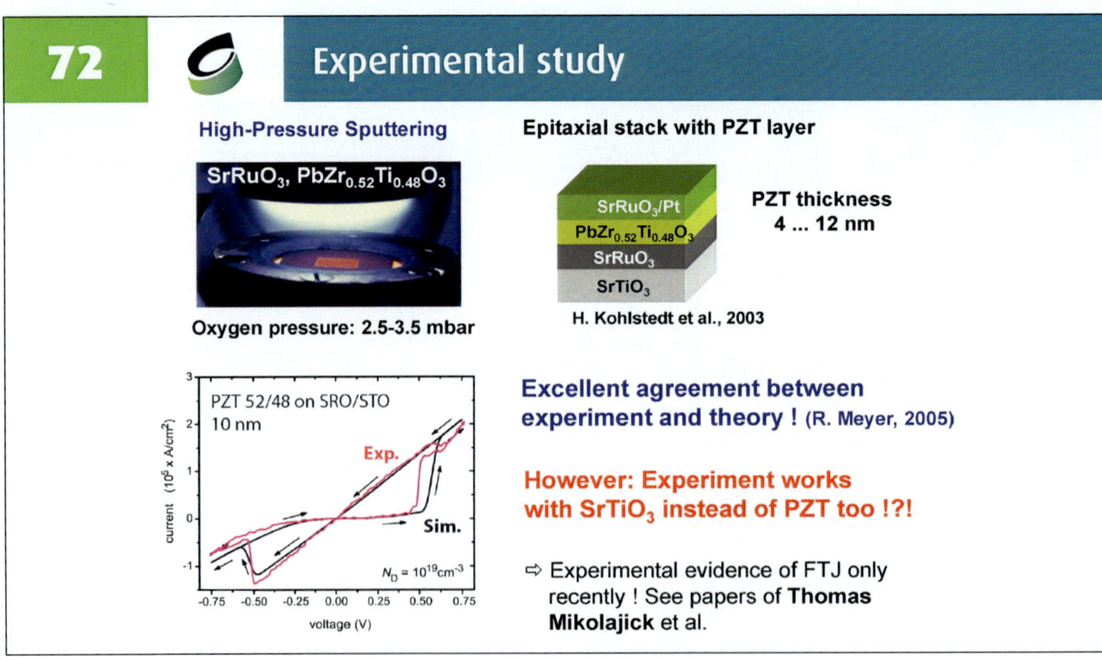

High-pressure sputtering, notably at oxygen pressures of 2.5-3.5 mbar as observed by H. Kohlstedt et al. in 2003, has enabled the experimental realization of ferroelectric tunnel junctions (FTJs) in materials such as SrRu03, SrTi03, and PbZr0.52Ti0.4803. Recent studies by Thomas Mikolajick et al. have provided experimental evidence for FTJs, with SrTiO3 demonstrating functionality comparable to PbZr0.52Ti0.4803, supported by excellent agreement between experimental results and theoretical predictions by R. Meyer in 2005 for epitaxial stacks incorporating PZT layers of 4 to 12 nm thickness.

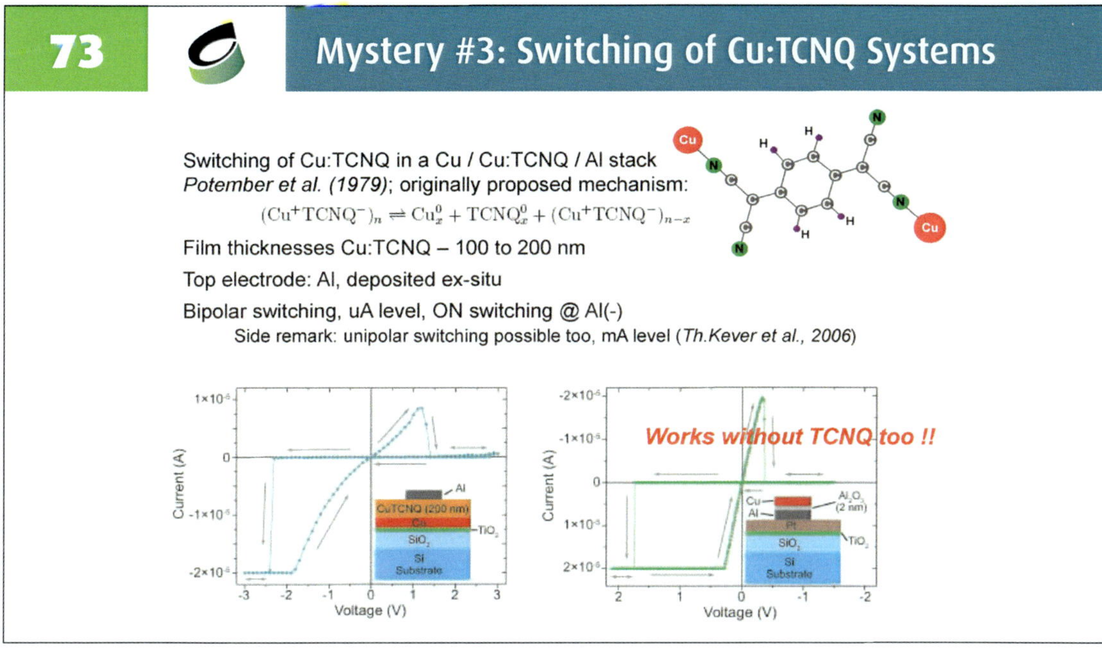

The mystery surrounding the switching of Cu:TCNQ systems, as investigated by Potember et al. in 1979, involves bipolar switching at the microampere level with Cu:TCNQ films of 100 to 200 nm thickness sandwiched between Cu and Al electrodes. Interestingly, unipolar switching, albeit at the milliampere level, has also been observed by Th. Kever et al. in 2006, suggesting potential variations in the underlying mechanisms or material configurations.

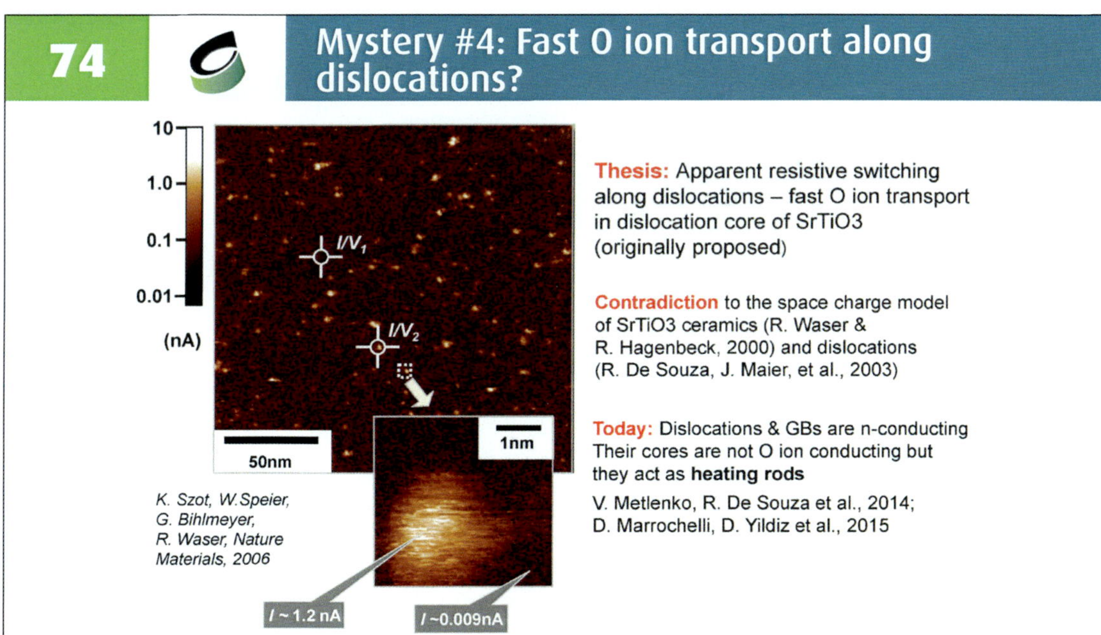

The mystery of apparent resistive switching along dislocations in SrTiO3, proposed by Szot et al. in 2006, initially suggested fast oxygen ion transport within the dislocation core, but it contradicts the space charge model of SrTiO3 ceramics. Recent studies by Metlenko, De Souza et al. in 2014 and Marrochelli, Yildiz et al. in 2015 indicate that dislocations and grain boundaries in SrTiO3 are electron-conducting, with their cores acting as heating elements rather than facilitating oxygen ion transport.

75 Operation mode of memristive elements

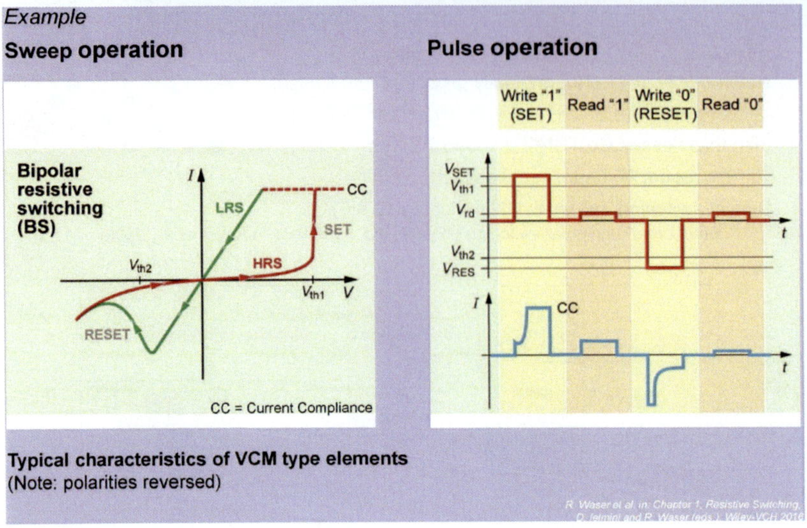

The operation modes of memristive elements, as outlined by R. Waser et al. in Wiley-VCH's 2016 publication, encompass sweep and pulse operations. In both modes, the typical characteristics of Voltage-Controlled Memory (VCM) type elements exhibit reversible changes in resistance, with polarities reversed depending on the direction of operation.

76 Operation modes of memristive elements

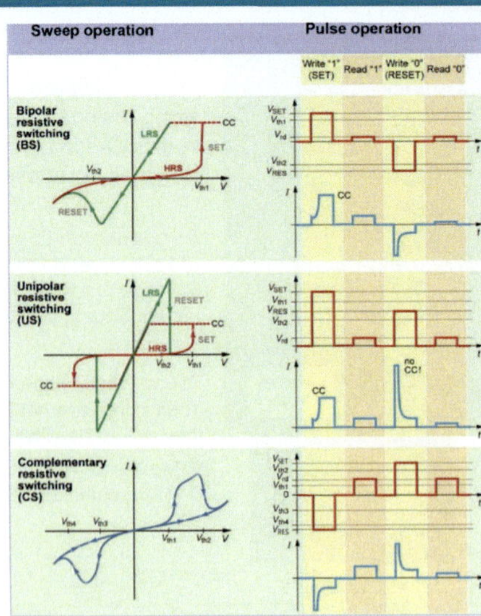

How memristors, devices with adjustable resistance, can be operated. Sweep mode uses gradual voltage changes for testing, while pulse mode uses quick voltage bursts for memory applications.

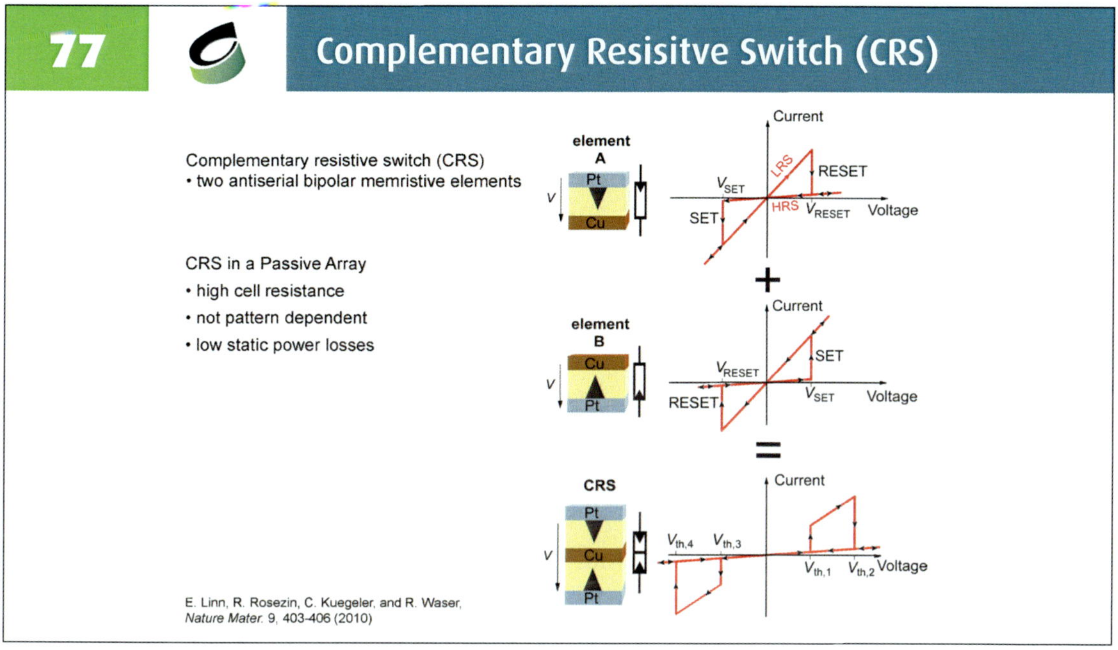

Complementary resistive switch (CRS) circuit in a passive array. with diagram it illustrates the voltage and current relationships between the two memristive elements that make up the CRS.

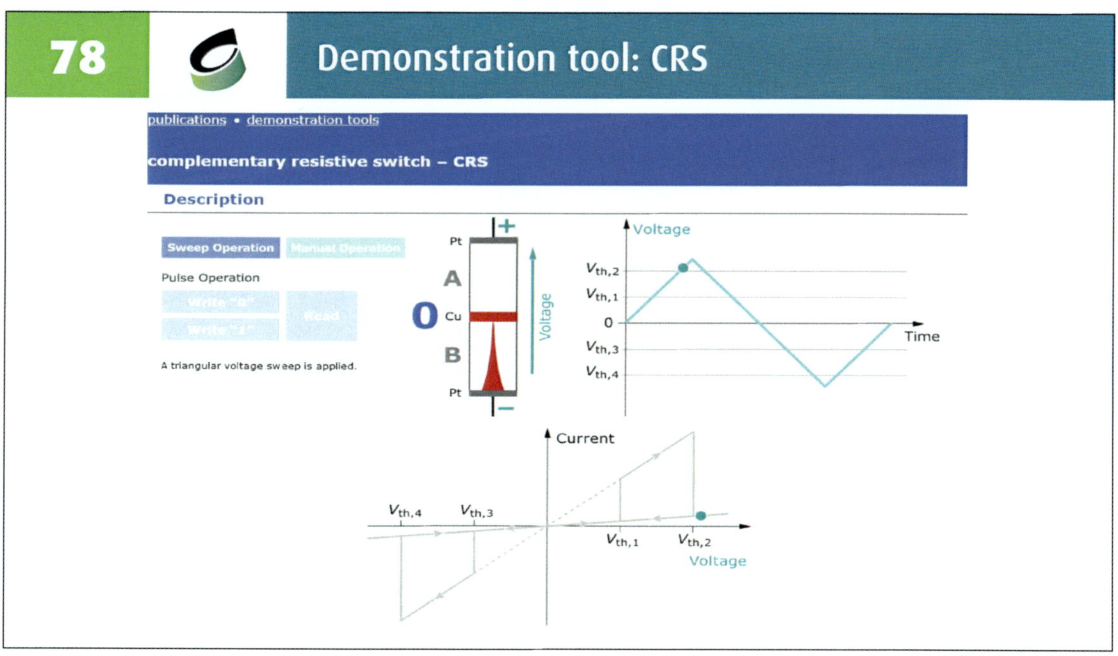

The CRS is a demonstration tool that can be used to regulate the voltage and current of a circuit. It can be operated in three different modes: sweep, pulse, and manual.

79 Architectures & Applications

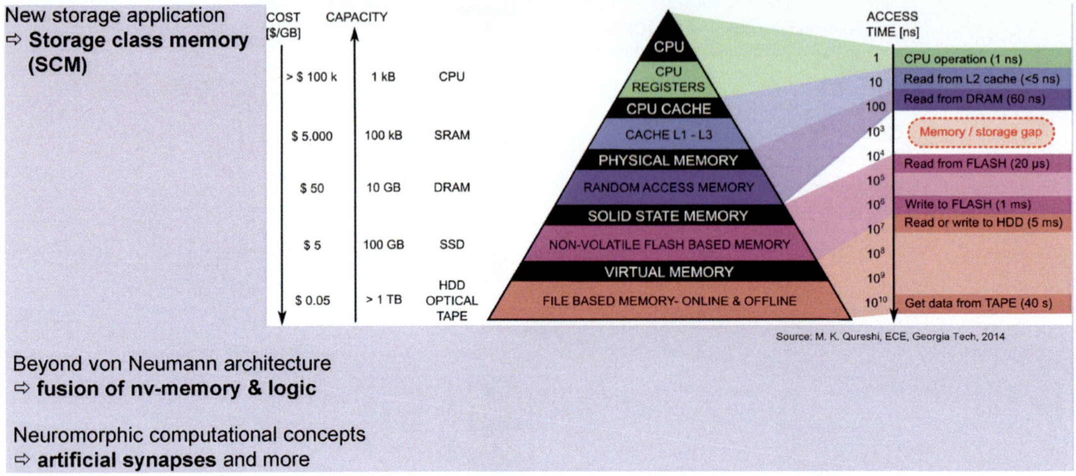

In this slide hierarchy of storage classes and applications, with increasing capacity and decreasing access time as you move down the pyramid is shown. Storage class memory (SCM) is the fastest and most expensive tier, while hard disk drives (HDD) are the slowest and least expensive tier.

80 Ultra-nonlinear switching kinetics ("Voltage time dilemma")

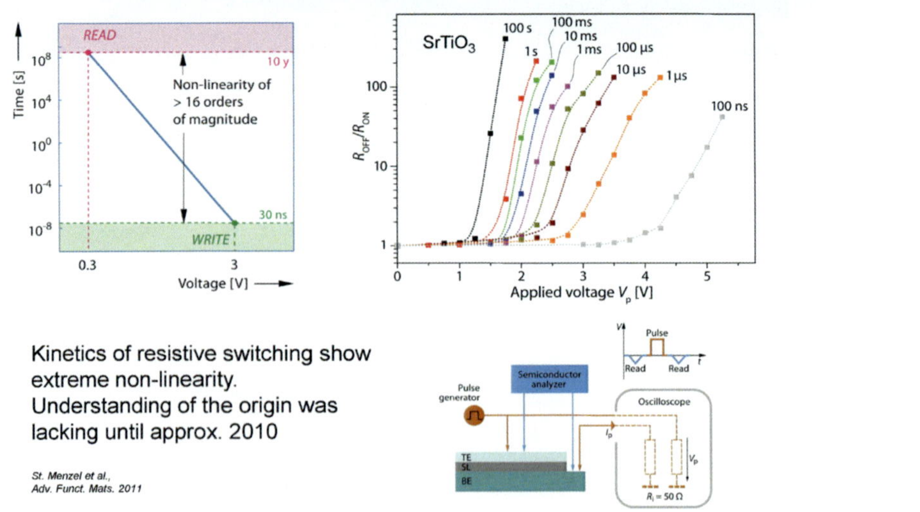

Diagram shows ultrafast non-linear switching kinetics of a resistive switching device, where the write and read times can vary over 16 orders of magnitude depending on the applied voltage.

81 Temperature- and field-enhanced ion mobility

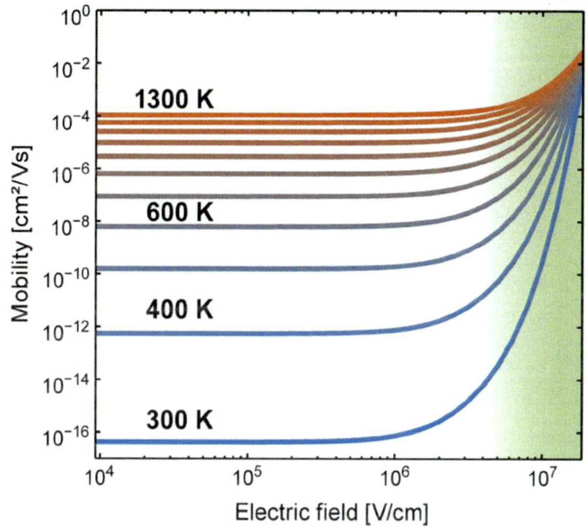

A scientific graph about the effect of temperature on ion mobility.

82 Temperature- and field-enhanced ion mobility

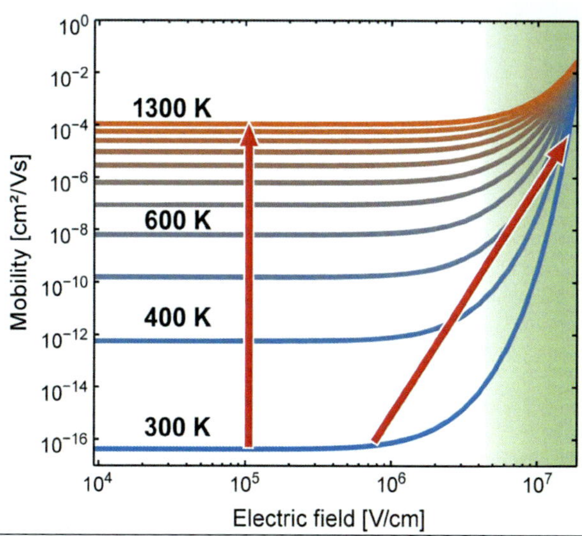

Ion mobility increases with electric field at all temperatures, and shows a stronger dependence at higher temperatures (300 K to 1300 K).

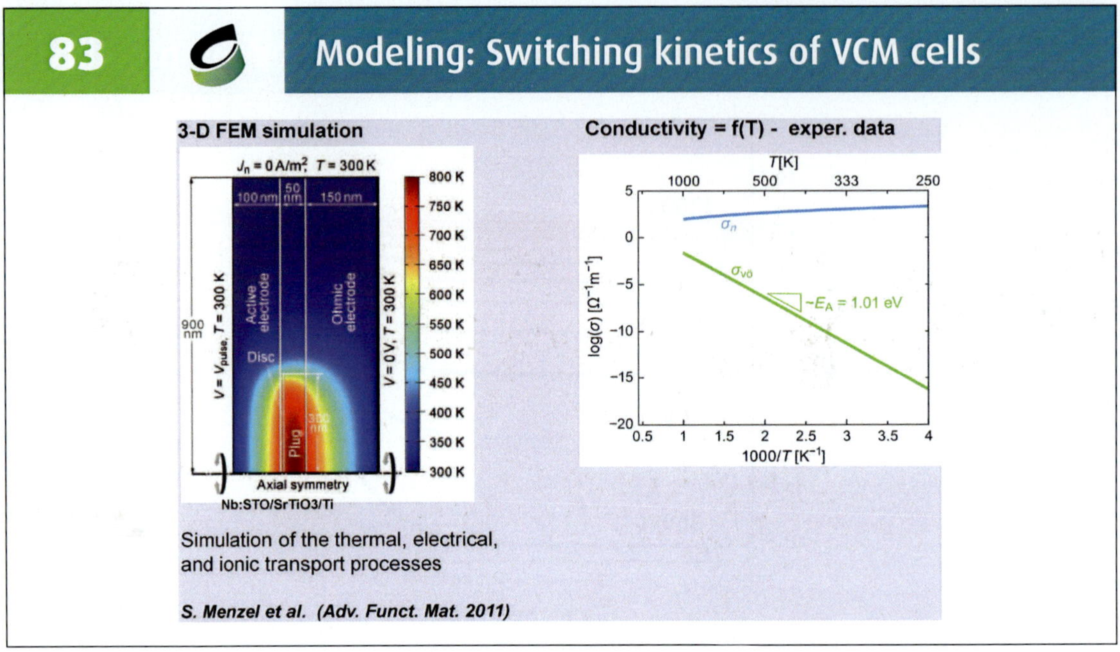

The presentation discussed a 3D FEM simulation of conductivity in VCM cells, finding a strong dependence on temperature.

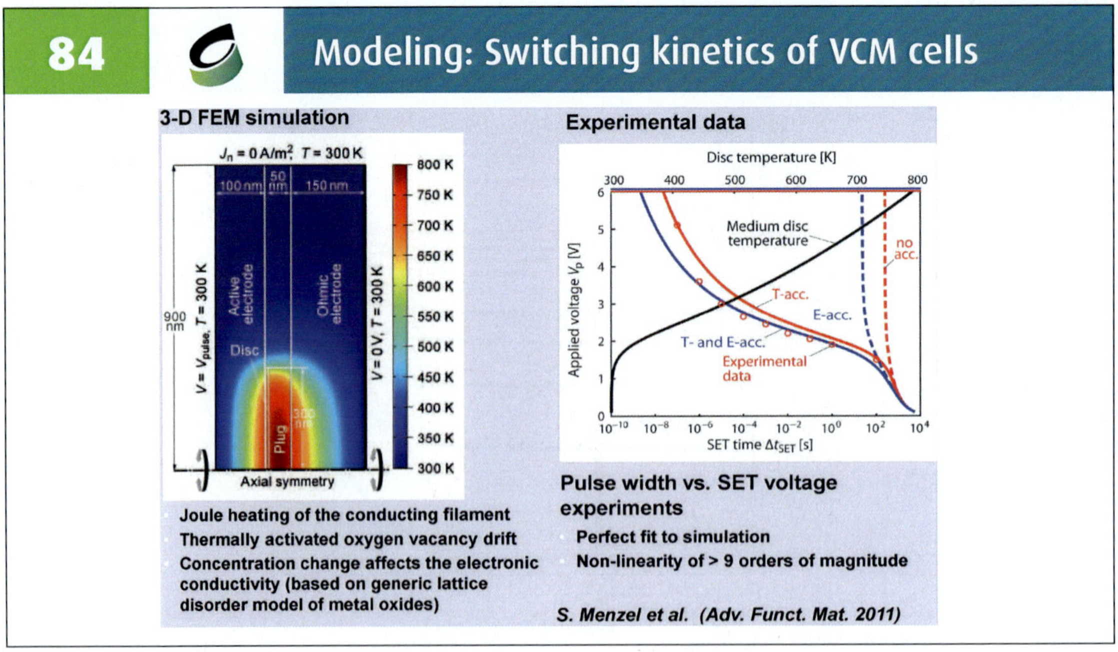

The discussion focused on modeling the switching kinetics of VCM cells, which showed good agreement between simulated and experimental data for conductivity changes due to oxygen vacancy concentration.

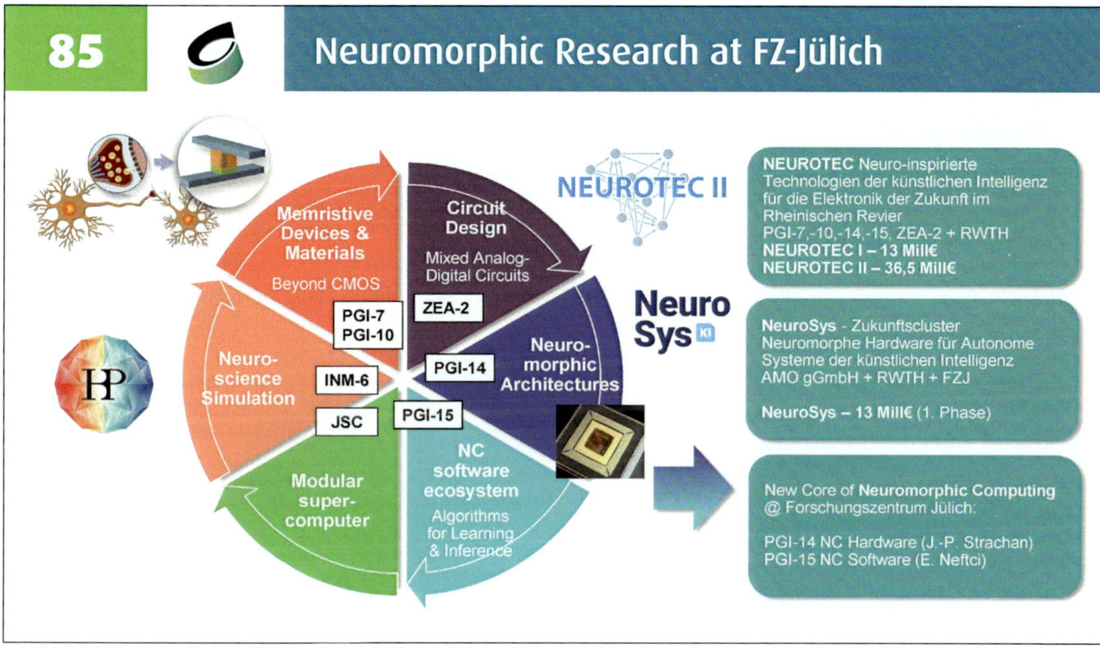

FZJ is conducting research on Neuromorphic computing, including memristive devices and circuits, mixed analog-digital circuits, and neuromorphic architectures, in collaboration with RWTH Aachen University and AMO GmbH.

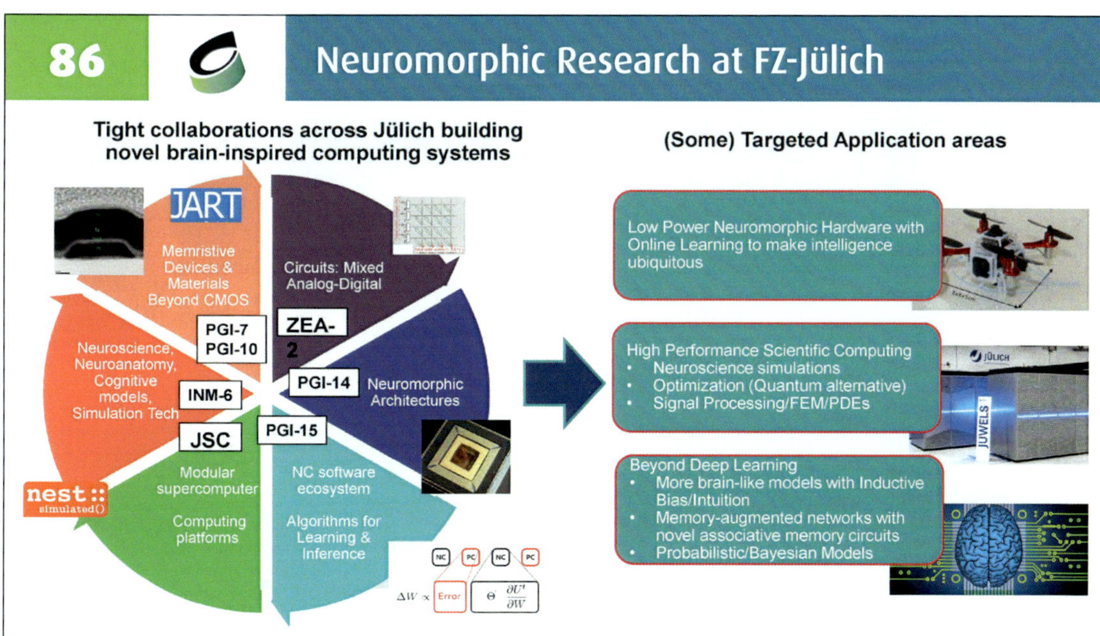

To develop novel brain-inspired computing systems through collaborations in neuroscience, cognitive models, and high-performance scientific computing.

87 Details of the forming process: Initial situation

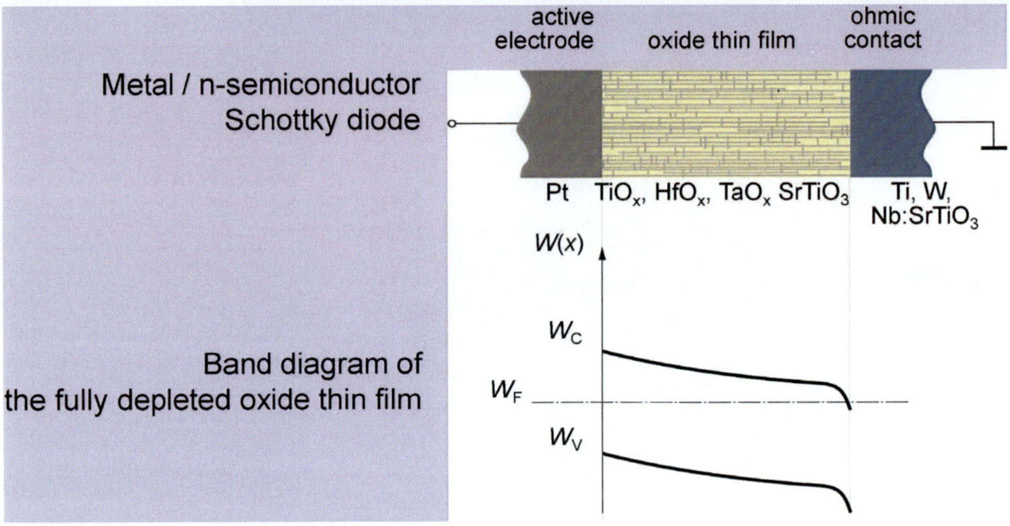

The initial stage of the forming process involves an oxide thin film and an ohmic contact.

88 Details of the forming process: electronic process

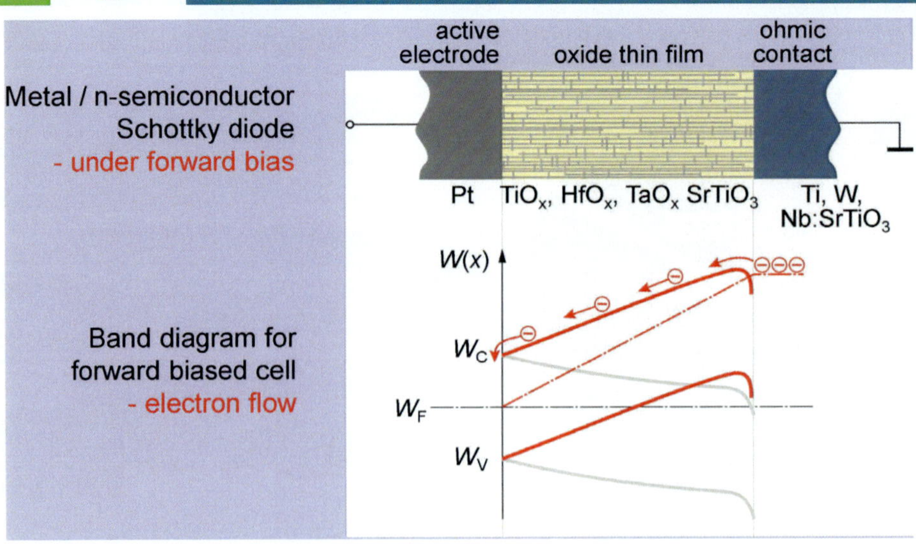

It discussed MIS devices, focusing on the formation of a Schottky diode at the interface between the metal and the n-type semiconductor.

89 Details of the forming process: ionic process

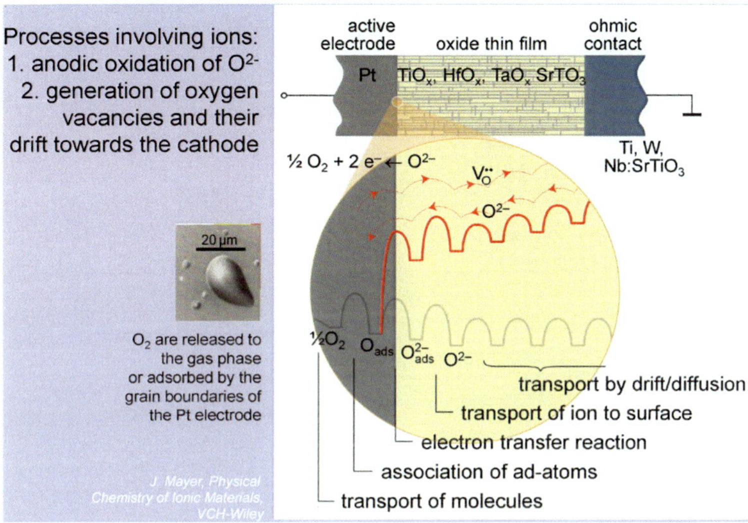

The discussion centered on the formation of an ionic oxide thin film through anodic oxidation and its impact on oxygen vacancy creation and transport.

90 Details of the forming process: overall process

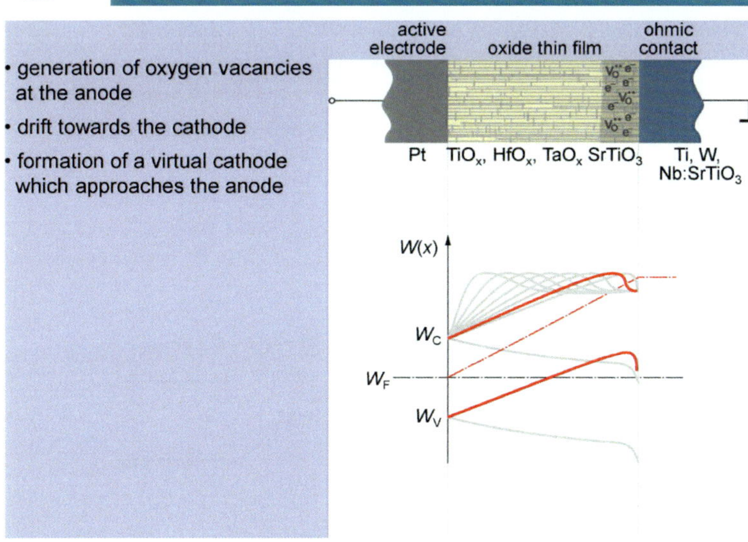

It discussed the overall process of forming a CRS device,Involves generation of oxygen vacancies at the anode and formation of a virtual cathode that approaches the anode.

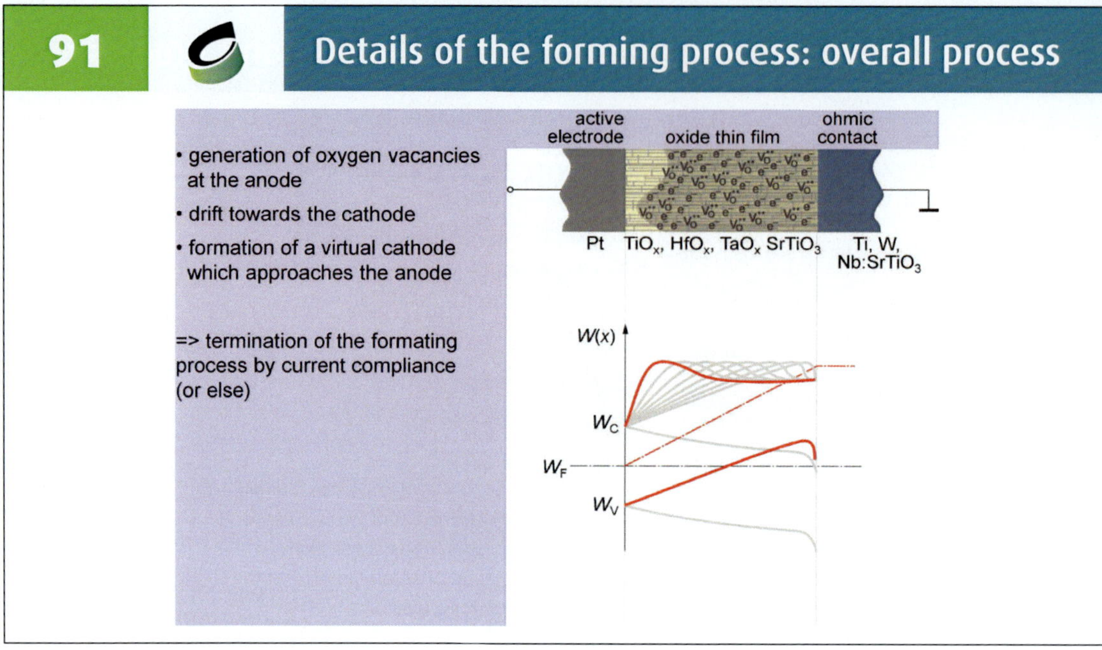

The discussion focused on the details of the forming process for an oxide thin film, which involve generation of oxygen vacancies at the anode and their migration towards the cathode.

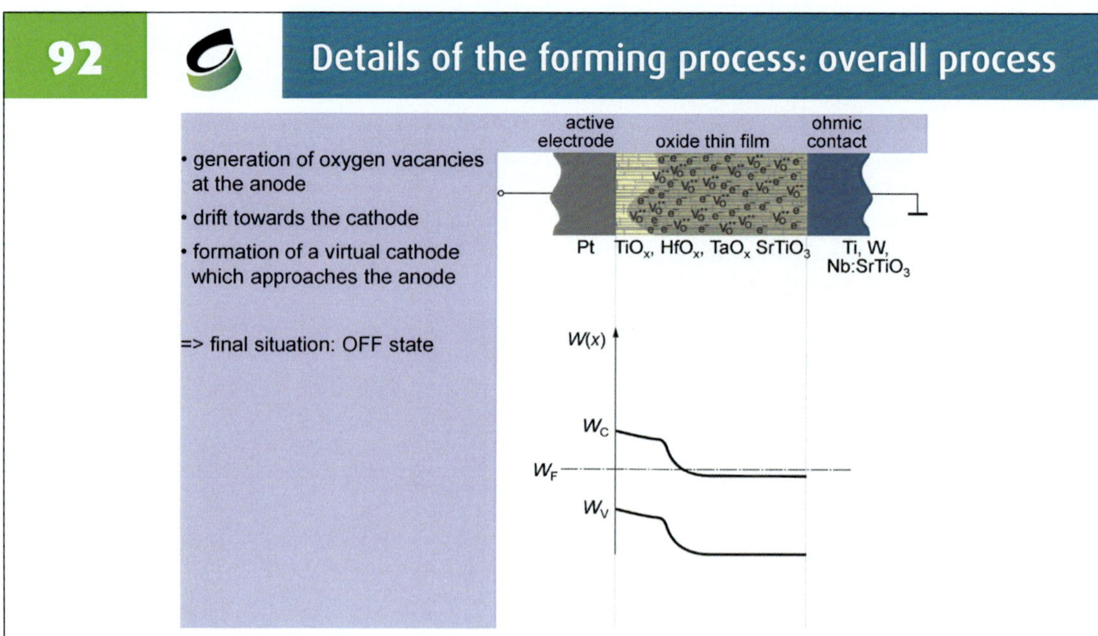

The presentation discussed the mechanism behind the formation of oxygen vacancies at the anode and their migration toward the cathode during the forming process of a resistive switching device.

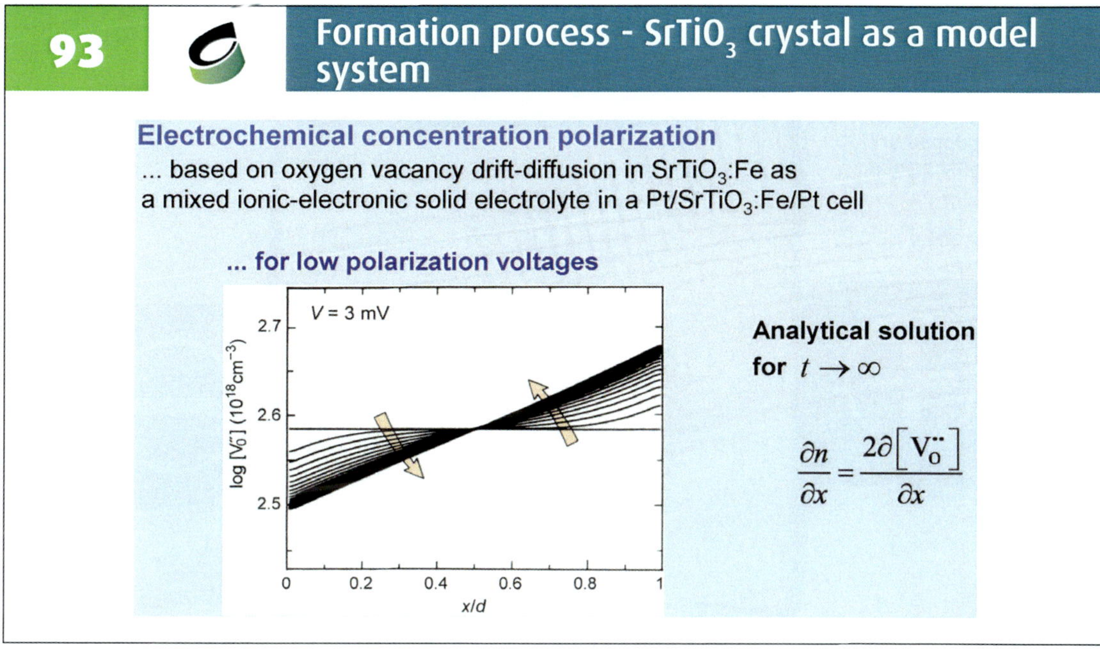

The presentation focused on electrochemical concentration polarization based on oxygen vacancy drift-diffusion in a strontium titanate (SrTiO3) crystal, used as a model system for a mixed ionic-electronic solid electrolyte.

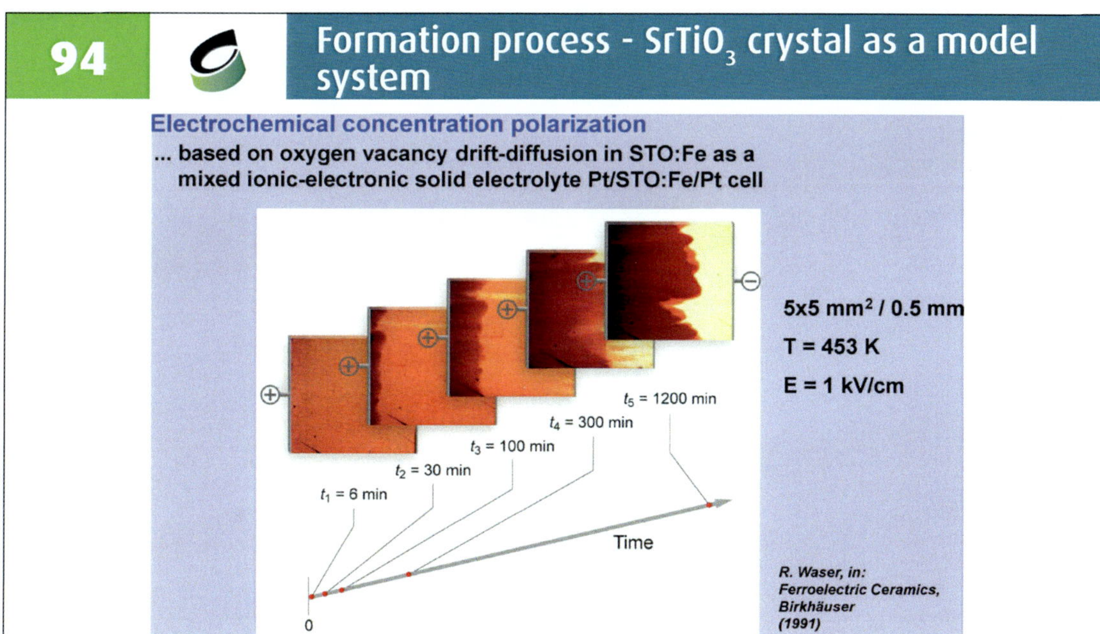

The presentation investigated the formation of SrTiO3 crystals as a model system for studying electrochemical concentration polarization caused by oxygen vacancy drift-diffusion.

95 Electrocoloration – the formation process made visible

A presentation on electrocoloration, where a voltage was applied to a single crystal (Fe-doped SrTiO3) to observe the formation process visualized by a color change (+ for formation, - for not forming).

96 Memristive devices

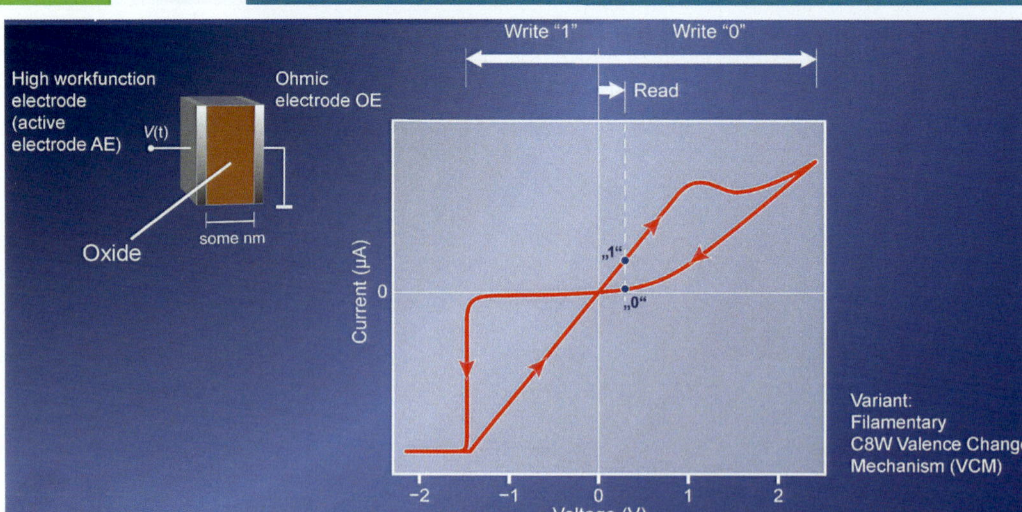

This presentation depicts a diagram illustrating the principle of memristive devices. The diagram shows a current vs voltage curve, where a high resistance state (HRS) and a low resistance state (LRS) are achieved by the formation and rupture of a conductive filament.

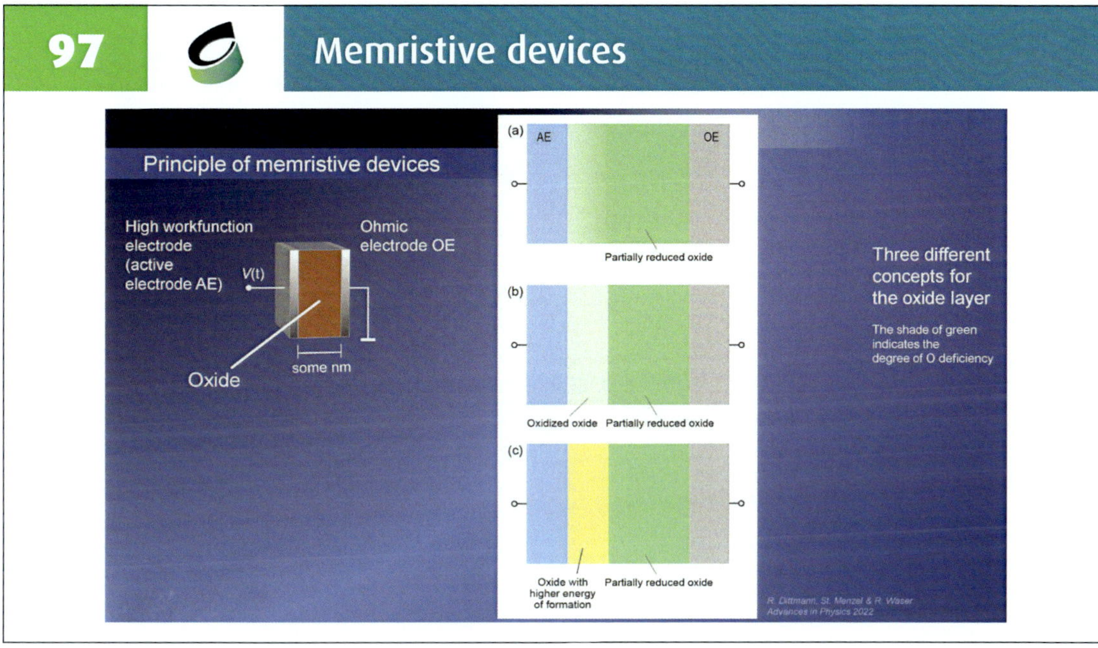

The discussion centered on the three different concepts for the oxide layer in memristive devices: an amorphous stoichiometric oxide, a partially reduced oxide, and an oxide with a higher energy of formation.

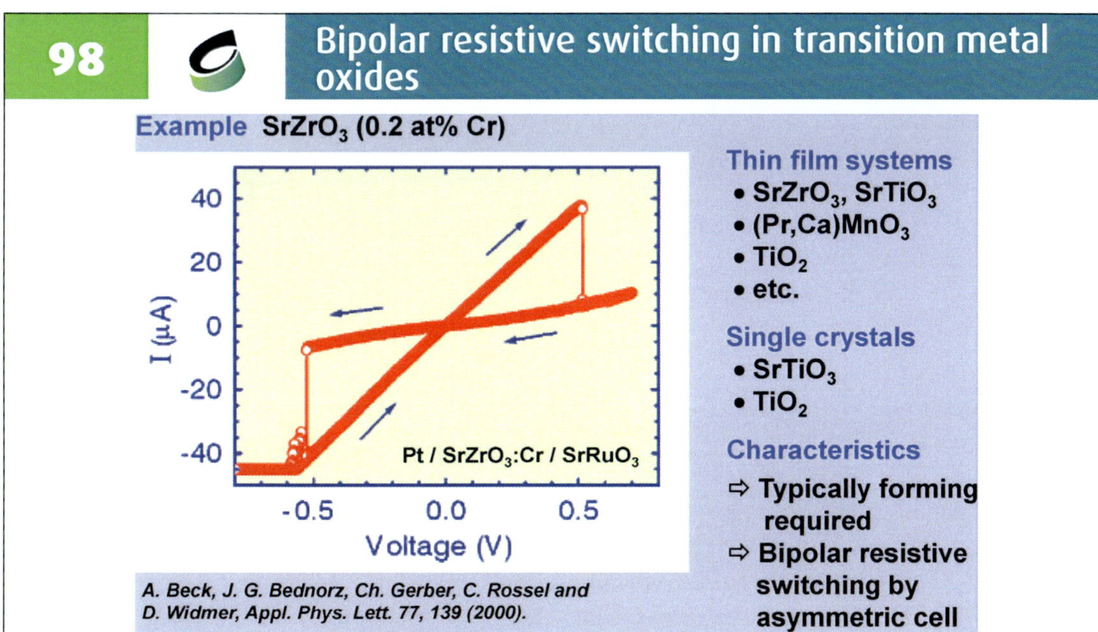

The discussion focused on bipolar resistive switching in transition metal oxides, noting that strontium zirconate (SrZrO3) doped with chromium (0.2%) is a typical example.

99 NEUROTEC I+II und NeuroSys

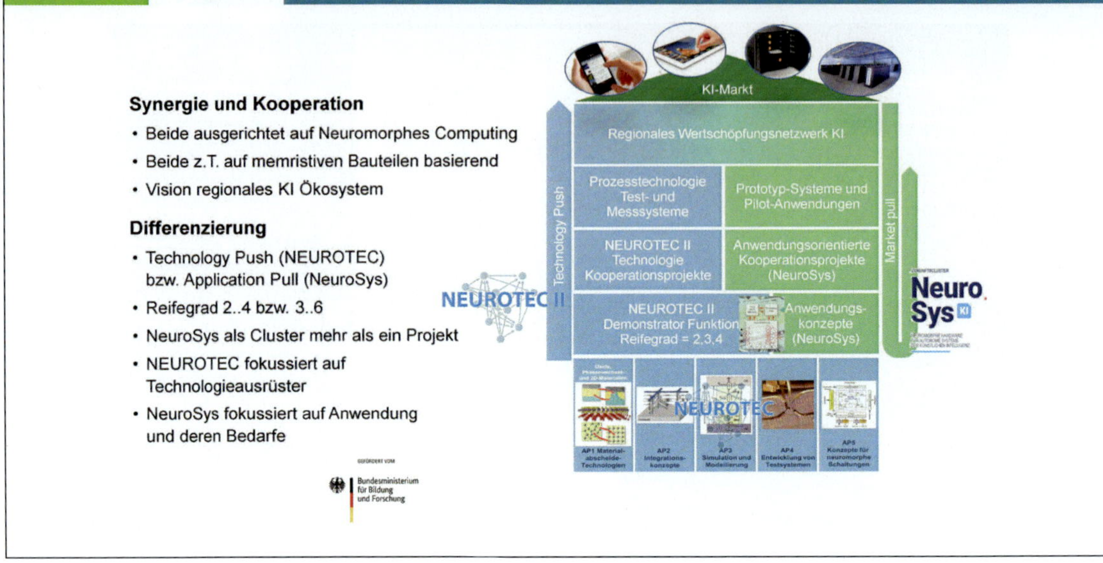

The discussion centered on NEUROTEC I+II and NeuroSys projects, which collaborate on developing a regional ecosystem for neuromorphic computing with a focus on memristive devices.

100 Observation of two VCM switching polarities

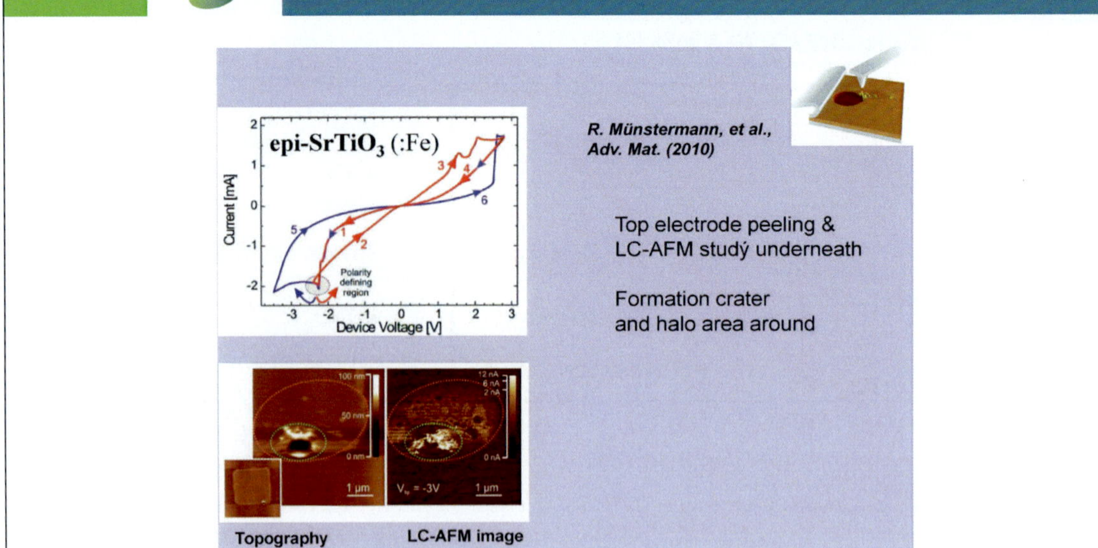

The presentation discussed observing two different polarities for resistive switching in epitaxial strontium titanate (SrTiO3) thin films grown on dysprosium scandate (DyScO3) substrates.

101 Observation of two VCM switching polarities

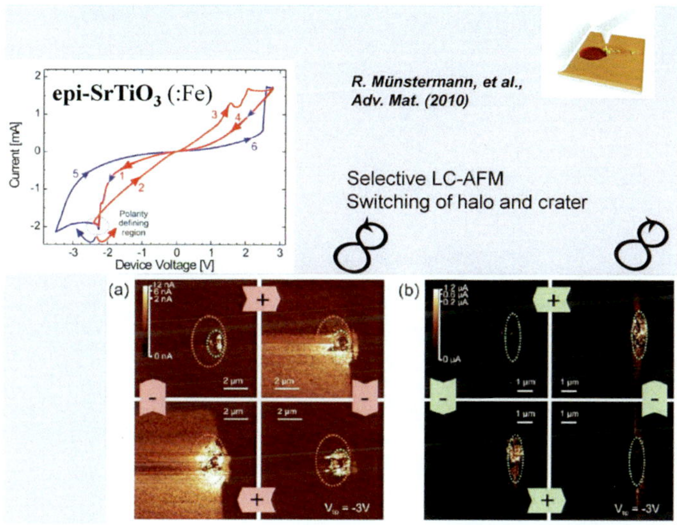

This presentation shows the observation of two distinct polarities for voltage control of magnetic states (LORA and LOVA) in epitaxial strontium titanate (SrTiO3) thin films.

102 Future of NAND Flash

NAND Flash properties:

	2011	2024
Cell area	$4F^2$	$4F^2$
Read time	100 µs	100 µs
Write time	1 ms	1 ms
Retention	10 years	10 years
Endurance (cycles)	10^4	$5*10^3$
Write operation voltage	15 V	15 V
Read operation voltage	1.8 V	1 V
Feature size 2D/3D	22 nm/-	8 nm/24 nm
MLC 2D/3D	3/-	4/2
Layers 3D	1	> 100

Cell properties do not improve much

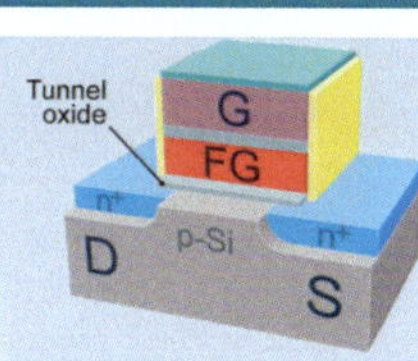

Source:
ITRS ERD 2011 / ORTC 2012

The future of NAND flash memory appears to be limited by endurance (write cycles) which haven't improved much since 2011, while other factors like cell area and read/write times have benefitted from scaling.

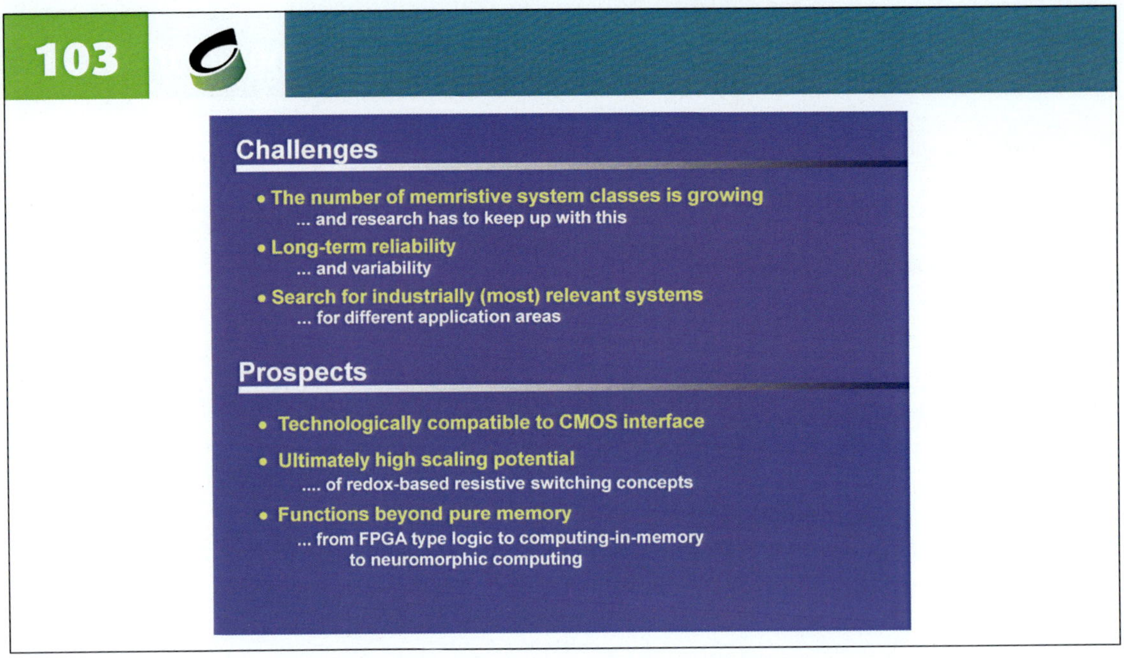

This presentation addressed the challenges facing the number of memristive system classes and how research needs to keep up with the growing number.

CHAPTER 02

Deep learning inference with in-memory computing

Abu Sebastian

IBM Research, Zurich

The emergence of in-memory computing (IMC) promises to revolutionize computational efficiency, with efforts aimed at achieving an unprecedented 1fJ/Operation energy efficiency. IMC leverages in-memory matrix-vector multiplication, particularly advantageous for deep learning inference, to address the plateau observed in existing architectures due to data transfer limitations. Recent applications of IMC span scientific computing, stochastic computing, and notably, deep learning, offering novel solutions to computational needs and data movement challenges. However, challenges such as conductance fluctuation, read noise, and temperature dependence in phase-change memory (PCM) pose obstacles to achieving precise computations. Innovative approaches, including additive noise training and nanoscale confinement of PCM materials, aim to mitigate these challenges. Additionally, the development of unique models for understanding 1/f noise in PCM and addressing interface resistance between material layers highlights ongoing research efforts. Despite these challenges, IMC shows promise in enhancing computational precision and efficiency, with potential applications across diverse fields. Experimental validations and technological advancements, including compact CMOS-based IMC cores and full-fledged accelerator architectures, underscore the transformative potential of IMC in advancing computational paradigms.

1 The dream of fJ/Op compute efficiency

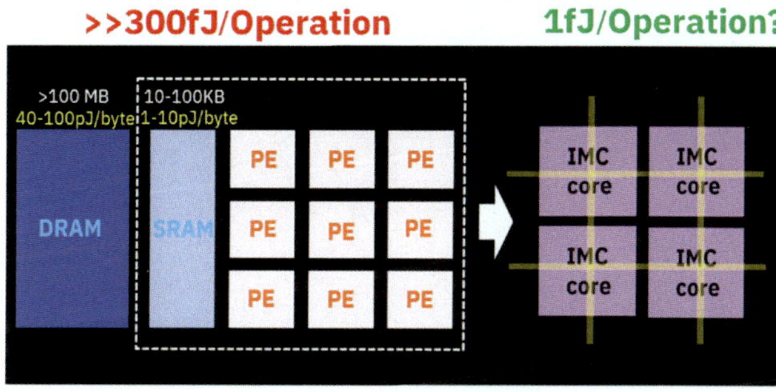

Murmann, IEEE TVLSI (2020)
Burr, Sebastian et al., IEEE Spectrum (2022)

- Improving the compute efficiency is one of the primary research goals
- Having a compute efficiency in the order of 1fJ/Operation is widely considered to be within reach but is still elusive
- Studies indicate that if we adhere to the existing computing architectures, the compute efficiency per operation is likely to plateau at the hundreds of fJ/Operation regime
- This is mostly arising from the need to shuttle data around between the memory and processing units
- One of the prominent computing approaches that is likely to keep alive the dream of fJ/Op is in-memory computing

Efforts to achieve 1fJ/Operation compute efficiency are ongoing, but existing architectures suggest a plateau around hundreds of fJ/Operation due to data transfer between memory and processing units. In-memory computing is an emerging computing paradigm that keeps alive the promise for 1fJ/Op energy efficiency.

2 In-memory computing (IMC)

- Perform "certain" computational tasks in place in memory
- Exploit the physical attributes of memory devices
- A sub-category of processing in memory (PIM) or compute in memory (CIM)
- Collective processing in memory

FOCUS | REVIEW ARTICLE

Memory devices and applications for in-memory computing

Abu Sebastian, Manuel Le Gallo, Riduan Khaddam-Aljameh and Evangelos Eleftheriou

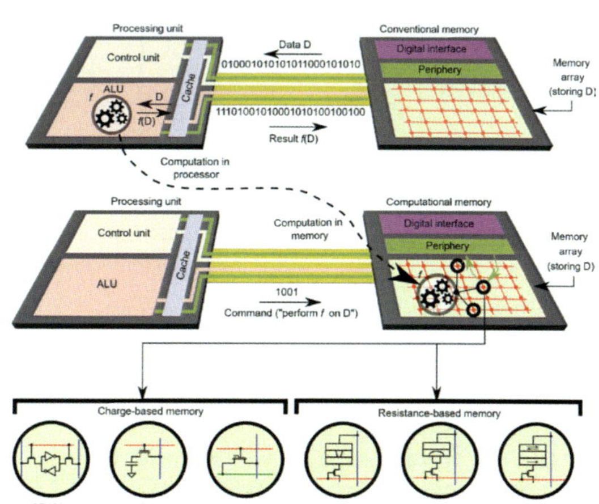

Deep learning inference is a killer application for IMC that exploits the compute primitive of in-memory matrix-vector multiplication.

3 The application landscape

Sebastian et al., Nature Nanotechnology (2020)

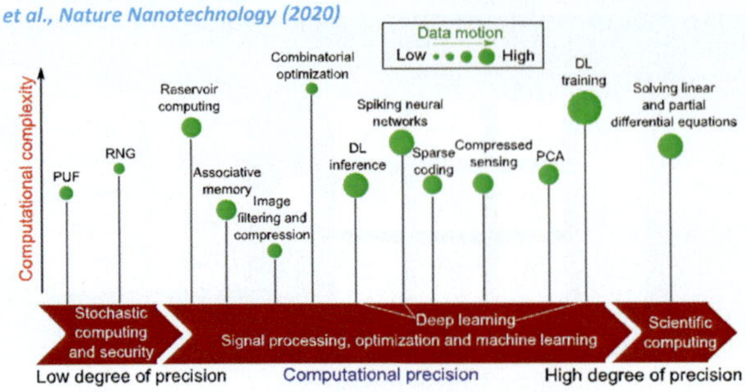

- In the last few years, in-memory computing has found applications in a wide range of areas
- These applications have varying degrees of computational complexity and requirements on data movement, both of which can be addressed by in-memory computing
- The application areas can be broadly classified in accordance with the computational precision that is required
- On one end of the spectrum, we have scientific computing where we require very high precision while on the other end, we have stochastic computing that relies on imprecision, in the middle we have areas such as SP, ML and optimization
- Within machine learning, deep learning considered to be particularly well suited for in-memory computing

In recent years, in-memory computing has been applied across diverse fields, addressing varying computational needs and data movement requirements. These applications span scientific computing, stochastic computing, and machine learning, with deep learning notably benefiting from in-memory processing.

4 In-memory matrix-vector multiplication

$$\begin{bmatrix} A_{11} & A_{12} \\ A_{21} & A_{22} \\ A_{31} & A_{32} \end{bmatrix} \begin{bmatrix} x_1 \\ x_2 \end{bmatrix} = \begin{bmatrix} b_1 \\ b_2 \\ b_3 \end{bmatrix}$$

MAP to conductance values | **MAP to read voltage** | **DECIPHER from the current**

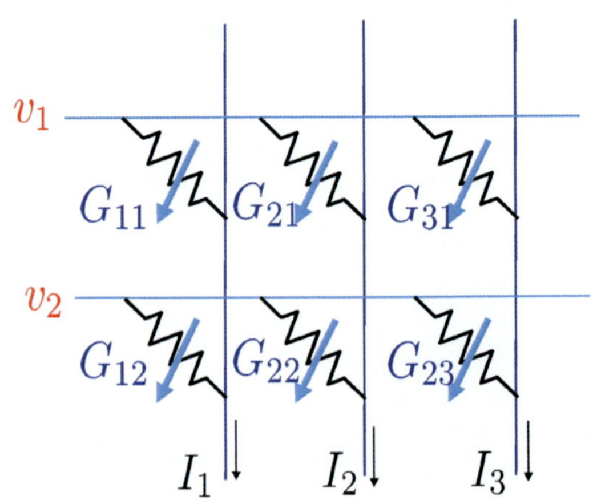

- In-place matrix-vector operation with O(1) time complexity
- Exploits analog storage capability of memristive devices and Kirchhoff's circuits laws

In-memory matrix-vector multiplication.
In-place matrix-vector operation with O(1) time complexity.
Exploits analog storage capability of memristive devices and Kirchhoff's circuits laws.

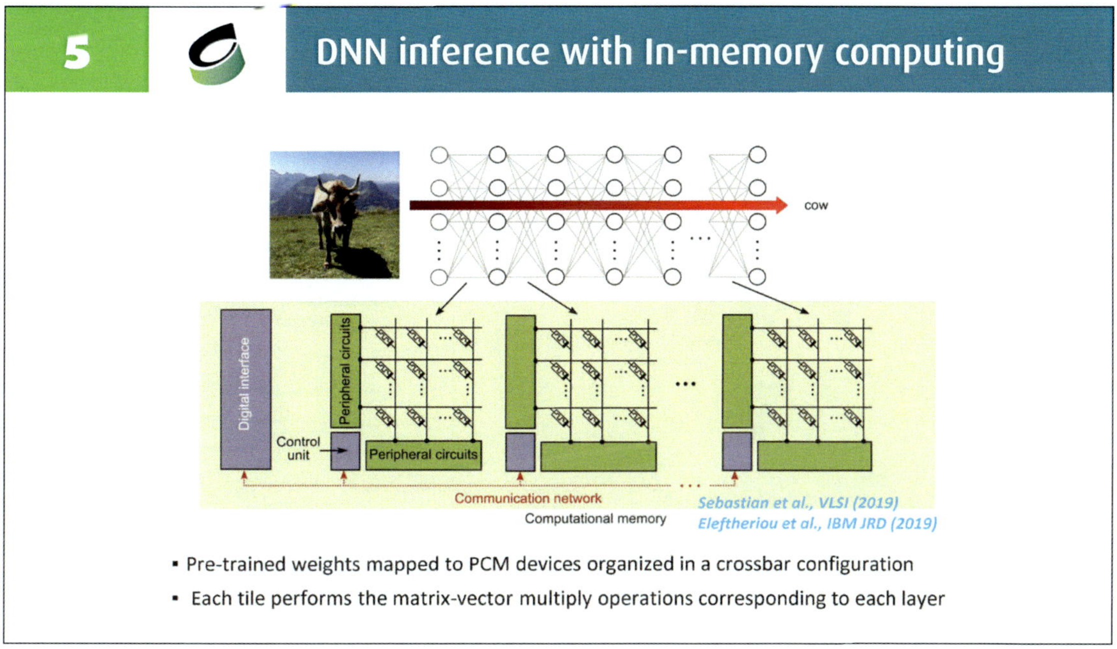

DNN inference enhanced by in-memory computing: pre-trained weights stored on PCM devices in crossbar configuration. Each tile executes matrix-vector multiplication for respective layers.

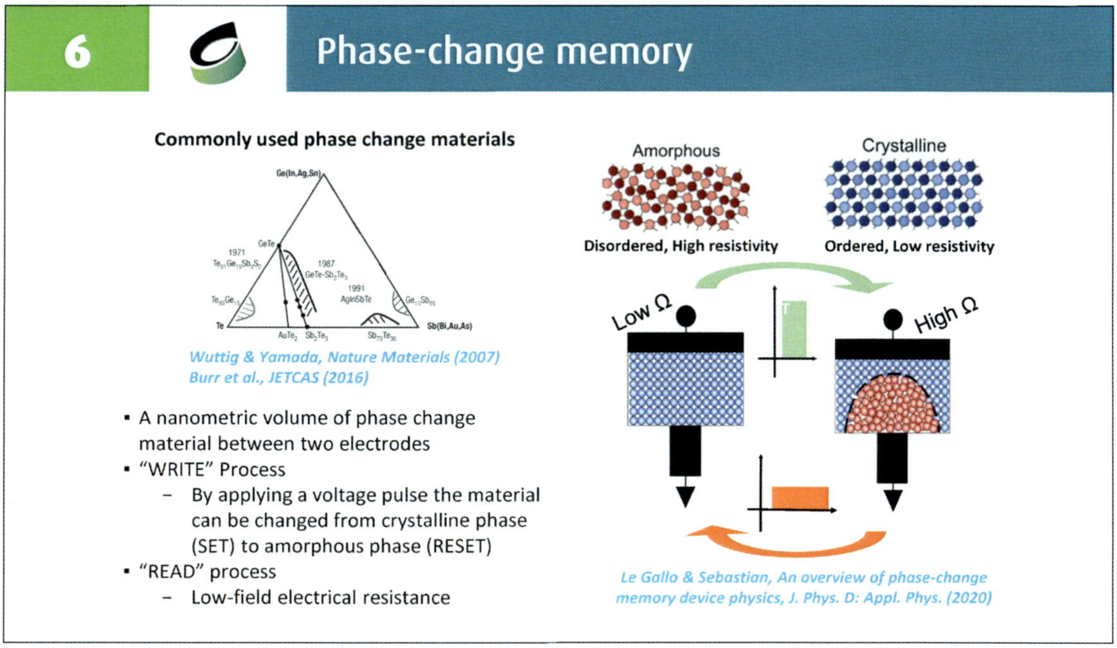

Exploring phase-change memory: Utilizing common phase change materials within nanometric volumes between electrodes. The WRITE process transitions material from crystalline (SET) to amorphous (RESET) phases via voltage pulses, while the READ process relies on low-field electrical resistance.

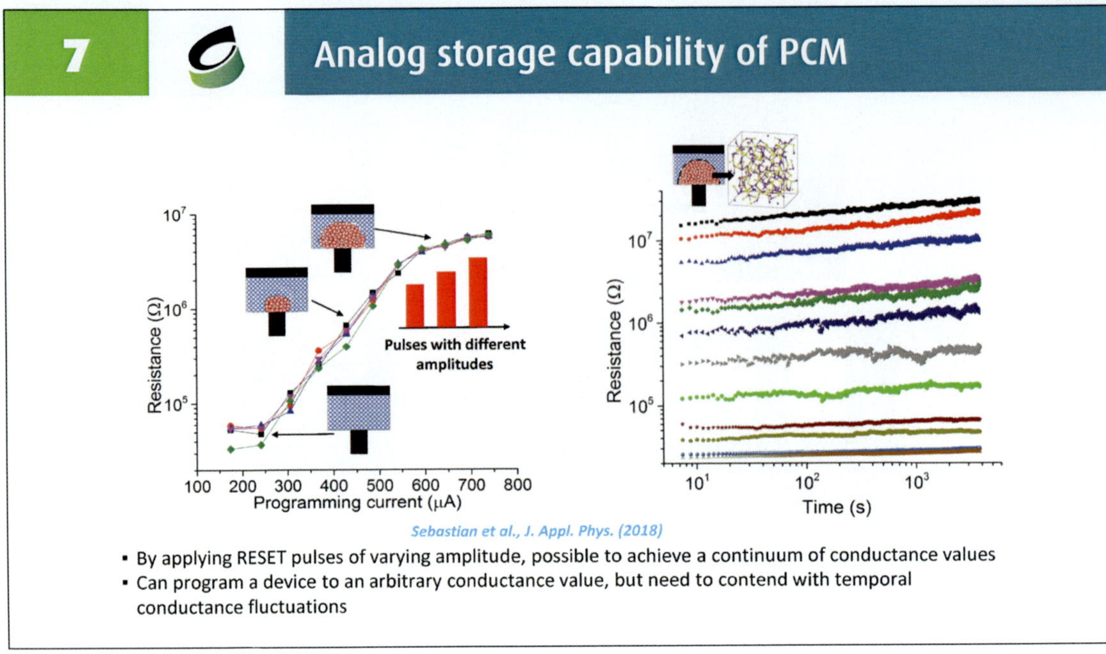

Unlocking analog storage in PCM: Varying amplitude RESET pulses enable a range of conductance values, albeit with temporal fluctuations. Programming devices to specific conductance levels necessitates the management of temporal conductance variations.

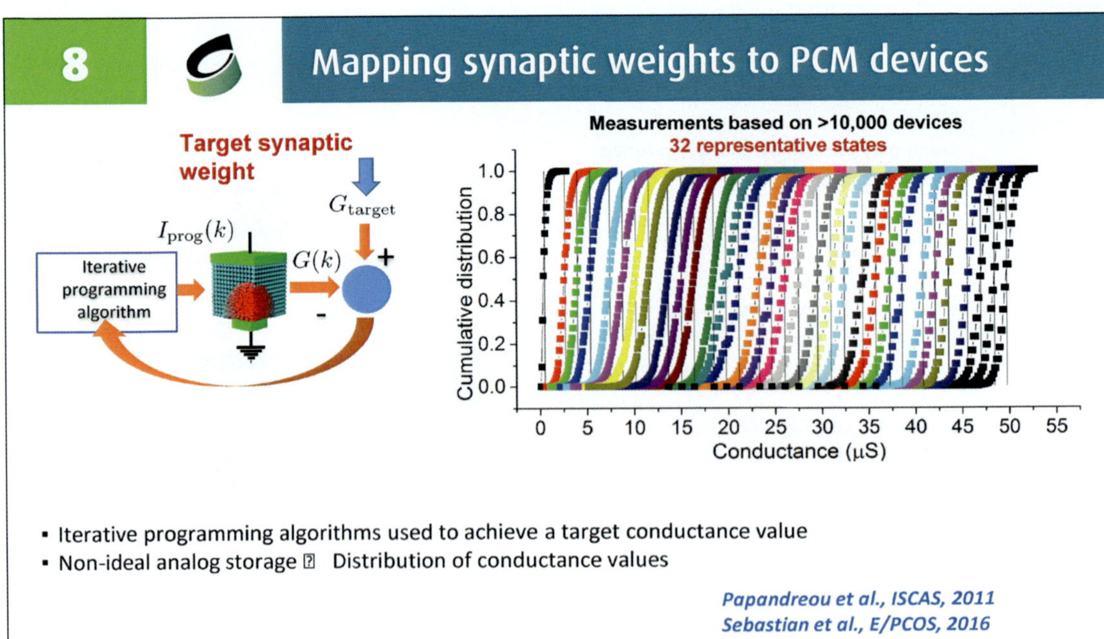

Mapping synaptic weights to PCM devices involves iterative programming for desired conductance values. Non-ideal analog storage results in conductance value distribution.

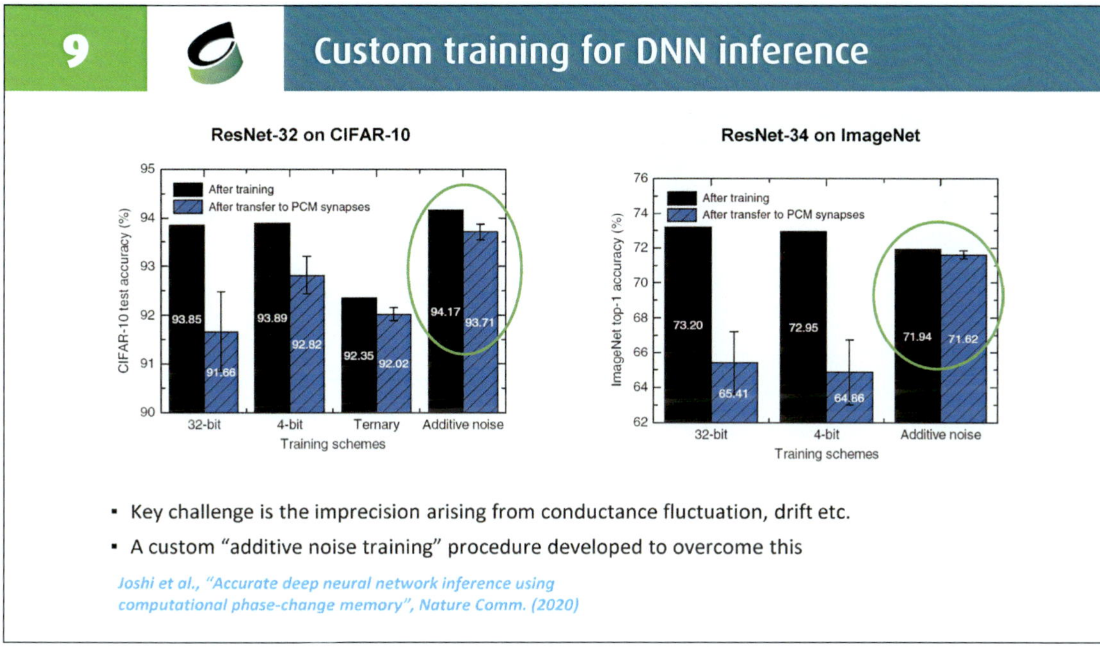

Custom training for DNN inference addresses challenges like conductance fluctuation and drift. An innovative 'additive noise training' technique is devised to mitigate these issues.

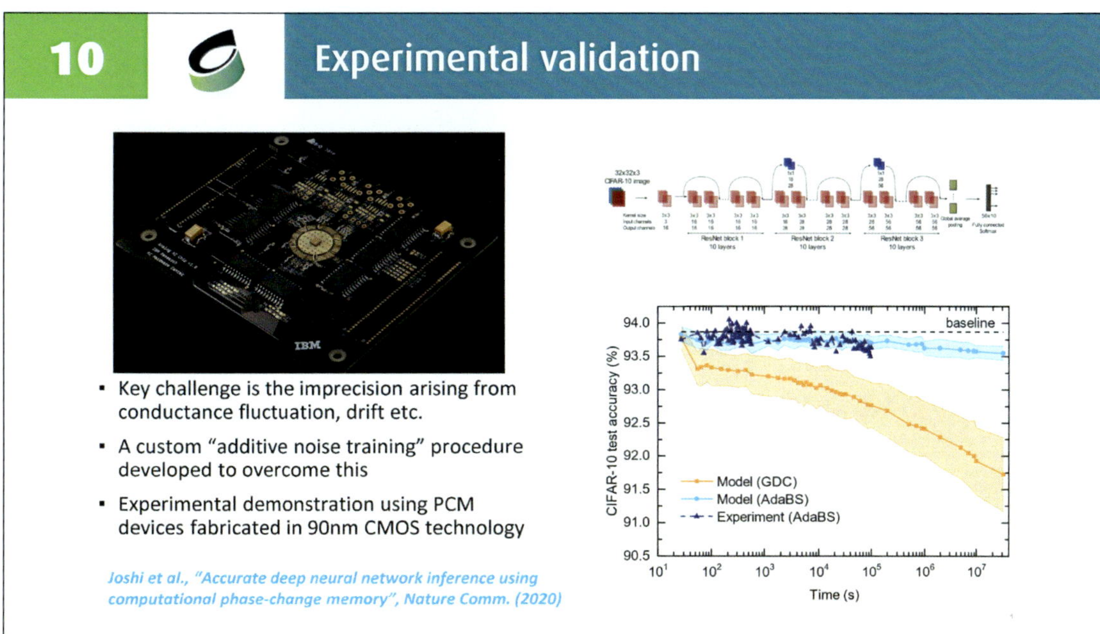

Experimental validation tackles challenges of conductance fluctuation and drift. A custom 'additive noise training' method is devised for mitigation. Demonstrations utilize PCM devices fabricated in 90nm CMOS technology.

11 IMC compute core in 14nm CMOS technology (single tile)

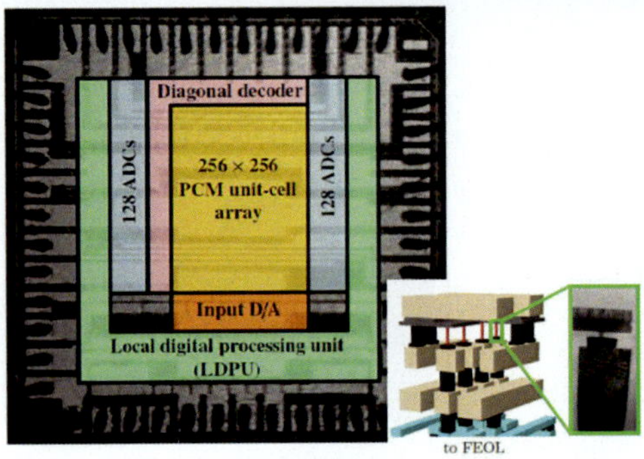

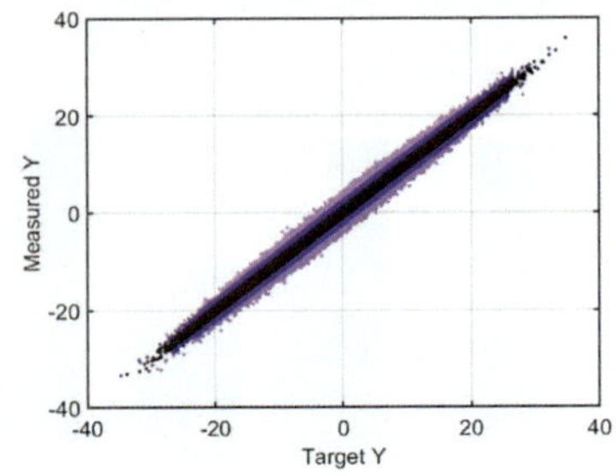

- 256x256 array of 8T4R unit cells
- PCM devices integrated in the back-end of 14nm CMOS chip
- Compact current controlled oscillator-based ADCs
- Local digital processing
- MVM performance: 10.5 TOPS/W and 1.6 TOPS/sq.mm.

Khaddam-Aljameh et al., Proc. VLSI (2021) (highlight paper)
Khaddam-Aljameh et al., JSSC (2022)

IMC compute core in 14nm CMOS technology features a 256x256 array of 8T4R unit cells with integrated PCM devices. Compact current-controlled oscillator-based ADCs and local digital processing enhance performance. Achieves 10.5 TOPS/W and 1.6 TOPS/sq.mm in matrix-vector multiplication.

12 Full-fledged IMC-based accelerator

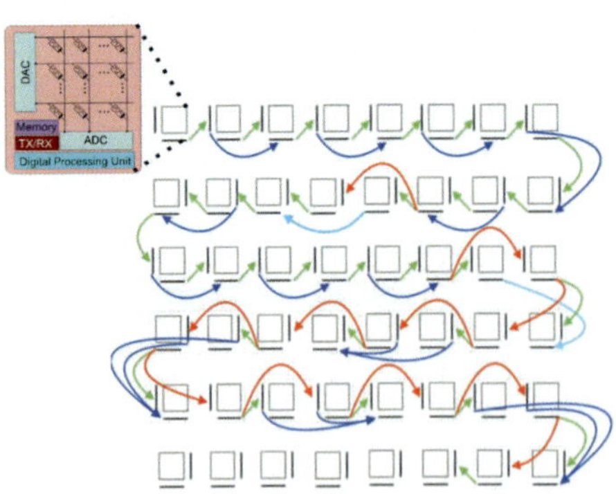

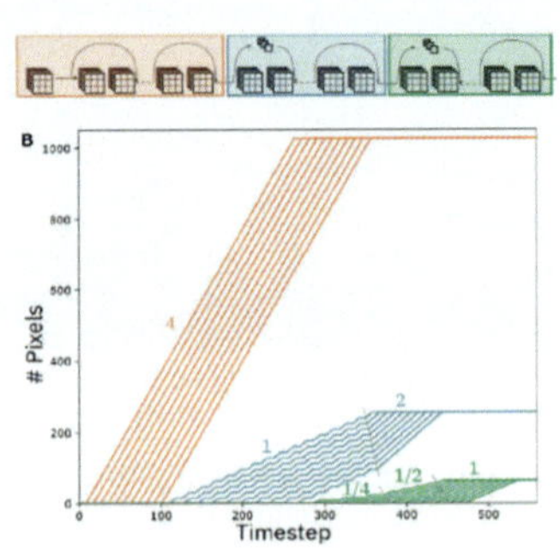

- Number of IMC tiles interconnected by a communication fabric (data-flow architecture)
- Amenable to highly spatially pipelined dataflows
- Could achieve unprecedented latency and throughput

Eleftheriou et al., IBM JRD (2019),
Dazzi et al., IEEE Trans. Computers (2021)
Dazzi et al., Front. Comp. Neuroscience (2021)

A full-fledged IMC-based accelerator features interconnected tiles via data-flow architecture. Enables highly spatially pipelined dataflows, leading to unprecedented latency and throughput.

13 Compute density

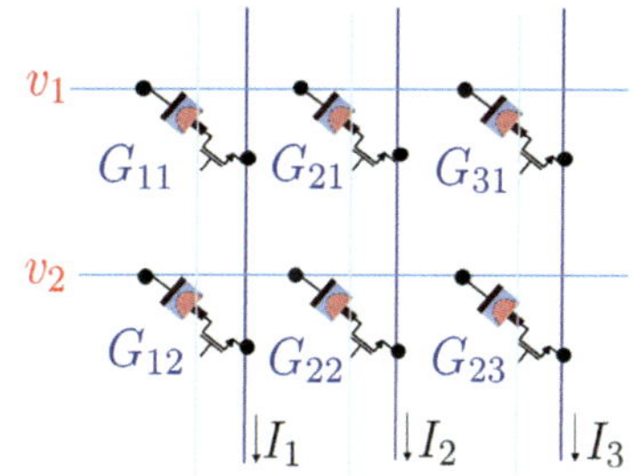

- Unprecedented parallelism at very low areal footprint
- Need to scale both the devices and the access transistors
- Key challenges
  - **High programming current**
  - **High threshold switching voltage**

Compute density highlights unprecedented parallelism with a minimal footprint. Challenges include scaling devices and access transistors due to high programming current and threshold switching voltage.

14 Programming current/energy

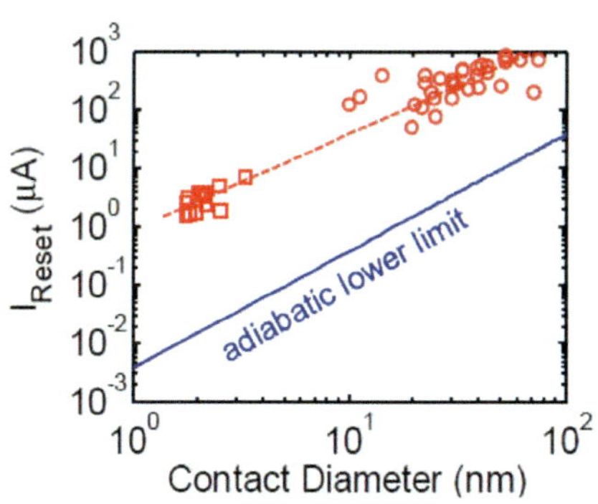

Xiong et al., IEDM (2016)

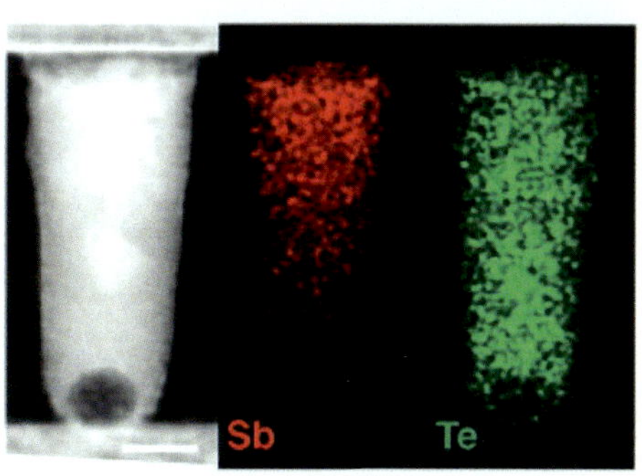

Xie et al., Adv. Mat. (2018)

- Need to reduce the volume of material that needs to be melted
- Scaling to small dimensions severely hampered by fabrication challenges and elemental segregation

Addressing programming current/energy involves minimising material volume for melting. Scaling to smaller dimensions faces fabrication challenges and elemental segregation hurdles.

15 Single-elemental phase-change memory

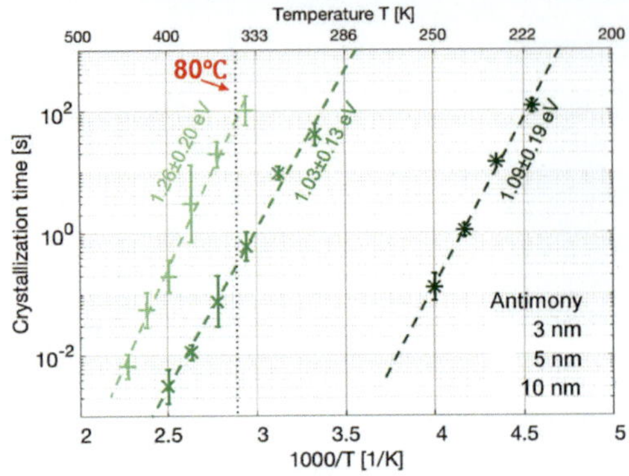

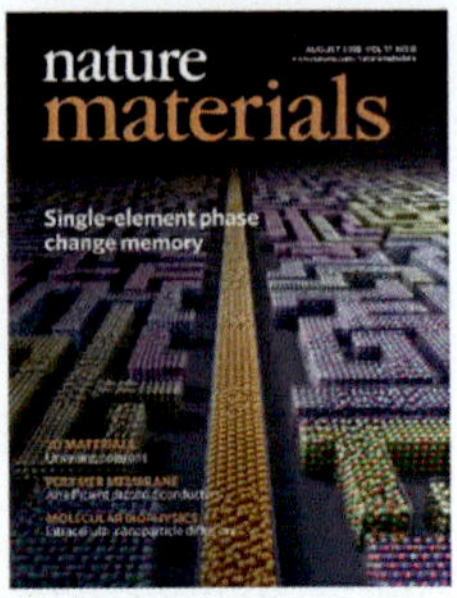

- **Single elemental Antimony** can be used for PCM when confined to nanoscale dimensions
- Directs research away from materials optimization to **nanoscale confinement**
- Could revive interest in several other PCM materials

Salinga et al., Nature Materials (2018)

They are utilizing single-elemental Antimony for PCM in nanoscale confinement. Shifts focus from material optimisation to nanoscale dimensions, potentially revitalizing interest in other PCM materials.

16 Compute precision

- Essential to achieve at least **4-bit precision**
- Accurate mapping of the operands
- Minimize the conductance fluctuations (time/temperature)
- Key challenges
  - **Read noise**
  - **Conductance drift**

Le Gallo et al., Nature Electronics (2018)

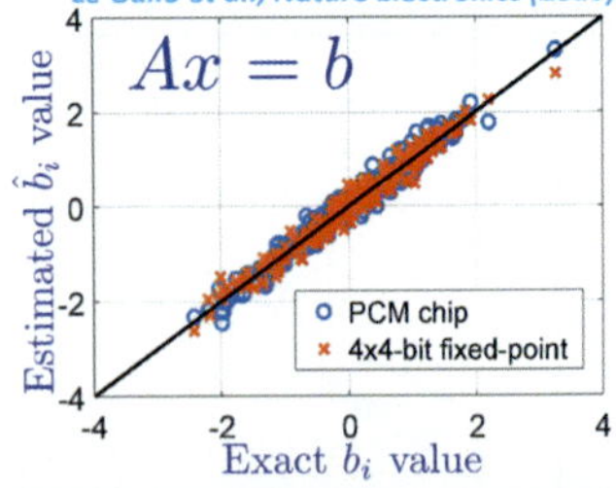

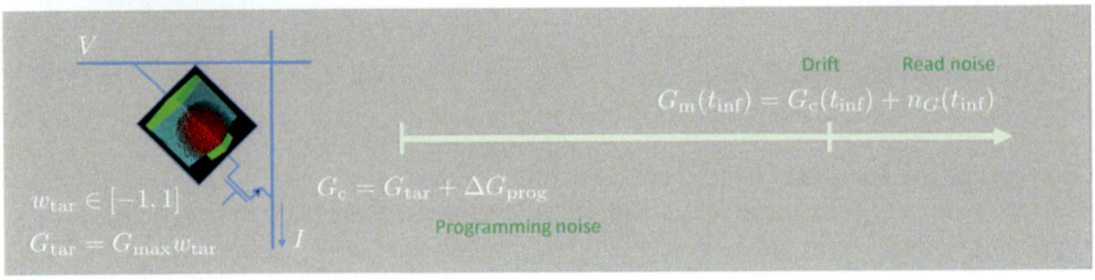

Compute precision crucial, aiming for at least 4-bit accuracy with accurate operand mapping. Challenges include minimising conductance fluctuations due to time/temperature, alongside addressing read noise and conductance drift.

17 Read noise

- A key source of computational imprecision
  - Fundamentally limits achievable programming precision
  - Determines the MVM accuracy

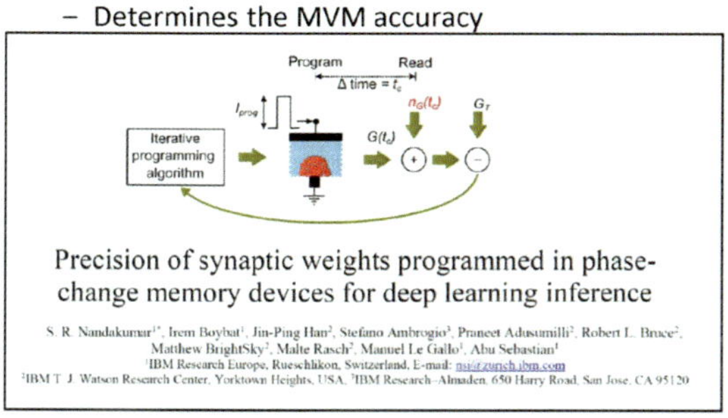

- **Research gaps**
  - **A unique model for 1/f noise in PCM**
  - **A thorough understanding of material dependence of 1/f noise**

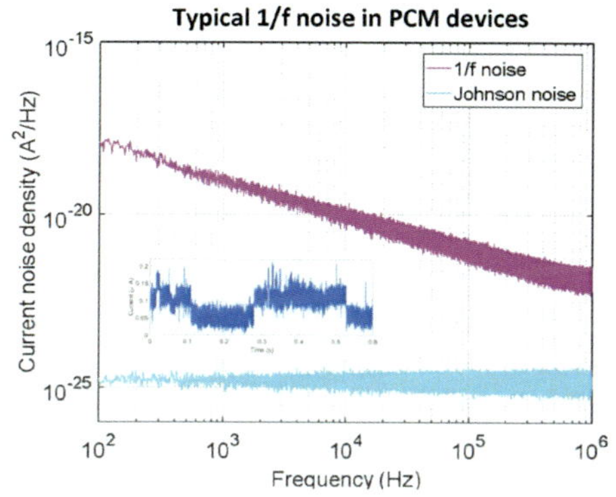

Fantini et al., APL (2006)
Nardone et al., PRB (2009)
Le Gallo et al., J. Appl. Physics D. (2020)

To achieve at least 4-bit precision, accurate mapping of the operands Is needed, and conductance fluctuations (time/temperature) are minimized. Key challenges are noise and Conductance drift.

Read noise, a primary source of computational imprecision, significantly limits achievable programming precision and impacts the accuracy of Matrix-Vector Multiplication). Bridging research gaps by developing a unique model for 1/f noise in Phase-Change Memory) and understanding material-dependent 1/f noise is crucial for effectively mitigating read noise in in-memory computing.

18 Conductance/Resistance drift

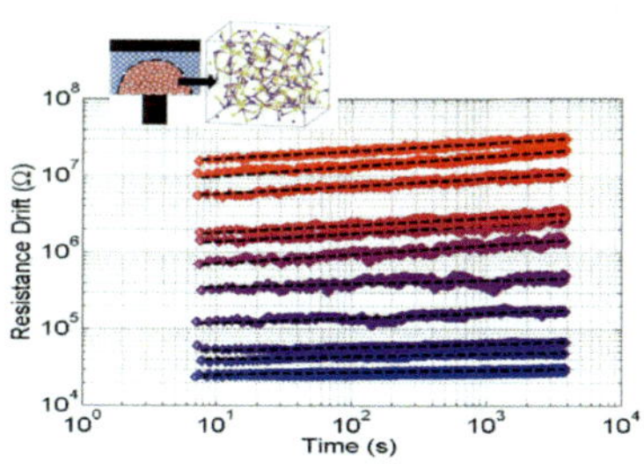

Ielmini et al., IEEE Trans. Elec. Dev (2009)
Fantini et al., Appl. Phys. Lett. (2012)

- Drift and drift variability pose key challenges to compute precision
- Research gaps
  - Microscopic origins of resistance drift
  - Kinetics of structural relaxation
  - A thorough understanding of material dependence of drift

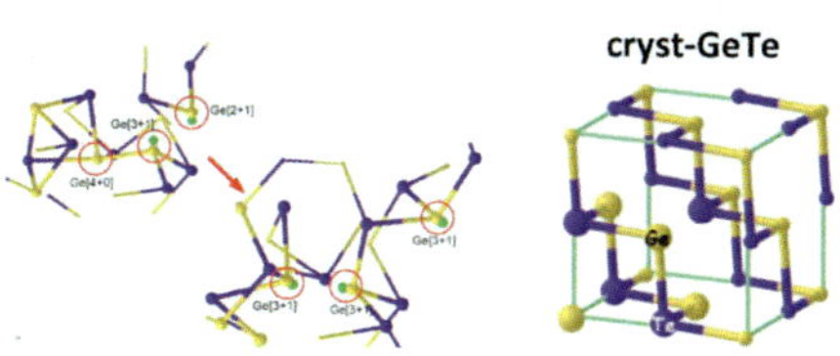

Raty et al., Nature Comm. (2015)
Zipoli et al., Phys. Rev. B (2016)

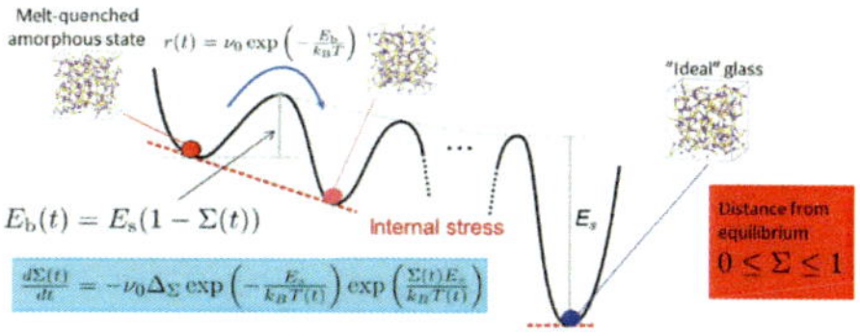

Le Gallo et al., Adv. Electr. Mat. (2018)
Kersting et al., Adv . Func. Mat. (2021)

Conductance or resistance drift presents significant challenges to computing precision due to its variability, necessitating research into the microscopic origins, structural relaxation kinetics, and drift material dependence. Bridging these gaps is crucial for developing effective strategies to mitigate drift in in-memory computing systems.

19 Temperature dependence

- Another challenge to maintaining sufficient compute precision is the temperature dependence of PCM conductance
- Although PCM exhibits substantial conductance-dependent temperature sensitivity, normalized distributions of the conductance states remain relatively constant across the applied time-temperature profile
- It was shown that it is possible to achieve and maintain high inference accuracies for a variety of networks under ambient temperature variations (33°C-80°C) by using a simple compensation scheme

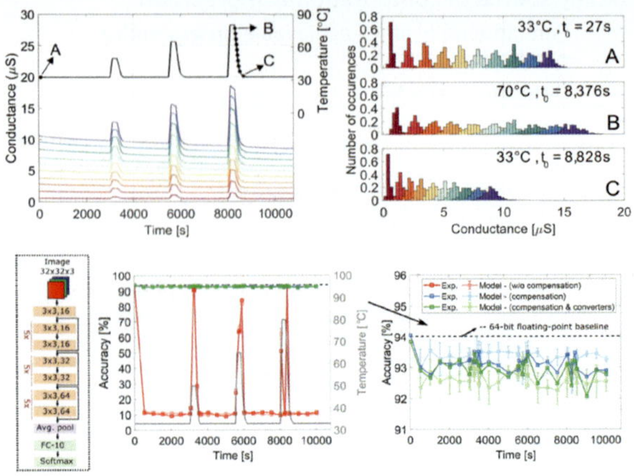

Boybat et al., "Temperature sensitivity of analog in-memory computing using phase-change memory", Proc. IEDM, 2021

The temperature dependence of Phase-Change Memory (PCM) conductance poses a challenge to compute precision, despite stable conductance state distributions across varied time-temperature profiles. However, studies have demonstrated the potential for maintaining high inference accuracies across ambient temperature variations (33°C-80°C) by implementing a straightforward compensation scheme.

20 Bipolar asymmetry

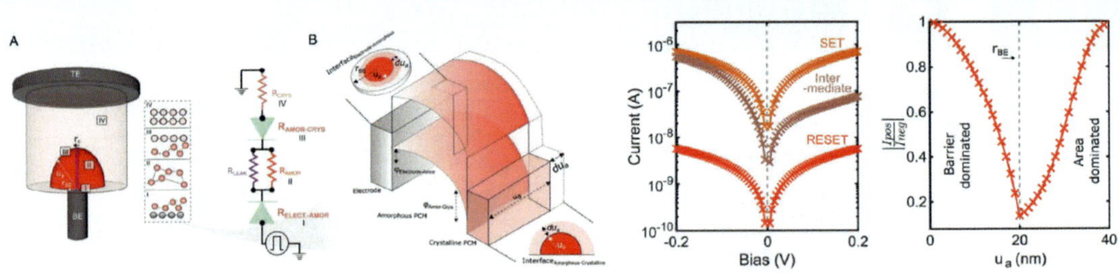

- Dependence on voltage polarity
- Back-to-back interfacial barriers
- Can be partially compensated
- Device-level solutions preferred

RESEARCH ARTICLE — ADVANCED MATERIALS www.advmat.de

Mechanism and Impact of Bipolar Current Voltage Asymmetry in Computational Phase-Change Memory

Syed Ghazi Sarwat, Manuel Le Gallo, Robert L Bruce, Kevin Brew, Benedikt Kersting, Vara Prasad Jonnalagadda, Injo Ok, Nicole Saulnier, Matthew BrightSky, and Abu Sebastian**

Bipolar asymmetry in computing, influenced by voltage polarity dependence and back-to-back interfacial barriers, presents a challenge that can be partially compensated for. Preferably, device-level solutions are sought to address this asymmetry in computing systems.

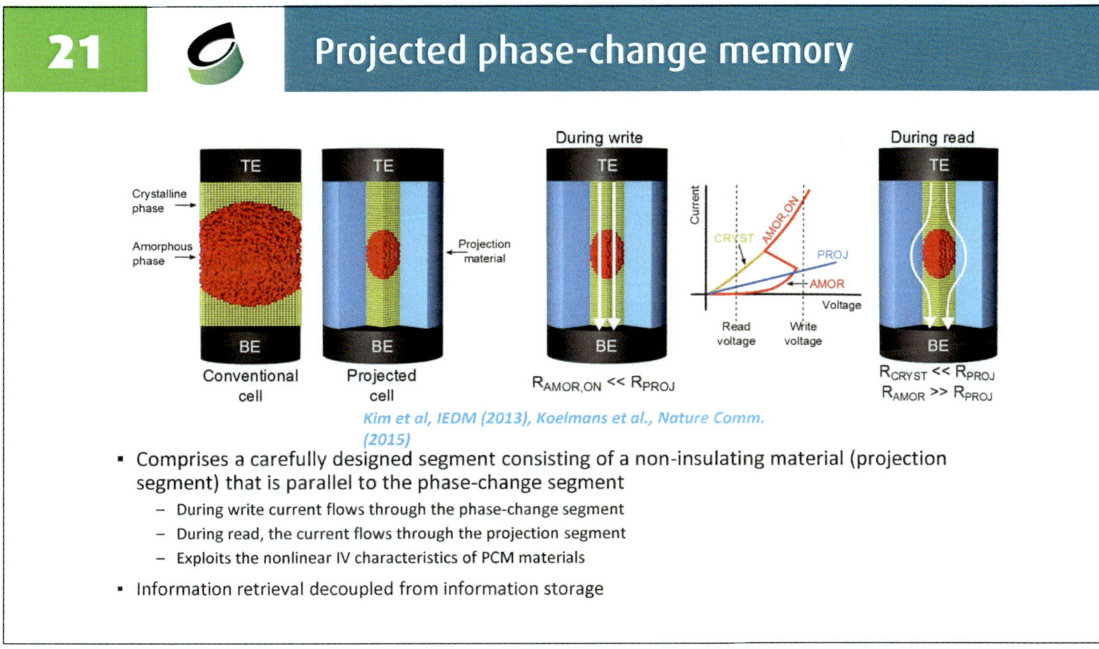

Projected phase-change memory utilises a distinct architecture comprising a phase-change segment and a non-insulating projection segment, with write and read operations flowing current through each segment, respectively. By leveraging the nonlinear IV characteristics of PCM materials, this design decouples information retrieval from storage, enhancing memory efficiency and performance.

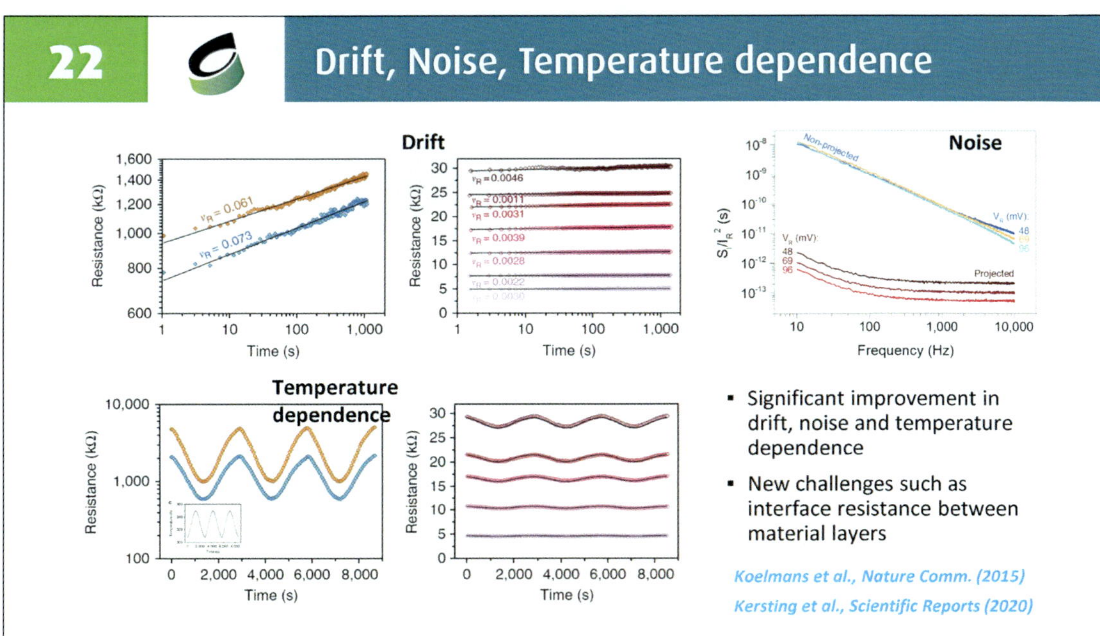

Drift, noise, and temperature dependence pose challenges in memory systems, but significant improvements have been achieved in addressing these issues. However, new challenges have emerged, such as interface resistance between material layers, which require further attention in advancing memory technology.

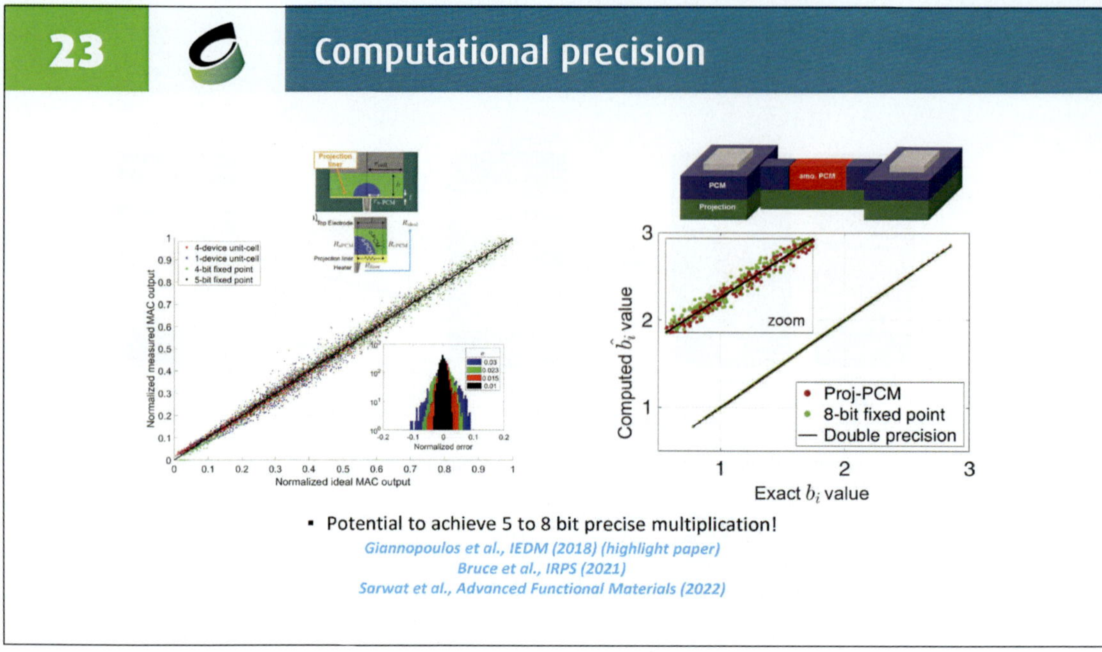

Figure: Potential to achieve 5 to 8-bit precise multiplication.

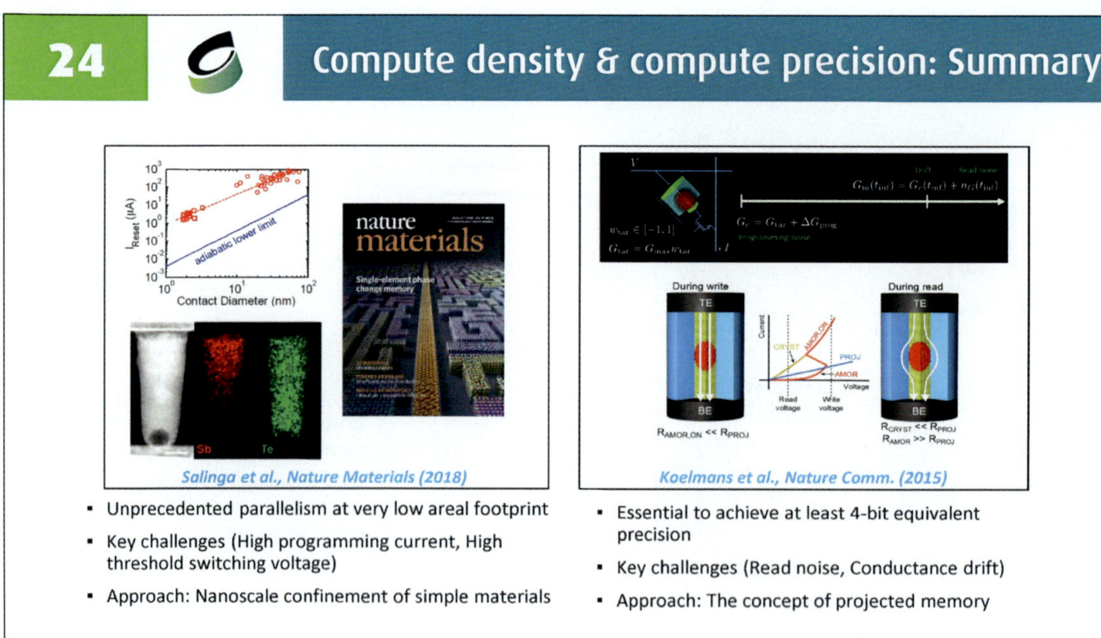

Figure 1: Unprecedented parallelism at minimal footprint faces challenges like high programming current and threshold switching voltage, addressed through nanoscale confinement of basic materials.

Figure 2: Achieving at least 4-bit equivalent precision is imperative, with challenges such as read noise and conductance drift, which are addressed through the concept of projected memory. This approach aims to mitigate these challenges and enhance precision in-memory operations.

25 Summary

- In-memory computing is an emerging computing paradigm that keeps alive the promise for 1fJ/Op energy efficiency
- Deep learning inference is a killer application for IMC that exploits the compute primitive of in-memory matrix-vector multiplication
- Phase-change memory (PCM) is particularly well suited for in-memory computing
- Iso-accuracy DNN inference is possible with custom training
- Recently fabricated mixed-signal IMC cores demonstrate the promise of this technology
- Materials research should primarily focus on increasing the compute density and compute precision
- Nanoscale confinement of materials such as Antimony a promising approach towards reducing the reset current
- Projected PCM has potential to address drift, noise and temperature dependence

CHAPTER 03

Modeling of Memristive Devices

Stephan Menzel

Peter-Grünberg-Insitut 7, Forschungszentrum Jülich GmbH

This talk presents a comprehensive exploration of memristive device modeling in this detailed presentation. The discussion covers various aspects of memristive systems, including the current-voltage relationship, redox-based devices, and multiscale modeling approaches. Detailed analyses are provided on topics such as electronic transport in VCM cells using DFT-NEGF studies, switching variability employing kinetic Monte Carlo modeling, and continuum modeling of switching dynamics. The presentation delves into the origins of switching variability, cycle-to-cycle variability, read noise, and the physical properties affecting filament behavior. Compact modeling of VCM cells is also discussed, highlighting electrical and ionic aspects, as well as insights into read instability and digital switching behaviors. The abstract concludes by summarizing the comprehensive understanding gained across various aspects of memristive device modeling, offering a framework for further research and optimization in this evolving field. The presentation serves as a valuable resource for researchers and practitioners in the realm of memristive devices.

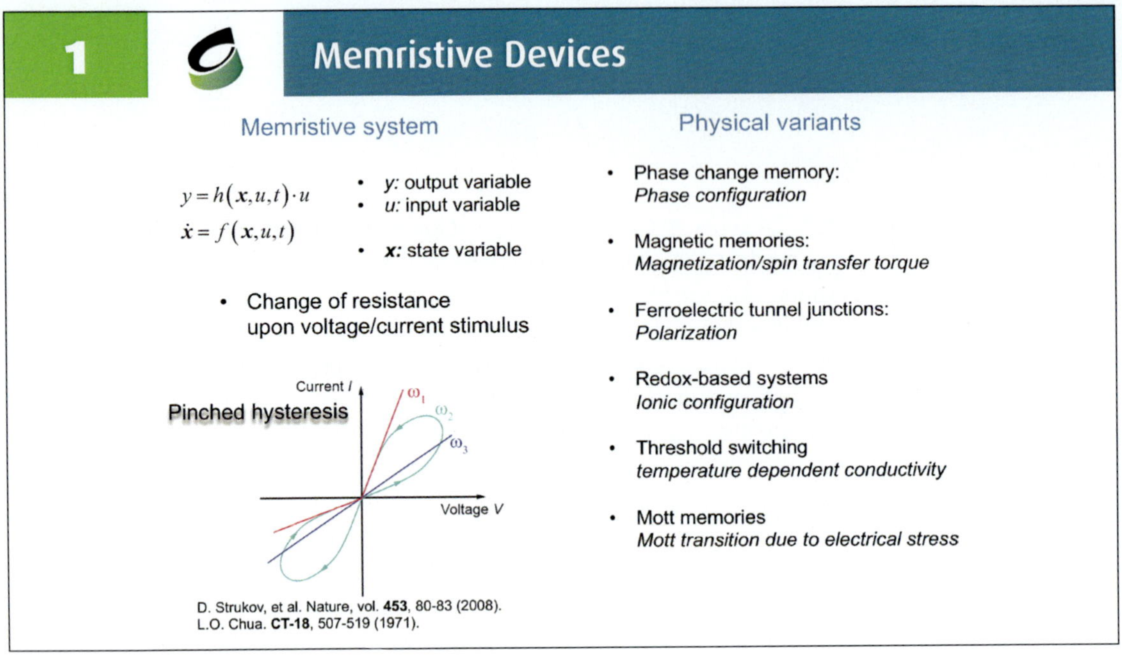

This slide discusses the current versus voltage relationship in a memristive system and its different physical variants.

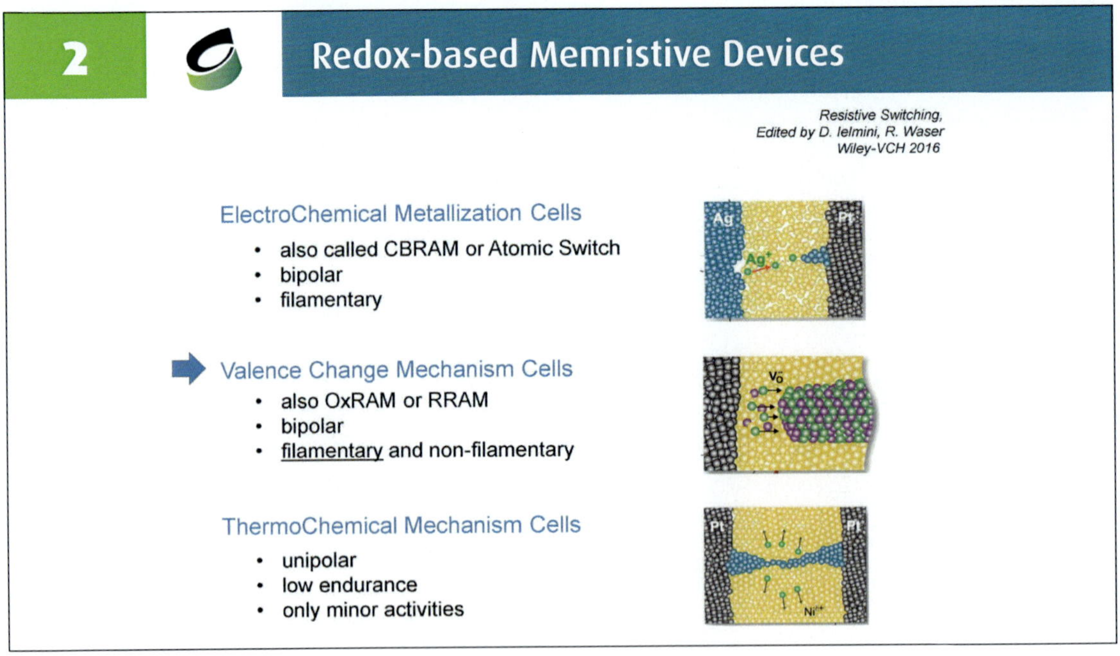

This slide focuses on the different Redox based memristive devices.

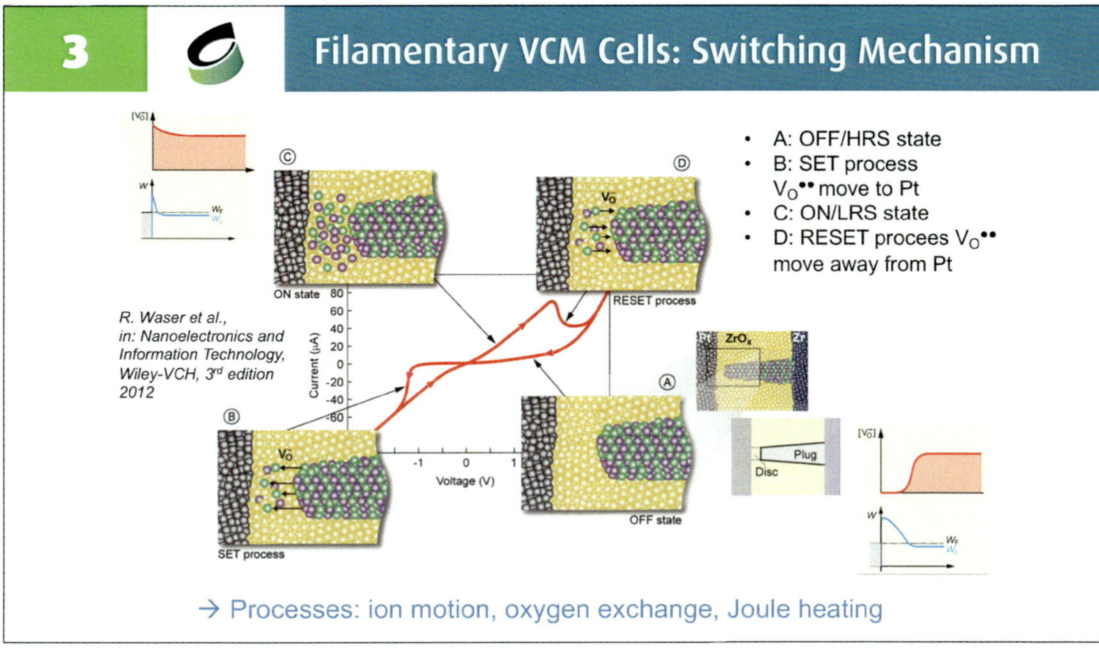

Explains the processes of ion motion, oxygen exchange, Joule heating in the switching mechanism of filamentary VCM cells.

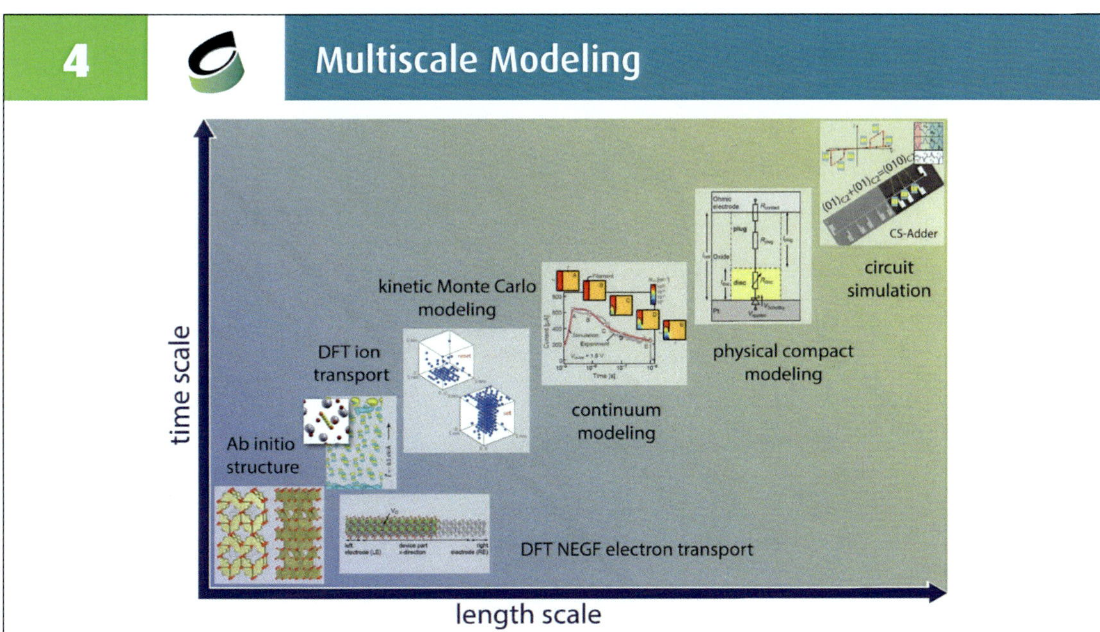

Gives an outline of multiscale modeling of memristive devices.

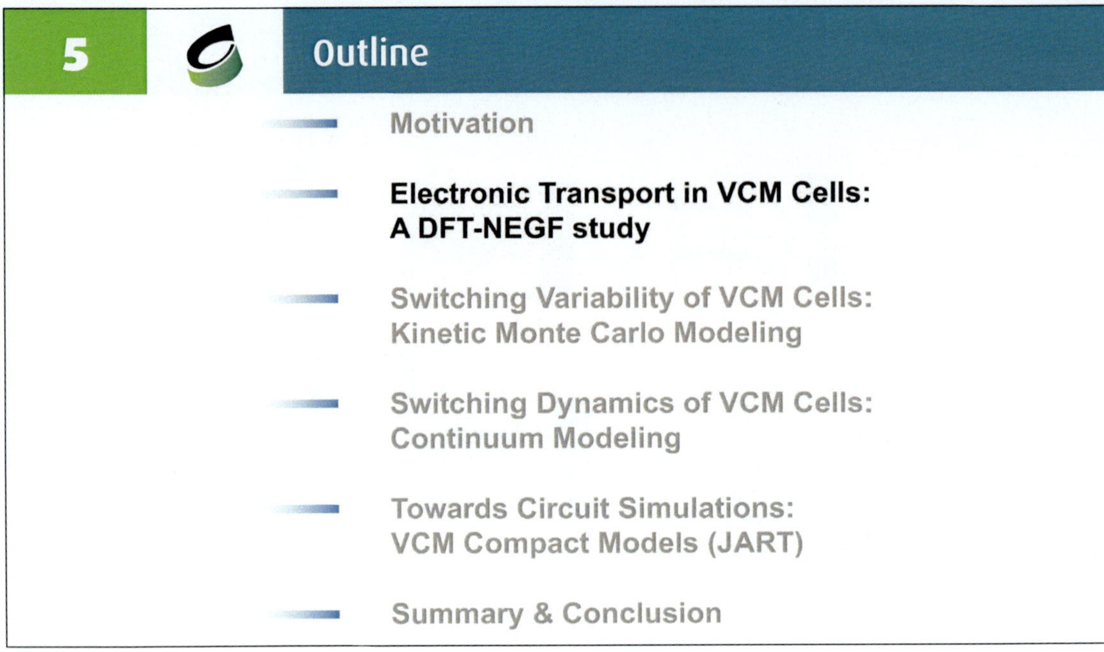

Introductory slide to the topic "Electronic Transport in VCM cells: A DFT-NEGF study".

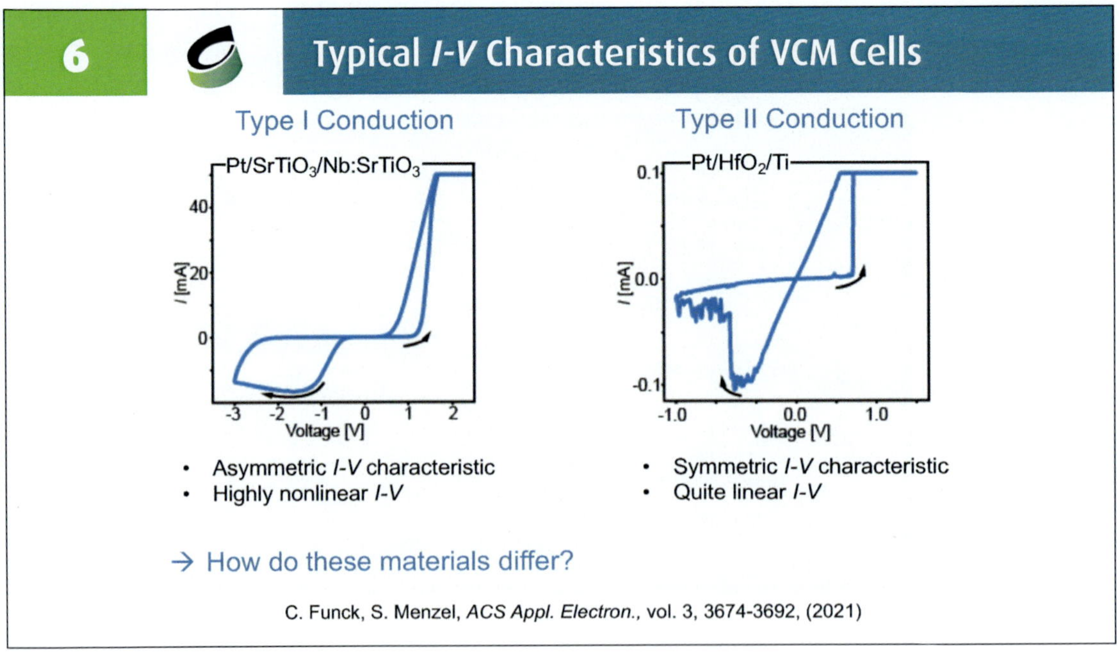

Discusses the IV characteristics of VCM cells.

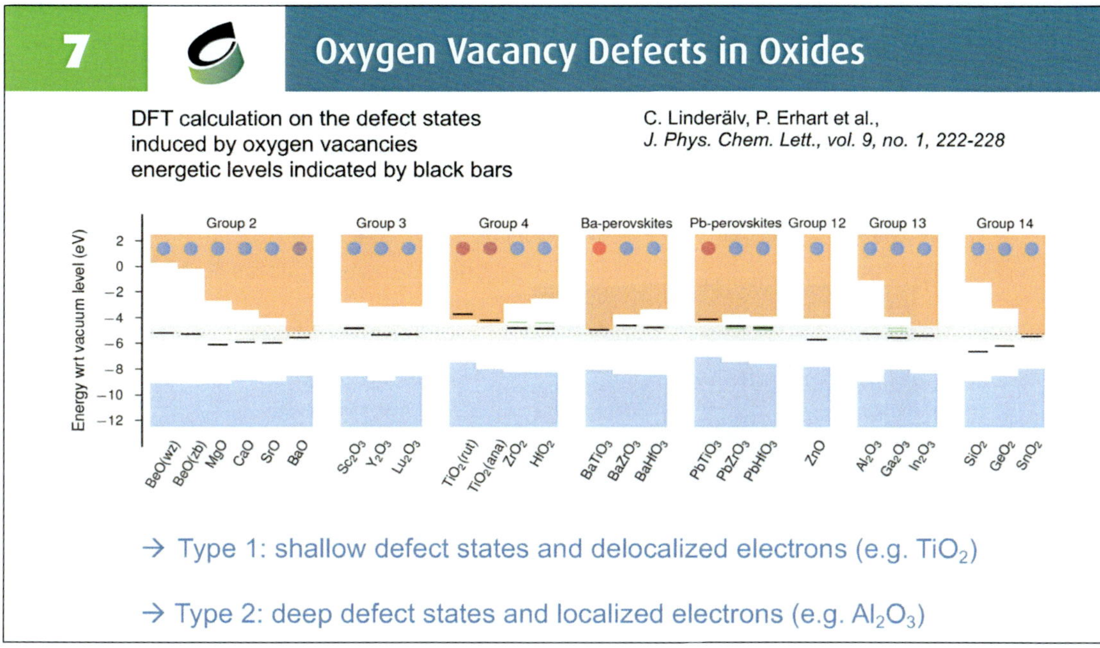

Mentions the different oxygen vacancy defects in oxides and its corresponding DFT calculations.

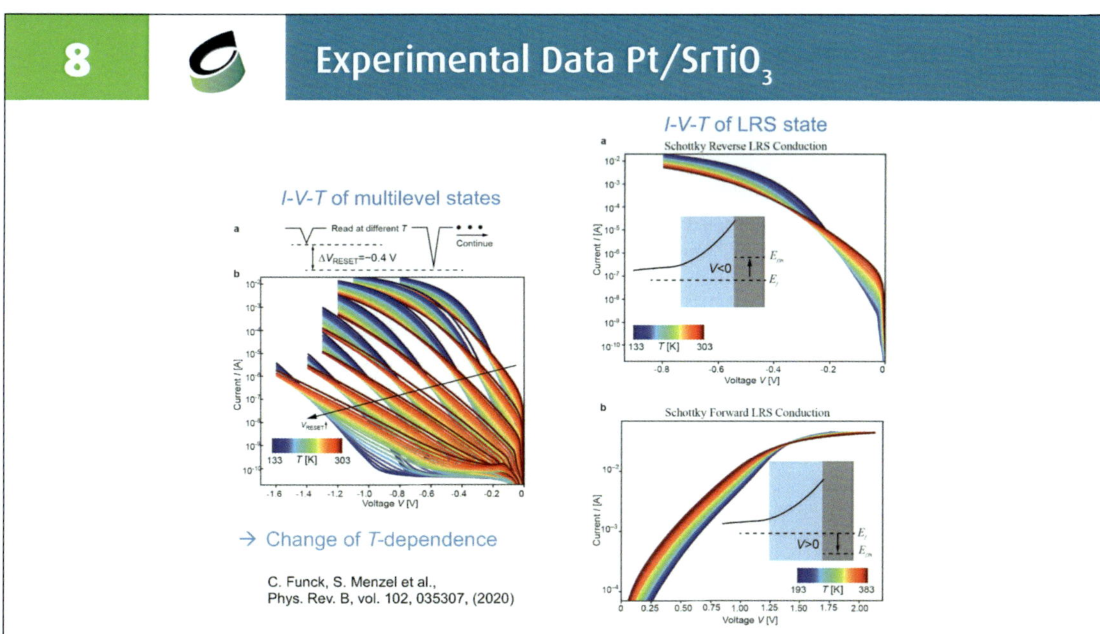

Illustrates the experimental data of Pt/$SrTiO_3$ and analyzes the I-V-T graphs for multilevel states and LRS states.

9 DFT-Nonequilibrium Green's Function (NEGF) Formalism

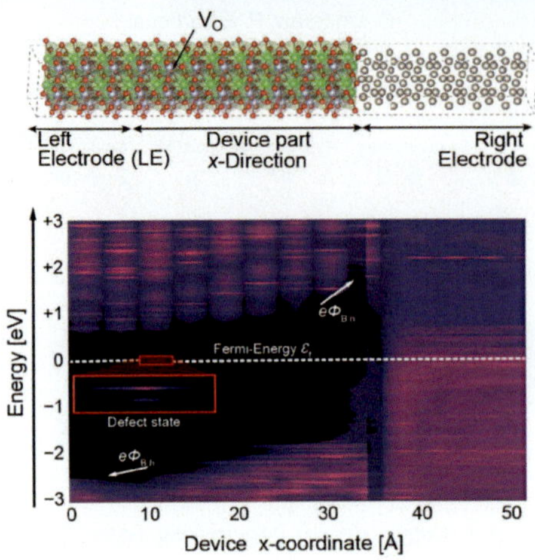

2×2×9 [100] $SrTiO_3$ supercell
2×2×4 [100] Pt supercell

↓ DFT PBE-GGA

Local density of states (LDOS)

DFT calculation details:

Code: ATK 2015
Exchange correlation: PBE

k-point Sampling for electrodes (3x3x100)

Local atomic basis set.

C. Funck, A. Marchewka, S. Menzel et al., *Adv. Electron. Mater.*, vol. 4, no. 7, 1800062, (2018)

Starts with the discussion of the theoretical framework "DFT-Non Equilibrium Green's Function (NEGF) Formalism" to study electronic transport properties in Pt/$SrTiO_3$ cell and its DFT calculation.

10 DFT-Nonequilibrium Green's Function (NEGF) Formalism

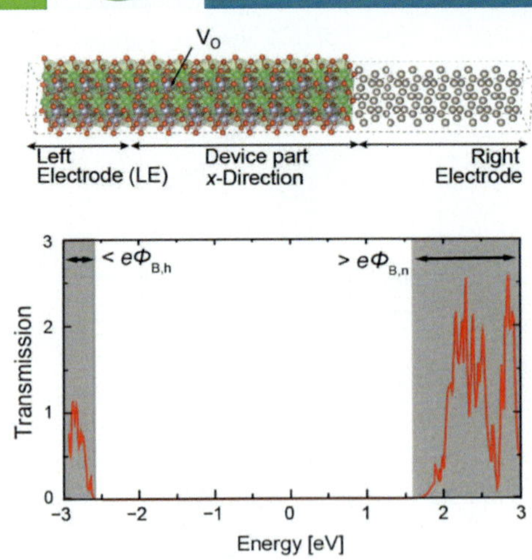

2×2×9 [100] $SrTiO_3$ supercell
2×2×4 [100] Pt supercell

↓ DFT PBE-GGA

Local density of states (LDOS)

↓ NEGF

Transmission spectrum

↓ Different T
No phonon scattering

Current characteristics

C. Funck, S. Menzel et al., *Adv. Electron. Mater.*, vol. 4, no. 7, 1800062, (2018)

$$I = \frac{e}{h}\int_{-\infty}^{\infty} T_R(E,\varepsilon_{LE},\varepsilon_{RE})\left[f(E-\varepsilon_{LE},T) - f(E-\varepsilon_{RE},T)\right] dE,$$

Calculates its transmission spectrum and current characteristics.

11 Spectral Current Analysis in Reverse Direction

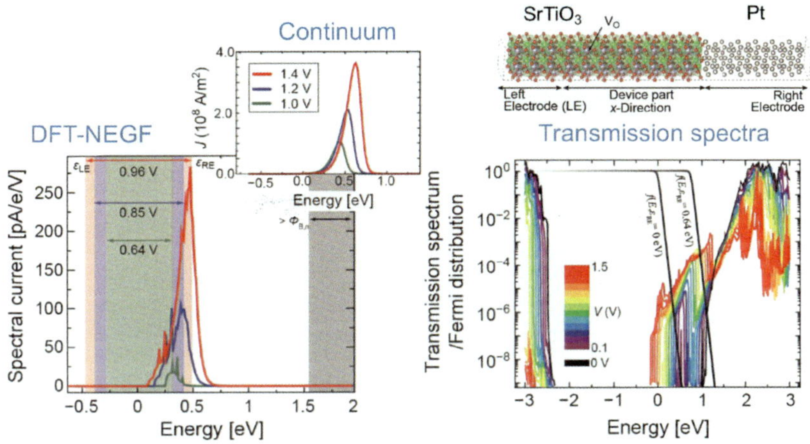

→ Spectral current reveals strong tunneling contribution

C. Funck, A. Marchewka, S. Menzel et al., *Adv. Electron. Mater.*, vol. 4, no. 7, 1800062, (2018)

Discusses the spectral current analysis in reverse direction and reveals its relationship with tunneling contribution.

12 Analysis of the Voltage Dependence

Low voltage regime

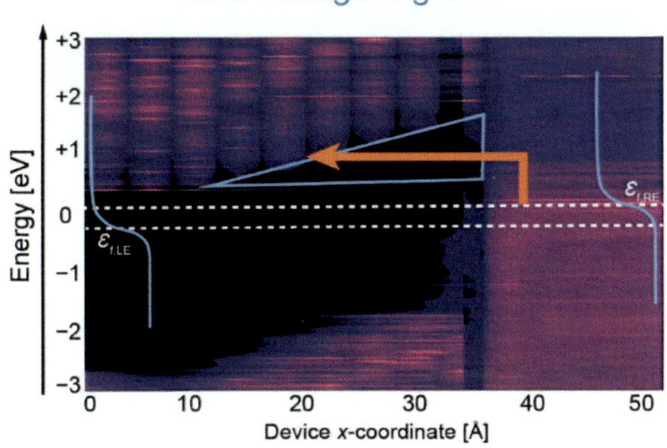

→ Only thermally excited electrons can tunnel (Boltzmann approximation valid)

→ Current increases with T

C. Funck, A. Marchewka, S. Menzel et al., *Adv. Electron. Mater.*, vol. 4, no. 7, 1800062, (2018)

Analyze the the tunneling effect of electrons with temperature in the low voltage regime of the device.

13 Analysis of the Voltage Dependence

Intersection voltage V_I

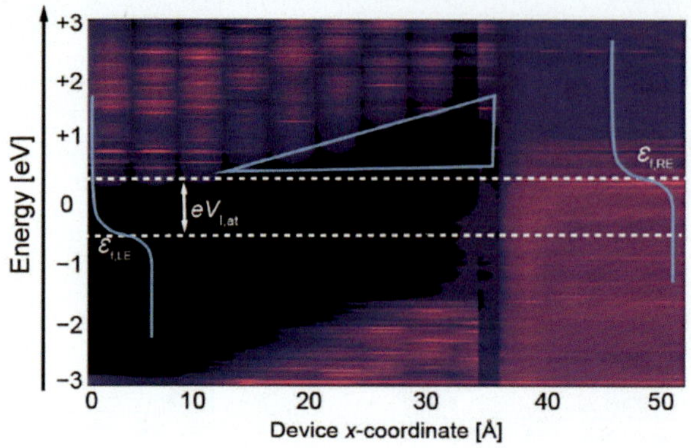

→ RE Fermi level and maximum of spectral current coincide
→ Boltzmann approximation invalid

C. Funck, A. Marchewka, S. Menzel et al., *Adv. Electron. Mater.*, vol. 4, no. 7, 1800062, (2018)

Continues the analysis and studies about the intersection voltage.

14 Analysis of the Voltage Dependence

High voltage regime

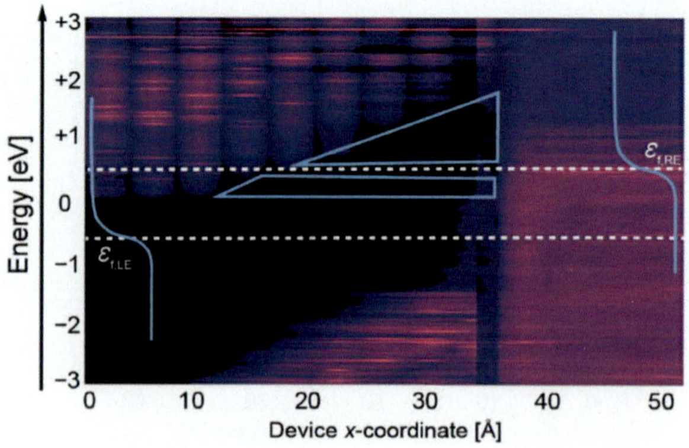

→ All electrons tunnel (no *T*-dep by tunneling)
→ *T*-dependence due to phonon scattering

C. Funck, A. Marchewka, S. Menzel et al., *Adv. Electron. Mater.*, vol. 4, no. 7, 1800062, (2018)

Switches the analysis to high voltage regime and the corresponding tunneling effect of electrons.

15 Reverse Direction: Multilevel States

Simulated *I-V-T* (Continuum Model)

Conduction Band Edge

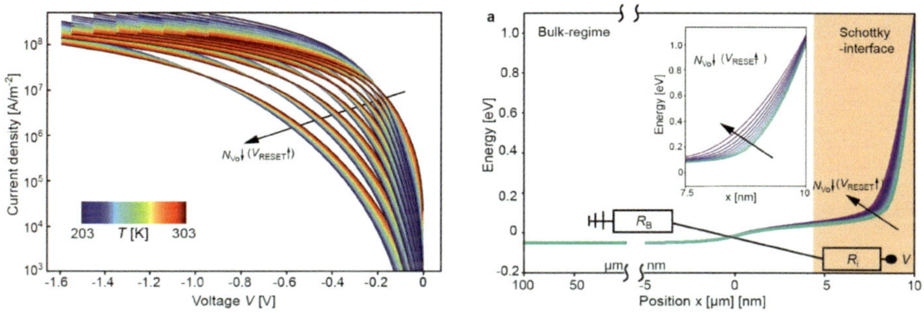

→ Shift of intersection voltage due increase of conduction band edge
→ ML states are described by change in the height and size of the barrier

C. Funck, S. Menzel et al., *Phys. Rev. B*, vol. 102, 035307, (2020)

Discusses the shift in intersection voltage with the increase in conduction band edge.

16 Summary of Type I Conduction

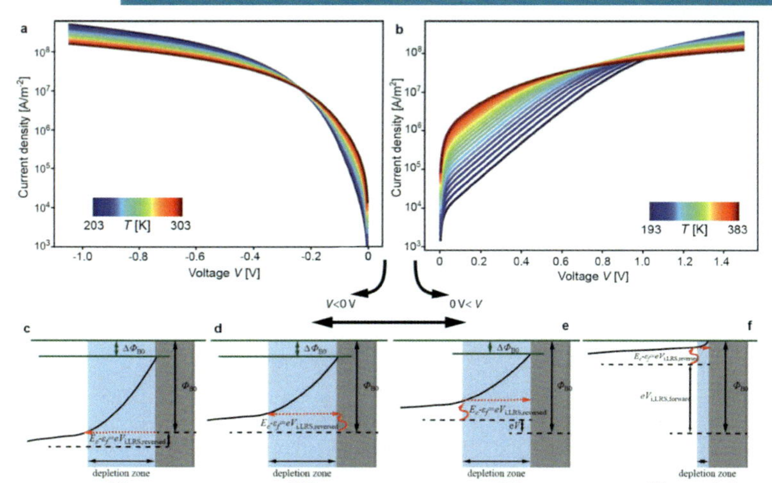

C. Funck, S. Menzel et al., *Phys. Rev. B*, vol. 102, 035307, (2020)

Summarizes the IV relationship in Type I conduction band materials.

17 DFT-NEGF Simulation of the HRS

C. Funck, S. Menzel, *ACS Appl. Electron.*, vol. 3, 3674-3692, (2021)

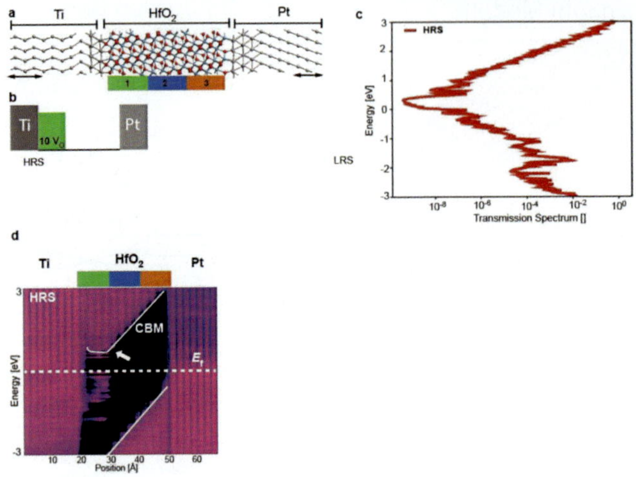

Discusses with the DFT-NEGF simulation of the HRS.

18 DFT-NEGF Simulation of an IRS

C. Funck, S. Menzel, *ACS Appl. Electron.*, vol. 3, 3674-3692, (2021)

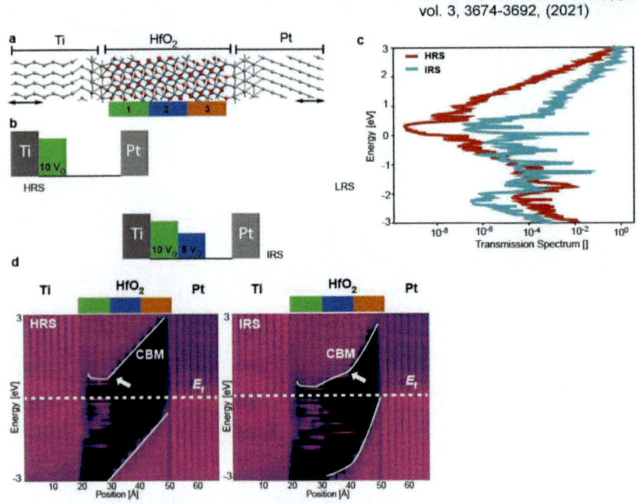

Discusses with the DFT-NEGF simulation of the IRS.

19 DFT-NEGF Simulation of the LRS

C. Funck, S. Menzel, *ACS Appl. Electron.*, vol. 3, 3674-3692, (2021)

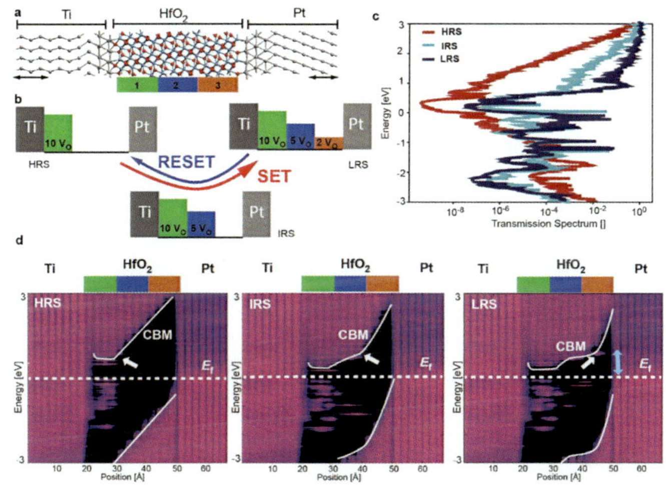

Discusses with the DFT-NEGF simulation of the LRS.

20 General Switching Mechanism

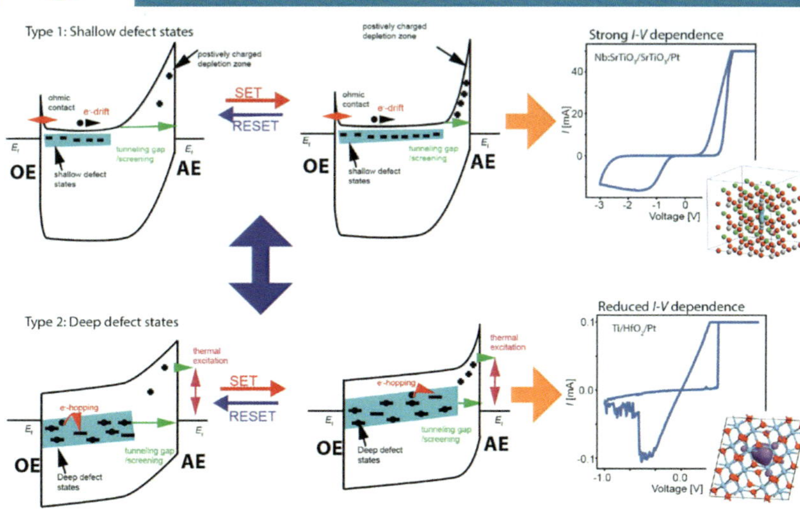

R. Dittman, S. Menzel, R. Waser, Advances in Physics, in press, DOI: 10.1080/00018732.2022.2084006

Illustrates the switching mechanisms in the shallow defect states and deep defect states and their IV relationships.

21 Outline

- Motivation
- Electronic Transport in VCM Cells: A DFT-NEGF study
- **Switching Variability of VCM Cells: Kinetic Monte Carlo Modeling**
- Switching Dynamics of VCM Cells: Continuum Modeling
- Towards Circuit Simulations: VCM Compact Models (JART)
- Summary & Conclusion

Introductory slide to the topic "Switching Variability of VCM Cells: Kinetic Monte Carlo Modeling".

22 KMC Model I: Electronic Model

Simulation geometry

- Oxide cube of 5 nm x 5 nm x 5 nm
- Possible defect sites every 0.5 nm (= a)
- Active (AE) and ohmic electrode (OE)

Defects

- Oxygen vacancies $V_O^{\cdot\cdot}$
- Reduced cations close to $V_O^{\cdot\cdot}$ act as electron traps

Conduction mechanism (Type II)

- Trap assisted tunnneling
- (1) e^- tunneling between electrode & traps
- (2) e^- hopping between traps
- At high defect densities band conduction

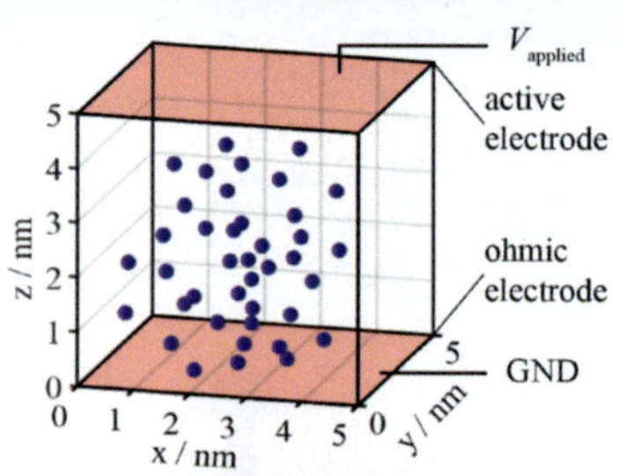

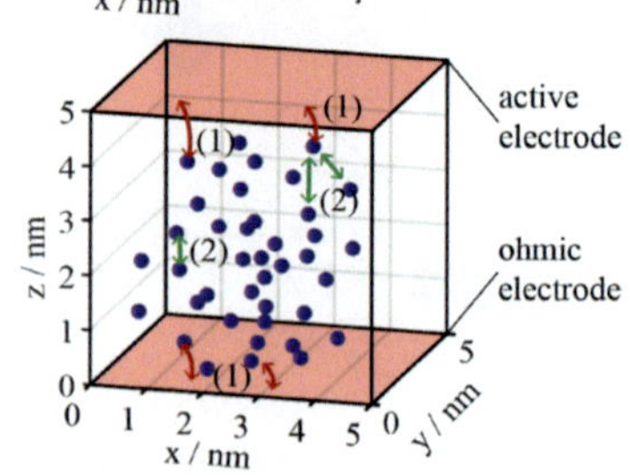

Discusses the simulation geometry, defects and conduction mechanism of the Electronic model based on KMC model I method.

23 KMC Model II: Ionic Model

Dynamic model

- Joule heating
- Oxygen exchange reaction at BE
- $V_O^{\cdot\cdot}$ hopping

$$\Gamma_{G,R,D} = \nu_0 \exp\left(-\frac{\Delta W_{G,R,D} - \alpha_i z E}{k_B T}\right)$$

- Weighted random selection of process
- Time increment

$$\Delta t = -\frac{\ln(\Gamma_i)}{\Gamma_{total}}$$

Simulation flow

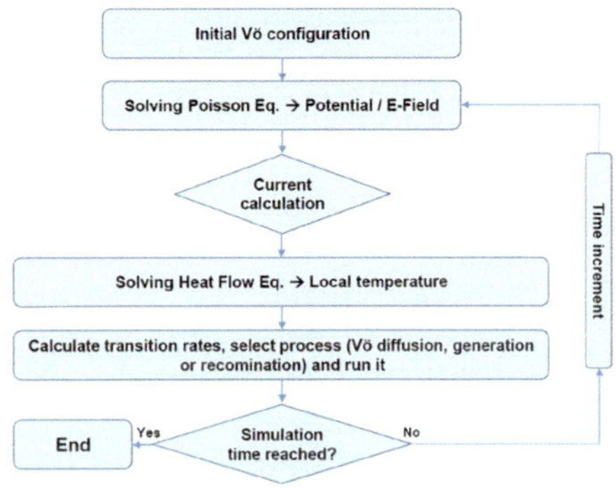

E. Abbaspour, S. Menzel, C. Jungemann et. al, *IEEE Trans. Nanotechnology,* vol. 17, no. 6, pp. 1181-1188 (2018)

E. Abbaspour, S. Menzel, C. Jungemann, *Journal of Computational Electronics,* vol. 19, pp. 1426-1432 (2020)

Discusses the dynamic model of Ionic model using KMC II method.

24 Origin of the Switching Variability: Stochasticity

Simulation of 100 *I*-*V* cycles

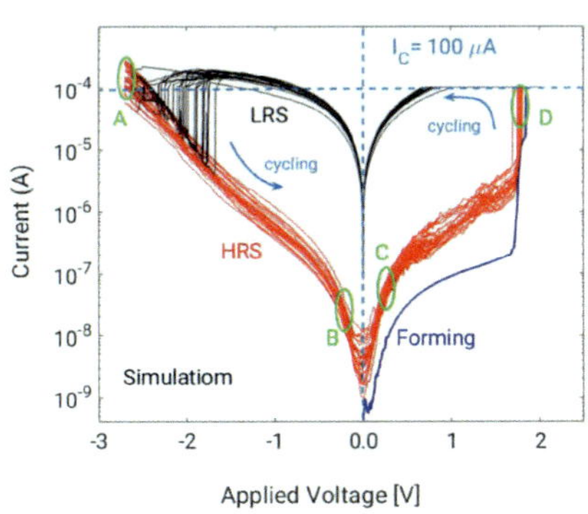

Filament evolution

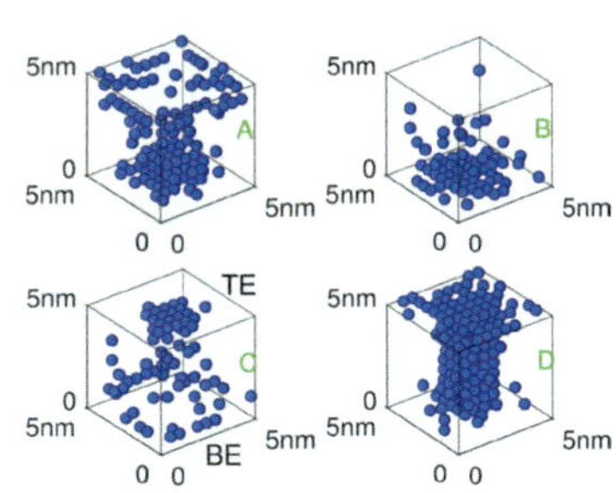

- O exchange at BE
- *V* applied to BE

→ Typical SET/RESET switching obtained

E. Abbaspour, S. Menzel, C. Jungemann et. al, *IEEE Trans. Nanotechnology,* vol. 17, no. 6, pp. 1181-1188 (2018)

Starts the discussion of the origin of Switching variability in Stochastic systems.

25 Origin of the Switching Variability: Stochasticity II

SET switching probability

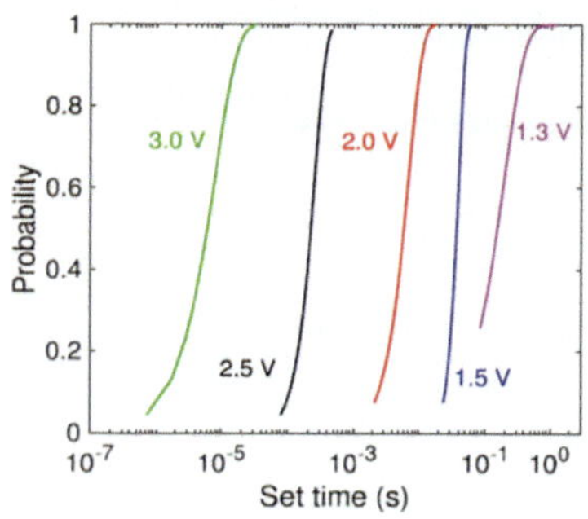

HRS/LRS distributions

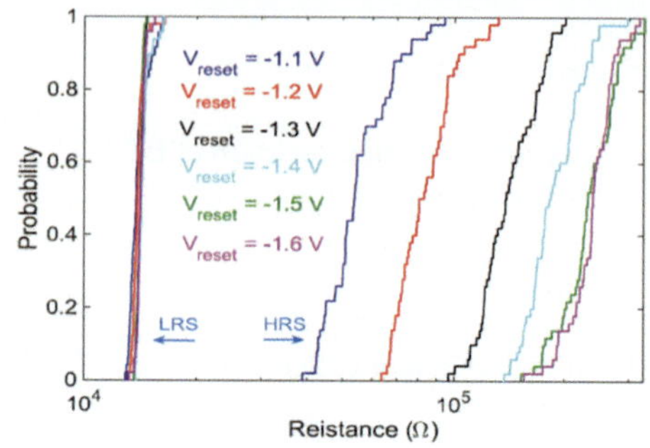

→ Distributions tighter than in experiment
→ Stochasticity is not sufficent to explain variability

E. Abbaspour, S. Menzel, C. Jungemann, *Journal of Computational Electronics*, vol. 19, pp. 1426-1432 (2020)

Summarizes the fact that stochasticity is not sufficient to explain variability by studying the relationship between set time and resistance with probabilities.

26 Relation between Read Instability & Retention

Retention (shifts & tilts)

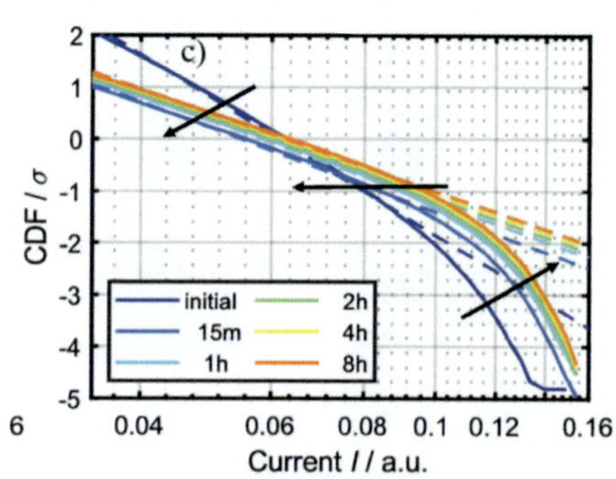

Read instability

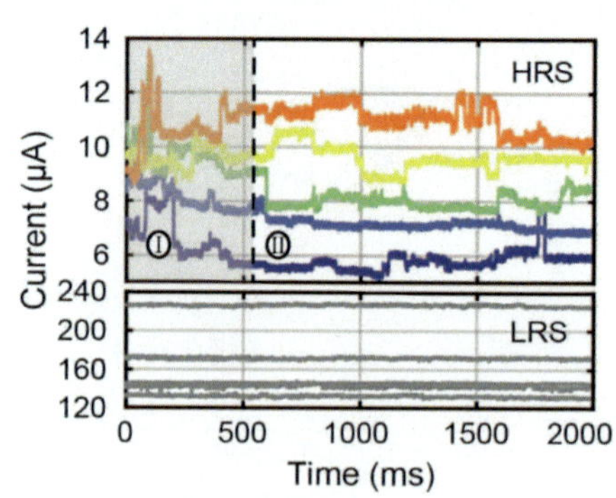

S. Wiefels, S. Menzel et al., *IEEE IMW* (2020)

S. Wiefels, S. Menzel et al., *IEEE T-ED*, vol. 67, no. 10, pp. 4208-4215 (2020)

→ Both processes depends on ionic motion, but occur on different timescales !?

Studies the relationship between read instability and retention.

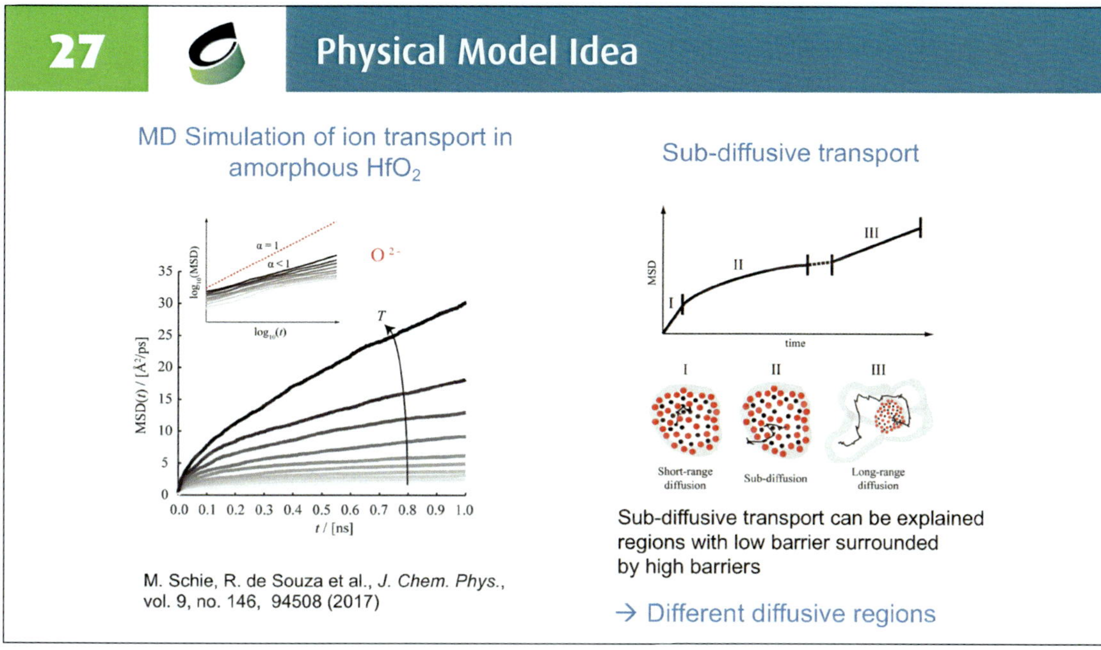

Mentions the physical model idea related to MD Simulation of ion transport in amorphous HfO_2.

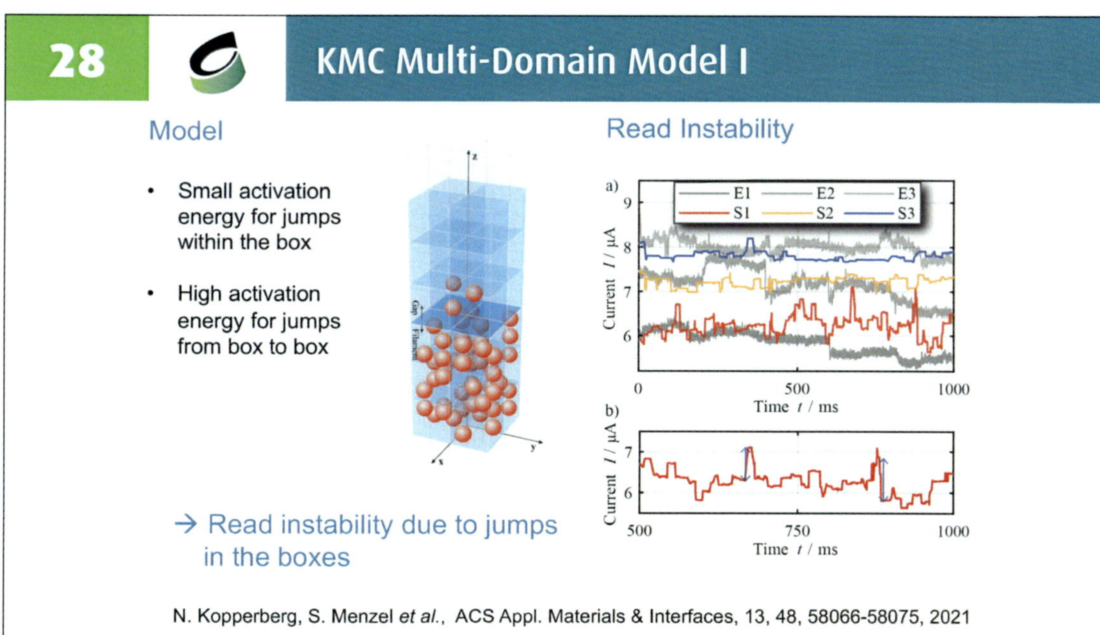

Read instability in VCM cells using KMC Multi-Domain Model I is discussed.

29 KMC Multi-Domain Model II

Tilting of the distribution

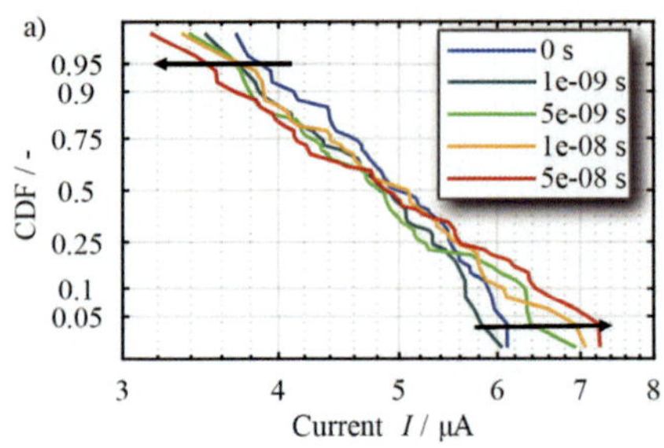

Shift of the distribution

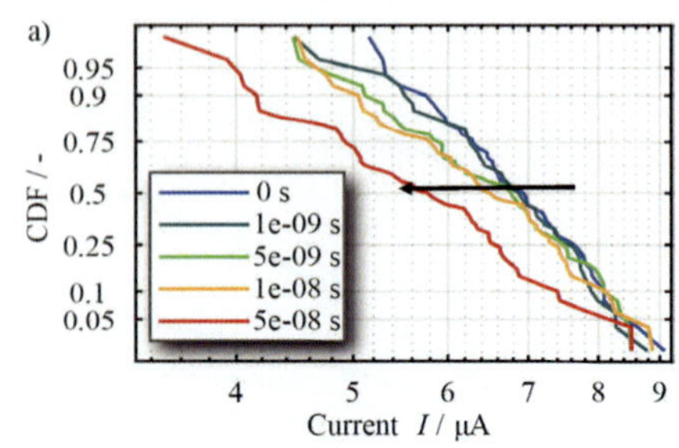

→ Retention due to jumps out of the boxes
→ depends on initial state

N. Kopperberg, S. Menzel *et al.*, ACS Appl. Materials & Interfaces, 13, 48, 58066-58075, 2021

Discusses the switching dynamics of VCM cells using KMC Multi-Domain Model II.

30 Note: Impact of Conduction Mechanism on Read Noise

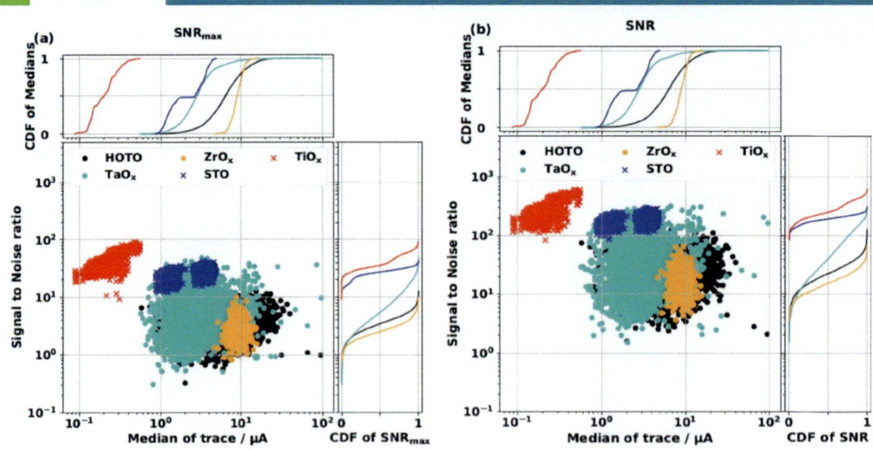

→ High read noise observed for Type II conduction materials

K. Schnieders, S. Menzel, R. Dittmann et al., *APL Materials* 10, 101114 (2022)

Summarizes the impact of conduction mechanism on read noise.

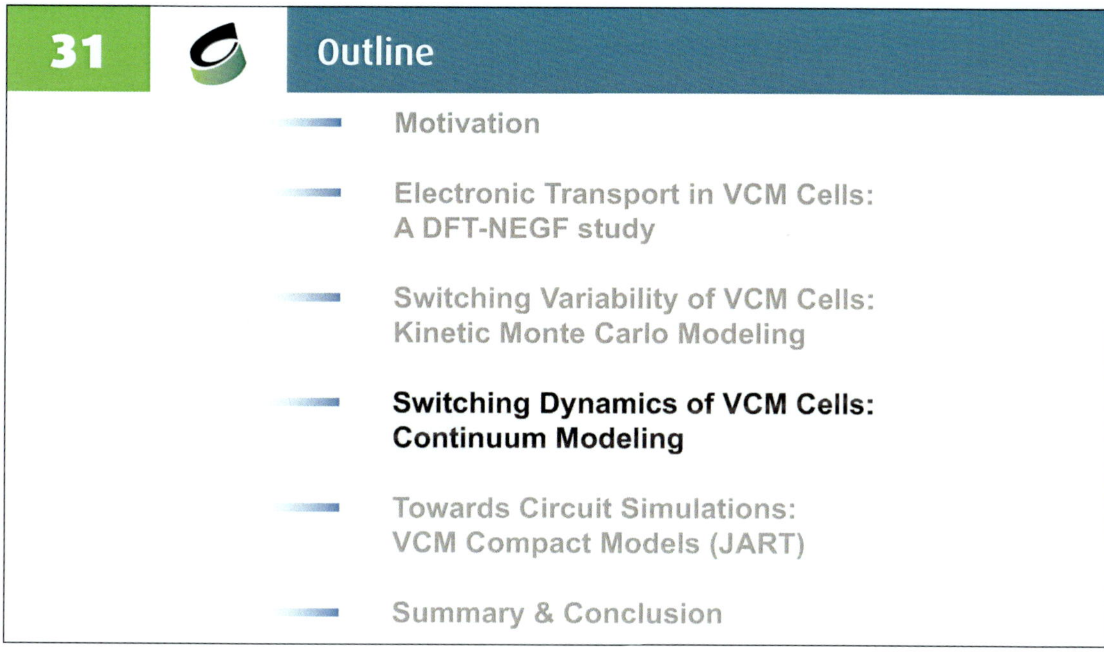

Introductory slide to the topic "Switching Dynamics of VCM Cells: Continuum Modeling".

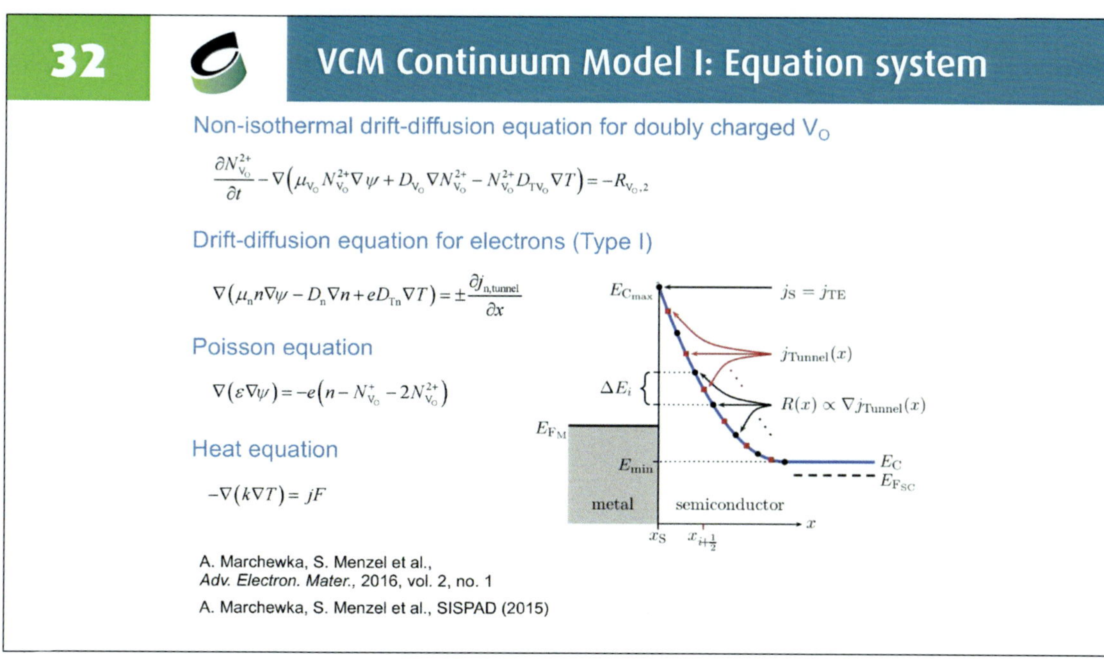

Describes the VCM Continuum Model I which includes a set of equations for modeling the electronics and thermal behavior of VCM cells.

33 VCM Continuum Model II: 2D Domains and Initial Conditions

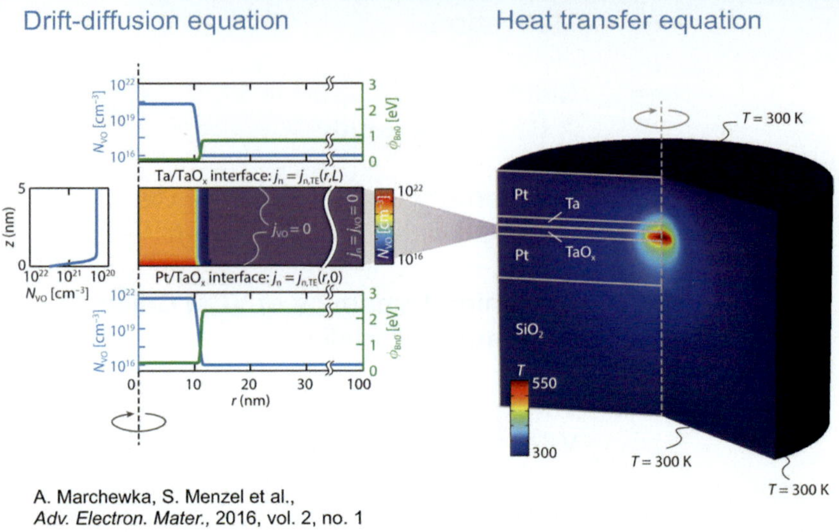

Gives the Drift-diffusion equation and heat-transfer equation of the VCM Continuum Model II.

34 Simulated *I-V* Characteristics

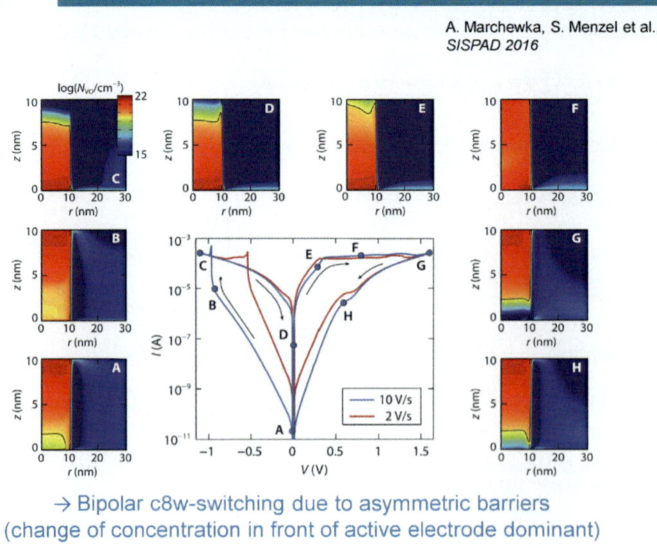

Shows the simulated IV characteristics accordingly.

35 Continuum Modeling of SET Dynamics

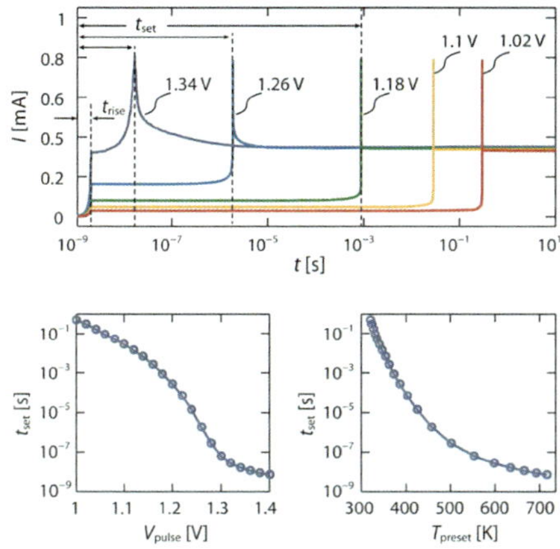

Comparison to Pt/SrTiO$_3$/Ti cell

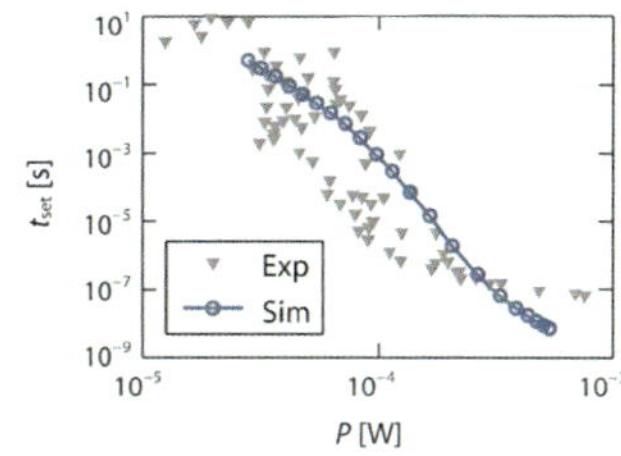

A. Marchewka, S. Menzel et al., SISPAD (2015)

→ Simulated current transients reproduce experimental behavior
→ Simulation model reproduces experimental power dependence

Compare the modeling of SET dynamics with Pt/SrTiO$_3$/Ti cell.

36 RESET Kinetics I: Simulation Results

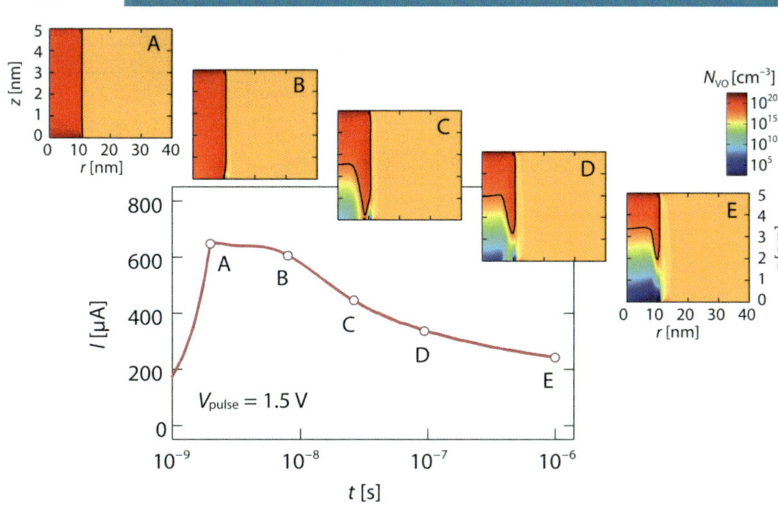

A. Marchewka, S. Menzel et al., *Adv. Electron. Mater.*, Vol. 2, No. 1, 2016

Gives the simulation results of RESET kinetics I.

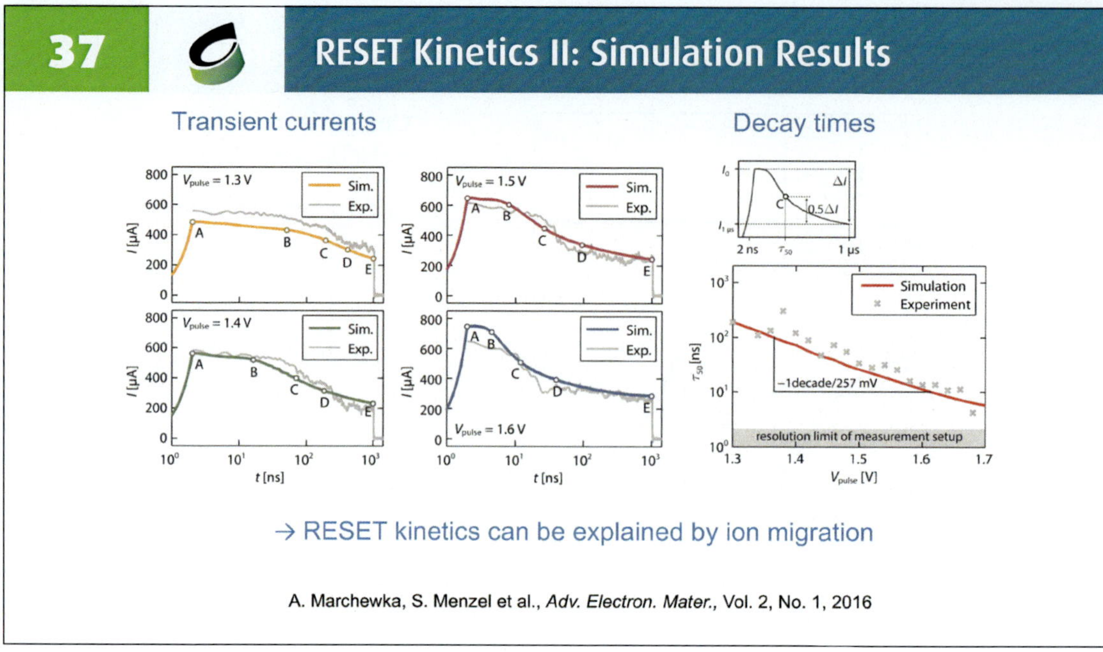

Gives the simulation results of RESET kinetics II.

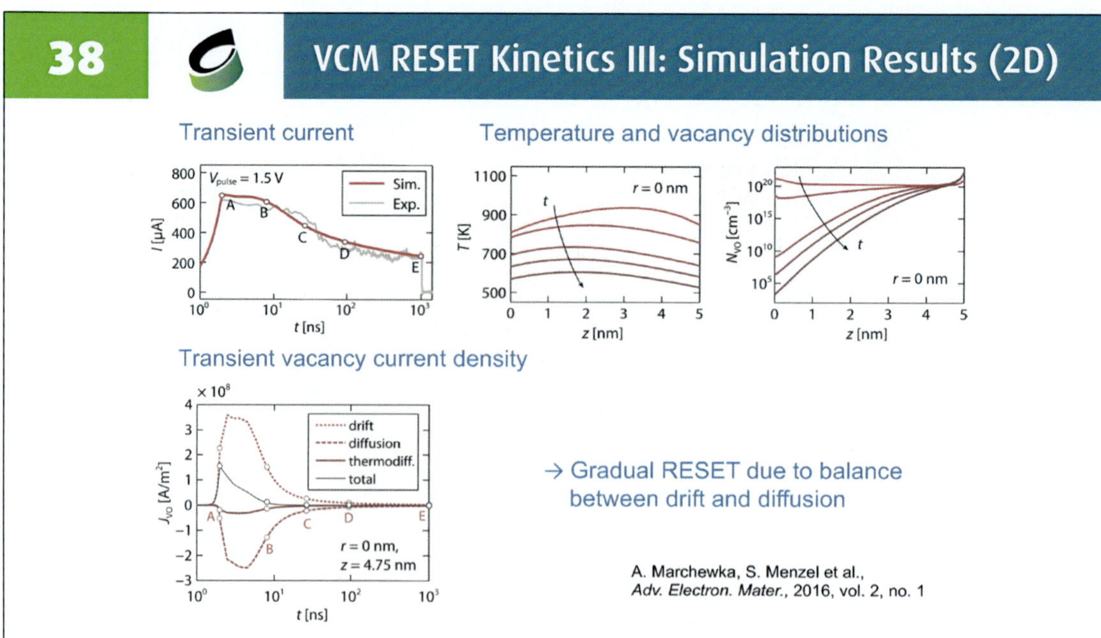

Gives the simulation results of RESET kinetics III.

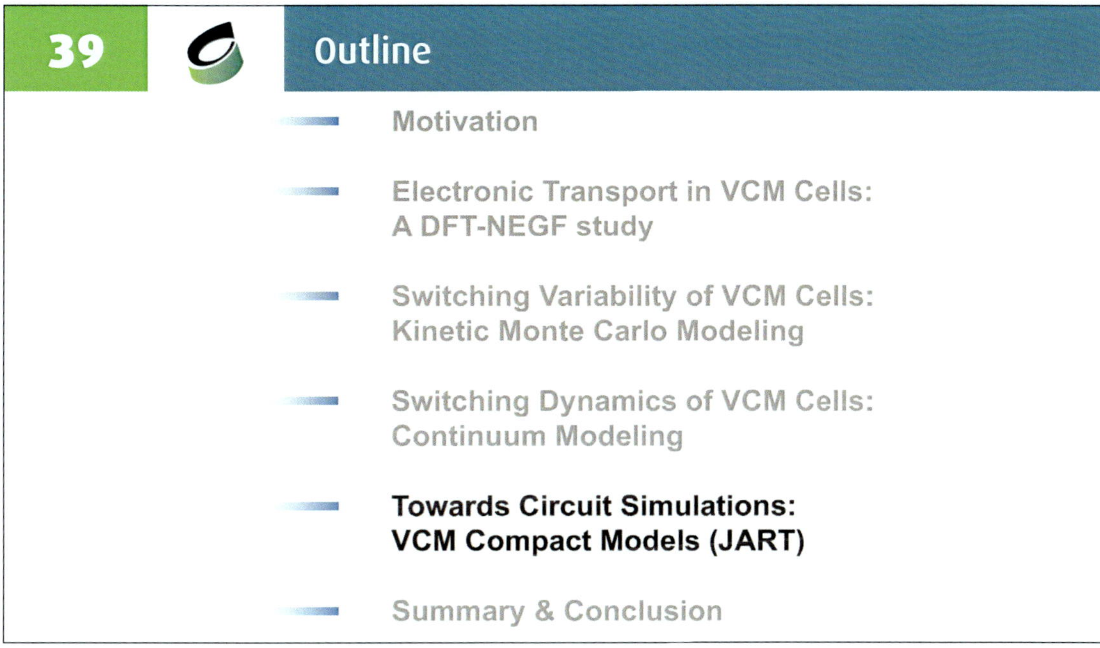

Introductory slide to the topic "Towards Circuit Simulations: VCM Compact Models (JART)".

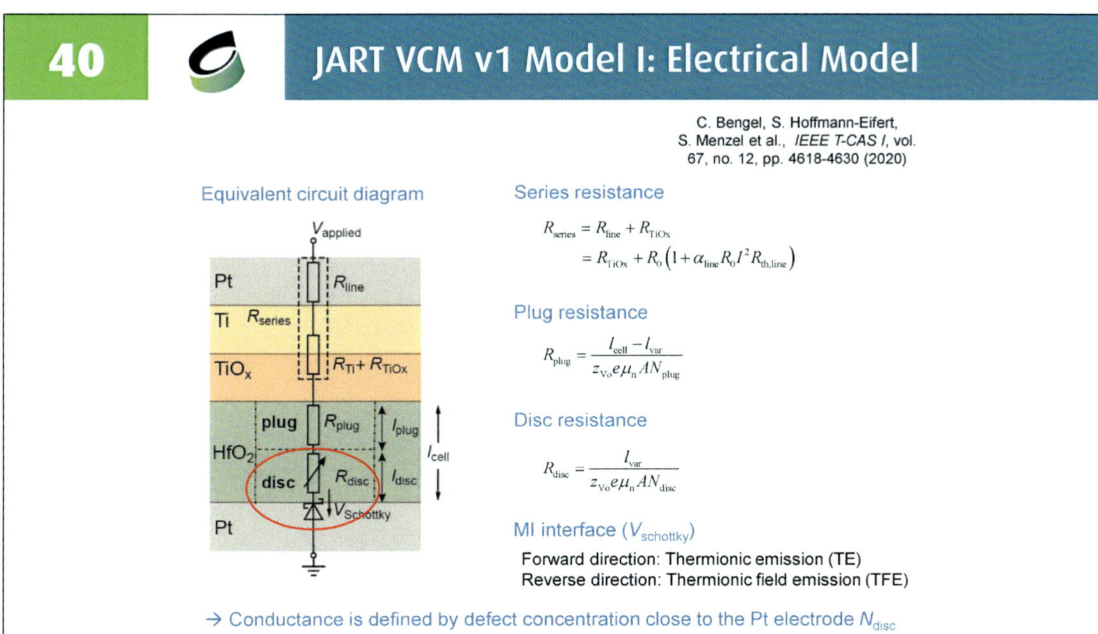

Focuses on the electrical aspects and behavior of different components of JART VCM v1 Model I.

41 JART VCM v1 Model II: Ionic Model

Driving force for is ionic current

$$\frac{dN_{\text{disc}}}{dt} = -\frac{I_{\text{ion}}}{e z_{\text{Vo}} A l_{\text{var}}}$$

$$I_{\text{ion}} = e z_{\text{Vo}} c_{\text{Vo}} a_{\text{Vo}} F_{\text{limit}} A \cdot \left(\exp\left(-\frac{\Delta W_{\text{min}}}{k_B T}\right) - \exp\left(-\frac{\Delta W_{\text{max}}}{k_B T}\right) \right) = f(E,T)$$

Electric field

$$\text{FB}: E = \left(V_{\text{disc}} + V_{\text{Schottky}} + V_{\text{plug}}\right) / l_{\text{cell}}$$

$$\text{RB}: E = V_{\text{disc}} / l_{\text{disc}}$$

Joule heating

$$T = T_0 + R_{\text{th,SET}} P_{\text{device}} \quad \text{SET}$$

$$T = T_0 + R_{\text{th,RESET}} P_{\text{device}} \quad \text{RESET}$$

Ion in periodic lattice

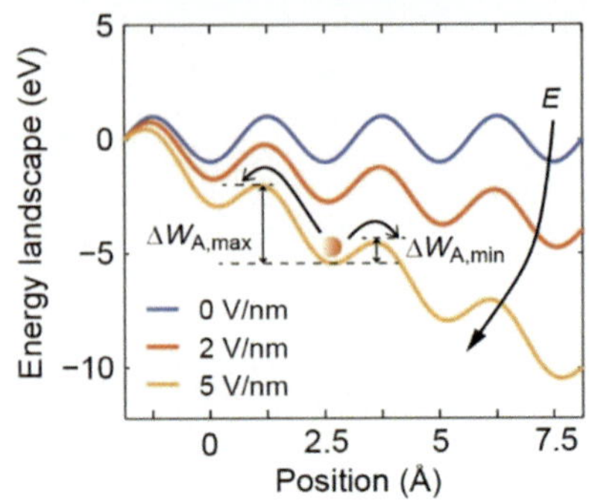

S. Menzel et al., *Faraday Discussions* vol. 213, pp. 197-213 (2018)

A. Genreith-Schriever, et al., *Phys. Rev. B* vol. 94, no. 22, pp. 224304 (2016)

Focuses on the ionic aspects of JART VCM v1 Model II.

42 Physical Origin of the Switching Variability

PEEM analysis of $SrTiO_3$-based cells

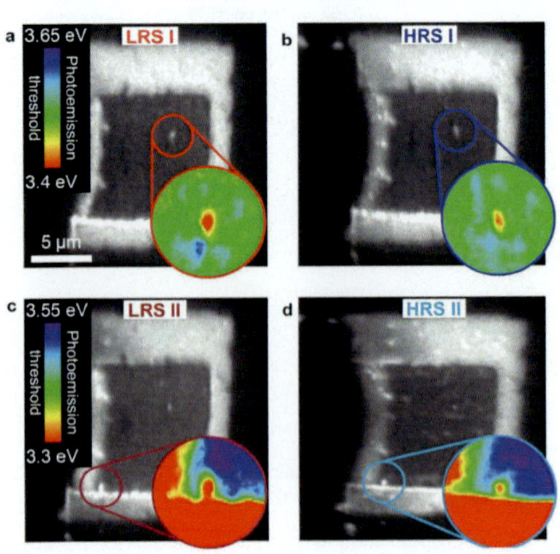

→ Parallel growth of multiple filaments

→ One filament „wins" during SET

→ Variability due to change of filament location and geometry

C. Baeumer, R.Dittmann, et al., *ACS Nano*, vol. 11, no. 7, pp. 6921-6929 (2017)

Discusses the physical origin of switching variability in $SrTiO_3$-based cells.

43 JART VCM v1: Modeling the Switching Variability I

→ Variability due change in filament properties

Device-to-device variability

Parameter are chosen randomly from a truncated Gaussian distribution

- Minimum disc concentration $N_{\mathrm{disc,min}}$
- Maximum disc concentration $N_{\mathrm{disc,max}}$
- Filament radius r
- Disc length l

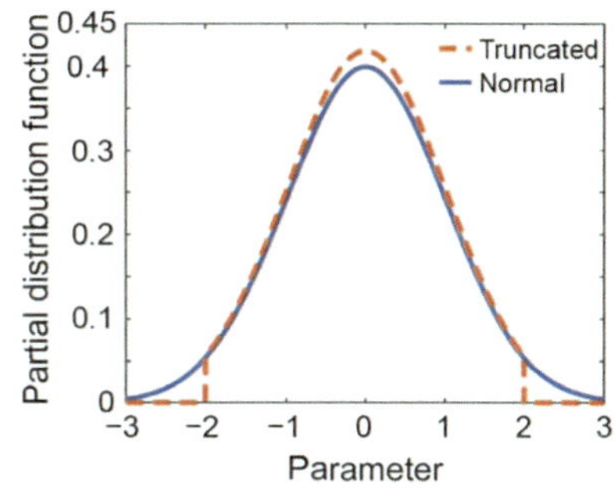

C. Bengel, S. Hoffmann-Eifert, S. Menzel et al., *IEEE T-CAS I*, vol. 67, no. 12, pp. 4618-4630 (2020)

Focuses the modeling of switching variability observed in devices attributed to changes in filament properties, disc concentrations etc.

44 JART VCM v1: Modeling the Switching Variability II

Cycle-to-cycle variability

Random walk algorithm is used to change the parameters from c2c

$$\xi_k = \xi_{k-1}\left(1 \pm \Delta\xi_{\max} \cdot P\right) \qquad P \in (0,1)$$

Change of r, and l depends on N_{disc}

SET

$$r_{\mathrm{var}} = r_{\mathrm{old}} + \left(r_{\mathrm{new}} - r_{\mathrm{old}}\right)\left(\frac{N_{\mathrm{disc}} - N_{\mathrm{disc,old}}}{N_{\mathrm{disc,max,var}} - N_{\mathrm{disc,old}}}\right)$$

$$l_{\mathrm{var}} = l_{\mathrm{old}} + \left(l_{\mathrm{new}} - l_{\mathrm{old}}\right)\left(\frac{N_{\mathrm{disc}} - N_{\mathrm{disc,old}}}{N_{\mathrm{disc,max,var}} - N_{\mathrm{disc,old}}}\right)$$

RESET

$$r_{\mathrm{var}} = r_{\mathrm{old}} + \left(r_{\mathrm{new}} - r_{\mathrm{old}}\right)\left(\frac{N_{\mathrm{disc,old}} - N_{\mathrm{disc}}}{N_{\mathrm{disc,old}} - N_{\mathrm{disc,min,var}}}\right)$$

$$l_{\mathrm{var}} = l_{\mathrm{old}} + \left(l_{\mathrm{new}} - l_{\mathrm{old}}\right)\left(\frac{N_{\mathrm{disc,old}} - N_{\mathrm{disc}}}{N_{\mathrm{disc,old}} - N_{\mathrm{disc,min,var}}}\right)$$

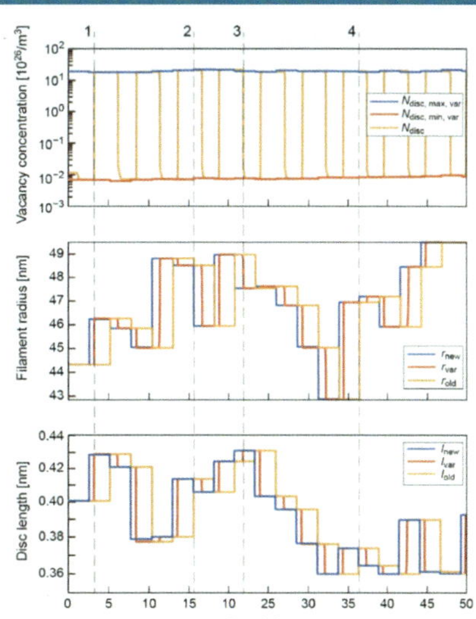

Studies the cycle-to-cycle variability in the JART VCM v1 model.

45 JART VCM v1: Modeling Read Noise

Physical Origin

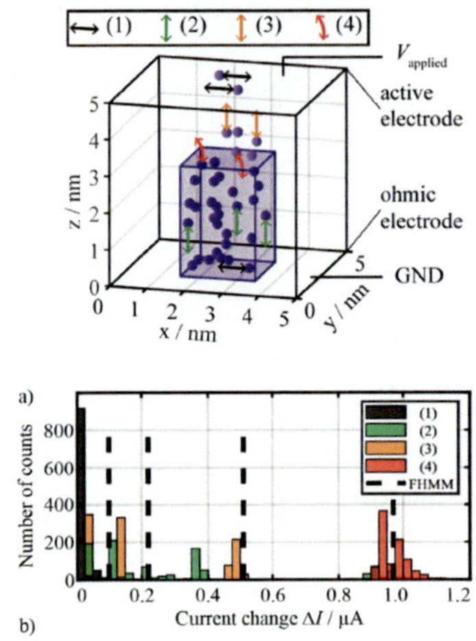

Read instability model

→ Most important jump between plug to disc

→ Change of N_{disc} by ±1, ±2 defects

Implemented using state machine

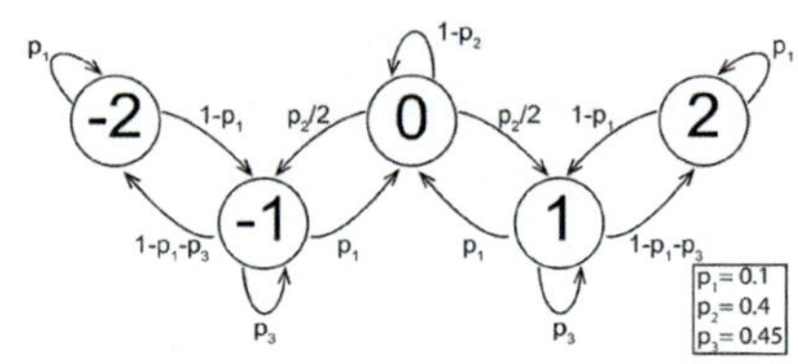

S. Wiefels, S. Menzel et al., *IEEE T-ED*, vol. 67, no. 10, pp. 4208-4215 (2020)

Describes the modeling of read noise in the JART VCM v1 model, with a focus on its physical origin and the associated read instability model.

46 Compact Modeling of SET Dynamics

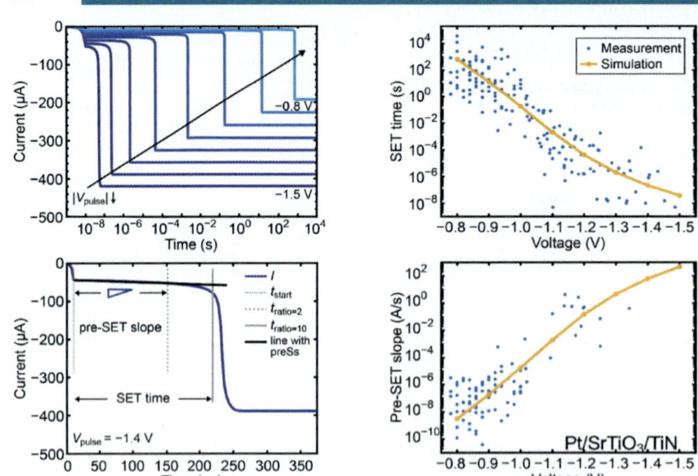

K. Fleck, S. Menzel et al., *Phys. Rev. Applied*, Vol. 6, No. 6, 064015, 2016
C. La Torre, PhD dissertation, 2019, RWTH Aachen University

Refers to the compact modeling of SET dynamics in a Pt/$SrTiO_3$/TiN device structure.

47 Compact Modeling of RESET Dynamics

RESET transients | RESET kinetics

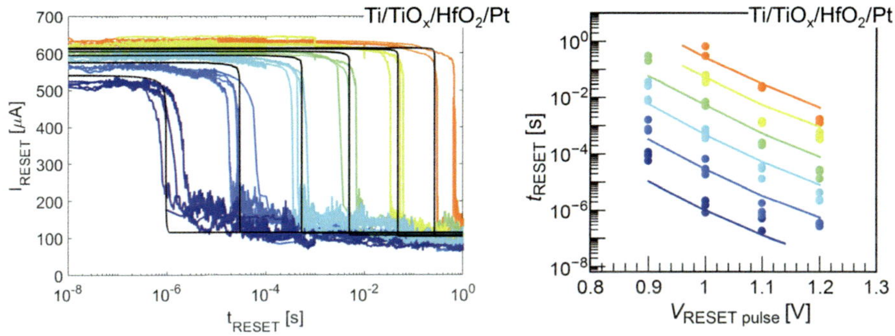

→ Voltage-divider between series resistance and actual device leads to strong state-dependence

F. Cüppers, S. Menzel et al., *APL Materials* vol. 7, no. 9, pp. 91105 (2019)

Refers to the compact modeling of RESET dynamics in a $Pt/SrTiO_3/TiN$ device structure.

48 Switching Variability I

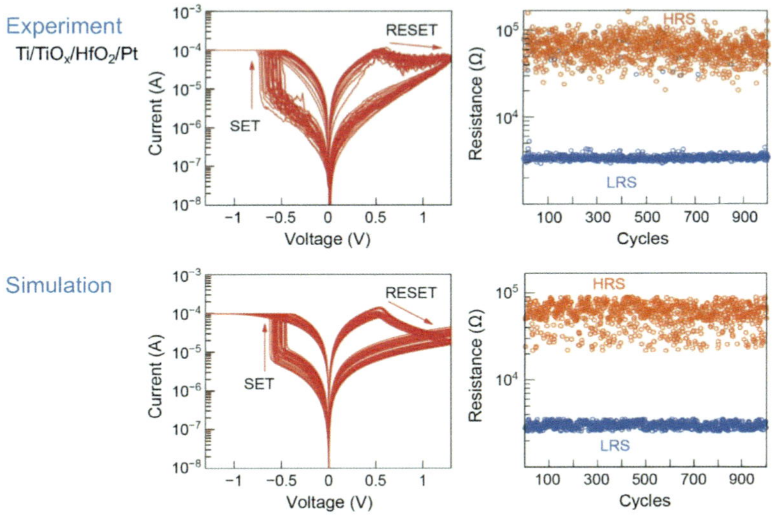

Discuss the current versus resistance simulation results related to variability experiments in resistive switching devices, such as memristors or RRAM, with a Ti/TiOx/HfO2/Pt structure.

49 Switching Variability II

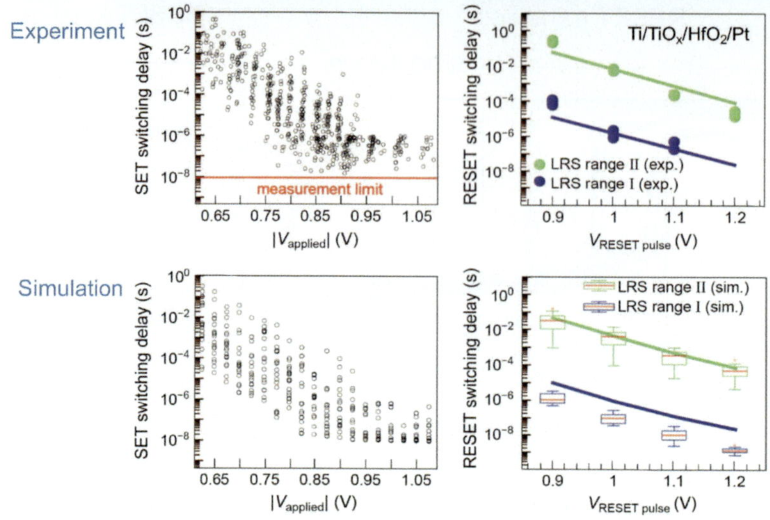

Discuss the SET and RESET switching delays versus voltage applied simulation results related to variability experiments in resistive switching devices, such as memristors or RRAM, with a Ti/TiOx/HfO2/Pt structure.

50 Switching Variability III

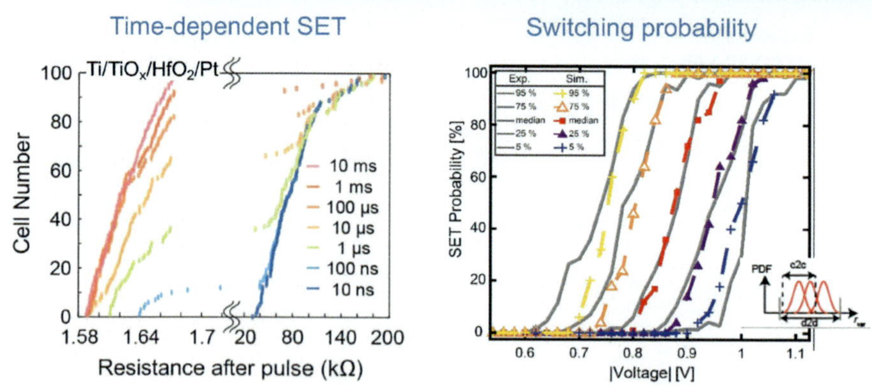

C. Bengel, S. Menzel et al., *IEEE T-CAS I*, vol. 67, no. 12, pp. 4618-4630 (2020)

C. Bengel, S. Menzel et al., *Frontiers in Neuroscience*, (2021)

→ „digital" switching observed

→ Different devices have similar c2c but different mean

Outlines the time-dependent SET behavior and switching probability in Ti/TiOx/HfO2/Pt devices.

51 Simulation of Read Instability: JART VCM 1b

Simulated *I-t* curve

Observed current jumps similar to experimental ones

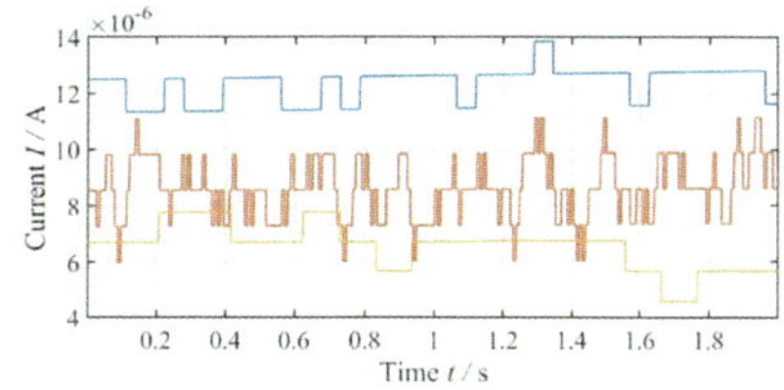

Simulated and experimental read distribution

- Log-normal distribution observed
- Log-normal distribution recovers after shaping experiment

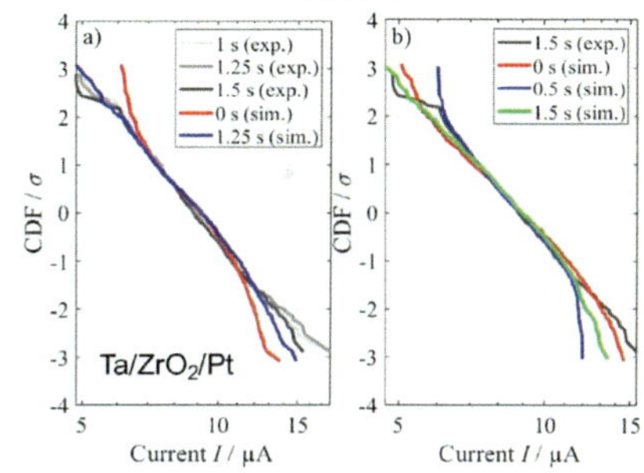

S. Wiefels, S. Menzel et al., *IEEE T-ED*, vol. 67, no. 10, pp. 4208-4215 (2020)

Discusses the read instability in Ta/ZrO2/Pt-based devices using the JART VCM 1b model.

52 Read Disturb with SET Polarity

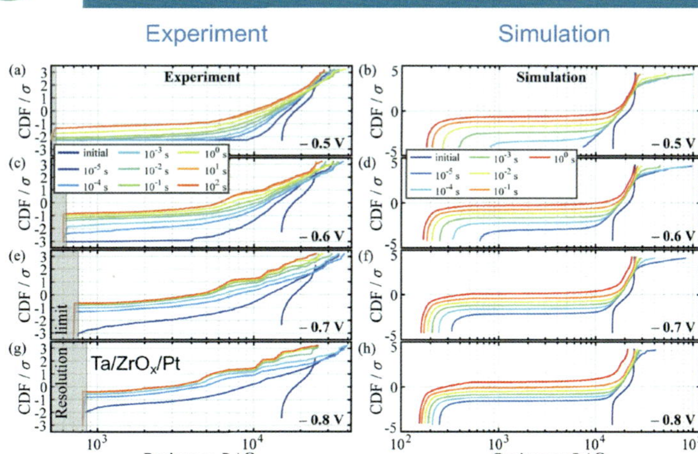

→ Digital switching observed

C. Bengel, S. Menzel et al., *Neuromorph. Comput. Eng.*, 2 034001 (2022)

Mentions the digital switching in the behavior of Ta/ZrOx/Pt-based devices as "READ disturb with SET polarity".

53 Read Disturb with RESET Polarity

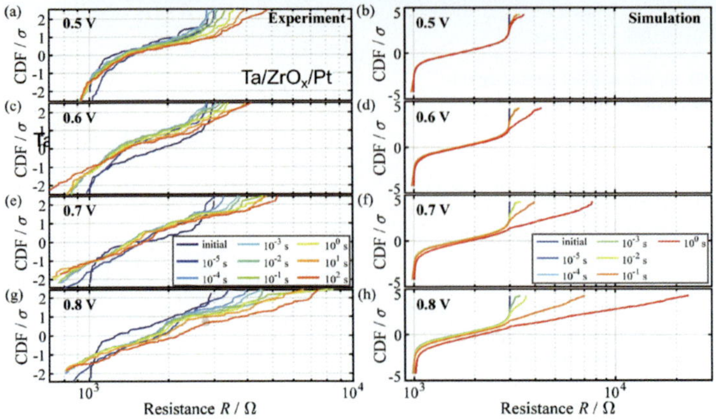

→ Gradual drift of LRS state

→ Very stable up to 0.5 V read voltage

C. Bengel, S. Menzel et al., *Neuromorph. Comput. Eng., 2 034001* (2022)

Investigates the read disturb with RESET polarity and the cumulative distribution function of SET time in Ta/ZrOx/Pt-based devices.

54 Outline

- Motivation
- Electronic Transport in VCM Cells: A DFT-NEGF study
- Switching Variability of VCM Cells: Kinetic Monte Carlo Modeling
- Switching Dynamics of VCM Cells: Continuum Modeling
- Towards Circuit Simulations: VCM Compact Models (JART)
- **Summary & Conclusion**

Introductory slide to "Summary and Conclusion".

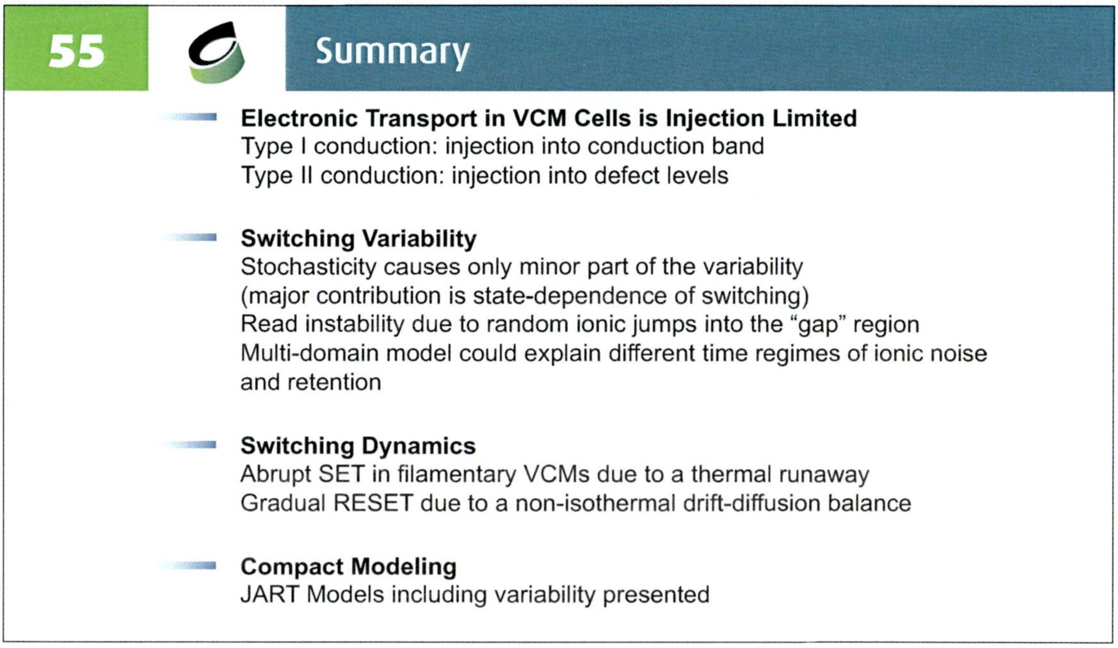

Summarizes the understanding of electronic transport, variability, dynamics, and compact modeling in VCM cells, offering insights into the underlying mechanisms and providing a framework for further research and optimization in this field.

Conclusion slide with the link to the JART website and the funding agency is mentioned.

SECTION 2: NONLINEAR MEMRISTIVE CIRCUITS

CHAPTER 04

Analog Neural Computing Under Variability Intense Memristor Arrays

Prof. Alex James

Professor of AI Hardware and Graphene Aurora Chair
Professor, Digital University Kerala

The presentation focuses on the utilization of analog neural computing through memristor arrays. It outlines the advantages of analog computing in neural networks, particularly in edge AI computing, emphasizing local data processing. By tracing the correlation between transistor scaling and AI literature growth, the pivotal role of hardware in AI development is underscored. Memristors, particularly in crossbars, are highlighted for their significance in analog computing. Various topics explored include the AI chip landscape, variability in analog neural computing, and backpropagation techniques. Additionally, the presentation delves into Hierarchical Temporal Memory (HTM) and its comparison with traditional neural network architectures. Novel concepts such as super-resolution crossbars and bridges using memristor technology are introduced, alongside techniques for expediting analog memristive chip design. The presentation concludes with a discussion on the challenges, recent advancements, and design considerations for achieving energy-efficient memristive chips. Overall, it offers a comprehensive overview of analog neural computing's potential applications and challenges.

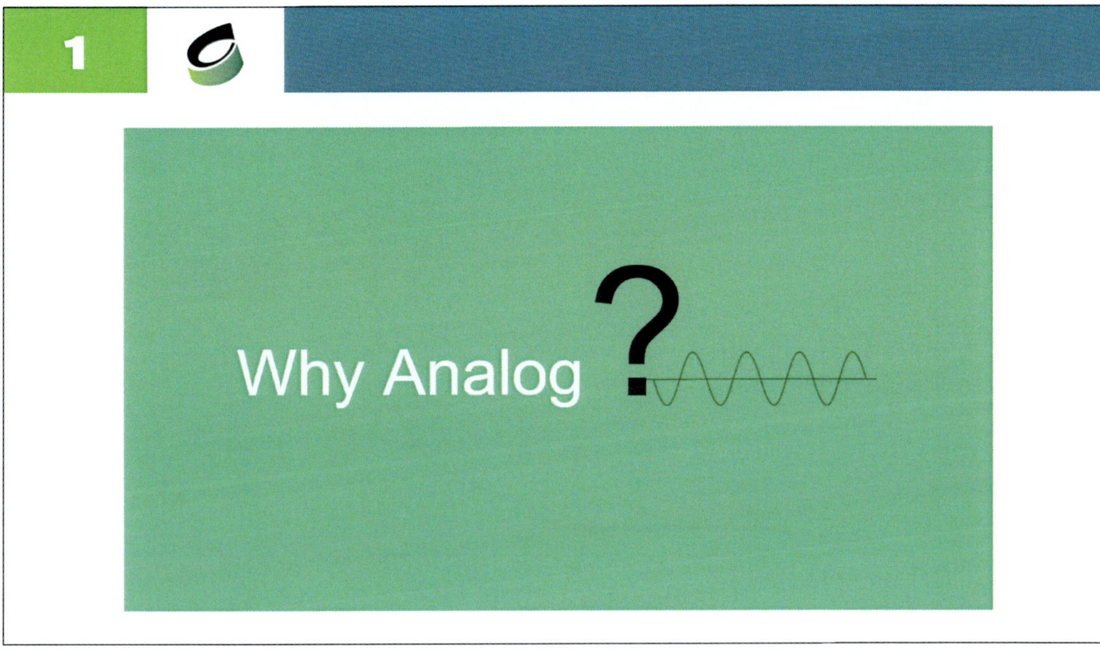

This slide discusses the rationale or advantages of utilizing analog computing in neural networks.

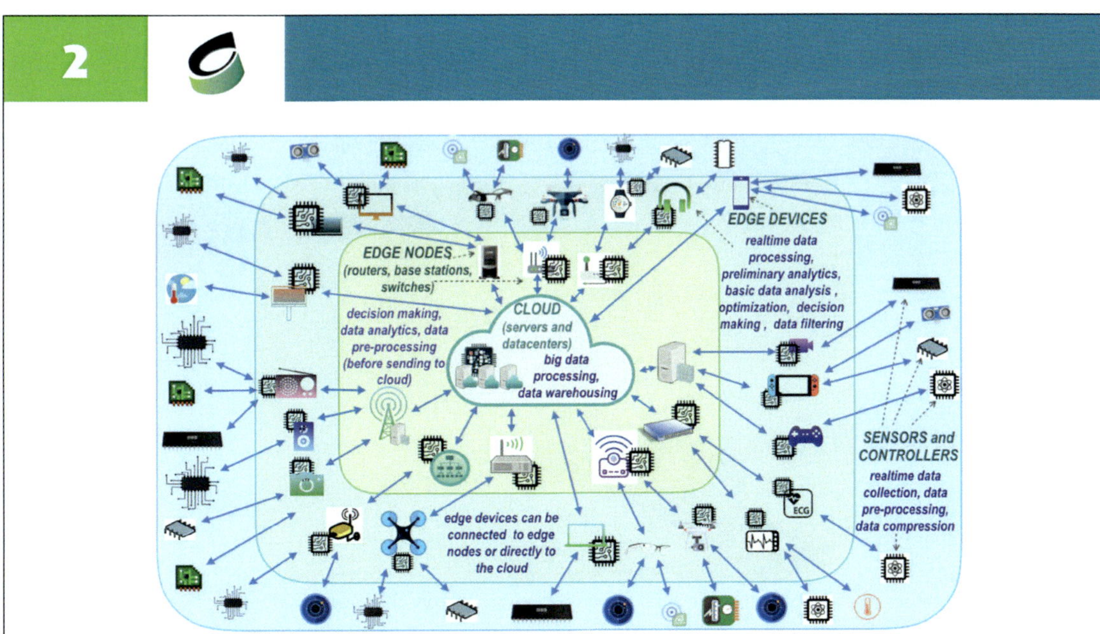

The focus is on edge artificial intelligence (AI) computing, which involves processing data locally on devices rather than relying on cloud-based servers.

3

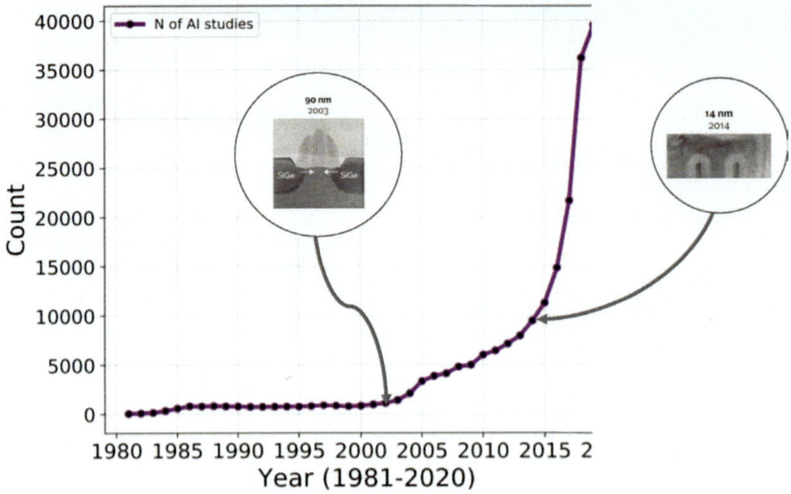

There is a strong correlation between the evolution of transistor scaling and that of the growth of AI literature, indicating that hardware plays a major role in the development of other domains.

4

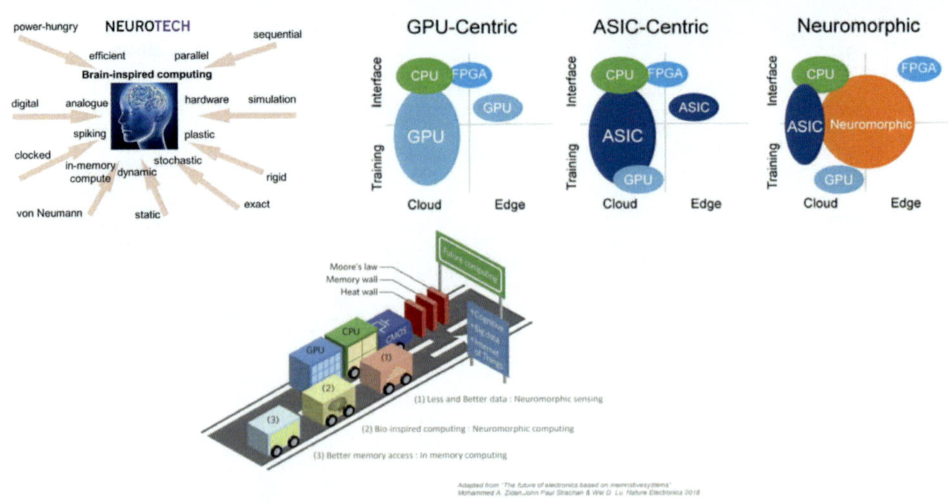

The neuromorphic computing approach has the potential to break through the limitations of conventional computing hardware, and extend the possibility of having learning in edge devices.

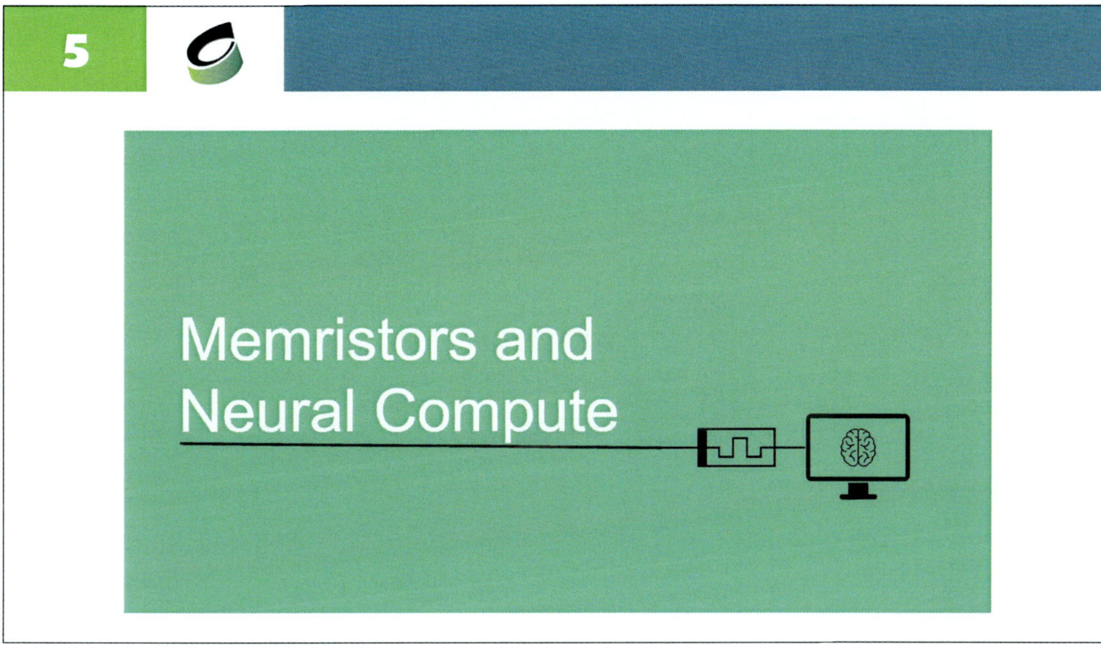

Discusses the relationship between memristors and neural computing, highlighting the significance of memristors in this context.

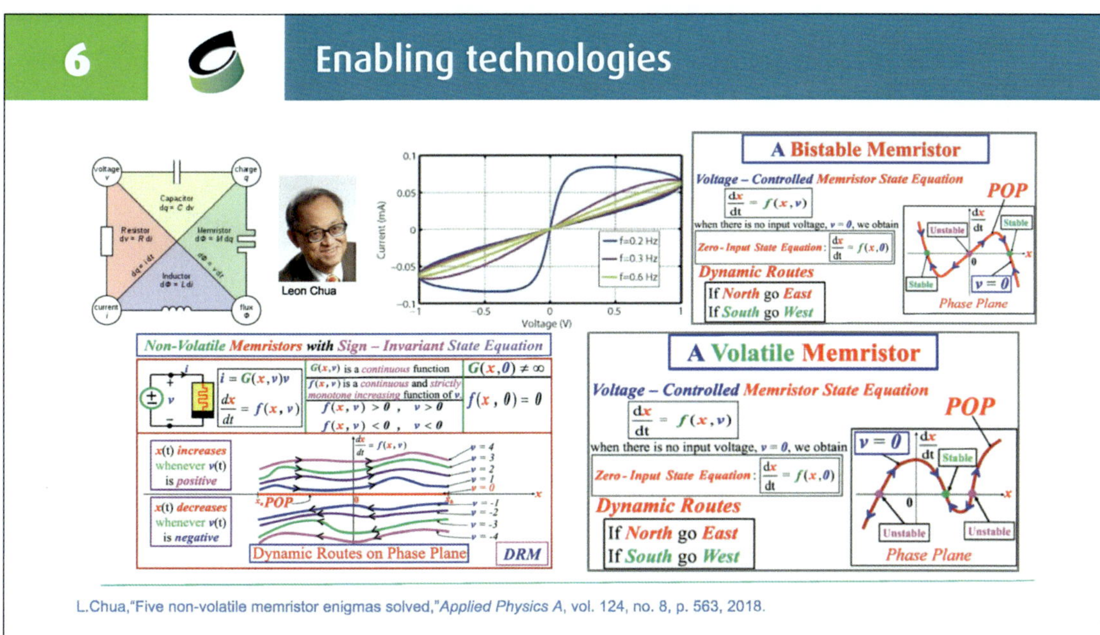

Mentions the work of Leon Chua and likely elaborates on the advancements in memristor technology that have enabled progress in neural computing.

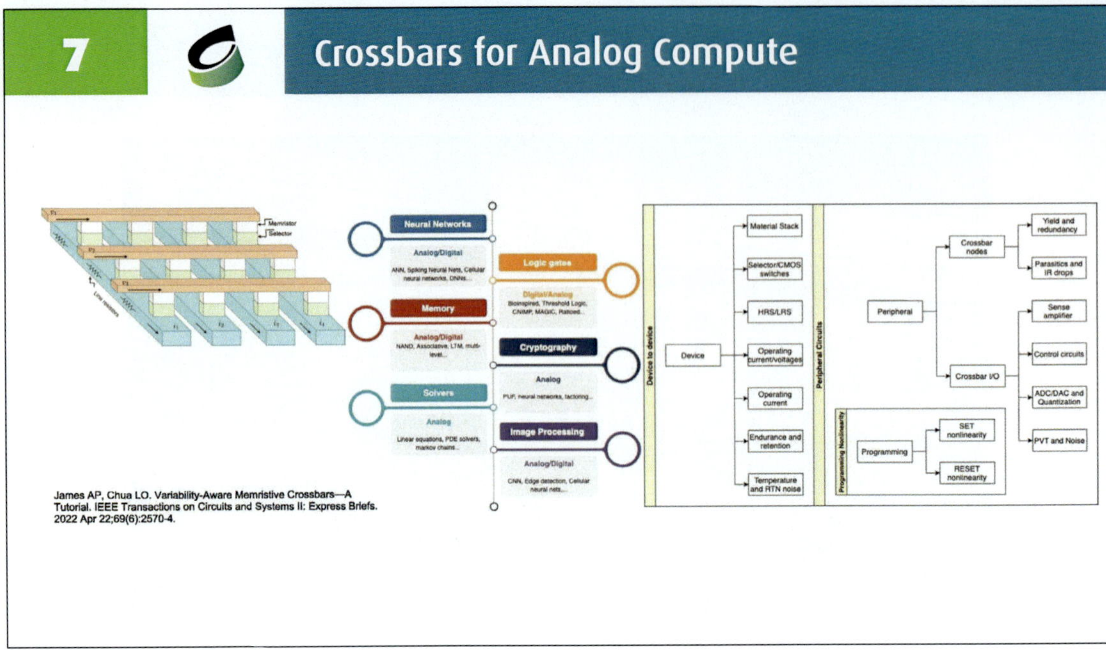

Discusses the use of crossbars in analog computing, particularly focusing on variability-aware memristive crossbars.

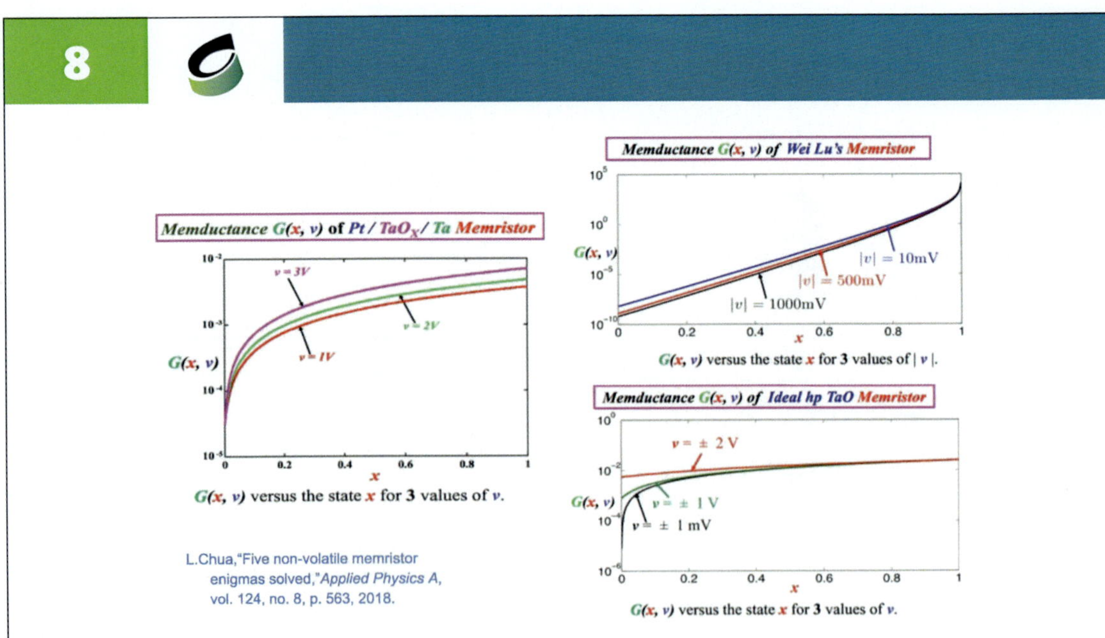

Reiterates the work of Leon Chua regarding memristor enigmas being solved, possibly in the context of analog computing.

9 Single layer neural network

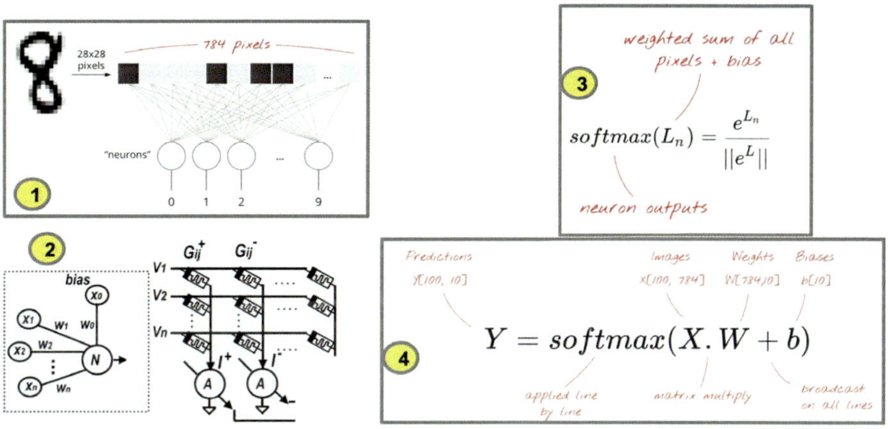

Krestinskaya O, James AP, Chua LO. Neuromemristive circuits for edge computing: A review. IEEE transactions on neural networks and learning systems. 2019 Mar 14;31(1):4-23.

Introduces a single-layer neural network and references a review on neuromemristive circuits for edge computing by Krestinskaya, James, and Chua for introduction to the topic.

10 Terms

Performance refers to the depth of an AI system's capabilities, i.e., how it compares to human- level performance for a given task.

Generality refers to the breadth of an AI system's capabilities, i.e., the range of tasks for which an AI system reaches a target performance threshold.

Defines and explains terms related to AI systems' performance and generality, providing clarity on these concepts.

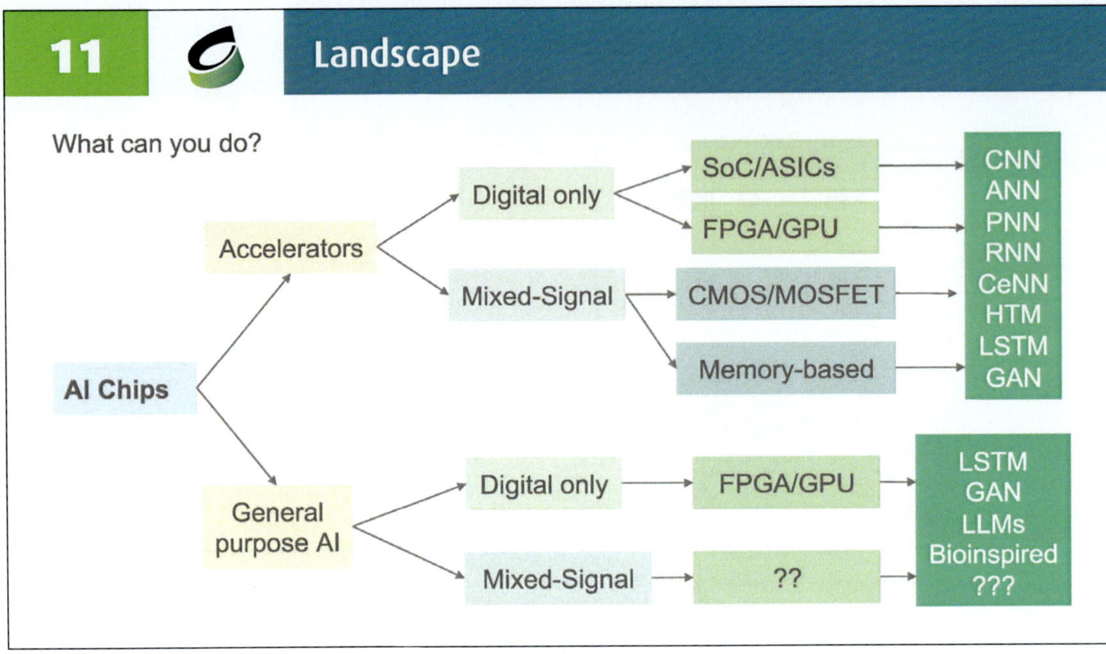

Provides an overview of the AI chip landscape, including different types of accelerators and architectures used in AI systems.

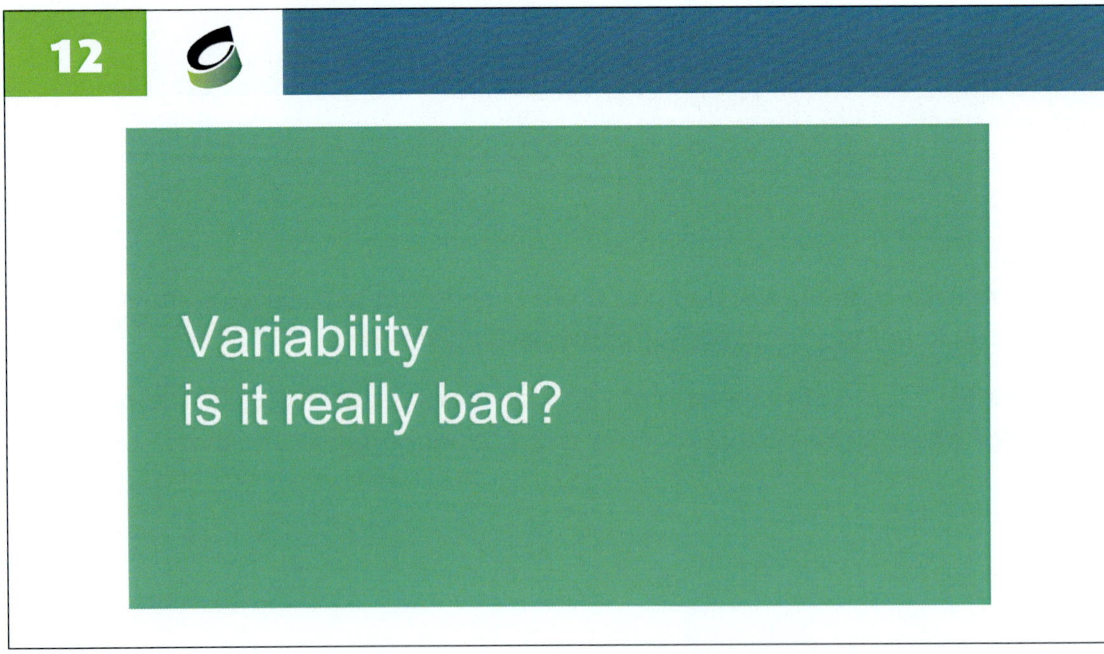

Discusses the concept of variability in the context of analog neural computing, exploring its implications and potential challenges.

13 Analog neural nets

Multiply and accumulate (MAC) units

Scaling MAC
- Sneek paths
- Smaller crossbars - lower sneak path errors
- Smaller crossbars - modular crossbar array(M2CA)

Non-idealities
- Devices
- Variability, aging, radiation
- Wire resistance and caps
- Interconnect issues
- Parasitics from selector devices and switches
- packaging issues, thermal, noise
- Read-out circuits
- Amplifiers

Failures
- Devices
- Rows/Columns

Energy

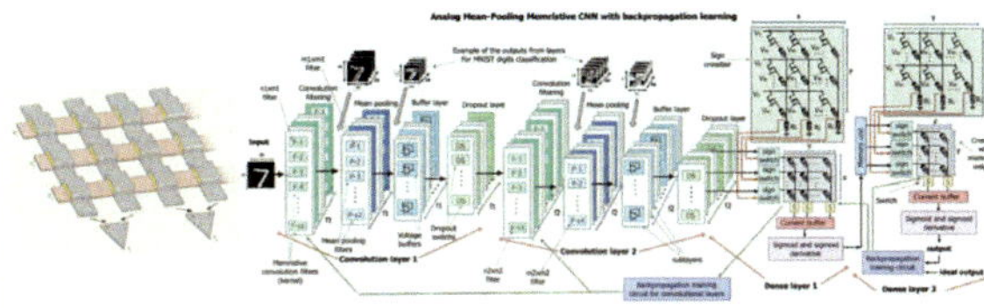

PERFORMANCE ACCURACY OF ANN, CNN AND BNN FOR DIFFERENT VARIABILITY OF MEMRISTIVE DEVICES

Architecture	Variation in memristive devices			
	10%	20%	30%	40%
	Simulation accuracy (%)			
ANN	98.4 ± 0.05	98.25 ± 0.1	98.05 ± 0.2	97.4 ± 0.4
CNN	93.3 ± 0.07	93.2 ± 0.13	93.05 ± 0.3	92.75 ± 1.2
BNN	97.70 ± 0.1	97.5 ± 0.2	97.25 ± 0.3	96.85 ± 0.4

VARIATION OF PERFORMANCE ACCURACY OF 2 LAYER ANN FOR DIFFERENCE NUMBER OF STABLE RESISTANCE LEVELS IN MEMRISTIVE DEVICES.

Levels	2	5	8	16	32	64	ideal
Accuracy (%)	9.80	75.43	88.40	95.38	95.43	95.44	95.49

Explores various aspects related to analog neural networks, including energy considerations, backpropagation, and challenges such as variability and non-idealities.

14 Backpropagation

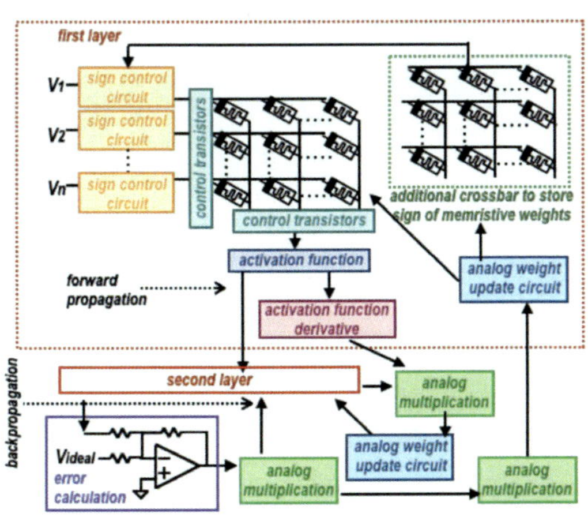

Derivative of Error with respect to weight

Old weight

$$^{*}W_x = W_x - a\left(\frac{\partial Error}{\partial W_x}\right)$$

New weight

Learning rate

O. Krestinskaya, K. N. Salama, and A. P. James, "Learning in memristive neural network architectures using analog backpropagation circuits," *IEEE Transactions on Circuits and Systems I: Regular Papers*, pp. 1–14, 2018.

Example of analog backpropagation implementation is shown.

15

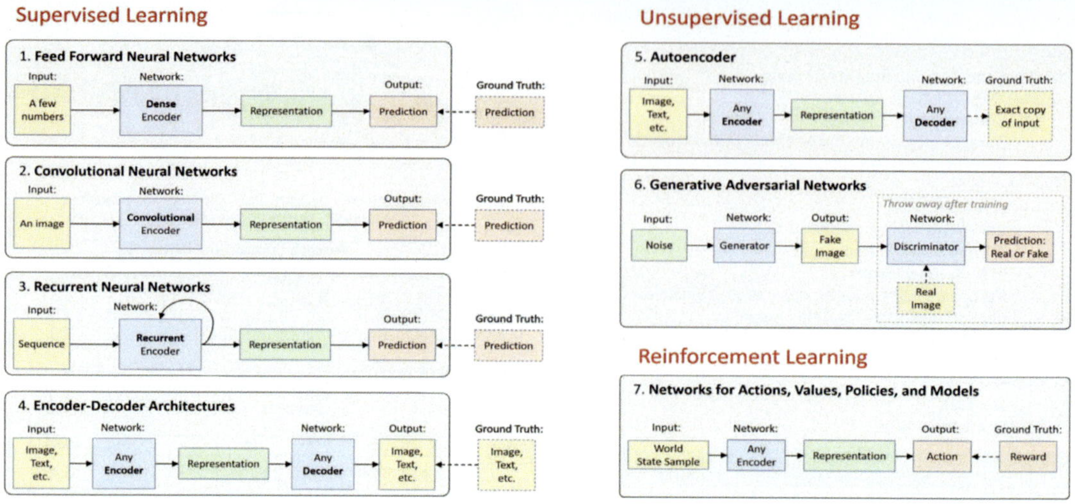

Different types of neural network configurations are graphically illustrated.

16 Hierarchical Temporal Memory (HTM)

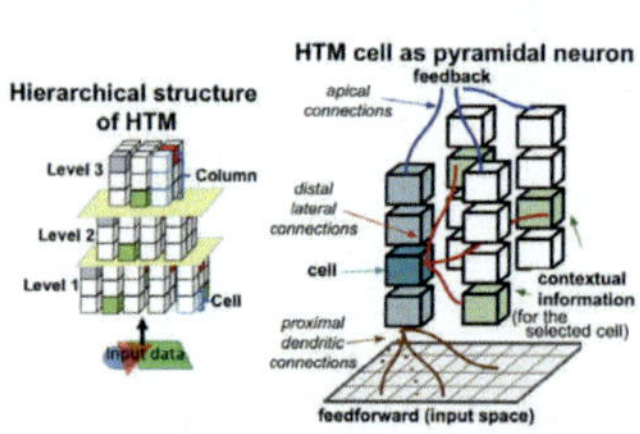

HTM is an algorithm is inspired by fundamental principles of human brain processing and mimics the processes in human neocortex

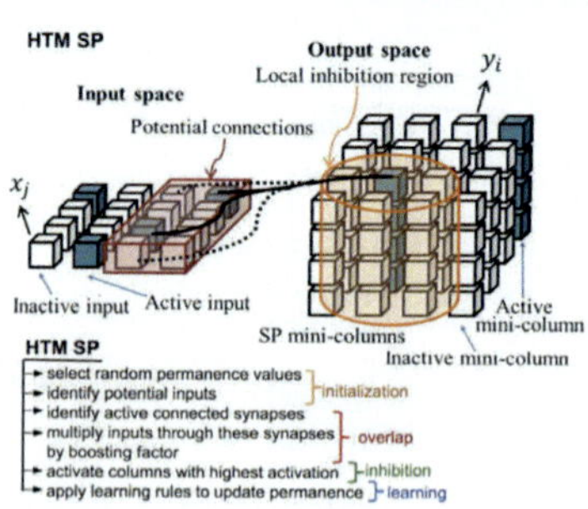

Spatial Pooler (SP): feature encoding, generating sparse distributed representation of input data

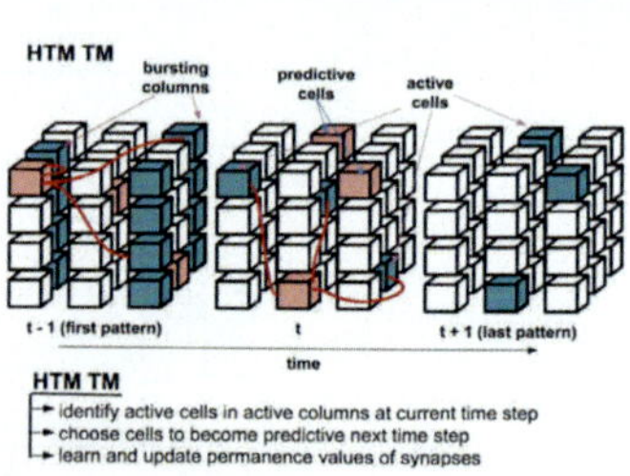

Temporal Pooler (TM): examines temporal changes of the input data and makes predictions based on previous experience

J. Hawkins and D. George, "Hierarchical temporal memory: Concepts, theory and terminology," Technical report, Numenta, Tech. Rep., 2006.
O. Krestinskaya, I. Dolzhikova and A. P. James, "Hierarchical Temporal Memory Using Memristor Networks: A Survey," in *IEEE Transactions on Emerging Topics in Computational Intelligence*, vol. 2, no. 5, pp. 380-395, Oct. 2018.

HTM is an algorithm is inspired by fundamental principles of human brain processing and mimics the processes in human neocortex Introduces Hierarchical Temporal Memory (HTM) as an algorithm inspired by human brain processing.

17 Hierarchical Temporal Memory (HTM)

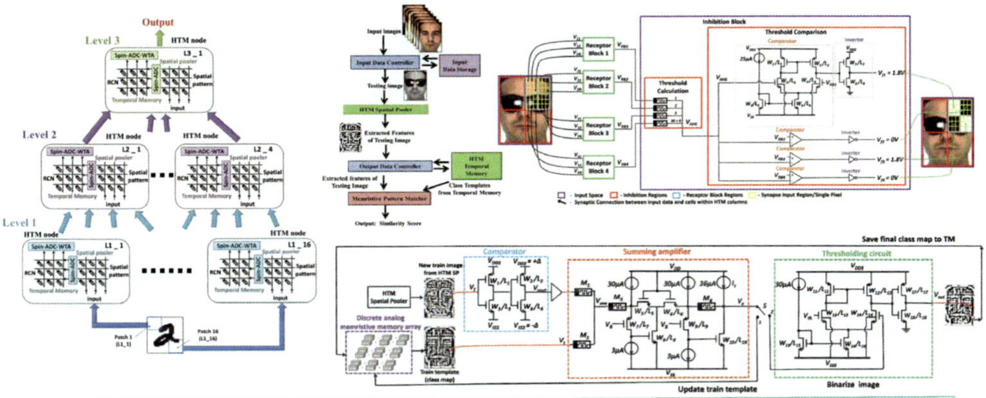

D. Fan, M. Sharad, A. Sengupta, and K. Roy, "Hierarchical temporal memory based on spin-neurons and resistive memory for energy- efficient brain-inspired computing," *IEEE transactions on neural networks and learning systems*, vol. 27, no. 9, pp. 1907–1919, 2016.
O. Krestinskaya, T. Ibrayev, and A. P. James, "Hierarchical temporal memory features with memristor logic circuits for pattern recognition,"*IEEE Transactions on Computer-Aided Design of Integrated Circuits and Systems*, vol. PP, no. 99, pp. 1–1, 2017.

Discusses HTM based on spin-neurons and resistive memory for brain-inspired computing, referencing works by Fan, Sharad, Sengupta, Roy, and Krestinskaya, Ibrayev, James.

18

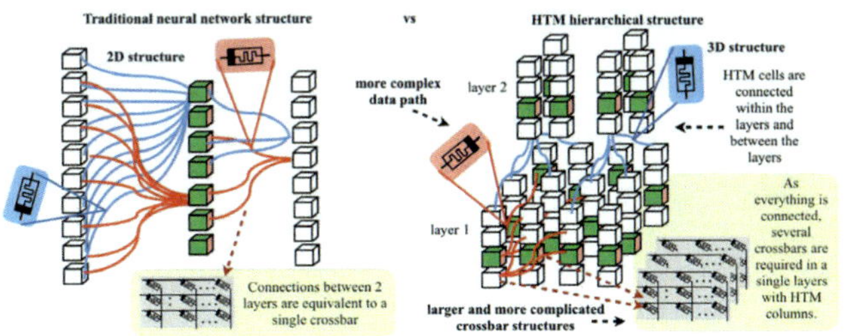

HTM versus CNN-LSTM

Smagulova K, Krestinskaya O, James AP. Who is the winner? memristive-cmos hybrid modules: Cnn-lstm versus htm. ***IEEE transactions on biomedical circuits and systems***. 2019 Nov 28;14(2):164-72.

HTM has high generalization ability
- Difficult to train

CNN-LSTM higher performance accuracy
- Easy to train

Compares HTM and CNN-LSTM in terms of generalization ability and training difficulty.

19 Dropouts - curious case

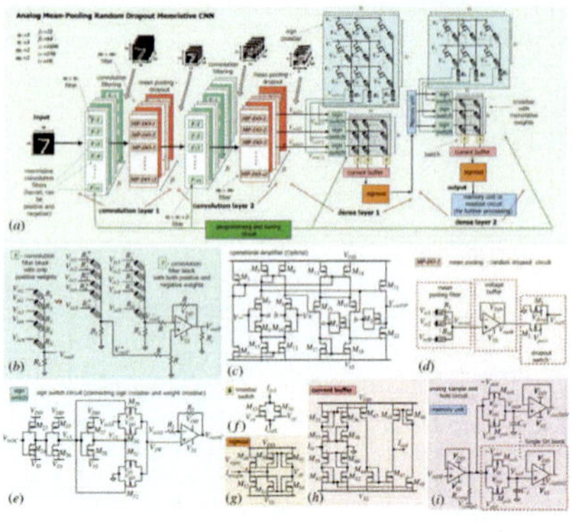

Krestinskaya O, James AP. Analogue neuro-memristive convolutional dropout nets. Proceedings of the Royal Society A. 2020 Oct 28;476(2242):20200210.

weights in conv. layer	accuracy MNIST ideal	MNIST hardware (64 levels)	fashion ideal	fashion hardware (64 levels)	Cifar10 ideal	Cifar10 hardware (64 levels)
only positive	99.6%	89.5%	97.6%	81.4%	90.6%	84.9%
positive and negative		99.5%		84.5%		89.9%

weights in conv. layer	maximum power per filter (3 × 3)	area per filter(3 × 3) *A*-memristor area	area from [34]	notes
only positive	0.043 mW*	$9A + 24.3\ \mu m^2$	$24.45\ \mu m^2$	can be parallel
positive and negative	38.26 mW**	$2(9A + 24.3\ \mu m^2) + 54.432\ \mu m^2$	$103.33\ \mu m^2$	requires sequential processing due to high power

*0.086 pJ per operation (or 0.86 pJ per a single convolution in 3 × 3 filter) for $f = 25$ MHz.
**0.0765 nJ per operation (or 0.765 nJ per a single convolution in 3 × 3 filter) for $f = 25$ MHz.

Delves into the concept of dropouts in neural networks, specifically in the context of analog neuro-memristive convolutional dropout nets.

20 Dropouts - curious case

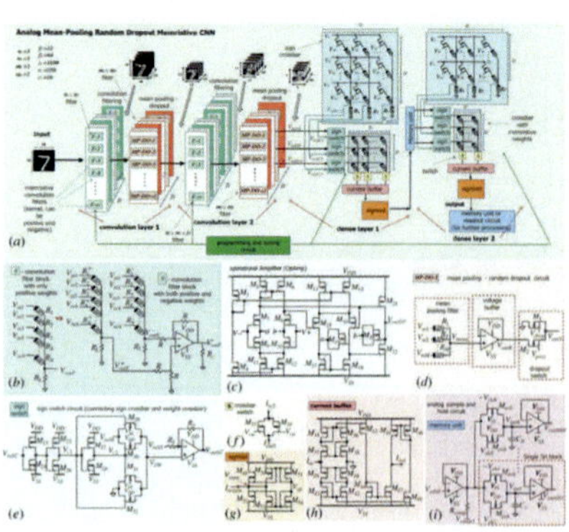

Krestinskaya O, James AP. Analogue neuro-memristive convolutional dropout nets. Proceedings of the Royal Society A. 2020 Oct 28;476(2242):20200210.

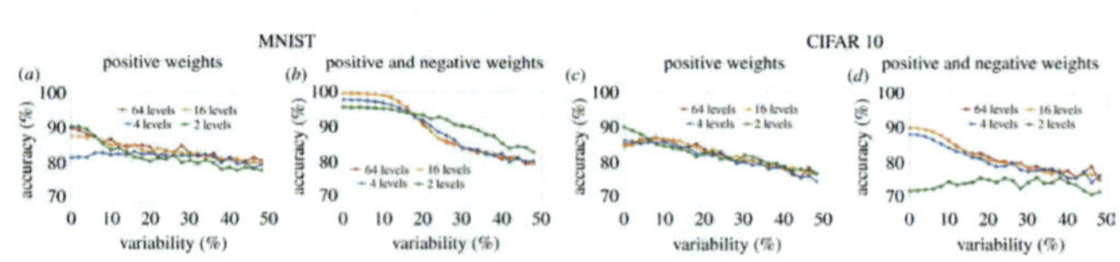

Main trends: –Accuracy decreases with memristor variability.
–for 'only positive' convolution weights, difference between different number of levels (stable resistive states) is not significant. But for 'positive/negative' case, 2–4 levels produce lower accuracy than 16–64 levels.

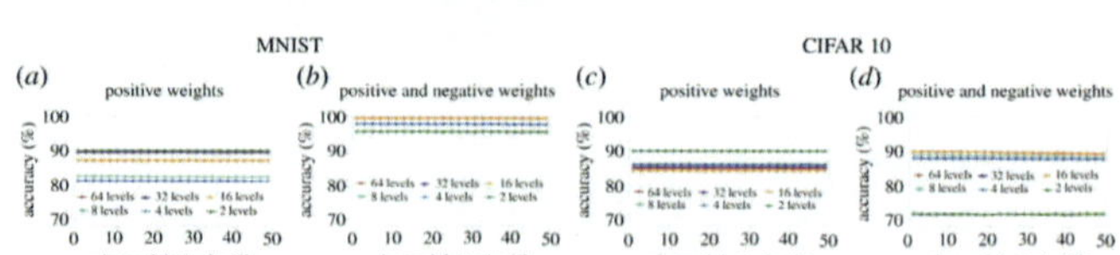

Main trends: –Adversarial attacks do not affect the performance (at least up to 50%).

Indicates that increase in variability can decrease the accuracy.

21 Dropouts - curious case

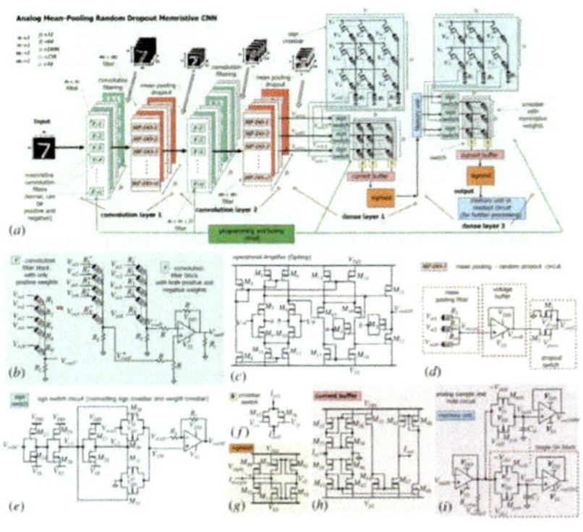

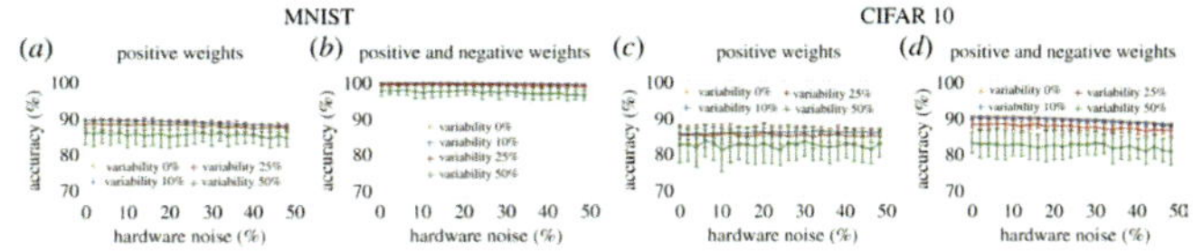

Main trends: –Hardware noise (up to 50%) has negligible effect on the performance.

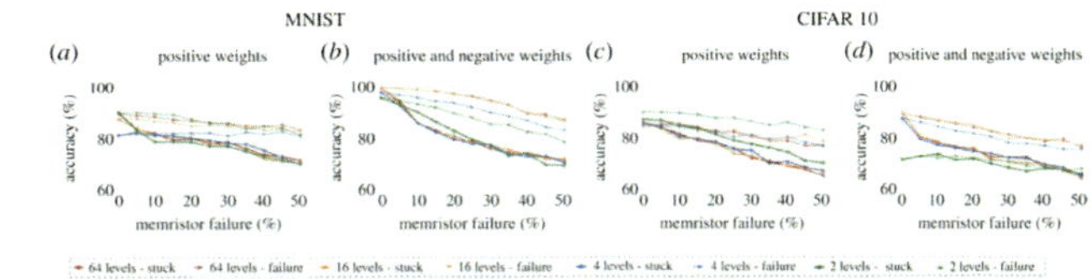

Main trends: –Memristor failure significantly reduces the accuracy.
–CNN can tolerate complete failure better than stuck devices.

Krestinskaya O, James AP. Analogue neuro-memristive convolutional dropout nets. Proceedings of the Royal Society A. 2020 Oct 28;476(2242):20200210.

Shows hardware noise has limited impact.

22 Dropouts - curious case

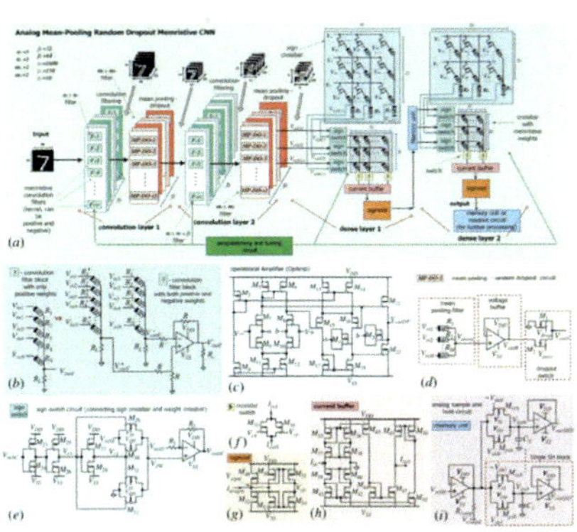

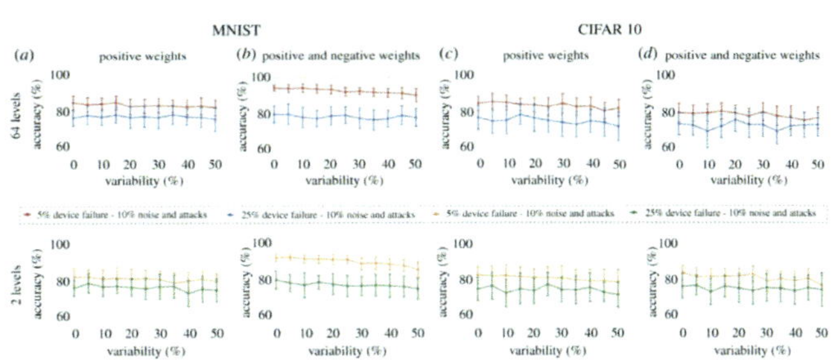

Main trends: –Accuracy is slightly reduced with memristor variability, when combined with other non-idealities.
–Device failure reduces performance accuracy significantly (worse than variability).
–When device failure is introduced, there is no difference between the cases of 64 and 2 levels.

Krestinskaya O, James AP. Analogue neuro-memristive convolutional dropout nets. Proceedings of the Royal Society A. 2020 Oct 28;476(2242):20200210.

When multiple non-idealities are combined, the reduction in accuracy is not significant.

23 Dropouts - curious case

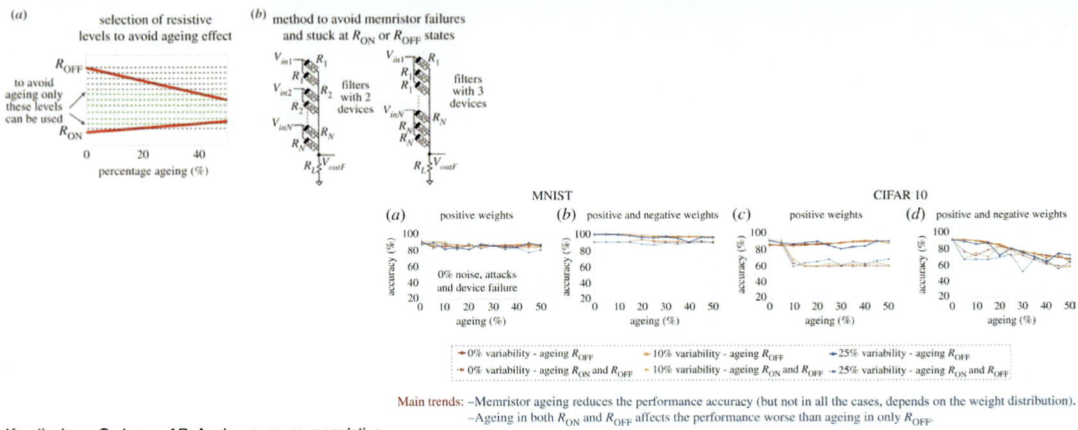

Krestinskaya O, James AP. Analogue neuro-memristive convolutional dropout nets. Proceedings of the Royal Society A. 2020 Oct 28;476(2242):20200210.

In some cases, the aging of memristor can have reduction in accuracy.

24 Superresolution crossbar

Convert a normal crossbar to superresolution by connecting the rows

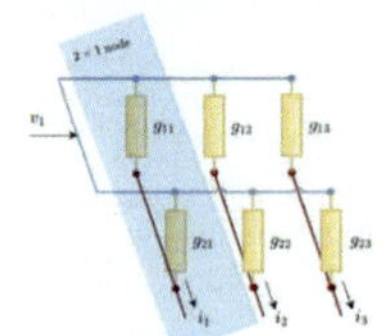

James AP, Chua LO. Analog Neural Computing With Super-Resolution Memristor Crossbars. ***IEEE Transactions on Circuits and Systems I: Regular Papers.*** 2021 May 21. (50th birthday of memristor)

EXAMPLE OF UNIQUE CONDUCTANCE VALUES OBTAINED FROM THE CROSSBAR NODES WITH COMBINATION OF 1, 2 AND 3 MEMRISTORS EACH HAVING $L = \{1, 2, 3\}$ CONDUCTANCE LEVELS.

L	1-M		2-M		3-M	
	Conductance value	$L_C = L$	Conductance value	L_C	Conductance value	L_C
1	g_1	1	$2g_1$	1	$3g_1$	1
2	g_1, g_2	2	$2g_1, 2g_2, (g_1 + g_2)$	3	$3g_1, 3g_2, g_1 + 2g_2, g_2 + 2g_1$	4
3	g_1, g_2, g_3	3	$2g_1, 2g_2, 2g_3, g_1 + g_2, g_1 + g_3, g_2 + g_3$	6	$3g_1, 3g_2, 3g_3, 2g_1 + g_2, 2g_1 + g_3, 2g_2 + g_1, 2g_2 + g_3, 2g_3 + g_1, 2g_3 + g_2, g_1 + g_2 + g_3$	10

Number of memristors (M)	Number of levels (L)	Unique conductance (L_C)	Sequence type
1	$\{1, 2, 3, 4, 5, 6, 7, 8, 9, 10, 11, 12..\}$	$\{1, 2, 3, 4, 5, 6, 7, 8, 9, 10, 11, 12..\}$	1-simplicial numbers
2	$\{1, 2, 3, 4, 5, 6, 7, 8, 9, 10, 11, 12..\}$	$\{1, 3, 6, 10, 15, 21, 28, 36, 45, 55, 66, 78, ..\}$	2-simplicial numbers
3	$\{1, 2, 3, 4, 5, 6, 7, 8, 9, 10, 11, 12..\}$	$\{1, 4, 10, 20, 35, 56, 84, 120, 165, 220, 286, 364, ..\}$	3-simplicial numbers
4	$\{1, 2, 3, 4, 5, 6, 7, 8, 9, 10, 11, 12..\}$	$\{1, 5, 15, 35, 70, 126, 210, 330, 495, 715, 1001, 1365, ..\}$	4-simplicial numbers
5	$\{1, 2, 3, 4, 5, 6, 7, 8, 9, 10, 11, 12..\}$	$\{1, 6, 21, 56, 126, 252, 462, 792, 1287, 2002, 3003, 4368, ..\}$	5-simplicial numbers
6	$\{1, 2, 3, 4, 5, 6, 7, 8, 9, 10, 11, 12..\}$	$\{1, 7, 28, 84, 210, 462, 924, 1716, 3003, 5005, 8008, 12376, ..\}$	6-simplicial numbers
7	$\{1, 2, 3, 4, 5, 6, 7, 8, 9, 10, 11, 12..\}$	$\{1, 8, 36, 120, 330, 792, 1716, 3432, 6435, 11440, 19448, 31824, ..\}$	7-simplicial numbers
8	$\{1, 2, 3, 4, 5, 6, 7, 8, 9, 10, 11, 12..\}$	$\{1, 9, 45, 165, 495, 1287, 3003, 6435, 12870, 24310, 43758, 75582, ..\}$	8-simplicial numbers

Convert a normal crossbar to superresolution by connecting the rows. Discusses the enhancement of a normal crossbar to achieve super-resolution by connecting the rows, referencing the work of James AP and Chua LO.

25 CNN

RECOGNITION ACCURACY WHEN USING A CNN ACCELERATOR WITH THE PROPOSED MEMRISTOR CROSSBAR NODES. THE TESTS ARE PERFORMED ON CIFAR-10 DATASET. THE IDEAL ACCURACY WITH INFINITE MEMRISTIVE STATES IS 91.26%.

1-M node		2-M node		3-M node		4-M node		5-M node		6-M node		7-M node		8-M node	
$L_C = L$	ACC (%)	L_C	ACC (%)	L_C	ACC (%)	L_C	ACC (%)	L_C	ACC (%)	L_C	ACC (%)	L_C	ACC (%)	L_C	ACC (%)
2	82.43	3	84.10	4	84.78	5	85.19	6	86.66	7	89.07	8	90.07	9	90.41
3	84.10	6	86.66	10	90.91	15	91.14	21	91.20	28	91.18	36	**91.26**	45	91.21
4	84.78	10	90.91	20	91.19	35	91.24	56	91.28	84	91.22	120	91.25	165	91.28
5	85.19	15	91.14	35	91.24	70	91.22	126	91.26	210	91.25	330	91.26	495	91.26
6	86.66	21	91.20	56	91.28	126	91.26	252	91.25	462	91.25	792	91.26	1287	91.26
7	89.07	28	91.18	84	91.22	210	91.25	462	91.25	924	91.26	1716	91.26	3003	91.26
8	90.07	36	91.26	120	91.25	330	91.26	792	91.26	1716	91.26	3432	91.26	6435	91.26
9	90.41	45	91.21	165	91.28	495	91.26	1287	91.26	3003	91.26	6435	91.26	12870	91.26
10	90.91	55	91.22	220	91.27	715	91.26	2002	91.26	5005	91.26	11440	91.26	24310	91.26
11	90.89	66	91.22	286	91.27	1001	91.25	3003	91.26	8008	91.26	19448	91.26	43758	91.26
12	91.02	78	91.23	364	91.27	1365	91.26	4368	91.26	12376	91.26	31824	91.26	75582	91.26

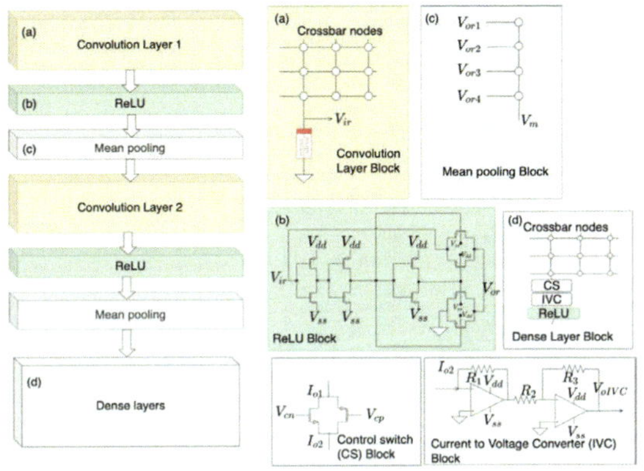

Introduces Convolutional Neural Networks (CNNs) that use the super-resolution crossbar, showing ideal accuracy.

26 Super-resolution bridge

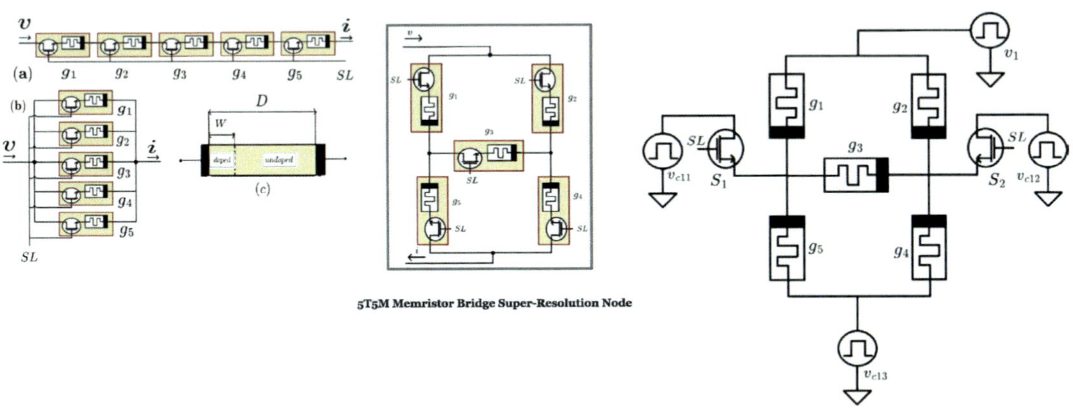

5T5M Memristor Bridge Super-Resolution Node

2T5M Memristor Bridge Super-Resolution Node

Radhakrishnan A, Pallathuvalappil S, Choubey B, James A. Bridge Memristor Super-Resolution Crossbars. IEEE Journal on Emerging and Selected Topics in Circuits and Systems. 2022 Nov 16;12(4):944-51.

Introduces the concept of a super-resolution bridge using memristor technology, referencing a specific study by Radhakrishnan, Pallathuvalappil, Choubey, and James.

27 Super-resolution bridge

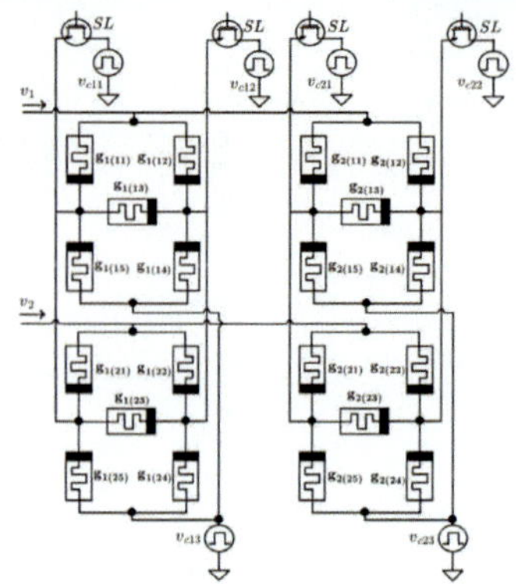

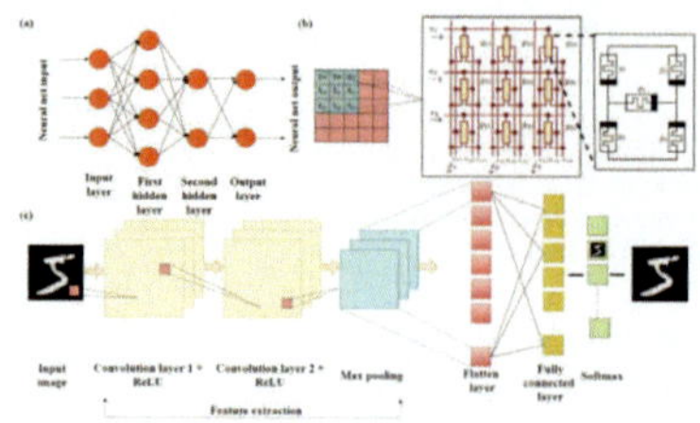

L	Conductance states per node		
	Bridge with equal interval	Bridge with unequal interval	Parallel [5]
1	1	1	1
2	11	11	6
3	57	60	21
4	202	210	56
5	560	585	126
6	1311	1359	252
7	2716	2800	462
8	5124	5268	792
9	9009	9225	1287
10	14965	15725	2002
11	23716	24156	3003
12	36150	36738	4368

L = Conductance states per memristor

Resolution states	Classification accuracy(%)			
	ANN		CNN	
	Equal interval	Not equal interval	Equal interval	Not equal interval
1	43.3	43.3	50	50
2	57	57	53	53
3	63	72.4	60	66.3
4	74	82	70	75.9
5	81	87	73	82.3
6	85	89.9	75.9	86.2
7	87	91.2	78	88.7
8	89	92	91.2	89.7
9	90	92.9	81.2	90.4
10	91	93	84.3	92.6
11	91.7	94	88	94.8
12	92.3	94.4	90	94.9
13	92.8	95.1	92.3	96
14	93.2	95.8	95	97.8
15	94.11	96.2	96.9	98
Ideal accuracy (infinity memristive states): ANN=96.8% CNN=98.20%				

Radhakrishnan A, Pallathuvalappil S, Choubey B, James A. Bridge Memristor Super-Resolution Crossbars. IEEE Journal on Emerging and Selected Topics in Circuits and Systems. 2022 Nov 16;12(4):944-51.

Shows the number of stable states created by the super-resolution bridge.

28 Super-resolution bridge

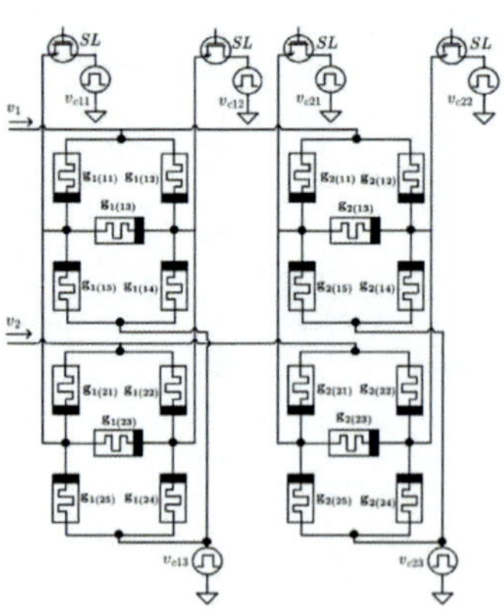

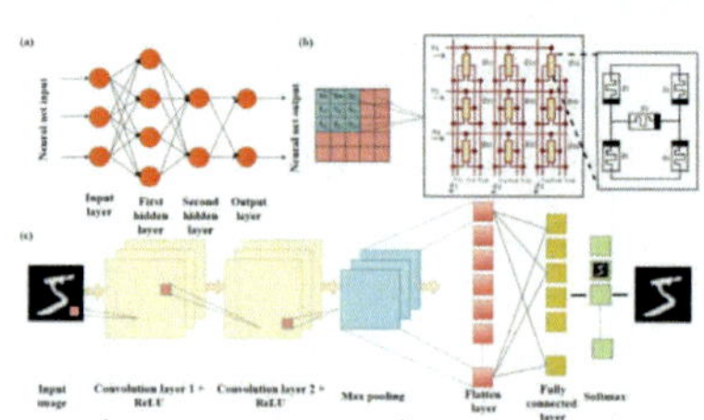

	Variability ratio/cases	RCE (%) for various L			
		3	5	7	12
E=2%	Case 1	0.97	0.34	0.20	0.038
	Case 2	1.31	0.49	0.24	0.08
	Case 3	0.14	0.17	0.08	0.02
	Case 4	0.62	0.42	0.17	0.33
E=10%	Case 1	1.6	2.1	0.71	0.78
	Case 2	3.6	2.0	1.10	0.89
	Case 3	2.28	0.8	0.89	0.4
	Case 4	1.11	1.89	0.7	0.66
E=20%	Case 1	9.0	8.88	5.6	3.6
	Case 2	9.22	9.0	7.3	7.1
	Case 3	3.3	3.0	2.0	1.89
	Case 4	8.9	9.0	4.9	4.8

L = Conductance states per memristor

Case 1: $R_{ON} + E * R_{ON}$ AND $R_{OFF} + E * R_{OFF}$

Case 2: $R_{ON} + E * R_{ON}$ AND $R_{OFF} - E * R_{OFF}$

Case 3: $R_{ON} - E * R_{ON}$ AND $R_{OFF} + E * R_{OFF}$

Case 4: $R_{ON} - E * R_{ON}$ AND $R_{OFF} - E * R_{OFF}$

Resolution states	Classification accuracy(%)			
	ANN		CNN	
	Equal interval	Not equal interval	Equal interval	Not equal interval
1	43.3	43.3	50	50
2	57	57	53	53
3	63	72.4	60	66.3
4	74	82	70	75.9
5	81	87	73	82.3
6	85	89.9	75.9	86.2
7	87	91.2	78	88.7
8	89	92	91.2	89.7
9	90	92.9	81.2	90.4
10	91	93	84.3	92.6
11	91.7	94	88	94.8
12	92.3	94.4	90	94.9
13	92.8	95.1	92.3	96
14	93.2	95.8	95	97.8
15	94.11	96.2	96.9	98
Ideal accuracy (infinity memristive states): ANN=96.8% CNN=98.20%				

Radhakrishnan A, Pallathuvalappil S, Choubey B, James A. Bridge Memristor Super-Resolution Crossbars. IEEE Journal on Emerging and Selected Topics in Circuits and Systems. 2022 Nov 16;12(4):944-51.

Impact of variability and its tolerance is reported.

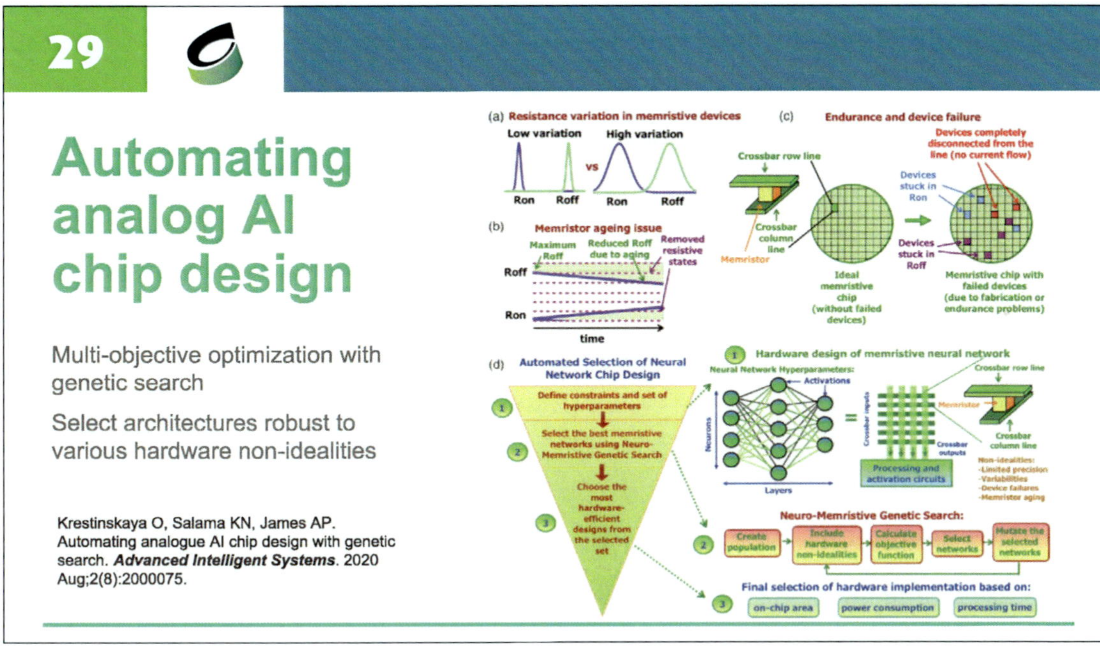

Presents a technique to speed up the design of analog memristive chips using multiobjective optimization taking into account both the hardware performance metrics and that of the system level performance metrics.

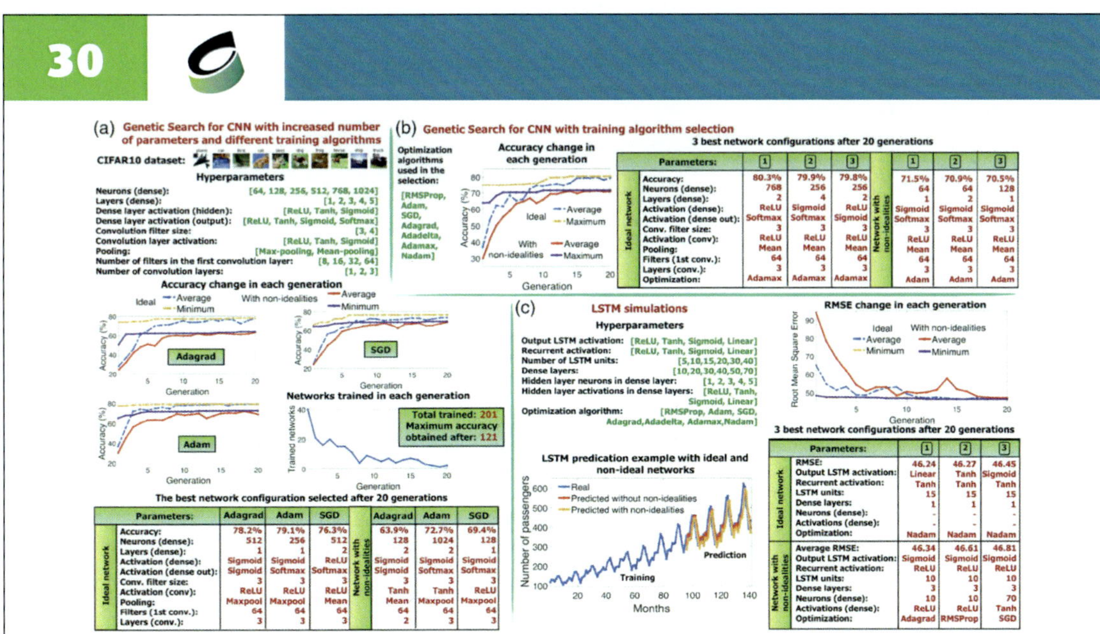

Presents the results of the generated combinations of standard cells and circuit architectures that can be implemented optimally for achieving high hardware and system performance.

31 Main challenges

(1) reliability and reconfigurability of the chip

(2) architecture design and scalability

(3) compatibility with older devices

(4) 3D integration challenges

(5) low power design challenges on multi-chip systems

(6) energy, area, and speed efficiency

(7) biocompatibility and multisensor integration issues

(8) design and test automation, and

(9) functional integration of multiple cognitive prior blocks required for AGI

Lists the main challenges in the analog memristive computing field.

32

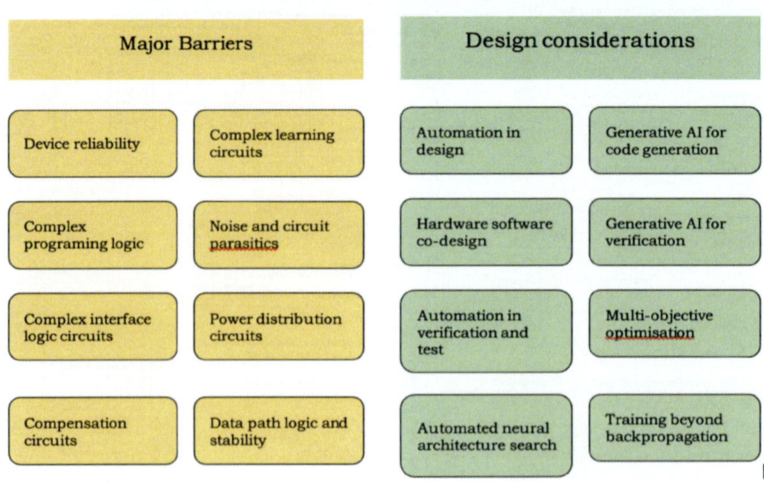

AP. James, Energy efficiency and design challenges in analog memristive chips, Nature Rev. Elec. Engg. 2024

Recent work highlighting the main barriers and design consideration to follow, to ensure an energy efficient design of memristive chips.

33 **Summary**

Analog computing requires hardware-software design thinking

Algorithms or hardware - what should come first?

Learning/error correction or primal knowledge

Automation will be key for scaling up the designs

Systems thinking will be inevitable for building practical applications

Lists the overall summary of the presentation.

CHAPTER 05

Edge of Chaos Clarifies the Origin for Yet-Unexplained Nonlinear Phenomena in Biological Systems

Alon Ascoli[1], *Ahmet Samil Demirkol*[2], *Leon Ong Chua*[3], and *Ronald Tetzlaff*[2]

[1]Politecnico di Torino
[2]Technische Universität Dresden
[3]University of California Berkeley

This pedagogical article introduces the simplest ever-reported bio-inspired circuit, which experiences symmetry-breaking instabilities under the effect of diffusion processes. Leveraging the capability of a NbO memristor to act as a local source of energy under suitable polarization, the proposed circuit reproduces the very same emerging phenomenon, known as destabilization of the homogeneous, first observed by Alan M. Turing in a two-cell array back in 1952, while employing half the number of degrees of freedom as compared to the original reaction-diffusion system discussed by the father of artificial intelligence.

The proposed circuit is a two-cell memristor cellular neural network, which employs just two dynamic circuit elements, specifically two globally-passive yet locally-active volatile resistance switching memories, based upon niobium oxide and fabricated at NaMLab, besides three linear and passive resistors, and two DC voltage sources.

A.M. Turing spent a huge amount of time to tune by trial and error the coefficients, appearing in the equations of the proposed fourth-order reaction-diffusion system, so as to demonstrate the emergence of dissipation-induced symmetry-breaking phenomena in a homogeneous cellular medium. On the other hand, the quantitative Theory of Local Activity, applicable universally to any physical system, allowed us to choose through a systematic methodology the circuit parameters of each of the two identical first-order memristive reaction cells so as to poise the composite second-order cell on the Edge of Chaos, which is the *conditio sine qua non* for the emergence of symmetry breaking phenomena, resulting in the steady-state formation of static patterns, across the bio-inspired system obtained after closing the composite cell itself across a suitable linear load resistance.

In summary, together with the other contributions, such as the article published in the first edition of this book [asc21], the pedagogical work, presented here, sheds light into the necessity to recur to the Physics Principle of the Edge of Chaos to gain a deep understanding of the mechanisms underlying emergent phenomena in natural systems.

[asc21] A. Ascoli, R. Tetzlaff, A.S. Demirkol, L.O. Chua, "Edge of Chaos Theory Explains Complex Phenomena in Memristor Circuits," chapter in book entitled "Selected Topics in Intelligent Chips with Emerging Devices, Circuits and Systems," River Publishers, Aalborg, Denmark, June 2023, e-ISBN: 9788770227643

This research study sheds light into the necessity to recur to the theoretical concepts from the Edge of Chaos Principle to understand the mechanisms underlying yet-unexplained nonlinear phenomena in biological systems.

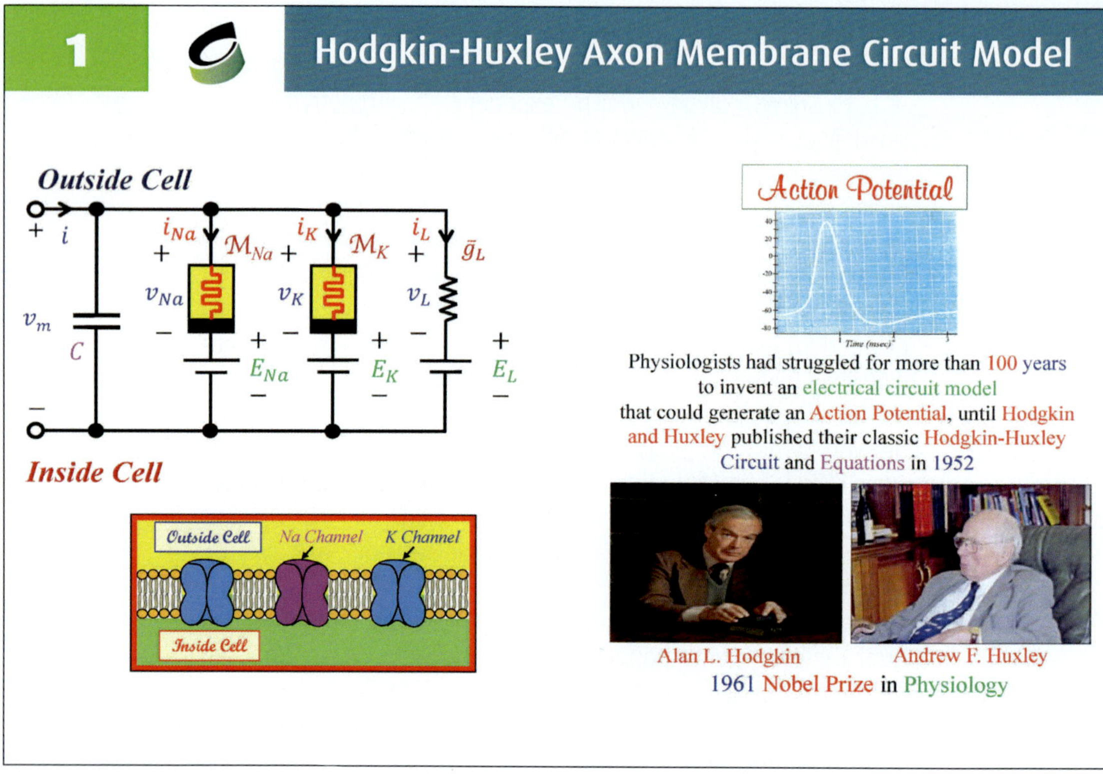

The Hodgkin-Huxley axon membrane model, developed in 1952 by Alan L. Hodgkin and Andrew F. Huxley, describes the main operating principles of a neuron under synaptic current excitation, including, especially, its ability to generate a train of electrical spikes, known as Action Potentials, in response to specific stimuli. It is composed of a set of four coupled nonlinear ordinary differential equations (ODEs) and admits a circuit-theoretic representation, including four parallel branches, which respectively host the membrane capacitance, the Potassium ion channel $\mathcal{M}_K$, the Sodium ion channel $\mathcal{M}_{Na}$, and a resistive channel of conductance $\bar{g}_L$, which accounts for the influence of other ions on the neuronal dynamics.

2 Hodgkin-Huxley Neuronal Axon Membrane Model Equations

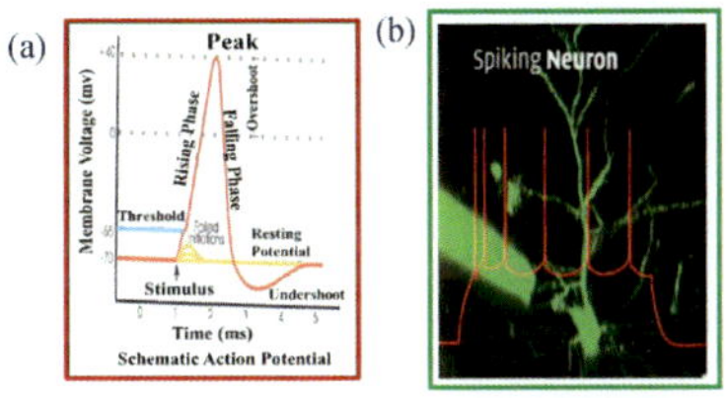

Sketch of an Action Potential (a) and of a Spiking Neuron (b)

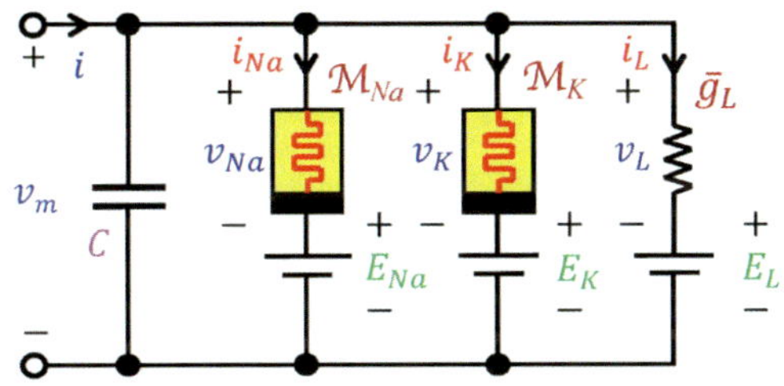

- In 1952 Hodgkin and Huxley introduced the following fourth-order ordinary differential equation (ODE) to model a neuron:

$$\frac{dv_m}{dt} = \frac{1}{C} \cdot i - \frac{1}{C} \cdot \bar{g}_K \cdot n^4 (v_m - E_K) - \frac{1}{C} \cdot \bar{g}_{Na} \cdot m^3 \cdot h \cdot (v_m - E_{Na}) - \frac{1}{C} \cdot \bar{g}_L \cdot (v_m - E_L) = f_1(n, m, h, v_m)$$

$$\frac{dn}{dt} = (1 - n) \cdot \alpha_n(v_m) - n \cdot \beta_n(v_m) = f_2(n, v_m)$$

$$\frac{dm}{dt} = (1 - m) \cdot \alpha_m(v_m) - m \cdot \beta_m(v_m) = f_3(m, v_m)$$

$$\frac{dh}{dt} = (1 - h) \cdot \alpha_h(v_m) - h \cdot \beta_h(v_m) = f_4(h, v_m)$$

where $v_m = v_{v_{Na}} - E_{Na} = v_K - E_K = v_L - E_L$ and $n, m, h \in \mathcal{D} \triangleq (0,1)$

A.L. Hodgkin, and A.F. Huxley, "A quantitative description of membrane current and its application to conduction and excitation in nerve," Journal of Physiology, vol. 117, pp. 500-544, 1952

The first equation descends from the application of the Kirchhoff's current law at the top node of the membrane capacitor C. The second equation governs the time evolution of the state variable n of the potassium ion channel $\mathcal{M}_K$, which is a first-order voltage-controlled generic memristor. The third (fourth) equation expresses the nonlinear dynamics of the state variable $m(h)$ of the sodium ion channel $\mathcal{M}_{Na}$, which falls in the class of second-order voltage-controlled generic memristors. The states of the memristive channels may only assume values within the open set (0,1). The quiescent state of the neuron under zero synaptic current input is set by three electrochemical gradients, modelled through DC bias sources E_K, E_{Na}, and E_L, appearing in series with $\mathcal{M}_K$, $\mathcal{M}_{Na}$, and $\bar{g}_L$, respectively.

3 Auxiliary functions in the Potassium (K) and Sodium (Na) Ion Channel Memristor Equations

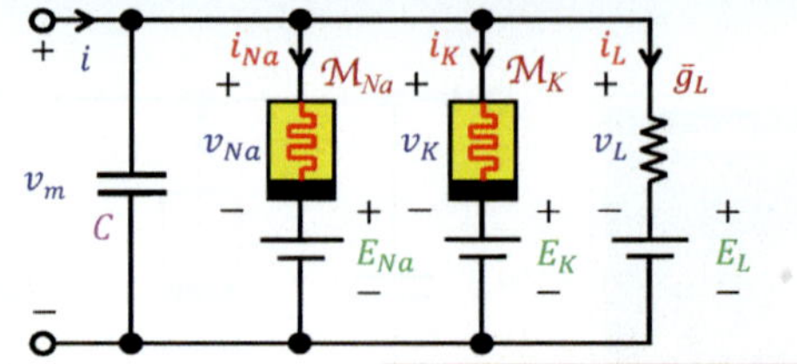

On the Na ion channel dynamics*

$$\alpha_m(v_m) = 0.1 \cdot 10^6 \cdot \frac{25 \cdot 10^{-3} - v_m}{\exp\left(\frac{25 \cdot 10^{-3} - v_m}{10 \cdot 10^{-3}}\right) - 1}$$

$$\beta_m(v_m) = 4 \cdot 10^3 \cdot \exp\left(\frac{-v_m}{18 \cdot 10^{-3}}\right)$$

$$\alpha_h(v_m) = 0.07 \cdot 10^3 \cdot \exp\left(\frac{-v_m}{20 \cdot 10^{-3}}\right)$$

$$\beta_h(v_m) = \frac{1 \cdot 10^3}{\exp\left(\frac{30 \cdot 10^{-3} - v_m}{10 \cdot 10^{-3}}\right) + 1}$$

On the K ion channel dynamics*

$$\alpha_n(v_m) = 0.01 \cdot 10^6 \cdot \frac{10 \cdot 10^{-3} - v_m}{\exp\left(\frac{10 \cdot 10^{-3} - v_m}{10 \cdot 10^{-3}}\right) - 1}$$

$$\beta_n(v_m) = 0.125 \cdot 10^3 \cdot \exp\left(\frac{-v_m}{80 \cdot 10^{-3}}\right)$$

** Physical units are omitted for the sake of simplicity*

Auxiliary voltage-dependent functions enter into the state equations of potassium and sodium ion channels, contributing to the process which regulates how they open or close under stimulation.

4 Train of Action Potentials

Parameters:
$E_K = -12 \cdot 10^{-3}$V
$\bar{g}_K = 36 \cdot 10^{-3}\Omega^{-1}$
$E_{Na} = 115 \cdot 10^{-3}$V
$\bar{g}_{Na} = 120 \cdot 10^{-3}\Omega^{-1}$
$E_L = 10.613 \cdot 10^{-3}$V
$\bar{g}_L = 0.3 \cdot 10^{-3}\Omega^{-1}$
$C = 1 \cdot 10^{-6}$F

Input: $i_{ext} = I_{ext} = 8\mu A$

Initial conditions: $v_C(0) = 0.01$ mV
$n(0) = 0.3$
$m(0) = 0.05$
$h(0) = 0.6$

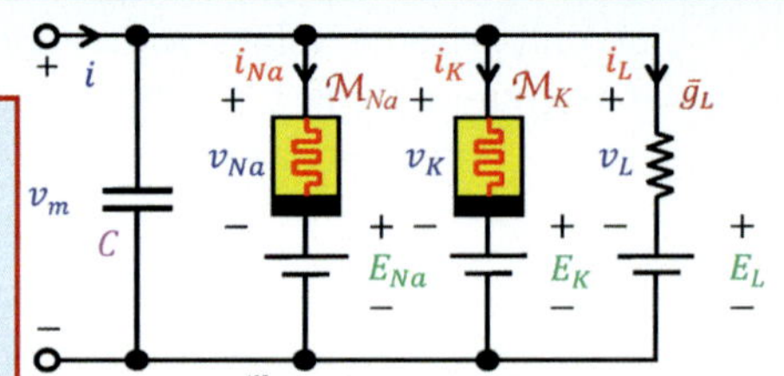

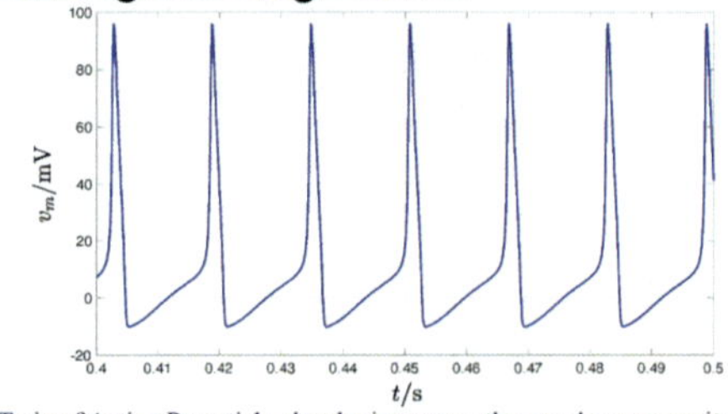

Train of Action Potentials, developing across the membrane capacitance C, as a net synaptic current i of constant value $I = 8\mu A$ enters the neuron

A. Ascoli, A.S. Demirkol, R. Tetzlaff, and L.O. Chua, "Edge of Chaos Theory Sheds Light Into the All-to-None Phenomenon in Neurons—Part I: On the Fundamental Role of the Sodium Ion Channel," IEEE TCAS–I: Regular Papers, vol. 71, no. 1, pp. 5-19, 2024, DOI: 10.1109/TCSI.2023.3339240

Setting the parameters of the Hodgkin-Huxley neuron model as reported in the seminal paper from 1952, a suitable choice for the constant net synaptic current I, flowing into the membrane, determines the emergence of a train of electrical spikes, also called Action Potentials, across the biological medium. This is an example of complexity par excellence.

5 The All-or-None Phenomenon

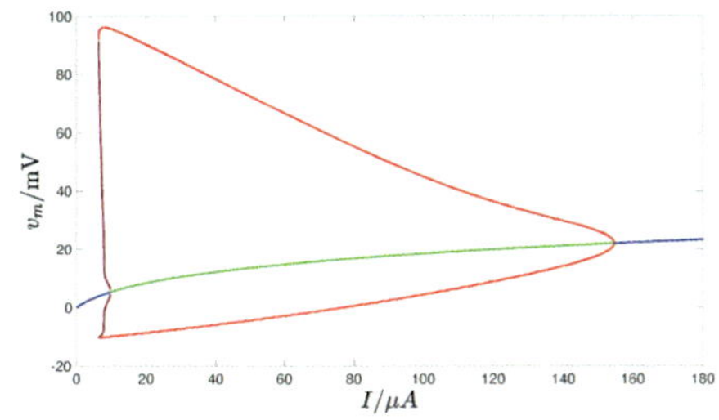

Blue trace: stable DC operating points for the neuron

Green trace: unstable DC operating points for the neuron

Top (bottom) red trace: maxima and minima in the stable oscillations

Top (bottom) brown trace: maxima and minima in the unstable oscillations

Bifurcation diagram of the Hodgkin-Huxley neuron model

- According to the Hodgkin-Huxley neuron model, the All-or-None phenomenon occurs in the neuron, when, as the net synaptic current I, entering its membrane, decreases below the fold limit cycle threshold, an outer limit cycle, corresponding to the train of stable electrical spikes (All), and an inner limit cycle, corresponding to the coexisting unstable oscillations, gently coalesce, annihilating each other other, in its four-dimensional state space, which hosts only the quiescent state (None) for lower net synaptic current values.
- The existence of unstable oscillations is thus fundamental for the appearance of the All-or-None phenomenon in neuronal axon membranes.
- The unstable limit cycle first appears in the four-dimensional state space of the Hodgkin-Huxley neuron model out of a *subcritical Hopf bifurcation*, when the neuron enters one of its two admissible *Edge of Chaos* domains about the operating point, enclosed by the unstable new-born cycle, and turning concurrently from an unstable focus to a stable focus.
- The stable cycle first emerges in the four-dimensional state space out of a *supercritical Hopf bifurcation*, when the neuron enters the other of its two *Edge of Chaos* domains at the operating point, enclosed by the stable new-born cycle, and turning concurrently from a stable focus into an unstable focus.

A. Ascoli, A.S. Demirkol, R. Tetzlaff, and L.O. Chua, "Edge of chaos theory sheds light into the all-to-none phenomenon in neurons—Part II: On the necessary and sufficient conditions for the observation of the entire life cycle of an action potential," IEEE TCAS–I: Regular Papers, to be published

The classical bifurcation diagram of the Hodgkin-Huxley neuron model, reported here, illustrates the entire life cycle of an Action Potential under net synaptic current sweep. Three are the fundamental bifurcations which control the onset, growth, and end of stable limit-cycle oscillations across the capacitance membrane. In the diagram the blue (green) curve represents the stable (unstable) branch of the neuron DC voltage V_m versus current I characteristic. Assuming I to decrease from some large value, say 180μA, initially the neuron sits at a globally asymptotically stable operating point. A supercritical Hopf bifurcation, spawning the birth of stable oscillations across the membrane capacitance, and a concurrent loss of stability for the neuron quiescent state, occurs when I descends to 155.731μA. The top and bottom red traces show the maxima and minima of the stable oscillations, respectively. As the net synaptic current keeps decreasing below this first bifurcation point, the stable oscillations, originally almost sinusoidal in shape, exhibit a larger and larger amplitude as well as a higher and higher degree of nonlinearity. When I descends down to 9.77003μA another Hopf bifurcation, this time from the subcritical class, emerges across the neuronal axon membrane, determining the birth of almost sinusoidal oscillations of unstable nature, while allowing the quiescent state to reacquire a stable nature. The top and bottom brown traces depict the maxima and minima of the unstable oscillations. As the net synaptic current descends below this second bifurcation point, the amplitude and nonlinearity of the unstable oscillations increase dramatically within a relatively short range of the control parameter. The growth rate of the inner unstable limit cycle is way larger than the growth rate of the outer stable one, and, as a result, when I decreases down to 6.256μA, as a result of a global phenomenon, known as fold limit cycle bifurcation, the two cycles gently coalesce one into the other, disappearing together from the state space, as the net synaptic current descends below this third bifurcation point. This global bifurcation determines the end of the life cycle of the Action Potential, with the neuron admitting a globally asymptotically stable operating point for any lower net synaptic current till the end of the input sweep. Importantly, for 6.256μA $< I <$ 9.77003μA the neuron features a bistable behavior. Depending upon the initial conditions of its four state variables v_m, n, m, and h, it converges asymptotically toward either of two admissible locally-stable attractors, namely the quiescent state or the outer limit cycle. Moreover, the fold limit cycle bifurcation is the mechanism explaining the All-to-None effect. In fact, for suitable initial conditions, as the net synaptic current descends below 6.256μA, the neuron turns from a condition, where large limit-cycle oscillations appear across its membrane capacitance (the All effect), to a scenario, where it sits at a quiescent state asymptotically (the None effect).

Since in a practical experiment, unstable steady states, also referred to as repellors in the theory of Nonlinear Dynamics, would not be observed, their presence in the state space of a system might go dangerously unnoticed. However, the onset of unstable oscillations across the neuronal axon membrane via subcritical Hopf bifurcation is of pivotal significance for the emergence of the All-to-None phenomenon.

Last but not least, it may be demonstrated that the neuron is poised on the Edge of Chaos when it undergoes each of the two local Hopf bifurcations, signaling the onset of stable and unstable limit-cycle oscillations across the membrane capacitance, respectively, which demonstrates that the epicenter of complex phenomena lies in the stable region of the Local Activity domain. Importantly, the possibility for the neuron to operate on the Edge of Chaos is enabled by the presence of the ion channels, whose current versus voltage loci include branches with negative slope, where the channels themselves admit a negative differential resistance (NDR), allowing them to amplify infinitesimal fluctuations in energy, an extraordinary property which confers them the locally active attribute.

6 Novel Nanotechnologies for Bio-Inspired Circuit Design

- *Electrical circuits, hosting **volatile memristors** with **Negative Differential Resistance (NDR)** effects, support **complex bio-mimetic phenomena**, e.g. the emergence and propagation of **All-or-None Spikes**, similarly as neuronal axon membranes, enabling the implementation of innovative bio-inspired signal processing paradigms, such as spike-based computing.*

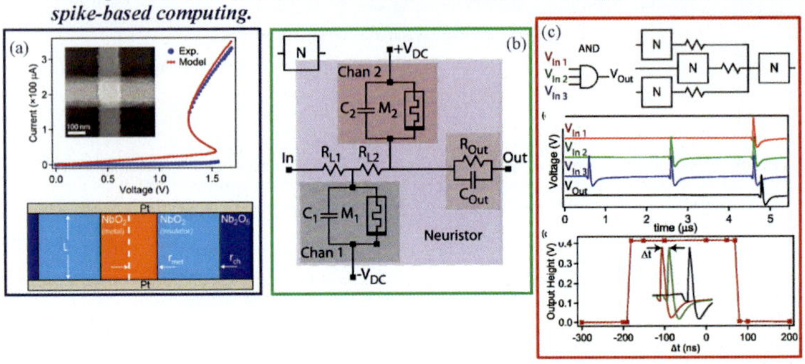

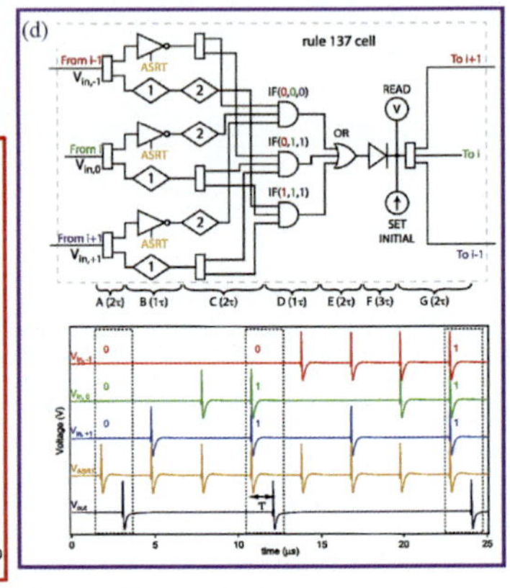

(a) DC characteristic of a NbO_2 memristor. Along the NDR branch the device is locally active. (b) A neuristor based upon two NbO_2 memristors. (c) A neuristor-based circuit, operating as a 3-input logic AND gate. (d) Spike-based computing in a network of neuristor logic gates, executing the universal cellular automaton rule 137.

M.D. Pickett, and R.S. Williams, "Sub-100 fJ and sub-nanosecond thermally driven threshold switching in niobium oxide crosspoint nanodevices," Nanotechnology, 23, 215202, 2012

M.D. Pickett, and R.S. Williams, "Phase transitions enable computational universality in neuristor-based cellular automata," Nanotechnology, 24, 384002, 2013

Disruptive nanotechnologies enable the fabrication of solid-state memristive devices, which, despite being unable to store data in a non-volatile form, feature *S*-shaped DC current versus voltage characteristics. When biased along the NDR branches of such loci, these devices acquire the sensational capability to act as local energy sources, similarly as the sodium and potassium ion channels in a biological neuron. This allows the design of innovative bio-inspired circuits and networks, which implementing spike-based computing paradigms, promise to improve the energy efficiency of traditional electronic systems.

7 Complexity in Reaction-Diffusion Systems

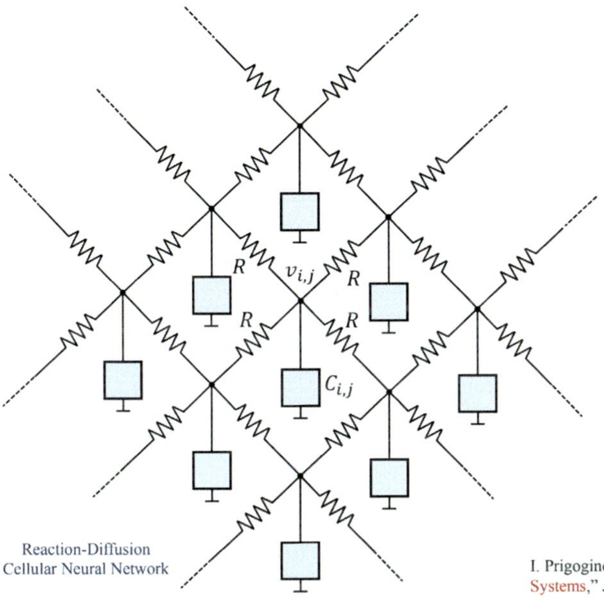

Reaction-Diffusion
Cellular Neural Network

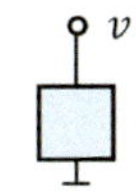

Isolated reaction cell C, consisting of an arbitrary nonlinear dynamic circuit

- Assume C to feature a globally asymptotically stable (GAS) operating point, such that, irrespective of the initial conditions assigned to its memory elements,

$$\lim_{t\to\infty} v = V$$

- An array of diffusively-coupled regularly-spaced identical reaction cells of this kind is an example of a homogeneous medium.
- It typically admits the homogeneous solution, i.e. perturbing slightly the initial condition of one cell relative to the common one, assigned to all the other cells,

$$\lim_{t\to\infty} v_{i,j} = V \quad \forall i \in \{1, \ldots, M\}, \forall j \in \{1, \ldots, N\}$$

- What are the necessary conditions for the destabilization of the homogeneous solution?
- Symmetry-breaking phenomena with spatio-temporal pattern formation in homogeneous cellular media of this kind are an example of complexity par excellence.

I. Prigogine, and G. Nicolis, "On Symmetry-Breaking Instabilities in Dissipative Systems," Journal of Chemical Physics, vol. 46, no. 9, pp. 3542-3550, 1967

Assume a physical system in the form of a reaction cell to feature a globally asymptotically stable operating point. Let us now consider a reaction-diffusion system, consisting of a two-dimensional infinitely-long and infinitely-wide array of regularly-spaced units, each of which is identical to the aforementioned reaction cell, and is coupled to each of its four direct neighbors, located in the north, south, east, and west directions, employing a very same linear and passive resistor R between one of its nodes and the corresponding one associated to the neighbor. It is reasonable to conjecture that such reaction-diffusion system would admit the homogeneous solution, after transients, emerging as the initial condition of at least one of the cells were displaced by a tiny amount relative to the common initial condition, corresponding to the quiescent state of the isolated cell, and assigned to all the others, would vanish. The homogeneous solution corresponds to the array steady state, where each of the cells asymptotically showcases the very same quiescent condition as it does while isolated from all the others. However, in some cases, this is not the case. If the isolated cell features suitable properties, symmetry-breaking phenomena may emerge across the reaction-diffusion array. In these circumstances the homogeneous solution of the array loses stability, and static or dynamic patterns form at steady state across the cellular medium. Symmetry-breaking effects occur frequently in physical systems, e.g. in electrical circuits, in biological media, and in chemical reactions.

8 The Turing Example-Part I: Isolated Identical Reaction Cells

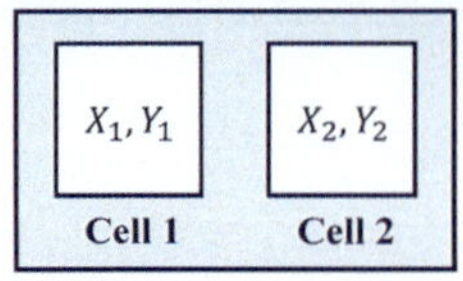

- Back in 1952, A.M. Turing, the Father of Artificial Intelligence, developed a mathematical model, showcasing the emergence of symmetry-breaking phenomena in a reaction-diffusion system.
- He considered two identical second-order reaction cells. Isolated one from the other, the cells were found to exhibit a common globally-asymptotially stable (GAS) operating point.

Cell 1:
$$\dot{X}_1 = 5 \cdot X_1 - 6 \cdot Y_1 + 1$$
$$\dot{Y}_1 = 6 \cdot X_1 - 7 \cdot Y_1 + 1$$

Cell 2:
$$\dot{X}_2 = 5 \cdot X_2 - 6 \cdot Y_2 + 1$$
$$\dot{Y}_2 = 6 \cdot X_2 - 7 \cdot Y_2 + 1$$

Unique steady-state solution for cell 1: $X_1 = 1, Y_1 = 1$

Unique steady-state solution for cell 2: $X_2 = 1, Y_2 = 1$

A.M. Turing, "The Chemical Basis of Morphogenesis," Philosophical Transactions of the Royal Society of London, Series B, Biological Sciences, vol. 237, no. 641, pp. 37-72, 1952

In a seminal paper from 1952 A.M. Turing presented a simple mathematical model capturing the emergence of diffusion-driven instabilities in a fourth-order two-cell array. The isolated cell, featuring two degrees of freedom, was found to exhibit one and only one stable operating point.

9 The Turing Example-Part II: Reaction-Diffusion System

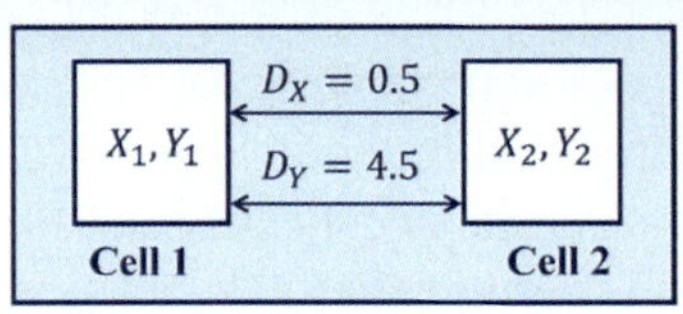

- The two identical cells are then let interact via diffusion processes. Cell-to-cell coupling is realized between both state variable pairs (X_1, X_2) and (Y_1, Y_2).
- The diffusion coefficient D_X (D_Y) for coupling state variables X_1 and X_2 (Y_1 and Y_2) is set to 0.5 (4.5).
- The two-cell reaction-diffusion array is a fourth-order ODE system.

Two-Cell Reaction-Diffusion Array:
$$\dot{X}_1 = 5 \cdot X_1 - 6 \cdot Y_1 + 1 + 0.5 \cdot (X_2 - X_1)$$
$$\dot{Y}_1 = 6 \cdot X_1 - 7 \cdot Y_1 + 1 + 4.5 \cdot (Y_2 - Y_1)$$
$$\dot{X}_2 = 5 \cdot X_2 - 6 \cdot Y_2 + 1 + 0.5 \cdot (X_1 - X_2)$$
$$\dot{Y}_2 = 6 \cdot X_2 - 7 \cdot Y_2 + 1 + 4.5 \cdot (Y_1 - Y_2)$$

- The homogeneous solution, i.e. $X_1 = 1, Y_1 = 1, X_2 = 1, Y_2 = 1$, satisfies the fourth-order ODE system, for which it represents an admissible operating point. In fact, according to common expectations, it should constitute the only possible steady state for the two-cell array.

However, connecting diffusively two identical copies of the aforementioned cell, while employing each of the two available state variable pairs, and assigning a suitable value to the coupling coefficient in each case, the homogeneous solution of the resulting reaction-diffusion system was found to be unstable. In fact, the two-cell array experienced symmetry-breaking phenomena, resulting in the divergence of its state variables over time.

Note that, adding some suitable nonlinearity into the model of each cell, the emergence of the blow-up phenomenon would be prevented, and the destabilization of the homogeneous solution would be accompanied by the progressive convergence of the two-cell array toward some inhomogeneous static pattern.

10 Turing's Two-Cell Reaction-Diffusion Array

- The model for the two-cell reaction-diffusion array may be recast as

$$\begin{bmatrix}\dot{X}_1\\\dot{Y}_1\\\dot{X}_2\\\dot{Y}_2\end{bmatrix}=\underbrace{\begin{bmatrix}5&-6&0&0\\6&-7&0&0\\0&0&5&-6\\0&0&6&-7\end{bmatrix}}_{\triangleq\,\boldsymbol{G}_1}\cdot\begin{bmatrix}X_1\\Y_1\\X_2\\Y_2\end{bmatrix}+\begin{bmatrix}1\\1\\1\\1\end{bmatrix}+\underbrace{\begin{bmatrix}-0.5&0&0.5&0\\0&-4.5&0&4.5\\0.5&0&-0.5&0\\0&4.5&0&-4.5\end{bmatrix}}_{\triangleq\,\boldsymbol{G}_2}\cdot\begin{bmatrix}X_1\\Y_1\\X_2\\Y_2\end{bmatrix}$$

- The eigenvalues of matrices $\boldsymbol{G}_1$ and $\boldsymbol{G}_2$, reported below, reveal their stable nature.

$$\lambda_1(\boldsymbol{G}_1)=-1 \qquad \lambda_1(\boldsymbol{G}_2)=-9$$
$$\lambda_2(\boldsymbol{G}_1)=-1 \qquad \lambda_2(\boldsymbol{G}_2)=-1$$
$$\lambda_3(\boldsymbol{G}_1)=-1 \qquad \lambda_3(\boldsymbol{G}_2)=0$$
$$\lambda_4(\boldsymbol{G}_1)=-1 \qquad \lambda_4(\boldsymbol{G}_2)=0$$

The system of state equations of the linear two-cell array from Turing may be recast as the sum between a reaction component, modelling the dynamics of the reaction cells, and a diffusive component, capturing the diffusive coupling effects. The eigenvalues of the matrix $\boldsymbol{G}_1$ ($\boldsymbol{G}_2$), characterizing the state-dependent part of the first component (the second component), are negative (are either negative or null). Thus both $\boldsymbol{G}_1$ and $\boldsymbol{G}_2$ are stable.

11 A Dissipative Vector Field

- Defining $\boldsymbol{V}\triangleq[X_1,Y_1,X_2,Y_2]^{\mathrm{T}}$, the fourth-order ODE reduces to

$$\dot{\boldsymbol{V}}=\boldsymbol{G}_1\cdot\boldsymbol{V}+\nabla^2\boldsymbol{V}$$

with

$$\nabla^2V\triangleq\boldsymbol{G}_2\cdot V$$

- The divergence of the diffusion part of the fourth-order ODE, i.e.

$$\mathrm{div}(\nabla^2V)\triangleq\boldsymbol{\nabla}\cdot(\nabla^2V)=\frac{\partial(0.5\cdot(X_2-X_1))}{\partial X_1}+\frac{\partial(4.5\cdot(Y_2-Y_1))}{\partial Y_1}+\frac{\partial(0.5\cdot(X_1-X_2))}{\partial X_2}+\frac{\partial(4.5\cdot(Y_1-Y_2))}{\partial Y_2}$$
$$=-0.5-4.5-0.5-4.5=-10$$

is strictly negative, indicating how, as time goes by, the diffusive coupling tends to contract the volume of the four-dimensional state-space, enclosing the ODE system trajectory initiated from any arbitrary starting point.

- ∇^2V is a dissipative vector field. Diffusion processes cannot be a source of instability.

The divergence of the diffusive component is negative, proving that the dissipative environment, where the two reaction cells are immersed, cannot provide the source for destabilizing the homogeneous solution of the cellular array.

12 Signs of Instability in Turing's Two-Cell Reaction-Diffusion Array

- The model for the two-cell reaction-diffusion array may be more compactly rewritten as

$$\begin{bmatrix}\dot{X}_1\\ \dot{Y}_1\\ \dot{X}_2\\ \dot{Y}_2\end{bmatrix} = \underbrace{\begin{bmatrix}5 & -6 & 0 & 0\\ 6 & -7 & 0 & 0\\ 0 & 0 & 5 & -6\\ 0 & 0 & 6 & -7\end{bmatrix}}_{\triangleq G_1} \cdot \begin{bmatrix}X_1\\ Y_1\\ X_2\\ Y_2\end{bmatrix} + \begin{bmatrix}1\\1\\1\\1\end{bmatrix} + \underbrace{\begin{bmatrix}-0.5 & 0 & 0.5 & 0\\ 0 & -4.5 & 0 & 4.5\\ 0.5 & 0 & -0.5 & 0\\ 0 & 4.5 & 0 & -4.5\end{bmatrix}}_{\triangleq G_2} \cdot \begin{bmatrix}X_1\\ Y_1\\ X_2\\ Y_2\end{bmatrix}$$

$$= \underbrace{\begin{bmatrix}4.5 & -6 & 0.5 & 0\\ 6 & -11.5 & 0 & 4.5\\ 0.5 & 0 & 4.5 & -6\\ 0 & 4.5 & 6 & -11.5\end{bmatrix}}_{\triangleq G = G_1 + G_2} \cdot \begin{bmatrix}X_1\\ Y_1\\ X_2\\ Y_2\end{bmatrix} + \begin{bmatrix}1\\1\\1\\1\end{bmatrix}$$

- Quite surprisingly, the eigenvalues of the matrix $\boldsymbol{G}$, reported below, reveal its unstable nature.

$$\lambda_1(\boldsymbol{G}) = -14$$
$$\lambda_2(\boldsymbol{G}) = -1$$
$$\lambda_3(\boldsymbol{G}) = -1$$
$$\lambda_4(\boldsymbol{G}) = +2$$

Summing up the matrices $\boldsymbol{G}_1$ and $\boldsymbol{G}_2$, characterizing the state-dependent part of the first component and the second component of the set of state equations, modelling Turing's reaction-diffusion system, one of the four eigenvalues of the resulting matrix $\boldsymbol{G}$ was found to be positive. The instability of $\boldsymbol{G}$ is an indication for the destabilization of the homogeneous solution of the two-cell array.

13 Understanding the Origin for Symmetry-Breaking Phenomena in Turing's Two-Cell Reaction-Diffusion Array

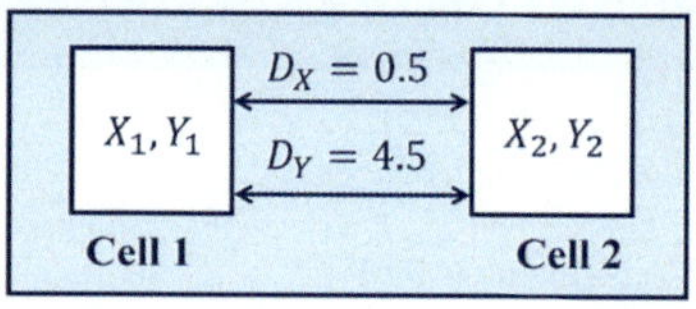

Two-Cell Reaction-Diffusion Array:

$$\begin{bmatrix}\dot{X}_1\\ \dot{Y}_1\\ \dot{X}_2\\ \dot{Y}_2\end{bmatrix} = \underbrace{\begin{bmatrix}5 & -6 & 0 & 0\\ 6 & -7 & 0 & 0\\ 0 & 0 & 5 & -6\\ 0 & 0 & 6 & -7\end{bmatrix}}_{\triangleq G_1} \cdot \begin{bmatrix}X_1\\ Y_1\\ X_2\\ Y_2\end{bmatrix} + \begin{bmatrix}1\\1\\1\\1\end{bmatrix} + \underbrace{\begin{bmatrix}-0.5 & 0 & 0.5 & 0\\ 0 & -4.5 & 0 & 4.5\\ 0.5 & 0 & -0.5 & 0\\ 0 & 4.5 & 0 & -4.5\end{bmatrix}}_{\triangleq G_2} \cdot \begin{bmatrix}X_1\\ Y_1\\ X_2\\ Y_2\end{bmatrix}$$

- Turing's diffusion-driven instability may not originate from the dissipative coupling between the two cells.
- So, what special property must $\boldsymbol{G}_1$ possess to explain the destabilization of the homogeneous solution in Turing's two-cell array?
- The symmetry-breaking phenomenon emerges as each of the two identical reaction cells is poised on the Edge of Chaos in the isolated case.
- As a final note, several nonlinear variants of Turing's equations have been proposed over the past decades in the literature for reproducing symmetry-breaking phenomena emerging in a wide class of biological reaction-diffusion systems.

Since the diffusion processes cannot be responsible for the destabilization of the homogeneous solution across the cellular medium, it is the cell, which must possess some fundamental property, which endows it with a highly-excitable nature. Thus, despite, while isolated, it would hide a high degree of excitability behind a quiet steady state, once set in diffusive interaction with an identical copy, it would be pushed away from the operating point it used to sit at.

The fundamental property, which the isolated cell must feature, for the destabilization of the homogeneous solution in Turing's reaction-diffusion model, refers to the locally-active nature of its stable quiescent state. In other words, diffusion-driven instabilities occur across the reaction-diffusion system if and only if the isolated cell operates on the Edge of Chaos around the respective operating point.

14

Local Activity and Edge of Chaos Theory

The missing Physics Law, explaining the origin for complex phenomena, such as the development of an Action Potential across a neuronal membrane, or the emergence of symmetry-breaking effects in reaction-diffusion cellular media, is Local Activity and its Crown Jewel Principle, referred to as the Edge of Chaos.

15

Schrödinger, Prigogine, Eigen, Gell-Mann, Turing, and Smale have all been searching for a missing new Physics Principle to explain Complexity in Physical Systems

Various luminaries, including Schrödinger, Gell-Mann, Smale, Prigogine, and Turing, have debated passionately for several decades about the origin for complex phenomena, without acquiring a comprehensive and conclusive understanding despite their intellectual calibre.

16 Local Activity is the Origin of Complexity

- *Local Activity, including its Crown Jewel, known as Edge of Chaos, is infact the Missing New Physics Principle*
- *It is a Universal Principle, grounded on solid theoretical foundations, which enable to derive Quantitative Conditions for a Physical System to support Emergent Phenomena, such as the appearance of an All-or-None Spike along a neuronal axon, or the occurrence of Symmetry-Breaking Effects in a Reaction-Diffusion Cellular Neural Network*

L.O. Chua, "Local Activity is the Origin of Complexity," International Journal of Bifurcation and Chaos, vol. 15, no. 11, pp. 3435-3456, 2005

The Universal Principle of the Edge of Chaos is grounded on solid theoretical foundations, which allow to define quantitative conditions for emergent phenomena to appear across any physical medium. In fact, recurring to this physics law of nature is a necessary step for explaining complexity in systems from fields as diverse as ecology, cellular biology, chemistry, physics, electronics, and even economics.

17 Definition of Local Activity

A system is said to be

locally active

if and only if it may

amplify

infinitesimal fluctuations in energy

under suitable polarization*

*There is no free lunch: in order for a locally-active system to act as local source of energy around some operating point, a DC source must concurrently supply it with the energy necessary for its polarization.

A system is said to be locally active if and only if it is able to act as an infinitesimal source of energy about some operating point, it is preliminarily biased into through a DC power supply.

18 Rigorous Condition for a Memristor to be Locally Active at a Bias Point

Extended memristor:
$v = R(x, i) \cdot i$
$\lim_{i \to 0} R(x, i) \neq \infty$
$\dot{x} = f(x, i)$
$x = [x_1, \ldots, x_n]^T \in \mathbb{R}^n$

- Let the nth-order current-controlled memristor* $\mathcal{M}$ be preliminarily biased on an operating point $\boldsymbol{Q} = \boldsymbol{X} = [X_1, \ldots, X_n]^T$ via a DC current source $I_S = I_P$
- $\boldsymbol{Q} = \boldsymbol{X}$ is unequivocally associated to a bias point $\boldsymbol{P} = (V_P, I_P)$ on the voltage V versus current I plane
- Let us superimpose a small-signal current $\delta i(t)$ on the input bias level I_P from time $t = t_0$

$$i_s(t) \equiv i(t) = I_P + \delta i(t) \longrightarrow \begin{cases} x(t) = X + \delta x(t) \\ v(t) = V_P + \delta v(t) \end{cases} \text{memristor response}$$

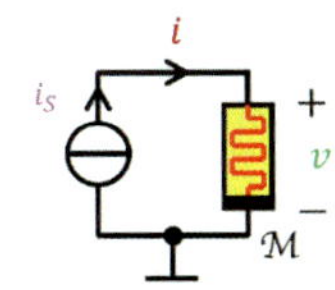

$i_S = I_S = I_P$ for $0 \leq t < t_0$
$i_S = I_P + \delta i$ for $t \geq t_0$

Definition:
A current-controlled memristor $\mathcal{M}$ is locally active at some operating point $\boldsymbol{Q}$ if and only if there exists at least one admissible small-signal current $\delta i(t)$, which, added up to the input bias level I_P from $t = t_0$, induces the one-port to act as <u>local source of energy</u> over a time frame $t \in [t_0, t]$, for some finite time $t = \bar{t} > t_0$, implying a negative sign for the net local energy, entering the one-port over the specified time frame, and computable as

$$\delta\mathcal{E}(t_0; \bar{t}) = \int_{t_0}^{\bar{t}} \delta p(t')dt,$$

where $\delta p(t) = \delta v(t) \cdot \delta i(t)$ is the local instantaneous power absorbed by $\mathcal{M}$

- Rigorous definition, yet impractical for testing purposes
- Fortunately, there exists a powerful theorem, which simplifies the testing procedure
- For the application of the theorem it is necessary for the one-port to admit a DC characteristic, which enables the derivation of its local transfer function about $\boldsymbol{Q}$

* While the definition is adapted to a current-controlled memristor, similar conclusions may be drawn *mutatis mutandis* for a voltage-controlled one.

Let us provide a rigorous definition for the condition, which a two-terminal memristive device must satisfy in order to act as a local source of energy about a given operating point. Without loss of generality, here we consider a nanodevice modelled by a differential algebraic equation (DAE) set falling in the class of current-controlled extended memristors. Similar conclusions may be drawn *mutatis mutandis* for a nanodevice modelled by a DAE set falling in the class of voltage-controlled extended memristors.

Assuming the current-controlled memristor to feature n degrees of freedom, the DAE set is composed of a nth-order ODE system, governing the time evolution of a n-dimensional state vector $\boldsymbol{x}$, and of one algebraic equation, referred to as Ohm's law, establishing a relationship between state vector $\boldsymbol{x}$, current i, and voltage v.

A DC current $I_s = I_p$ is injected into the memristor from the time instant $t = 0$, assuming a given initial condition $x_0 = x(0)$ for the memory state x. After transients decay to zero, the memristor is found to sit at some operating point $Q = X$, while the voltage across it attains a certain value V_p. Afterwards, from the time instant $t = t_0$, an infinitesimally-small or local signal δi is superimposed on top of the DC input level I_p, inducing small-signal responses δx and δv in the device state and voltage, respectively. The memristor is said to be locally active about the operating point Q if and only if there exists at least one local input perturbation δi such that the infinitesimal energy $\delta\varepsilon(t_0; t)$, the device absorbs over the time interval $[t_0, t]$, with $t > t_0$, becomes negative for at least one finite time instant $t = \bar{t}$. Despite its rigor, this condition is not usable for verification purposes, since before ruling out the possibility that the device may ever operate in the local activity regime about some operating point Q, after superimposing each of the possible infinitely-many local current perturbations upon any of the admissible infinitely-many DC current stimuli, one should compute the integral of the local instantaneous power δp(t) = δv(t) · δi(t), absorbed by the device, from the time instant $t = t_0$, at which the small-signal current stimulus is first let flow through its physical medium, to any future time instant $t > t_0$. A local activity test of this kind is rather impractical. Fortunately, there exists a powerful theoretical result, known as Theorem of Local Activity, which simplifies the procedure aimed at inferring whether the memristor is locally-active about some operating point. The application of this theorem requires the preliminary derivation of the device local transfer function.

19 DC Voltage versus Current Locus of a First-Order Current-Controlled Memristor

- First-order current-controlled extended memristor model: $\begin{cases} v = R(x,i)\cdot i \\ \lim_{i\to 0} R(x,i) \neq \infty \\ \dot{x} = f(x,i) \end{cases}$
- Insert a DC current I_S into the memristor $\Longrightarrow$ Let $I_S \equiv I = I_P$, where I_P is any real number

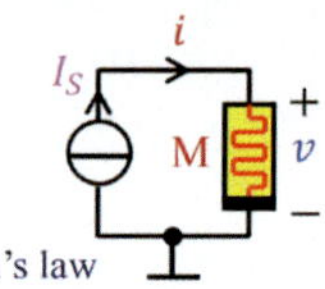

$\Longrightarrow$ Solve $\dot{x} = f(x, I_P) = 0 \Longrightarrow$ State solutions: $X^{(1)}(I_P), \ldots, X^{(k)}(I_P)$

$\Downarrow$

Calculate the corresponding voltage values from Ohm's law

$V_P^{(1)}(I_P) = R(X^{(1)}(I_P), I_P)\cdot I_P, \quad \ldots, \quad V_P^{(k)}(I_P) = R(X^{(k)}(I_P), I_P)\cdot I_P$

$\Downarrow$

Mark the following points on the V versus I plane (in the left figure k is let equal to 3)

$P^{(1)} = (I_P, V_P^{(1)}(I_P)), \quad \ldots, \quad P^{(k)} = (I_P, V_P^{(k)}(I_P))$

$\Downarrow$

Repeat the above procedure for each value of $I_P \in (-\infty, \infty)$

$\Downarrow$

Interpolate the current-voltage pairs derived in all the iterations

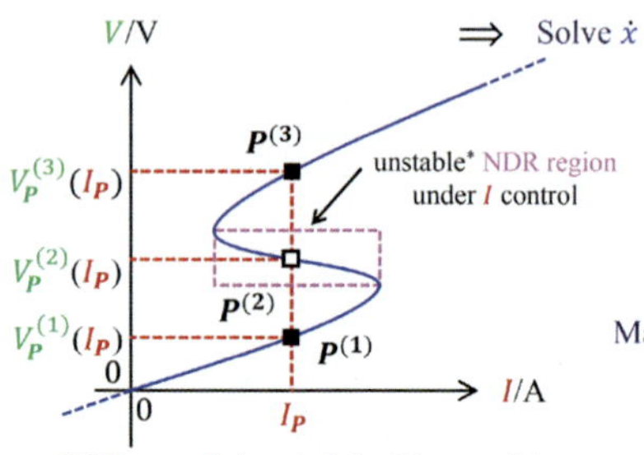

DC V versus I characteristic of the memristor (a purely-positive sweep in I_P was carried out here)

Note: A NDR branch on the DC V versus I locus of a one-port is a signature for local activity

* If the DC V versus I locus has positive (negative) slope at a bias point $P = (V_P, I_P)$, the one-port is said to have positive (negative) differential resistance, PDR (NDR) for short, therein. In the example shown here, the lower and upper PDR branches (the NDR branch) are stable (is unstable) under current control, as clarified later.

Before calculating the local transfer function of the current-controlled memristor, its DC voltage versus current characteristic needs to be determined. In order to achieve this goal, the following three-step procedure is iterated for any admissible bias current, which may be injected into the device.

1) All the stable or unstable operating points, which the device may ever admit, when a given DC current is let flow continuously through its physical medium, are derived through the computation of the zeros of the state evolution function.
2) The Ohm law is then employed to obtain the bias voltage, which would fall across the device, for each of the operating points derived in the previous step.
3) Points, featuring as abscissa the DC current stimulus under consideration, and as ordinates the bias voltages, which may consequently fall across the device, as recorded in the previous step, are marked in the voltage versus current plane.

A suitable interpolation method is finally employed to find the curve, which best fits the set of points marked in the voltage versus current plane over the various iterations of the three-step procedure. This finally delivers a good approximation for the device DC voltage versus current characteristic. A locally-active volatile memristor physical realization typically features a S-shaped DC V versus I locus. Along the branches, characterized by a negative (positive) slope, a device of this kind is said to admit a negative (positive) differential resistance, NDR (PDR) for short. The presence of a NDR branch along the DC V versus I characteristic of a device is a signature for its local activity.

20 Local Model of a First-Order Current-Controlled Memristor

$$\begin{cases} v = v(x,i) = R(x,i)\cdot i \\ \lim_{i\to 0} R(x,i) \neq \infty \\ \dot{x} = f(x,i),\ x \in \mathbb{R} \end{cases}$$ DAE set of a first-order extended memristor

- *Linearize the DAE set* about an operating point $Q = X$, which is associated unequivocally to a bias point $\boldsymbol{P} = (I_P, V_P)$ in the voltage versus current plane.

$$\begin{cases} \dfrac{d\delta x}{dt} = a(Q)\cdot\delta x + b(Q)\cdot\delta i \\ \delta v = c(Q)\cdot\delta x + d(Q)\cdot\delta i \end{cases} \quad \text{where} \quad \begin{cases} a(Q) \triangleq \left.\dfrac{\partial f(x,i)}{\partial x}\right|_Q & b(Q) \triangleq \left.\dfrac{\partial f(x,i)}{\partial i}\right|_Q \\ c(Q) \triangleq \left.\dfrac{\partial v(x,i)}{\partial x}\right|_Q & d(Q) \triangleq \left.\dfrac{\partial v(x,i)}{\partial i}\right|_Q \end{cases}$$

- Transform the linearized system in the *Laplace domain* (with $\delta x(0) = 0$)

$$s\cdot\mathcal{L}\{\delta x(t)\} = a(Q)\cdot\mathcal{L}\{\delta x(t)\} + b(Q)\cdot\mathcal{L}\{\delta i(t)\} \longrightarrow \mathcal{L}\{\delta x(t)\} = \frac{b(Q)}{s-a(Q)}\cdot\mathcal{L}\{\delta i(t)\}$$

$$\mathcal{L}\{\delta v(t)\} = c(Q)\cdot\mathcal{L}\{\delta x(t)\} + d(Q)\cdot\mathcal{L}\{\delta i(t)\} \longrightarrow \mathcal{L}\{\delta v(t)\} = c(Q)\cdot\frac{b(Q)}{s-a(Q)}\cdot\mathcal{L}\{\delta i(t)\} + d(Q)\cdot\mathcal{L}\{\delta i(t)\}$$

- The local impedance Z_Q of the memristor about Q is thus computed via

$$Z_Q(s) \triangleq \left.\frac{\mathcal{L}\{\delta v(t)\}}{\mathcal{L}\{\delta i(t)\}}\right|_Q = K_Z(Q)\cdot\frac{s - z_Z(Q)}{s - p_Z(Q)}$$

where

$$K_Z(Q) = d(Q),\quad z_Z(Q) = \frac{a(Q)\cdot d(Q) - b(Q)\cdot c(Q)}{d(Q)},\quad p_Z(Q) = a(Q)$$

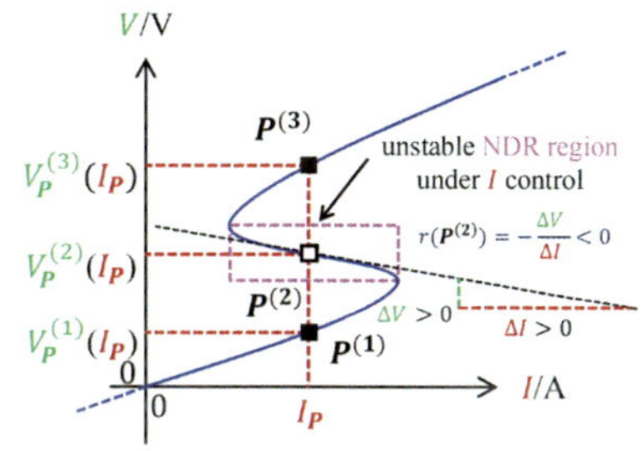

DC V versus I characteristic of the memristor (a purely-positive sweep in I_P was carried out here)

$$R_2(Q) = \frac{K_Z(Q)\cdot z_Z(Q)}{p_Z(Q) - z_Z(Q)} \qquad L(Q) = \frac{K_Z(Q)}{z_Z(Q) - p_Z(Q)}$$

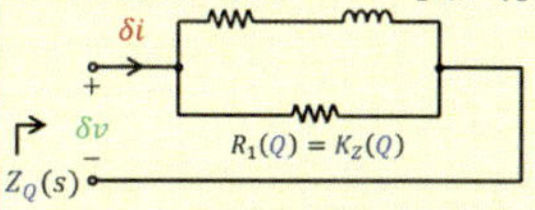

Memristor* small-signal equivalent circuit model at Q

*The small-signal resistance of the memristor at an operating point Q corresponds to the slope of its DC V versus I locus at the corresponding bias point $\boldsymbol{P} = (V_P, I_P)$ (see the upper figure on the right). With reference to the memristor local equivalent circuit model, this quantity is simply computed as $Z_Q(0) = r(\boldsymbol{P}) = R_1(Q) \parallel R_2(Q)$.

The local transfer function of a two-terminal current-controlled device, such as the memristor under the zooming lens in this pedagogical article, is its small-signal impedance. Its computation about a given operating point is obtained through a four-step procedure, as outlined below:

1) The state equation and the Ohm law are first linearized by developing the functions, appearing on their right-hand sides, in Taylor series about the operating point of interest, while disregarding terms of order equal to or higher than 2.
2) The small-signal model is then transformed in the Laplace domain.
3) The Laplace transform of the local state response is expressed as a function of the Laplace transform of the infinitesimal current perturbation employing the representation of the small-signal state equation in the Laplace domain.
4) Using the formula for the Laplace transform of the local state response, as derived in the previous step, inside the representation of the small-signal Ohm's law in the Laplace domain allows to determine a closed-form expression for the local impedance $Z_Q(s)$ of the memristor about the given operating point.

A one-port, hosting a suitable combination of linear circuit elements within its core, and featuring as impedance the local impedance of the memristor about the operating point under consideration, may then be synthesized. The values of the linear elements within the circuit representation of the memristor small-signal model depend upon the operating point about which its local impedance was earlier evaluated. The choice of the small-signal equivalent circuit model of the current-controlled memristor is not unique. For the choice made here, the small-signal resistance $r(Q)$ of the device at a given operating point $Q = X$, denoting the slope of its DC voltage V versus current I characteristic at the corresponding bias point $\boldsymbol{P} = (I, V)$, may be computed as the parallel combination between the two resistances $R_1(Q)$ and $R_2(Q)$ evaluated at the operating point Q.

21 Local Activity Theorem

- A current-controlled n^{th}-order one-port* with state vector $x = [x_1, \ldots, x_n]^T$ is said to be locally active at $Q = X = [X_1, \ldots, X_n]^T$ if and only if at least one of the following 4 conditions applies:

1. $Z_Q(s)$ has a pole p on the RHP, such that $\Re\{p\} > 0$
2. $Z_Q(s)$ has a simple pole of the form $p = j\omega_P$, whose residue, namely
$$\mathrm{res}(Z_Q(s), p) = \lim_{s \to j\omega_P} (s - j\omega_P) \cdot Z_Q(s),$$
is either a complex number or a negative real number
3. $Z_Q(s)$ has a multiple pole on the imaginary axis, reading as $p = j\omega_P$
4. $\Re\{Z_Q(j\omega)\}$ is negative for at least one real-valued angular frequency $\omega = \omega_0 \in (-\infty, \infty)$

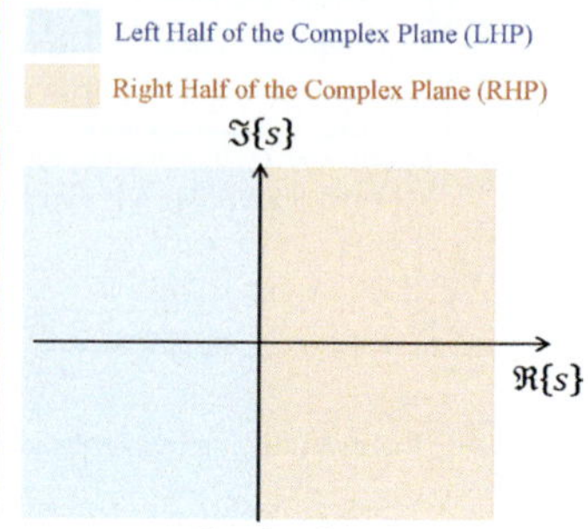

Complex plane

Notes:

- Conditions 2. and 3. refer to marginal cases
- Condition 4. holds true at an operating point Q for the one-port, corresponding to a bias point $P = (I_P, V_P)$, which lies on a NDR branch of the one-port DC V versus I characteristic, since the following inequality holds therein:
$$\Re\{Z_Q(j\omega)\}\Big|_{\omega=\omega_0=0} \equiv r(P) < 0$$

*Despite it is invoked here to test a nth-order one-port under current control, using the local admittance $Y \triangleq Z^{-1}$ in place for the local impedance Z as one-port local transfer function, the Local Activity Theorem may be applied *mutatis mutandis* to any n^{th}-order one-port under voltage control. Moreover, the Theorem has been extended for testing multi-ports for Local Activity. For more details, the interested reader is invited to consult the seminal work cited here:

L.O. Chua, "Local Activity is the Origin of Complexity," International Journal of Bifurcation and Chaos, vol. 15, no. 11, pp. 3435-3456, 2005

The Theorem of Local Activity asserts that a current-controlled memristor is locally active about an operating point if at least one of four conditions holds true, i.e. if 1) its local impedance about the operating point of interest admits at least one pole with positive real part, or if 2) its local impedance about the operating point of interest admits a simple pole on the imaginary axis while the associated residue is either a complex number or a negative real number, or if 3) its local impedance about the operating point of interest admits a multiple pole on the imaginary axis, or even if 4) the real part of its local impedance, evaluated about the operating point of interest, and expressed in terms of the Fourier variable $j\omega$, is negative for at least one finite value of the angular frequency.

Importantly, the small-signal resistance $r(\boldsymbol{P})$ of a current-controlled memristor about a given bias point $\boldsymbol{P} = (I, V)$, lying along the respective DC V versus I characteristic, and associated unequivocally to one of its admissible operating points, say $Q = X$, may be determined by evaluating its local impedance $Z_Q(j\omega)$ about Q for $\omega = 0$. It follows that the current-controlled memristor is locally active at a given operating point Q provided its small-signal resistance $r(\boldsymbol{P})$, evaluated at the corresponding bias point $\boldsymbol{P} = (I, V)$, is negative, in view of condition 4) from the Local Activity Theorem.

22 Edge of Chaos Corollary

- A current-controlled n^{th}-order one-port* with state vector $\boldsymbol{x} = [x_1, \ldots, x_n]^{\mathrm{T}}$ is said to be on the Edge of Chaos at $\boldsymbol{Q} = \boldsymbol{X} = [X_1, \ldots, X_n]^{\mathrm{T}}$ if and only if the following 2 conditions concurrently hold true:

  (i) All the poles of $Z_Q(s)$ lie on the LHP

  (ii) $\Re\{Z_Q(j\omega)\}$ is negative for at least one real-valued angular frequency $\omega = \omega_0 \in (-\infty, \infty)$

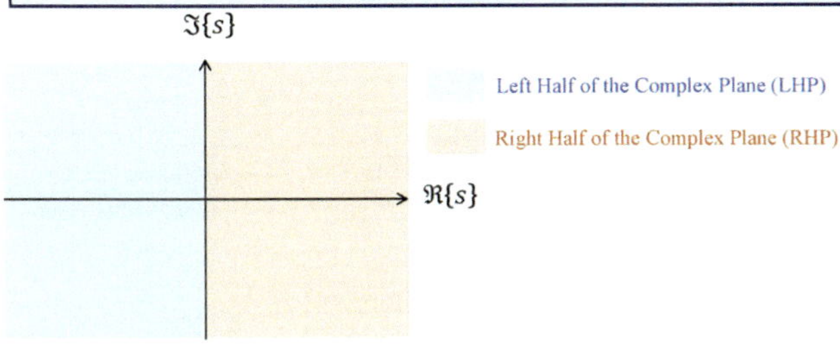

Complex plane

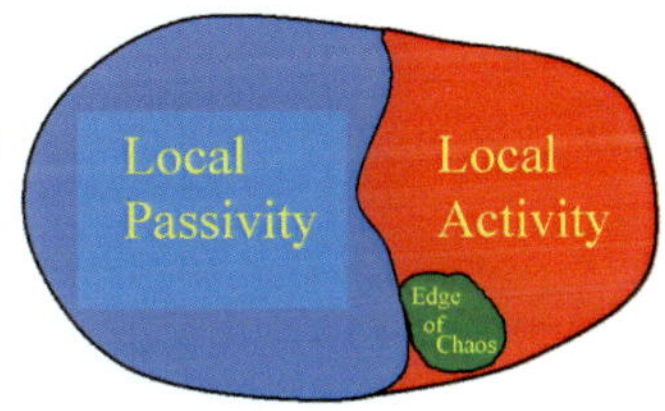

Edge of Chaos: the "Pearl" Embedded within Local Activity

* Despite it is invoked here to test a nth-order one-port under current control, using the local admittance $Y \triangleq Z^{-1}$ in place for the local impedance Z as one-port local transfer function, the Edge of Chaos Corollary may be applied *mutatis mutandis* to any n^{th}-order one-port under voltage control. Moreover, the Corollary has been extended for testing multi-ports for Edge of Chaos. For more details, the interested reader is invited to consult the seminal work cited here:

L.O. Chua, "Local Activity is the Origin of Complexity," International Journal of Bifurcation and Chaos, vol. 15, no. 11, pp. 3435-3456, 2005

According to the Edge of Chaos Corollary, two are the conditions, ensuring a current-controlled memristor to be poised on the Edge of Chaos at a given operating point *Q*. This is the case if 1) the poles of its local impedance $Z_Q(s)$ about *Q* lie on the left half of the complex plane (LHP), and if 2) the real part $R\{Z_Q(j\omega)\}$ of its local impedance $Z_Q(j\omega)$ about *Q* is negative for at least one real-valued angular frequency ω. This means that the memristor local impedance $Z_Q(s)$ at *Q* satisfies only condition 4) from the Local Activity Theorem. All in all, the memristor is poised on the Edge of Chaos at *Q* if and only if it is asymptotically stable yet locally active at *Q*.

23 DC Characteristics of a Volatile Locally-Active Niobium Oxide Memristor

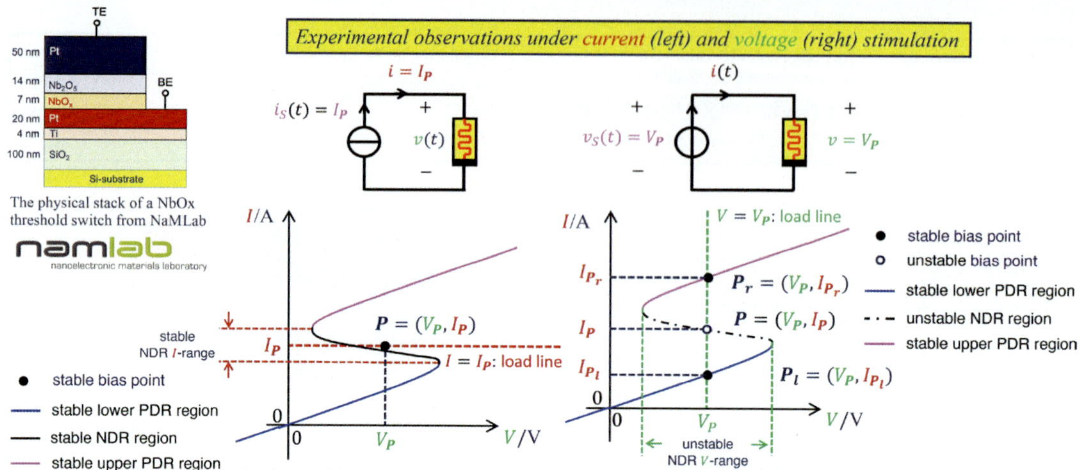

The physical stack of a NbOx threshold switch from NaMLab

Sketch of the DC I versus V characteristic of the current-driven device. The whole locus is stable.

Sketch of the DC I versus V locus of the voltage-driven device. The solid branches (dash-dotted branch) are (is) stable (unstable)

- Condition 4. of the Local Activity Theorem holds true at DC, i.e. for $\omega = 0$, at any NDR bias point.
- However the NDR branch is stable only under current sweep. On the other hand, the PDR branches are stable under both voltage and current sweeps.

The DC current I versus voltage V characteristic of the niobium oxide memristor, fabricated at NaMLab, is odd symmetric. In the first quadrant of the current versus voltage plane the DC characteristic features one NDR branch, sandwiched between two PDR branches, the left (right) of which, being associated to high (low) volatile resistance states, is referred to as high (low) resistance state PDR branch, or HRS (LRS) PDR branch for short, in the remainder of this article. Recording the DC I versus V locus in the lab by changing the current, flowing through the device, in small steps, and recording for each value, assigned to the current stimulus, the voltage, which correspondingly falls between the device terminals after transients decay to zero, it is possible to record all the admissible device bias points. On the other hand, sweeping the voltage, let fall across the device physical stack, and recording for each input value the current, which correspondingly flows through the device at steady state, the device bias points, lying along the NDR branch, are unobservable, which reveals their instability under voltage stimulation. In fact, in case the value of the voltage stimulus assumes the lower (upper) bound of the NDR range during a monotonic increase (decrease) from smaller (larger) values, the device is found to converge to a bias point sitting along the HRS (LRS) PDR branch. Differently from the current stimulation scenario, where the device exhibits a monostable response to any DC input value, when a bias voltage from the NDR range is applied across it in the voltage excitation case, the memristor acts as a bistable system.

24 A generic memristor model for the NaMLab Threshold Switch

- To explain the experimental observations, we developed the following model for the device:

$$\begin{cases} \dot{x} = g(x, v) \text{ and} \\ i = G(x) \cdot v \end{cases}$$

with state evolution and memductance functions respectively expressed by

$$g(x, v) = a_0 + a_1 \cdot x + b_2 \cdot v^2 + c_{21} \cdot v^2 \cdot x + c_{22} \cdot v^2 \cdot x^2 + c_{23} \cdot v^2 \cdot x^3 + c_{24} \cdot v^2 \cdot x^4 + c_{25} \cdot v^2 \cdot x^5 \text{ and}$$

$$G(x) = d_0 + d_1 \cdot x + d_2 \cdot x^2 + d_3 \cdot x^3 + d_4 \cdot x^4$$

- Chua's Unfolding Principle was invoked to develop ad hoc polynomial series expansions for $g(x, v)$ and $G(x)$.

Table 1 Parameter setting of the first-order volatile locally-active memristor model

a_0	a_1	b_2	c_{21}	c_{22}	c_{23}	c_{24}
$5.19 \cdot 10^9$	$-2.05 \cdot 10^7$	$7.21 \cdot 10^9$	$-0.07 \cdot 10^9$	$2.27 \cdot 10^5$	$-2.40 \cdot 10^2$	$1.25 \cdot 10^{-1}$

c_{25}	d_0	d_1	d_2	d_3	d_4
$-2.69 \cdot 10^{-5}$	$6.50 \cdot 10^{-3}$	$-6.66 \cdot 10^{-5}$	$2.14 \cdot 10^{-7}$	$-2.14 \cdot 10^{-10}$	$1.19 \cdot 10^{-13}$

L.O. Chua, "Resistance Switching Memories are Memristors," Applies Physics A, vol. 102, pp. 765-783, 2011

A. Ascoli, S. Slesazeck, H. Mähne, R. Tetzlaff, and T. Mikolajick, "Nonlinear Dynamics of a Locally-Active Memristor," IEEE Trans. Circuits and Systems-I (TCAS-I): Regular Papers, vol. 62, no. 4, pp. 1165–1175, Apr. 2015, DOI: 10.1109/TCSI.2015.2413152

In order to analyze these experimental observations to gain a deep understanding of the mechanisms behind the stability (instability) of the NDR branch of the device DC characteristic under current (voltage) stimulation, a simple yet accurate mathematical model of the niobium oxide device was then developed. Invoking Chua's Unfolding Principle, state evolution function and Ohm's law were shaped as polynomial series expansions in state and voltage, while their coefficients were fitted to experimental data, acquired from a device sample through a number of DC and AC tests, employing a standard optimization algorithm. The proposed DAE set falls in the class of first-order generic memristors. While the voltage (current) appears here as the input (output) variable, the memristor is strictly speaking a current-controlled device in Circuit Theory, since its DC characteristic is single-valued if and only if the current is taken as the independent variable.

25 Memristor Dynamic Route Map (DRM) under Current Excitation

- The model may be cast in a form suitable under current stimulation: $\begin{cases} \dot{x} = g(x, R(x) \cdot i) \\ v = R(x) \cdot i \quad \text{with} \quad R(x) = G^{-1}(x) \end{cases}$
- A DRM is a family of State Dynamic Routes (SDRs), each of which is the $\dot{x}$ versus x locus, which is derived by assigning one DC value to the stimulus.

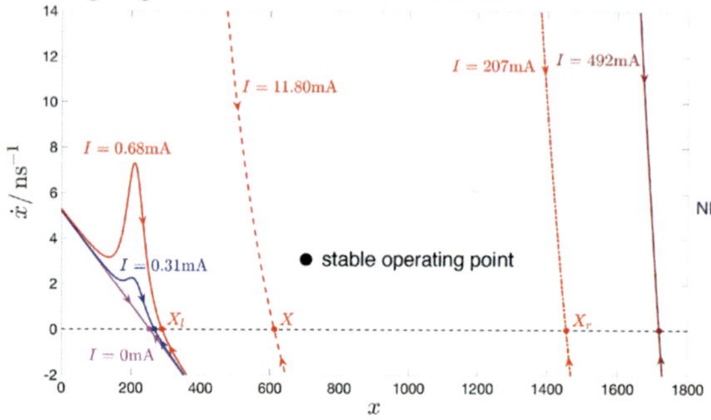

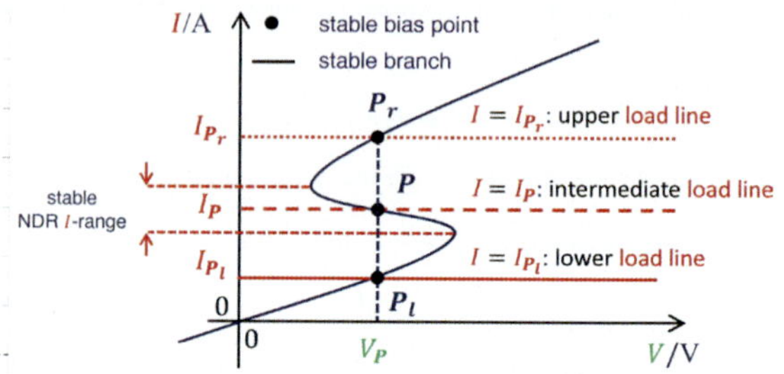

Memristor DRM under a range of DC current values. The 3 red SDRs are associated to the same NDR DC voltage, i.e. $V = 0.875\text{V}$ (see the 3 load lines on the right sketch).

Sketch of the device DC I versus V characteristic under current sweep. The whole locus is stable.

- Irrespective of the DC current, there exists one and only one globally asymptotically stable (GAS) operating point $Q = X$ for the device

A. Ascoli, A.S. Demirkol, R. Tetzlaff, S. Slesazeck, T. Mikolajick, and L.O. Chua, "On Local Activity and Edge of Chaos in a NaMLab Memristor", *Frontiers in Neuroscience*, vol. 15, no. 651452, 30pp., 2021, DOI: 10.3389/fnins.2021.651452

Since the memductance function depends upon the state only, the model may be easily recast as a current-controlled generic memristor DAE set. This allows to investigate the response of the device to a DC current stimulus. Plotting the time derivative of the state, i.e. the state evolution function, against the state itself, for any DC value, assigned to the current stimulus, the resulting trace, representing a State Dynamic Route (SDR), provides insights into time evolution and asymptotic behavior of the state x from any initial condition x_0. Arrows, pointing to the east (west), are superimposed on top of the SDR, while it visits the upper (lower) half of the $\dot{x}$ versus x plane, to reveal a progressive increase (decrease) in the state when the state evolution function is positive (negative). Any intersection between a SDR and the horizontal axis, where $\dot{x} = 0$, represents an equilibrium $x = X$ for the state equation. An equilibrium $x = X$ is asymptotically stable if and only if the slope of the SDR is negative therein. The family of all the possible SDRs represents the Dynamic Route Map (DRM), which allows to draw a comprehensive picture for the device response to DC stimulation. Under any DC current stimulus, irrespective of the initial condition, the state x of the NaMLab device is found to converge toward a specific operating point $Q = X$, associated unequivocally to a globally asymptotically stable bias point $\boldsymbol{P} = (V, I)$ on the DC characteristic, after transients vanish. Particularly, this occurs also for $I = 0$, when the SDR, referred to as Power Off Plot (POP), shows that only one memory state, specifically $X = -a_0/a_1$, is available in the device for data storage, when the power is switched off, confirming its volatile nature. All in all, recurring to the system-theoretic concept referred to as DRM, a powerful graphic tool for analyzing first-order systems, allows to explain the stability of all the bias points of the device DC characteristic under current stimulation.

26 Memristor DRM under Voltage Excitation

- Differential algebraic equation (DAE) set suitable under voltage stimulation: $\begin{cases} \dot{x} = g(x, v) \\ i = G(x) \cdot v \end{cases}$

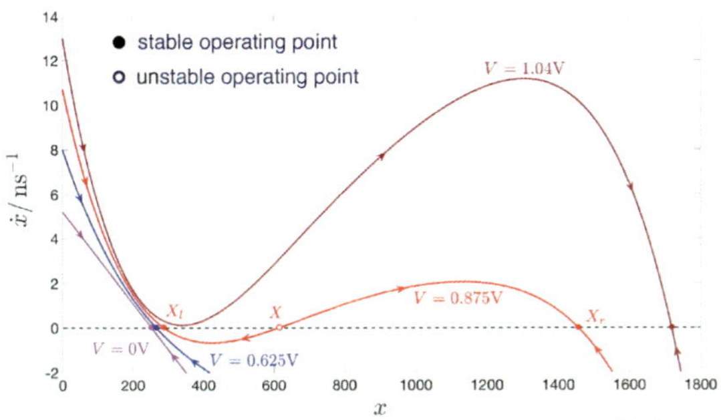

Memristor DRM under a range of DC voltage values. For $V = 0.875$V three are the possible DC currents for the device (see the load line on the right sketch)

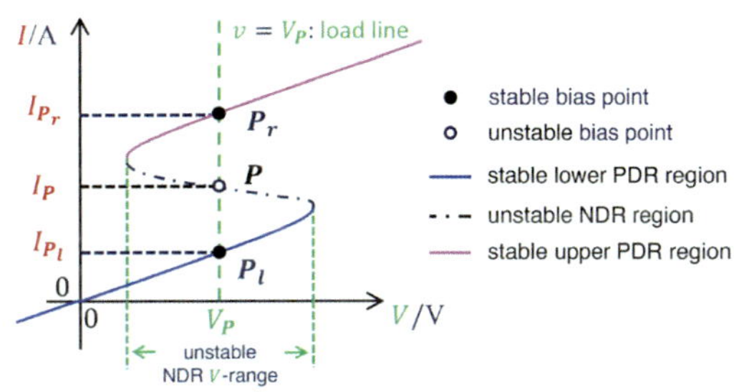

Sketch of the device DC I versus V characteristic under voltage sweep. The NDR branch – refer to the dashed-dotted black trace – is unstable.

- If the DC voltage is chosen from the NDR range, there exists 3 admissible operating points for the device, namely $Q_l = X_l$, $Q = X$, and $Q_r = X_r$, of which the intermediate NDR one $Q = X$ is unstable.

A. Ascoli, A.S. Demirkol, R. Tetzlaff, S. Slesazeck, T. Mikolajick, and L.O. Chua, "On Local Activity and Edge of Chaos in a NaMLab Memristor", *Frontiers in Neuroscience*, vol. 15, no. 651452, 30pp., 2021, DOI: 10.3389/fnins.2021.651452

Considering now the voltage-controlled representation of the DAE set, and drawing the DRM of the device for this excitation scenario, each of the resulting SDRs crosses with negative slope the horizontal axis in a single point only if and only if the stimulus is chosen outside of the NDR range. In this case, similarly as upon current stimulation, the device is found to approach asymptotically a globally asymptotically stable operating point, irrespective of the initial condition. In particular, this is the case also for $V = 0$, when, as may be inferred from the POP, the NaMLab memristor may only sit at the memory state $X = -a_0/a_1$ after transients vanish, as expected of a volatile device. Different is what happens in case the voltage stimulus is chosen within the NDR range. In this case a SDR intersects the state axis in three different locations. The left (right) one corresponds to a locally stable bias point lying along the HRS (LRS) PDR branch of the device DC characteristic. However, $\partial\dot{x}/\partial x$ is negative at the intermediate crossing, which explains the instability nature of the associated state equation equilibrium, which corresponds to a bias point located on the NDR branch of the I versus V locus. All in all, the graphical analysis of the state behaviour, enabled by the DRM technique, explains the reason for the instability of the NDR branch of the device DC characteristic under voltage stimulation.

27 Bias circuit for stabilizing a NDR operating point of the voltage driven device

$$v = \hat{v}(x) = \frac{V_S}{1 + G(x) \cdot R_S}$$ memristor voltage

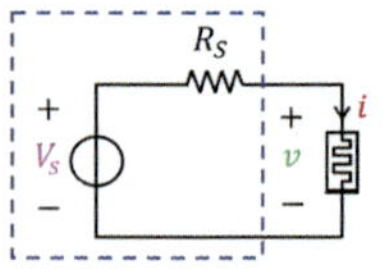

DC bias circuit

$$\begin{cases} \dot{x} = g(x, \hat{v}(x)) \\ i = G(x) \cdot \hat{v}(x) \end{cases}$$ memristor differential algebraic equation (DAE) set

$$I = \hat{I}(V) = \frac{V_S - V}{R_S}$$ load line formula

- The crossings betwen the state evolution function versus state locus and the state axis provide the admissible device operating points
- The intersections betwen the load line and the DC I versus V locus identify the admissible device bias points

A. Ascoli, R. Tetzlaff, A.S. Demirkol, L.O. Chua, "Edge of Chaos Theory Explains Complex Phenomena in Memristor Circuits," chapter in book entitled "Selected Topics in Intelligent Chips with Emerging Devices, Circuits and Systems," River Publishers, Aalborg, Denmark, June 2023, e-ISBN: 9788770227643

A simple bias circuit adaptation may be developed for stabilizing a NDR bias point $\boldsymbol{P} = (V_P, I_P)$ of the device under voltage stimulation. It is sufficient to insert an appropriate linear resistance R_S between the memristor and the DC voltage source. While it generated a bias voltage $V = V_P$, when no resistor was included in the design, the battery is set here so as to source a suitable constant stimulus V_S, as explained shortly. The bias circuit adaptation modifies the load line equation, earlier reading as $V = V_P$, and now expressed via $I = (V_S - V)/R_S$. The bias parameter pair (V_S, R_S) must be chosen in such a way that the load line $I = (V_S - V)/R_S$ crosses the device DC I versus V locus in the NDR bias point $\boldsymbol{P}$ under stabilization, only. The insertion of the series resistor between DC voltage source and memristor also changes the state evolution function, earlier formulated as $g(x, V_{\boldsymbol{P}})$, and now defined via $g(x, v(x))$, with $v(x) = V_S/(1 + G(x) \cdot R_S)$.

28 Stabilization of a NDR Bias Point on the NDR Branch of the DC Characteristic of the Voltage-Driven Device

- Apply a DC voltage V_S across the R_S-$\mathcal{M}$ series one-port.
- Choose three (R_S, V_S) pairs, so that the resulting load lines cross the DC I-V locus in one NDR bias point $\boldsymbol{P} = (V_{\boldsymbol{P}}, I_{\boldsymbol{P}})$:
- In order to stabilize the NDR bias point, the modulus R_S^{-1} of the slope $-R_S^{-1}$ of the load line $I = \hat{I}(V)$ has to be smaller than the modulus $-r^{-1}(\boldsymbol{P})$ of the slope $r^{-1}(\boldsymbol{P})$ of the DC I-V locus at $\boldsymbol{P}$, i.e.

$$R_S > -r(\boldsymbol{P}) \triangleq -\frac{1}{\left.\frac{di}{dv}\right|_{\boldsymbol{P}}}$$ condition for the stabilization of a NDR bias point $\boldsymbol{P}$

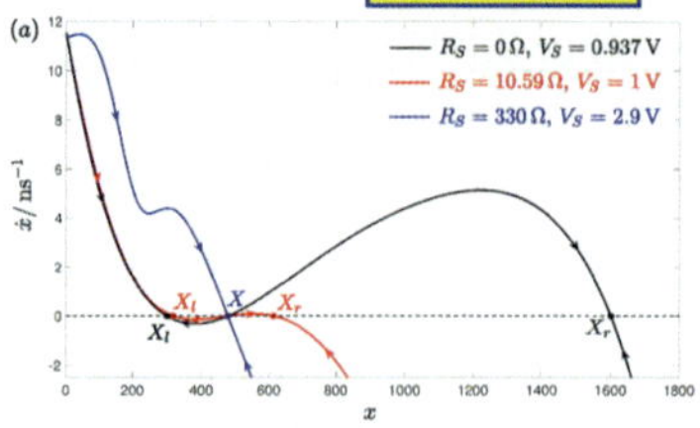

Three $\dot{x} = g(x, \hat{v}(x))$ versus x loci, sharing NDR operating point $Q = X$

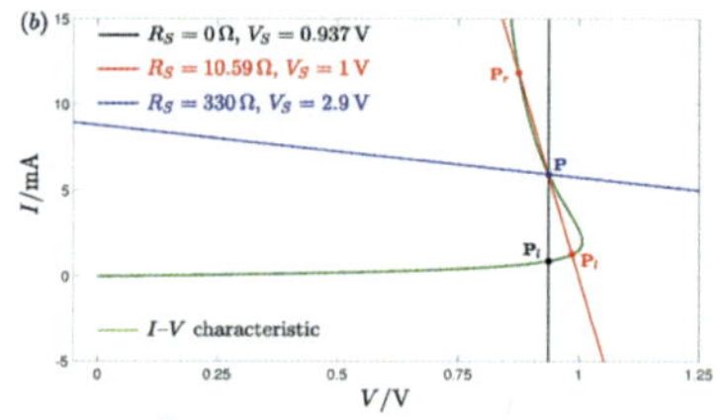

Load line $I = \hat{I}(V)$ for each (R_S, V_S) pair in (a) and DC I-V locus

A. Ascoli, S. Slesazeck, H. Mähne, R. Tetzlaff, and T. Mikolajick, "Nonlinear Dynamics of a Locally-Active Memristor," IEEE Trans. Circuits and Systems-I (TCAS-I): Regular Papers, vol. 62, no. 4, pp. 1165–1175, Apr. 2015, DOI: 10.1109/TCSI.2015.2413152

If and only if the bias parameter (V_S, R_S) is selected in such a way to induce one and only one intersection, specifically the NDR bias point $\boldsymbol{P} = (V_P, I_P)$ under stabilization, between load line and device DC I versus V characteristic, does the resulting memristor SDR, namely the $g(x, v(x))$ versus x locus, cross with negative slope the horizontal axis in a single point, featuring as abscissa the memristor operating point $Q = X$ associated to $\boldsymbol{P}$, which endows the device with the desired globally asymptotically stable quiescent state. Under these circumstances the series resistance is found to satisfy the inequality $R_S > -r(\boldsymbol{P})$, implying the modulus of the slope of the load line to be smaller than the modulus of the slope $g(\boldsymbol{P}) = r^{-1}(\boldsymbol{P})$ of the device DC I versus V characteristic at $\boldsymbol{P}$.

29 Small-Signal Equivalent Circuit Model of the NaMLab Threshold Switch

- Memristor DAE set under voltage stimulation:

$$\begin{cases} \dot{x} = g(x,v) \\ i = \hat{\imath}(x,v) = G(x)\cdot v \end{cases} \quad \text{with} \quad \begin{aligned} g(x,v) &= a_0 + a_1\cdot x + b_2\cdot v^2 + c_{21}\cdot v^2\cdot x + c_{22}\cdot v^2\cdot x^2 + c_{23}\cdot v^2\cdot x^3 + c_{24}\cdot v^2\cdot x^4 + c_{25}\cdot v^2\cdot x^5 \\ G(x) &= d_0 + d_1\cdot x + d_2\cdot x^2 + d_3\cdot x^3 + d_4\cdot x^4 \end{aligned}$$

- Local admittance $Y(s)$ of the voltage-driven memristor about an operating point $Q = X$ associated to a bias point $\boldsymbol{P} = (V_P, I_P)$:

$$Y_Q(s) = \left.\frac{\mathcal{L}\{\delta i(t)\}}{\mathcal{L}\{\delta v(t)\}}\right|_Q = K_Y(Q)\cdot\frac{s - z_Y(Q)}{s - p_Y(Q)} \quad \text{where} \quad \begin{cases} K_Y(Q) = d(Q) \\ z_Y(Q) = \dfrac{a(Q)\cdot d(Q) - b(Q)\cdot c(Q)}{d(Q)} \\ p_Y(Q) = a(Q) \end{cases} \quad \text{and} \quad \begin{aligned} a(Q) &\triangleq \left.\frac{\partial g(x,v)}{\partial x}\right|_Q & b(Q) &\triangleq \left.\frac{\partial g(x,v)}{\partial v}\right|_Q \\ c(Q) &\triangleq \left.\frac{\partial \hat{\imath}(x,v)}{\partial x}\right|_Q & d(Q) &\triangleq \left.\frac{\partial \hat{\imath}(x,v)}{\partial v}\right|_Q \end{aligned}$$

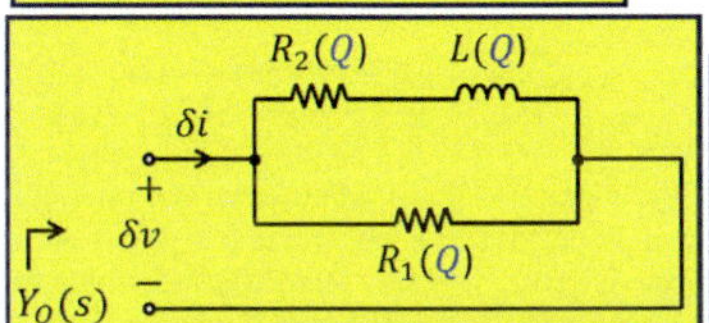

$$R_1(Q) = \frac{1}{K_Y(Q)}$$
$$R_2(Q) = -\frac{p_Y(Q)}{K(Q)\cdot(p_Y(Q) - z_Y(Q))}$$
$$L(Q) = \frac{1}{K(Q)\cdot(p_Y(Q) - z_Y(Q))}$$

Memristor small-signal equivalent circuit about Q

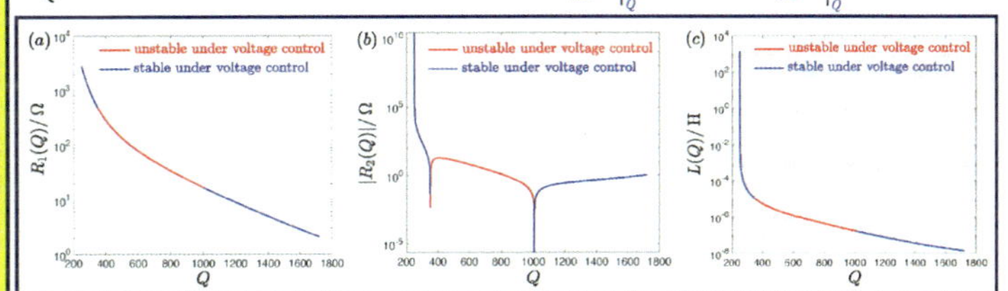

Dependence of the memristor small-signal equivalent circuit parameters upon its DC operating point $Q = X$

$R_1(Q) = \dfrac{1}{G(X)} > 0$ memristance at $Q = X$ $\quad r(\boldsymbol{P}) = R_1(Q) \parallel R_2(Q) = \dfrac{1}{\left.\frac{di}{dv}\right|_{\boldsymbol{P}}}$ small-signal resistance at $\boldsymbol{P}$

A. Ascoli, A.S. Demirkol, R. Tetzlaff, S. Slesazeck, T. Mikolajick, and L.O. Chua, "On Local Activity and Edge of Chaos in a NaMLab Memristor", *Frontiers in Neuroscience*, vol. 15, no. 651452, 30pp., 2021, DOI: 10.3389/fnins.2021.651452

In order to apply the Local Activity Theorem to the NaMLab device, its small-signal model is derived here on the basis of its voltage-controlled DAE set, following the theoretical methodology presented at slide 21. The circuit-theoretic representation of the memristor local model is a linear one-port, consisting of the parallel combination between a first resistance R_1, and a two-element branch, composed of the series connection between an inductance L and a second resistance R_2. The admittance of this one-port coincides with the memristor local admittance $Y(s)$. The values of the three circuit elements depend upon the operating point Q about which the device local admittance is evaluated. The inductance $L(Q)$ and the first resistance $R_1(Q)$ are found to be positive at any operating point. On the other hand, the second resistance $R_2(Q)$ is positive at any HRS or LRS PDR bias point, while it gets negative throughout the NDR branch of the device DC characteristic. Importantly, defining the memristor local impedance as $Z(s) = Y^{-1}(s)$, since the first-order DAE set of the device falls in the class of generic memristors, $R_1(Q) = \lim_{\omega\to\infty} Z_Q(j\omega)$ is found to coincide with the device memristance $R(Q) = G^{-1}(Q)$ at the operating point of interest. Moreover, the device small-signal resistance $r(Q) = Z_Q(j\omega)|_{\omega=0}$ at Q may be computed via $R_1(Q) \parallel R_2(Q)$.

30 Pole-Zero Diagram of the Memristor Local Admittance

- The memristor local admittance at Q reads as

$$Y_Q(s) = \left.\frac{\mathcal{L}\{\delta i(t)\}}{\mathcal{L}\{\delta v(t)\}}\right|_Q = K_Y(Q) \cdot \frac{s - z_Y(Q)}{s - p_Y(Q)}$$

where

$$K_Y(Q) = \frac{1}{R_1(Q)} \quad > 0 \text{ for all } Q = X$$

$$p_Y(Q) = -\frac{R_2(Q)}{L(Q)} \quad > 0 \text{ if } Q = X \text{ lies along the NDR branch}$$

$$z_Y(Q) = -\frac{R_1(Q) + R_2(Q)}{L(Q)} \quad < 0 \text{ for all } Q = X$$

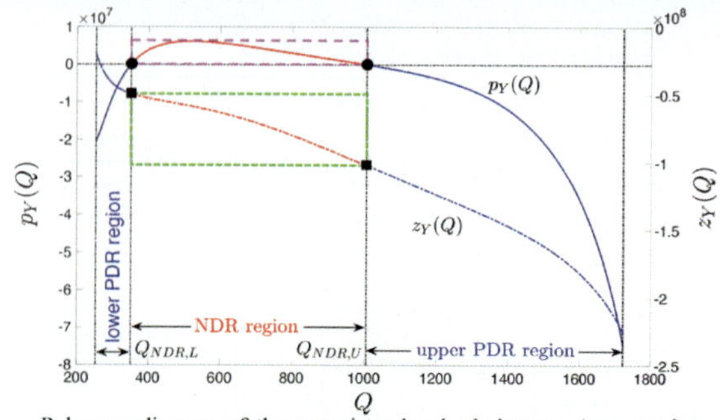

Pole-zero diagram of the memristor local admittance. An operating point Q for the device is associated to a bias point $\boldsymbol{P} = (V, I)$ along the NDR branch of the device DC V versus I locus if it lies within the range $(Q_{NDR,L}, Q_{NDR,U})$, where $Q_{NDR,L} = 351$, and $Q_{NDR,U} = 1006$. Circles (Squares) show the poles (zeros) at the NDR range endpoints.

- The memristor is locally active and unstable at any NDR bias point Q under voltage stimulation (see magenta box)
- Zero on the LHP at any NDR bias point Q (see green box)
- Under current stimulation, the device is on the *Edge of Chaos* at any NDR bias point Q, since Q is stable, and*

$$\Re\{Z_Q(0)\} = r(\boldsymbol{P}) = R_1(Q) \parallel R_2(Q) < 0$$

* The local impedance of a one-port is the inverse of its local admittance.

The device local admittance $Y(s)$ features one real zero and one real pole, only. While the zero is strictly negative, the polarity of the pole becomes positive at any NDR operating point, whereas it is negative elsewhere. The positive polarity of the pole across the NDR domain was expected as any device bias point along the branch of the DC I versus V locus, admitting a negative slope, is unstable under voltage excitation. Noting that the zero of $Y(s)$ coincides with the pole of $Z(s)$, the negative sign, which the zero of the local admittance features throughout the NDR region, is not surprising as well, given that the whole NDR branch of the device DC characteristic is stable under current stimulation. Additionally, since $r(Q)$ is negative at any NDR operating point, the current-controlled device is not solely stable but also locally active therein, in view of condition 4) of the Local Activity Theorem. In other words, the memristor is on the Edge of Chaos at any NDR bias point under current stimulation.

31 Memristor on Edge of Chaos along the NDR region under current stimulation

- The local input impedance of the current-driven memristor about Q reads as

$$Z_Q(s) = \left.\frac{\mathcal{L}\{\delta v(t)\}}{\mathcal{L}\{\delta i(t)\}}\right|_Q = K_Z(Q)\cdot\frac{s - z_Z(Q)}{s - p_Z(Q)} = \frac{1}{K_Y(Q)}\cdot\frac{s - p_Y(Q)}{s - z_Y(Q)}$$

- $\Re\{Z_Q(j\omega)\} < 0$ for*

$$\omega_{l,Z_Q} \triangleq -\frac{1}{L(Q)}\cdot\sqrt{-R_2(Q)\cdot(R_1(Q)+R_2(Q))} < \omega < \omega_{r,Z_Q} \triangleq +\frac{1}{L(Q)}\cdot\sqrt{-R_2(Q)\cdot(R_1(Q)+R_2(Q))}$$

  which holds true only in the NDR region, where $R_2(Q) < 0$, since $R_1(Q) + R_2(Q) > 0$ for all Q
- Since, additionally, $p_Z(Q) < 0$ for all Q, the memristor is on the *Edge of Chaos* at any NDR operating point Q under current excitation
- Note: irrespective of the voltage or current form of stimulation, the device is *locally-passive* in either of the two PDR regions

* For a one-port, the conditions which satisfy the inequality $\Re\{Z_Q(j\omega)\} < 0$ also satisfy the inequality $\Re\{Y_Q(j\omega)\} < 0$, and viceversa, irrespective of Q

In fact, the fourth condition of the Local Activity Theorem applies for a range of angular frequencies, centered around $\omega = 0$, if and only if the memristor is poised somewhere along the NDR branch of its DC characteristic. Considering, furthermore, that the conditions satisfying $R\{Y_Q(j\omega)\} < 0$ coincide with those satisfying $R\{Z_Q(j\omega)\} < 0$, we may conclude that

1) under voltage control the device is locally active yet unstable at any NDR operating point, while being locally passive at any HRS or LRS PDR operating point;
2) under current control the device is on the Edge of Chaos at any NDR operating point, while being locally passive at any HRS or LRS PDR operating point.

32 Satisfaction of Condition 4. from the Local Activity Theorem at a given NDR bias point

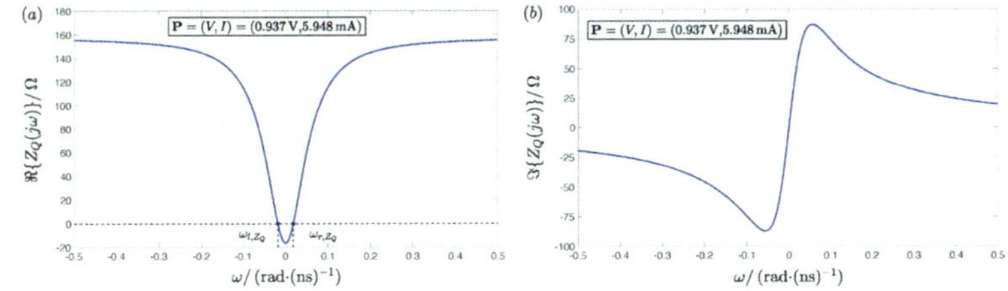

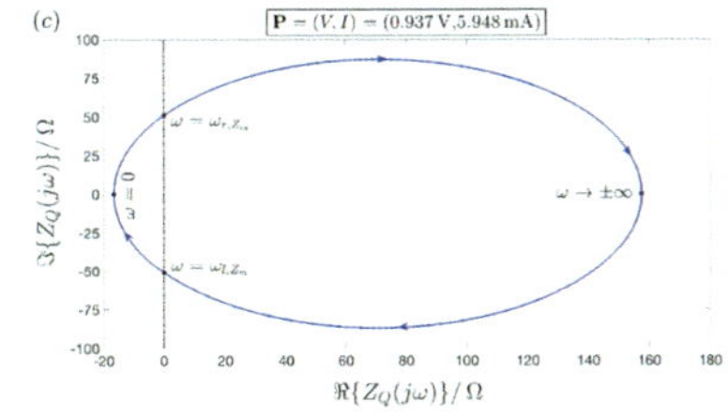

- A NDR operating point Q satisfies the constraint $Q_{NDR,L} < Q < Q_{NDR,U}$, where $Q_{NDR,L} = 351$, and $Q_{NDR,U} = 1006$
- If $Q = 478$, the bias point $\boldsymbol{P} = (V, I)$ is numerically* found to lie at $(0.937\ \mathrm{V}, 5.948\ \mathrm{mA})$, while $\omega_{r,Z_Q} = -\omega_{l,Z_Q} = 18.305\ \mathrm{Mrad}\cdot\mathrm{s}^{-1}$

* As both the DC $Q = X$ versus I and the DC V versus I loci are single-valued, first the ordinate and then the abscissa of $\boldsymbol{P}$ are determined for each DC state.

From the analysis of the NaMLab memristor response to DC stimulation, reported in slides 26 and 27, the device operating point Q is associated to a bias point $\boldsymbol{P}$, lying on the NDR branch of the DC I versus V characteristic, if and only if it falls within the range $(Q_{NDR,L}, Q_{NDR,U}) = (351,1006)$.

For example, stabilizing the device on the NDR operating point $Q = 478$ by injecting a DC current I = 5.948mA into its physical stack, the Nyquist diagram of its local impedance $Z(j\omega)$, evaluated about Q, is found to visit the LHP for a range of angular frequencies centered around $\omega = 0$. This provides firm evidence for the satisfaction of condition 4) from the Local Activity Theorem at the NDR operating point $Q = 478$, where, as a result, the NaMLab device is poised on the Edge of Chaos.

33 The NaMLab Memristor is a Local Energy Source

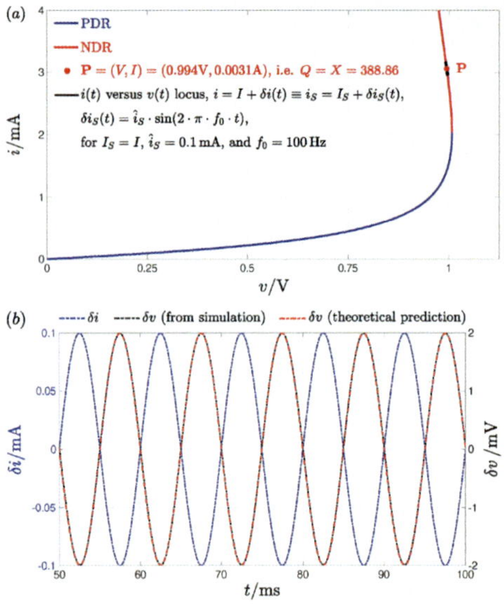

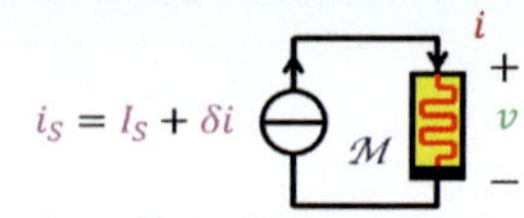

- A bias current I_S polarizes the device in a specific NDR bias point ***P***
- An infinitesimal current perturbation δi_S is added to the DC input in the form

$$\delta i_S(t) = \hat{\imath}_S \cdot \sin(2 \cdot \pi \cdot f_0 \cdot t)$$

- The device local voltage response δv to the local current stimulus δi_S is cast as*

$$\delta v(t) = |Z_Q(j \cdot 2 \cdot \pi \cdot f_0)| \cdot \hat{\imath}_S \cdot \sin\left(\omega_0 \cdot t + \arg(Z_Q(j \cdot 2 \cdot \pi \cdot f_0))\right),$$

where

$$|Z_Q(j \cdot 2 \cdot \pi \cdot f_0)| = 19.92, \quad \text{and} \quad \arg\{Z_Q(j \cdot 2 \cdot \pi \cdot f_0)\} = 179.99°$$

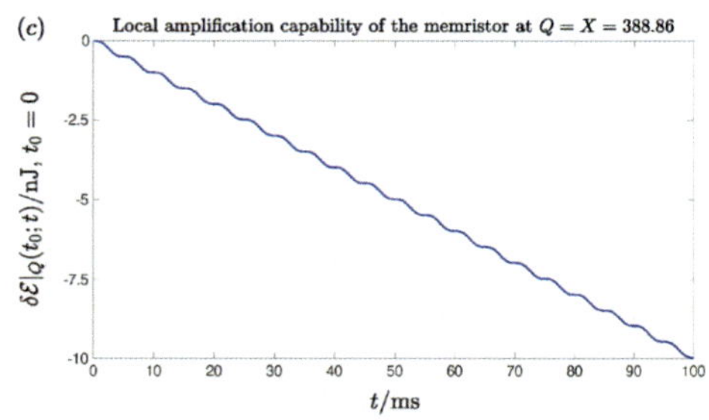

* Physical units are omitted for simplicity.

With the aim to demonstrate that the NaMLab device may act as a local source of energy under suitable polarization, a DC current I = 3.1mA is let flow continuously through its physical stack from t = 0 to stabilize it into the bias point Q = 389, which falls within the NDR range $(Q_{NDR,L}, Q_{NDR,U})$ = (351,1006). At some later time, say from $t = t_0$, an infinitesimal current perturbation ∂i is added on top of the DC stimulus I, and the local voltage δv, which is found to fall across the device, as a result, is recorded thereafter. Computing the time integral of the local instantaneous power $\delta p(t) = \delta v(t) \cdot \delta i(t)$, absorbed by the memristor, over a closed time interval $[t_0, t]$, with $t > t_0$, the small-signal energy $\delta\mathcal{E}(t_0; t)$, entering the device over this temporal range, is found to decrease monotonically from zero. Thus, the NaMLab device, biased around the NDR operating point Q = 389, delivers local energy to the current source it is coupled to! Note however that, including also the power, supplied to the memristor for polarization purposes, in the argument of the integrand function, the total energy, absorbed by the NaMLab device over the time frame $[0, t]$, with $t > t_0$, would never go negative. In fact, as any other memristor physical realization, the niobium oxide volatile resistance switching memory, fabricated at NaMLab, is a globally passive device.

34 Complete DC current versus voltage locus of the NaMlab Memristor

- Let us derive the DC current-voltage locus of the device for both positive and negative voltages
- As mentioned earlier, plotting the DC voltage data against the DC current data, a single-valued curve is obtained, revealing that, strictly speaking, the NaMLab memristor falls in the circuit-theoretic class of current-controlled devices.
- Moreover, as anticipated above, the DC Q versus I locus is also found to be single-valued.

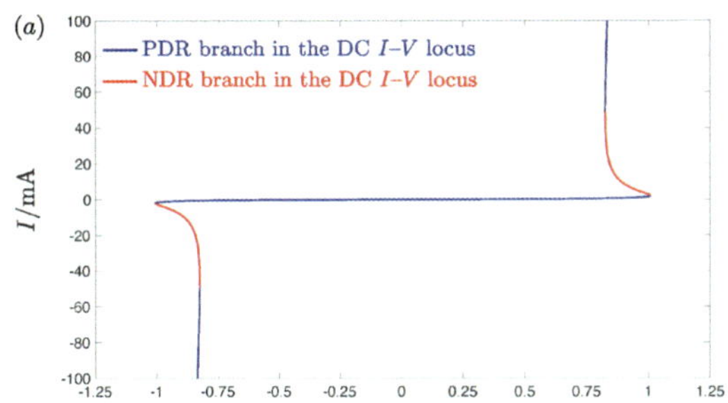

DC I-V characteristic of the NaMLab memristor

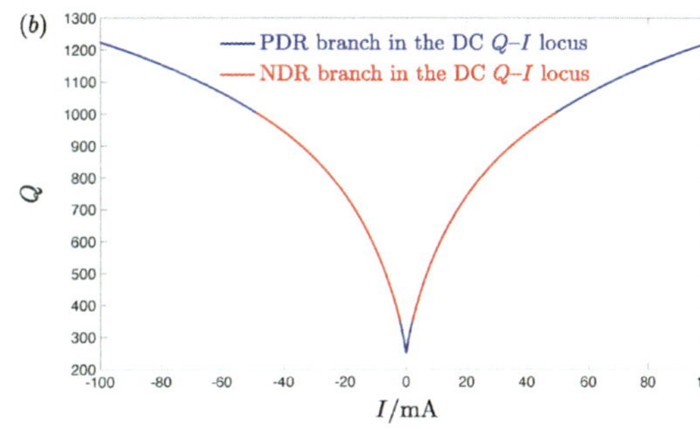

DC Q-I characteristic of the NaMLab memristor

From a circuit-theoretic viewpoint the NaMLab memristor is a current-controlled device, given that plotting its bias voltage V against its bias current I, the resulting trace is single-valued. As discussed in slide 26, each DC current stimulus I poises the device into a specific operating point Q, which is unequivocally associated to one and only one DC voltage V.

Considering also negative input values in the analysis of the device response to DC current stimuli, the particular forms of state evolution and memductance functions in the DAE set of the NaMLab memristor – refer to slide 25 – endow the Q versus I locus with even symmetry and the V versus I locus with odd symmetry.

35 Toward the Design of the Bio-Inspired Two-Cell Array: Circuit, and Large-Signal Model of the Basic Cell

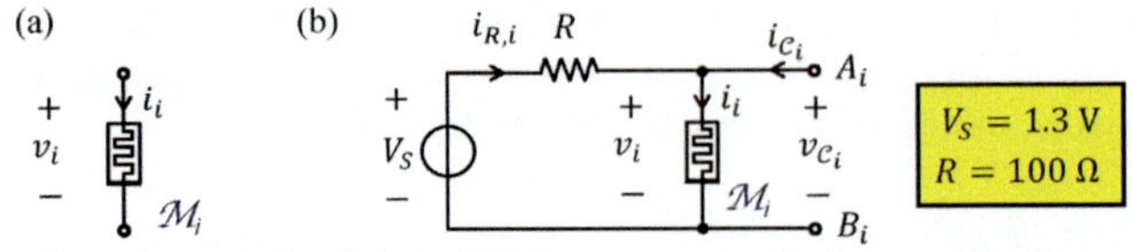

(a) Symbol of memristor $\mathcal{M}_i$ in the cell $\mathcal{C}_i$, $i \in \{1,2\}$. Its state, current, and voltage are referred to as x_i, i_i, and v_i. (b) Circuit topology of the cell $\mathcal{C}_i$. While studying the dynamics of the cell in its isolated state, i_{C_i} is set to 0. The DC parameters V_S and R polarize $\mathcal{M}_i$ in the NDR bias point $\boldsymbol{P_i} = (V_i, I_i) = (0.994\text{V}, 0.0031\text{A})$, associated to the operating point $Q_i = X_i = 388.86$.

- The cell $\mathcal{C}_i$ is used in the design of a bio-inspired two-cell *Memristor Cellular Neural Network* (M-CNN) ($i \in \{1,2\}$), capable to reproduce the symmetry-breaking phenomenon in Turing's two-cell reaction-diffusion array by leveraging only *half the number of states* as compared to the original system.

- DAE set model for the cell $\mathcal{C}_i$ under $i_{C_i} = 0$:

$$\dot{x}_i = g(x_i, v_i) \qquad g(x_i,v_i) = a_0 + a_1 \cdot x_i + b_2 \cdot v_i^2 + c_{21} \cdot v_i^2 \cdot x_i + c_{22} \cdot v_i^2 \cdot x_i^2 + c_{23} \cdot v_i^2 \cdot x_i^3 + c_{24} \cdot v_i^2 \cdot x_i^4 + c_{25} \cdot v_i^2 \cdot x_i^5$$

with

$$v_i = V_S \cdot \frac{1}{1 + R \cdot G(x_i)} \qquad G(x_i) = d_0 + d_1 \cdot x_i + d_2 \cdot x_i^2 + d_3 \cdot x_i^3 + d_4 \cdot x_i^4$$

- The current through $\mathcal{M}_i$ can be computed via $i_i = G(x_i) \cdot v_i$

A. Ascoli, A.S. Demirkol, L.O. Chua, and R. Tetzlaff, "Edge of Chaos is Sine Qua Non for Turing Instability," IEEE Trans. Circuits and Systems-I (TCAS-I): Regular Papers, vol. 69, no. 11, pp. 4596-4609, Nov. 2022, DOI: 10.1109/TCSI.2022.3194465

The basic cell, to be employed shortly for the design of the simplest ever-reported bio-inspired electrical network, specifically a second-order two-cell array, reproducing the symmetry-breaking instabilities emerging in Turing's fourth-order reaction-diffusion model, is a first-order electrical system, composed of the parallel combination between one NaMLab memristor, and a branch, consisting of the series connection between a DC voltage source and a linear and passive resistance. For each index i in the set {1,2}, the bias parameters $V_{s,i}$ and R_i are chosen in such a way to stabilize the NaMLab memristor $\mathcal{M}_i$, belonging to the cell $\mathcal{C}_i$, in a specific bias point $\boldsymbol{P_i} = (V_i, I_i)$, located in the NDR branch of the respective DC I_i versus V_i characteristic, and associated to a particular operating point $Q_i = X_i$. As the two cells of the proposed cellular array are identical one to the other, $R_1 = R_2 = R$ and $V_{s,1} = V_{s,2} = V_s$, which implies $\boldsymbol{P_1} = \boldsymbol{P_2} = \boldsymbol{P} = (V,I)$ and $Q_1 = Q_2 = Q = X$. Importantly, for each index i in the set {1,2}, $A_i - B_i$ is chosen as coupling port in the circuit of the cell $\mathcal{C}_i$.

36 Toward the Design of the Bio-Inspired Two-Cell Array: DC Locus and Small-Signal Model of the Basic Cell

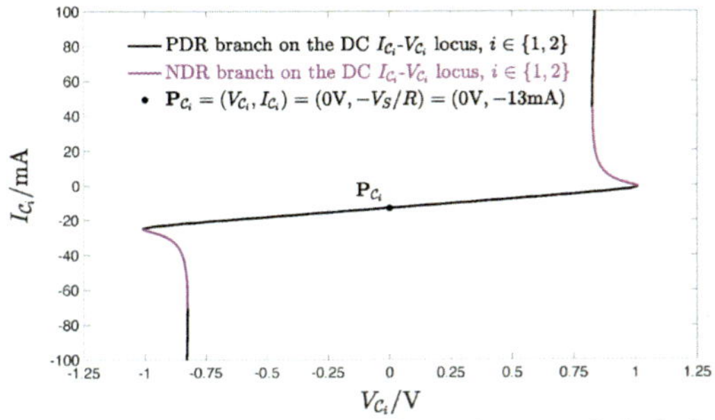

DC I_{C_i} versus V_{C_i} characteristic of the cell C_i. Interestingly, for $I_{C_i} = 0$, the cell bias point $\boldsymbol{P}_{C_i}$ lies on the upper magenta-colored NDR branch.

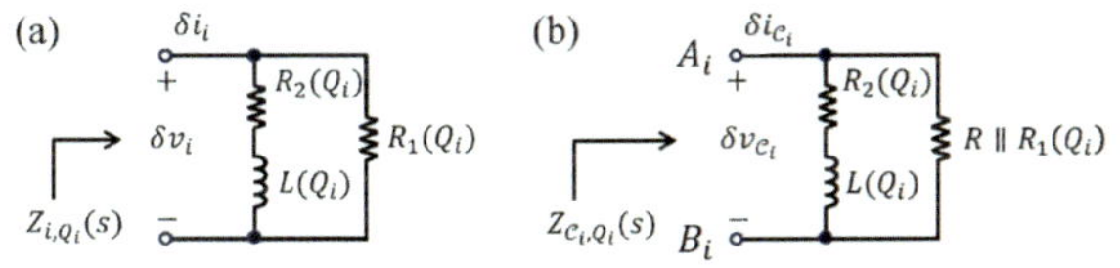

Small-signal of the memristor $\mathcal{M}_i$ (a) and of the cell C_i (b) at $Q_i = X_i$.

- The local input impedance Z_i of the memristor $\mathcal{M}_i$ about Q_i reads as

$$Z_{i,Q_i}(s) = \frac{\mathcal{L}\{\delta v_i(t)\}}{\mathcal{L}\{\delta i_i(t)\}}\bigg|_{Q_i} = K_{Z_i}(Q_i) \cdot \frac{s - z_{Z_i}(Q_i)}{s - p_{Z_i}(Q_i)}$$

- An operating point Q_{C_i} of the cell C_i is equivalent to an operating point $Q_i = X_i$ of the respective memristor $\mathcal{M}_i$.
- The local input impedance Z_{C_i} of the cell C_i about Q_i reads as

$$Z_{C_i,Q_i}(s) = \frac{\mathcal{L}\{\delta v_{C_i}(t)\}}{\mathcal{L}\{\delta i_{C_i}(t)\}}\bigg|_{Q_i} = K_{Z_{C_i}}(Q_i) \cdot \frac{s - z_{Z_{C_i}}(Q_i)}{s - p_{Z_{C_i}}(Q_i)}$$

where

$$K_{Z_{C_i}}(Q_i) = R \parallel R_1(Q_i)$$

$$z_{Z_{C_i}}(Q_i) = -\frac{R_2(Q_i)}{L(Q_i)}$$

$$p_{Z_{C_i}}(Q_i) = -\frac{R \parallel R_1(Q_i) + R_2(Q_i)}{L(Q_i)}$$

Applying the theoretical methods, discussed for the memristor earlier on, to the one-port, featuring positive and negative terminals A_i and B_i, respectively, and identifiable in the circuit of the ith cell C_i ($i \in \{1,2\}$), the DC current I_{Ci} versus voltage V_{Ci} characteristic is found to feature a similar shape as the I_i versus V_i locus of its NaMLab device $\mathcal{M}_i$. Notably, due to the presence of the bias voltage source V_s within the circuit topology of the cell C_i, part of the DC characteristic of the bipole with port terminals A_i and B_i visits the fourth quadrant of the I_{Ci} versus voltage V_{Ci} plane. The small-signal equivalent circuit of the cell C_i may be retrieved from the local circuit model of its memristor $\mathcal{M}_i$ by replacing the resistance $R_1(Q_i)$ with the parallel combination $R \parallel R_1(Q_i)$ between the resistances R and $R_1(Q_i)$. A closed-form expression for the local impedance $Z_{ci}(s)$ of the cell C_i at the port $A_i - B_i$ may then be determined through the application of basic circuit-theoretic principles to the circuit of the linearized model of the cell C_i itself. Finally, the only dynamic element in the circuit of the cell C_i is the memristor $\mathcal{M}_i$. It follows that the operating point Q_{Ci} of the cell C_i, about which its local impedance $Z_{Ci}(s)$ at the port $A_i - B_i$ is evaluated, coincides with the operating point Q_i of its memristor $\mathcal{M}_i$.

37 Insights on Bias and Operating Points for the Cell and Its Memristor

- The only state of the cell $\mathcal{C}_i$ is provided by its memristor $\mathcal{M}_i$.
- Thus, the operating point $Q_{\mathcal{C}_i}$ for $\mathcal{C}_i$ coincides with the operating point $Q_i = X_i$ for $\mathcal{M}_i$.
- However, the corresponding bias points $\boldsymbol{P}_{\mathcal{C}_i}$ and $\boldsymbol{P}_i$, lying respectively on the DC $I_{\mathcal{C}_i}$ - $V_{\mathcal{C}_i}$ and I_i - V_i loci, differ one from the other.
- This is the case since, while $V_{\mathcal{C}_i} = V_i$, however $I_{\mathcal{C}_i} = I_i - (V_S - V_i)/R \neq I_i$.
- Therefore, given a certain PDR or NDR operating domain for the memristor $\mathcal{M}_i$, $Q_{\mathcal{C}_i}$ ($V_{\mathcal{C}_i}$) exhibits the same variation range as Q_i (V_i), whereas the two intervals, across which $I_{\mathcal{C}_i}$ and I_i are respectively allowed to assume values, are different one from the other.

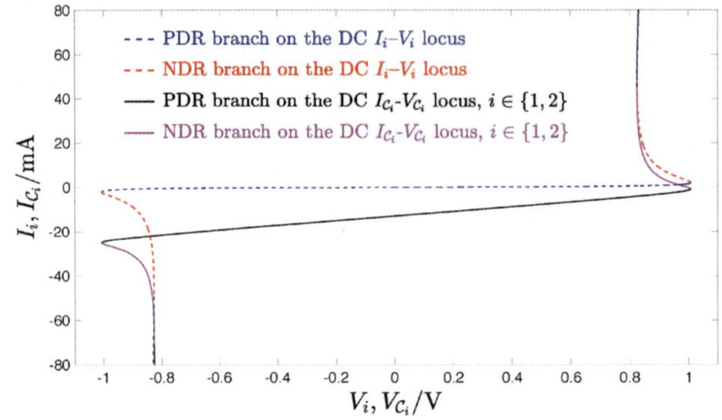

DC current versus voltage loci of the cell $\mathcal{C}_i$ and of its memristor $\mathcal{M}_i$, for $i \in \{1,2\}$

The cell $\mathcal{C}_i$ and its memristor $\mathcal{M}_i$ share the same operating point, i.e. $Q_{ci} = Q_i$. However, the I_{ci}- versus voltage V_{ci} locus differs from the I_i versus V_i locus. In fact, while $V_{ci} = V_i$, however $I_{ci} = I_i - (V_s - V_i)/R \neq I_i$. For example, when $V_{ci} = V_i = 0$, we have $I_i = 0$, whereas $I_{ci} = -V_s/R$. Thus, considering any PDR or NDR operating domain for the memristor $\mathcal{M}_i$, Q_{ci} (V_{ci}) features the same variation range as Q_i (V_i), being the two variables identical one to the other, whereas I_{ci} and I_i assume values across different regions.

38 Testing Memristor and Basic Cell for Local Activity and Edge of Chaos

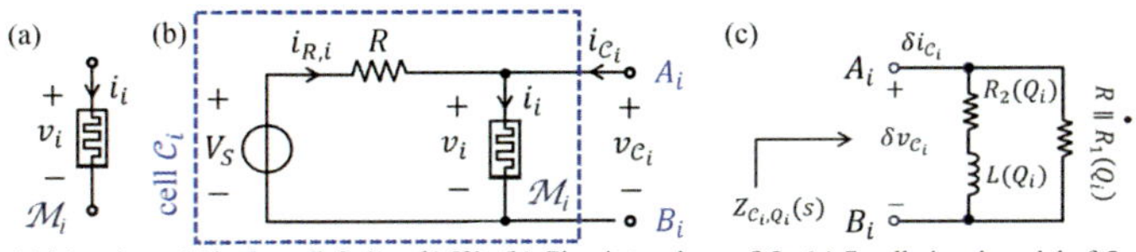

(a) Memristor $\mathcal{M}_i$ in the cell C_i ($i \in \{1,2\}$). (b) Circuit topology of C_i. (c) Small-signal model of C_i.

- Range of frequencies, across which condition 4. from the Local Activity Theorem applies:

$$\Delta\omega\big|_{\Re\{Z_{i,Q_i}\}<0} = 2\cdot\sqrt{-z_{Z_i}(Q_i)\cdot p_{Z_i}(Q_i)} \quad \text{for } Z_i$$

$$\Delta\omega\big|_{\Re\{Z_{C_i,Q_i}\}<0} = 2\cdot\sqrt{-z_{Z_{C_i}}(Q_i)\cdot p_{Z_{C_i}}(Q_i)} \quad \text{for } Z_{C_i}$$

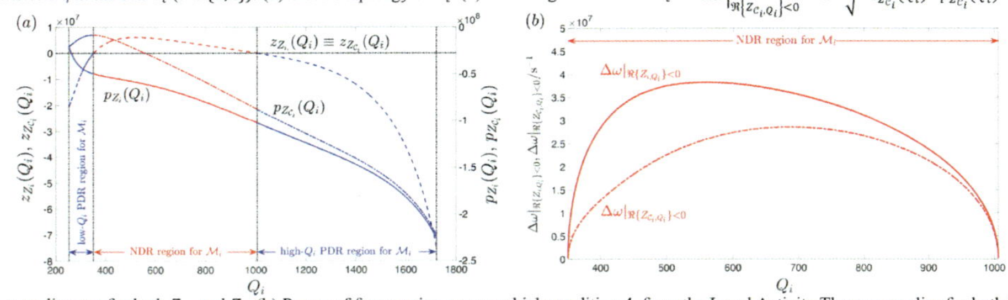

(a) Pole-zero diagram for both Z_{C_i} and Z_i. (b) Range of frequencies, across which condition 4. from the Local Activity Theorem applies for both Z_{C_i} and Z_i

- The pole of Z_i and the pole of Z_{C_i} are negative irrespective of Q_i.
- The common zero of Z_i and of Z_{C_i} becomes positive at each NDR operating point $Q_i = X_i$ of the memristor $\mathcal{M}_i$
- Condition 4. from the Local Activity Theorem holds true for both Z_i and Z_{C_i} throughout the NDR region of the memristor $\mathcal{M}_i$
- Thus*, under current stimulation, both M_i and C_i are on the *Edge of Chaos* around each NDR operating point $Q_i = X_i$ of the memristor $\mathcal{M}_i$

* If a one-port is locally-active under voltage stimulation, it is locally-active under current stimulation as well, and viceversa.

Given that the operating point Q_{C_i} of the cell C_i coincides with the operating point Q_i of its memristor $\mathcal{M}_i$, in the remainder of this article we shall refer to the latter one only. Employing results derived in slide 37, the zero $z_{Z_{C_i}}$ of the local impedance $Z_{C_i}(s)$ of the cell C_i at the port $A_i - B_i$ is found to be real and positive at each memristor NDR operating point Q_i, while it assumes some real and negative value elsewhere. On the other hand, the pole $p_{Z_{C_i}}$ of the same transfer function is found to be a real and negative value at any memristor operating point Q_i. Moreover, $\Re\{Z_{C_i}(j\omega)\}$ goes negative for at least one finite angular frequency ω throughout the memristor NDR domain. It follows that

1) under current control the cell C_i is on the Edge of Chaos at each memristor NDR operating point $Q_i \in (Q_{NDR,L}, Q_{NDR,U}) = (351,1006)$, whereas it is locally passive elsewhere;
2) under voltage control the cell C_i is locally active yet unstable at each memristor NDR operating point $Q_i \in (Q_{NDR,L}, Q_{NDR,U}) = (351,1006)$, whereas it is locally passive elsewhere.

39 Evidence for the "Sleeping State" of the Basic Cell

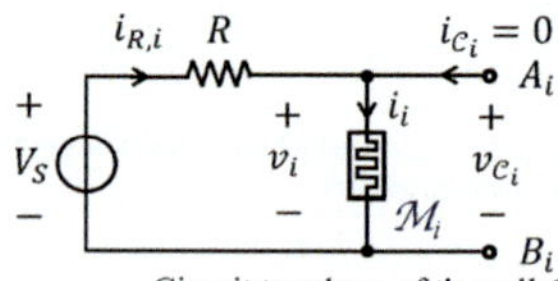

Circuit topology of the cell C_i $i \in \{1,2\}$

$V_s = 1.3$ V
$R = 100\ \Omega$

- In the isolated case, where $i_{C_i} = 0$, the cell C_i admits a globally asymptotically stable operating point, specifically $Q_i = X_i = 388.86$, which belongs to the NDR region of the respective memristor $\mathcal{M}_i$.
- Moreover, *under current stimulation the isolated cell is on the Edge of Chaos at* Q_i.
- Thus, behind an apparently "*sleeping state*", it hides a *high degree of excitability*.

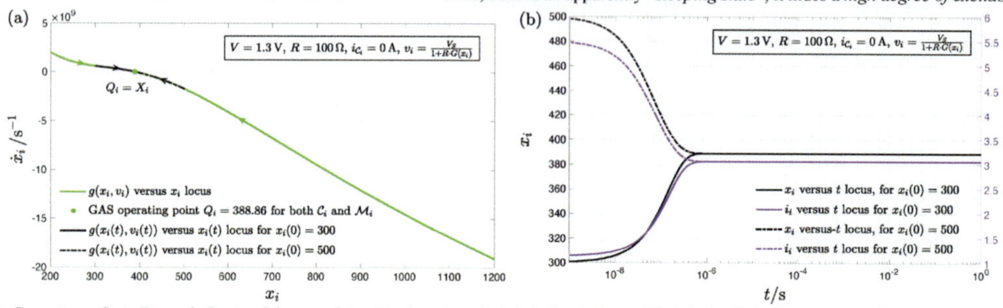

(a) Green trace: State Dynamic Route of the memristor $\mathcal{M}_i$ when the cell C_i is isolated. Green filled circle: Globally asymptotically stable (GAS) operating point Q_i for both C_i and $\mathcal{M}_i$.

(b) Black/Violet solid (dash-dotted) trace: time waveform of the state x_i/current i_i of the memristor $\mathcal{M}_i$, resulting from a simulation of the isolated cell C_i for $x_i(0) = 300$ ($x_i(0) = 500$). The black solid (dash-dotted) trace in (a) shows the dynamic evolution of the trajectory point $(x_i(t), \dot{x}_i(t))$ over the course of the simulation from plot (b) for $x_i(0) = 300$ ($x_i(0) = 500$).

With reference to the circuit of the cell $\mathcal{C}_i$, setting the bias parameter pair (V_s, R) to (1.3V, 100Ω) stabilizes the NaMLab memristor $\mathcal{M}_i$ at the NDR bias point $\boldsymbol{P} = (V, I) = (0.994\text{V}, 0.0031\text{A})$. In this case, irrespective of the initial condition $x_{i,0} = x_i(0)$ assigned to the state x_i of the memristor $\mathcal{M}_i$, the solution $x_i = x_i(t)$ to the DAE set, which describes the cell $\mathcal{C}_i$, as reported at slide 36, is found to converge progressively toward the globally asymptotically stable NDR operating point $Q_i = 388.86$, associated unequivocally to the memristor NDR bias point $\boldsymbol{P}$ mentioned above. As expected for $v_i = V_s/(1 + R \cdot G(x_i))$, the $g(x_i, v_i)$ versus x_i locus crosses only once with negative slope the horizontal axis at $X_i = 388.86$.

The cell $\mathcal{C}_i$ is thus found to be in a dormant condition, asymptotically. However, as descends from the analysis in slide 39, $\mathcal{C}_i$ is poised on the Edge of Chaos at Q_i, and consequently hides a high degree of excitability behind the apparently sleeping state. As shall be clarified shortly, coupling two identical cells of this kind through a sufficiently small linear and passive resistance shall excite them away from their common sleeping state, triggering symmetry-breaking effects across the reaction-diffusion system, which shall converge asymptotically to one of two admissible static inhomogeneous solutions, depending upon the initial conditions assigned to its two memristors.

40 **The Two-Cell M-CNN: the Simplest Ever-Reported Circuit Reproducing Turing Instability**

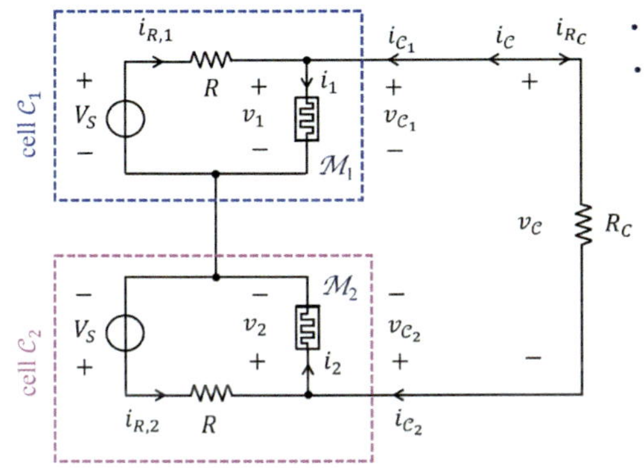

An All-Memristor Bio-Inspired Circuit

$V_S = 1.3\text{V}, R = 100\ \Omega$

- The circuit consists of 2 batteries, 3 resistors, and 2 NaMLab memristors
- Model of the second-order bio-inspired array:

$$\dot{x}_1 = g(x_1, v_1)$$
$$\dot{x}_2 = g(x_2, v_2)$$

where

$$v_1 = V_S \cdot \frac{2 \cdot R + R_C \cdot (1 + R \cdot G(x_2))}{R_C \cdot (1 + R^2 \cdot \eta(x_1, x_2)) + 2 \cdot R + R \cdot (R + R_C) \cdot \xi(x_1, x_2)}$$

$$v_2 = V_S \cdot \frac{2 \cdot R + R_C \cdot (1 + R \cdot G(x_1))}{R_C \cdot (1 + R^2 \cdot \eta(x_1, x_2)) + 2 \cdot R + R \cdot (R + R_C) \cdot \xi(x_1, x_2)}$$

and

$$\eta(x_1, x_2) \triangleq G(x_1) \cdot G(x_2)$$
$$\xi(x_1, x_2) \triangleq G(x_1) + G(x_2)$$

A. Ascoli, A.S. Demirkol, L.O. Chua, and R. Tetzlaff, "Edge of Chaos is Sine Qua Non for Turing Instability," IEEE Trans. Circuits and Systems-I (TCAS-I): Regular Papers, vol. 69, no. 11, pp. 4596-4609, Nov. 2022, DOI: 10.1109/TCSI.2022.3194465 41

The proposed second-order bio-inspired two-cell array, in which two identical copies of the basic first-order battery-resistor-memristor circuit, shown at slide 36, are coupled via a linear and passive resistance R_C, employs just two dynamic elements, particularly two NaMLab memristors, two DC voltage sources, and three resistances to replicate the complex symmetry-breaking phenomenon observed by Turing in a reaction-diffusion system with twice the number of dynamical states.

Our reaction-diffusion system is an *All-Memristor Array*, in the sense that it includes no other dynamic circuit element except for the two memristors.

The model of the two-cell array, shown in this slide, consists of a pair of ODEs, in which the memristors' states and voltages are constrained to comply with a couple of algebraic relations at all times. The proposed bio-inspired reaction-diffusion system is a simple *Memristor Cellular Neural Network* (M-CNN).

41 Numerical Simulations of the M-CNN: Homogeneous Solution

- For any R_C value larger than $R_{C,P} = 49.7\Omega$, the two-cell array converges to the homogeneous solution, i.e.

$$\lim_{t\to\infty}(x_1(t), x_2(t)) = \boldsymbol{Q_h} = (Q_h, Q_h), \text{ with } Q_h = X_h = 388.86,$$

irrespective of $x_1(0)$ and $x_2(0)$.

- No current is found to flow through R_C at steady state, i.e.

$$\lim_{t\to\infty} i_{\mathcal{C}}(t) = I_{\mathcal{C},h} = 0$$

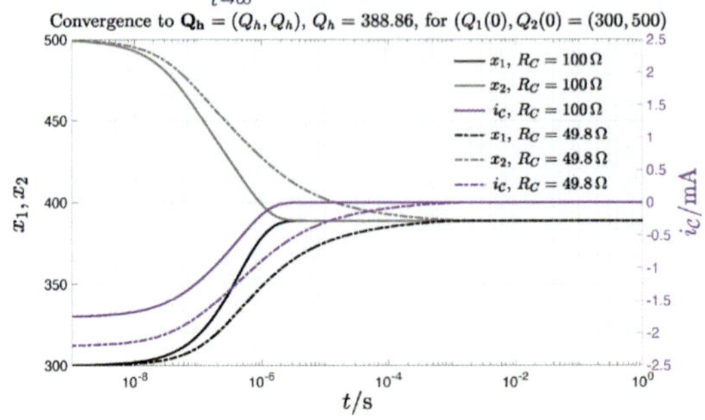

Time waveforms of x_1, x_2 and $i_{\mathcal{C}}$, when each value from the set $\{49.8\Omega, 100\Omega\}$ is assigned to the resistor R_C, for $(x_1(0), x_2(0)) = (300{,}500)$

- In the isolated case, the current-driven cell $\mathcal{C}_i$ is poised on the Edge of Chaos at $Q_i = X_i = 388.86$ for each i value in the set $\{1{,}2\}$.
- This is the *Conditio Sine Qua Non* for symmetry-breaking phenomena to appear in the M-CNN, where a linear and passive resistance R_C, inserted between the positive terminals of $\mathcal{M}_1$ and $\mathcal{M}_2$, let the identical cells $\mathcal{C}_1$ and $\mathcal{C}_2$ interact via diffusion processes.

The components of the operating point $\boldsymbol{Q} = (Q_1, Q_2)$ of the second-order bio-inspired two-cell array are the operating points Q_1 and Q_2 of the memristors M_1 and M_2, respectively. Keeping the same bias parameter pair, as specified in slide 40, specifically $(V_s, R) = (1.3\text{V}, 100\boldsymbol{\Omega})$, for each of the two identical cells of the array, and decreasing the coupling resistance, inserted between the positive terminals of the two memristors, from a very large value, irrespective of the initial conditions $x_{1,0}$ and $x_{2,0}$, assigned to the states x_1 and x_2 of the memristors $\mathcal{M}_1$ and $\mathcal{M}_2$ respectively, the system is found to converge asymptotically to the operating point $\boldsymbol{Q_h} = (Q_h, Q_h)$, with $Q_h = 388.86$, as long as $R_c > R_{c,P} = 49.7\boldsymbol{\Omega}$. In these circumstances, each of the two cells progressively approaches the very same quiescent state it would sit asymptotically on its own. $\boldsymbol{Q_h}$ is in fact the *homogeneous solution* of the two-cell array. Importantly, here the current i_c, which flows through the coupling resistance R_c, vanishes as transients decay to zero, when time tends to positive infinity, i.e. converges asymptotically to the level $I_{c,h} = 0$ associated to the homogeneous solution of the two-cell array.

42 Numerical Simulations of the M-CNN: Inhomogeneous Static Solutions

- For $R_{C,P} = 49.7\Omega$ the two-cell array undergoes a *supercritical pitchfork bifurcation*, at which the homogeneous solution loses stability, while the two memristor states converge toward distinct levels as transients vanish
- In fact, for all $R_C \leq R_{C,P}$ the bio-inspired circuit acts as a bistable dynamical system, supporting the asymptotic formation of one of two possible R_C-dependent inhomogeneous static solutions, specifically
  $\lim_{t\to\infty}(x_1(t), x_2(t)) = \boldsymbol{Q_{inh_j}} = (Q_{1,inh_j}, Q_{2,inh_j})$, with $Q_{1,inh_j} = X_{1,inh_j} \neq Q_{2,inh_j} = X_{2,inh_j}$, and $X_{1,inh_j}, X_{2,inh_j} \neq X_h$ for $j \in \{1,2\}$
- The current through R_C asymptotically approaches either of two possible R_C-dependent values at steady state, i.e.

$$\lim_{t\to\infty} i_C(t) = I_{C,inh_j} \text{ with } I_{C,inh_j} \neq I_{C,h} = 0 \quad \text{for } j \in \{1,2\}$$

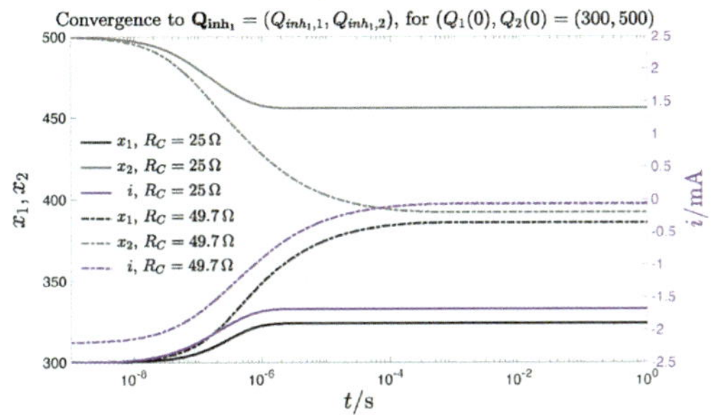

Time waveforms of x_1, x_2, and i_C for each R_C value in the set $\{25\Omega, 49.7\Omega\}$ when $(x_1(0), x_2(0)) = (300{,}500)$

- The initial conditions $x_1(0)$ and $x_2(0)$ determine which of the two possible *Turing patterns** shall emerge across the two-cell array
- The figure on the left shows the asymptotic convergence of the M-CNN toward one of the two coexisting inhomogeneous static solutions.
- The relations to follow link corresponding variables of the coexisting attractors:

$$Q_{1,inh_2} = Q_{2,inh_1}$$
$$Q_{2,inh_2} = Q_{1,inh_1}$$
$$I_{C,inh_2} = -I_{C,inh_1}$$
$$V_{C,inh_2} = -V_{C,inh_1}$$

* An inhomogeneous static solution, developing across a homogeneous medium, is typically referred to as a *Turing pattern*.

When R_C is decreased down to $R_{C,P} = 49.7\Omega$ *symmetry-breaking effects* set in across the two-cell array. Its homogeneous solution $\boldsymbol{Q_h}$ loses stability via a *supercritical pitchfork bifurcation*, which spawns two locally stable *inhomogeneous static solutions*, also referred to as *Turing patterns*, namely $\boldsymbol{Q_{inh_1}} = (Q_{1,inh_1}, Q_{2,inh_1})$ and $\boldsymbol{Q_{inh_2}} = (Q_{1,inh_2}, Q_{2,inh_2})$, with $Q_{1,inh_2} = Q_{2,inh_1}$ and $Q_{2,inh_2} = Q_{1,inh_1}$. Importantly the current i_C through the coupling resistor R_C is found to converge toward some finite non-zero value asymptotically, specifically I_{C,inh_1} and I_{C,inh_2} with $I_{C,inh_2} = -I_{C,inh_1}$, when the two-cell array approaches the *inhomogeneous static solution* $\boldsymbol{Q}_{inh_1}$ and $\boldsymbol{Q}_{inh_2}$, respectively.

In fact, the bio-inspired reaction-diffusion system acquires a *bistable behaviour*, converging to one of the two coexisting solutions $\boldsymbol{Q_{inh_1}}$ and $\boldsymbol{Q_{inh_2}}$, depending upon the initial conditions $x_{1,0}$ and $x_{2,0}$ assigned preliminarily to the states x_1 and x_2 of its two memristors $\mathcal{M}_1$ and $\mathcal{M}_2$.

The two-cell array keeps acting as a bistable dynamical system for any other lower coupling resistance value, but the constant levels Q_{1,inh_1} and Q_{2,inh_1} ($Q_{1,inh_2} = Q_{2,inh_1}$ and $Q_{2,inh_2} = Q_{1,inh_1}$), respectively attained by the memristors' states x_1 and x_2 asymptotically for initial conditions leading to the first (second) Turing pattern, change accordingly.

43 Insights into the Pre-Bifurcation Monostability

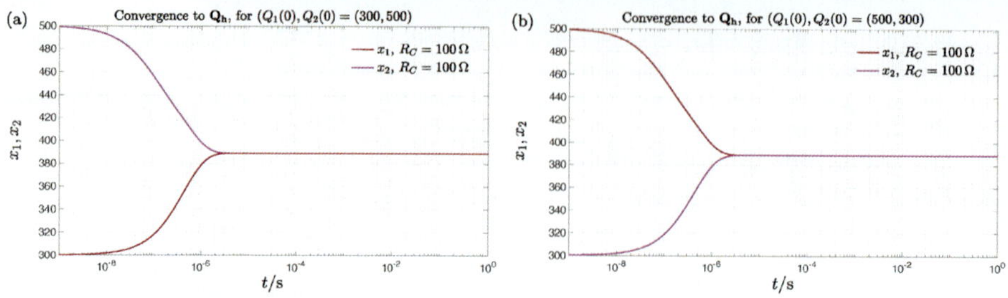

Convergence of the two-cell array toward the homogeneous solution $\boldsymbol{Q_h}$ for $R_C = 100\Omega$ when $(x_1(0), x_2(0))$ is set to (300,500) (a) and to (500,300) in (b)

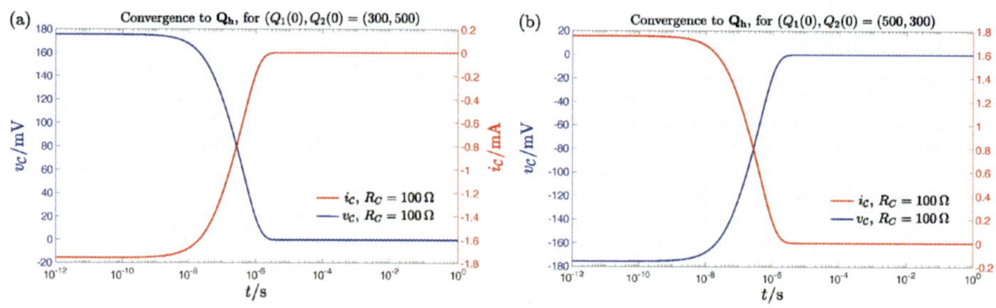

Convergence of v_C and i_C toward $V_{C,h} = 0$ and $I_{C,h} = 0$, respectively, for $R_C = 100\Omega$, when $(x_1(0), x_2(0))$ is set to (300,500) in (a) and to (500,300) in (b)

In a scenario, preceding the *supercritical pitchfork bifurcation*, when $R_C = 100\Omega$, the memristors' states converge progressively toward the same constant value, specifically $Q_h = 388.86$ whereas the current (voltage) signal i_C (v_C), flowing through (falling across) the coupling resistance, approaches asymptotically the level $I_{C,h} = 0$ ($V_{C,h} = 0$), associated to the homogeneous solution of the two-cell array, irrespective of the initial conditions assigned to x_1 and x_2 e.g., as shown in this slide, for $(x_{1,0}, x_{2,0}) = (300,500)$ and for $(x_{1,0}, x_{2,0}) = (500,300)$.

44 Insights into the Post-Bifurcation Bi-stability

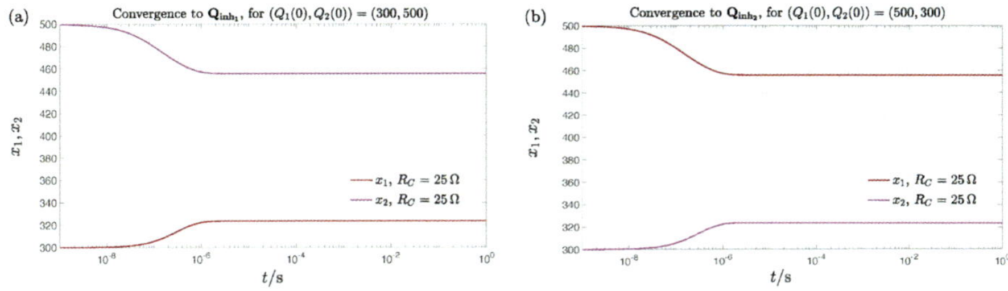

Convergence of the M-CNN toward $\boldsymbol{Q}_{inh_1}$ (a) and toward $\boldsymbol{Q}_{inh_2}$(b) for $R_C = 25\Omega$, when $(x_1(0), x_2(0))$ is respectively set to (300,500) and to (500,300).

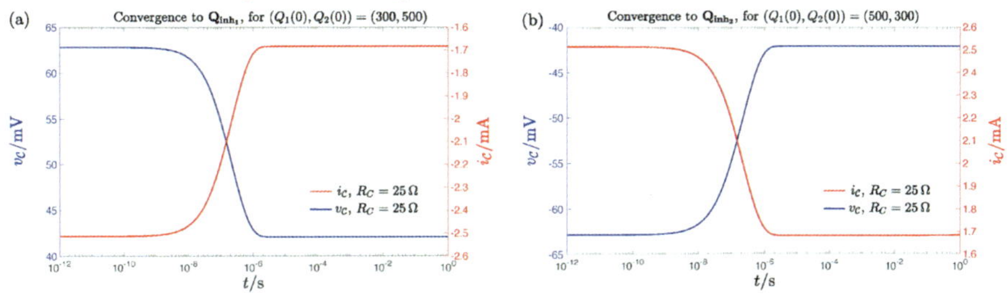

Convergence of v_C (i_C) toward I_{C,inh_1} (V_{C,inh_1}) in (a) and toward I_{C,inh_2} (V_{C,inh_2}) in (b) for $R_C = 25\Omega$, with $(x_1(0), x_2(0))$ set in turn to (300,500) and to (500,300)

In a scenario, following the *supercritical pitchfork bifurcation*, when $R_C = 25\Omega$, the memristors' states tend toward different asymptotic values, and, concurrently, the current and voltage across the coupling resistance attains a non-zero level after transients vanish. In particular, choosing a set of initial conditions, belonging to the basin of attraction of the first (second) inhomogeneous static solution, specifically $(x_{1,0}, x_{2,0}) = (300{,}500)$ ($(x_{1,0}, x_{2,0}) = (500{,}300)$), the M-CNN is found to approach progressively the operating point $\boldsymbol{Q}_{inh_1} = (Q_{1,inh_1}, Q_{2,inh_1}) = (323.7079{,}455.9304)$ ($\boldsymbol{Q}_{inh_2} = (Q_{1,inh_2}, Q_{2,inh_2}) = (Q_{2,inh_1}, Q_{1,inh_1}) = (455.9304{,}323.7079)$). In the first (latter) case the coupling resistor current and voltage respectively converge toward the non-zero values $I_{c,inh_1} = -1.684$mA and $V_{c,inh_1} = 42.091$mV ($I_{c,inh_2} = -I_{c,inh_1} = 1.684$mA and $V_{c,inh_2} = -V_{c,inh_1} = -42.091$mV).

Very importantly, diffusion-driven instabilities may occur across the proposed bio-inspired two-cell array when and only when, as is the case here, each of its two identical cells, let evolve on its own from any initial condition, is found to sit on a stable and locally active operating point, asymptotically.

45 Investigation of the Local and Global Behaviour of the All-Memristor Array via Phase Portrait Analysis

- Important conclusions can be drawn from the numerical investigations:

1. Pre-bifurcation: irrespective of the initial condition, the memristor states x_1 and x_2 converge to the same value $Q_h = X_h$ as time goes to infinity. The asymptotic value $i_{C,h}$ for i_C is always 0. Similarly, the asymptotic value $V_{C,h}$ for v_C is always 0.
2. Post-bifurcation: the state vector may converge toward one of two coexisting R_C-dependent inhomogeneous static solutions, namely $\boldsymbol{Q_{inh_1}} = (Q_{1,inh_1}, Q_{2,inh_1})$ or $\boldsymbol{Q_{inh_2}} = (Q_{1,inh_2}, Q_{2,inh_2})$, with $Q_{1,inh_2} = Q_{2,inh_1}$ and $Q_{2,inh_2} = Q_{1,inh_1}$. Correspondingly, the current i_C would converge either to I_{C,inh_1} or to I_{C,inh_2}, with $I_{C,inh_2} = -I_{C,inh_1}$, respectively. Similarly, the voltage v_C would converge in turn to V_{C,inh_1} or to V_{C,inh_2}, with $V_{C,inh_2} = -V_{C,inh_1}$. The initial conditions determine which of the two inhomogeneous static solutions appears across the M-CNN.

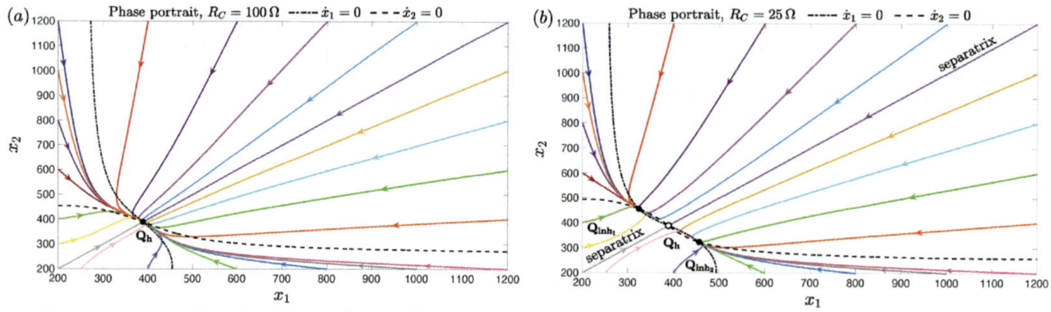

Phase portrait in a pre-bifurcation mono-stability scenario, i.e. for $R_C = 100\Omega$ (a), and in a post-bifurcation bi-stability scenario, i.e. for $R_C = 25\Omega$ (b)

The application of the Local Activity Theory sheds light into the behavior of a system around one of its operating points. Such analysis needs to be complemented with the study of the dynamics of the system away from its operating points in order to draw a comprehensive picture of its evolution from any admissible initial condition.

In this respect, acquiring the *phase portrait* of the system is necessary. For a second-order system, as is the case here, the phase portrait is a family of trajectories, each of which is obtained by plotting the data series of one of the two state variables, say x_2, against the data series of the other state variable, say x_1, as it results from the numerical simulation of the mathematical model of the system from a specific initial condition $(x_{1,0}, x_{2,0})$. The admissible operating points of the system lie at the intersections between two special curves, namely the loci $\dot{x}_1 = 0$ and $\dot{x}_2 = 0$, referred to as x_1- and x_2- *nullclines*, respectively.

The phase portrait of the proposed bio-inspired M-CNN is obtained by integrating numerically the second-order DAE set, which is reported at slide 41, for a large number of initial conditions. The phase portrait of the dynamical system in a pre-bifurcation scenario, specifically for $R_C = 100\Omega$, reveals the *mono-stability* of the two-cell array, which converges toward the homogeneous solution $\boldsymbol{Q_h}$ from any initial condition $(x_{1,0}, x_{2,0})$ across the x_2 versus x_1 plane, which is also referred to as *phase plane*. However, as revealed by the phase portrait of the system in a post-bifurcation scenario, specifically for $R_C = 25\Omega$, the two-cell array exhibits *bi-stability*, approaching asymptotically the first (second) inhomogeneous static solution $\boldsymbol{Q_{inh_1}}$ ($\boldsymbol{Q_{inh_2}}$) if the initial condition $(x_{1,0}, x_{2,0})$ lies to the left (right) of the *stable manifold* of a *saddle point*, which represents the unstable homogeneous solution $\boldsymbol{Q_h}$ of the M-CNN. The stable manifold of this saddle point is referred to as *separatrix*, since it physically divides the basins of attraction of the two coexisting attractors $\boldsymbol{Q_{inh_1}}$ and $\boldsymbol{Q_{inh_2}}$.

46 Zoom-in View of a Pre- and Post-Bifurcation Phase Portrait

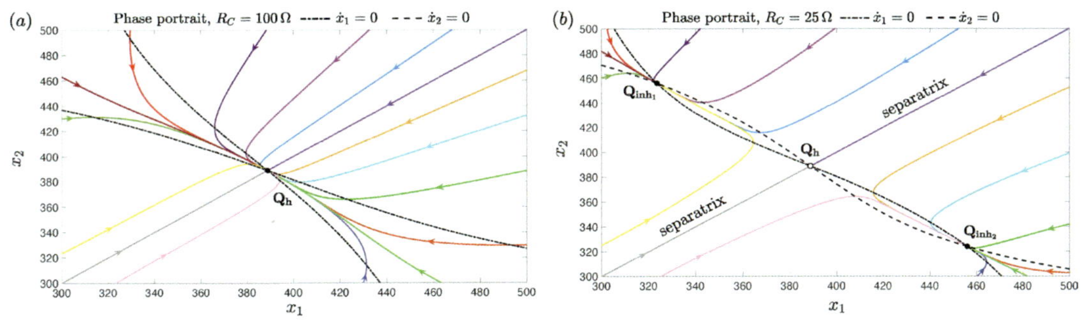

Close-up view of the phase portrait for $R_C = 100\Omega$ around Q_h (a) and for $R_C = 25\Omega$ around point attractors Q_{inh_1} and Q_{inh_2} and point repellor Q_h (b)

The left (right) illustration in this slide is a close-up view of the phase portrait of the reaction-diffusion system around the region hosting the globally asymptotically stable homogeneous solution (the two locally asymptotically stable inhomogeneous static solutions and the unstable homogeneous solution) for $R_C = 1000\Omega$ ($R_C = 25\Omega$).

47 DC Current-Voltage Locus of the Basic Cell with Inverted Port Voltage and Inverted Port Current

- Reconsidering the analysis of the second cell $\mathcal{C}_2$ from a different perspective, as discussed in this slide, allows to gain a deeper insight into the emergent phenomena observed across the two-cell array .
- While the approach is illustrated here in general for the ith cell $\mathcal{C}_i$ ($i \in \{1,2\}$), it shall be applied to $\mathcal{C}_2$ only, in the next slide.

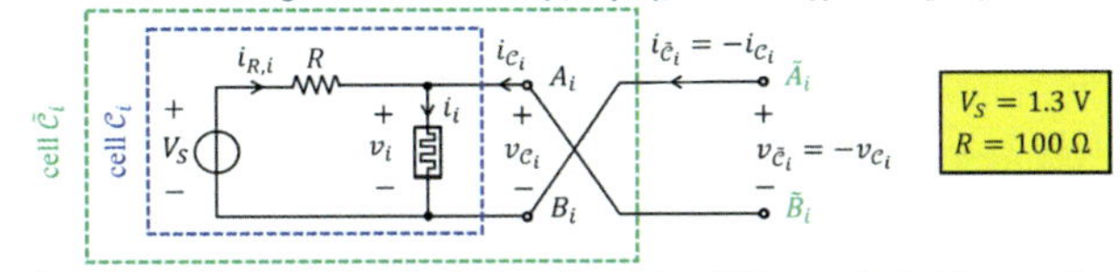

Circuit topology of the cell $\tilde{\mathcal{C}}_i$, obrained from the circuit topology of the cell $\mathcal{C}_i$ by reversing both its port voltage and its port current.

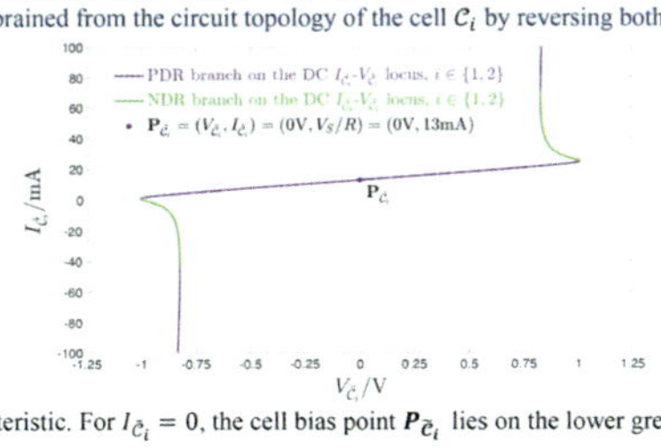

DC $I_{\tilde{\mathcal{C}}_i}$ versus $V_{\tilde{\mathcal{C}}_i}$ characteristic. For $I_{\tilde{\mathcal{C}}_i} = 0$, the cell bias point $P_{\tilde{\mathcal{C}}_i}$ lies on the lower green-colored NDR branch.

Reconsidering the second cell $\mathcal{C}_2$ in the proposed bio-inspired reaction-diffusion network from a different perspective allows to draw a deeper insight into the complexity emerging across the cellular array. Despite the approach is introduced here with reference to the ith M-CNN cell $\mathcal{C}_i$ ($i \in \{1,2\}$), it will be applied to $\mathcal{C}_2$ only, in slide 49. Inverting the polarities of both port voltage and port current of the cell $\mathcal{C}_i$, which implies $i_{\tilde{\mathcal{C}}i} = -i_{\mathcal{C}i}$ and $v_{\tilde{\mathcal{C}}i} = -v_{\mathcal{C}i}$, results in a new cell, referred to as $\tilde{\mathcal{C}}_i$, and admitting the same operating point as its memristor $\mathcal{M}_i$, i.e. $Q\tilde{\mathcal{C}}_{ii} = Q_i$. Applying the theoretical methods, discussed for the memristor $\mathcal{M}_i$ as well as for the cell $\mathcal{C}_i$ previously, a new one-port, featuring positive and negative terminals $\tilde{A}_i$ and $\tilde{B}_i$ respectively, is identifiable in the circuit of the cell $\tilde{\mathcal{C}}_i$. While admitting a similar shape as the DC characteristic of its memristor $\mathcal{M}_i$, the DC current $I_{\tilde{\mathcal{C}}i}$ versus voltage $V_{\tilde{\mathcal{C}}i}$ locus of this one-port differs from the current I_i versus voltage V_i locus, since $V_{\tilde{\mathcal{C}}i} = -V_i$, and $I_{\tilde{\mathcal{C}}i} = -I_i + (V_s - V_i)/R \neq I_i$. In fact, it is straightforward to realize that the DC characteristic of the new one-port with terminals $\tilde{A}_i$ and $\tilde{B}_i$ visits also the second quadrant of the $I_{\tilde{\mathcal{C}}i}$ versus $V_{\tilde{\mathcal{C}}i}$ plane, besides the first and third quadrants. For example, when $V_{\tilde{\mathcal{C}}i} = V_i = 0$, we have $I_i = 0$, whereas $I_{\tilde{\mathcal{C}}i} = +V_s/R$.

48 DC Current-Voltage Locus of the Composite Cell

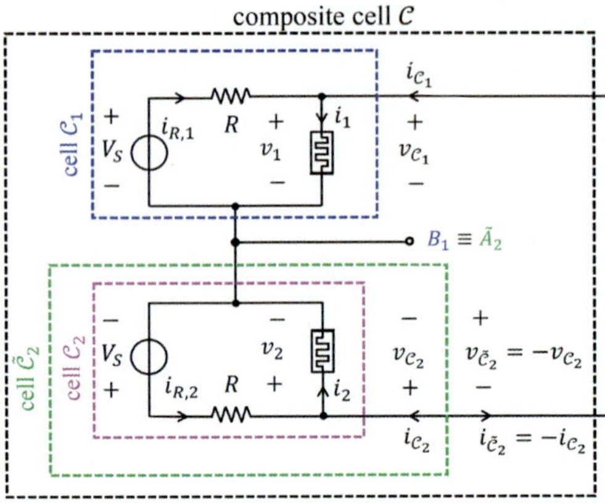

Composite Cell of the proposed *All-Memristor* Bio-Inspired Circuit, where $\mathcal{C}$ is closed across a linear and passive resistance $R_{\mathcal{C}}$

$V_S = 1.3\text{V}, R = 100\ \Omega$

- Port voltage and current of the composite cell $\mathcal{C}$:

$$v_{\mathcal{C}} = v_{\mathcal{C}_1} + v_{\tilde{\mathcal{C}}_2} = v_{\mathcal{C}_1} - v_2$$
$$i_{\mathcal{C}} = i_{\mathcal{C}_1} = i_{\tilde{\mathcal{C}}_2} = -i_{\mathcal{C}_2}$$

- The $I_{\mathcal{C}}$-$V_{\mathcal{C}}$ locus may be obtained graphically by summing the voltages $V_{\mathcal{C}_1}$ and $V_{\tilde{\mathcal{C}}_2}$ for each $I_{\mathcal{C}} = I_{\mathcal{C}_1} = I_{\tilde{\mathcal{C}}_2}$ value

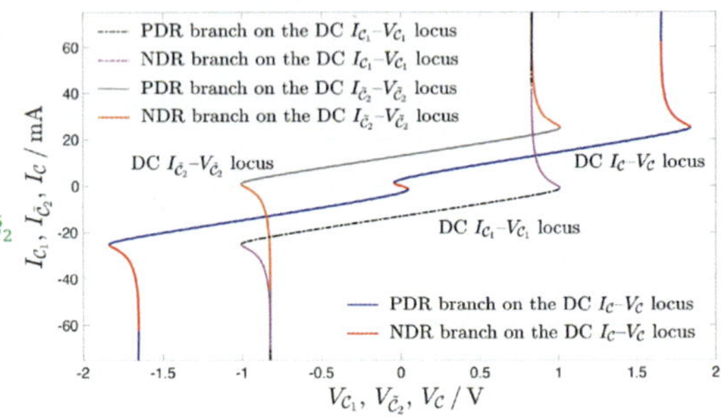

DC current versus voltage loci of the cells $\mathcal{C}_1$, $\tilde{\mathcal{C}}_2$, and $\mathcal{C}$

A. Ascoli, A.S. Demirkol, L.O. Chua, and R. Tetzlaff, "Edge of Chaos is Sine Qua Non for Turing Instability," IEEE Trans. Circuits and Systems-I (TCAS-I): Regular Papers, vol. 69, no. 11, pp. 4596-4609, Nov. 2022, DOI: 10.1109/TCSI.2022.3194465

Let us focus our attention on the proposed M-CNN, assuming it is closed across an open circuit, first. Considering the variant of the second cell $\mathcal{C}_2$ with reversed port voltage and reversed port current, which implies $i_{\tilde{\mathcal{C}}2} = -i_{\mathcal{C}2}$ and $v_{\tilde{\mathcal{C}}2} = -v_{\mathcal{C}2}$, i.e. focusing on its variant topology, referred to as $\tilde{\mathcal{C}}_2$, it is possible to combine the DC $I_{\mathcal{C}1}$ versus $V_{\mathcal{C}1}$ characteristic of the cell $\mathcal{C}_1$ with the $I_{\tilde{\mathcal{C}}2}$ versus $V_{\tilde{\mathcal{C}}2}$ characteristic of the cell $\tilde{\mathcal{C}}_2$ to derive the DC $I_{\mathcal{C}}$ versus $V_{\mathcal{C}}$ characteristic of a *composite cell,* which features positive and negative terminals A_1 and $\tilde{B}_2$. The composite cell is indicated as $\mathcal{C}$ in the remainder of this article. In fact, it is worthy to observe that $i_{\tilde{\mathcal{C}}} = i_{\tilde{\mathcal{C}}_1} = i_{\tilde{\mathcal{C}}2}$, while $v_{\tilde{\mathcal{C}}} = v_{\tilde{\mathcal{C}}_1} + v_{\tilde{\mathcal{C}}2}$. This allows to derive graphically the current-voltage characteristic of the composite cell. In fact, for each value assigned to the current $I_{\mathcal{C}}$, the corresponding voltage $V\mathcal{C}$ on the DC characteristic of the composite cell $\mathcal{C}$ may be obtained by summing the abscissa of the point $\boldsymbol{P}_{\mathcal{C}1}$, sitting on the $I_{\mathcal{C}1}$ versus $V_{\mathcal{C}1}$ locus, and featuring the ordinate $I_{\mathcal{C}1} = I_{\mathcal{C}}$, to the abscissa of the point $\boldsymbol{P}_{\tilde{\mathcal{C}}2}$, sitting on the $I_{\tilde{\mathcal{C}}2}$ versus $V_{\tilde{\mathcal{C}}2}$ loci, and featuring the ordinate $I_{\tilde{\mathcal{C}}2} = I_{\mathcal{C}}$. Therefore, from a circuit-theoretic viewpoint the cell $\mathcal{C}$ is a current-controlled one-port.

As a very insightful remark, the DC locus of the cell $\mathcal{C}$ admits three NDR branches. This shows that, combining simple memristive one-ports, as done in this slide, it is possible to obtain two-terminal circuit elements, endowed with richer dynamics than the basic one-ports they are composed of. This may inspire the development of a novel circuit-theoretic approach to memristor device modelling in the years to come.

49 DC Behaviours of the Constitutive Units of the Composite Cell

- The DC current-voltage locus of the composite cell $\mathcal{C}$ has *three NDR branches!*
- For $I_{\mathcal{C}} = 0$, implying $I_{\mathcal{C}_1} = I_{\tilde{\mathcal{C}}_2} = 0$, the cell bias point $\boldsymbol{P}_{\mathcal{C}} = (V_{\mathcal{C}}, I_{\mathcal{C}})$ lies on the inner red-colored NDR branch.
- For each $I_{\mathcal{C}}$ value, cells $\mathcal{C}_1$ and $\tilde{\mathcal{C}}_2$ must respectively comply with the following DC constraints:

$$V_1 = \frac{V_S + R \cdot I_{\mathcal{C}}}{1 + R \cdot G(Q_1)}$$

$$V_2 = \frac{V_S - R \cdot I_{\mathcal{C}}}{1 + R \cdot G(Q_2)}$$

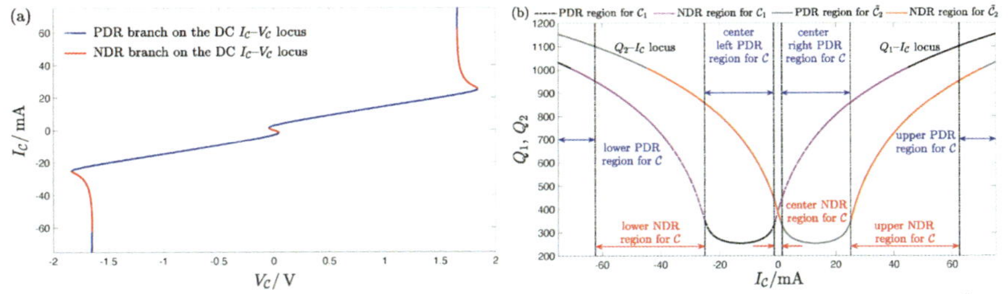

(a) DC $I_{\mathcal{C}}$ versus $V_{\mathcal{C}}$ locus of the composite cell $\mathcal{C}$. (b) DC Q_1 versus $I_{\mathcal{C}}$ and Q_2 versus $I_{\mathcal{C}}$ loci, with indication of PDR and NDR regions of cells $\mathcal{C}_1$, $\tilde{\mathcal{C}}_2$, and $\mathcal{C}$.

A. Ascoli, A.S. Demirkol, L.O. Chua, and R. Tetzlaff, "Edge of Chaos is Sine Qua Non for Turing Instability," IEEE Trans. Circuits and Systems-I (TCAS-I): Regular Papers, vol. 69, no. 11, pp. 4596-4609, Nov. 2022, DOI: 10.1109/TCSI.2022.3194465

Recalling that the memristor $\mathcal{M}_i$, for each index i in the set {1,2}, is a current-controlled device, a single-valued function expresses its state $Q_i = X_i$ in terms of the current $I_{\mathcal{C}}$, flowing through the composite cell $\mathcal{C}$. Moreover, an algebraic relation relates the DC current $I_{\mathcal{C}}$ to the corresponding bias levels, assumed by the voltage V_i and by the state $Q_i = X_i$ of the memristor $\mathcal{M}_i$ within the circuitry of the composite cell.

Importantly, on the basis of the results, collected in this analysis, given a certain bias current $I_{\mathcal{C}}$, flowing through the composite cell $\mathcal{C}$, it is possible to specify whether the one-ports, featuring positive and negative terminals A_1 and B_1, $\tilde{A}_2$ and $\tilde{B}_2$, and A_1 and $\tilde{B}_2$, respectively, admit a negative or a positive small-signal resistance.

50 Local Impedance of the Composite Cell

- The local input impedance* $Z_{\mathcal{C}}$ of the composite cell $\mathcal{C}$ about a given operating point $\boldsymbol{Q} = (Q_1, Q_2) = (X_1, X_2)$ reads as

$$Z_{\mathcal{C},\boldsymbol{Q}}(s) = \left.\frac{\mathcal{L}\{\delta v_{\mathcal{C}}(t)\}}{\mathcal{L}\{\delta i_{\mathcal{C}}(t)\}}\right|_{\boldsymbol{Q}} = K_{Z_{\mathcal{C}}}(\boldsymbol{Q}) \cdot \frac{(s - z_{1,Z_{\mathcal{C}}}(\boldsymbol{Q})) \cdot (s - z_{2,Z_{\mathcal{C}}}(\boldsymbol{Q}))}{(s - p_{1,Z_{\mathcal{C}}}(\boldsymbol{Q})) \cdot (s - p_{2,Z_{\mathcal{C}}}(\boldsymbol{Q}))}$$

where

$$K_{Z_{\mathcal{C}}}(\boldsymbol{Q}) = K_{Z_{\mathcal{C}_1}}(Q_1) + K_{Z_{\mathcal{C}_2}}(Q_2)$$

$$p_{1,Z_{\mathcal{C}}}(\boldsymbol{Q}) = p_{Z_{\mathcal{C}_1}}(Q_1)$$

$$p_{2,Z_{\mathcal{C}}}(\boldsymbol{Q}) = p_{Z_{\mathcal{C}_2}}(Q_2)$$

$$z_{1(2),Z_{\mathcal{C}}}(\boldsymbol{Q}) = \frac{v_{Z_{\mathcal{C}}}(\boldsymbol{Q}) + (-)\sqrt{v_{Z_{\mathcal{C}}}^2(\boldsymbol{Q}) - 4 \cdot p_{Z_{\mathcal{C}}}(\boldsymbol{Q})}}{2}$$

$$v_{Z_{\mathcal{C}}}(\boldsymbol{Q}) = \frac{K_{Z_{\mathcal{C}_1}}(Q_1) \cdot \left(z_{Z_{\mathcal{C}_1}}(Q_1) + p_{Z_{\mathcal{C}_2}}(Q_2)\right) + K_{Z_{\mathcal{C}_2}}(Q_2) \cdot \left(z_{Z_{\mathcal{C}_2}}(Q_2) + p_{Z_{\mathcal{C}_1}}(Q_1)\right)}{K_{Z_{\mathcal{C}_1}}(Q_1) + K_{Z_{\mathcal{C}_2}}(Q_2)}$$

$$p_{Z_{\mathcal{C}}}(\boldsymbol{Q}) = \frac{K_{Z_{\mathcal{C}_1}}(Q_1) \cdot z_{Z_{\mathcal{C}_1}}(Q_1) \cdot p_{Z_{\mathcal{C}_2}}(Q_2) + K_{Z_{\mathcal{C}_2}}(Q_2) \cdot z_{Z_{\mathcal{C}_2}}(Q_2) \cdot p_{Z_{\mathcal{C}_1}}(Q_1)}{K_{Z_{\mathcal{C}_1}}(Q_1) + K_{Z_{\mathcal{C}_2}}(Q_2)}$$

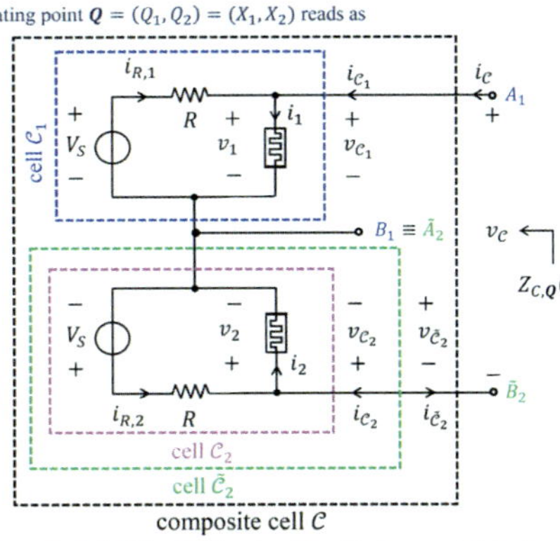

Composite Cell of the *All–Memristor* Circuit to be presented shortly

* The small-signal model of $\tilde{\mathcal{C}}_2$ is equivalent to the small-signal model of $\mathcal{C}_2$.

A. Ascoli, A.S. Demirkol, L.O. Chua, and R. Tetzlaff, "Edge of Chaos is Sine Qua Non for Turing Instability," IEEE Trans. Circuits and Systems-I (TCAS-I): Regular Papers, vol. 69, no. 11, pp. 4596-4609, Nov. 2022, DOI: 10.1109/TCSI.2022.3194465

Replacing the cells $\mathcal{C}_1$ and $\tilde{\mathcal{C}}_2$ with their small-signal equivalent circuit models allows to derive the local circuit model of the composite cell $\mathcal{C}$. Applying basic circuit-theoretic principles to the resulting linear circuit, the local impedance $Z_{\mathcal{C},\boldsymbol{Q}}(s)$ of the composite cell $\mathcal{C}$ at the port $A_1 - \tilde{B}_2$, or, equivalently, of the one-port with positive and negative terminals A_1 and $\tilde{B}_2$, about a given operating point $\boldsymbol{Q} = (Q_1, Q_2)$, is found to feature two zeros and two poles, whose closed-form expressions depend upon the memristors' operating points.

51 Testing the Composite Cell for Local Activity and Edge of Chaos

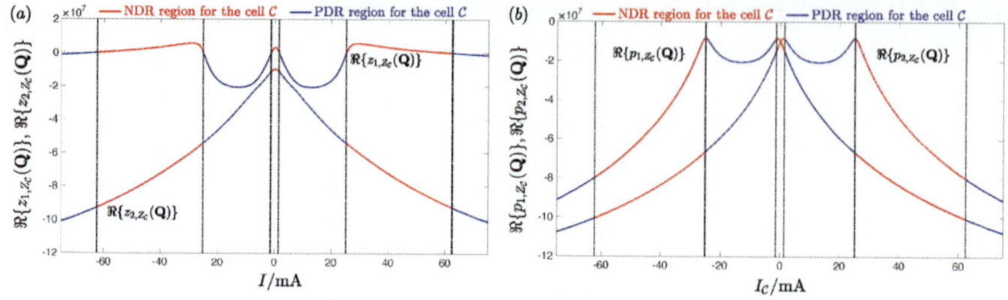

(a) ((b)) Real part of each of the two zeros (poles) of the local impedance $Z_{\mathcal{C}}$ of the cell $\mathcal{C}$ versus the DC value $I_{\mathcal{C}}$ of the respective port current

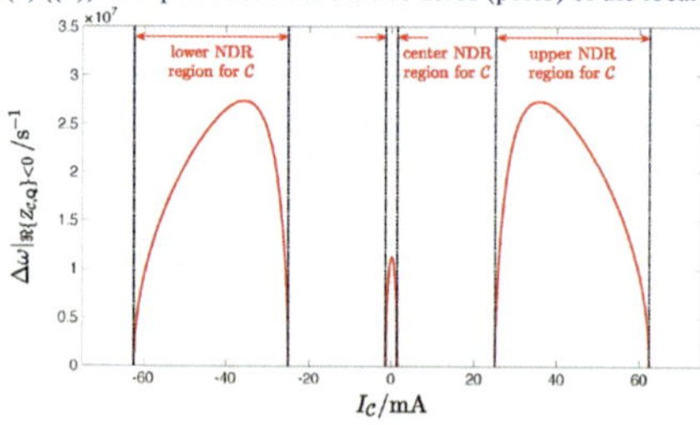

Range of frequencies, across which condition 4. from the Local Activity Theorem applies for $Z_{\mathcal{C}}$

- Range of frequencies, across which condition 4. from the Local Activity Theorem applies for $Z_{\mathcal{C}}$:

$$\Delta\omega\Big|_{\Re\{Z_{\mathcal{C},\boldsymbol{Q}}\}<0} = 2\cdot\sqrt{\frac{\lambda_{Z_{\mathcal{C}}}(\boldsymbol{Q}) - (+)\sqrt{\lambda^2_{Z_{\mathcal{C}}}(\boldsymbol{Q}) - 4\cdot\mu_{Z_{\mathcal{C}}}(\boldsymbol{Q})}}{2}}$$

where

$$\lambda_{Z_{\mathcal{C}}}(\boldsymbol{Q}) = \alpha_{Z_{\mathcal{C}}}(\boldsymbol{Q}) + \beta_{Z_{\mathcal{C}}}(\boldsymbol{Q}) - \gamma_{Z_{\mathcal{C}}}(\boldsymbol{Q})\cdot\delta_{Z_{\mathcal{C}}}(\boldsymbol{Q}) \qquad \beta_{Z_{\mathcal{C}}}(\boldsymbol{Q}) = p_{1,Z_{\mathcal{C}}}(\boldsymbol{Q})\cdot p_{2,Z_{\mathcal{C}}}(\boldsymbol{Q})$$

$$\mu_{Z_{\mathcal{C}}}(\boldsymbol{Q}) = \alpha_{Z_{\mathcal{C}}}(\boldsymbol{Q})\cdot\beta_{Z_{\mathcal{C}}}(\boldsymbol{Q}) \qquad \gamma_{Z_{\mathcal{C}}}(\boldsymbol{Q}) = z_{1,Z_{\mathcal{C}}}(\boldsymbol{Q}) + z_{2,Z_{\mathcal{C}}}(\boldsymbol{Q})$$

$$\alpha_{Z_{\mathcal{C}}}(\boldsymbol{Q}) = z_{1,Z_{\mathcal{C}}}(\boldsymbol{Q})\cdot z_{2,Z_{\mathcal{C}}}(\boldsymbol{Q}) \qquad \delta_{Z_{\mathcal{C}}}(\boldsymbol{Q}) = p_{1,Z_{\mathcal{C}}}(\boldsymbol{Q}) + p_{2,Z_{\mathcal{C}}}(\boldsymbol{Q})$$

- $\mathcal{C}$ is poised on the Edge of Chaos at each DC port current $I_{\mathcal{C}}$ along either of the 3 NDR branches of the respective DC locus, in particular for $I_{\mathcal{C}} = 0$, when $\boldsymbol{Q} = \boldsymbol{Q_h}$

We now have all the information necessary to test the composite cell $\mathcal{C}$ for Local activity and Edge of Chaos. The real parts of the two poles of the local impedance $Z_{\mathcal{C},\boldsymbol{Q}}(s)$ of the composite cell $\mathcal{C}$ at the port $A_1 - \tilde{B}_2$ are found to be negative at any admissible operating point $\boldsymbol{Q}$, i.e. for any of the possible DC values assumed by its control signal, which is the bias current stimulus $I_{\mathcal{C}}$. On the other hand, the real parts of the two zeros of $Z_{\mathcal{C},\boldsymbol{Q}}(s)$ are found to be positive at each operating point $\boldsymbol{Q}$, such that the composite cell $\mathcal{C}$ admits a negative small-signal resistance, whereas they keep negative elsewhere. Finally, the real part of the local frequency response $Z_{\mathcal{C},\boldsymbol{Q}}(j\omega)$ of the composite cell $\mathcal{C}$ at the port $A_1 - \tilde{B}_2$ is found to go negative for at least one finite angular frequency ω about each operating point $\boldsymbol{Q}$ associated to a bias point $\boldsymbol{P_{\mathcal{C}}} = (V_{\mathcal{C}}, I_{\mathcal{C}})$, lying along either of the three NDR branches of the DC $I_{\mathcal{C}}$ versus $V_{\mathcal{C}}$ characteristic.

We may thus conclude that

1) under current control the composite cell $\mathcal{C}$ is poised on the Edge of Chaos about any operating point $\boldsymbol{Q}$ associated to a bias point $\boldsymbol{P_{\mathcal{C}}} = (V_{\mathcal{C}}, I_{\mathcal{C}})$, lying along either of the three NDR branches of its DC $I_{\mathcal{C}}$ versus $V_{\mathcal{C}}$ characteristic, while it is locally passive elsewhere;
2) under voltage control the composite cell $\mathcal{C}$ is locally active yet unstable about any operating point $\boldsymbol{Q}$ associated to a bias point $\boldsymbol{P_{\mathcal{C}}} = (V_{\mathcal{C}}, I_{\mathcal{C}})$, lying along either of the three NDR branches of its DC $I_{\mathcal{C}}$ versus $V_{\mathcal{C}}$ characteristic, while it is locally passive elsewhere.

It is worth to observe that, under current control, the composite cell $\mathcal{C}$ is on the Edge of Chaos for $I_{\mathcal{C}} = 0$, i.e. when it is closed across an open circuit, or, equivalently, for $R_c \to \infty$, which implies $\boldsymbol{Q} = \boldsymbol{Q_h}$, since the two cells $\mathcal{C}_1$ and $\tilde{\mathcal{C}}_2$ do not interact one with the other via diffusion processes in such a scenario. The stable and locally active operation of the composite cell $\mathcal{C}$ for $I_{\mathcal{C}} = 0$ is the *conditio sine qua non* for the emergence of symmetry-breaking effects across the cellular array obtained by loading the port $A_1 - \tilde{B}_2$ by means of a linear and passive resistor of suitable finite resistance R_c.

52 Transition from Mono- to Bi-stability via Pitchfork Supercritical Bifurcation

- Each crossing between the load line $I_{\mathcal{C}} = -V_{\mathcal{C}}/R_{\mathcal{C}}$ and the DC $I_{\mathcal{C}}$ versus $V_{\mathcal{C}}$ locus represents a possible bias point $\boldsymbol{P}_{\mathcal{C}} = (V_{\mathcal{C}}, I_{\mathcal{C}})$ for the cell $\mathcal{C}$ when closed across a load resistance $R_{\mathcal{C}}$
- For all $R_{\mathcal{C}} > R_{\mathcal{C},P} = 49.75\Omega$ the cell $\mathcal{C}$ admits a single GAS bias point, i.e. $\boldsymbol{P}_{\mathcal{C},h} = (V_{\mathcal{C},h}, I_{\mathcal{C},h}) = (0,0)$
- For all $R_{\mathcal{C}} \leq R_{\mathcal{C},P} = 49.75\Omega$, beside the bias point $\boldsymbol{P}_{\mathcal{C},h} = (V_{\mathcal{C},h}, I_{\mathcal{C},h}) = (0,0)$, which is however unstable, the cell $\mathcal{C}$ admits two locally-stable load resistance-dependent bias points, i.e. $\boldsymbol{P}_{\mathcal{C},inh_1} = (V_{\mathcal{C},inh_1}, I_{\mathcal{C},inh_1})$ and $\boldsymbol{P}_{\mathcal{C},inh_2} = (V_{\mathcal{C},inh_2}, I_{\mathcal{C},inh_2}) = -\boldsymbol{P}_{\mathcal{C},inh_1} = (-V_{\mathcal{C},inh_1}, -I_{\mathcal{C},inh_1})$
- At $R_{\mathcal{C}} = R_{\mathcal{C},P} = 49.75\Omega$ the load line becomes tangent to the DC locus of $\mathcal{C}$. This is a signature of a ***Pitchfork Supercritical Bifurcation***

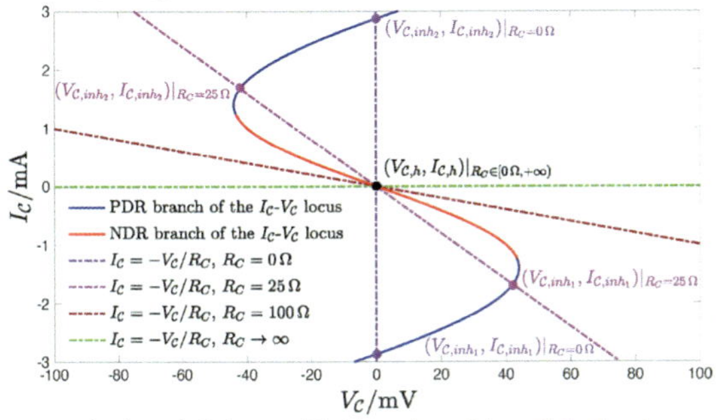

Determination of all the possible bias points of the cell $\mathcal{C}$, closed across a load resistance $R_{\mathcal{C}}$, for a number of pre- and post-bifurcation scenarios

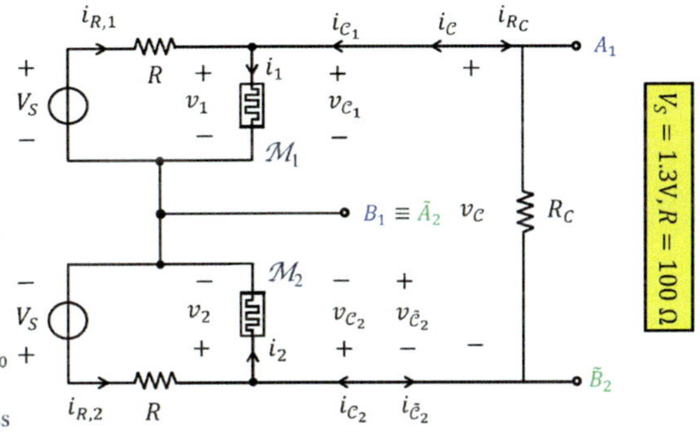

An All-Memristor Bio-Inspired Circuit

A. Ascoli, A.S. Demirkol, L.O. Chua, and R. Tetzlaff, "Edge of Chaos is Sine Qua Non for Turing Instability," IEEE Trans. Circuits and Systems-I (TCAS-I): Regular Papers, vol. 69, no. 11, pp. 4596-4609, Nov. 2022, DOI: 10.1109/TCSI.2022.3194465

When the port $A_1 - \tilde{B}_2$ is closed across a resistor with a given finite resistance $R_{\mathcal{C}}$, an admissible operating point $\boldsymbol{Q}$ for the composite cell $\mathcal{C}$ corresponds to one intersection between its DC $I_{\mathcal{C}}$ versus $V_{\mathcal{C}}$ characteristic and the load line, whose formula $I_{\mathcal{C}} = -V_{\mathcal{C}}/R_{\mathcal{C}}$ expresses the current $I_{\mathcal{C}}$ in terms of the current $I_{R_{\mathcal{C}}}$ flowing through the coupling resistance in the bio-inspired all-memristor array. Decreasing the load resistance $R_{\mathcal{C}}$ from positive infinity, as long as its value is larger than a critical level, specifically $R_{\mathcal{C},P} = 49.7\Omega$, there exists one and only one crossing between the $I_{\mathcal{C}}$ versus $V_{\mathcal{C}}$ characteristic and the load line, namely $\boldsymbol{P}_{\mathcal{C},h} = (V_{\mathcal{C},h}, I_{\mathcal{C},h}) = (0,0)$, which is unequivocally associated to the composite cell operating point $\boldsymbol{Q} = \boldsymbol{Q}_h$. In these circumstances the proposed second-order reaction-diffusion system converges to the homogeneous solution, irrespective of the initial conditions assigned to its memristors' states. As $R_{\mathcal{C}}$ is reduced toward the critical level, the modulus of the slope of the load line progressively increases, until it matches the modulus of the slope of the DC $I_{\mathcal{C}}$ versus $V_{\mathcal{C}}$ characteristic when $R_{\mathcal{C},P} = 49.7\Omega$. Here the load line is found to be tangent to the DC locus of the composite cell $\mathcal{C}$, which, as a result, undergoes a *Supercritical Pitchfork Bifurcation,* which destabilizes its bias point $\boldsymbol{P}_{\mathcal{C},h} = (V_{\mathcal{C},h}, I_{\mathcal{C},h}) = (0,0)$, associated to the homogeneous solution $\boldsymbol{Q} = \boldsymbol{Q}_h = (Q_h, Q_h)$ of the two-cell array, while concurrently endowing it with two additional locally stable bias points, specifically $\boldsymbol{P}_{\mathcal{C},inh_1} = (V_{\mathcal{C},inh_1}, I_{\mathcal{C},inh_1})$ and $\boldsymbol{P}_{\mathcal{C},inh_2} = (V_{\mathcal{C},inh_2}, I_{\mathcal{C},inh_2}) = (-V_{\mathcal{C},inh_1}, -I_{\mathcal{C},inh_1})$, associated to the coexisting inhomogeneous static solutions $\boldsymbol{Q} = \boldsymbol{Q}_{inh_1} = (Q_{1,inh_1}, Q_{2,inh_1})$ and $\boldsymbol{Q} = \boldsymbol{Q}_{inh_2} = (Q_{1,inh_2}, Q_{2,inh_2}) = (Q_{2,inh_1}, Q_{1,inh_1})$, respectively. Decreasing the resistance further below the bifurcation point, the M-CNN keeps acting as a *bistable system,* admitting two locally-stable inhomogeneous static solutions, which change with the value assigned to $R_{\mathcal{C}}$. Importantly, recalling the right illustrations in slides 46 and 47, depicting in turn a global and a local view of the phase portrait of the second-order DAE set from slide 41 for $R_{\mathcal{C}} = 25\Omega$, in the regime of bi-stability, i.e. for $R_{\mathcal{C}} \leq R_{\mathcal{C},P}$, the two-cell array shall progressively converge toward the first (second) inhomogeneous static solution $\boldsymbol{Q}_{inh_1}$ ($\boldsymbol{Q}_{inh_2}$) in case the initial condition $(x_{1,0}, x_{2,0})$ is chosen to the left (right) of the stable manifold of the saddle point, which represents the unstable homogeneous solution $\boldsymbol{Q}_h$ of the M-CNN.

53 Bifurcation Diagrams

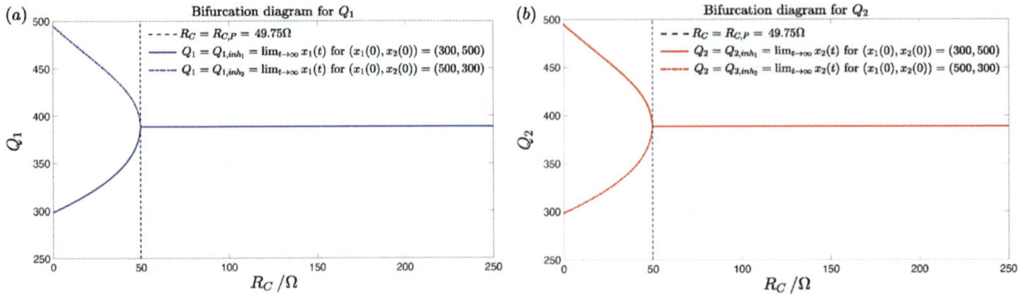

(a) ((b)) Q_1 (Q_2) versus R_C locus. At $R_C = R_{C,P} = 49.75\Omega$ the homogeneous solution of the M-CNN loses stability. For each $R_C \leq R_{C,P}$ the array supports the asymptotic formation of one of two Turing patterns. The initial conditions determine which of them emerges across the M-CNN.

- The bifurcation diagrams illustrate all the possible operating points for each memristor in the two-cell array for any R_C value in the set $[0,250\ \Omega]$

Decreasing monotonically the coupling resistance R_C from 250Ω to 0Ω, and recording all the asymptotic values, each of the two memristors' states x_1 and x_2 may ever admit for any possible choice of the initial condition ($x_{1,0}$, $x_{2,0}$) across the x_2 versus x_1 phase plane, allows to draw the bifurcation diagrams of the DAE set from slide 41. The ith diagram ($i \in \{1,2\}$) shows any admissible operating point $Q_i = X_i$ of the memristor $\mathcal{M}_i$ for each value of the bifurcation parameter R_C. For each i value from the set $\{1,2\}$, decreasing the coupling resistance from 250Ω, the only possible operating point Q_i for the memristor $\mathcal{M}_i$ is found to be equal to Q_h = 388.86 as long as R_C keeps larger than $R_{C,P}$ = 49.7Ω. As the coupling resistance descends down to this critical point, the homogeneous solution is no longer observable across the two-cell array, which is however found to support the formation of one of two possible inhomogeneous static solutions, also referred to as *Turing patterns.* In fact, assigning any of the two values in the set $\{1,2\}$ to the index i, for each coupling resistance equal to or smaller than the bifurcation threshold $R_{C,P}$, the state x_i of the memristor $\mathcal{M}_i$ is found to converge asymptotically toward the point attractor Q_{i,inh_1} (Q_{i,inh_2}) if the initial condition ($x_{1,0}$, $x_{2,0}$) across the x_2 versus x_1 phase plane is chosen to the left (right) of the stable manifold of the saddle point, whose i^{th} coordinate represents the unstable solution Q_h for x_i. Importantly, comparing the bifurcation diagrams relative to the first and second state variables in the DAE set from slide 41, it may be realized that $Q_{1,inh_2} = Q_{2,inh_1}$ and that $Q_{2,inh_2} = Q_{1,inh_1}$, which confirms what was anticipated a number of times earlier on.

54 Local Impedance at the Load Resistor Port in the Two-Cell Array

- The two-cell array features the same state vector as the composite cell $\mathcal{C}$, thus one of its operating points may still be denoted as $\boldsymbol{Q} = (Q_1, Q_2)$
- The local impedance Z_O, computed at the port A_1-$\tilde{B}_2$ across R_C in the M-CNN, about an operating point $\boldsymbol{Q}$ reads as

$$Z_{O,\boldsymbol{Q}}(s) = \left.\frac{\mathcal{L}\{\delta v_O(t)\}}{\mathcal{L}\{\delta i_O(t)\}}\right|_{\boldsymbol{Q}} = R_C \parallel Z_{\mathcal{C},\boldsymbol{Q}}(s)$$

$$= K_{Z_O}(\boldsymbol{Q}) \cdot \frac{(s - z_{1,Z_O}(\boldsymbol{Q})) \cdot (s - z_{2,Z_O}(\boldsymbol{Q}))}{(s - p_{1,Z_O}(\boldsymbol{Q})) \cdot (s - p_{2,Z_O}(\boldsymbol{Q}))}$$

where

$$K_{Z_O}(\boldsymbol{Q}) = \frac{R_C \cdot K_{Z_\mathcal{C}}(\boldsymbol{Q})}{R_C + K_{Z_\mathcal{C}}(\boldsymbol{Q})}$$

$$z_{1,Z_O}(\boldsymbol{Q}) = z_{1,Z_\mathcal{C}}(\boldsymbol{Q})$$

$$z_{2,Z_O}(\boldsymbol{Q}) = z_{2,Z_\mathcal{C}}(\boldsymbol{Q})$$

$$p_{1(2),Z_O}(\boldsymbol{Q}) = \frac{\sigma_{Z_O}(\boldsymbol{Q}) + (-)\sqrt{\sigma_{Z_O}^2(\boldsymbol{Q}) - 4 \cdot \chi_{Z_O}(\boldsymbol{Q})}}{2}$$

$$\sigma_{Z_O}(\boldsymbol{Q}) = \frac{\left(z_{1,Z_\mathcal{C}}(\boldsymbol{Q}) + z_{2,Z_\mathcal{C}}(\boldsymbol{Q})\right) \cdot K_{Z_\mathcal{C}}(\boldsymbol{Q}) + \left(p_{1,Z_\mathcal{C}}(\boldsymbol{Q}) + p_{2,Z_\mathcal{C}}(\boldsymbol{Q})\right) \cdot R_C}{K_{Z_\mathcal{C}}(\boldsymbol{Q}) + R_C}$$

$$\chi_{Z_O}(\boldsymbol{Q}) = \frac{z_{1,Z_\mathcal{C}}(\boldsymbol{Q}) \cdot z_{2,Z_\mathcal{C}}(\boldsymbol{Q}) \cdot K_{Z_\mathcal{C}}(\boldsymbol{Q}) + p_{1,Z_\mathcal{C}}(\boldsymbol{Q}) \cdot p_{2,Z_\mathcal{C}}(\boldsymbol{Q}) \cdot R_C}{K_{Z_\mathcal{C}}(\boldsymbol{Q}) + R_C}$$

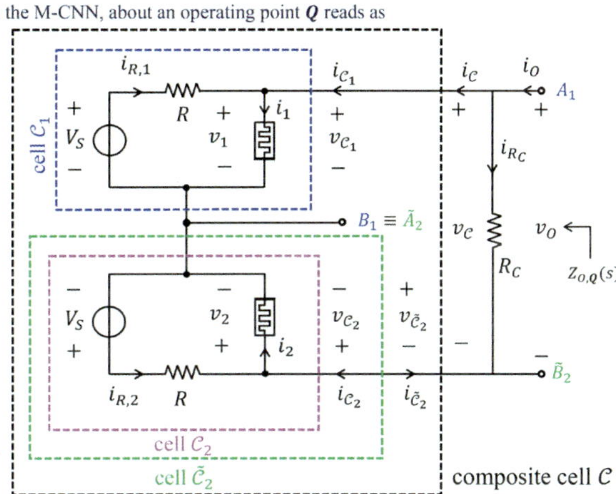

The simplest ever-reported circuit supporting symmetry-breaking phenomena

A. Ascoli, A.S. Demirkol, L.O. Chua, and R. Tetzlaff, "Edge of Chaos is Sine Qua Non for Turing Instability," IEEE Trans. Circuits and Systems-I (TCAS-I): Regular Papers, vol. 69, no. 11, pp. 4596-4609, Nov. 2022, DOI: 10.1109/TCSI.2022.3194465

The local impedance $Z_O(s)$ of the proposed bio-inspired M-CNN at the port $A_1 - \tilde{B}_2$ may be easily derived as the parallel combination between the local impedance $Z_\mathcal{C}(s)$ of the composite cell $\mathcal{C}$ at the same port and the load resistance R_C. Since the two-cell array admits the very same two-dimensional state vector as the composite cell $\mathcal{C}$, its operating point $\boldsymbol{Q_O}$ coincides with the operating point $\boldsymbol{Q} = (Q_1, Q_2)$ of the composite cell. Monitoring how the poles of $Z_O(s)$, evaluated about the two-cell array operating point $\boldsymbol{Q} = \boldsymbol{Q_h}$, evolve across the complex plane, under a monotonic decrease in the coupling resistance from positive infinity, allows to study the local stability properties of the homogeneous solution of the reaction-diffusion network.

55 Further Insights into the Pitchfork Supercritical Bifurcation

- As may be inferred via numerical simulations, for $(x_1(0), x_2(0)) = (300,500)$, if $R_C > R_{C,P}$ $(R_C \le R_{C,P})$, then $\boldsymbol{Q}$ is found to be equal to $\boldsymbol{Q_h}$ $(\boldsymbol{Q_{inh_1}})$. In the first (latter) case $\boldsymbol{Q}$ is independent of (depends upon) the load resistance.
- The real part of one of the two poles of Z_O about $\boldsymbol{Q} = \boldsymbol{Q_h}$ becomes zero (positive) when R_C is decreased down to (below) the bifurcation point

$$R_{C,P} = R_{C,P}(Q_h) = -\frac{2}{\frac{1}{R_2(Q_h)} + \frac{1}{R \parallel R_1(Q_h)}},$$

which is equal to 49.75Ω for $Q_h = 388.86$, as follows from the cell parameter setting, envisaging $V_S = 1.3\text{V}$ and $R = 100\ \Omega$, considered here.

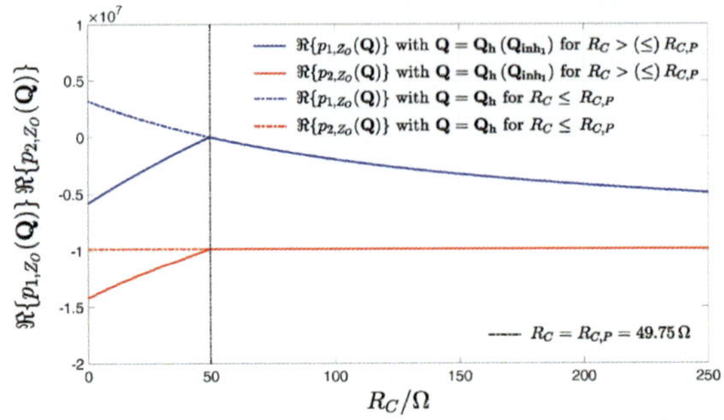

Solid traces: Loci of the real parts of the two poles of the local impedance Z_O, seen at the port A_1-$\tilde{B}_2$ in the bio-inspired array, and computed about the operating point $\boldsymbol{Q}$, which the cell $\mathcal{C}$, closed on the load resistor, is found to admit asymptotically for $(x_1(0), x_2(0)) = (300,500)$, versus the bifurcation parameter R_C. Dash-dotted traces: real parts of p_{1,Z_O} and p_{2,Z_O} about the unstable operating point $\boldsymbol{Q} = \boldsymbol{Q_h}$ for $R_C \le R_{C,P}$.

The critical value $R_{C,P}$, to be assigned to the resistance R_C, for the emergence of symmetry-breaking phenomena across the all-memristor two-cell array, may be determined by studying the condition, under which the real part of at least one of the two poles of the transfer function $Z_O(s)$, evaluated at the M-CNN operating point $\boldsymbol{Q}$, associated to the homogeneous solution $\boldsymbol{Q_h}$, first goes positive.

This analysis results in a closed-form expression for the critical threshold $R_{C,P}$ of the coupling resistance in terms of the cell DC voltage V_S and of the linear resistance R, and in terms of the memristor local circuit model parameters $R_1(Q)$ and $R_2(Q)$, evaluated at the operating point $Q = Q_h$, which each cell admits in the isolated case. Setting V_S and R respectively to 1.3V and to 100Ω, which implies $Q_h = 388.86$, the formula for the critical coupling resistance $R_{C,P}$ provides a value of 49.7Ω, in agreement with the numerical simulations described earlier on.

In the pole diagram, illustrated in this slide, the solid traces show the evolution of the real parts of the two poles of the local impedance $Z_O(s)$ of the bio-inspired two-cell array at the load resistor port about the operating point $\boldsymbol{Q}$, which the two-cell array itself is found to admit asymptotically when the initial condition is set to a phase-plane point $(x_{1,0}, x_{2,0}) = (300,500)$, sitting on the basin of attraction of the first inhomogeneous static solution $\boldsymbol{Q_{inh_1}}$, as R_C is decreased from 250Ω to 0Ω. On the other hand, the dashed lines illustrate the dependence of the real parts of the two poles of $Z_O(s)$, evaluated at the homogeneous solution $\boldsymbol{Q_h}$ of the cellular network, upon coupling resistance values smaller than the bifurcation threshold $R_{C,P}$.

All in all, when, over the course of its monotonic decrease from 250Ω, R_C is reduced down to $R_{C,P}$, the strenght of the diffusive interaction between the two identical reaction cells, poised on the Edge of Chaos on their own, is sufficient to induce the destabilization of the homogeneous solution of the second-order bio-inspired network, which stands as the simplest ever-reported reaction-diffusion system, undergoing symmetry-breaking phenomena, which, furthermore, endow it with a bistable character, allowing the asymptotic formation of two locally-stable Turing patterns across its physical medium.

56 Conclusions

- The Simplest Ever-Reported Bio-Inspired Circuit, undergoing Diffusion-Driven Symmetry-Breaking Instabilities, was presented.
- The proposed circuit leverages half the number of degrees of freedom as compared to Turing's Two-Cell Reaction-Diffusion Array.
- It is a Two-Cell Memristor Cellular Neural Network, and employs only two dynamic elements, specifically two globally-passive yet locally-active resistance switching memories from NaMLab, besides three linear and passive resistors, and two DC voltage sources.
- While A.M. Turing must have spent a huge amount of time to develop equations, revealing the diffusion-induced destabilization of the homogeneous solution in a homogeneous cellular medium, the Quantitative Theory of Local Activity allows us to choose the circuit parameters of each of the two identical memristive reaction cells so as to poise the composite cell on the Edge of Chaos, before a suitable load resistance, across which it is closed, may result in static pattern formation across the overall bio-inspired circuit.
- The two Take-Home-Messages are

1. *Edge of Chaos is Sine Qua Non for Turing Instability*
2. *Locally-Active Memristors Enable to Reproduce Complex Phenomena in High-Order Biological Systems through Lower-Order Circuits*

A. Ascoli, A.S. Demirkol, L.O. Chua, and R. Tetzlaff, "Edge of Chaos is Sine Qua Non for Turing Instability," IEEE Trans. Circuits and Systems-I (TCAS-I): Regular Papers, vol. 69, no. 11, pp. 4596-4609, Nov. 2022, DOI: 10.1109/TCSI.2022.3194465

A. Ascoli, A.S. Demirkol, L.O. Chua, and R. Tetzlaff, "Edge of Chaos Resolves Smale Paradox," IEEE Trans. Circuits and Systems-I (TCAS-I): Regular Papers, vol. 69, no. 3, pp. 1252-1265, March 2022, DOI: 10.1109/TCSI.2021.3133627

Much time was devoted by A.M. Turing, the Father of Artificial Intelligence, to find by trail and error the proper set of values to assign to the coefficients of a set of fourth-order ODEs, so that its solutions could reproduce the emergence of symmetry-breaking phenomena across a reaction-diffusion system, when its two identical second-order reaction cells, sitting at a common globally asymptotically stable quiescent state on their own, were let interact via diffusion processes. On the other hand, the quantitative theory of Local Activity facilitates the selection of a suitable set of bias parameters for the stabilization of a NaMLab memristor on a negative differential resistance bias point, allowing to poise the resulting battery-resistor-memristor circuit on the Edge of Chaos about the respective operating point, which is the *conditio sine qua* for the emergence of symmetry-breaking phenomena across the proposed bio-inspired cellular array, composed of two identical resistively-coupled copies of the battery-resistor-memristor circuit. Remarkably, the adoption of the NaMLab niobium oxide volatile memristor, capable to act as a local source of energy when polarized on a NDR bias point, allows to design an energy-efficient electrical network, which supports diffusion-driven instabilities while employing half the number of dynamical states as compared to Turing's fourth-order reaction-diffusion ODE system.

CHAPTER 06

Memristor Computing Systems: at the crossroad between Circuit Theory and Artificial Intelligence

Fernando Corinto

Politecnico di Torino

This talk covers various facets, including analog computing platforms for solving complex problems, the utilization of coupled oscillators, and advancements in memristor-based technologies. The presentation delves into circuit analysis methods like the Flux-Charge Analysis Method (FCAM), emphasizing its application in modeling memristors and simplifying differential equations. It further discusses memristor-based oscillators, their dynamics, bifurcations, and experimental validations. The application of memristive systems extends to associative memories, where concepts like Hebbian learning and equilibrium propagation are elucidated. Challenges such as synchronization of memristor arrays and mapping learning algorithms into circuit equations are addressed. Additionally, Weakly-Coupled Oscillatory Neural Networks and the architecture of Memristive Oscillatory Associative Memories (MOAMS) are explored. The presentation concludes with recommendations for further reading and gratitude. Overall, it offers a comprehensive overview of the intersection between circuit theory and artificial intelligence, showcasing the potential and challenges of memristor computing systems.

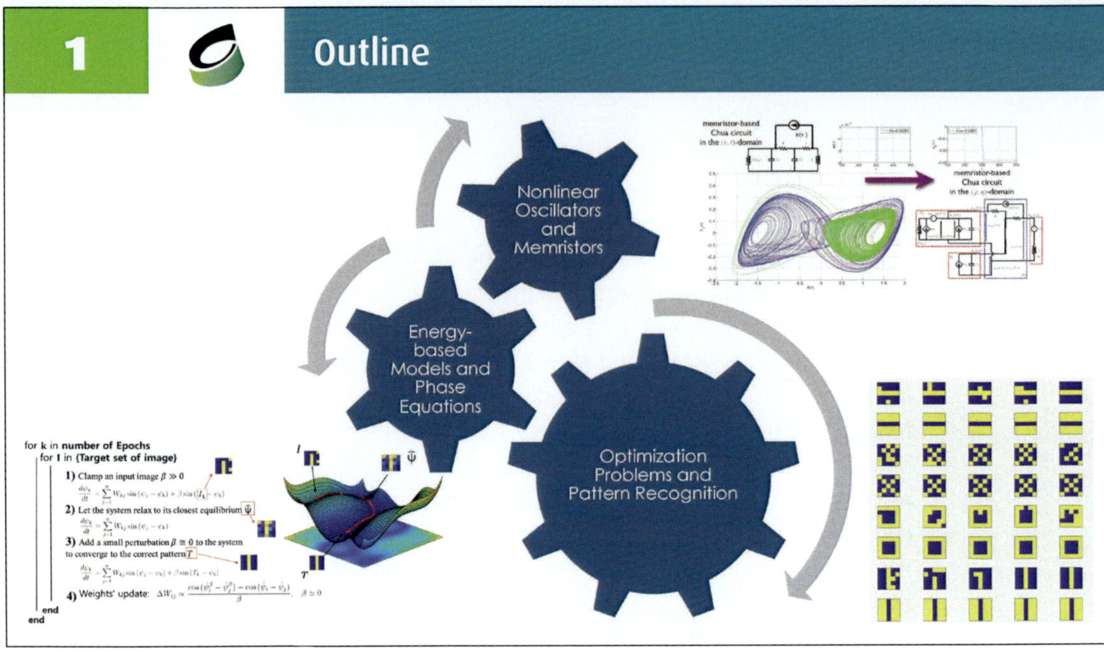

It delves into various aspects, including nonlinear oscillators and memristors, energy based models and phase equations, optimization problems and pattern recognition.

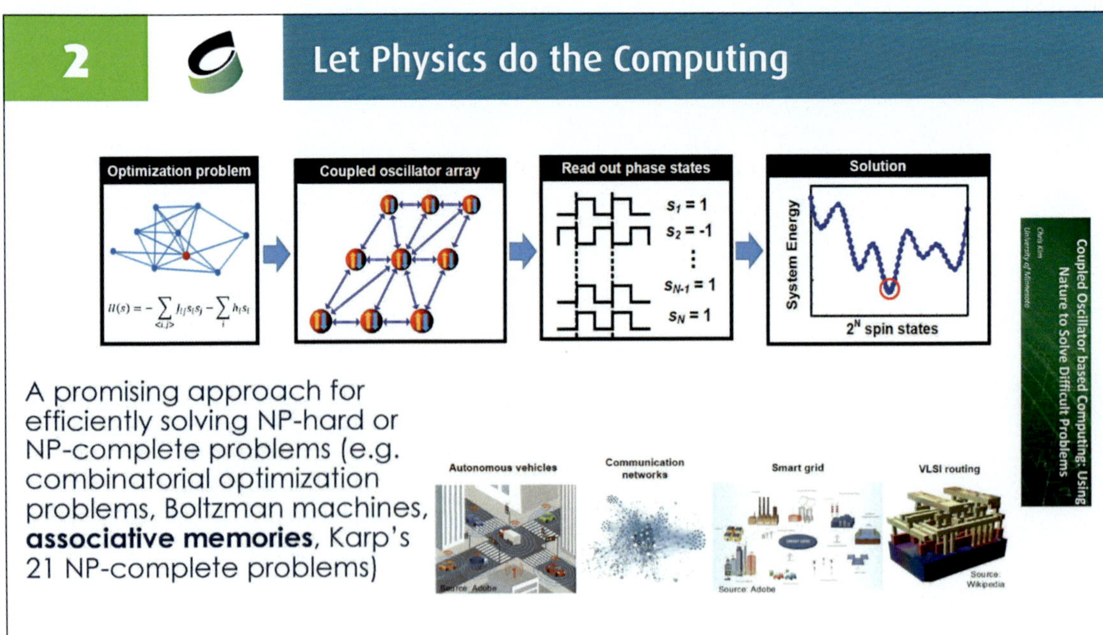

Analog computing platforms offer a promising avenue for effectively tackling NP-hard or NP-complete problems, including combinatorial optimization problems, Boltzmann machines, associative memories, and Karp's 21 NP-complete problems. For instance, the utilization of a coupled oscillator array enables the extraction of diverse phase states of an optimization problem to discern the solution.

3 Devices for Analog Computing

Comparison of Coupled Oscillator Technologies

	Qubit	Optical	Phase transition	Spintronic	CMOS
Conceptual figure	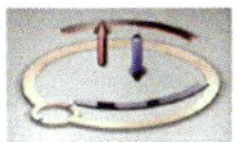	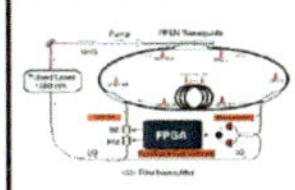	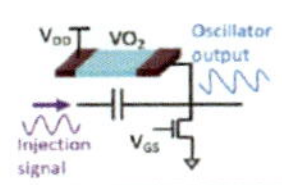	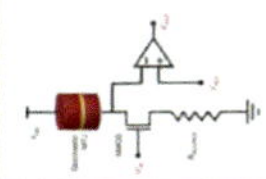	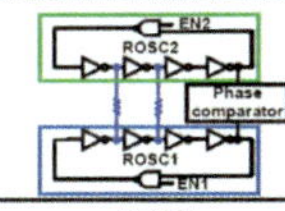
# of oscillators	~2000 (single chip)	100-2000 (lab setup)	4 (probe station)	8 (board level)	2000+ (1.3mm² chip in 65nm)
Advantages	Under debate	Room temperature	Room temperature	Room temperature	Room temperature, leverages CMOS, cloud/edge computing
Disadvantages	Cryogenic cool, 25kW power, premature tech. cloud only	1km optical fiber, FPGA chip, complex setup	Premature device, no real area advantage over CMOS	Premature device, no real area advantage over CMOS	Will it outperform GPUs and software solvers?
Integrated system in 10 yrs?	No	No	No	No	Yes
Target applications	NP-hard and NP-complete combinatorial optimization problems (e.g. supply chain, AI/ML, transportation, smart grid, communication, IC design, bioinformatics, computer vision, and robotics)				

To date, a variety of coupled oscillator technologies within analog computing have been aimed at addressing NP-hard and NP-complete combinatorial optimization problems. These include implementations such as qubit, optical, phase transition, spintronic, and CMOS technologies.

4 Memristor-based Oscillators

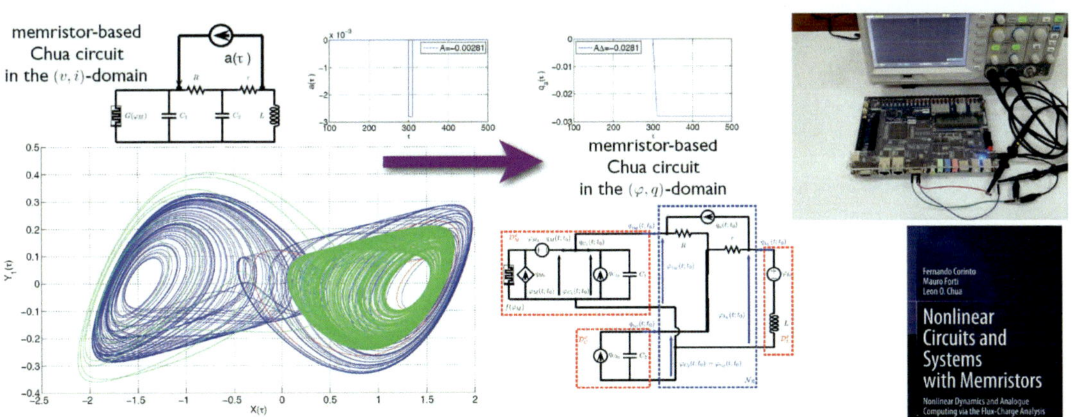

Pulse-controlled Tunable Periodic/Chaotic Attractors in Memristor-based Oscillators

Discusses the ability to switch between periodic and chaotic attractors in memristor-based oscillators which can offer new opportunities for designing and implementing advanced computing systems. The significance of pulse-controlled tunable periodic/chaotic attractors in memristor-based oscillators lies in their ability to provide a means of controlling the behavior of the oscillators.

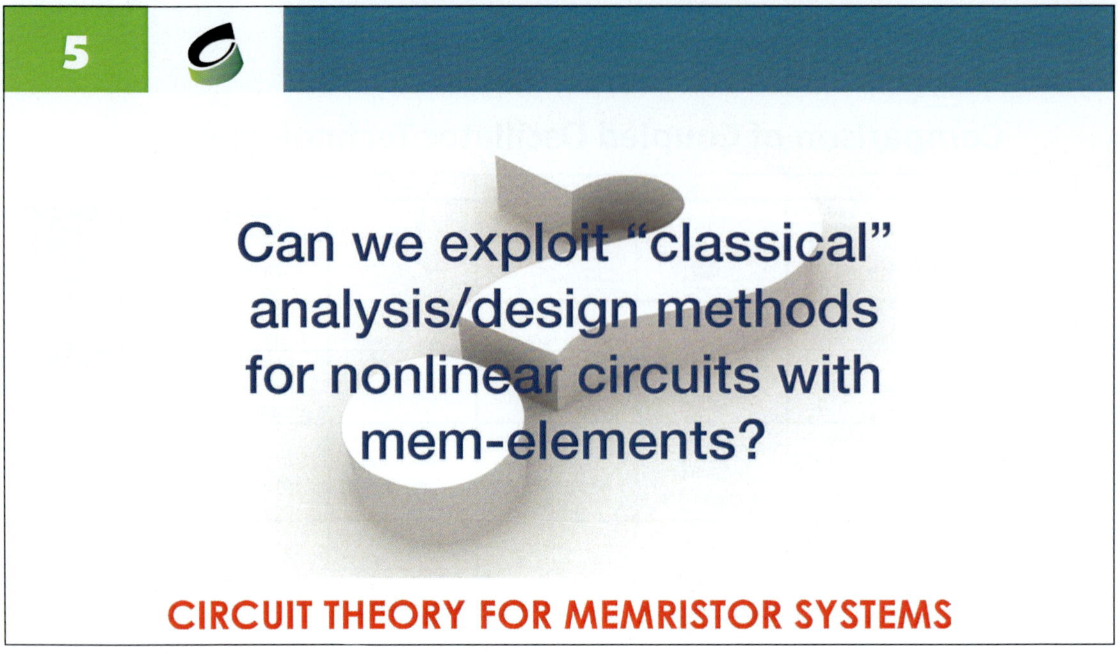

Investigates the potential to leverage traditional analysis and design techniques for nonlinear circuits incorporating memristive elements.

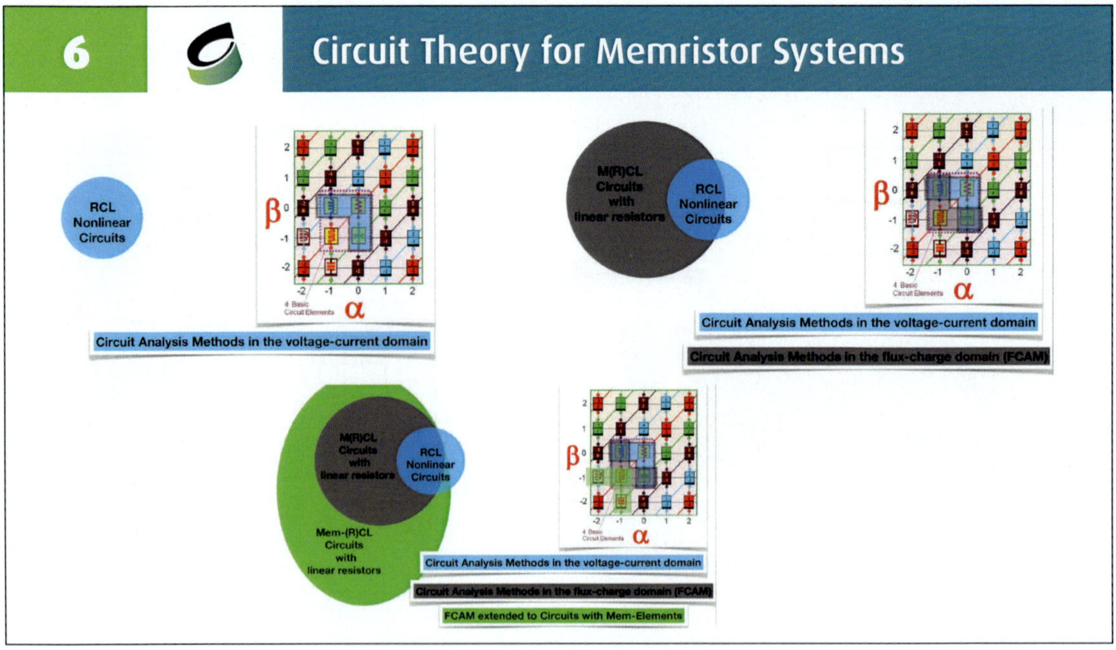

Three primary memristor system-based computing circuits are emphasized: RCL nonlinear circuits, utilized as methods for circuit analysis in the voltage-current domain; M(RC)L circuits, incorporating linear resistors for circuit analysis in the flux-charge domain; and Mem-(RC)L circuits, which extend FCAM circuits to incorporate memristive elements.

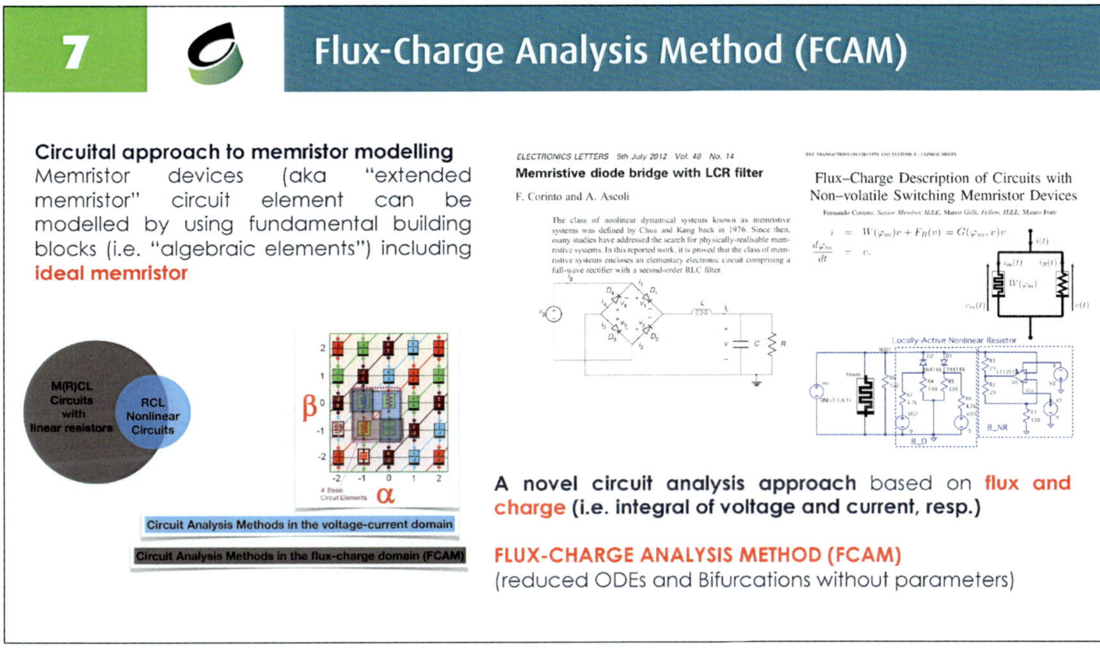

Discusses the circuit-based method for modeling memristors, employing fundamental building blocks specific to memristive devices, and introduces a novel circuit analysis approach grounded in flux and charge considerations.

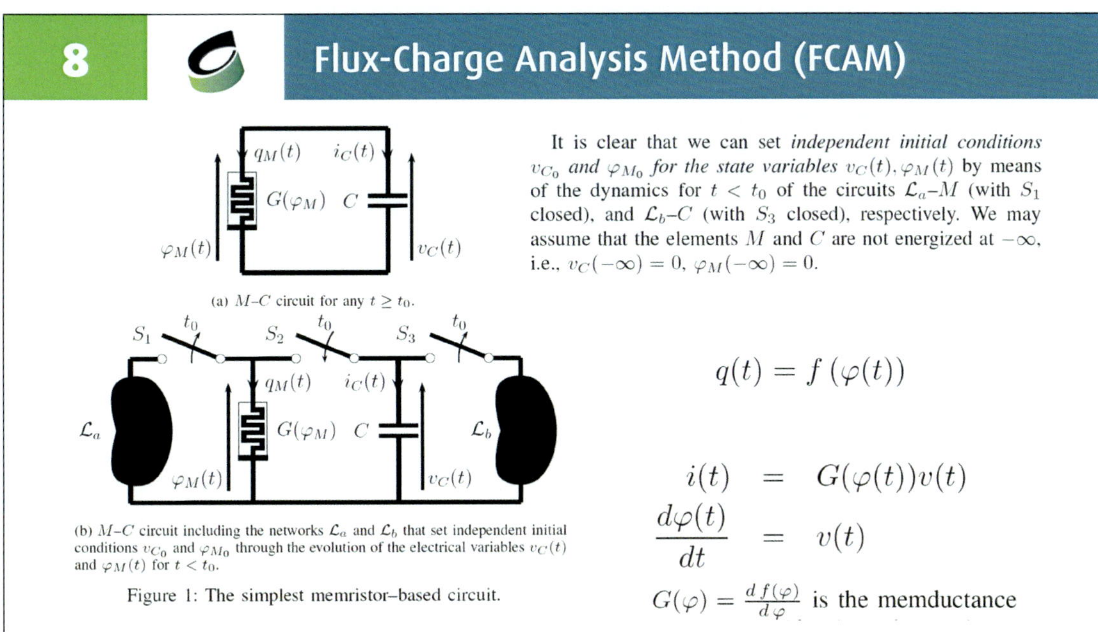

The Flux-Charge Analysis Method (FCAM) is examined, presenting a fresh approach to circuit analysis centered on memristor modeling through fundamental building blocks, notably the ideal memristor. Its core components involve leveraging flux and charge—integral representations of voltage and current—as fundamental elements for analysis.

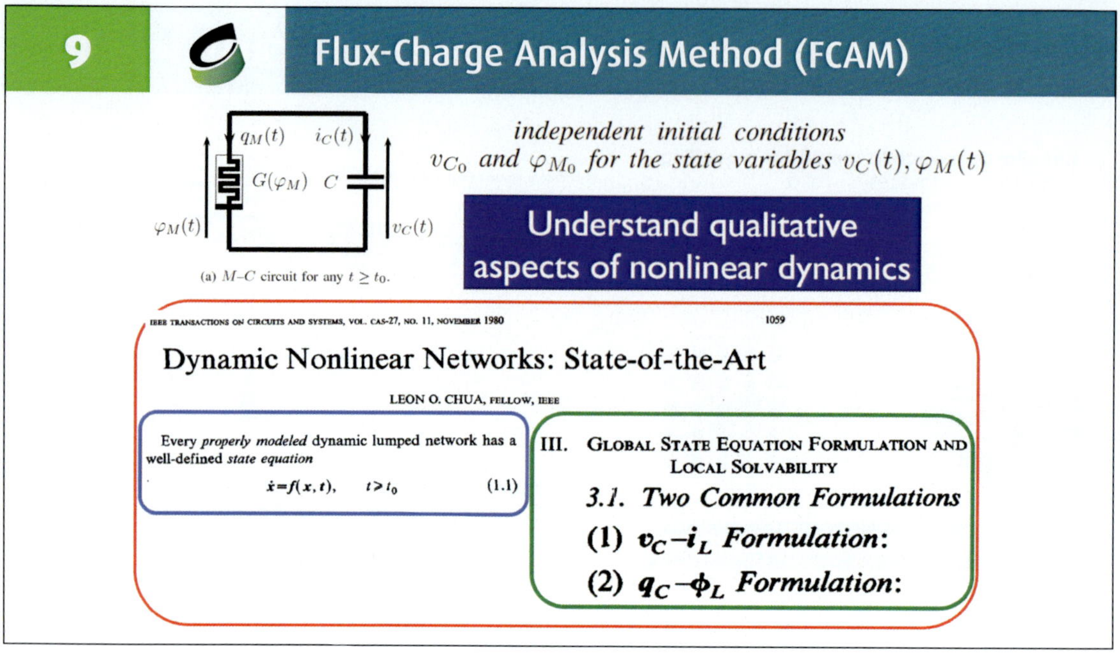

Focuses on further details of Flux-Charge Analysis Method (FCAM), for memristor modeling. This methodology enables the reduction of ordinary differential equations and facilitates bifurcations without parameters.

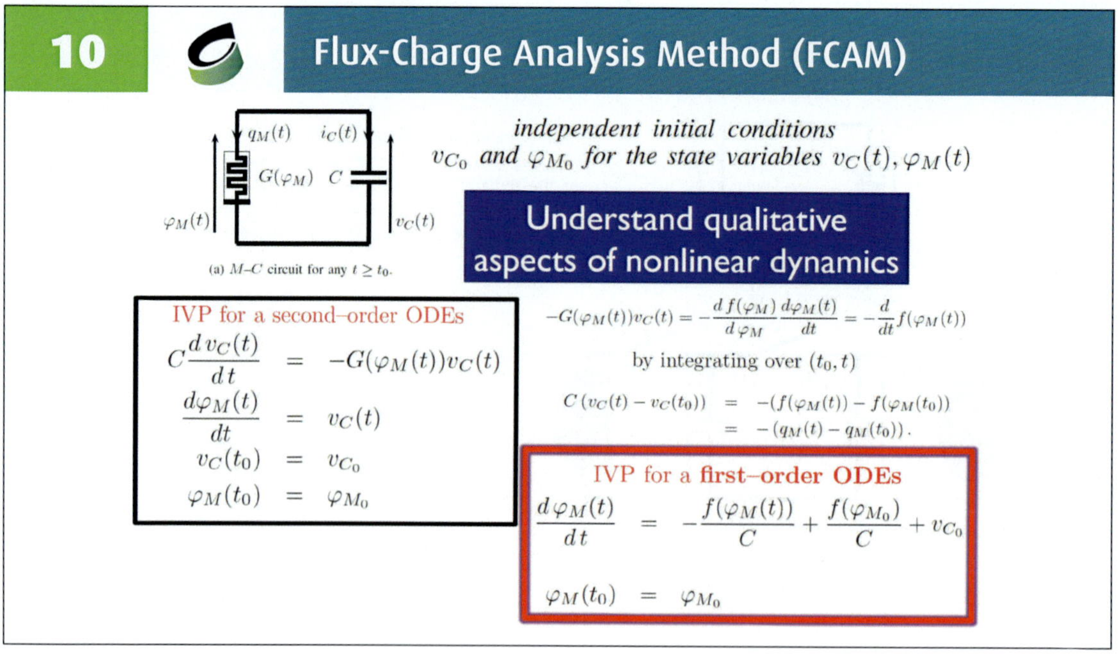

Discusses the initial value problems (IVPs) of first and second-order ordinary differential equations (ODEs).

11 Flux-Charge Analysis Method (FCAM)

The pillars of the **new flux–charge method** are:

- the use of KqL and KφL in terms of the incremental flux and charge:

$$\begin{aligned} \varphi_k(t;t_0) &\doteq \varphi_k(t) - \varphi_k(t_0) = \int_{t_0}^{t} v_k(\tau)d\tau \\ q_k(t;t_0) &\doteq q_k(t) - q_k(t_0) = \int_{t_0}^{t} i_k(\tau)d\tau \end{aligned}$$

  for any $t \geq t_0$, where $v_k(t)$ is the voltage across and $i_k(t)$ is current through any two-terminal element in $\mathcal{LM}$, respectively;

- the use of Constitutive Relations (CRs) expressed according to the same electrical variables, that is, CRs of resistors, capacitors, inductors and memristors specified in terms of incremental charges and fluxes.

IEEE TRANSACTIONS ON CIRCUITS AND SYSTEMS—I: REGULAR PAPERS, VOL. 63, NO. 11, NOVEMBER 2016 1907

Memristor Circuits: Flux—Charge Analysis Method

Fernando Corinto, *Senior Member, IEEE*, and Mauro Forti

Defines and explains technique based on the Kirchhoff Flux and Charge Laws, along with the constitutive relations of circuit elements expressed in terms of incremental flux and charge. Key benefits of formulating circuit equations in the (φ, q)-domain, as opposed to the traditional voltage-current-domain, include: Streamlined analysis of nonlinear dynamics and bifurcations through a reduced set of ODEs and Enhanced clarity regarding the impact of initial conditions.

12 Flux-Charge Analysis Method (FCAM)

$$\begin{cases} C\frac{d\varphi_C(t;t_0)}{dt} = -f(\varphi_C(t;t_0) + \varphi_{M_0}) + Q_0 \\ \varphi_C(t_0;t_0) = 0 \end{cases}$$

$$Q_0 = f(\varphi_M(t_0)) + Cv_C(t_0)$$

total charge $Q(t)$ for $t \geq t_0$

$$Q(t) = f(\varphi_M(t)) + Cv_C(t)$$

Property 2: The total charge $Q(t) = Q_0$ for all $t \geq t_0$.

$$q_M = f(\varphi_M) = -a\varphi_M + b\varphi_M^3, \qquad a, b > 0$$

Three equilibrium points for

$$\begin{aligned} \bar{\varphi}_M^{\alpha} &= -\frac{(1+\sqrt{3}\imath)a}{\sqrt[3]{12}\sigma} - \frac{(1-\sqrt{3}\imath)\sigma}{2\sqrt[3]{18}b} & \bar{\varphi}_C^{\alpha} &= \bar{\varphi}_M^{\alpha} - \varphi_{M_0} \\ \bar{\varphi}_M^{\beta} &= -\frac{(1-\sqrt{3}\imath)a}{\sqrt[3]{12}\sigma} - \frac{(1+\sqrt{3}\imath)\sigma}{2\sqrt[3]{18}b} & \bar{\varphi}_C^{\beta} &= \bar{\varphi}_M^{\beta} - \varphi_{M_0} \\ \bar{\varphi}_M^{\gamma} &= \frac{\sqrt[3]{\frac{2}{3}}a}{\sigma} + \frac{\sigma}{\sqrt[3]{18}b} & \bar{\varphi}_C^{\gamma} &= \bar{\varphi}_M^{\gamma} - \varphi_{M_0} \end{aligned}$$

where $\sigma = \sqrt[3]{\sqrt{3(27b^4Q_0^2 - 4a^3b^3)} + 9b^2Q_0}$.

$$\varphi_M(t) = \varphi_C(t;t_0) + \varphi_{M_0}$$

Analyses property related to FCAM method which is based upon the physical principle of charge conservation. As per the principle: the state variables evolve while the total charge stays constant at its initial value. The solution for the flux involving equilibrium points also discussed.

13 Flux-Charge Analysis Method (FCAM)

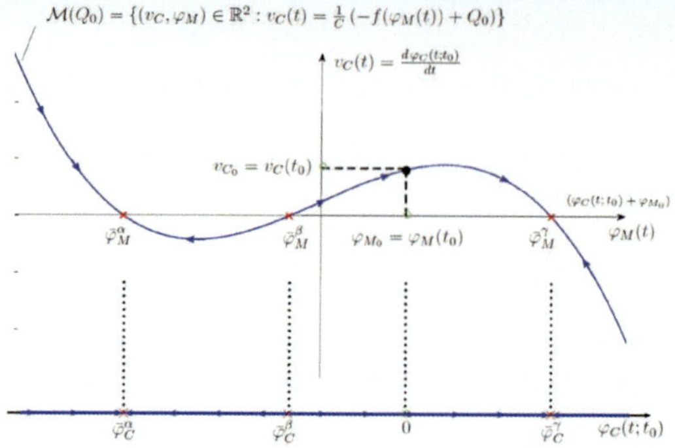

Figure 12: Invariant manifold $\mathcal{M}(Q_0)$ and equilibrium points

A *saddle–node bifurcation without parameters* occurs if Q_0 is such that $\bar{\varphi}_M^\beta$ and $\bar{\varphi}_M^\alpha$ (or else $\bar{\varphi}_M^\beta$ and $\bar{\varphi}_M^\gamma$) collide and annihilate each other. The conditions $\bar{\varphi}_M^\beta = \bar{\varphi}_M^\alpha$ and $\bar{\varphi}_M^\beta = \bar{\varphi}_M^\gamma$ provide

$$Q_0^{(sn)} = \pm\frac{2}{3}\frac{\sqrt[3]{a^2}}{\sqrt{3b}}.$$

Defines the invariant manifold concept to investigate nonlinear dynamics and bifurcations in the M – C circuit.

14 2nd order Memristor-based Oscillators

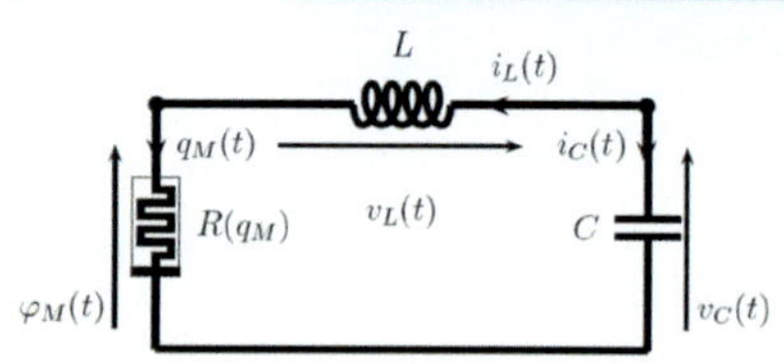

Figure 12: The M–L–C obtained from the M–C circuit in Fig. 8 by inserting a parasitic inductance L in series with the memristor.

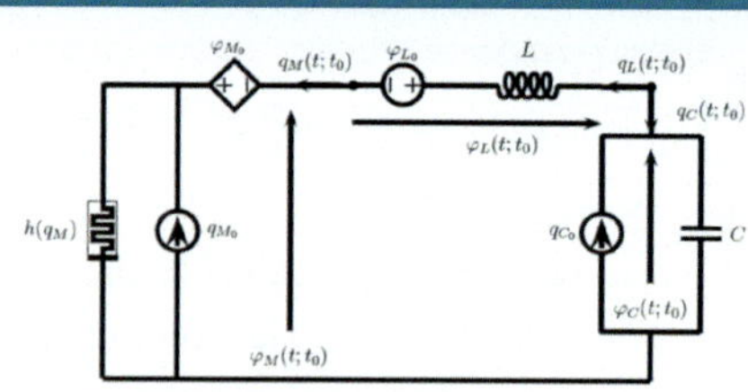

Figure 13: Equivalent circuit in the (φ, q)–domain of the M–L–C circuit in Fig. 12 for $t \geq t_0$.

$$\begin{aligned}
C\frac{d}{dt}\varphi_C(t;t_0) &= -q_L(t;t_0) + q_{C_0} \\
L\frac{d}{dt}q_L(t;t_0) &= \varphi_C(t;t_0) - h(q_L(t;t_0) + q_{M_0}) \\
&\quad + h(q_{M_0}) + \varphi_{L_0} \\
\varphi_C(t_0;t_0) &= 0 \\
q_L(t_0;t_0) &= 0.
\end{aligned}$$

$$\begin{aligned}
\frac{dx(t)}{dt} &= \frac{-y(t) + Q_0}{C} \\
\frac{dy(t)}{dt} &= \frac{x(t) - h(y(t))}{L} \\
x(t_0) &= h(q_{M_0}) + \varphi_{L_0} \\
y(t_0) &= q_{M_0}
\end{aligned}$$

for $t \geq t_0$, where

$$Q_0 = q_{C_0} + q_{M_0} = Cv_{C_0} + q_{M_0}$$

Introduces and examines second-order memristor-based oscillators, elucidating the dynamic pathway of initial value problems that depict the equilibrium points.

15 2nd order Memristor-based Oscillators

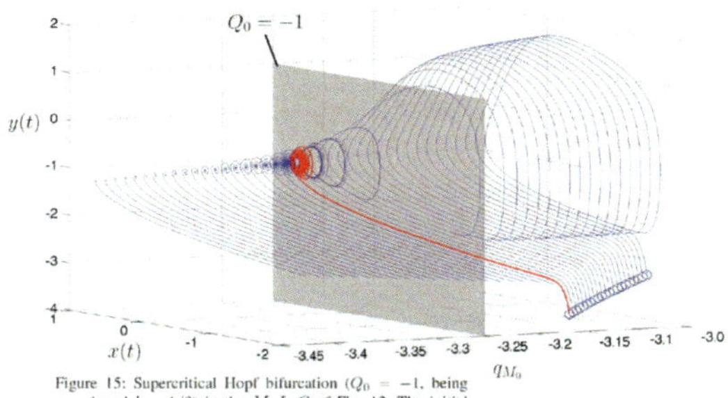

Figure 15: Supercritical Hopf bifurcation ($Q_0 = -1$, being $a = 1$ and $b = 1/3$) in the M–L–C of Fig. 12. The initial condition q_{M_0} is such that, according to (16), $Q_0 = Cv_{C_0} + q_{M_0} = -1$ for $q_{M_0} = -3.25$ (being $C = 9/2$ and $v_{C_0} = 0.5$).

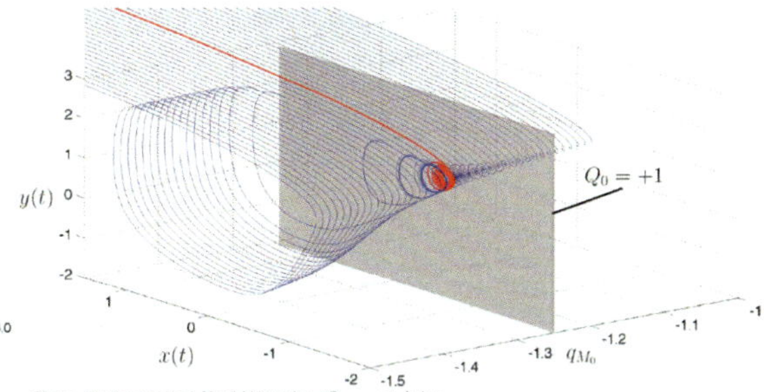

Figure 16: Supercritical Hopf bifurcation ($Q_0 = +1$, being $a = 1$ and $b = 1/3$) in the M–L–C of Fig. 12. The initial condition q_{M_0} is such that, according to (16), $Q_0 = Cv_{C_0} + q_{M_0} = +1$ for $q_{M_0} = -1.25$ (being $C = 9/2$ and $v_{C_0} = 0.5$).

Hopf Bifurcation without Parameters

Analyses the second-order memristor-based oscillators using Hopf bifurcations without parameters.

16 3rd order Memristor-based Oscillators

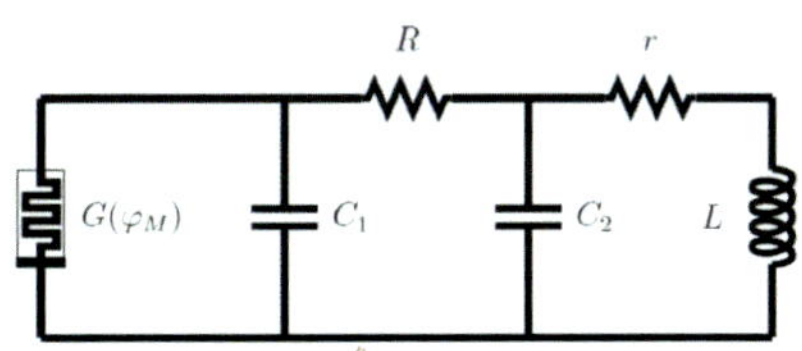

Figure 17: Memristor–based Chaotic Oscillator.

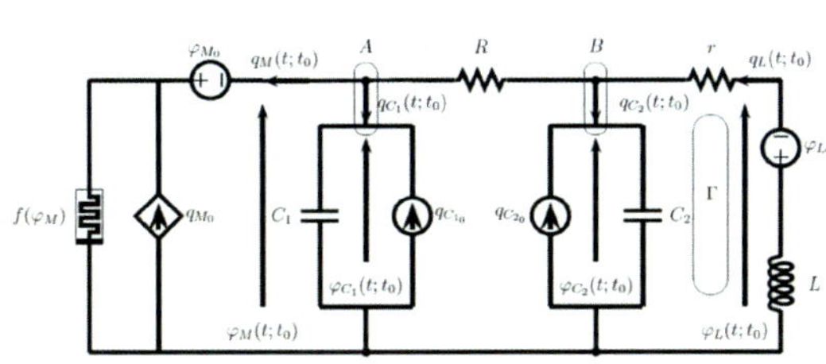

Figure 18: Equivalent circuit of the Memristor–based Chaotic Oscillator in the (φ, q)–domain.

$$
\begin{aligned}
\frac{dx(t)}{dt} &= \alpha\left[-x(t) + y(t) - n(x(t)) + X_0\right] \\
\frac{dy(t)}{dt} &= x(t) - y(t) + z(t) \\
\frac{dz(t)}{dt} &= -\beta y(t) \\
x(t_0) &= \varphi_{M0} \\
y(t_0) &= \varphi_{L0} \\
z(t_0) &= -\varphi_{M0} + \varphi_{L0} + q_{C_{2_0}}
\end{aligned}
$$

and

$$X_0 = \varphi_{M_0} - Li_{L_0} + n(\varphi_{M_0}) + C_1 v_{C_{1_0}}$$

Examines the third-order memristor-based chaotic oscillators and their equivalent circuits.

17 3rd order Memristor-based Oscillators

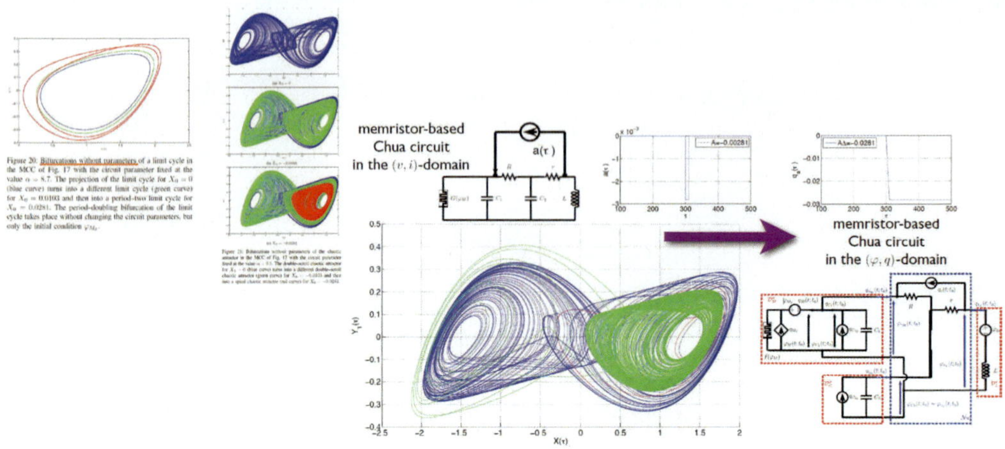

Emphasizes the development of third-order memristor-based chaotic oscillators, incorporating the transition from the memristor-based Chua's circuit in the voltage-current domain to its counterpart in the flux-charge domain.

18 MemElements-based Oscillators

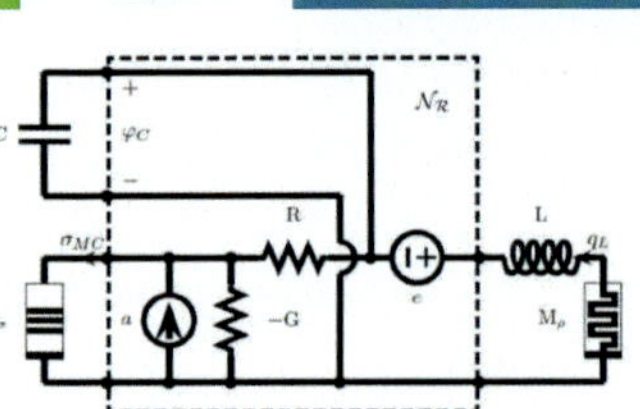

Figure 7: Chaotic network with memcapacitor and memristor.

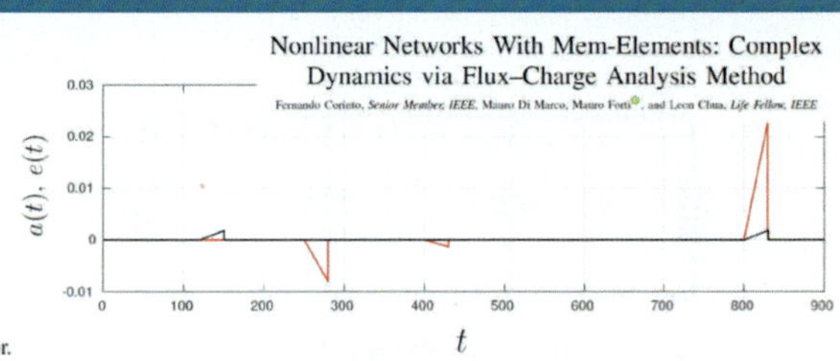

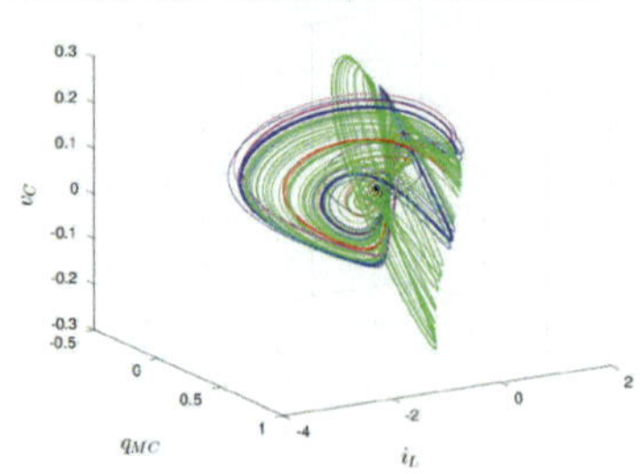

Figure 12: 3D view of the trajectory evolution.

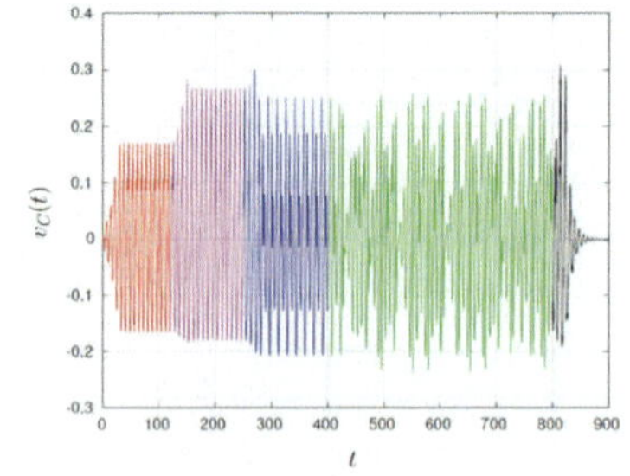

Figure 14: Time-domain evolution of capacitor voltage $v_C(t)$.

Introduces dynamics within nonlinear networks featuring emerging nanoscale mem-elements. Expands the flux-charge analysis method (FCAM) from nonlinear circuits with memristors to encompass a wider range of nonlinear networks N, which also include memcapacitors and meminductors.

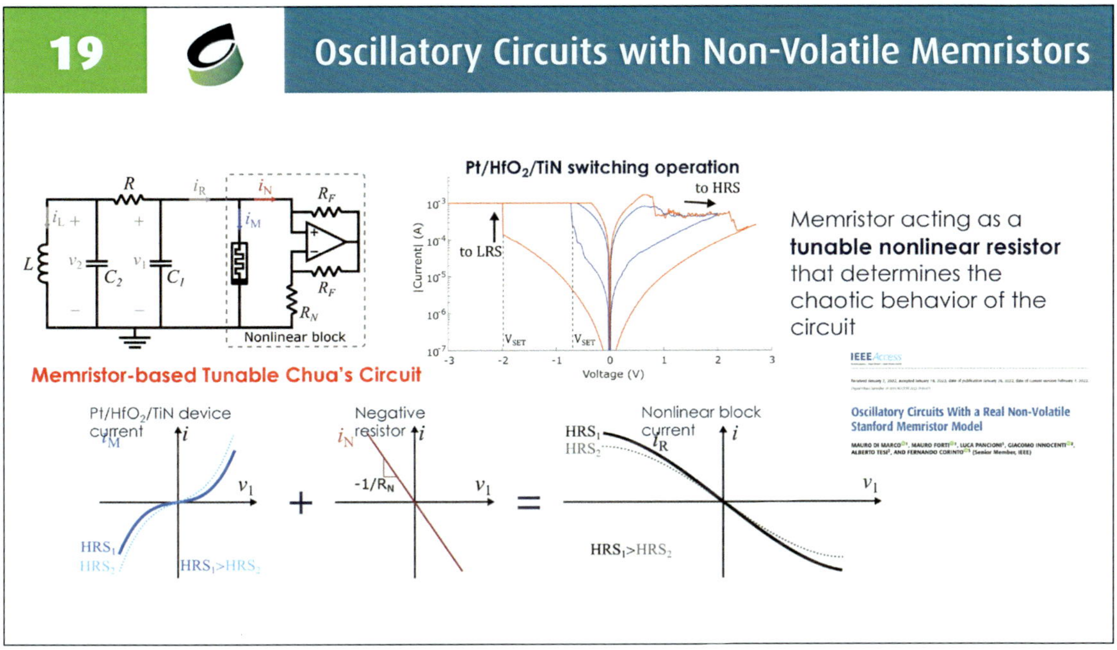

Presents the dynamics and bifurcations in a class of nonlinear oscillators with real non-volatile memristor devices obeying Stanford model.

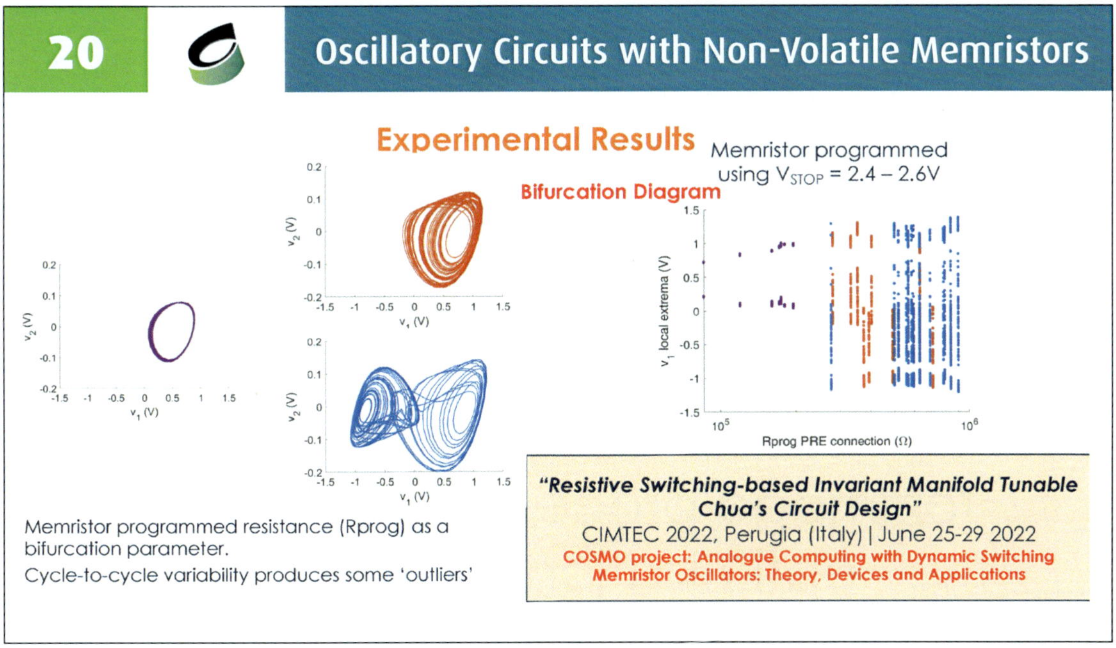

Presents the experimental findings of oscillatory circuits employing non-volatile memristors, where the programmed resistance (Rprog) of the memristor serves as a bifurcation parameter. Additionally, depicted "outliers" arising from cycle-to-cycle variability.

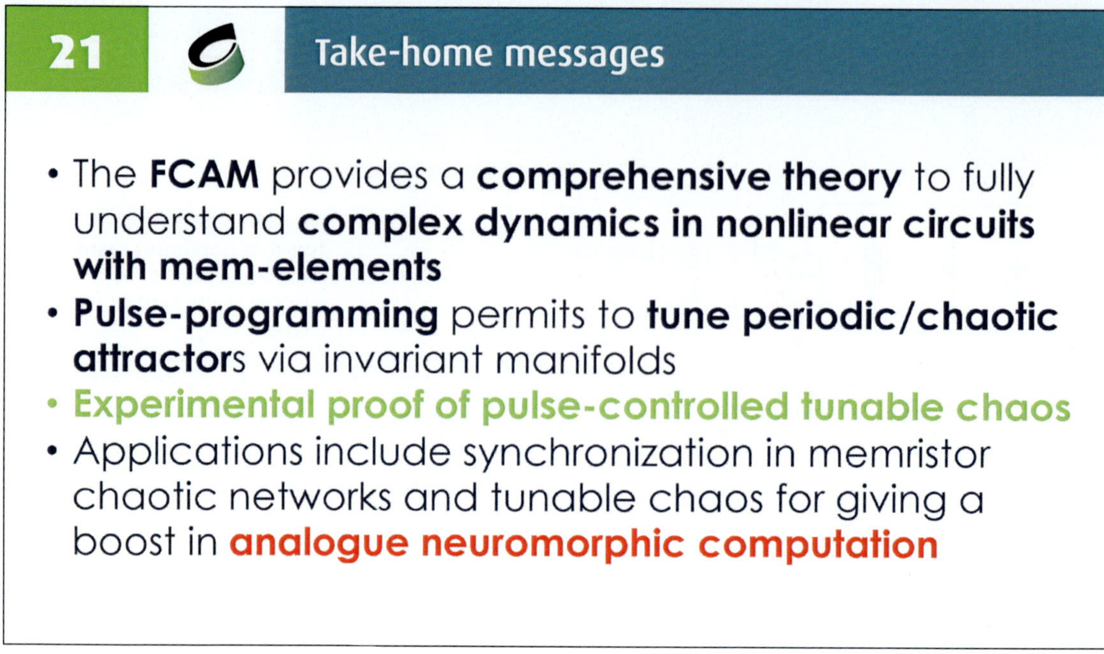

Summarizes FCAM as a comprehensive framework for gaining a deep understanding of the intricate dynamics within nonlinear circuits featuring mem-elements. Demonstrates adjustment of periodic or chaotic attractors through invariant manifolds utilizing pulse-programming. Experimental evidence demonstrates the efficacy of pulse-controlled tunable chaos.

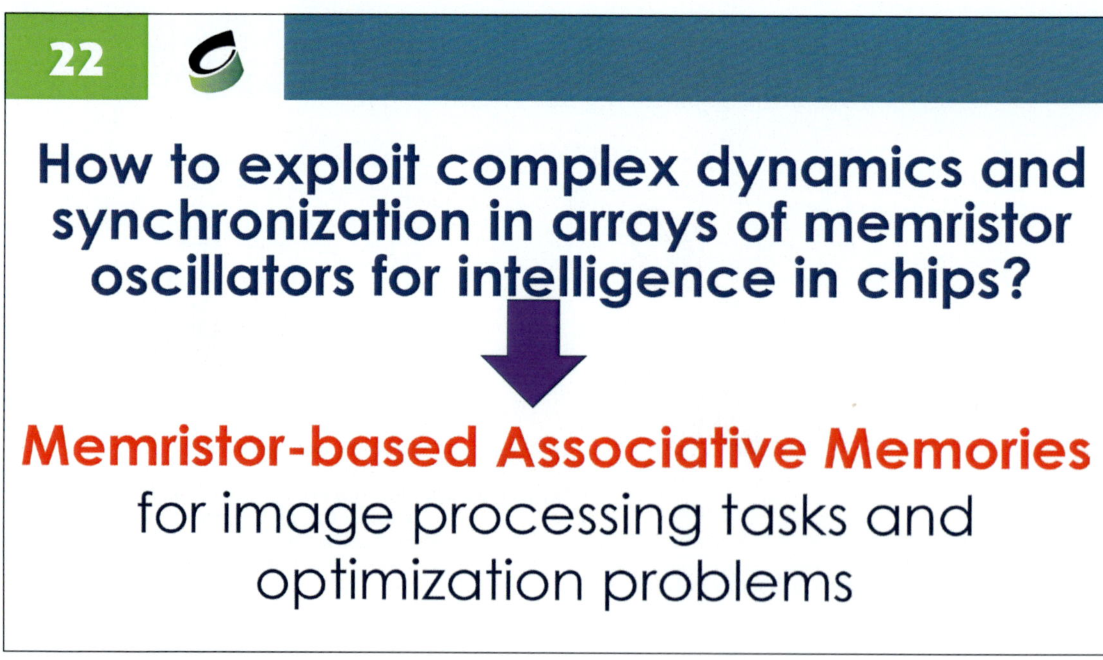

Presents the challenge of intricate dynamics and synchronization of arrays of memristor oscillators to enhance chip intelligence. Suggests utilizing memristor-based associative memories for tasks like image processing and optimization problems.

23 Oscillatory Associative Memories

Energy-based Models provide one well-grounded foundation for associative memory models:

1) Hopfield Neural Network

$$\dot{x}_i = -x_i + \sum_{j=1}^{N} W_{ij} g(x_j) + \theta_i \quad \forall i = 1, \ldots, N \qquad E(\mathbf{x}, \mathbf{W}) = \sum_{i=1}^{N} \int_0^{x_i} s g'(s) ds - \frac{1}{2} \sum_{i,j=1}^{N} W_{ij} g(x_i) g(x_j) - \theta_i g(x_i)$$

2) Kuramoto Model

$$\dot{\psi}_i = \sum_{j=1}^{N} w_{ij} \sin(\psi_j - \psi_i) \quad \forall i = 1, \ldots, N \qquad E(\psi, \mathbf{W}) = -\frac{1}{2} \sum_{i,j=1}^{N} w_{ij} \cos(\psi_i - \psi_j)$$

Two important **key points**:

1) Patterns = attractors of the dynamical network

2) Inference = provide a corrupted pattern to the system and let it converge to the closest equilibrium

Introduces Oscillatory Associative Memories, based on energy-based models like the Hopfield neural network, Kuramoto model, etc., which offer a solid foundation for associative memory frameworks.

24 Oscillatory Associative Memories

Suppose to have the following set of patterns:

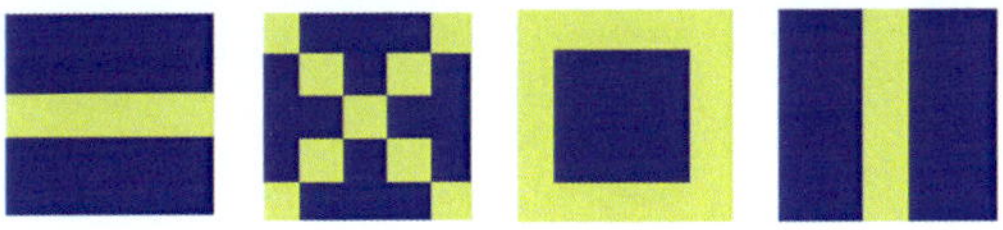

A yellow square corresponds to the **phase difference** between each oscillator and a reference $\Delta\psi =$ 0ual to $\Delta\psi = \pi$

, and a blue square corre: 'ifference equal to

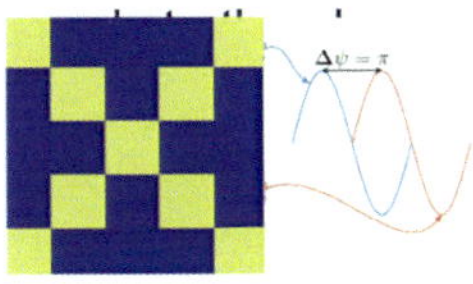

Hebbian learning rule builds the weight matrix **W** which stores the $m = 4$ targets $\mathbf{T}_i$:

$$\mathbf{W} = \frac{1}{m} \sum_{i=1}^{m} [\cos(\mathbf{T}_i)\cos(\mathbf{T}_i)^T - \mathbb{I}_{n \times n}]$$ ⟶ Prone to local minima!

Explains patterns that may arise within associative memories are discussed, along with the definition of the Hebbian learning rule responsible for constructing the weight matrix.

25 Memristor Oscillatory Associative Memories (MOAMs)

frontiers in Neuroscience

ORIGINAL RESEARCH
published: 24 March 2020
doi: 10.3389/fnins.2020.00240

Equilibrium Propagation for Memristor-Based Recurrent Neural Networks

Gianluca Zoppo*, Francesco Marrone and Fernando Corinto

Equilibrium Propagation is a **two-phase local learning rule** for Energy-based Models.

The method is a **supervised** technique that avoids the high computational cost characterizing the backpropagation algorithm.

The weights update computes the gradient of an associated cost function $\mathcal{C}(\mathbf{T},\psi)$ that measures the distance between the target and the current equilibrium of the system.

GOAL: adapt this learning rule to oscillatory neural networks with memristors defined by the Kuramoto model and optimize the cost function

$$\mathcal{C}(\mathbf{T},\psi) = N - \sum_{i=1}^{N} \cos(T_i - \hat{\psi}_i)$$

Delves into Memristive Oscillatory Associative Memories, introducing the equilibrium propagation, a two-phase local learning rule for energy-based models. Notably, the method's advantage lies in its ability to circumvent the computational expenses associated with the backpropagation algorithm.

26 Memristor Oscillatory Associative Memories (MOAMs)

for **k** in **number of Epochs**
  for **I** in **{Target set of image}**

1) Clamp an input image $\beta \gg 0$

$$\frac{d\psi_k}{dt} = \sum_{j=1}^{n} W_{kj} \sin(\psi_j - \psi_k) + \beta \sin(I_k - \psi_k)$$

2) Let the system relax to its closest equilibrium $\hat{\Psi}$

$$\frac{d\psi_k}{dt} = \sum_{j=1}^{n} W_{kj} \sin(\psi_j - \psi_k)$$

3) Add a small perturbation $\beta \cong 0$ to the system to converge to the correct pattern T

$$\frac{d\psi_k}{dt} = \sum_{j=1}^{n} W_{kj} \sin(\psi_j - \psi_k) + \beta \sin(T_k - \psi_k)$$

4) Weights' update: $\Delta W_{ij} \propto \frac{\cos(\hat{\psi}_i^\beta - \hat{\psi}_j^\beta) - \cos(\hat{\psi}_i - \hat{\psi}_j)}{\beta}, \quad \beta \simeq 0$

  end
end

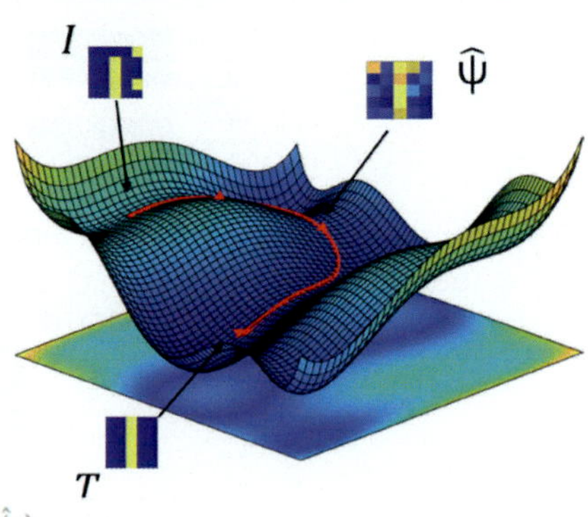

Explains that the system converges to the nearest equilibrium upon receiving a corrupted pattern, with the learning process entailing the adjustment of weights to specify the desired equilibria.

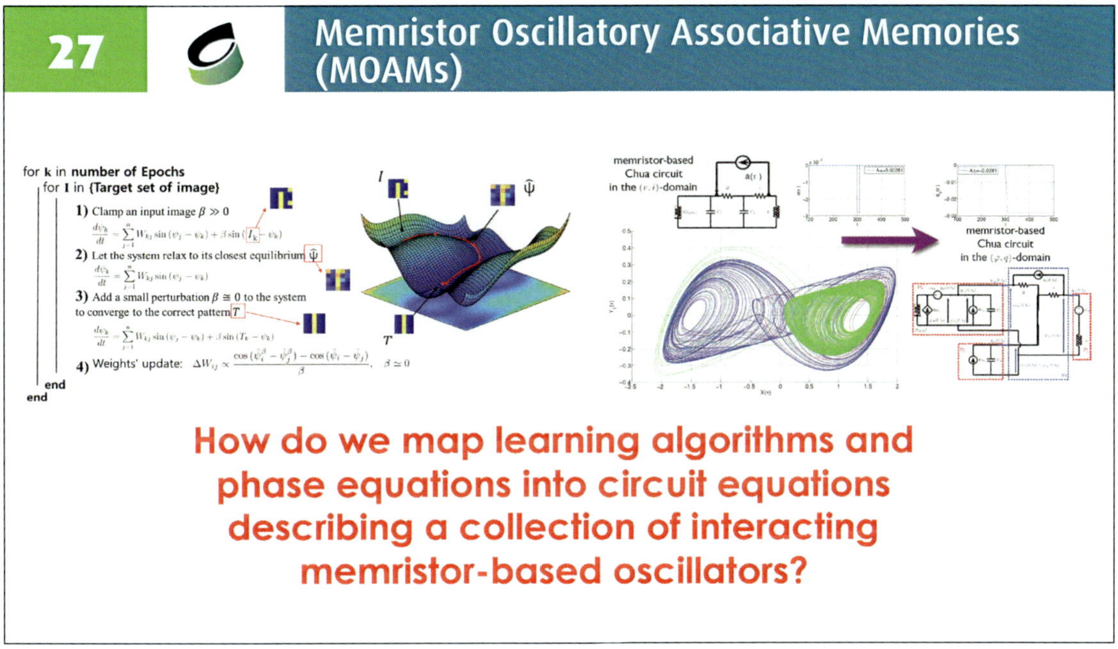

Presents the challenge of mapping learning algorithms and phase equations into circuit equations describing a collection of interacting memristor-based oscillators.

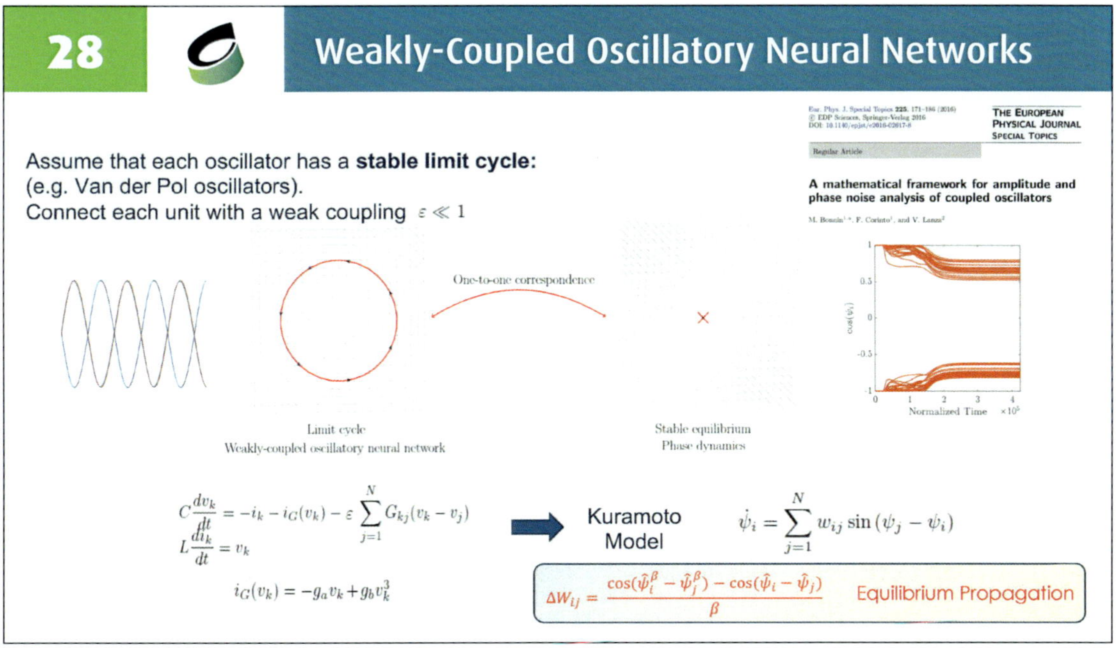

Explores the basics and mathematical formulations of Weakly-Coupled Oscillatory Neural Networks.

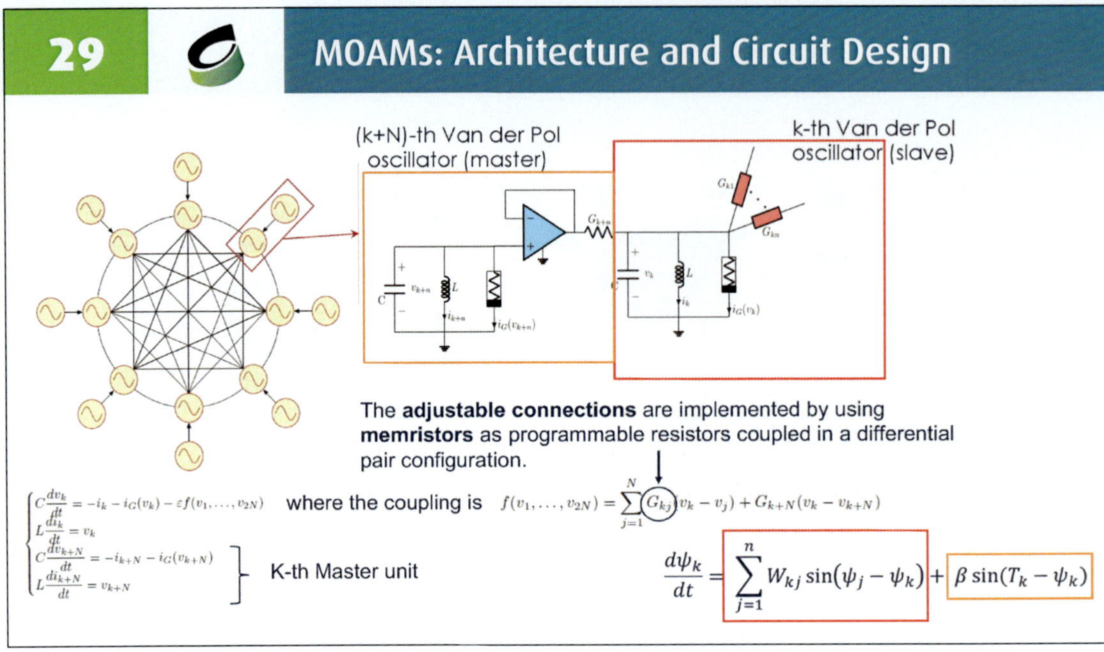

Elucidates the architecture and circuit design of MOAMS, detailing the involvement of memristors as programmable resistors coupled in a configuration of differential pairs.

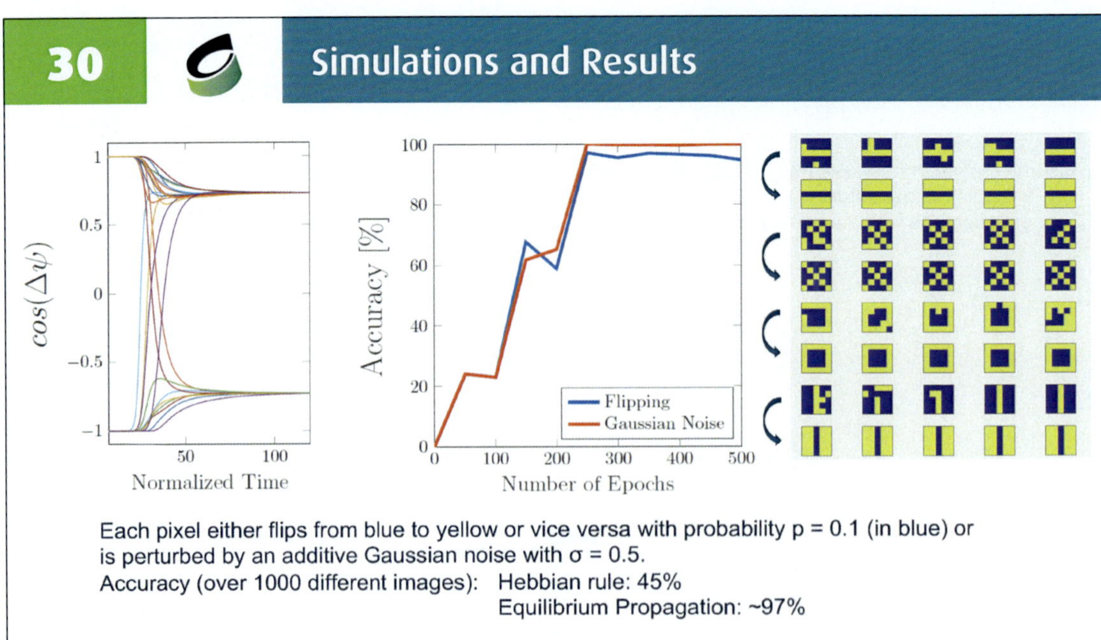

Examines simulations and outcomes, offering insights into pixels transitioning or encountering perturbations from additive Gaussian noise.

31 Key Takeaways

A holistic approach to analogue computing with memristor oscillators:

- **Physical and circuit models of memristor devices**
- **Tools (FCAM, Harmonic Balance) for understanding nonlinear dynamics and bifurcation to control complex behaviour**
- **Experimental results of tuneable periodic/chaotic behavior via memristor programming**
- **Comprehensive mathematical frameworks for a one-to-one mapping nonlinear oscillators including memristor into phase models**
- **Advanced learning algorithms to solve optimization problems via collective phase dynamics:** Inference lies in letting the system converge to its equilibria. Learning consists in finding the correct weights defining the required equilibria. Equilibrium Propagation is a two-phase local learning rule that requires a low computational cost. Equilibrium Propagation outperforms conventional approaches in solving associative memories and is well suited for analogue hardware implementation. Device-to-device variability and systematic bias in the weight updates degrade the system performance. This motivates further studies on the robustness of Equilibrium Propagation.

Lists the overall summary of the presentation.

32 Suggested Readings

- Corinto, Fernando, Mauro Forti, and L. Chua. "Nonlinear Circuits and Systems with Memristors." Nonlinear Dynamics and Analogue Computing via the Flux-Charge Analysis Method. Springer Verlag. Available online at: https://link.springer.com/book/10.1007/978-3-030-55651-8
- Corinto, Fernando, and Mauro Forti. "Memristor circuits: Flux—Charge analysis method." IEEE Transactions on Circuits and Systems I: Regular Papers 63.11 (2016): 1997-2009.
- Corinto, Fernando, and Mauro Forti. "Memristor circuits: Bifurcations without parameters." IEEE Transactions on Circuits and Systems I: Regular Papers 64.6 (2017): 1540-1551.
- Corinto, Fernando, and Mauro Forti. "Memristor circuits: Pulse programming via invariant manifolds." IEEE Transactions on Circuits and Systems I: Regular Papers 65.4 (2017): 1327-1339.
- Corinto, Fernando, et al. "Nonlinear networks with mem-elements: Complex dynamics via flux–charge analysis method." IEEE Transactions on Cybernetics 50.11 (2019): 4758-4771.
- Di Marco, Mauro, et al. "Oscillatory Circuits With a Real Non-Volatile Stanford Memristor Model." IEEE Access 10 (2022): 13650-13662.
- Scellier, Benjamin, and Yoshua Bengio. "Equilibrium propagation: Bridging the gap between energy-based models and backpropagation." Frontiers in computational neuroscience 11 (2017): 24.
- Zoppo, Gianluca, Francesco Marrone, and Fernando Corinto. "Equilibrium propagation for memristor-based recurrent neural networks." Frontiers in neuroscience 14 (2020): 240.
- Ascoli, Alon, et al. "Graph Coloring via Locally-Active Memristor Oscillatory Networks." Journal of Low Power Electronics and Applications 12.2 (2022): 22.

Provides a list of recommended readings for gaining familiarity with memristive oscillatory architectures.

Wraps up the presentation with gratitude.

SECTION 3: MEMRISTIVE CIRCUIT APPLICATIONS

CHAPTER 07

Aspects of Digital Circuit Design using Memristor Ratioed Logic (MRL)

Spiros Nikolaidis

Aristotle University of Thessaloniki

The presentation on digital circuit design using Memristor Ratioed Logic (MRL) offers a comprehensive exploration of memristor-based logic gates and their integration into digital architectures. Beginning with an introduction to memristors as the fourth basic circuit element, the presentation covers their functional characteristics, device structure, and operational principles. It delves into the challenges and advantages of memristor-based logic, discussing design families like Memristor Aided Logic (MAGIC), Material Implication (IMPLY), and Memristor Ratioed Logic (MRL). MRL, in particular, is highlighted for its compatibility with CMOS technology and its potential for single-step computations. The presentation elucidates the implementation of MRL in crossbar structures, showcasing its application in logic gates and combinational circuits. Through examples like full adders, encoders, decoders, and multiplexers, the presentation demonstrates the efficiency and scalability of memristor-based designs compared to traditional CMOS approaches. It concludes by emphasizing the need for further research to fully exploit the benefits of hybrid MRL-CMOS architectures in real-world applications. Overall, the presentation provides valuable insights into the emerging field of memristor-based digital circuit design and its potential for future computing paradigms.

1 Memristor – the missing element

- Resistor relates Voltage - Current
- Capacitor relates Voltage - Charge
- Inductor relates Current – Flux
- ? relates Charge - Flux

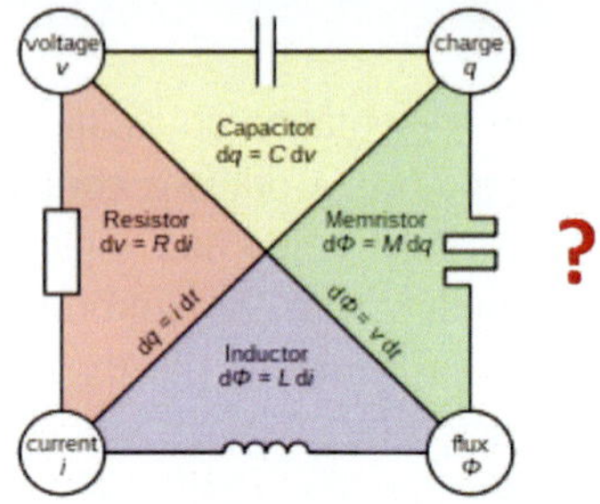

- Memristor – the missing element
- It was described and named in 1971 by Leon Chua, University of California, Berkeley.
- In 2008, Stanley Williams (HP labs) developed a practical model

This slide discusses the functional relation between voltage, current, flux and charge, along with the circuit components connecting those relations. The missing element, memristor, which functionally relates charge and flux, was discovered by Prof. Leon Chua, and Stanley Williams developed the practical model at HP labs.

2 Memristor function

Device	Differential equation
Resistor	R=dV/dI
Capacitor	C=dq/dV
Inductor	L=dΦ/dI
Memristor	M=dΦ/dq

Substituting the flux as the time integral of the voltage, and charge as the time integral of current

$$M(q(t)) = \frac{d\Phi/dt}{dq/dt} = \frac{V(t)}{I(t)}$$

Memristor is a resistor with memory and memristance is expressed in Ohms (Ω)

It is a voltage-controlled device whose memristance depends on its voltage history

It retains its resistance level even after power has been shut down – non-volatile device

Introduces memristor as the fourth basic circuit element. Memristor is a resistor with non-volatile memory whose memristance depends on its voltage history, and its resistance level is retained even when the power is shut down.

3 Memristor hysteresis loop

- This two-terminal device is a thin semiconductor film sandwiched between two metal contacts.
- Typical i-v characteristic of memristor

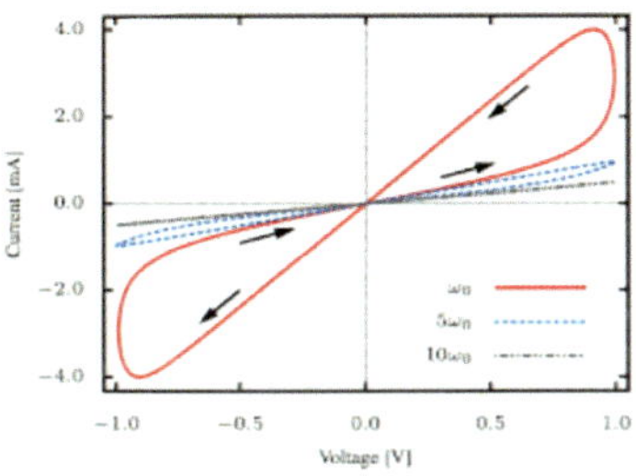

- Two stable cases for the resistance value
- Use for memory
- Low resistance value corresponds to logical 1
- High resistance value corresponds to logical 0

Discusses the device structure as a thin semiconductor film sandwiched between two metal contacts and the i-v characteristic of the memristor with its low and high resistance states. As frequency changes, the hysteresis loop also transforms.

4 Memristor: a resistor with memory

- Memristor is basically a resistor with varying resistance, dependent on the history of the device.
- It can be used for memory, where the data is stored as a resistance.
- **This varying resistance can also be used for producing logic**
- **The resistance of the device increases due to current flow in one direction and decreases due to current flow in the other direction.**
- This resistance depends on the direction, magnitude and duration of the applied voltage across its terminals
- It is bounded between a minimum value (R_{ON}) and a maximum value (R_{OFF}).

Provides an overview of using the memristor for producing logic whose resistance value is bounded between a minimum(R_{ON}) and maximum(R_{OFF}) value. The resistance increases due to the current flow in one direction and decreases due to the current flow in the other.

5 Memristor model

- Memristor HP TiO_2 basic model

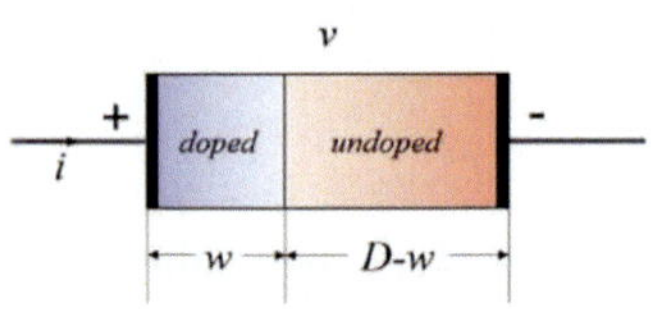

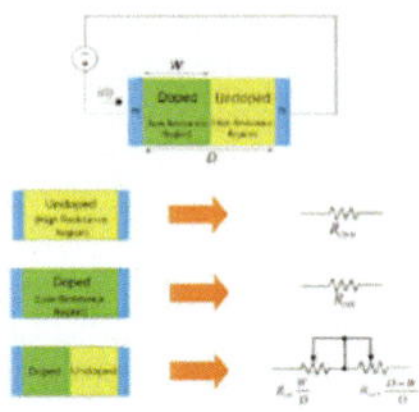

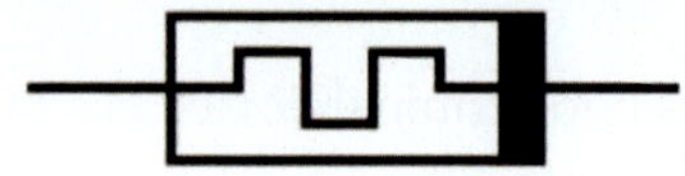

Memristor Symbol

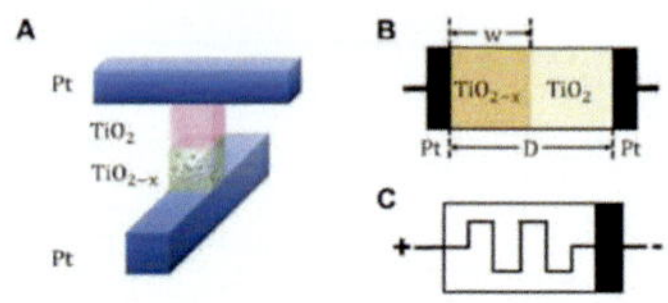

▪The resistance of the device increases due to current flow in one direction and decreases due to current flow in the other direction.

MRL is based on this and to Roff>>Ron

Discusses the HP TiO2 basic model of memristor. The relation between the Memristor Ratioed Logic (MRL) and the two resistance states R_{ON} and R_{OFF} is also covered.

6 Fabrication advantages

- Memristors can be fabricated between two metal layers
- **In this way, memristors devices can be fabricated above the CMOS transistors.**
- The size of a typical memristor is relatively small, similar to the processing of a via between metal layers.
- **Hence, memristor-based circuits may be smaller than transistor-only CMOS circuits.**
- **Therefore, they can exhibit high density and good scalability.**

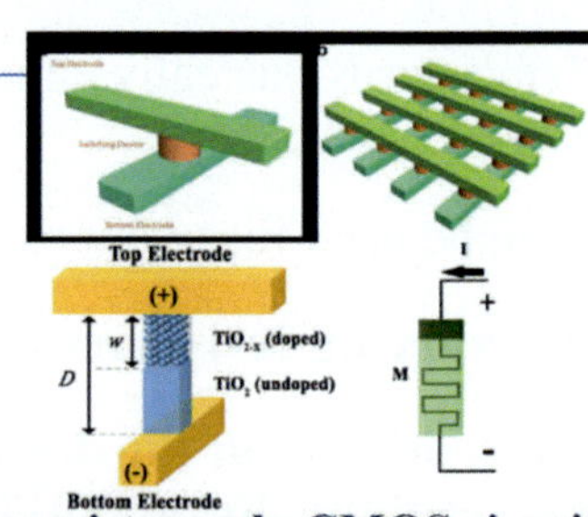

Mentions the fabrication advantages of memristors. Memristor devices are flexible to fabricate above the CMOS transistors and may be smaller than transistor-only CMOS circuits with high density and good scalability.

7 Challenges

- The read and write time for these devices can be as fast as one nanosecond.
- The switching speed can be as fast as hundreds of picoseconds.
- The dynamic power dissipation is relatively low
- On the other hand, deeply scaled CMOS logic suffers from high leakage current
- Memristors may solve many major problems in the semiconductor industry, providing nonvolatile, dense, fast, and power efficient memory and logic.

Provides an overview of challenges of the memristive devices like read and write time, switching speed, power dissipation, leakage currents, etc.

8 Memristor design families

- Several memristive logic design families have emerged in the literature, each with its own characteristics, capabilities and usage:
- Memristor Aided Logic (MAGIC)
- Material Implication (IMPLY)
- Memristor Ratioed Logic (MRL)

Introduces memristor design families Memristor Aided Logic (MAGIC), Material Implication (IMPLY) and Memristor Ratioed Logic (MRL).

9 Material Implication (IMPLY) Gates

- In IMPLY logic, the memristor states (ROFF, RON) represent the logical state variables (0, 1), respectively.

IMPLY gate

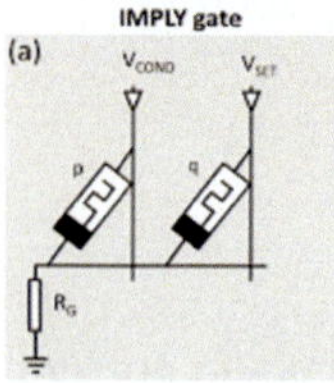

Truth table

(b)

p	q	p→q
0	0	1
0	1	1
1	0	0
1	1	1

The IMPY gate consists of the two memristors p and q and the resistor RG.

The initial memristances of p and q represent the input to the gate

- The output is written into memristor q after applying VCOND and VSET simultaneously.
- IMPLY logic function together with FALSE (a function that always yields the value zero as an output) comprises a computationally complete logic structure

Discusses Material Implication (IMPLY) gates comprising two memristors and a resistor in which (RON, ROFF) is interpreted as the state variables (1,0). The initial memristances represent input to the gate.

10 IMPLY Logic

IMPLY gate

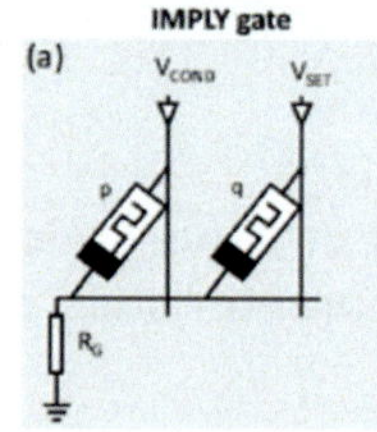

Truth table

(b)

p	q	p→q
0	0	1
0	1	1
1	0	0
1	1	1

- IMPLY gate implements the function q = p'+q and can be used as a basis for any logic function.
- The same memristors are used to store the logical state and/or perform a logical operation.
- All available designs with IMPLY logic require several time steps to accomplish the target computations.
- **This fact leads to an overhead in terms of time delay compared to other logic implementation techniques.**
  - Also, it requires complex control logic
- The IMPLY logic gate can be integrated easily inside a crossbar

Explains the IMPLY logic, in which the same two memristors are used to perform logical operations and to store logical states. Since it demands several time steps to accomplish target computations, it leads to an overhead in terms of time delay. Also, requires complex control logic.

11 Memristor Aided Logic (MAGIC)

- The logic state is represented by the resistance stored in the utilized memristors, where ROFF represents logical "0" and RON the logical "1".
- Unlike IMPLY, MAGIC uses separate memristors to store the input bits, and an additional memristor to store the output bit.

MAGIC-NOR gate

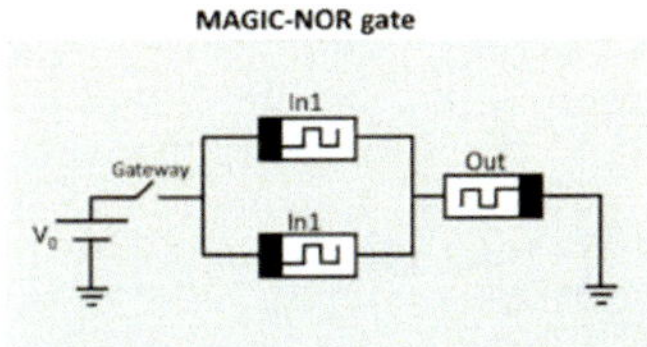

NOR MAGIC gate where in1 and in2 serve as the input memristors and out serves as the output memristor.

The initial execution step includes writing a logical 1 into the output memristor

- For evaluation we drive the gate with voltage Vo and the result of the NOR operation of in1 and in2 is written simultaneously into out.
  - ➢ It requires more steps than one to execute the computation

Explains Memristor Aided Logic (MAGIC) that uses separate memristors to store the input bits and additional memristor to store the output bit. It requires more steps than one to execute the computation.

12 MAGIC Logic

- The MAGIC cannot realize cascade connection between multiple logic gates and multiple fan-outs due to its structure.
- Using MAGIC NOR, the applications of MAGIC in memristor-based crossbars are straightforward
- MAGIC NOR can be used as the basis to perform logic computation inside the memory, thus adding processing capabilities.

Discusses that MAGIC cannot realise the cascade connection between multiple logic gates and multiple fan-outs. Application in memristor-based crossbar and as MAGIC NOR to perform logic computation inside the memory is also covered.

13 Memristor Ratioed Logic (MRL)

- MRL uses voltage signals as state variables: **compatible with standard CMOS logic.**
- **The primitive computational operators (OR, AND) are based on memristors, and CMOS inverters are added to provide a complete logic structure and signal restoration.**

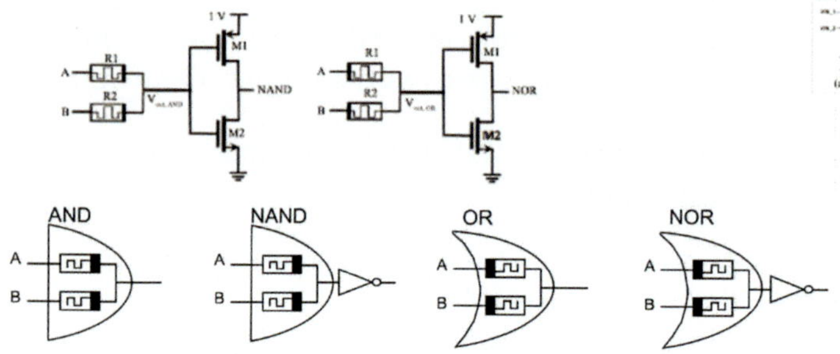

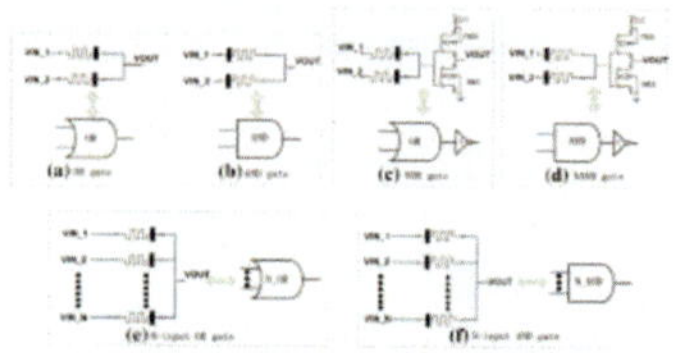

Both OR and AND gates consist of two anti-serial memristors, whereas for NOR and NAND, a CMOS inverter is added at the output.

Explores Memristor Ratioed Logic (MRL) that uses voltage signals as state variables. The OR and AND are based on memristors, and CMOS inverters are added to provide a complete logic structure and signal distortion.

14 MRL compatible with CMOS

- The memristive devices are utilized solely for logic computation and not for storing a logical state
  - ➤ The initial state only affects the computational time
- The main idea behind MRL is to replace as many transistors with memristors as possible while keeping the role of the intended digital architecture the same.
- Of the mentioned logic design styles, MRL is the only approach that uses voltage as state variable, like the conventional CMOS.
- The topology of the circuit determines the logical function, similar to CMOS.
- The computation in MRL is accomplished in a single step.
- **Thus, MRL is the most qualified for integration in current CMOS designs.**

Discusses the compatibility of MRL with CMOS in which as many transistors are replaced with memristors, keeping the intended digital architecture the same. MRL is the most qualified for integration in current CMOS designs, and computations can be accomplished in a single step.

15 MRL by using crossbar

- This integration should be performed in a way that efficiently exploits the promising characteristics of memristive devices, such as density and scalability.
- This can be achieved through the **use of the crossbar structure**, which is a highly adapted topology for arranging memristors at the top of the CMOS layer.

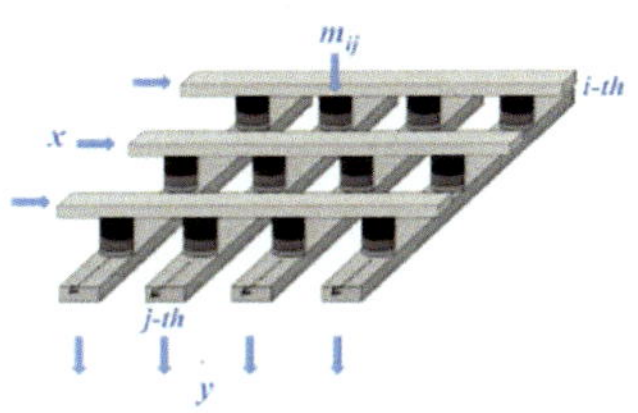

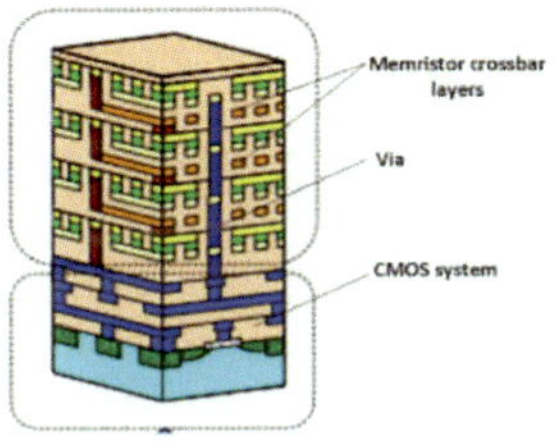

Explores the integration of MRL in the crossbar, which can be achieved by using the crossbar structure for arranging memristors at the top of the CMOS layer.

16 MRL Logic

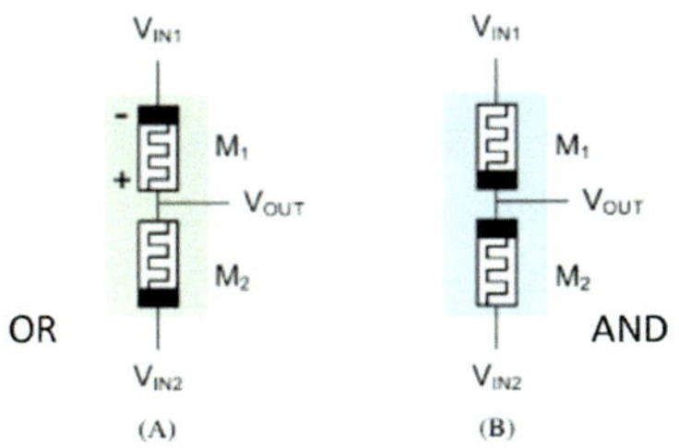

Both MRL OR and AND gates react similarly when both inputs are set either to logic "1" or "0".

- ➢ no current flows through the anti-serial memristors, leading to the transfer of the input voltage to the output.

- In the case one input is set to "0" and the other to "1", a current flows from the input with higher potential (logic "1") to the input with lower potential (logic "0").
  - ➢ Static power dissipation

Explains MRL logic in which MRL OR and AND gates react similarly when both inputs are set to logic 1 or 0. If one input is 0 and the other to 1, a current flows from the input with a higher potential(logic 1) to the input with a lower potential(logic 0).

17 MRL Logic (2)

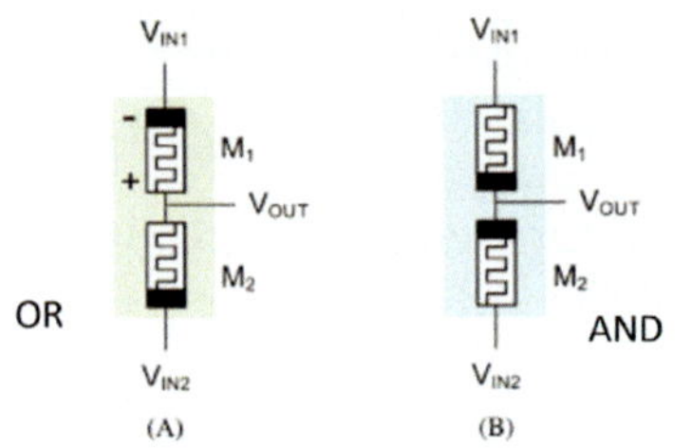

- The resulting potential difference (for inputs "1" and "0") changes the internal state of both memristors in an opposite manner.

- One memristor tends to attain the RON state while the other tends to attain the ROFF state

- The connected memristors form a **voltage divider circuit.**
- Assuming ROFF >> RON (VIN1 =1 , VIN2=0)

$$\text{Vout}, OR = \frac{Ron}{Ron+Roff}VCC \approx 0 \qquad \text{Vout}, AND = \frac{Roff}{Ron+Roff}VCC \approx VCC$$

Illustration of MRL Logic (2) wherein the connected memristors form a voltage divider circuit. Here, one memristor tends to attain the RON state, and the other tends to attain ROFF state.

18 MRL AND

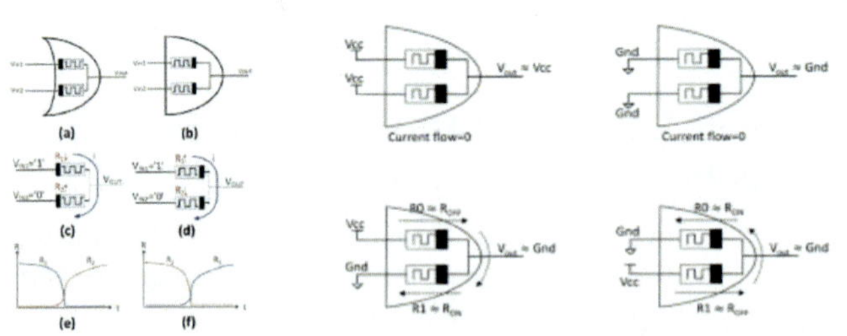

Logical operations of the MRL AND gate corresponding to all input combinations.

- The delay of the logic gates depends mostly of the time required for the memristive devices to be fully switched
- Cascading gates leads to floating outputs because of voltage degradation
- **Therefore, CMOS inverters can be used as buffers after several stages to restore the attained logical state.**

Discusses MRL AND, in which the delay of logic gates depends mainly on the time required for the memristive devices to be fully switched and cascading gates lead to floating outputs because of voltage degradation. Hence, CMOS inverters can be used as buffers after several stages to restore the attained logic state.

19 Crossbar for circuits

- A memristive crossbar is a two-dimensional grid of memristors distributed along vertical and horizontal nanowires.
- A memristor is allocated between a column and a row.
- A memristive crossbar is characterized by its simple and dense structure and could be fabricated on the top of a CMOS layer.
- **We target in the implementation of Boolean functions on a crossbar**

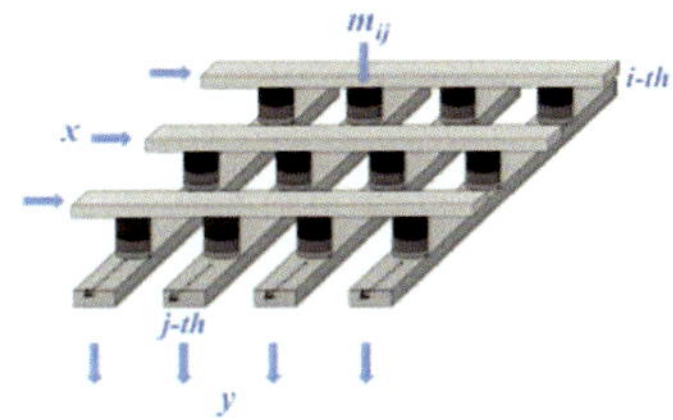

Explores the implementation of boolean functions on a memristive crossbar. Crossbars are two-dimensional grid of memristors distributed along vertical and horizontal nanowires and memristors are allocated between each row and column.

20 Implementation in a crossbar topology

- Any Boolean function can be written in the form of the sum of products.
- It can be implemented using MRL-AND and MRL-OR with the aid of CMOS inverters.

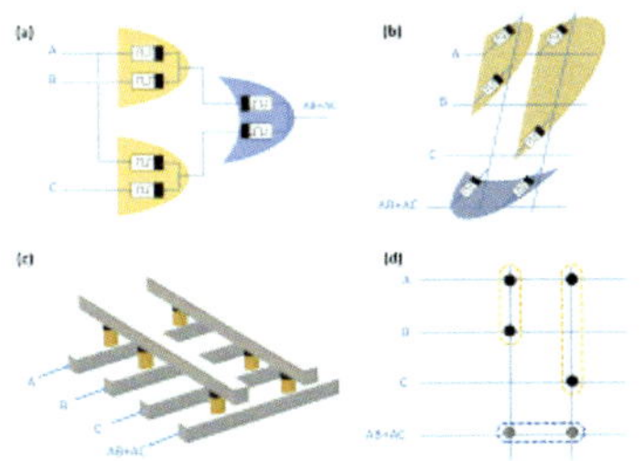

Example: Design and implementation of the simple function F = AB + AC

Although the array includes AND and OR structures the positive poles of the memristors lay on the same planar

Khaled Alhaj Ali, Mostafa Rizk, Amer Baghdadi, Jean-Philippe Diguet, Jalal Jomaah, **"Hybrid Memristor-CMOS Implementation of Combinational Logic Based on X-MRL,"** Electronics 2021, 10, 1018, https://doi.org/10.3390/electronics10091018

Illustrates the implementaiopn of boolean function in a crossbar technology which can be written in the form of sum of products. The implementation can be done using MRL-AND and MRL-OR with th eaid of CMOS inverters.

21 Design of a Full-Adder

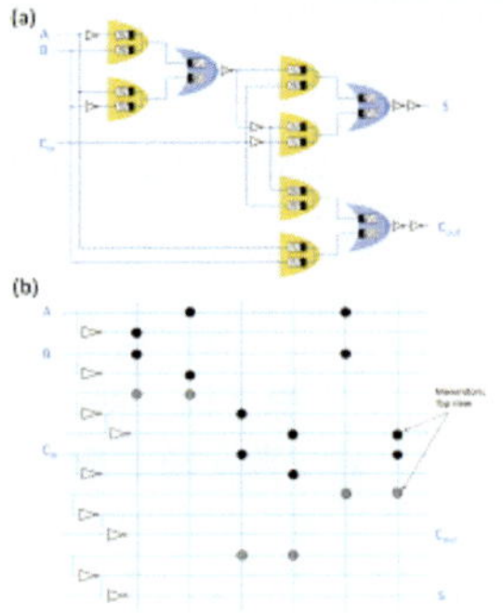

$S = A \oplus B \oplus Cin = Cin(A'B+AB')' + C'in(A'B+AB')$

$Cout = (A \oplus B)Cin + AB = AB + ACin + BCin$

The CMOS inverters are responsible for either

- ➢ inverting (NOT operation) and/or
- ➢ performing signal restoration

There are cascaded MRL gates where the current can flow from the output node into the input of the next logic gate. **This may create problems!**

Khaled Alhaj Ali, Mostafa Rizk, Amer Baghdadi, Jean-Philippe Diguet, Jalal Jomaah, "**Hybrid Memristor-CMOS Implementation of Combinational Logic Based on X-MRL,**" Electronics 2021, 10, 1018, https://doi.org/10.3390/electronics10091018

Discusses the design of a full adder in which the CMOS inverters are responsible for either inverting and/or performing signal restoration. The cascaded MRL gates can cause the current flow from the output node into the input of the next logic gate which may create problems.

22 Circuit characteristics

- The proposed array is different from conventional crossbar arrays as a certain number of crosspoints are vacant.
- In this array, all memristors are accessed simultaneously, leading to deterministic current paths.
- Thus, there are no unexpected current paths, and consequently it is **sneak-path free**.

Explains the circuit characteristics like the simultaneous access of all current paths. Since there are no unexpected current paths , crossbar will be sneak-path free.

23 Layout of full adder – Evaluation by the authors

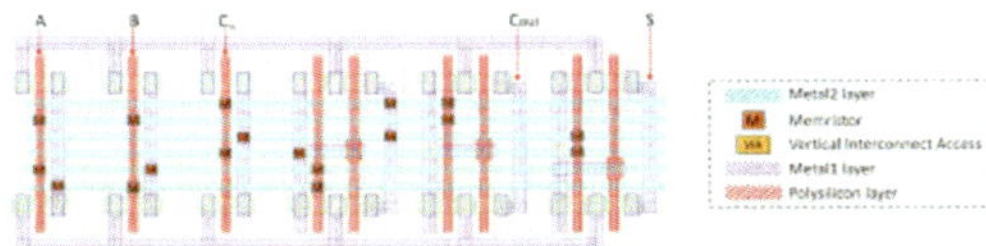

The positions of the allocated memristors have been assigned virtually

- The allocated memristors in the proposed design do not add any overhead in terms of the implementation area. **Area is defined only by the transistors**
- **The total area of the design is 8.16 μm² compared to the area of 14.78 μm² utilized in the case of a pure CMOS implementation, leading to a 44.79% area saving.**
- **One step is needed for the calculation: needed time 2.5ns, needed energy 0.69pJ**
- **The authors of the corresponding paper propose the use of hafnium oxide (HfOx) memristor which has a switching delay of 300ps at 1.4V and ratio Roff/Ron>100.**
- **Max frequency: 200 MHz**

Illustrates the layout of full adder using Hafnium Oxide memristor having a switching delay of 300ps, maximum frequency of 200 MHZ. The area is 8.16um2 defined and is defined only by the transistors. One step is needed for the calculation and the time required is 2.5ns. Energy needed is 0.69pJ.

24 Memristor based PLA structure

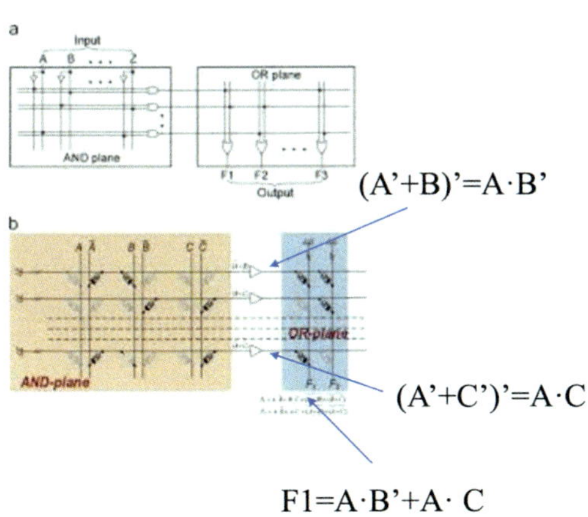

Use for Sum of Product Boolean functions

Programming means the existence of memristors at the appropriate cross-points

Kyoungrok Cho, Sang-Jin Lee, Kamran Eshraghian, "Memristor-CMOS logic and digital computational components," Microelectronics Journal, 46, pp. 214-220, 2015

Discusses the alternative approach in MRL-CMOS implementation. In the memristor based PLA structure, programming means the existence of memristors at the appropriate cross-points.

25 New basic logic circuits based on MRL (Modified MRL)

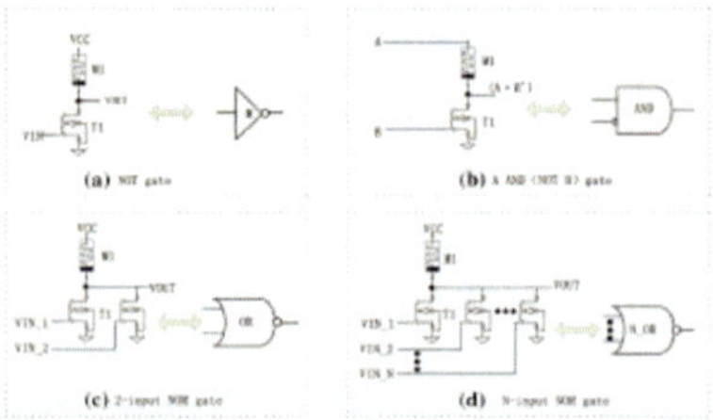

Pseudo nMOS logic

Memristor is biased forward and R_M=Ron

For the transistor when conducts $R_T \approx 0$ otherwise R_T>>Ron

- For the inverter: When Vin=0 , Vout= $\frac{RT}{Ron+RT} VCC \approx 1$ When Vin=1 , Vout= $\frac{RT}{Ron+RT} VCC \approx 0$
- New logic gate: when the DC bias voltage *VCC* is used as the input signal A, we can obtain the logic **F = *A* ·*B'***
- However, there is static power dissipation

Explores the modified MRL in memristor is biased forward and $R_M = R_{ON}$. There will be a static power dissipation.

26 Memristive XOR functionality

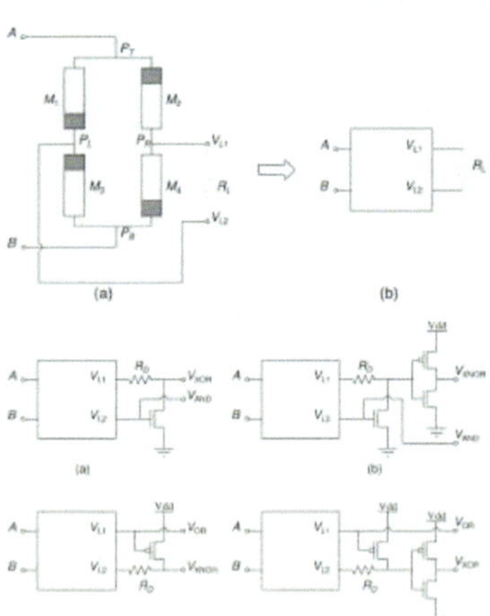

$V_{L1} = A + B, \quad V_{L2} = A \cdot B, \quad V_{L1} \geq V_{L2}$

If we interpret the current flowing from V_{L1} to V_{L2} as logical 1, and the absence of this current as logical 0, then clearly this is a XOR functionality

Memristive XOR and XNOR gates with dual functionality

The use of resistor may be a problem (large area)

An inverter buffer can be added at the output to restore logic levels

Xiaohan Yang, Adedotum Adeyemo, Anu Bala, Abusaleh jabir, "**Novel Techniques for Memristive Multifunction Logic Design**," Integration, Vol. 65, pp.219-230, March 2019

Explains the memristive XOR functionality. An efficient XOR can lead to an efficient implementation of many circuit structures.

27 A multifunctional logic gate (XOR, AND)

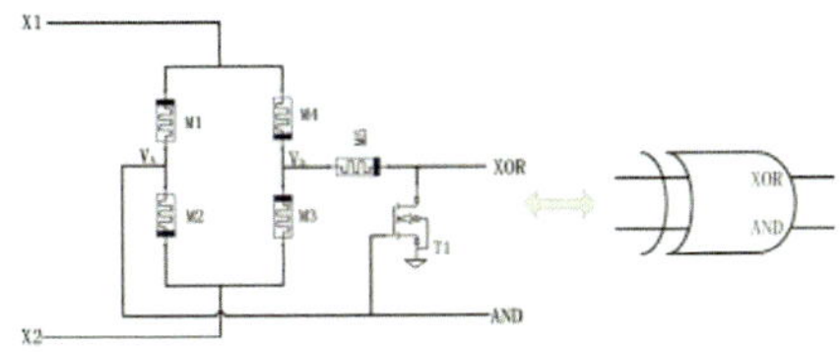

$V_A = X_1 \cdot X_2$, $V_B = X_1 + X_2$

$$Vxor = V_A' \cdot V_B = (X_1 \cdot X_2)' \cdot (X_1 + X_2) = X_1 \oplus X_2$$

This gate can be used as basic logic circuit to design some combinational circuits.

Gongzhi Liu, Shuhang Shen, Peipei Jin, Guangyi Wang, Yan Liang, **"Design of Memristor-Based Combinational Logic Circuits,"** Circuits, Systems and Signal Processing, 40:5825-5846, June 2021, DOI:10.1007/s00034-021-01770-1

Explains a multifunctional logic gate (XOR, AND). The gate can be used as basic logic circuit to design some combinational circuits. In the structure, dividing cascading memristors will be a problem. It depends on the supply voltage levels.

28 Memristor based combinational logic circuits

- Conventional MRL is based on logic defined by two memristors and two transistors
- **Modified MRL is based on logic defined by one memristor and one transistor**
- The output signal of the new MRL is obtained not by dividing the input voltage, but by the saturation voltage and cut-off voltage of the NMOS transistor.
- **Therefore, the modified MRL can achieve the accurate output of the signal without adding the inverter at the output to boost the signal.**
- **Significant decrease of the required transistors**

Presents memristor-based combinational logic circuits. Modified MRL is based on the logic defined by one memristor and one transistor. Hence it achieves the accurate output of the signal without adding the inverter at the output to boost the signal. The number of required transistors are significantly decreased.

29 Memristor based binary encoder (3-bit)

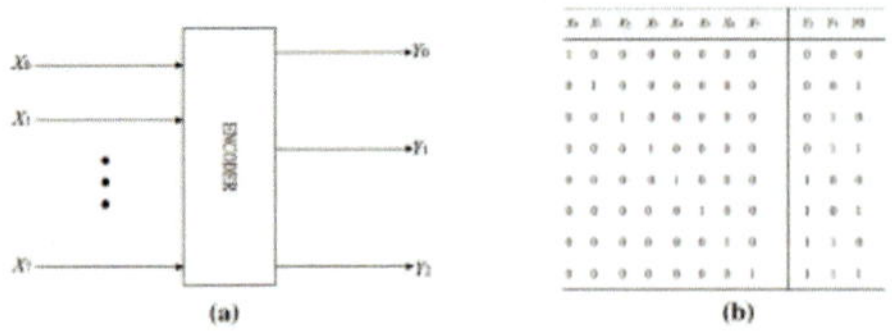

(a) (b)

$$Y_2 = X_4 + X_5 + X_6 + X_7$$

$$Y_1 = X_2 + X_3 + X_6 + X_7$$

$$Y_0 = X_1 + X_3 + X_5 + X_7$$

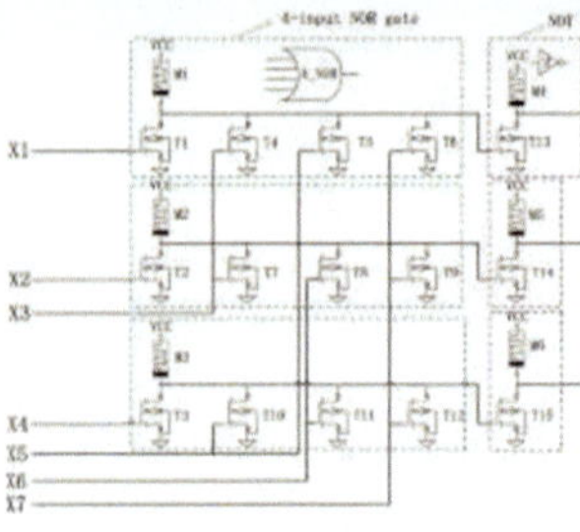

Use of 15 nMOS and 6 memristors while the traditional CMOS requires 30 transistors

Gongzhi Liu, Shuhang Shen, Peipei Jin, Guangyi Wang, Yan Liang, **"Design of Memristor-Based Combinational Logic Circuits,"** Circuits, Systems and Signal Processing, 40:5825-5846, June 2021, DOI:10.1007/s00034-021-01770-1

Shows the benefit of using modified MRL. The memristor based binary 3-bit encoder uses 15 NMOS transistors and 6 memristors while the traditional CMOS requires 30 transistors.

30 Memristor based binary decoder (3-bit)

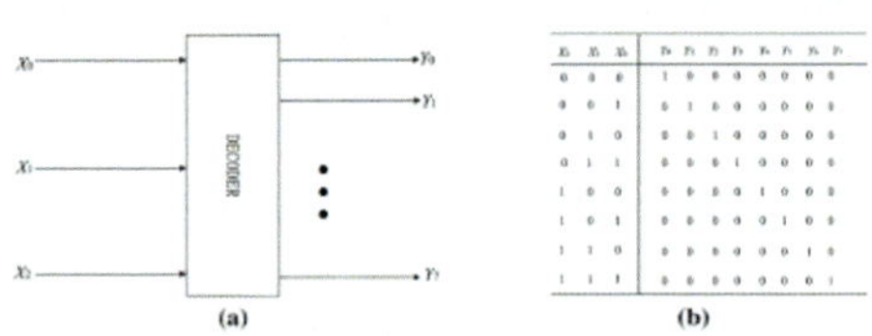

(a) (b)

$Y_0 = X_2'X_1'X_0'$ $Y_1 = X_2'X_1'X_0$ $Y_2 = X_2'X_1X_0'$

$Y_3 = X_2'X_1X_0$ $Y_4 = X_2X_1'X_0'$ $Y_5 = X_2X_1'X_0$

$Y_6 = X_2X_1X_0'$ $Y_7 = X_2X_1X_0$

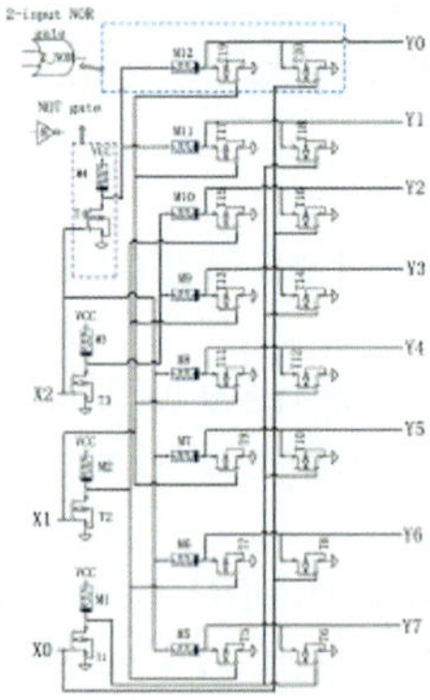

$$Y_0 = X_2'(X_1 + X_0)' = X_2'X_1'X_0'$$

$$Y_1 = X_2'(X_1 + X_0')' = X_2'X_1'X_0$$

$$Y_2 = X_2'(X_1' + X_0)' = X_2'X_1X_0'$$

<<

Use of 20nMOS and 12 memristors while the traditional CMOS requires 54 transistors

Gongzhi Liu, Shuhang Shen, Peipei Jin, Guangyi Wang, Yan Liang, **"Design of Memristor-Based Combinational Logic Circuits,"** Circuits, Systems and Signal Processing, 40:5825-5846, June 2021, DOI:10.1007/s00034-021-01770-1

Explains the benefit of using modified MRL. The memristor based binary 3-bit decoder uses 20 NMOS transistors and 12 memristors while the traditional CMOS requires 54 transistors.

31 Memristor-based multiplexer

- 4x1 multiplexer

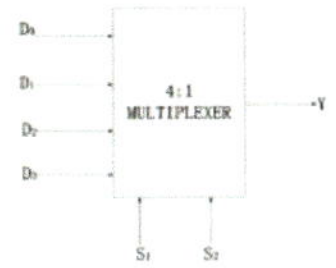

(a)

INPUT		OUTPUT
S1	S2	Y
0	0	D0
0	1	D1
1	0	D2
1	1	D3

(b)

$$D_3(S_1' + S_2')' = D_3S_2S_1$$

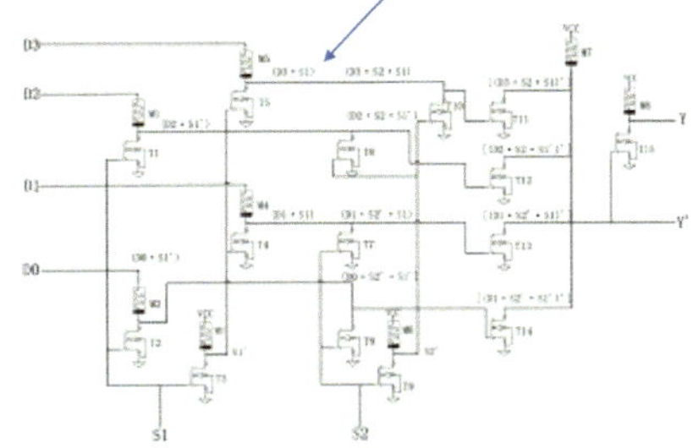

$$Y = D_0S_2'S_1' + D_1S_2'S_1 + D_2S_2S_1' + D_3S_2S_1$$

- Use of 15 nMOS transistors and 8 memristors while the traditional CMOS requires 36 transistors

Gongzhi Liu, Shuhang Shen, Peipei Jin, Guangyi Wang, Yan Liang, "**Design of Memristor-Based Combinational Logic Circuits,**" Circuits, Systems and Signal Processing, 40:5825-5846, June 2021, DOI:10.1007/s00034-021-01770-1

Explains the benefit of modified MRL. The memristor based 4x1 multiplexer uses 15 NMOS transistors and 8 memristors while the traditional CMOS requires 36 transistors.

32 Memristor-based full adder

- Full adder with 6 nMOS transistors and 16 memristors
- Use of the multifunctional gate

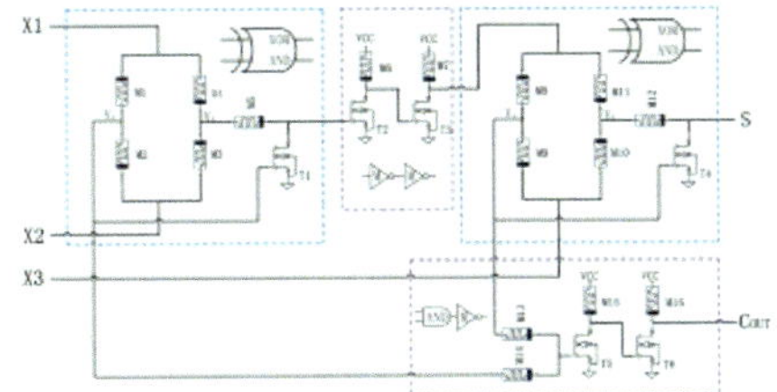

S=X1⊕X2⊕X3

Cout=(X1·X2)+X3·(X1⊕X2)

- The previously presented full adder uses 18 transistors and 18 memristors
- Conventional CMOS design requires 28 transistors

Gongzhi Liu, Shuhang Shen, Peipei Jin, Guangyi Wang, Yan Liang, "**Design of Memristor-Based Combinational Logic Circuits,**" Circuits, Systems and Signal Processing, 40:5825-5846, June 2021, DOI:10.1007/s00034-021-01770-1

Presents memristor-based full adder with 6 NMOS transistors and 16 memristors where conventional CMOS design requires 28 transistors.

33 Memristor based numerical comparator

- 1-bit numerical comparator – the basis for all numerical comparators

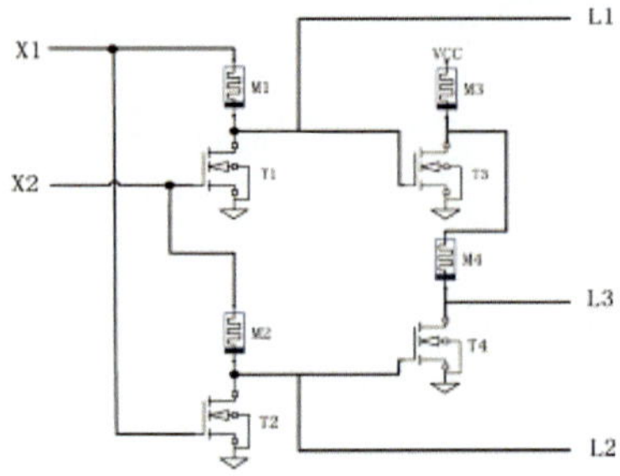

$$L_1 = X_1X_2' \quad (X1 > X2)$$

$$L_3 = (X_1'X_2 + X_1X_2')' \quad (X1 = X2)$$

$$L_2 = X_1'X_2 \quad (X1 < X2)$$

- Use of 4 NMOS transistor and 4 memristors while the traditional CMOS requires 16 transistors
- **Using general memristor models, all the presented circuits have been simulated (by the authors of the corresponding references) and their correct functionality has been verified.**

Shows memristor based numerical comparator using 4 NMOS transistors and 4 memristors where the traditional CMOS requires 16 transistors.

34 Other multifunctional logic gates (XOR, AND, OR)

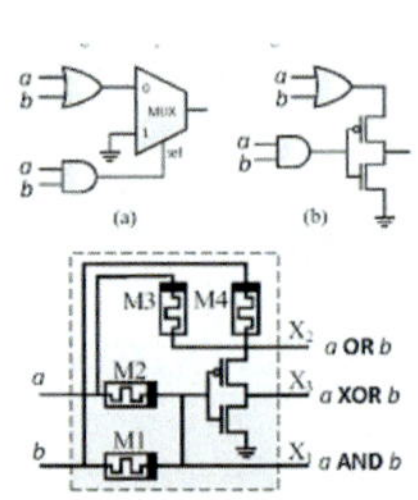

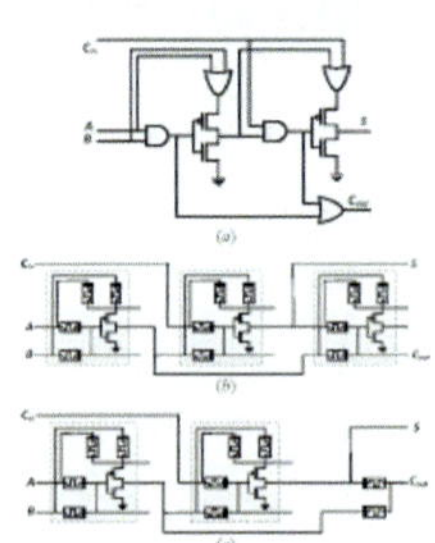

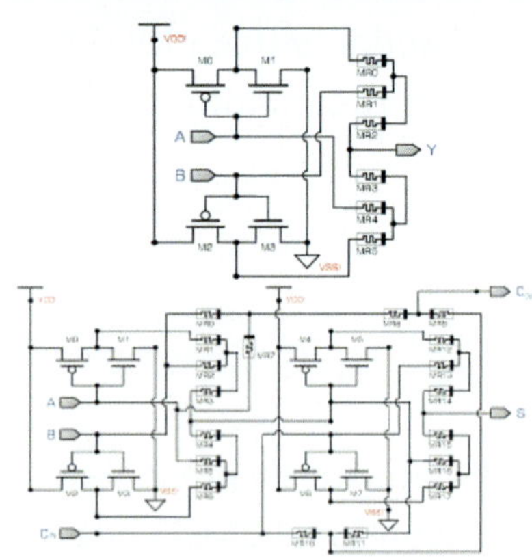

Mehri Teimoori, Arash Ahmadi, Shahpour Alirezaee, Majid Ahmadi, "**A Novel Hybrid CMOS-Memristor Logic Circuit Using Memristor Ratioed Logic**," IEEE Canadian Conference on Electrical and Computer Engineering, 2016.

Tejinder Singh, "**Hybrid Memristor-CMOS (MeMOS) based Logic Gates and Adder Circuits**", https://doi.org/10.48550/arXiv.1506.06735

Presents the alternative structures of the multifunctional logic gate have been proposed which can be used as promites in the design of more complicated circuit structures.

35 Conclusions

- With the appropriate memristive device a hybrid combination of the MRL and CMOS seems to provide substantial benefits compared to the pure CMOS logic
- A memristive device with no or low threshold with its resistance to increase or decrease to a value depending on the current direction is preferred.
- Simulation results based on appropriate values for the memristor model parameters demonstrates that MRL circuits can be effectively used to design power efficient low complexity systems
- **However, real implementations need to make clear possible benefits of using this hybrid MRL – CMOS approach**

Lists the overall summary of the presentation. Concludes that the real implementations need to make clear possible benefits of using the hybrid MRL-CMOS approach.

CHAPTER 08

Neuromorphic Computing with Memristors: From Circuits to Algorithms

Cory Merkel

Rochester Institute of Technology

The presentation delves into the intricate landscape of machine learning, artificial intelligence (A.I.), and data mining, with a focus on the advancements and challenges witnessed in 2021. It begins by highlighting the resource-intensive nature of deep learning and the ongoing quest to bridge the gap between A.I. systems and the human brain's remarkable capabilities. This leads to explorations in neuromorphic devices, brain-inspired architectures, and trustworthy algorithms, aiming to enhance A.I.'s efficiency and reliability. The agenda encompasses topics ranging from memristors and memristor synapses to training memristor-based neuromorphic systems and addressing adversarial robustness in memristor crossbar circuits. Detailed explanations cover the mathematical underpinnings of memristors, the implementation of artificial neural networks, and the intricacies of current-mode neuron designs. Additionally, the presentation sheds light on stochastic representation, gradient descent learning, and the challenges posed by adversarial attacks in neural networks. By leveraging insights from neuroscience, materials engineering, and data science, the presentation underscores the interdisciplinary efforts required to advance A.I. technologies and mitigate security risks. Through discussions on energy-aware neural networks, model robustness, and decision region analysis, the presentation offers a comprehensive exploration of the complexities and potential avenues for innovation in the field of A.I.

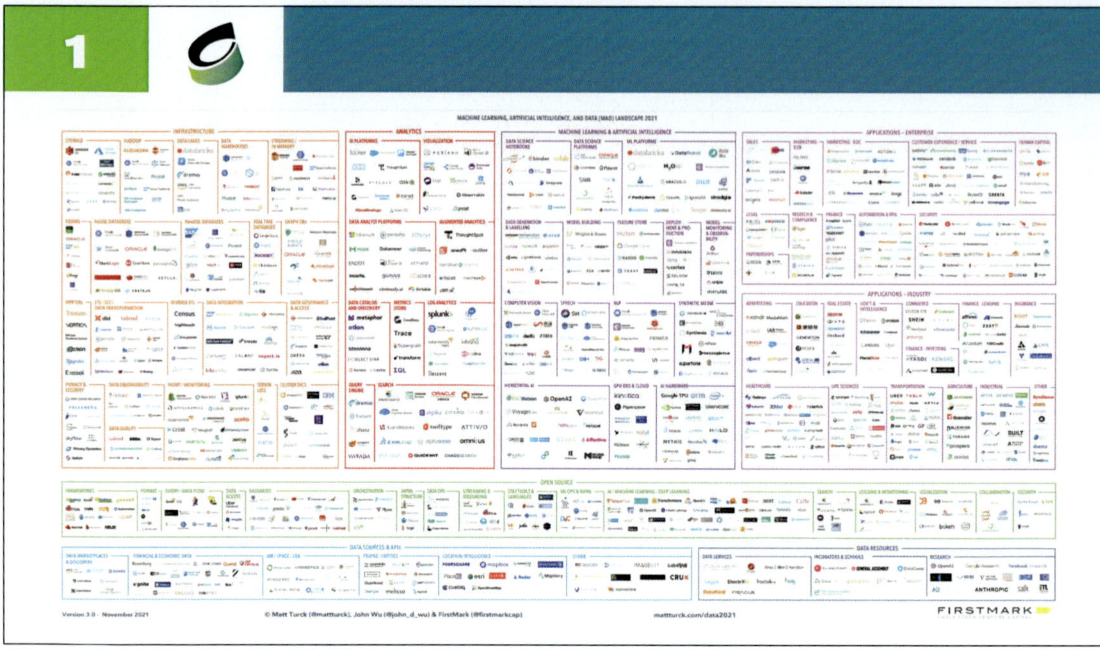

Machine learning, artificial intelligence and data mining Landscape 2021.

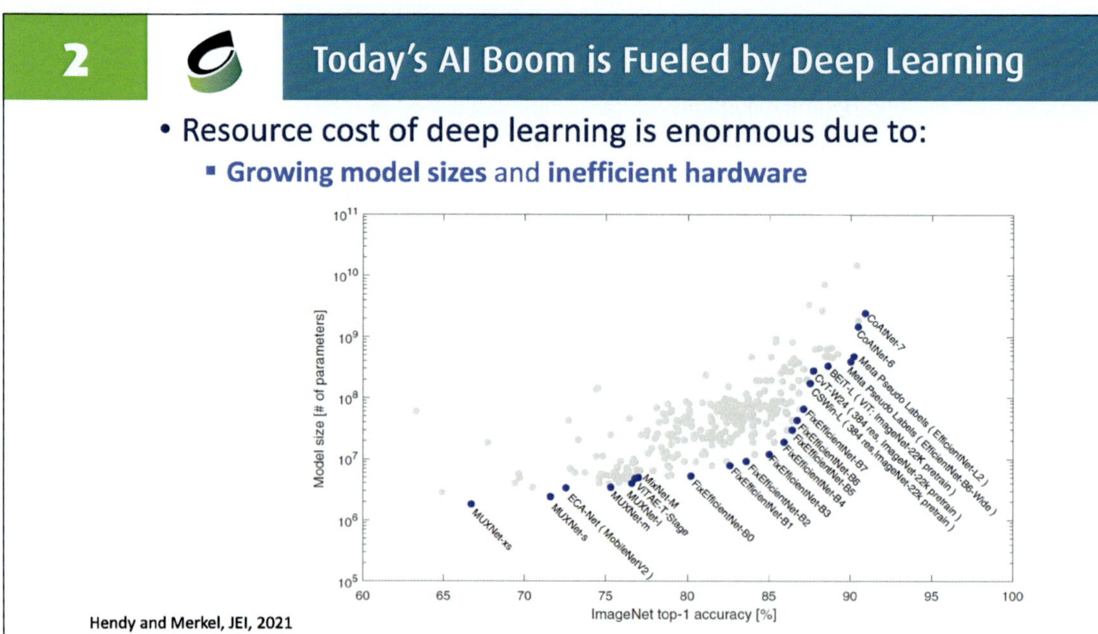

Today's A.I. Boom is Fueled by Deep Learning. The resource cost of deep learning is enormous due to growing model sizes and inefficient hardware.

3 Deep Learning vs. Biology

- Human brain still vastly outperforms AI even in simple tasks like image classification

	Computer[1]	Brain
Signals	Digital	Mixed signal
Precision	High	Low
Parallelism	Low	Very high
Info Density	Low	High
CPU-Memory	Separated	Coupled
Mutability	Constant	Plastic
Size	204 m²	2.17 liters
Weight	6 tons	3 lbs
Power	87 kW	10-20 W

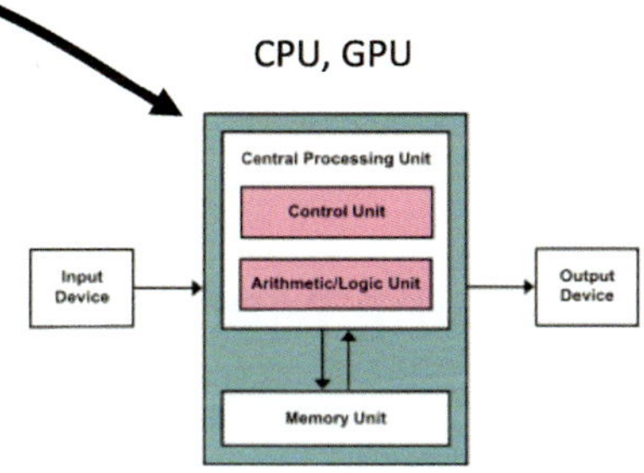

- What principles can we borrow from the brain to make AI better?

[1]Based on Google's 16,000 core machine (Le et al., 2012)

The human brain still vastly outperforms A.I. even in simple tasks like image classification. What principles can we borrow from the brain to make A.I. better?

4 Brain Lab

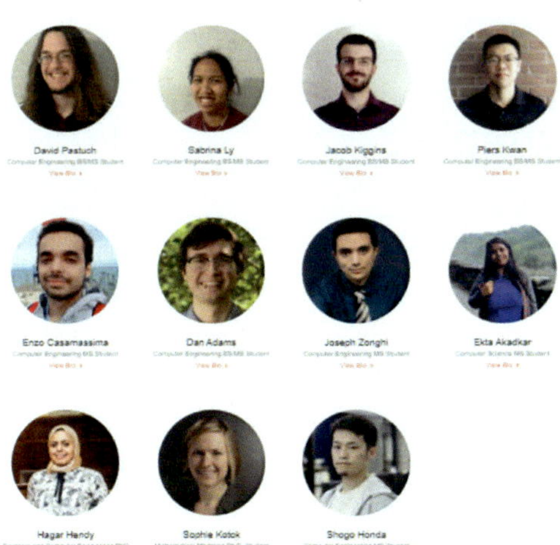

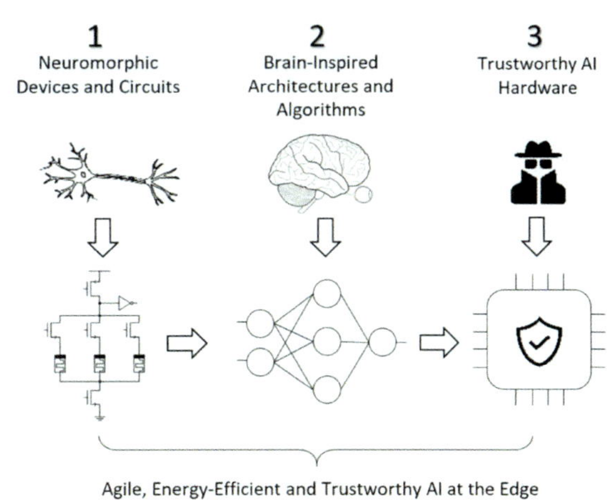

Neuromorphic devices and circuits, brain-inspired architecture and algorithms, Trustworthy algorithms.

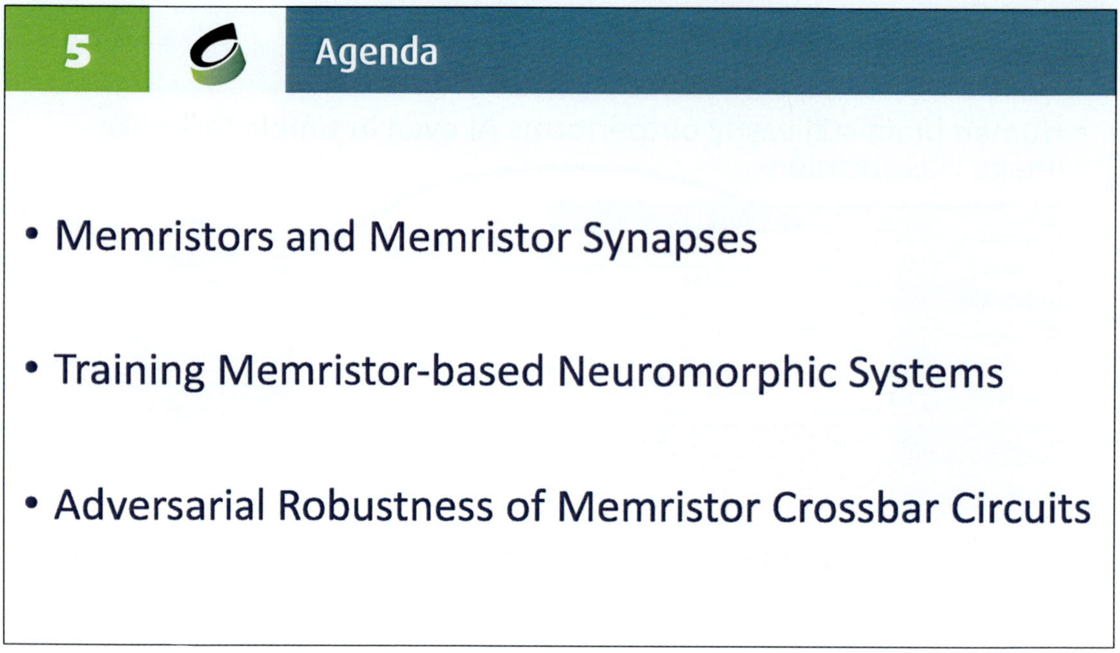

The Agenda for the presentation are Memristors and Memristor Synapses, Training Memristor-based Neuromorphic Systems and Adversarial Robustness of Memristor Crossbar Circuits.

Agenda - Memristors and Memristor Synapses.

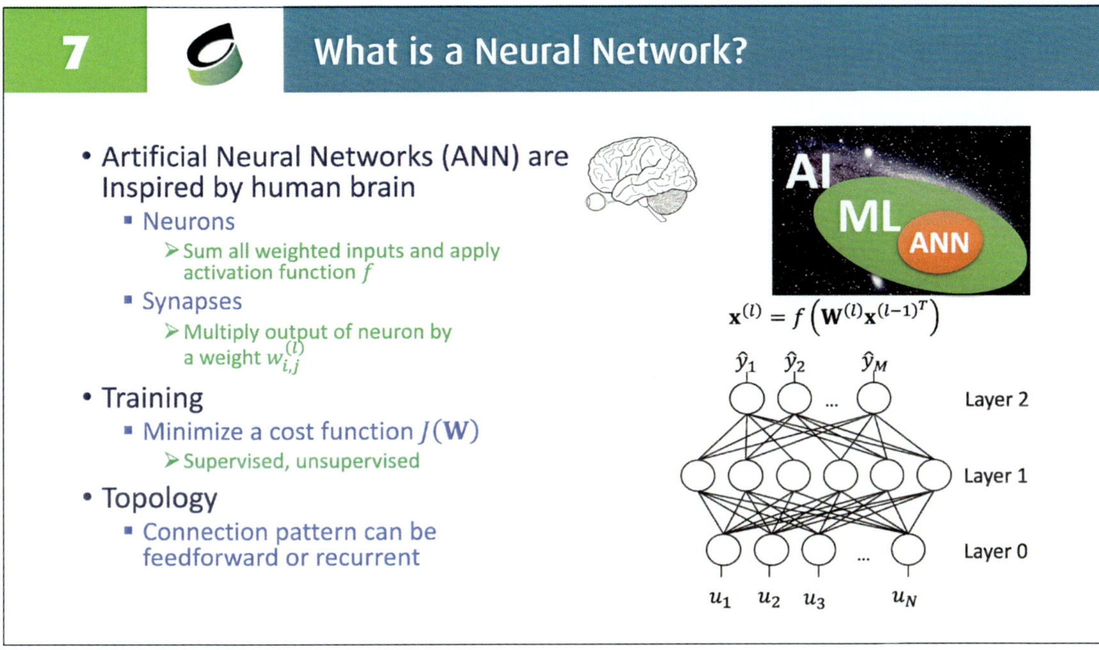

Artificial Neural Networks (ANN) is inspired by the human brain neurons and synapses.

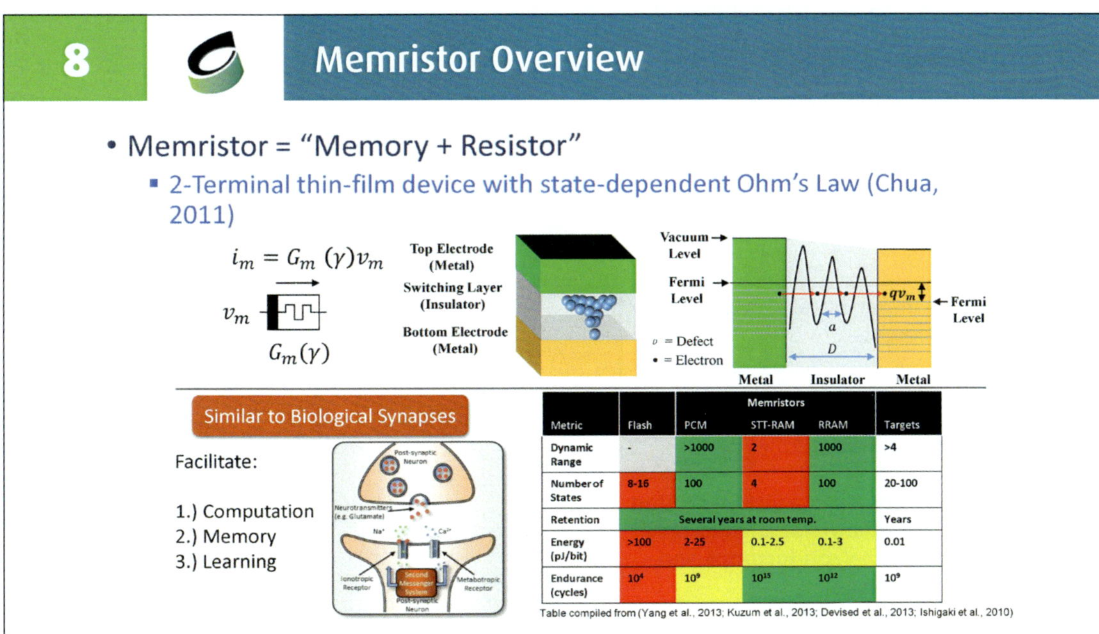

Metric	Flash	Memristors PCM	STT-RAM	RRAM	Targets
Dynamic Range	-	>1000	2	1000	>4
Number of States	8-16	100	4	100	20-100
Retention	Several years at room temp.				Years
Energy (pJ/bit)	>100	2-25	0.1-2.5	0.1-3	0.01
Endurance (cycles)	10^4	10^9	10^{15}	10^{12}	10^9

Table compiled from (Yang et al., 2013; Kuzum et al., 2013; Devised et al., 2013; Ishigaki et al., 2010)

Overview of Memristor, Memristor = "Memory + Resistor".

9 Semi-Empirical Memristor Model

- **Linear ion drift model:**

$$\frac{dx}{dt} = \mu_I \frac{R_{on}}{D^2} i(t) = \frac{\frac{\mu_I}{D^2} v(t)}{\sqrt{r^2 + 2(r-1)\left(B - \frac{\mu_I}{D^2}\phi(t)\right)}}$$

$$B \equiv \frac{x_0^2(r-1)}{2} - r x_0$$

$$r \equiv R_{off}/R_{on}$$

$$i(t) = \frac{v(t)}{R_{on}\sqrt{r^2 + 2(r-1)\left(B - \frac{\mu_I}{D^2}\phi(t)\right)}} = W(\phi)v(t)$$

- **Home-brewed semi-empirical model:**

$$\frac{dx}{dt} = \begin{cases} \xi_4^+ \sinh\left(\xi_5^+ v(t) - \xi_6^+ V_{tp}\right) f_{win}(x), & v > V_{tp} \\ \xi_4^- \sinh\left(\xi_5^- v(t) - \xi_6^- V_{tp}\right) f_{win}(x), & v < V_{tn} \\ 0, & \text{otherwise} \end{cases}$$

$$i = \begin{cases} xG_{on}v + (1-x)G_{off}\xi_1^+ \sinh\left(\frac{v}{\xi_1^+}\right) & v \geq 0 \\ xG_{on}v + (1-x)G_{off}\xi_1^- \sinh\left(\frac{v}{\xi_1^-}\right) & v < 0 \end{cases}$$

These models are easy to implement in SPICE/Verilog-AMS

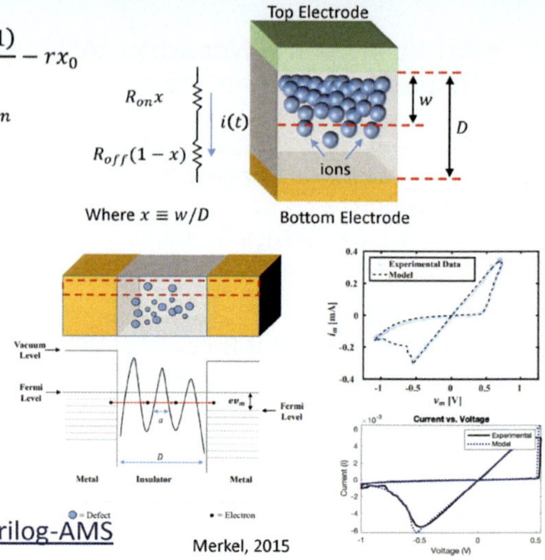

Merkel, 2015

Explains the math behind the Semi-Empirical Memristor Model.

10 Memristor Synapse Circuits

- A single memristor can be used as a synapse
- Multiple memristors usually used for bipolar weights
- Example: Current-mode design:

$$s_{ij} = \left(2\frac{G_{m1j}}{G_{m1j} + G_{m2j}} - 1\right)x_j = w_{ij}x_j$$

- **Many designs need a virtual ground so that the memristor current only depends on the conductance and the input voltage**

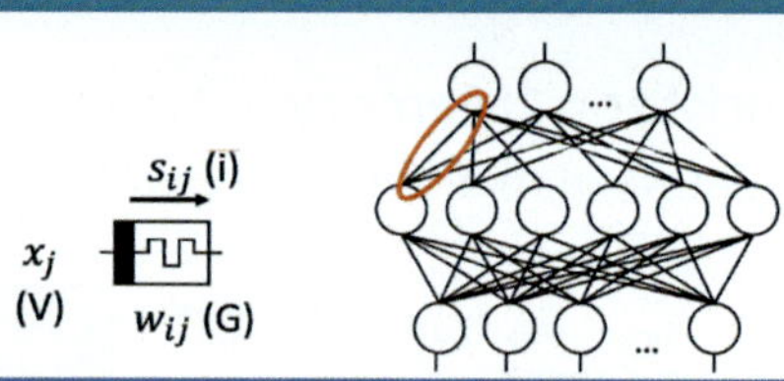

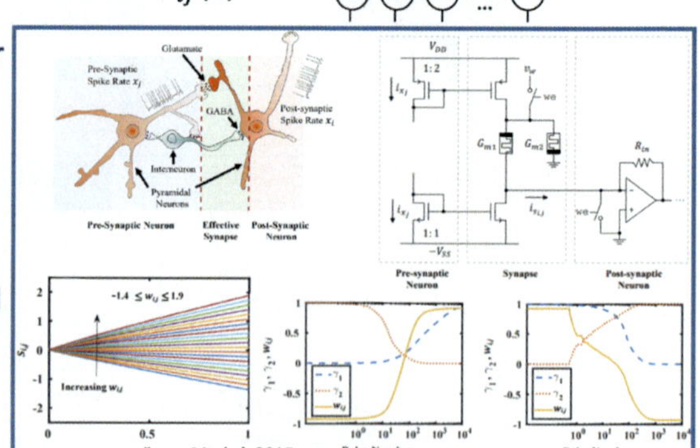

Merkel, 2015

Memristors mimic synapses in neural network designs due to their ability to mimic synaptic behaviour. Bipolar weights in these networks often employ multiple memristors; implementing a virtual ground ensures that the memristor current is solely influenced by its conductance and the input voltage, facilitating accurate weight adjustments.

11 Amplifier Gain Requirements

- If op amps are used, then they need fairly high gain (depending on how much linearity is assumed in dot product operation):

$$s_{ij} = \left(2\frac{G_{m1j}}{G_{m1j} + G_{m2j}} - 1\right)x_j = w_{ij}x_j$$

$$w_{ij} = \left(\frac{2G_{m1j}}{G_{m1j} + G_{m2j}}\right)\left(\frac{1 + A_0 + \frac{G_{m2j}R_{in}i_{in}^*}{2i_{inj}}}{1 + A_0 + \sum_k \frac{G_{m1k}G_{m2k}R_{in}}{G_{m1k} + G_{m2k}}}\right) - 1$$

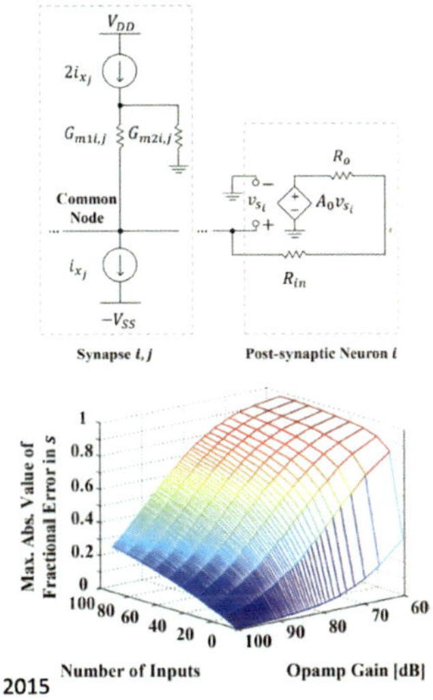

Merkel, 2015

Amplifier Gain Requirements: If op amps are used, they need relatively high gain.

12 Memristor Crossbars

- Memristor crossbars compute the matrix-vector product: $\mathbf{s} = \mathbf{W}\mathbf{x}$

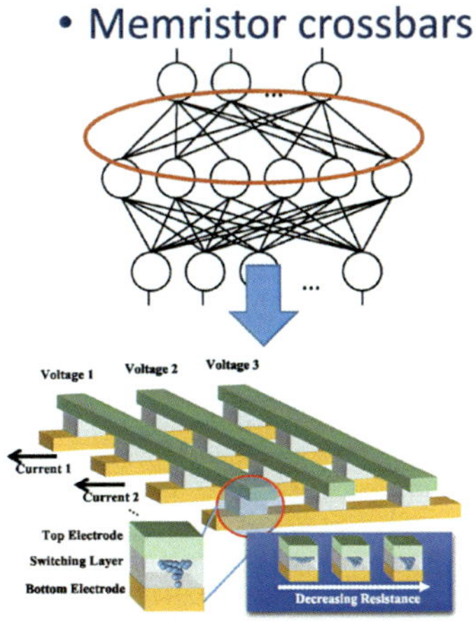

Merkel, *Nature Electronics News and Views*, 2018.

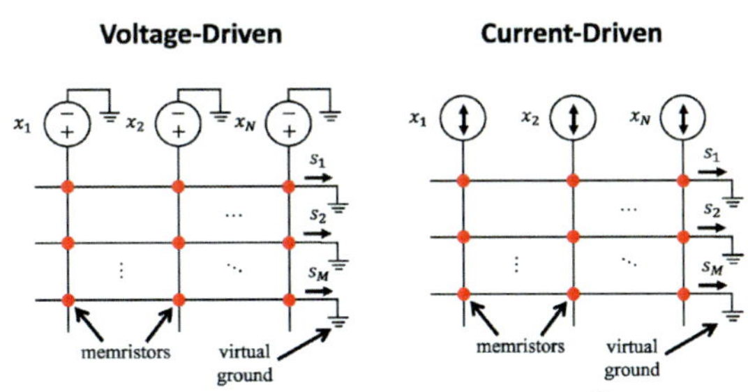

Voltage-Driven: $w_{i,j} = G_{i,j}$

Current-Driven: $w_{i,j} = \frac{G_{i,j}}{\sum_{k=1}^{M} G_{k,j}}$

Merkel, "Current-mode memristor crossbars for neuromorphic computing," *NICE*, 2019.

- Current-driven crossbars can be used in subthreshold current-mode designs

Explains the working of memristor crossbars.

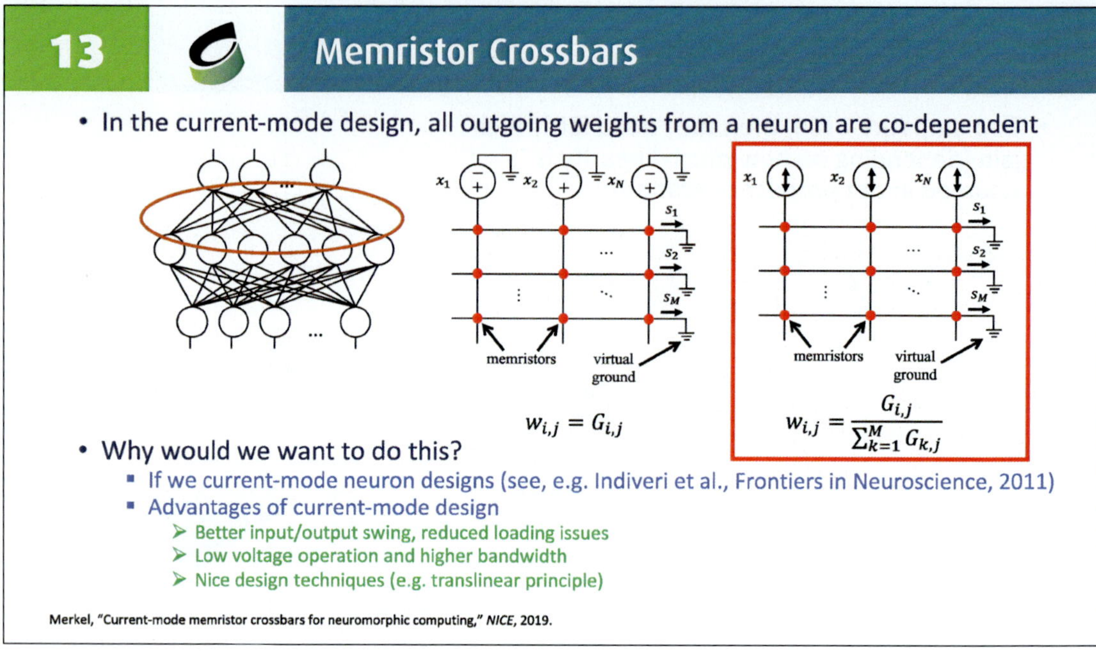

Current-mode neuron designs implement interdependent outgoing weights to achieve benefits such as improved input/output swing, mitigated loading issues, low voltage operation, increased bandwidth, and the facilitation of elegant design techniques like the translinear principle.

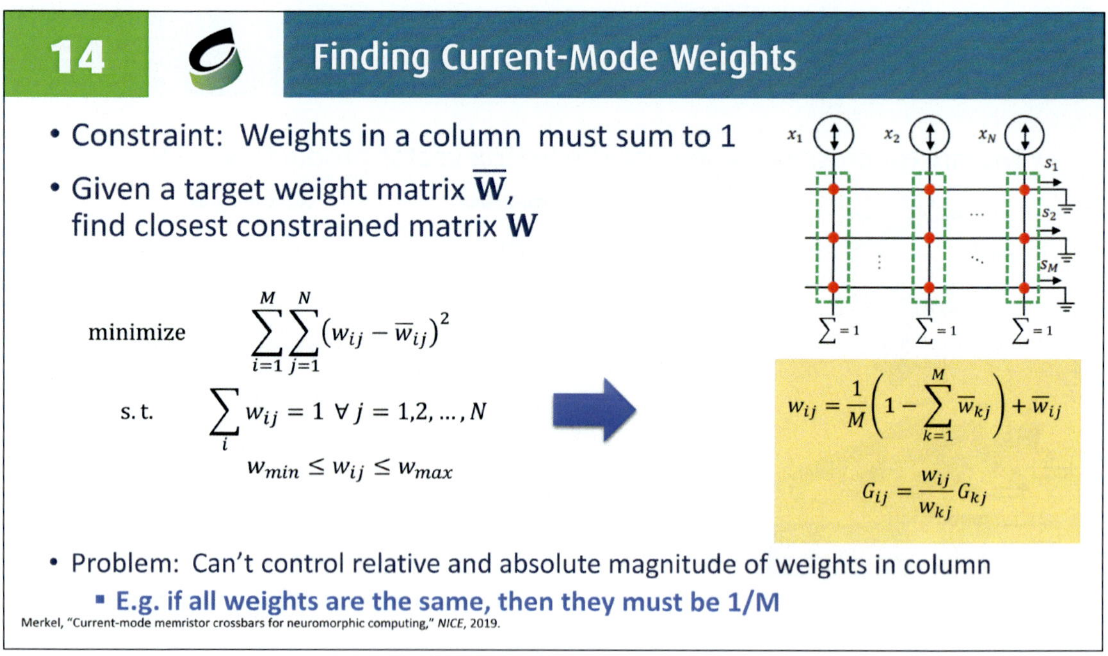

Explains how to find Current-Mode Weights.

15 A Better Way

- Dummy row allows magnitude of entire column to be adjusted
- Now,
  - Constrain voltage-mode weights as:

$$\max\left(\frac{1-w_{max}}{M-1}, w_{min}\right) \leq \overline{w}_{ij} \leq \min\left(\frac{1-w_{min}}{M-1}, w_{max}\right)$$

  - Set current-mode weights as

$$w_{ij} = \overline{w}_{ij} \ \forall\, i < M$$

$$w_{Mj} = 1 - \sum_{k=1}^{M} \overline{w}_{kj}$$

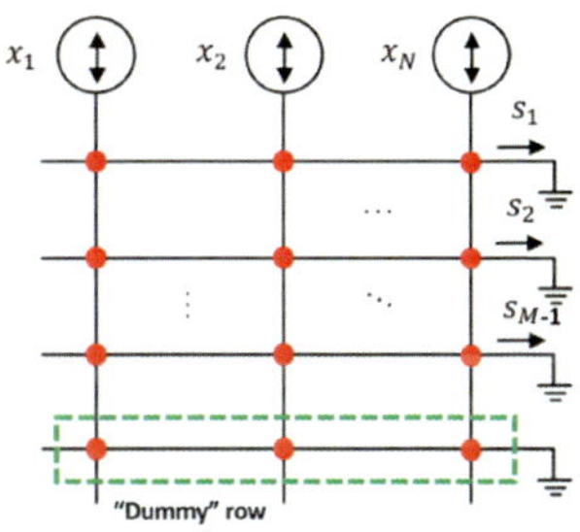

Merkel, "Current-mode memristor crossbars for neuromorphic computing," *NICE*, 2019.

Explains a better way to use dummy rows to adjust the magnitude of the entire column.

16 Current-mode Crossbar Training

- **We can train offline using the proposed method, but online training is more desirable**
- **Gradient descent can be implemented as**

$$\Delta G_{ij}^{l} = -\alpha \frac{\partial J}{\partial G_{ij}^{l}} = \alpha \sum_{k=1}^{M^l} \delta_k^l x_j^{l-1} \frac{\partial w_{kj}^{l}}{\partial G_{ij}^{l}}$$

$$\approx \begin{cases} \alpha \delta_i^l x_j^l & i \neq M \\ -\langle \Delta G_{1:M-1,j}^{l} \rangle & i = M \end{cases}$$

MLP trained on MNIST

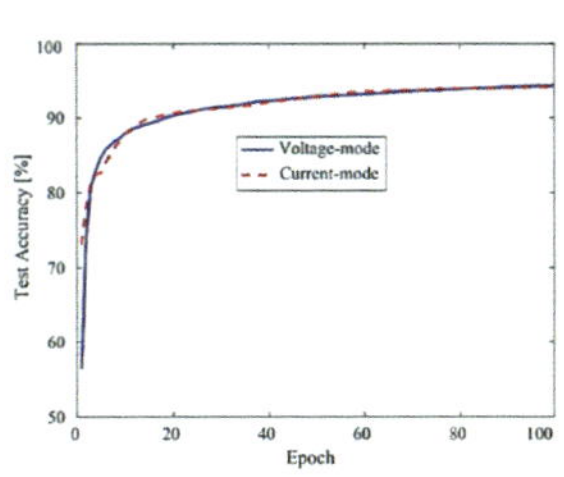

Just use delta rule for non-dummy rows and average the updates for dummy row

Merkel, "Current-mode memristor crossbars for neuromorphic computing," *NICE*, 2019.

While offline Current-mode Crossbar Training is feasible, online training is preferred.

17 Inspiration from Spiking Neurons

- Memristor crossbars consume constant power
- Can we borrow ideas from real biological neurons?
  - Active ion pumps setup concentration gradients

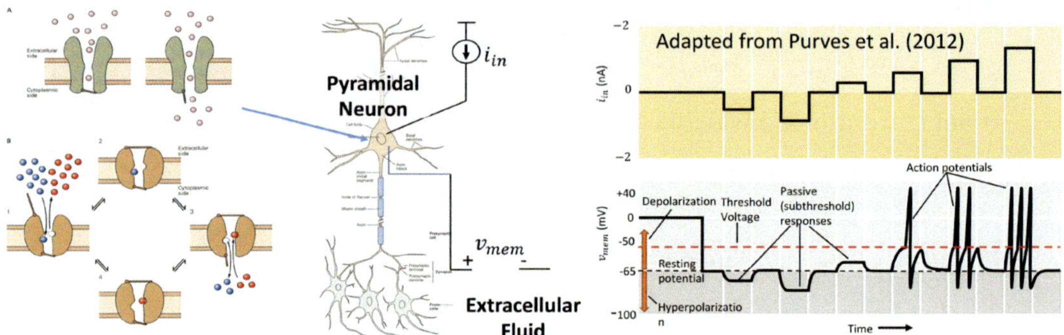

Drawing inspiration from Active ion pumps set concentration gradients in real biological neurons.

18 Domino Logic Neuron

- Encodes dot product $\mathbf{w}_i^l \mathbf{x}^{l-1}$ with pulse width of v_x^l

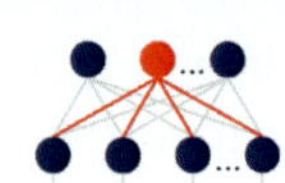

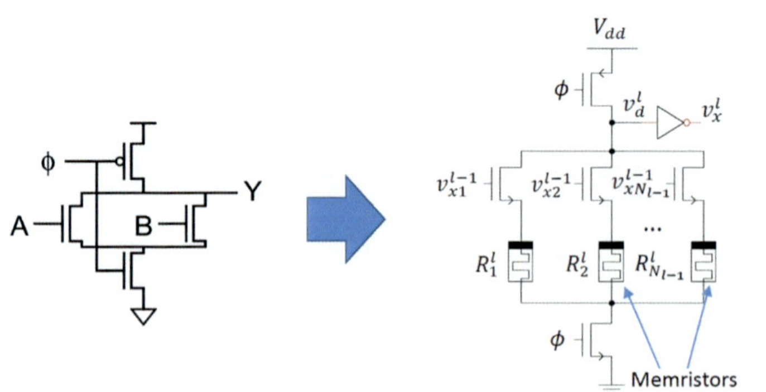

$$v_d^l(t) = V_{dd} \exp\left(-\int_0^t \frac{G_i^l(\tau)}{C} \mathrm{d}\tau\right)$$

$$G_i^l(t) = \frac{1}{V_{dd}} \sum_{j=1}^{N_{l-1}} v_{xj}^{l-1}(t) G_{ij}^l$$

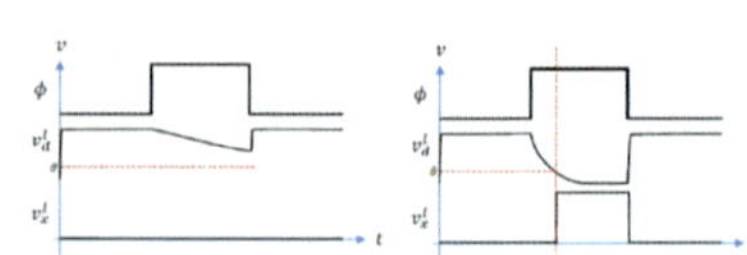

Merkel and Nikam, *ICONS*, 2020.

Explains Domino Logic Neuron.

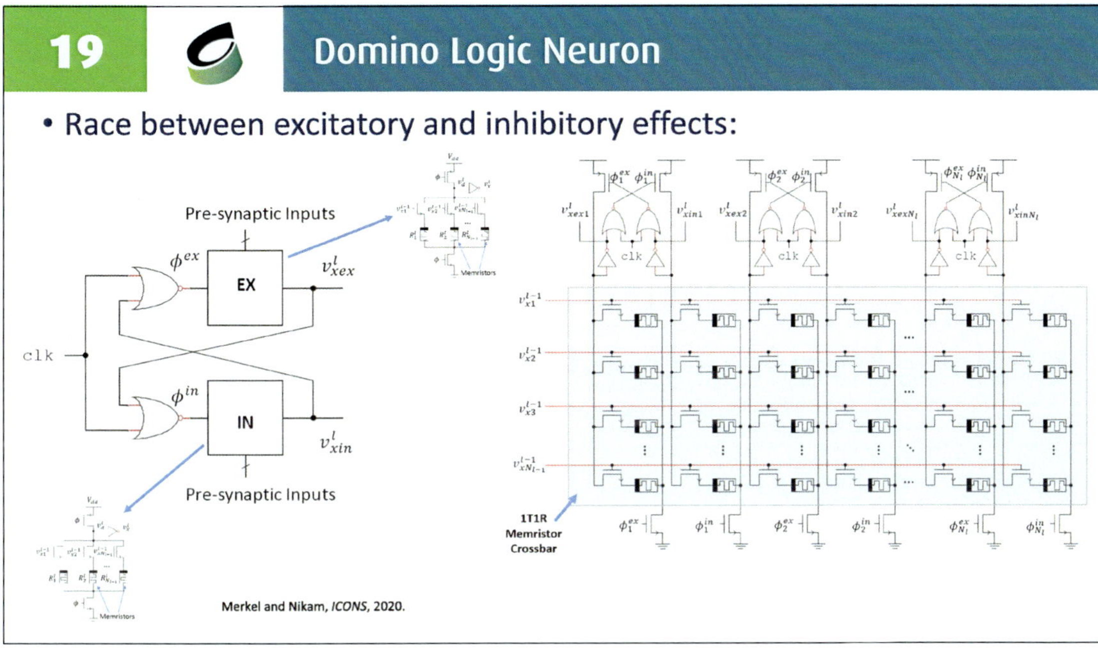

Explains the Race between excitatory and inhibitory effects of Domino Logic Neuron.

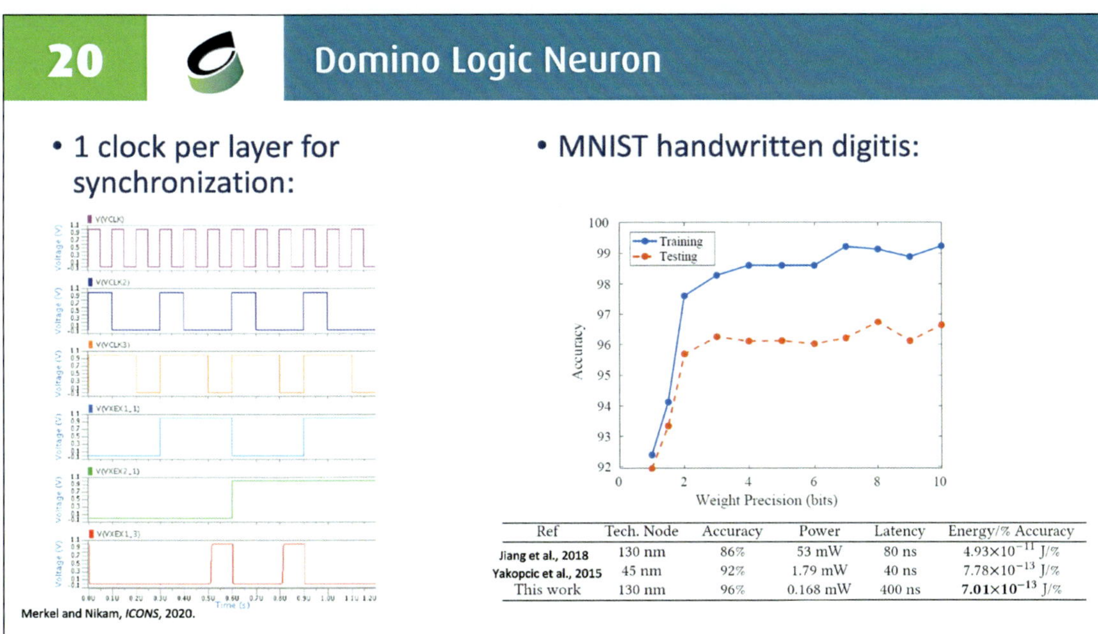

Ref	Tech. Node	Accuracy	Power	Latency	Energy/% Accuracy
Jiang et al., 2018	130 nm	86%	53 mW	80 ns	4.93×10^{-11} J/%
Yakopcic et al., 2015	45 nm	92%	1.79 mW	40 ns	7.78×10^{-13} J/%
This work	130 nm	96%	0.168 mW	400 ns	**7.01×10^{-13}** J/%

Chart representing accuracy vs weight precision(bits) in M.N.I.S.T. handwritten digits and clock synchronisation per layer is explained.

21 Agenda

- Memristors and Memristor Synapses
- Training Memristor-based Neuromorphic Systems
- Adversarial Robustness of Memristor Crossbar Circuits

Highlights the Training Memristor-based Neuromorphic Systems in the Agenda.

22 Stochastic Representation

- Represent data as length L Bernoulli process X with probability p:

$$p = \frac{value}{max.value}$$

- Example: Stochastic representation of 011_2

$$p = \frac{3}{7} \rightarrow X = 10010101001001$$

$\longrightarrow t$

$$\Pr(X_t = 1) = p$$

- Bipolar representations also available for representation of negative numbers

Explains the stochastic representation, representing data as length L Bernoulli process X with probability P. An example of stochastic representation is given.

23 Stochastic Logic

- Stochastic representations enable complex arithmetic with simple logic gates:

Unipolar Representation | **Bipolar Representation**

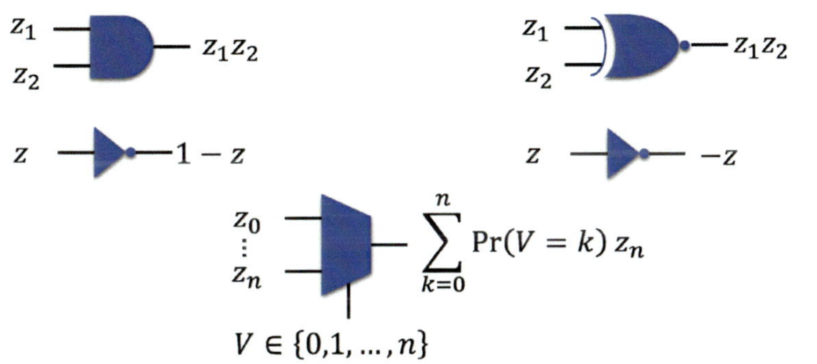

Explains the ability of stochastic logic to express complex arithmetic with simple logic gates such as Unipolar and Bipolar representation.

24 Gradient Descent Learning

- ANNs are trained by minimizing a cost function:

$$\Delta w_{i,j}^{(l)} = -\alpha \frac{\partial J\left(w_{i,j}^{(l)}\right)}{\partial w_{i,j}^{(l)}} = \alpha \frac{\partial J\left(g_1\left(g_2\left(\ldots g_n\left(w_{i,j}^{(l)}\right)\right)\right)\right)}{\partial w_{i,j}^{(l)}}$$

- Using chain rule for derivatives:

$$\Delta w_{i,j}^{(l)} = -\alpha \underbrace{\frac{\partial J}{\partial g_1}\frac{\partial g_1}{\partial g_2}\cdots\frac{\partial g_n}{\partial w_{i,j}^{(l)}}}$$

This computation is expensive in hardware!

Explain the Gradient Descent Learning, which contains ANN and its equation derived using the chain rule.

25 Gradient Descent with Stochastic Logic

- Least mean squares learning rule for single network layer:

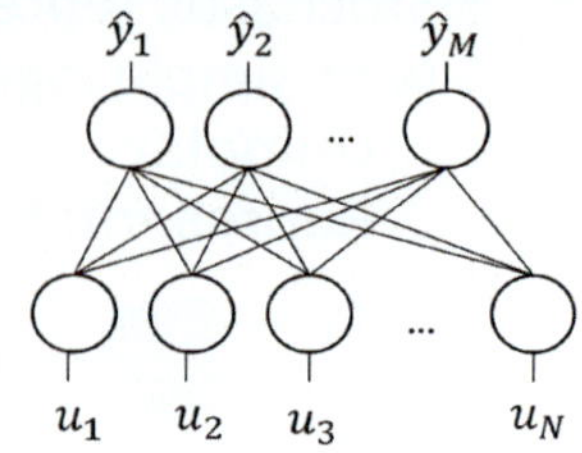

$$J \equiv MSE \rightarrow \Delta w_{i,j} = \alpha u_j(y_i - \hat{y}_i)$$

$$\Delta W_{i,j} = \alpha \mathrm{sign}(y_i - \hat{y}_i) U_j |Y_i - \hat{Y}_i|$$

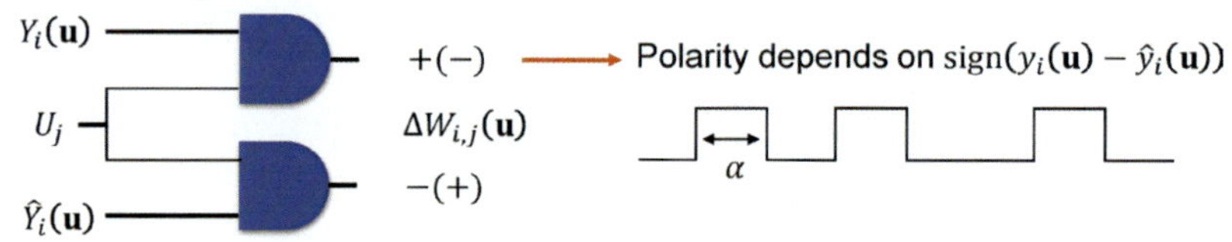

Expected value of $\Delta W_{i,j} \approx \Delta w_{i,j}$

Explains the Least mean squares learning rule for a single network layer, which is obtained by integrating gradient descent with stochastic logic.

26 Deterministic Weight Update vs. Expected Stochastic Weight Update

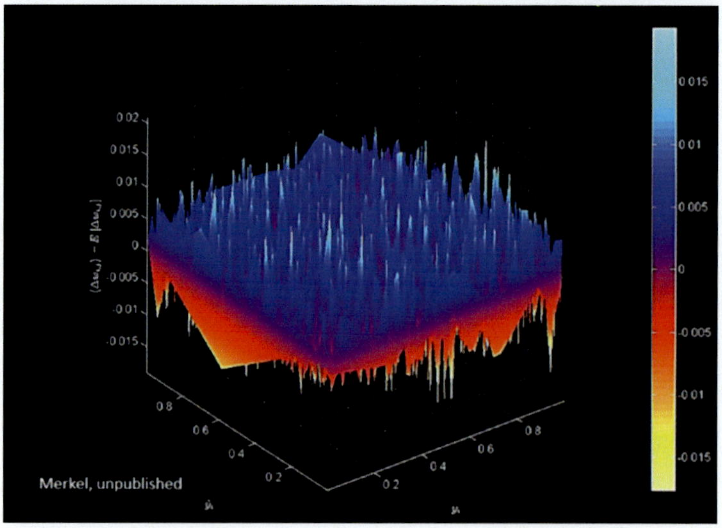

Shows a 3d representation of Deterministic Weight Update vs Expected Stochastic Weight Update.

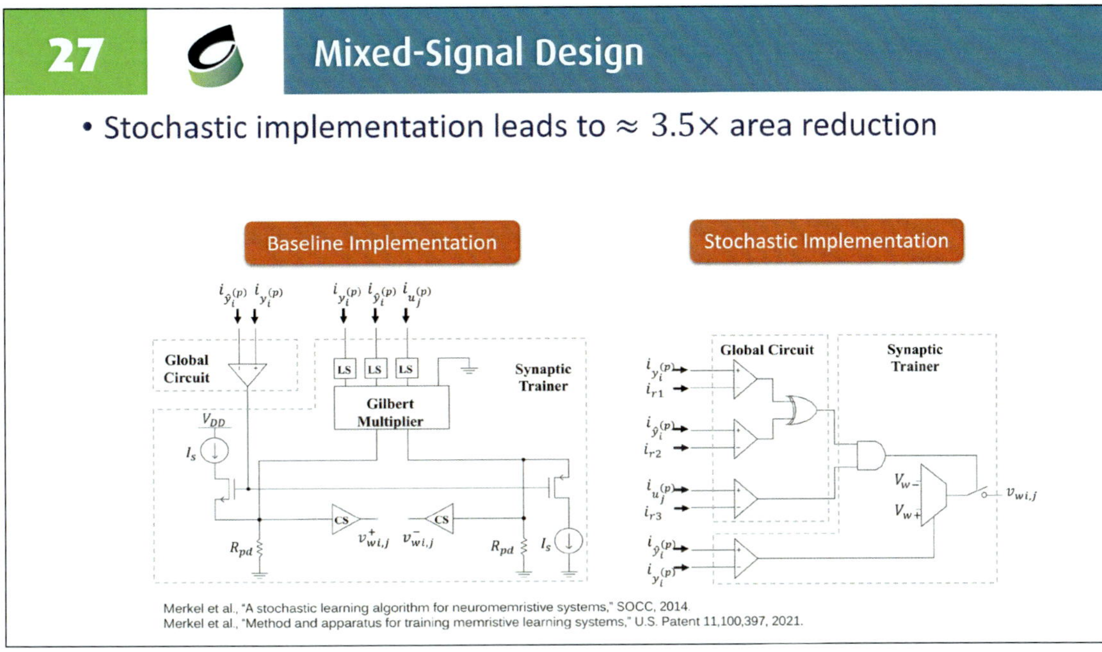

Explains the mixed-signal design to stochastic implementation that leads to area reduction, which can be implemented by baseline and stochastic methods.

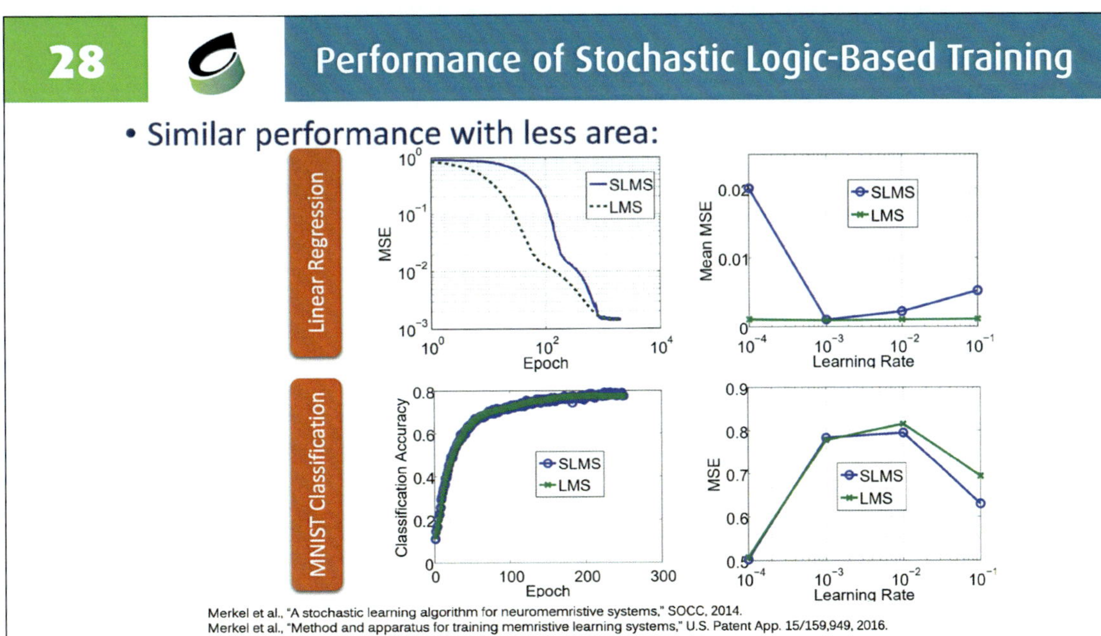

Charts explaining the comparison between Linear regression and M.N.I.S.T. classification, which determines the performance of stochastic logic-based training.

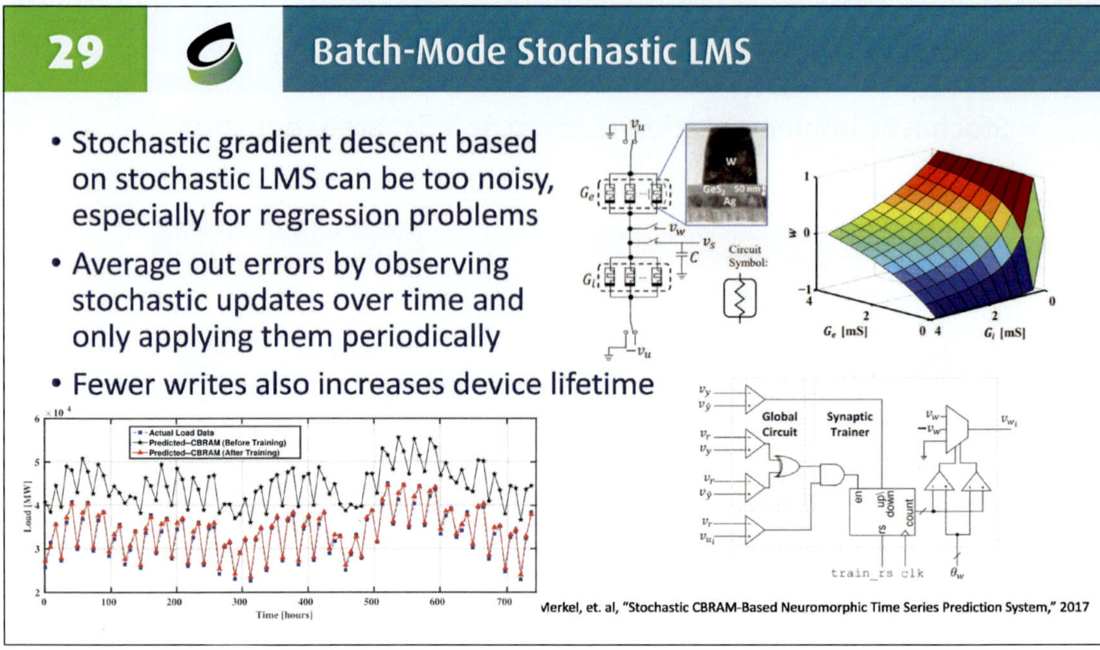

Stochastic gradient descent using stochastic L.M.S. can introduce excessive noise in regression tasks. To mitigate this, averaging errors by observing stochastic updates over time and applying them periodically can reduce noise and extend the device's lifetime.

Highlights the heading: Adversarial Robustness of Memristor Crossbar Circuits.

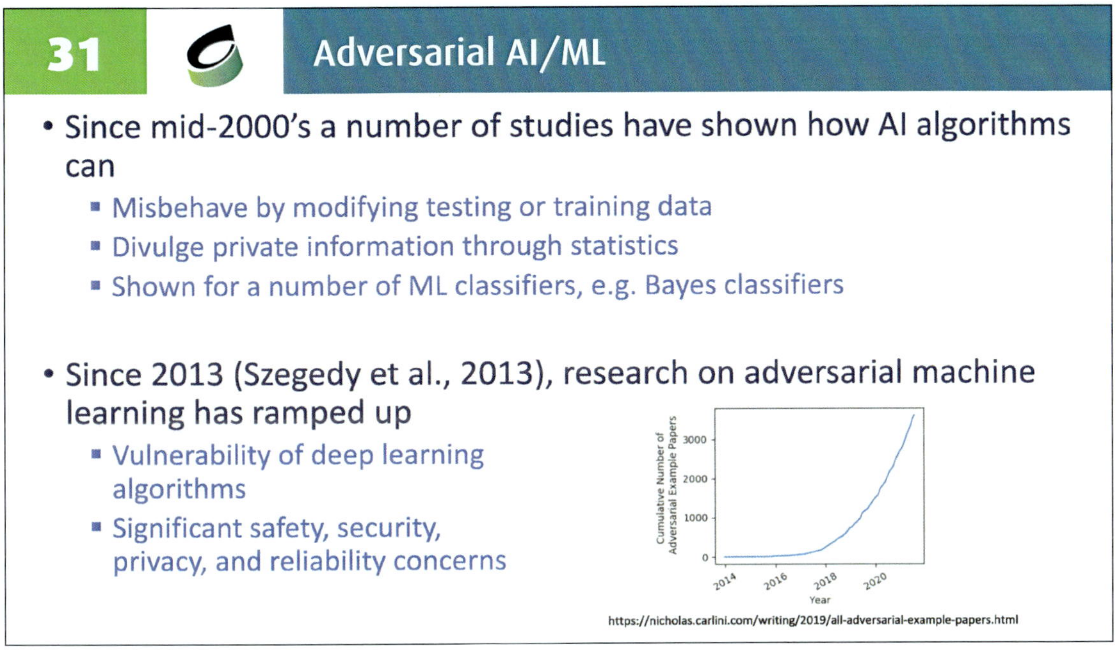

Since the mid-2000s, A.I. algorithms have demonstrated misbehaviour by altering data and leaking private information, mainly seen in classifiers like Bayes. Research since 2013 has intensified on adversarial machine learning, revealing deep learning's vulnerability and raising concerns about safety, security, and privacy.

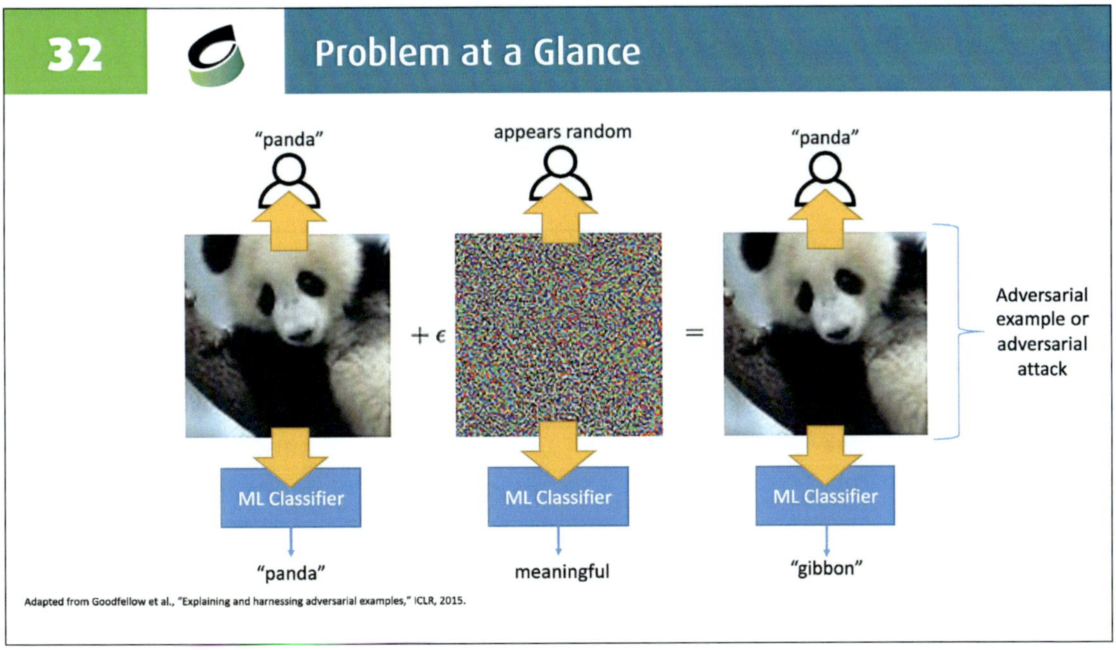

Figure.

33 Adversarial Evasion Attacks

- Exploit small-margin decision boundaries
  - Modify model inputs "just enough" to cause misclassification
    - For images, modifications are usually imperceptible to humans

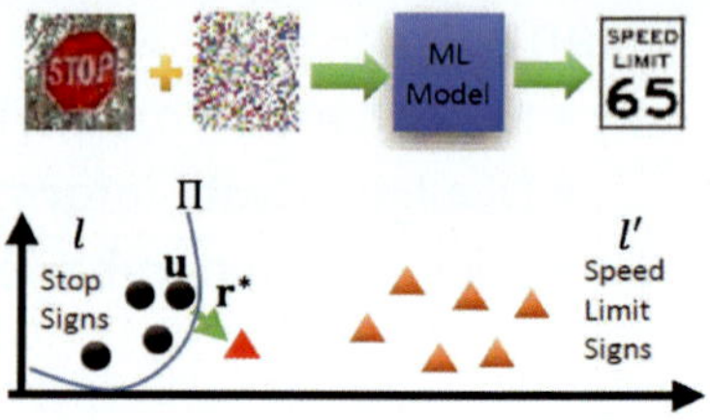

$$\mathbf{r}^* = \underset{\mathbf{r}\in\mathcal{R}}{\arg\max}\ \mathcal{L}_{\Pi}(\mathbf{u}+\mathbf{r}, l)$$
$$\text{s.t. } l' \neq l$$

- The max. size of the perturbation is usually bounded by an ℓ^p norm
  - $p = 0, 1, 2, \infty$ are common
- Attacks can be white-box, black-box, targeted, untargeted
- How susceptible is neuromorphic hardware to these attacks?

Explains the evasion of adversarial attacks, small-margin boundaries size of the perturbation, attacks can be white-box, black-box, targeted, untargeted.

34 How Susceptible is Neuromorphic Hardware to These Attacks?

- How to specific hardware characteristics manifest themselves into potential model vulnerabilities?

Manifestation of Fault	Underlying Fault/Possible Sources
Untrainable weights (stuck at value)	Stuck-at 0/1 fault Defective/worn out device (e.g. memristor)
Weight drift	Memory state decay Destructive memory reads
Random changes in weights/activations	Single event upsets caused by radiation
Non-linear matrix-vector products	Sneak paths in crossbar-based implementations Insufficient gain in analog neuron circuits
Variations in weights/activations	Random process variations
Quantized weights and activations	Limited hardware precision

Merkel, unpublished

Explains how specific hardware characteristics manifest into potential model vulnerabilities.

35 Using Power Information to Attack Neuromorphic Systems

- Attacker can measure current/power consumption information from the crossbar
  - This will be related the sum of the 1-norms of all of the weight matrix columns

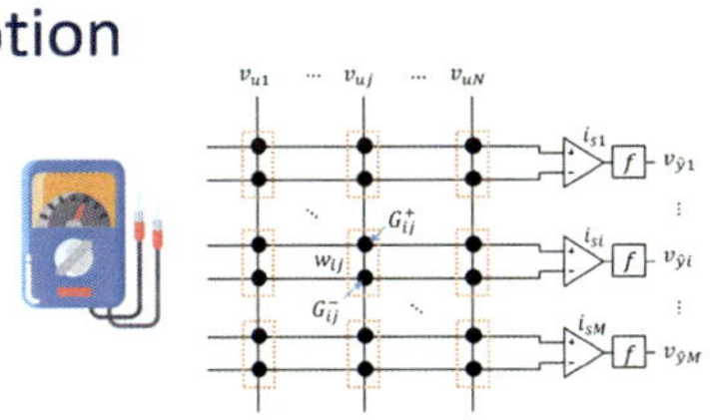

- Sensitivity of the loss w.r.t. to an input is

$$\frac{\partial \mathcal{L}}{\partial u_j} = \sum_{i=1}^{M} \frac{\partial \mathcal{L}}{\partial \hat{y}_i} f'(s_i) w_{ij} \quad \rightarrow \quad \left|\frac{\partial \mathcal{L}}{\partial u_j}\right| \leq \sum_{i=1}^{M} \left|\frac{\partial \mathcal{L}}{\partial \hat{y}_i} f'(s_i)\right| |w_{ij}|$$

- f' is positive, so equality will hold when $\frac{\partial \mathcal{L}}{\partial \hat{y}_i}$ has same sign as w_{ij}
- How often will this be the case?

Attackers exploit power consumption information from crossbars by measuring the current, which correlates with the sum of the first norm of all matrices. This enables adversarial strategies targeting neuromorphic systems, posing security risks that need mitigation.

36 Correlation Between 1-Norm and Sensitivity

Dataset	Activation	Mean Correlation Train	Mean Correlation Test	Correlation of Mean Train	Correlation of Mean Test
MNIST	Linear	0.70	0.70	0.99	0.98
	Softmax	0.52	0.52	0.92	0.92
CIFAR-10	Linear	0.26	0.26	0.87	0.87
	Softmax	0.33	0.33	0.91	0.91

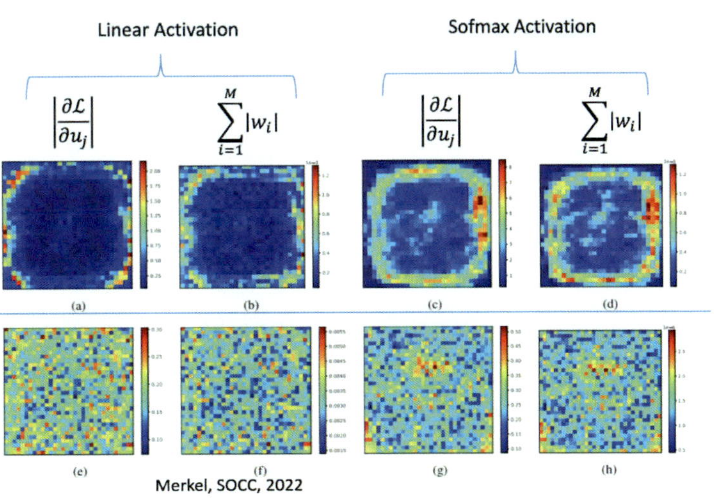

Merkel, SOCC, 2022

Correlation between 1-Norm and sensitivity.

37 Single-Pixel Attacks Using only Power Information

- We use the 1-norm of weight matrix columns to infer the most sensitive pixel
  - Adjust most sensitive pixel by *attack strength*
    - + → Add to pixel, - → Subtract from pixel, RD → Randomly choose to add or subtract

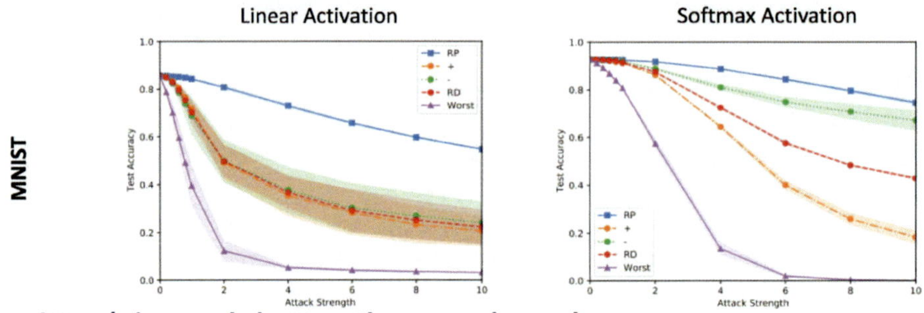

- Attack is much better than random chance

Merkel, SOCC, 2022

Single-pixel attacks leveraging power information utilise the 1-norm of weight matrix columns to identify the most sensitive pixel, then adjust it based on attack strength using addition, subtraction, or random perturbation. This approach enables adversarial manipulation of neural networks using only power consumption data, presenting novel security challenges.

38 Single-Pixel Attacks Using only Power Information

- We use the 1-norm of weight matrix columns to infer the most sensitive pixel
  - Adjust most sensitive pixel by *attack strength*
    - + → Add to pixel, - → Subtract from pixel, RD → Randomly choose to add or subtract

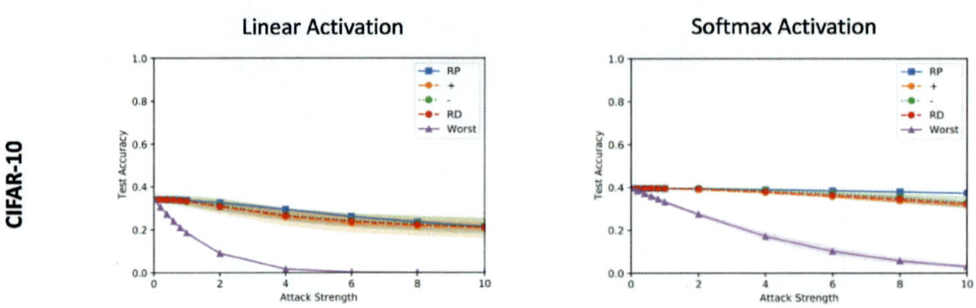

- Attack not as effective: Likely due to low clean accuracy

Explains how we can infer the most sensitive pixel by using the 1-norm of weight matrix columns.

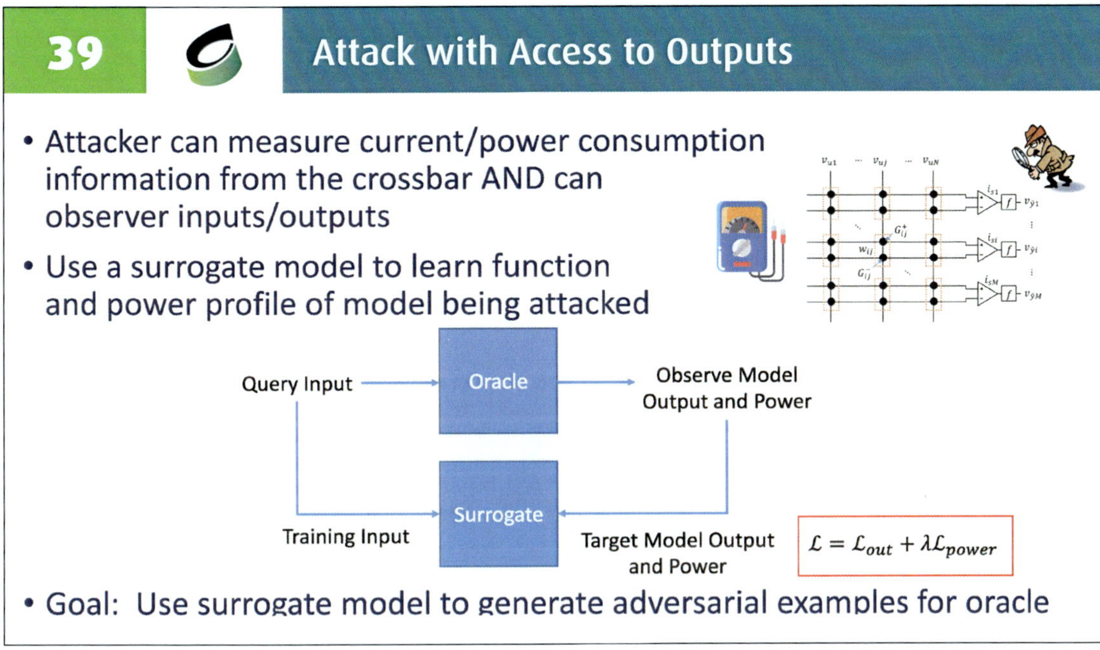

Attackers leverage current/power consumption data and observe inputs/outputs to train a surrogate model, enabling sophisticated attacks on neuromorphic systems by learning the targeted model's function and power profiles.

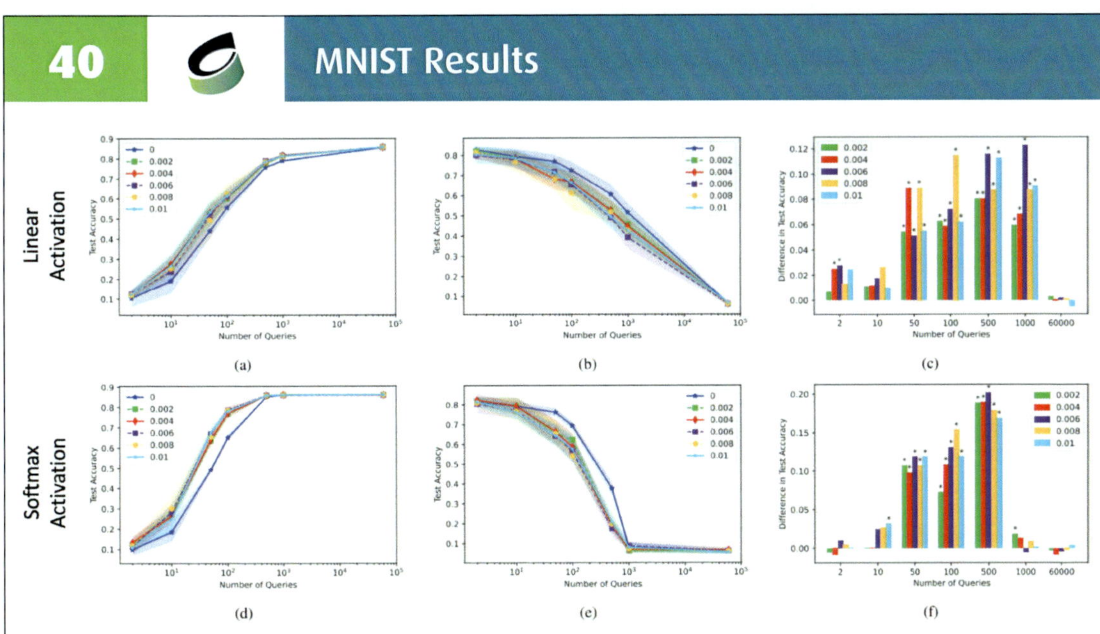

M.N.I.S.T. Results.

41 CIFAR-10 Results

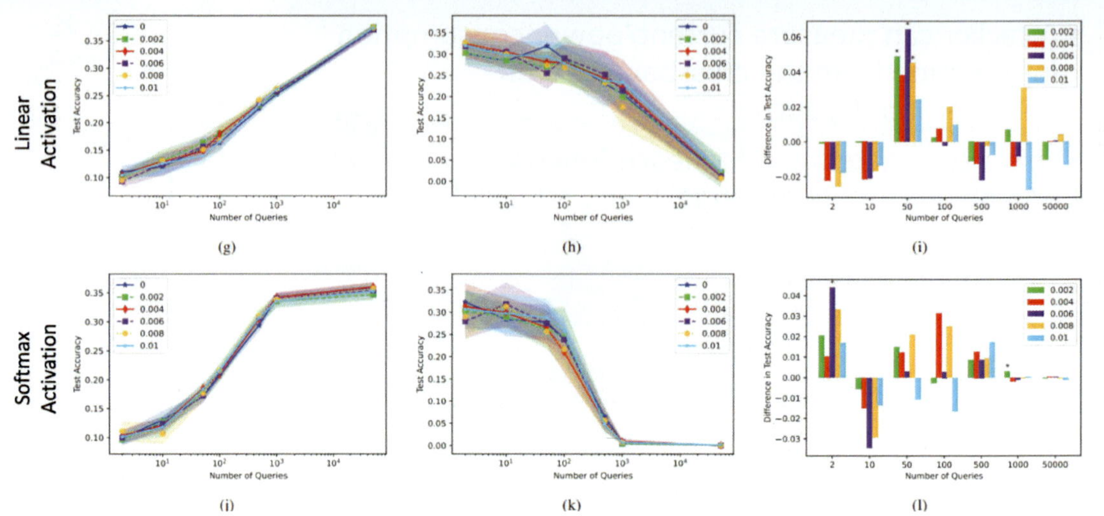

CIFAR-10 Results

42 Multi-pixel Attacks Based on Power Consumption in MLPs

- Using power consumption information to perturb more than one pixel has diminishing effect:

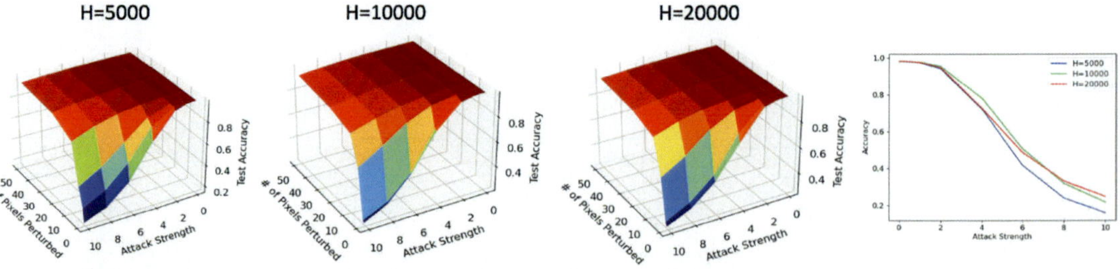

- Ongoing work is exploring other methods of incorporating power consumption information into attack formulation, especially in multi-layer networks

Multi-pixel attacks leveraging power consumption in M.L.P.s exhibit diminishing effects, prompting exploring alternative methods of integrating power consumption information into attack strategies, particularly in multi-layer networks. This ongoing work aims to enhance adversarial strategies by considering more comprehensive power-related vulnerabilities..

43 Single Layer Results with MSTAR

	Mean Corr. Coef.		Corr. Coef. Of Mean	
	Train	Test	Train	Test
Linear	0.44	0.44	0.94	0.94
Softmax	0.60	0.60	0.96	0.96

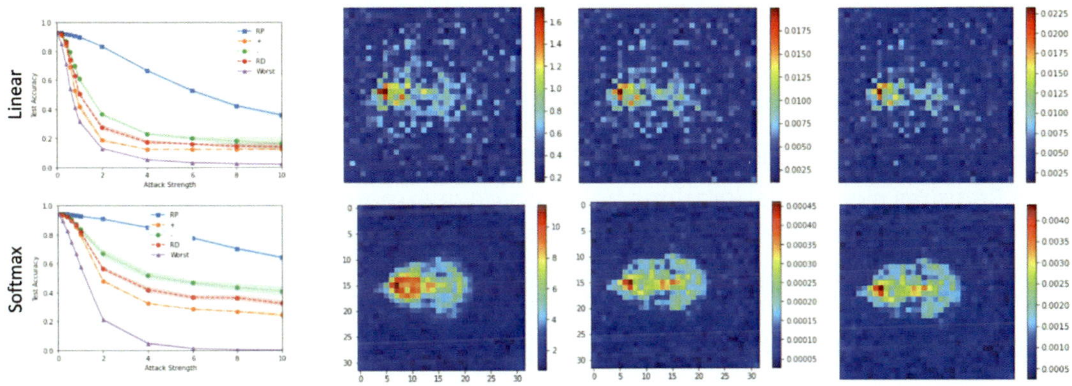

Single Layer Results with MSTAR.

44 MLP with MSTAR

	Mean Corr. Coef.		Corr. Coef. Of Mean	
	Train	Test	Train	Test
Linear	0.49	0.49	0.90	0.90
Softmax	0.61	0.61	0.82	0.86

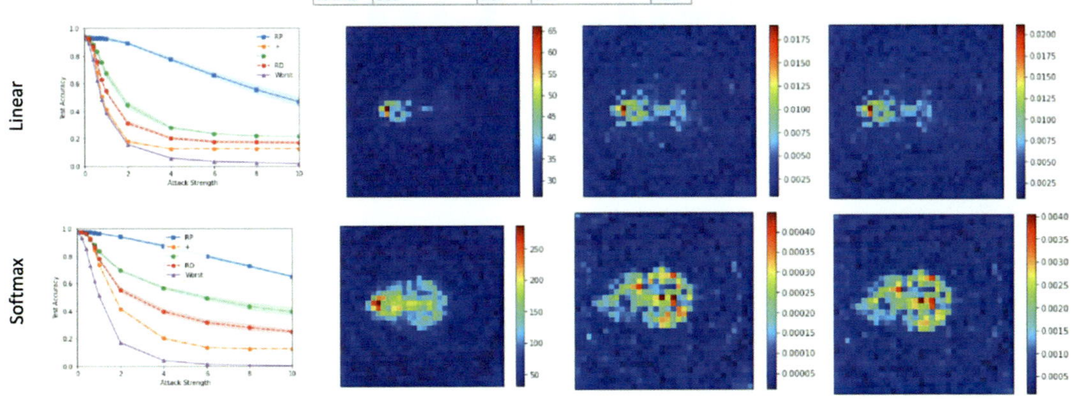

M.L.P. with MSTAR.

45 Conclusions

- Current trajectory of AI is not sustainable and will not work for resource-constrained application domains
  - Innovative technologies and algorithmic breakthroughs are needed to push AI to the edge
  - **New memory technologies are one of the keys**
- Neuromorphic computing is a team sport!
  - Requires close collaborations between device/materials engineers, data scientists, electrical/computer engineering, and neuroscientists, to name a few
- Innovations will come at the intersections between these disciplines

The current path of A.I. is unsustainable for resource-constrained domains, necessitating innovative technologies and algorithmic breakthroughs to bring A.I. to the edge. Neuromorphic computing underscores the importance of interdisciplinary collaboration between device/materials engineers, data scientists, electrical/computer engineers, and neuroscientists, with future advancements expected at the intersections of these disciplines.

46 Thank You!

47 A Better Way

- Dummy row allows magnitude of entire column to be adjusted
- Now,
  - Constrain voltage-mode weights as:

$$\max\left(\frac{1-w_{max}}{M-1}, w_{min}\right) \le \overline{w}_{ij} \le \min\left(\frac{1-w_{min}}{M-1}, w_{max}\right)$$

  - Set current-mode weights as

$$w_{ij} = \overline{w}_{ij} \ \forall\, i < M$$

$$w_{Mj} = 1 - \sum_{k=1}^{M} \overline{w}_{kj}$$

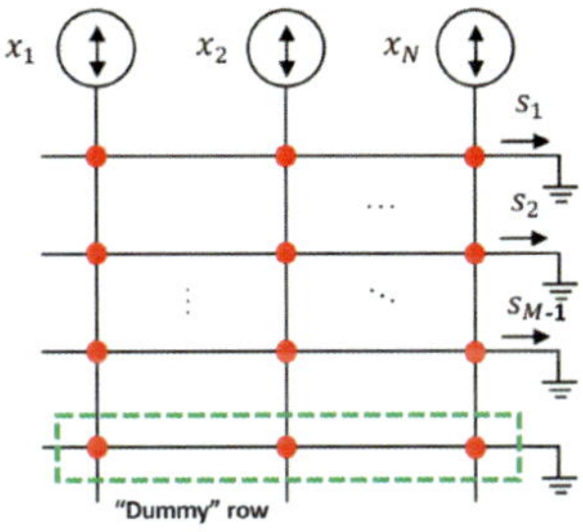

Merkel, "Current-mode memristor crossbars for neuromorphic computing," *NICE*, 2019.

Explains a better way to find current mode weights using dummy rows.

48 Current-mode Crossbar Training

- We can train offline using the proposed method, but online training is more desirable
- Gradient descent can be implemented as

$$\Delta G_{ij}^{l} = -\alpha \frac{\partial J}{\partial G_{ij}^{l}} = \alpha \sum_{k=1}^{M^l} \delta_k^l x_j^{l-1} \frac{\partial w_{kj}^{l}}{\partial G_{ij}^{l}}$$

$$\approx \begin{cases} \alpha \delta_i^l x_j^l & i \neq M \\ -\langle \Delta G_{1:M-1,j}^{l} \rangle & i = M \end{cases}$$

MLP trained on MNIST

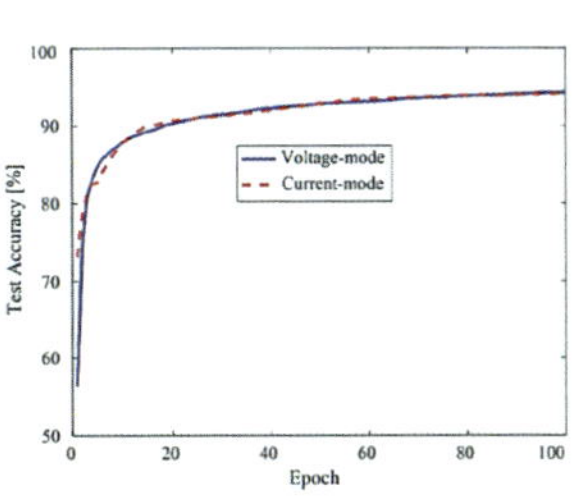

Just use delta rule for non-dummy rows and average the updates for dummy row

Merkel, "Current-mode memristor crossbars for neuromorphic computing," *NICE*, 2019.

The proposed method makes Offline training feasible, but online training is preferable in current-mode crossbar training. Gradient descent serves as an effective optimisation technique in this training approach.

49 You Can Read Memristor Crossbars with Time

- Memristor crossbars consume constant static current

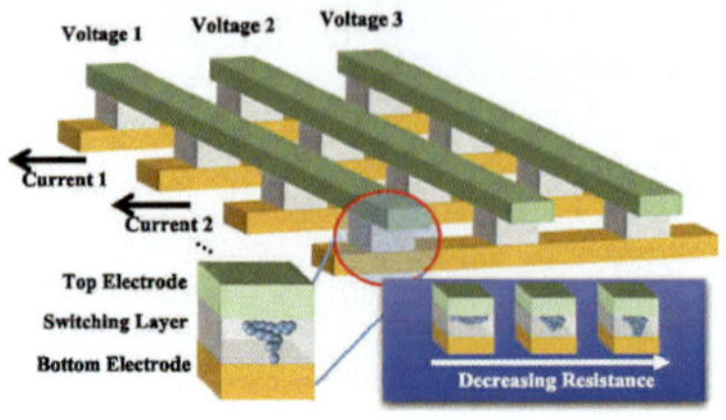

- We can borrow a potential solution from digital logic design: **dynamic logic**

Explains how to read memristor crossbars with time. dynamic logic.

50 Example: Binary Decision Problem

- Consider a binary decision problem (e.g. autonomous vehicle turing right of left at fork)
  - Energy cost of executing good decision is $E_{X\min}$
  - For bad decision, cost is $\gamma E_{X\min}$
- Choosing # of layers can be found using optimal energy-accuracy point:

$$A^* = \frac{1}{b}\log\left(\frac{(\gamma-1)E_{X\min}}{E_{D\max}ab}\right)$$

∝ **MACs for 6-layer network**

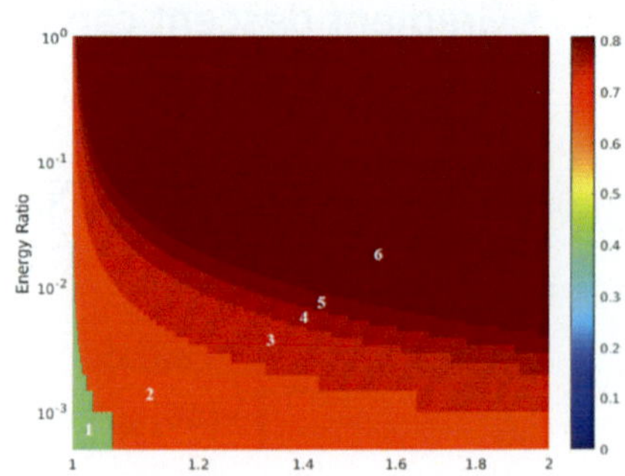

Example: Example: Binary Decision Problem.

51 Dynamic Energy-Accuracy Scaling

- Since α may not be fixed (e.g. dynamic amounts of available energy, changing mission parameters), we need several energy-accuracy points available
- For our 6-layer CNN example, one option:
  - Have separate 1-layer, 2-layer, 3-layer, 4-layer, 5-layer, and 6-layer networks available
    - ➢ Big overhead
- Could also combine into a single network

Dynamic Energy-Accuracy Scaling involves accommodating varying energy availability and mission parameters by providing multiple energy-accuracy points. For example, in a 6-layer CNN, options include maintaining separate networks for each layer configuration or consolidating them into a single network, each with its associated overhead considerations.

52 From 6 Networks to 1 Network

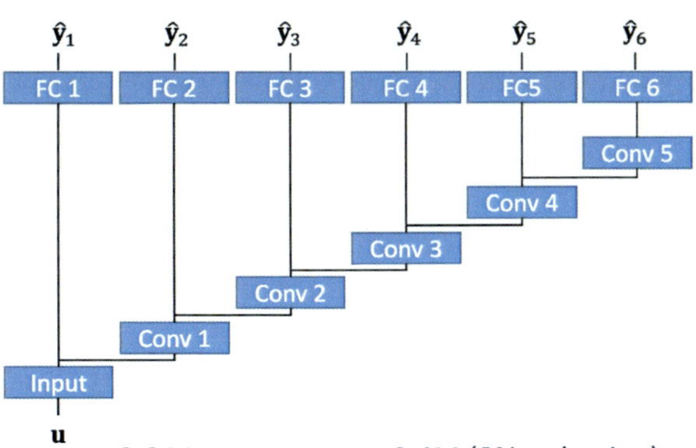

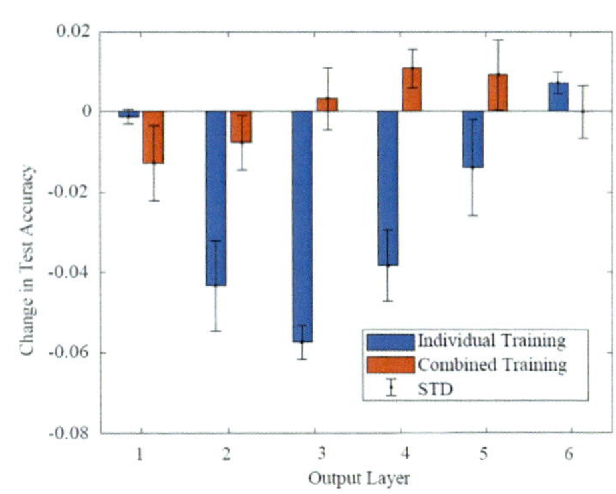

- ≈2.6 M parameters to 2.4M (8% reduction)
  - Could reduce more with weight sharing?
- FC layers can be trained individually with Conv layers fixed
  - Better performance with combined training: $\mathcal{L} = \sum_i \mathcal{L}_i$
- Same $C_D(A)$ as before assuming unused portions are gated, but fewer parameters

Explains the reduction of 6 networks to 1 network.

53 Accuracy vs. Desired Accuracy

- As desired accuracy increases, more layers are used
  - Use a layer if desired accuracy < min accuracy or confidence > desired accuracy

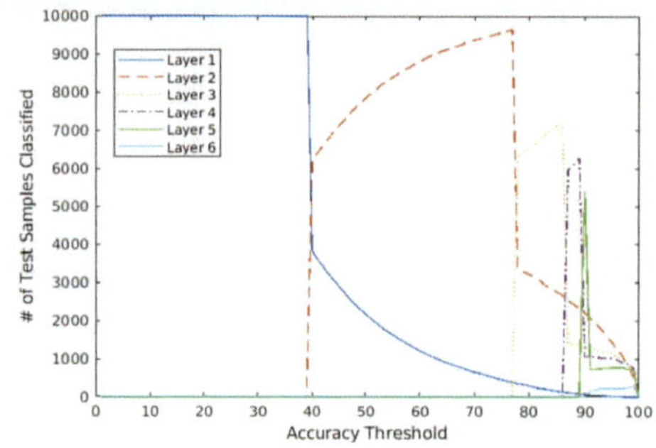

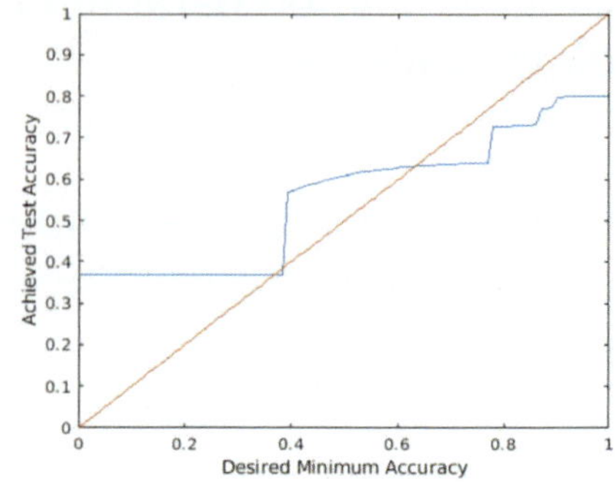

- Could shift achieved accuracy curve left by modifying policy for using a layer

The number of layers used increases as the desired accuracy rises, with each layer employed if the desired accuracy exceeds a minimum threshold or confidence surpasses the desired accuracy. Adjusting the layer utilisation policy can shift the achieved accuracy curve to the left, influencing the network's performance in meeting desired accuracy levels.

54 Accuracy vs. Desired Accuracy

- Optimization problem:

$$\text{minimize } l \in \{1,2,\ldots,L\}$$
$$s.t. \quad \beta_1 A_{train_l} \geq \tilde{A} \vee C_l \geq \beta_2 \tilde{A}$$

Where A_{train_l} is the training accuracy for layer l, $\tilde{A}$ is the desired accuracy level, C_l is the confidence level for layer l, and β_1, β_2 are hyperparameters

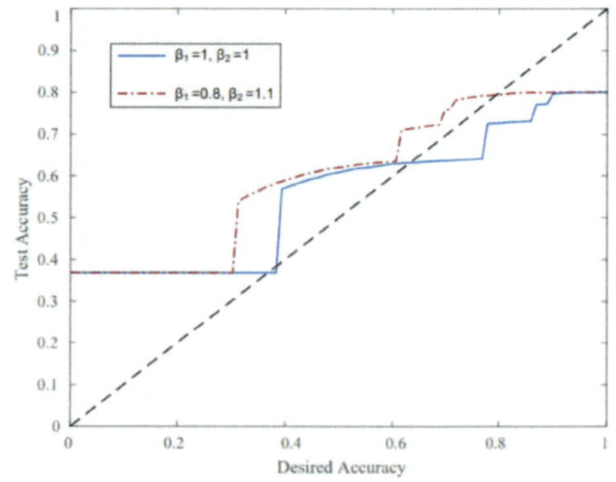

Explains more about Accuary VS Desired accuracy.

55 Energy Cost

- The shape of the new energy-accuracy curve has more flat parts, leading to significant energy savings, depending on the desired accuracy

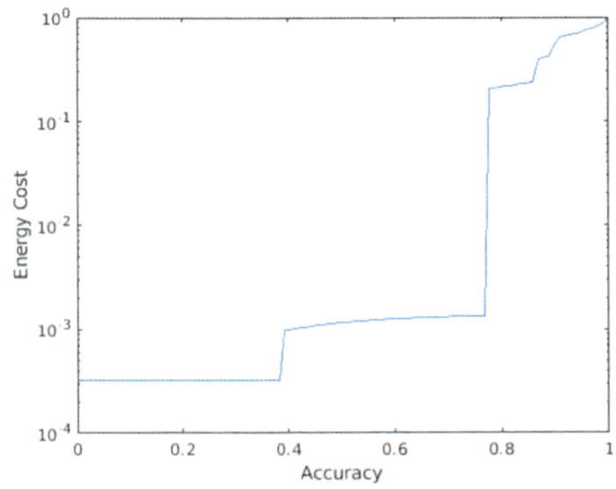

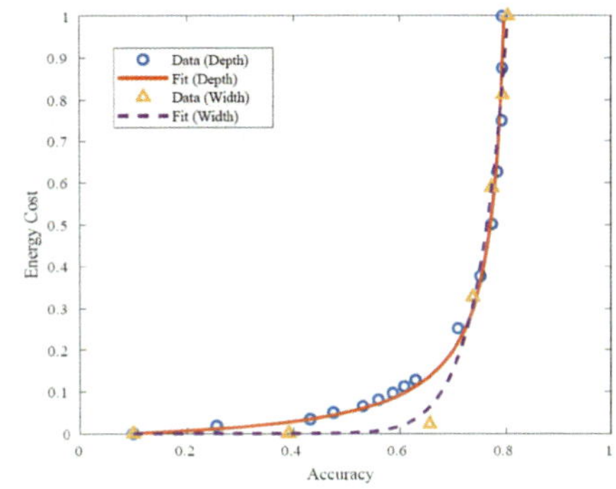

The revised energy-accuracy curve features flatter regions, resulting in notable energy savings contingent on the desired accuracy.

56 Energy Constraints

- Pools of energy are shared by different resources (e.g. neurons)
  - Inspired by work of Burroni et al., Front. Neurosci., 2017
- Pools are depleted when energy-consuming operations occur
  - e.g. action potential generation
  - Pools replenish at a fixed rate

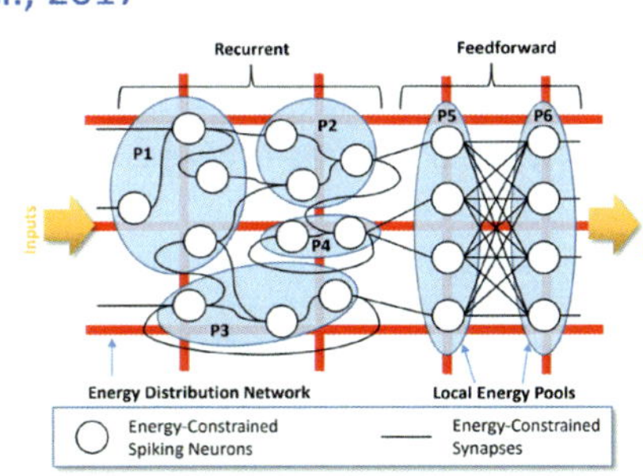

- A certain level of energy is required for each operation

Energy constraints in neural networks involve shared energy pools among various resources, such as neurons, following the framework inspired by Burroni et al. (2017). These pools deplete during energy-intensive operations like action potential generation and replenish at a fixed rate, with specific energy requirements for each operation.

57 Using Power Analysis to Improve Transferability

- What if we force a surrogate model to have the same power consumption as a black box model?

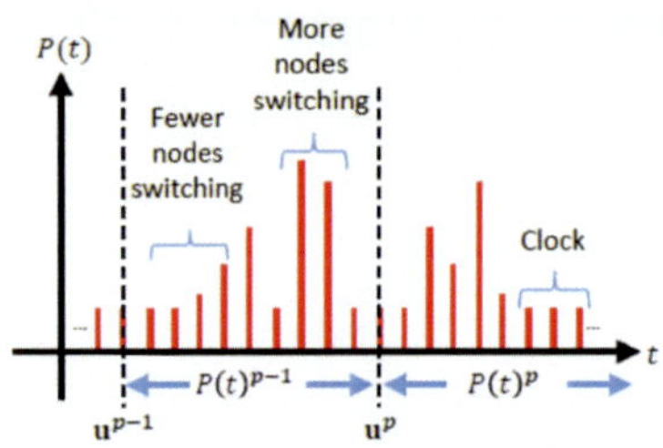

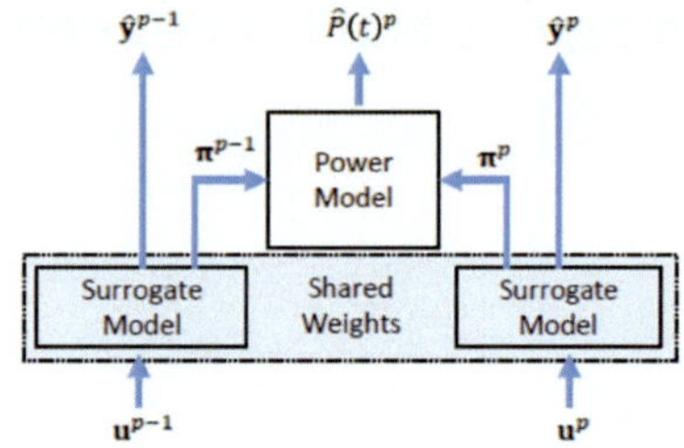

- Does power information help constrain solution space of the surrogate so that the surrogate model has increased similarity with the black box model?

$$\mathcal{L} = \alpha \mathcal{L}_{CE} + \beta \underbrace{\mathcal{L}_{power}}_{\text{MSE}}$$

Explains the use of Power Analysis to Improve Transferability.

58 Leveraging Switching Power

- Total power consumption is sum of static and dynamic power:

$$P = P_{static} + P_{dynamic}$$

Circuit is "idle" (P_{static}) — Nodes are switching ($P_{dynamic}$)

- First, we consider an MLP with hidden layer composed of neurons that have step activation
  - For each new input, hidden neurons will either stay at 0/1, switch from 0 to 1, or switch from 1 to 0
  - Dynamic power (more specifically, switching power) is consumed during the 0-to-1 transition
  - If we can measure the switching power, we can determine how many neurons switched from 0 to 1:

$$\hat{P}^p = (\boldsymbol{\pi}^p - \boldsymbol{\pi}^{p-1})(\boldsymbol{\pi}^p)^\top$$

Leveraging Switching Power: total power consumption is the sum of static and dynamic power.

59 Leveraging Switching Power

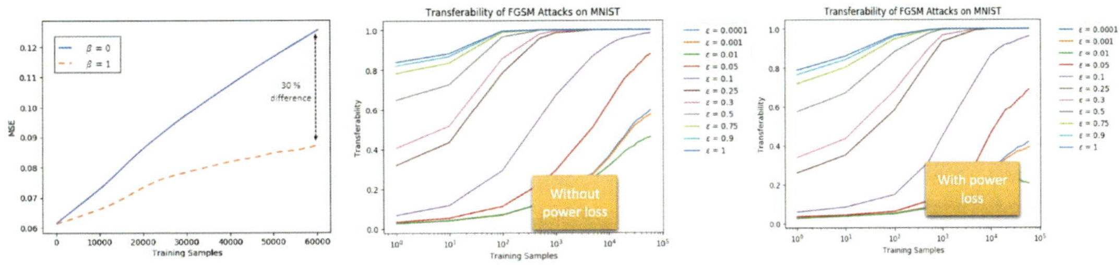

- Switching power information reduces the MSE between the weights of surrogate and black box model
- So far, switching power information is not observed to improve transferability
  - Why this is happening?
    - Power loss is reaching a local minimum that is far from the oracle
    - There are power loss minima that lead to very different hidden states from the oracle
- Currently exploring smarter querying to get more useful information (e.g. gradient information) from power data

Li and Merkel, ICANN, 2021

Switching power information reduces M.S.E. between surrogate and black box model weights but doesn't enhance transferability, likely due to power loss reaching distant local minima and disparities in hidden states. More brilliant querying methods are being explored to address this issue, like utilising gradient information from power data.

60 Leveraging Power Consumption from Matrix-Vector Dot Product in Crossbars

- In crossbar-like implementations of neural networks, the majority of power will come from the matrix-vector product:

$$\mathbf{s} = \mathbf{W}\mathbf{x}$$

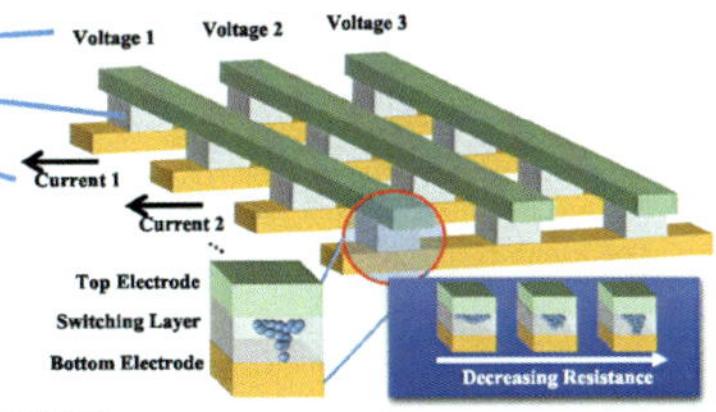

- The power consumption will be related to the sum of the crossbar outputs:

$$P = \sum_j v_j^2 \sum_i G_{ij} \propto \sum_j x_j^2 \sum_i W_{ij}$$

- Constraining surrogate to match power consumption or oracle forces the sum of weights over the rows to be the same

In crossbar-like neural network implementations, the primary power consumption arises from the matrix-vector product, with power consumption directly linked to the sum of crossbar outputs. To ensure consistency, constraints are imposed on the surrogate to match power consumption, or the oracle enforces exact sums of weights over the rows.

61 Leveraging Power Consumption from Matrix-Vector Dot Product in Crossbars

- Linear regression results for MNIST on an MLP
  - Assumed dual crossbar implementation (1 xbar holds positive component of weights while other holds negative component)
  - Loss is weighted sum of cross entropy and MSE of power consumption
- Initial results show that addition of power loss can improve accuracy when fewer queries are used
  - Linear regression is convex: Power loss will be more useful in multi-layer net

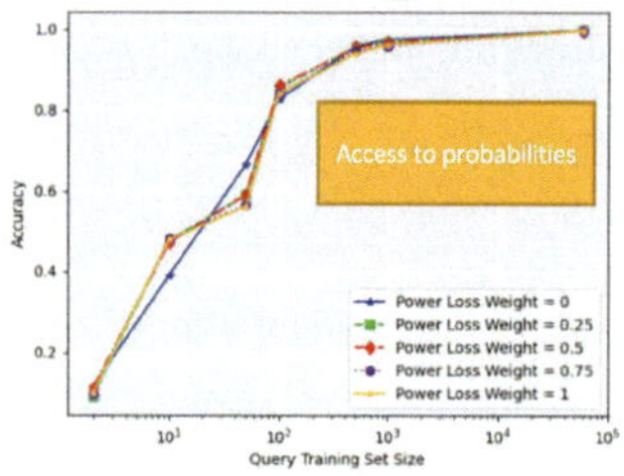

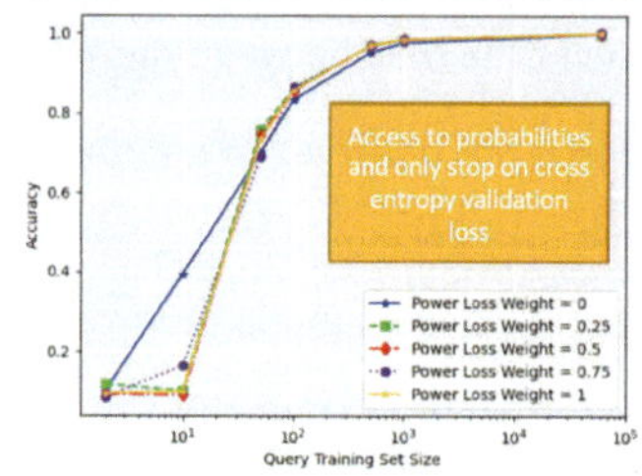

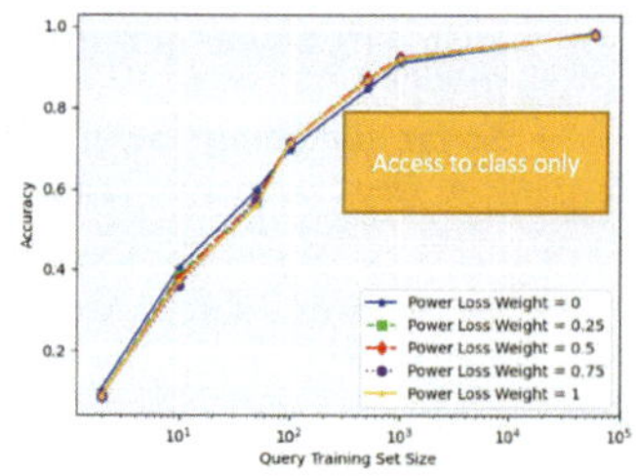

Leveraging power consumption from the matrix-vector dot product enhances accuracy in M.N.I.S.T. MLPs, especially with fewer queries, suggesting its value. Initial results indicate that integrating power loss, a combination of cross entropy and mean squared error of power consumption, can improve accuracy, particularly in multi-layer networks with convex linear regression.

62 Leveraging Power Consumption from Matrix-Vector Dot Product in Crossbars

- MLP trained on MNIST:
  - Transferability of adversarial examples can be improved by incorporating power consumption information into the surrogate training process

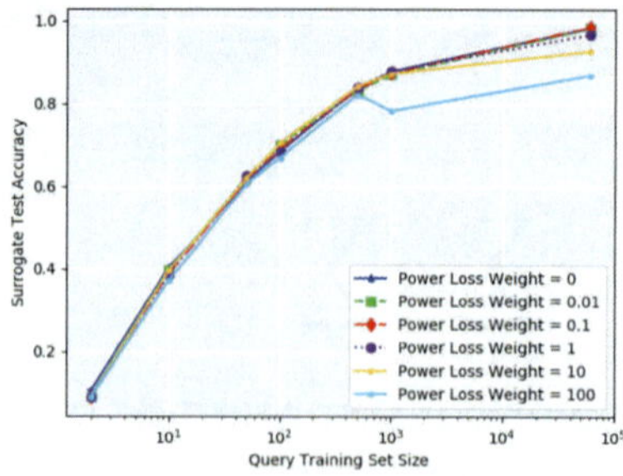

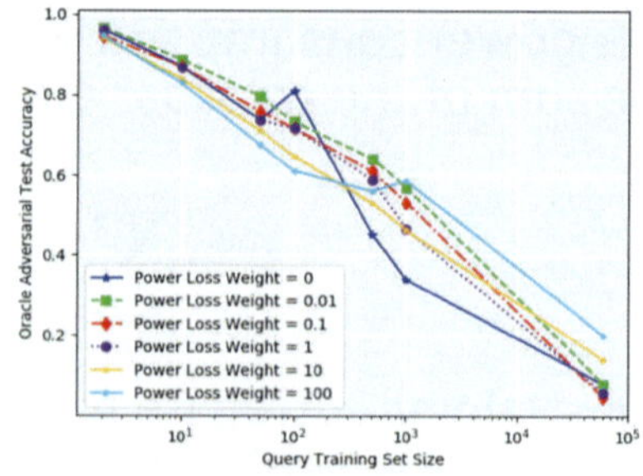

Transferability of adversarial examples can be improved by incorporating power consumption information into the surrogate training process.

63 Agenda

• Memristor circuits for artificial synapses

• **Energy-aware neural networks**

Energy-aware neural networks

64 Cost of Making a Decision

- Cost of using an AI system to make an inference:

$$C = \alpha C_D + (1 - \alpha) C_X$$

- C_D is cost of making decision
- C_X is cost of executing decision
- α is the relative cost
  - High α: e.g. lots of energy to make decision with little payoff
    - ➢ Time sensitive (publishing a paper before someone else does)
  - Low α: e.g. very important to make a good decision
    - ➢ Disease diagnosis and treatment plan

Merkel, *IGSC*, 2020.

Explains the cost of making and executing the decision.

65 Cost in a Neural Network

- In a neural network, we can relate the cost to energy and accuracy
- Number of multiply-accumulate (MAC) operations used as a proxy for energy:

$$MAC_{conv} \approx (W_i - W_f + 1)(H_i - H_f + 1)N_c W_f H_f N_f$$

$$MAC_{fc} \approx N_u N_y$$

- More accurate energy models will be implementation-dependent
  - Digital, analog, memristor, etc.

Merkel, *IGSC*, 2020.

In a neural network, we can relate the cost to energy and accuracy; more accurate energy models will be implementation-dependent.

66 Varying Network Depth and Width

- CNN on CIFAR-10
  - Each convolutional layer is 64 3x3 filters except when network width is adjusted
  - Softmax output
  - Dropout, L2 regularization, and early stopping used

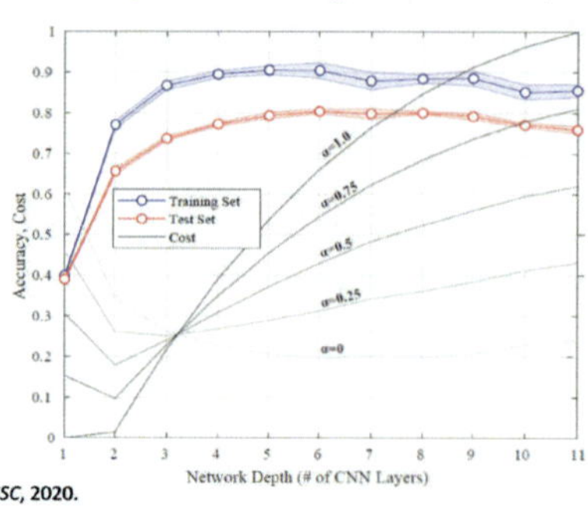

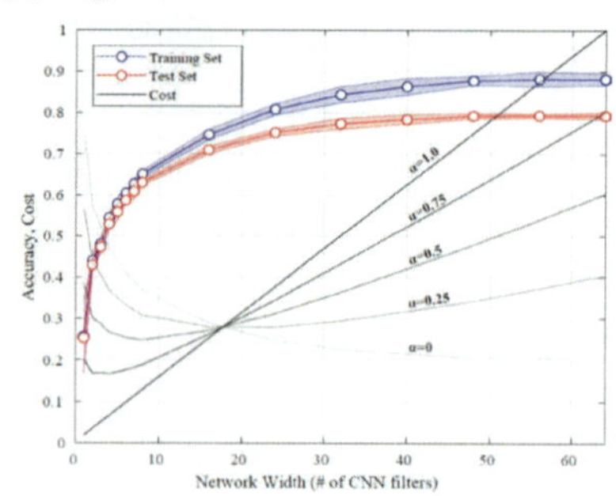

Merkel, *IGSC*, 2020.

The study investigates different CNN architectures on CIFAR-10, employing varying depths and widths with standard techniques like dropout, L2 regularisation, and early stopping for optimisation. Each convolutional layer features 64 3x3 filters, except when adjusting network width, ultimately leading to softmax output for classification.

67 Optimal Energy-Accuracy Point

- Can find optimal energy-accuracy point as

$$\frac{\partial C}{\partial A} = \alpha \frac{\partial}{\partial A} C_D(A) + (1-\alpha)\frac{\partial}{\partial A} C_X(A)$$
$$= \alpha \frac{\partial}{\partial A} C_D(A) + (1-\alpha)\frac{\partial}{\partial A}(1-A) = 0$$

$$\frac{\partial}{\partial A} C_D(A) = \frac{1-\alpha}{\alpha}$$

$C_X \propto$ chance of misclassification

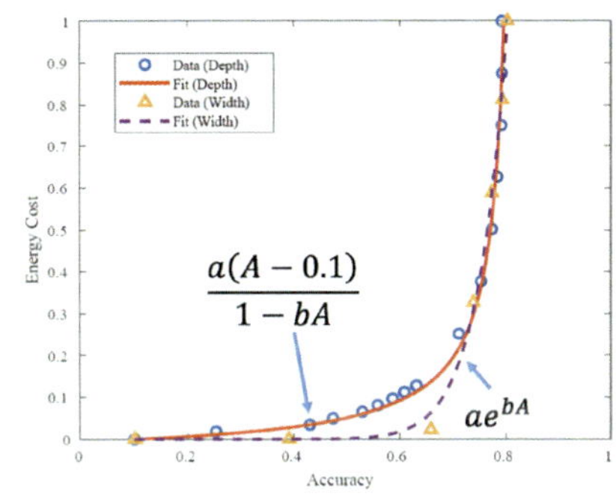

- Analytical solution for optimal A may be complicated depending on the form of $C_D(A)$
  - $C_D(A) = ae^{bA}$ works well for modeling # of layers:

$$A^* = \frac{1}{b}\log\left(\frac{1-\alpha}{\alpha ab}\right)$$

Finding the optimal energy accurate point.

68 Can We Do Better?

- Training set confidence distribution over layers:
  - Confidence is taken as largest softmax output
  - Mean confidence scales with accuracy/layer

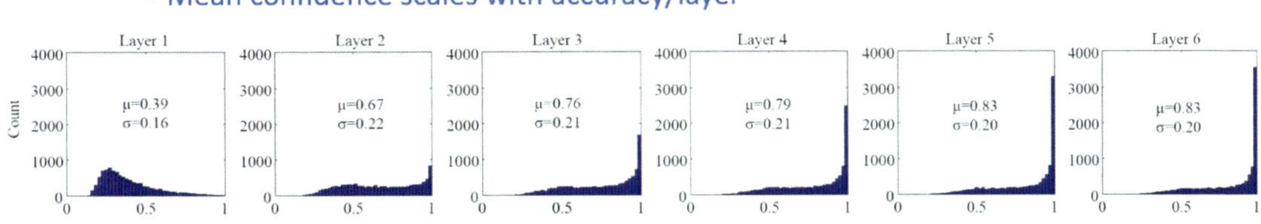

- It may be possible to use high-confidence prediction in lower layers to achieve an accuracy that normally requires larger # of layers

The confidence distribution in the training set across layers is determined by selecting the most considerable softmax output, where mean confidence correlates with accuracy per layer.

69 Accuracy and Confidence

- Accuracy of samples above a particular confidence level:

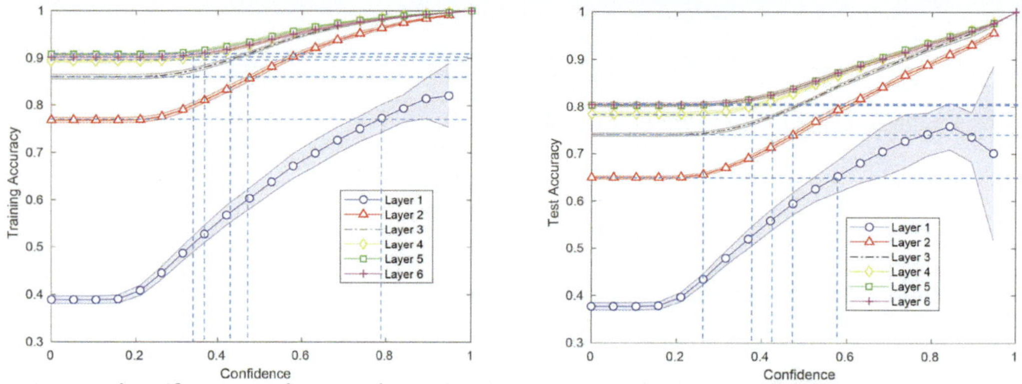

- Example: If we can live with an 85% accuracy, then we might use a 2-layer network with confidence above ≈50% even though on average, they give ≈78% accuracy

Explains accuracy and confidence.

70 Energy Cost

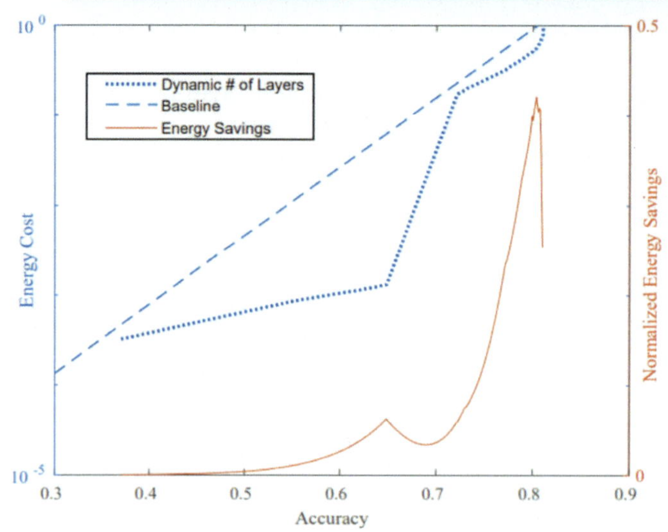

Explains about energy cost.

71 Bio-Inspired Energy Constraints on Neural Networks

- In biology, energy and function were co-optimized
- Astrocytes closely regulate neuronal energy consumption in brains
  - Maintain a reserve of energy to supply to neurons (Boison, 2018)
  - Astrocytes are thought to retrieve glucose from blood vessels, where they can metabolize it to lactate and transport to neurons for use as energy (Jha, 2018)
  - Astrocytes are generally connected to several neuron soma (Halassa, 2007)

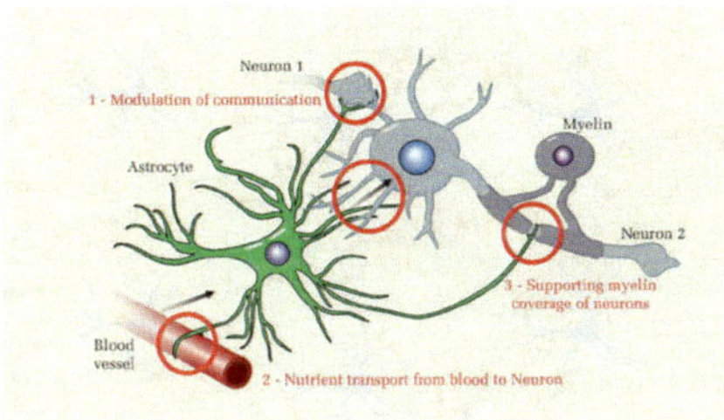

www.wingsforlife.com/en/latest/astrocytes-the-star-cells-1637

Astrocytes in the brain optimise energy and function by regulating neuronal energy consumption, supplying lactate derived from glucose metabolism, and connecting to multiple neuron soma for efficient energy transfer.

72 Liquid State Machines

- Spiking neural network with random input and recurrent layer (reservoir) weights
- Pools of energy are shared by different resources (e.g. neurons)
  - Inspired by work of Burroni et al., Front. Neurosci., 2017
- Pools are depleted when energy-consuming operations occur
  - e.g. action potential generation
  - Pools replenish at a fixed rate
- A certain level of energy is required for each operation

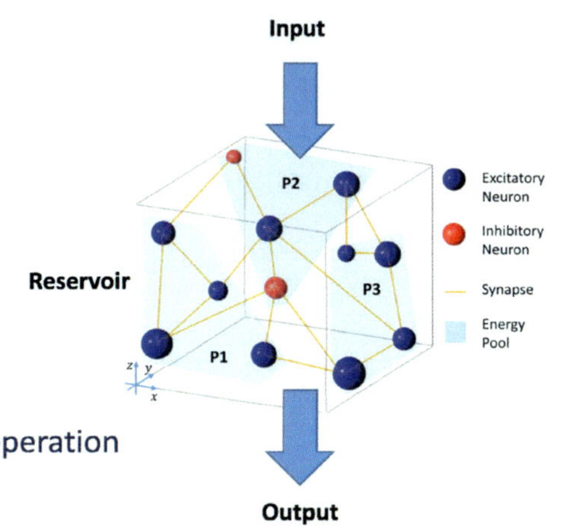

Fountain and Merkel, 2020

Liquid State Machines (L.S.M.s) are spiking neural networks comprising random input and recurrent layer weights, akin to a reservoir, with energy pools shared among resources.

73 One-channel EEG Seizure Detection Task Results

- Maximum accuracy on the task is seen at non-zero energy constraint settings for all pool settings
  - An average boost of 5.4% in testing accuracy (75.0% at base, improved to 80.4%) was observed
    - ➢ Hypothesis testing results in a p-score of 0.00029
    - ➢ This occurs at an energy consumption of 0.2 with 4 neurons per pool
- Data collected over 20 random initializations of the reservoir and testing/training split

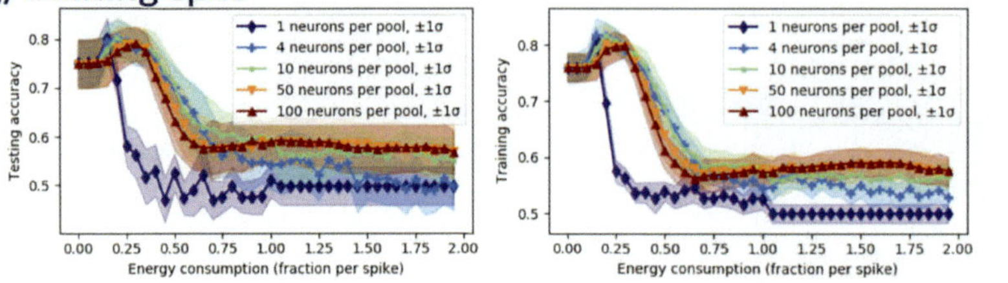

Fountain and Merkel, 2020

In the one-channel E.E.G. seizure detection task, optimal accuracy is attained with non-zero energy constraints across all pool configurations, resulting in a 5.4% average accuracy boost (from 75.0% to 80.4%), supported by a significant p-score of 0.00029 at an energy consumption of 0.2 with four neurons per pool, based on 20 random reservoir initialisations and testing/training splits.

74 One-channel EEG seizure detection task results

- Lyapunov exponent estimate monotonically decreases
  - Correlation coefficient drops with increasing number of neurons per pool
- Separation ratio reaches its peak at the same point as the maximum in testing accuracy
  - Correlation coefficient is consistently ~0.85

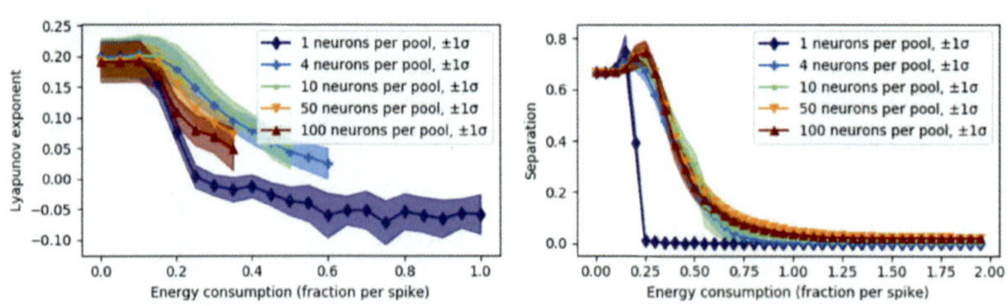

Fountain and Merkel, 2020

The Lyapunov exponent estimate shows a consistent monotonic decrease, while the correlation coefficient decreases with an increasing number of neurons per pool. Additionally, the separation ratio peaks alongside the maximum testing accuracy, with the correlation coefficient maintaining a consistent value of approximately 0.85 throughout the analysis.

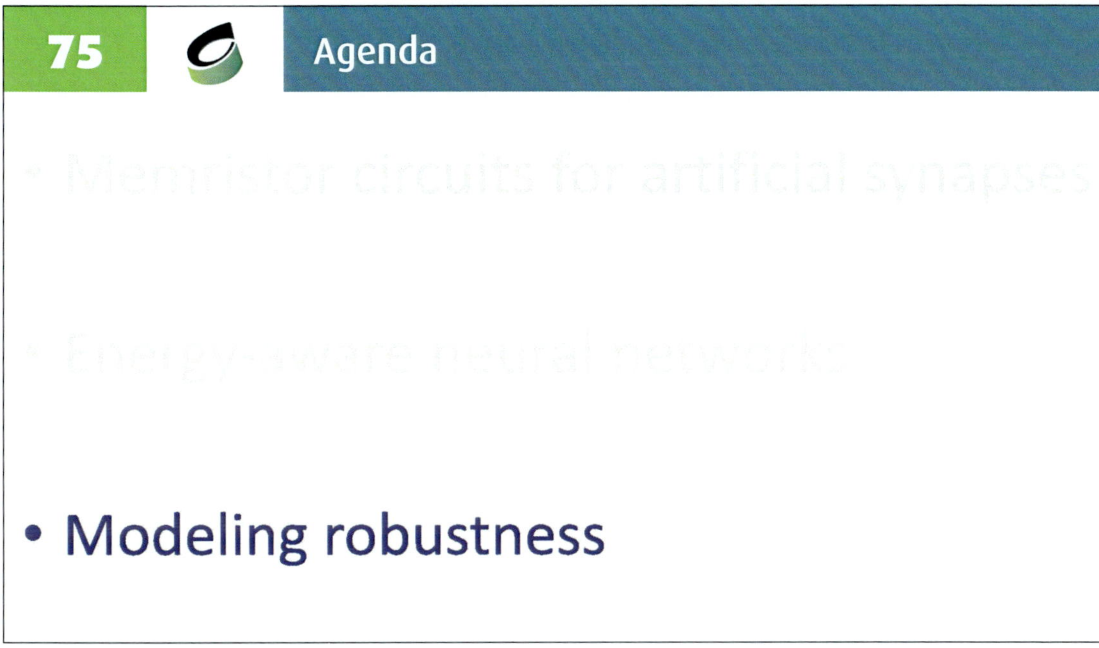

Agenda Modeling robustness.

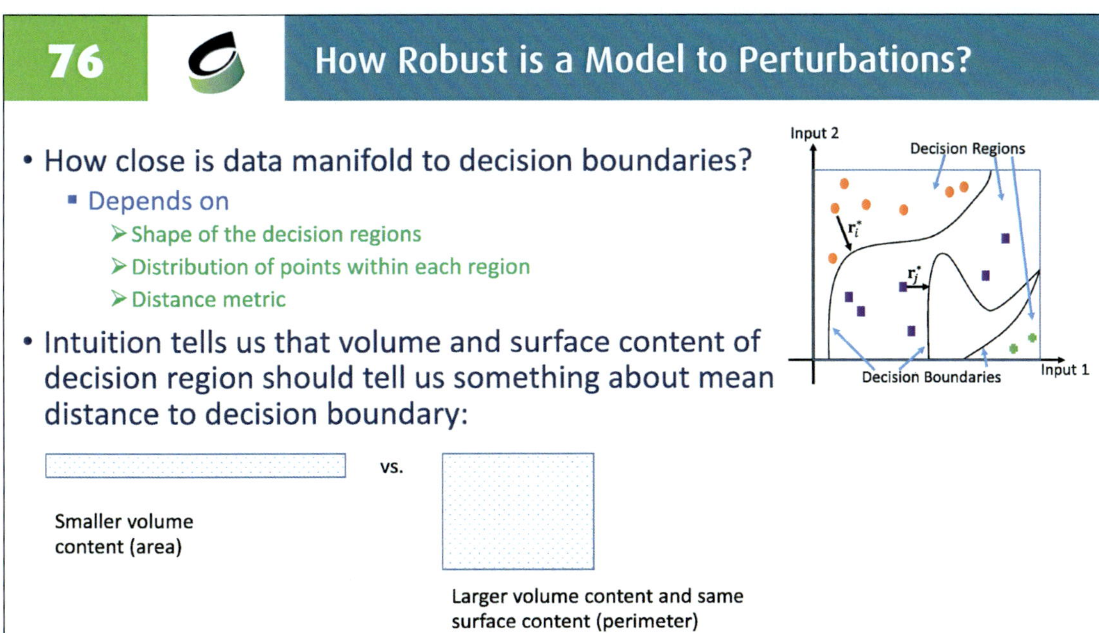

The closeness of the data manifold to decision boundaries depends on the decision regions' shape, point distribution within each region, and the chosen distance metric. Intuitively, decision regions' volume and surface area reflect the mean distance to decision boundaries, indicating class separation in the feature space.

77 How Robust is a Model to Perturbations?

- Approach:
  - Model the geometries of decision regions and the effect of hardware faults on those geometries
- To start, assume 1.) $V = cr^n$, 2.) S=dV/dr, 3.) Data are uniformly distributed
  - V and S are volume and surface content, c is a constant, r is a characteristic length, n is the dimension of the input
    - ➤ Hyperspheres, hypercubes, simplices, etc.

$$\bar{\epsilon} \equiv \mathbb{E}_{\mathbf{x}\in\mathcal{S}}\left[\min_{\mathbf{y}\in\mathcal{S}'}\|\mathbf{x}-\mathbf{y}\|_2\right] = \frac{1}{V}\int_0^r \epsilon S(\epsilon)d\epsilon = \cdots = \frac{n}{n+1}\frac{V}{S} \approx \frac{V}{S}$$

$$f(\epsilon) = \frac{S(\epsilon)}{V} = \frac{nc(r-\epsilon)^{n-1}}{cr^n} = \frac{n(r-\epsilon)^{n-1}}{r^n}$$

$$F(\epsilon) = 1 - \frac{1}{r^n}(r-\epsilon)^n = 1 - \left(1 - \frac{\epsilon}{r}\right)^n$$

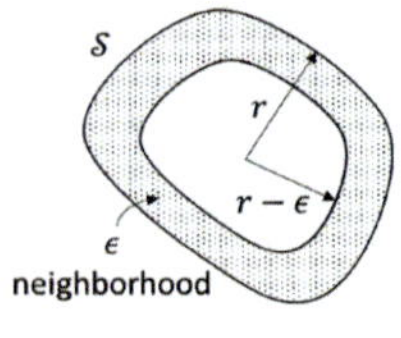

$$A_r(\epsilon) \equiv 1 - F(\epsilon) = \left(1 - \frac{\epsilon}{r}\right)^n$$

Relative accuracy: Adversarial accuracy expressed as fraction of clean accuracy.

Merkel, unpublished.

Explains more about the robustness of a model to perturbations.

78 How Robust is a Model to Perturbations?

- We have applied this model to determine the lower bound relative accuracy of neural networks with quantized weights
- A neural network with step activation divides input space into polytope decision regions
  - Minimum decision region size is inversely proportional to weight precision
    - ➤ Minimum size of decision region (partially) determines lower bound of robustness
    - ➤ Simplest polytope is simplex (generalization of triangle)

Draghici, 2002

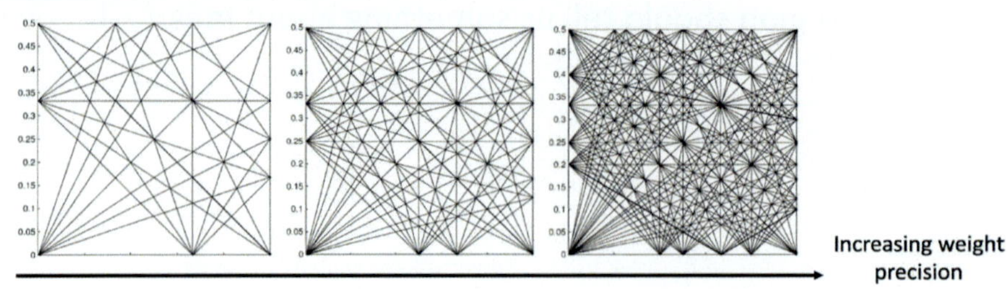

The model establishes the minimum achievable accuracy of neural networks employing quantised weights, where step activation divides the input space into polytope decision regions. The minimum size of these decision regions, partially determined by weight precision inversely, is a crucial factor influencing the lower bound of robustness, with the simplex, a generalisation of a triangle, representing the simplest polytope.

79 Mean Distance to Simplex Face

- For large n and q-bit weights, a hypercube of length $l = 2^{1-q}$ can be shattered by a set of simplices with longest edge $s \leq \sqrt{n}2^{-q}$ (Draghici, 2002)
- Volume and surface area of regular n-simplex (Rabinowitz et al., 1989):

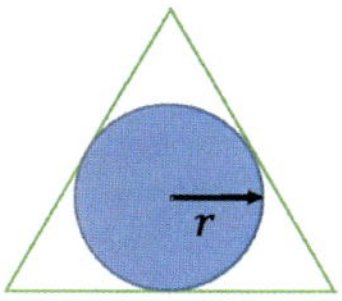

$$V = \frac{s^n}{n!}\sqrt{\frac{n+1}{2^n}} \qquad S = \frac{(n+1)s^{n-1}}{(n-1)!}\sqrt{\frac{n}{2^{n-1}}}$$

- With $s = \sqrt{n}2^{-q}$:

$$r = n\frac{V}{S} = \frac{2^{-q}}{\sqrt{2n}}$$

$$A_r(\epsilon) = \left(1 - \frac{\epsilon}{h}\right)^n = \left(1 - \epsilon\sqrt{2n}\,2^q\right)^n$$

Quantization should provide some protection to adversarial attacks.

Explains the mean distance to simplex face.

80 MNIST Results on MLP (784-100-10)

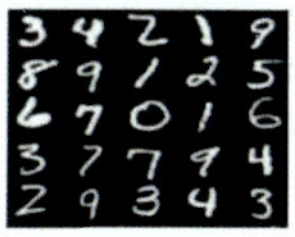

- Attack strengths range from 0 to 0.5 in 0.005 increments

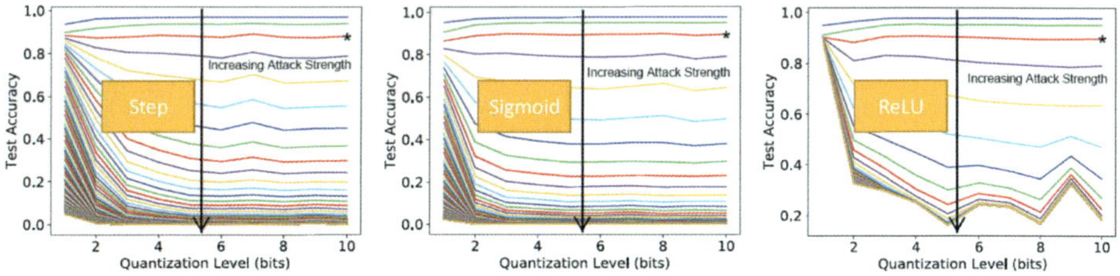

- ReLU is most robust at extreme quantization
- Effect of quantization depends on attack strength
- *Critical Attack Strength*: Attack strength where quantization has minimal effect on accuracy

In M.L.P. experiments on M.N.I.S.T., varying attack strengths from 0 to 0.5 in 0.005 increments reveal ReLU's highest robustness at extreme quantisation levels, while the impact of quantisation varies with attack strength. A critical attack strength emerges where quantisation minimally affects accuracy, indicating a pivotal point in the trade-off between quantisation and model robustness.

81 2 Spiral Results (2-100-1)

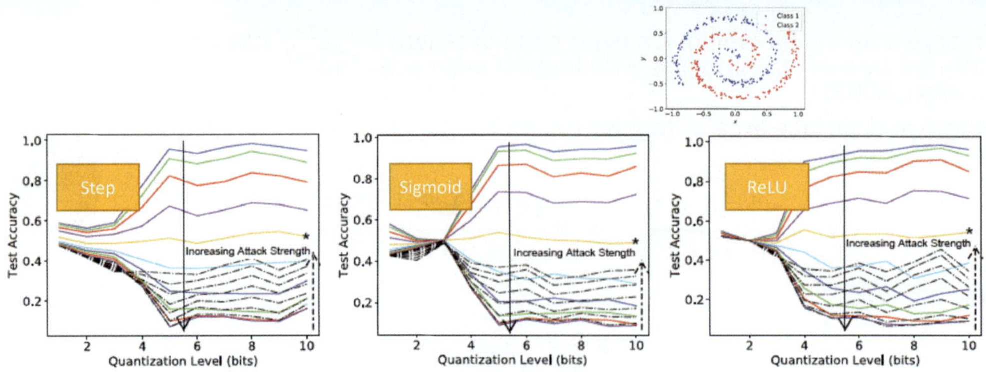

- Quantization has significant impact on clean accuracy
- Concept of *critical attack strength* holds even with lower dimensional inputs
- As attack strength increases, test accuracy eventually stops decreasing and approaches random chance

Quantization significantly affects clean accuracy, and critical attack strength persists even with lower dimensional inputs. As attack strength rises, test accuracy eventually plateaus and approaches random chance, indicating a threshold beyond which further attacks have diminishing impact.

82 Lower Bound A_r Model vs. simulation Results

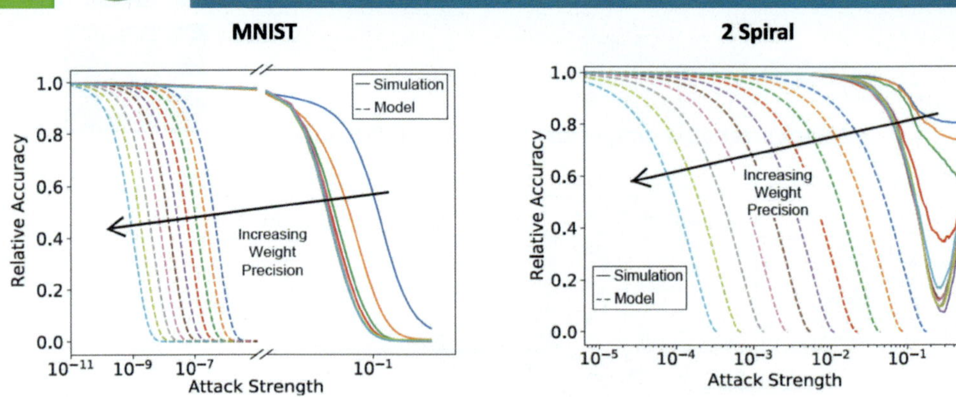

- **Bounds could be tightened by revisiting assumptions about**
  - Decision region shape – Likely not convex and more complex than simplex
  - Distribution of data in decision region

To refine accuracy and robustness assessments of neural networks, it's essential to reassess assumptions about decision region shape, which are likely non-convex and more intricate than a simplex, alongside understanding the distribution of data within decision regions. These considerations offer avenues for tightening bounds and improving the fidelity of model evaluations.

CHAPTER 09

In-memory computing meets Neuromorphic Event-Driven Vision

Arindam Basu

City University of Hong Kong

The intersection of in-memory computing (IMC) and neuromorphic event-driven vision represents a significant frontier in the Internet of Video Things, as highlighted in the collaborative efforts at the Brain Lab @ CityU HK. This presentation highlights how two aspects of neuromorphic systems--event-driven perception and in-memory computing can be combined to develop very efficient visual perception systems. In particular, event-based image sensors enable data reduction at the source by being sensitive to change in temporal contrast. In this work, the image buffer used to store these data is used to also process it. It is shown that using SRAM or combinations of SRAM and DRAM as the image buffer, basic operations such as noise removal and region proposal may be accomplished within the memory itself. Hence, the following recognition system may be turned off in blank frames and also focus only on salient regions, thus saving significant energy at the system level. Overall, the convergence of neuromorphic event-driven vision and in-memory computing heralds a paradigm shift in low-power, high-performance computing, particularly in applications such as surveillance and intelligent transportation systems of smart cities.

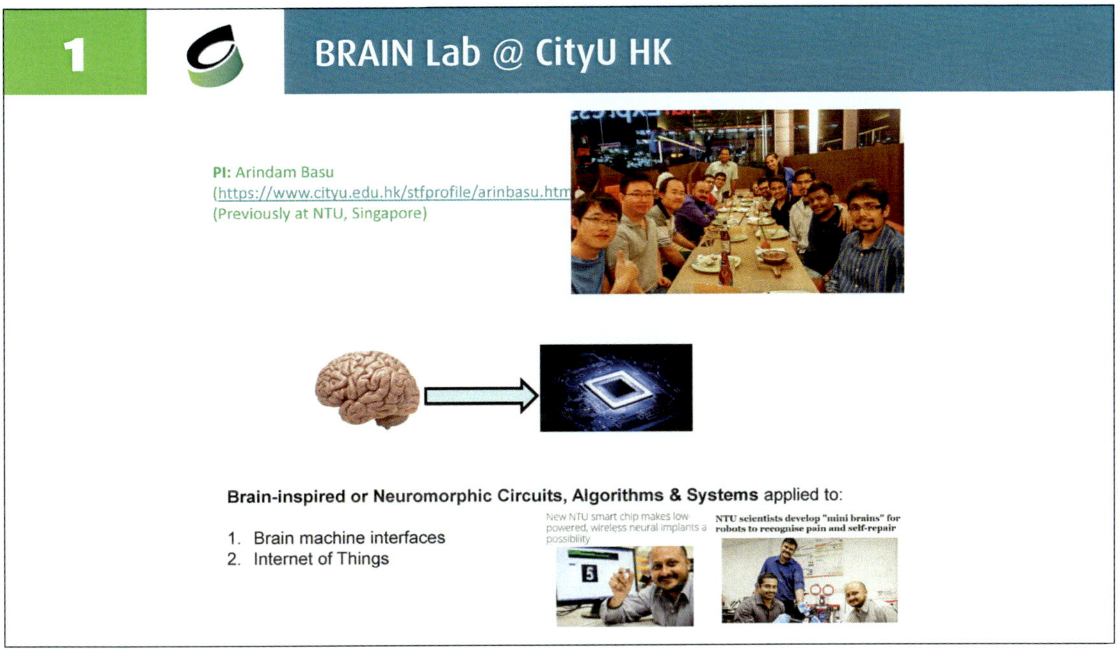

Activities at the Brain lab @ City U HK where they apply neuromorphic circuits, algorithms and systems to brain machine interfaces and IoT. Notable works by the team are mentioned wherein they developed a smart chip for wireless neural implants and mini-brains for robots.

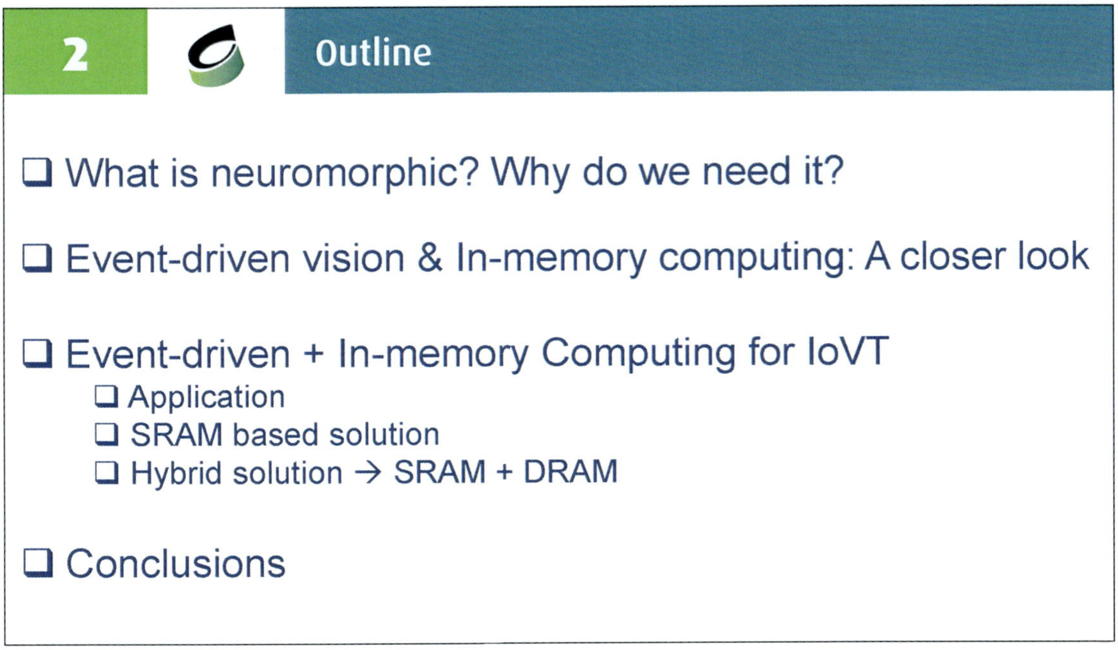

The presentation outlines neuromorphic systems, event driven in-memory computing and their applications. It also briefs about SRAM and DRAM+SRAM based solutions for in-memory computing.

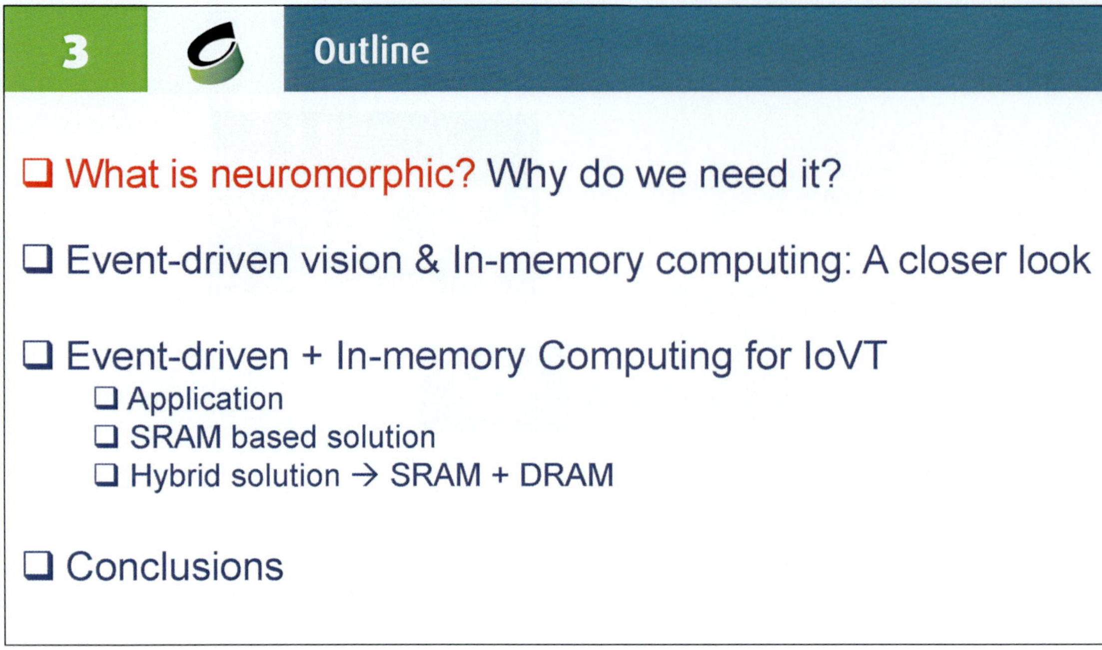

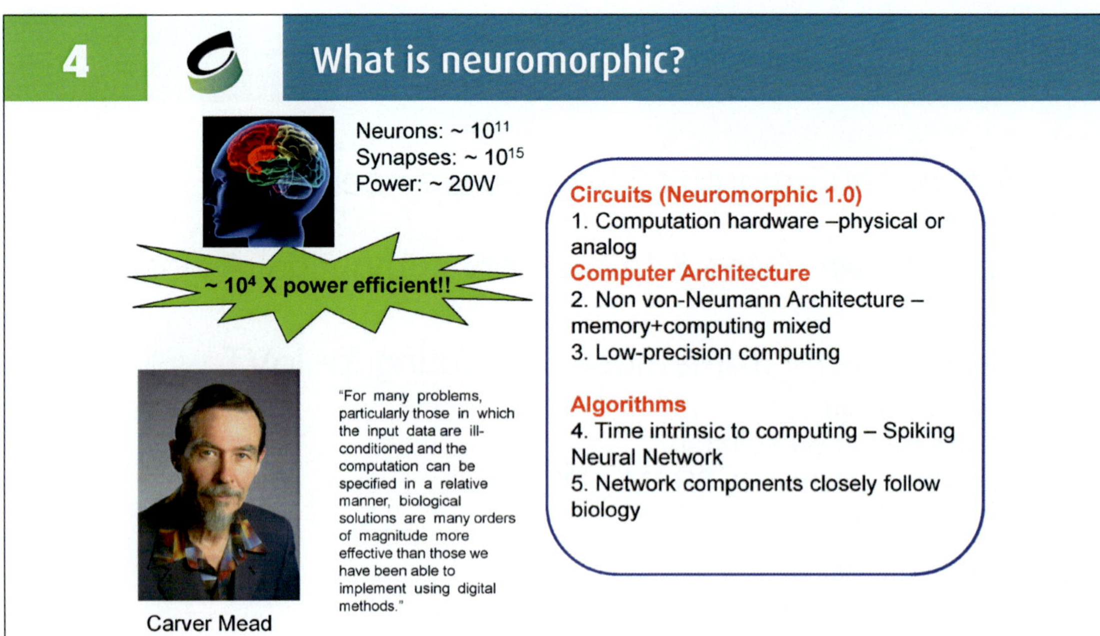

Biological methods are more effective in computing data. While the early definition of Neuromorphic systems involved only analog circuit based processing, modern connotation of the term has broadened in meaning to different communities to cover In-memory computing and Spiking Neural Networks.

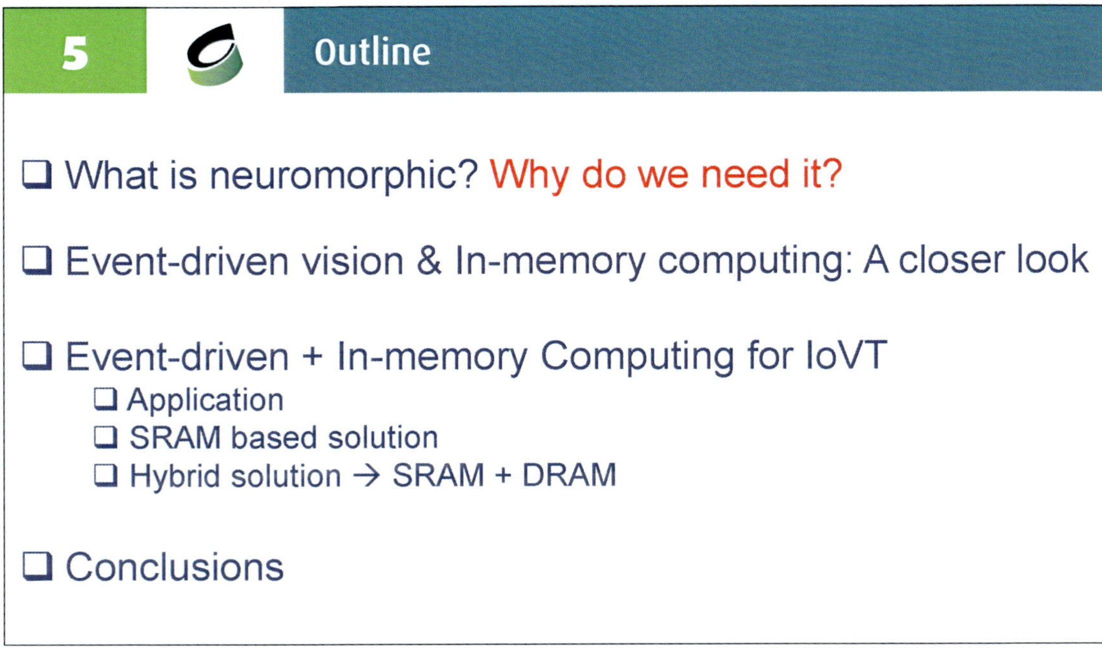

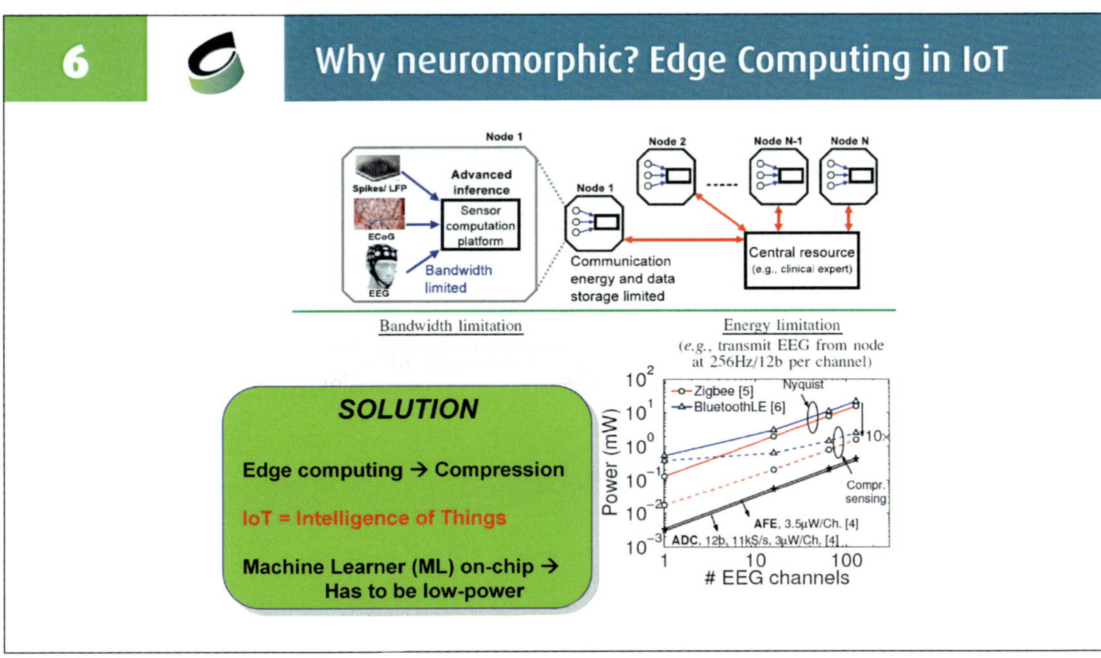

Edge computing in IoT is a better way for efficient and energy efficient computing. Considering EEG signal processing as an example, machine learning on chip can be a plausible solution to limitations that are faced currently.

7 How to create low-energy ML ASIC?

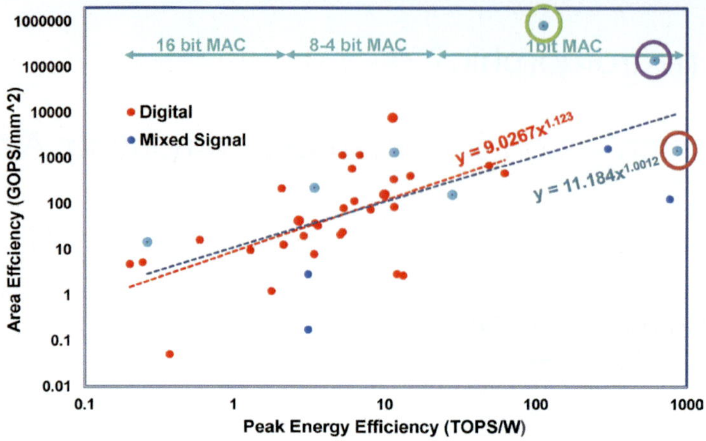

H. Valavi, P. J. Ramadge, E. Nestler and N. Verma, "A 64-Tile 2.4-Mb In-Memory-Computing CNN...", *JSSC*, 2019
W. Khwa *et al.*, "A 65nm 4Kb algorithm-dependent computing-in-memory SRAM...", ISSCC, 2018
Q. Dong *et al.*, "15.3 A 351TOPS/W and 372.4GOPS Compute-in-Memory SRAM Macro...", ISSCC, 2020

Comparison of digital and mixed signal systems is done in machine learning application specific integrated chips (ML-ASIC). It is studied that on an average mixed signal systems give better area and power optimization in ML-ASIC. This provides the motivation to use neuromorphic concepts to reduce energy in edge computing systems.

8 Outline

- ❑ What is neuromorphic? Why do we need it?
- ❑ Event-driven vision & In-memory computing: A closer look
- ❑ Event-driven + In-memory Computing for IoVT
  - ❑ Application
  - ❑ SRAM based solution
  - ❑ Hybrid solution → SRAM + DRAM
- ❑ Conclusions

9 Neuromorphic Sensor—Event-driven vision

Asynchronous change detecting pixels—Simulating human retina!!

- High dynamic range
- Data reduction
- Fast reaction
- Ideal sampling

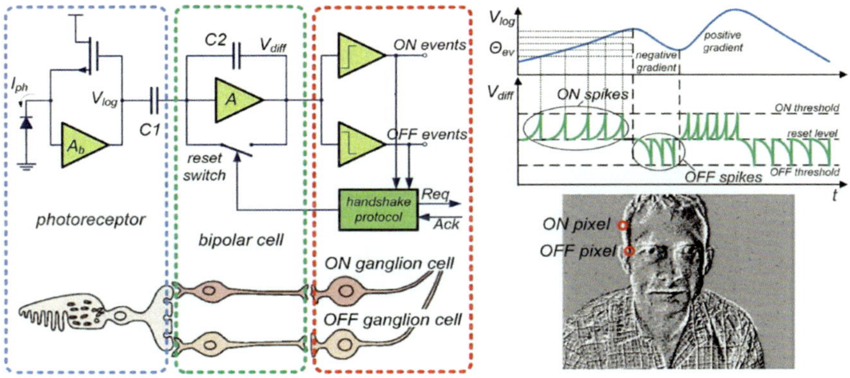

Application in simulating a human retina by building pixels that could detect asynchronous changes by designing systems that could mimic biosystems like ganglion cells, photoreceptors and bipolar cells. The objective is to attain

1) High dynamic range
2) Data reduction
3) Fast reaction
4) Ideal sampling

10 In-memory Computing (IMC): Needs

Integer	
Add	
8 bit	0.03pJ
32 bit	0.1pJ
Mult	
8 bit	0.2pJ
32 bit	3.1pJ

FP	
FAdd	
16 bit	0.4pJ
32 bit	0.9pJ
FMult	
16 bit	1.1pJ
32 bit	3.7pJ

Memory	
Cache	(64bit)
8KB	10pJ
32KB	20pJ
1MB	100pJ
DRAM	1.3-2.6nJ

Memory access energy dominates!

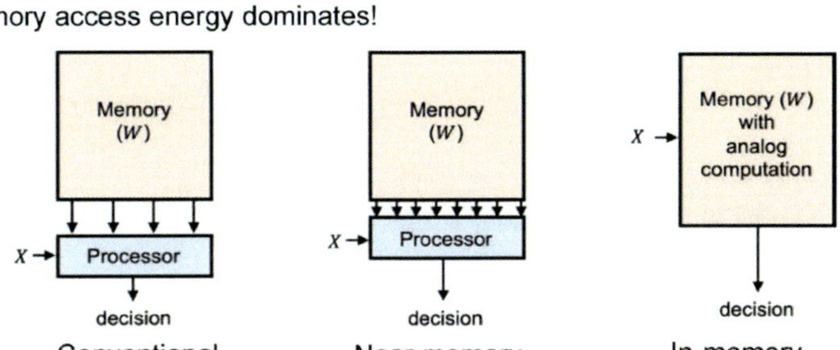

By comparing memory accessing methods, a conclusion is reached where in memory computing(IMC) with analog computation is the key.

11 Volatile memory based IMC

- **Pros**
  - Ubiquitously available on-chip
  - Stable technology
  - Low write energy, high endurance compared to NVM–good for learning as well
- **Cons**
  - Only 1-bit per cell → Area overhead for multi-bit
  - Produce multi-bit output also difficult
  - Volatile problem (e.g. refresh)

Pros and cons of volatile in memory computing, e.g. SRAM or DRAM is discussed. While it has Low write energy and high endurance, it is volatile, meaning it loses data if power is absent. It is stable and available on chip, but is only one bit per cell adding to area occupied and complexity in multi-bit processing.

12 Volatile memory based IMC: Example in this talk

SRAM/DRAM
Image storage

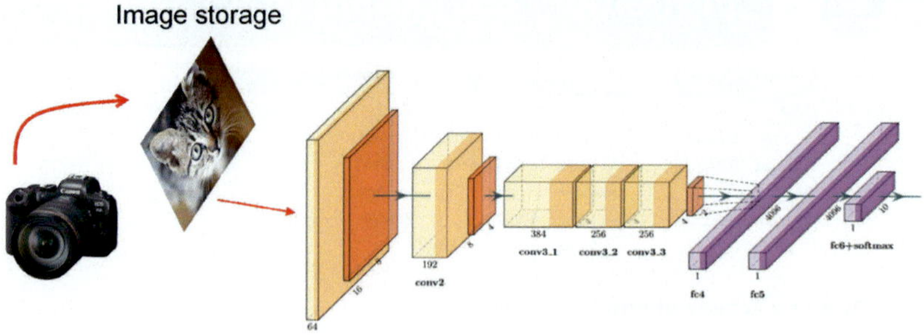

RRAM for DNN

- What if memory storing image can process it?
  - Remove noise, Region proposal
  - Trigger ON the following DNN

Scopes the Application of RRAM in DNN to remove noise and reduce area for image processing.

13 Outline

- ❑ What is neuromorphic? Why do we need it?
- ❑ Event-driven vision & In-memory computing: A closer look
- ❑ Event-driven + In-memory Computing for IoVT
  - ❑ Application
  - ❑ SRAM based solution
  - ❑ Hybrid solution → SRAM + DRAM
- ❑ Conclusions

14 Internet of Video Things (IoVT)

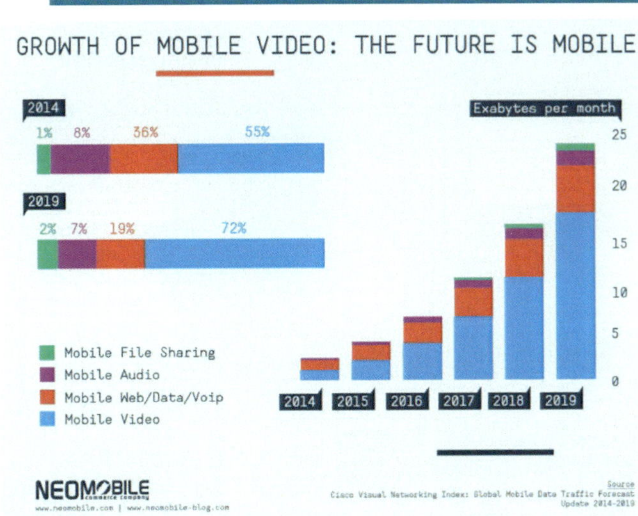

15 Unattended Ground Sensor (UGS)

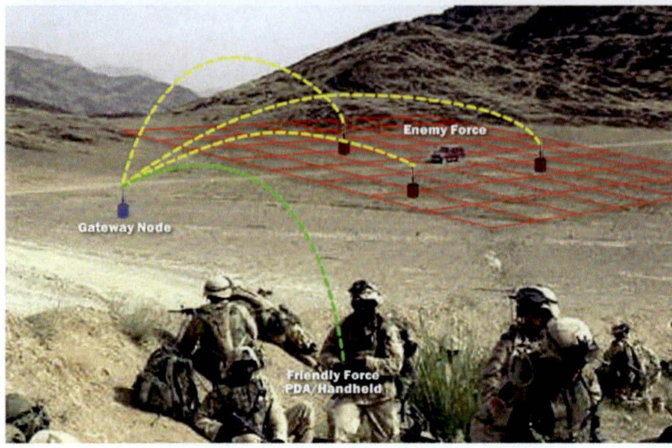

- Battery operated
- Wireless Tx
- Challenge for Visual UGS

16 Intelligent Transportation System (ITS)

Our neuromorphic solution integrates at RSU

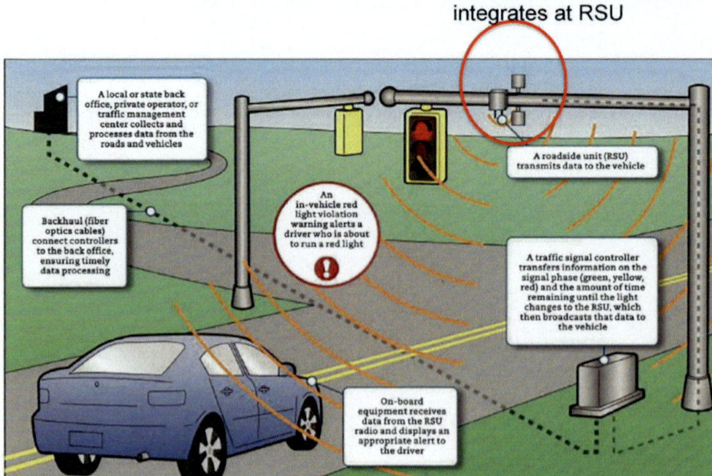

17 Demonstration

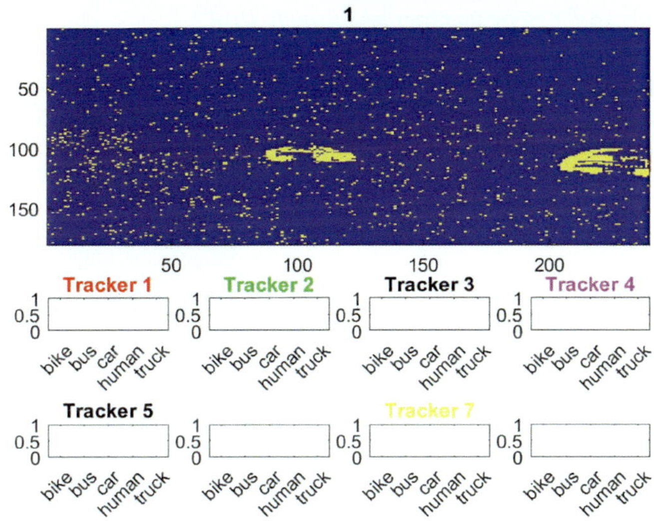

18 Outline

- ❑ What is neuromorphic? Why do we need it?
- ❑ Event-driven vision & In-memory computing: A closer look
- ❑ Event-driven + In-memory Computing for IoVT
  - ❑ Application
  - ❑ SRAM based solution: Temporal sparsity
  - ❑ Hybrid solution → SRAM + DRAM
- ❑ Conclusions

19 Temporal Sparsity & Denoise

- Is event-based processing always frugal?
  - ➢ Noise causes problems
  - ➢ Temporal sparsity maybe hard to exploit?
  - ➢ Cannot duty cycle? IoT energy saving?

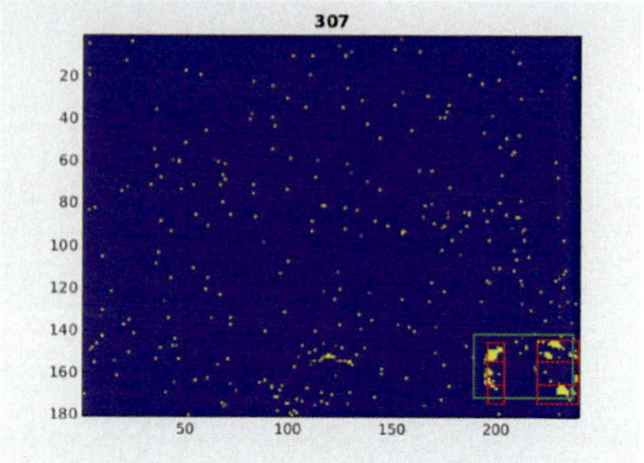

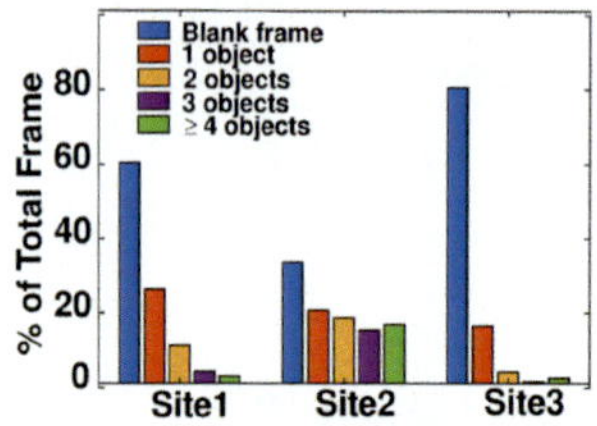

51% frames are empty

Solution: Remove noise to detect Blank frames

20 Event-based binary image (EBBI) comparison

- Alternative to storing timestamp for each pixel → store 1/0 if any event happened/not in frame duration t_F (ignore polarity)
- Hybrid → Frame + Binary event
- Binary representation → Event based binary image (EBBI)
  - ➢ Blank frame detection (BFD) → 2X energy savings at system level!
  - ➢ What about BFD detection overhead?
  - ➢ BFD entirely In-memory → Huge energy/time savings compared to von-Neumann

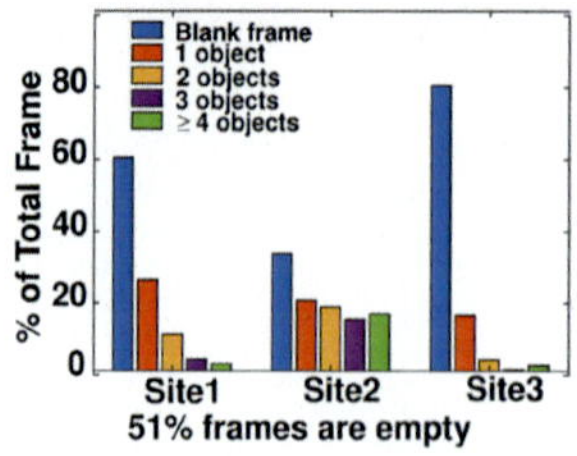

51% frames are empty

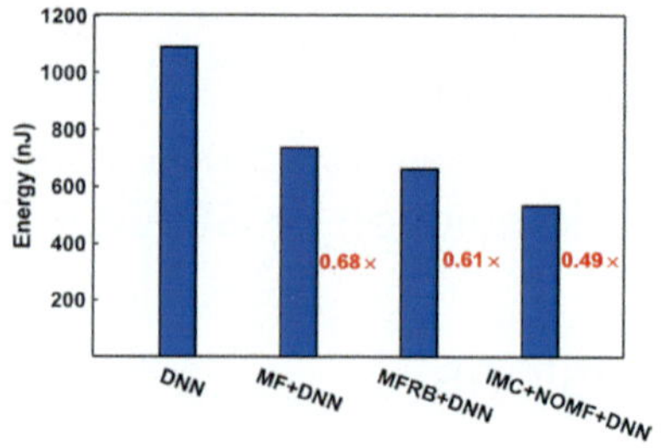

[1]S. K. Bose, D. Singla and A. Basu, "A 51.3-TOPS/W, 134.4-GOPS In-Memory Binary Image Filtering in 65-nm CMOS," in *IEEE Journal of Solid-State Circuits*, doi: 10.1109/JSSC.2021.3098539.

21 Denoise—Non-overlap Median Filter (NOMF)

- Conventional NVS → Event domain filtering (nearest neighbour filter etc.)
- For EBBI frames → Median filtering is good!

$$P_{mf}(i,j) = \text{median}\left(\left\{P(i+k, j+l) \mid k,l \subset \mathbb{Z} \text{ and } \in \left[-\frac{n-1}{2}, \ldots, \frac{n-1}{2}\right]\right\}\right)$$

- For binary images → $P_{mf}(i,j) = \begin{cases} 1, & \text{if } \Sigma P(i+k, j+l) \geq \left\lceil \frac{n^2}{2} \right\rceil \\ 0, & \text{otherwise} \end{cases}$

- Propose Approximate median filter→ Can be mapped to IMC
- Regular nxn Median filter → Stride = 1
- Proposed nxn NOMF → Stride = n

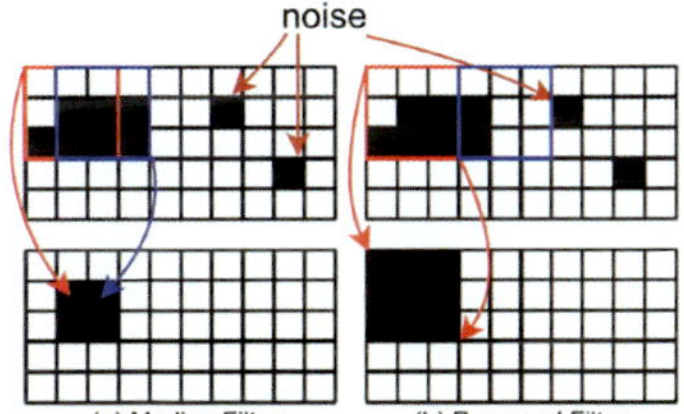

22 Top level Architecture

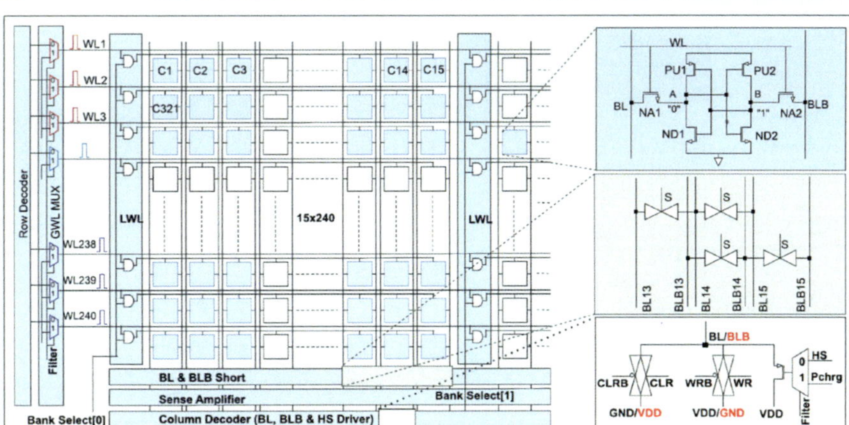

- 320x240 SRAM macro → QVGA
- Filter size, n = 3,5 supported
- Banks have 15 columns → 22 banks (~320/15)

23 NOMF Operation—Exploit Read Disturb

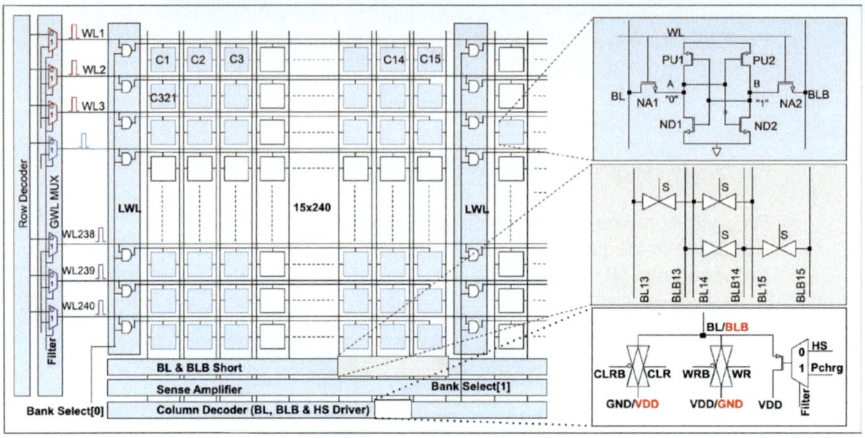

- Pre-charge bit-lines, WL Enable "n" rows, short "n" column BLs (BLBs)
- Latches fight – majority wins and flips bits of minority
- All banks filter in parallel → need only 240/n cycles!

24 Results-I (Characterization)

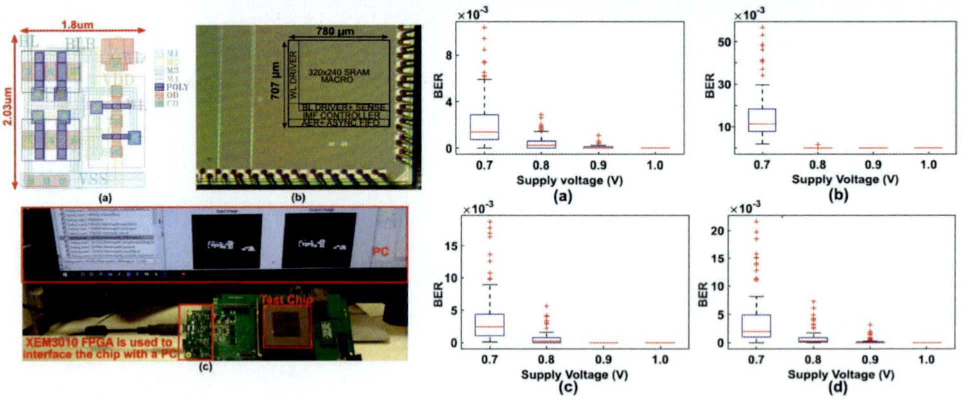

- 65nm CMOS
- FPGA for interfacing with PC

- Results from 4 IC
- ~ 8K "3x3" patches tested
- Worst case patch → 126 5"1"4"0" patterns

25 Results-II (Video)

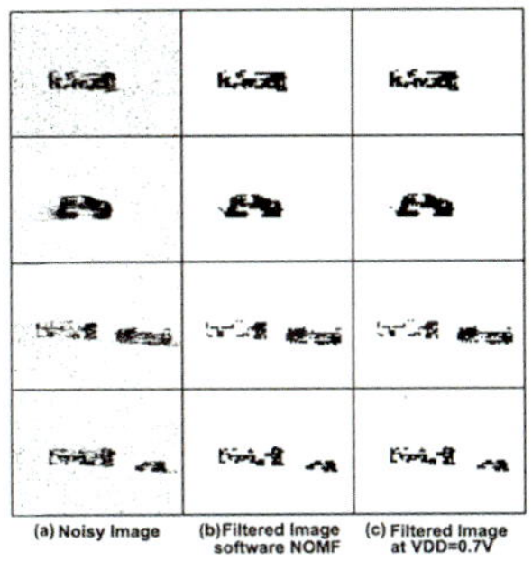

- 1000 frames
- 240x180 images from DAVIS

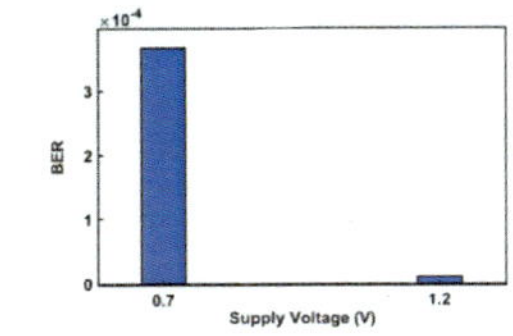

- ~50X lower than worst case

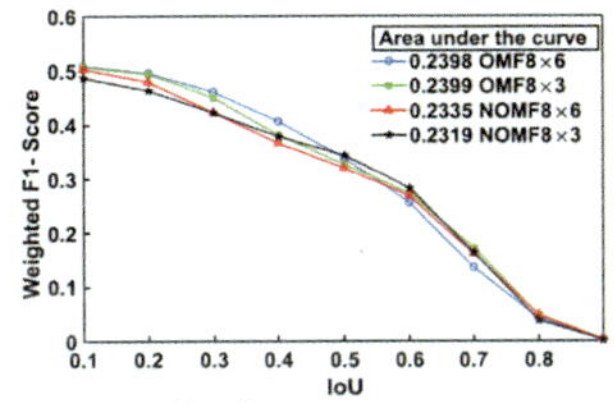

$$F1 = 2\frac{P \times R}{P + R}$$

P → Precision = TP/(TP+FP)
R → Recall = TP/(TP+FN)

- A_{GP} → Area of ground truth
- A_P → Area of the proposed region

$$IoU = \frac{AGT \cap Ap}{AGT \cup Ap}$$

26 Performance Comparison

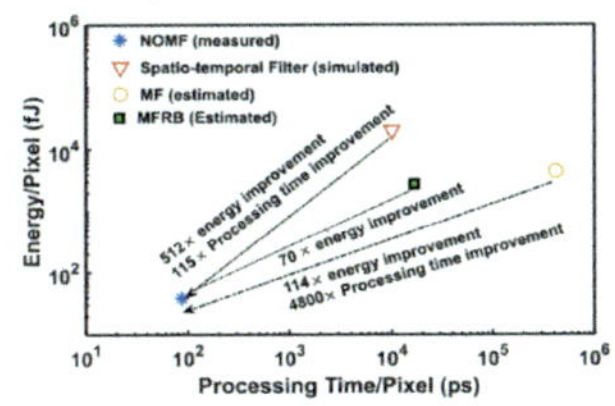

COMPARISON WITH DIFFERENT IMAGE FILTERING WORKS

	This Work	ISCAS'15 [36]	TCAS-II'20 [37]	ASP-DAC'20 [38]
Implementation & Technology	ASIC, 65nm	ASIC, 180nm	ASIC, 65nm	FPGA Artix-7
Method	Spatial filter, in-memory Computing	Spatio-temporal filter	Spatial filter, in-memory Computing	Spatio-temporal filter
Memory Size	9.375kB	-	9.375kB	0.25kB
Area/ cell (μm^2)	3.6	400	3.54	-
Latency/ Pixel (ns)	0.086 at 24MHz	10	0.11	10
Power	450μW at 0.7V, 24MHz	1mW at 1Meps	-	0.471W

COMPARISON OF DIFFERENT PUBLISHED IMC WORKS

	This Work	ISSCC'19 [19]	TCAS-I'19 [39]	JSSC'18 [10]	SoVC'19 [40]	JSSC'19 [16]
Technology	65nm	55nm	65nm	65nm	28nm	65nm
SRAM type	6T	Twin-8T	6T (split WL)	6T	6T (split WL)	10T
Algorithm	Image denoising	CNN	BNN	Versatile DNN	BNN	BWN
SRAM capacity	9.375kB	0.468kB	2kB	102kB	2kB	2kB
Operating Voltage (V)	0.7-1.2	1	1	0.55-1	1	0.8-1
Input/ Weight precision	1[b] / 1[b]	1-4 / 2-5	1 / 1	1 / 1	1 / 1	6 / 1
Peak Throughput (GOPS)	134.4		278.2	1380	615	8
Peak Energy Efficiency (TOPS/W)	51.3	18.37-72.02 [13.15-51.5][a]	30.49-55.8	6.3	300 [55.6][a]	51.3
Area Efficiency (TOPS/mm^2)	0.243	-	33.13	0.089-0.365	-	0.127

[a]Assuming energy efficiency to 65nm implementation following energy efficiency ∝ 1/(Tech. node)^2
[b]Assume the filter kernel is convoluted over the binary image and all the filter weights are 1.

27 Outline

- ❑ What is neuromorphic? Why do we need it?
- ❑ Event-driven vision & In-memory computing: A closer look
- ❑ Event-driven + In-memory Computing for IoVT
  - ❑ Application
  - ❑ SRAM based solution: Spatial sparsity
  - ❑ Hybrid solution → SRAM + DRAM
- ❑ Conclusions

28 Spatial sparsity—Need for region proposal

- Exploit temporal redundancy → 2x energy savings
- Frame size → 240x180
- Average object size → 71x32→ 5% of frame size
- Spatial redundancy → Always on dedicated Region of Interest (RoI) Module
- RoI module → Calculates bounding boxes & Triggers DNN

Mean Object Size (Height x Width)

	Car/Van	Bus	Bike	Truck
Site1	16x42	31x94	15x21	22x50
Site2	25x47	52x107	17x22	35x61
Site3	34x84	64x180	26x44	50x104

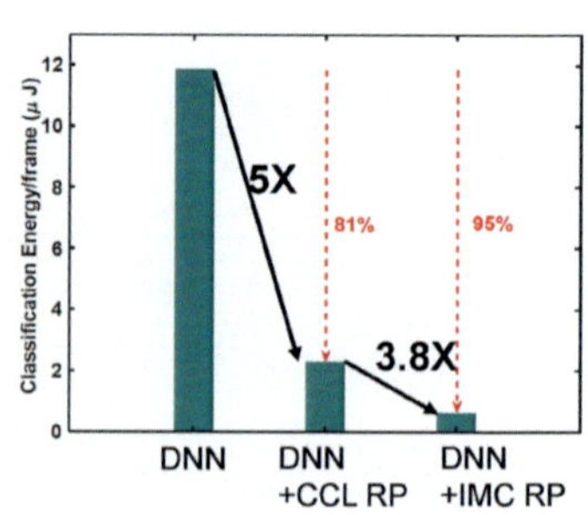

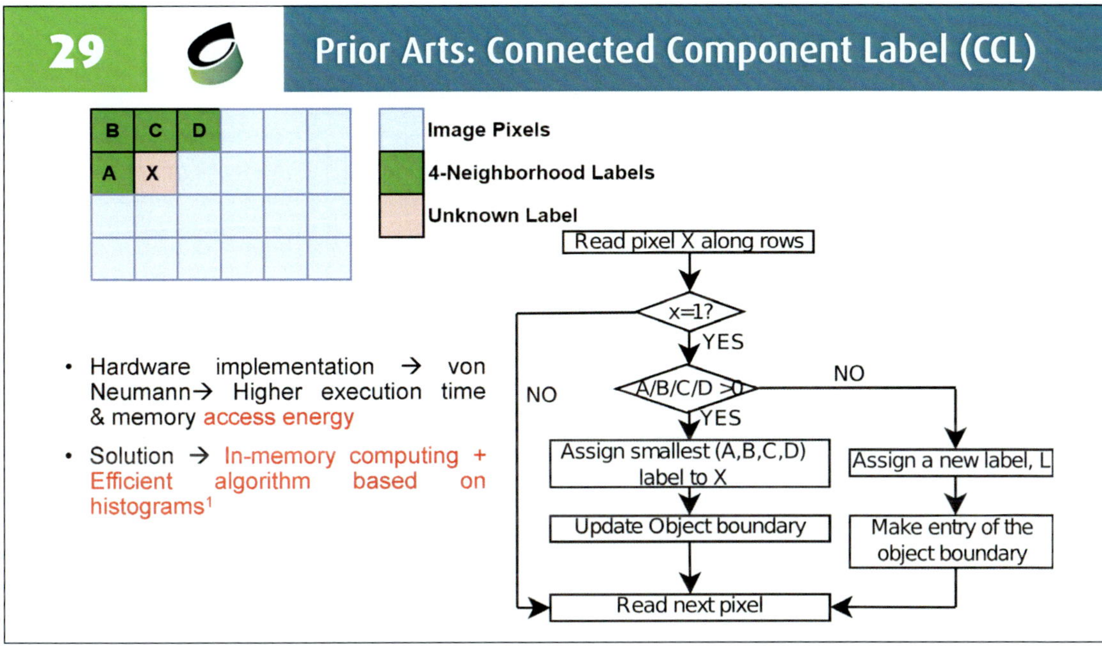

Connected component labeling is an image processing technique and its algorithm is presented in the flowchart. This algorithm prioritizes efficiency in memory computations and saving execution time.

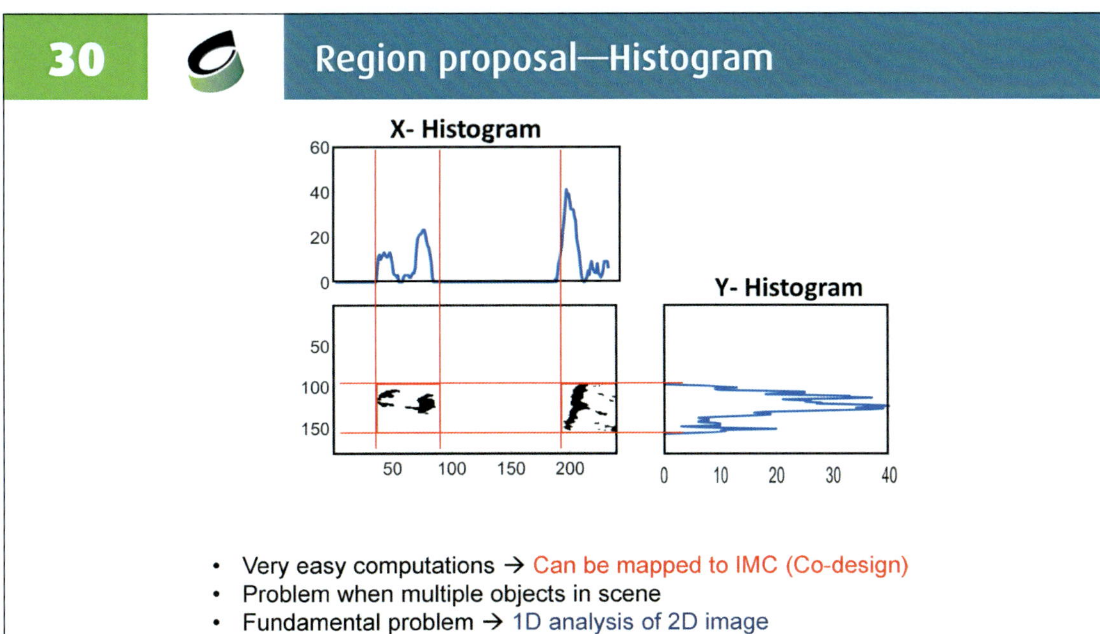

A method is proposed to Identifying the potential region of interest within image using a histogram. The potential Challenges for this method are multiple objects in scene and 1D analysis of 2D image.

31 Proposed Method: 9T-SRAM based RoI

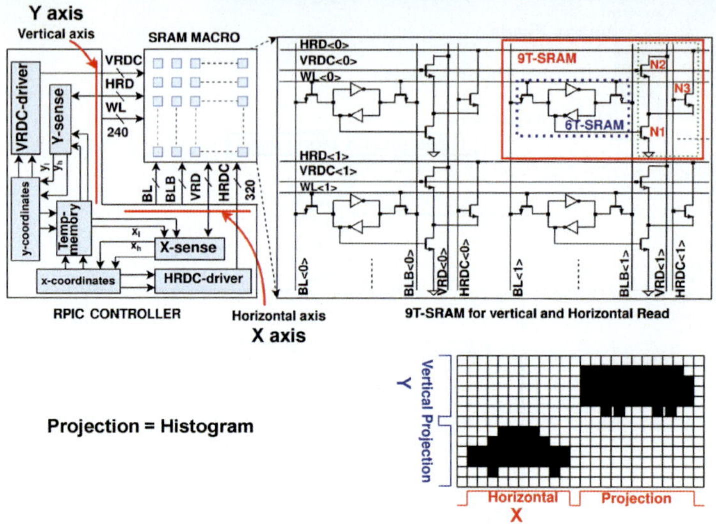

- 320x240 SRAM cells
- Unit cell → 9T-SRAM
- N1 & N2→ Vertical Read
- N1 & N3 → Horizontal Read
- H read/ Y projection→ Precharge HRD lines → Enable HRDC lines → sense HRD lines
- V read/X projection→ Precharge VRD lines → Enable VRDC lines → sense VRD lines

9T-SRAM based RoI is a method based on 9 transistor SRAM. SRAM is a volatile memory that store data using transistor.

32 Iterative and Selective Search

- Horizontal (X) Search→ Rough estimate of object locations
- Vertical (Y) Search → enable individual objects→ Accurate y-coordinates
- Horizontal (X) Search → enable individual objects→ Accurate x-coordinates

The flowchart represents the algorithm for iterative and selective search. The algorithm is for finding rough estimation of objects in an image.

33 Test setup & RP demonstration

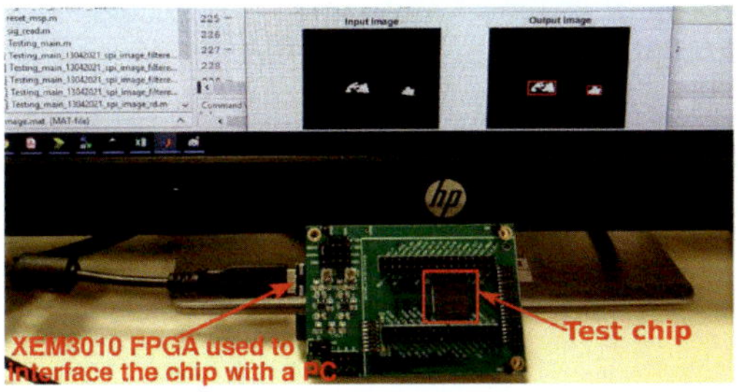

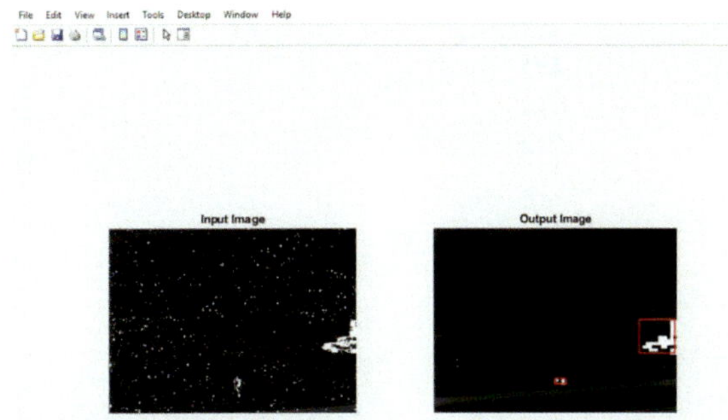

- Noisy images → MATLAB and FPGA → Chip
- RoI → red rectangle

For testing the chip, noisy images from MATLAB are fed to it through an FPGA. The processed output is displayed on a connected computer monitor.

34 Performance-I

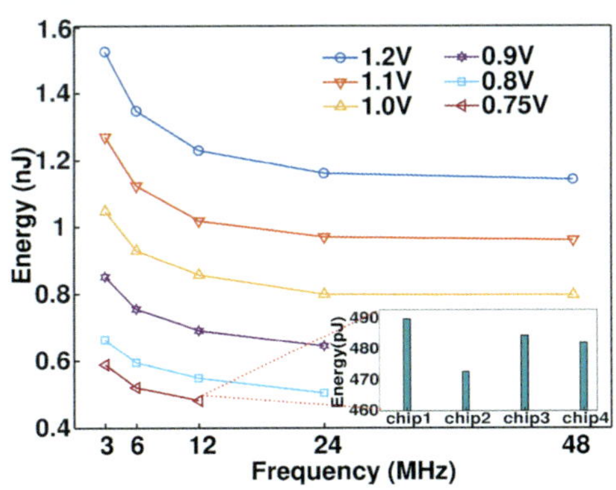

ASSCC 21
JSSC 22

- Minimum energy/object detection → 0.482nJ
- RP & blank frame detection → sub-nJ energy at 1V

The graph shows how much energy it takes to find objects in a picture. The key point is that this system uses very little energy (sub-nanojoules) to do this, especially when working at 0.75 volt

35 Performance-II

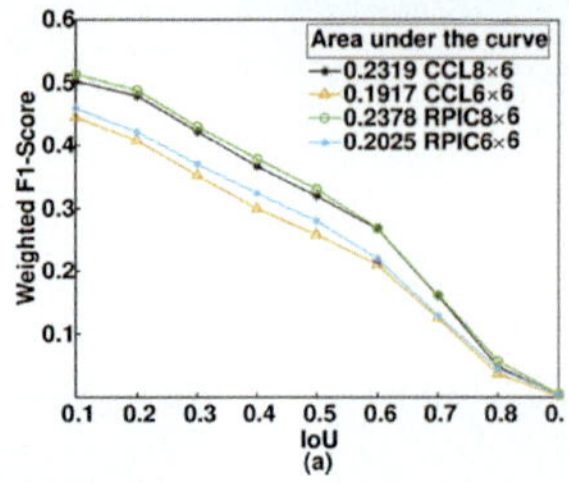

- A_{GP} → Area of ground truth
- A_P → Area of the proposed region

$$IoU = \frac{AGT \cap Ap}{AGT \cup Ap}$$

$$F1 = 2\frac{P \times R}{P + R}$$

P → Precision = TP/(TP+FP)
R → Recall = TP/(TP+FN)

Performance at part with CCL

	This Work	CICC'20 [26]	TCAS-I'19 [27]	JSSC'18 [28]	VLSI'19 [29]
Technology	65nm	65nm	65nm	65nm	28nm
SRAM type	9T	8T	6T (split WL)	6T	6T (split-WL)
Algorithm	RoI Calculation	ML	BNN	Versatile DNN	BNN
SRAM capacity (kB)	9.375	2	2	102	2
Operating volatge (V)	0.75-1.2	0.45-0.8	1	0.55-1	1
input/weight precision	1 / 1[b]	1	1 / 1	1 / 1	1 / 1
Peak Throughput (GOPS)	409	-	278.2	1380	615
Peak Energy Efficiency (TOPS/W)	389	490	30.49-55.8	6.3	300 [55.6][a]

[a] Assuming energy efficiency to 65nm implementation following energy efficiency $\propto \frac{1}{(Tech.node)^2}$.
[b] Assume the read bit line discharging is similar to BNN (input and weight precision is 1) works.

The slide compares different ways to find objects in images using a special memory (SRAM). Their method uses less space and energy than other methods to find objects.

36 Outline

- ❑ What is neuromorphic? Why do we need it?
- ❑ Event-driven vision & In-memory computing: A closer look
- ❑ Event-driven + In-memory Computing for IoVT
  - ❑ Application
  - ❑ SRAM based solution: Temporal sparsity
  - ❑ Hybrid solution → SRAM + DRAM
- ❑ Conclusions

37 Hybrid In-memory computing (IMC)

- Current IMC uses volatile SRAM or other NVM
- What about mixing modes?

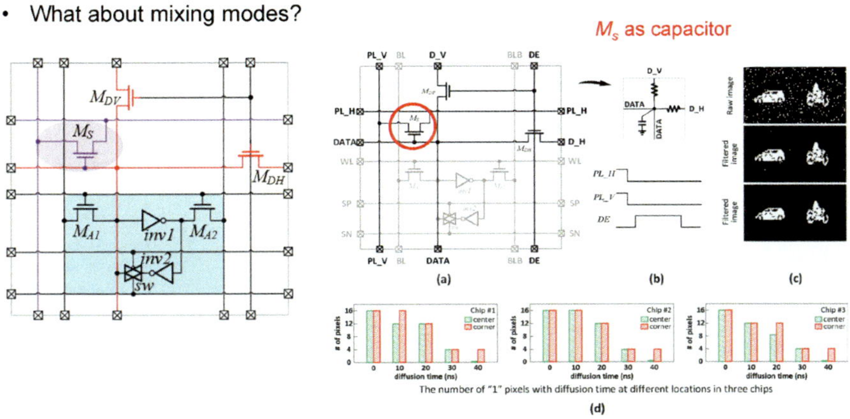

- HRAM = DRAM + SRAM → Denoise by diffusion + RP by projection

Introducing an new concept of IMC, HRAM by mixing memory modes , SRAM and DRAM modes.it aim to leverage the strengths of both SRAM (speed) and DRAM (analog storage) for optimized performance in IMC applications.

38 Future: Hybrid In-memory computing (IMC)

- Additional near-memory processor to merge fragmented objects

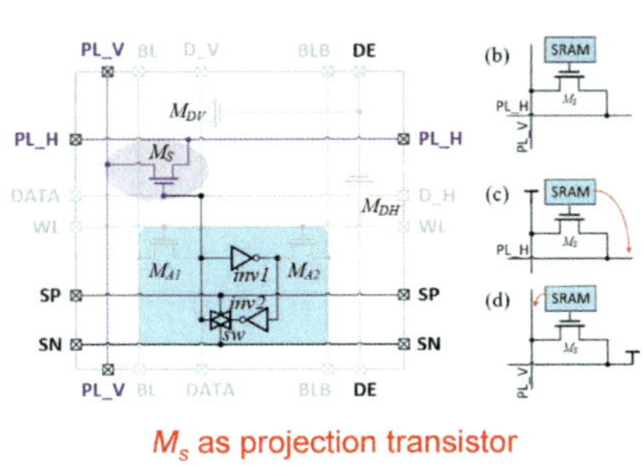

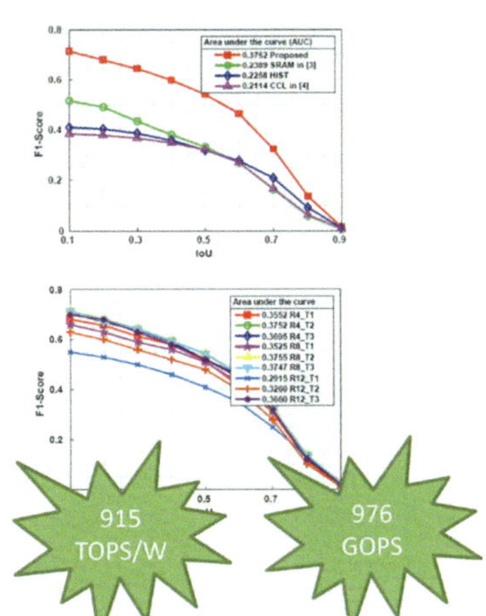

- Other options → SRAM + RRAM/Flash?

Hybrid IMC as the potential approach for in-memory computing. Hybrid IMC combines different memory types like SRAM and DRAM, potentially offering advantages in terms of speed, capacity, and data movement reduction for IMC tasks.

39 Outline

- ❑ What is neuromorphic? Why do we need it?
- ❑ Event-driven vision & In-memory computing: A closer look
- ❑ Event-driven + In-memory Computing for IoVT
  - ❑ Application
  - ❑ SRAM based solution
  - ❑ Hybrid solution → SRAM + DRAM
- ❑ Conclusions

40 Conclusion: Neuromorphic 2.0 has big opportunities

- Event-driven vision a great fit for IoVT
  - ✓ Static camera → Background removal, Motion sensing
  - ✓ Dynamic camera → Fast reaction, High DR
- In-memory computing (IMC) key to Ultra low-energy & high throughput computing
- Volatile IMC → good for frequent rewrites
- Hybrid IMC is an exciting new direction

The potential of Neuromorphic 2.0, particularly its advantages in low-power, high-performance computing for applications like computer vision. Event-driven vision and In-Memory Computing are presented as key enabling technologies for this advancement.

41 Questions

THANK YOU

BRAIN Lab

PI: Arindam Basu (https://www.cityu.edu.hk/stfprofile/arinbasu.htm)

Collaborators

Nripan Mathews (NTU), Chang Chip Hong (NTU), Shih-Chii Liu (ETH), Nitish Thakor (JHU), Camilo Libedinsky (NUS), Rosa So (I^2R), J Hasler (GaTech), Debdeep Mukhopadhyay (IIT KGP)

SECTION 4: MEMRISTIVE SYSTEMS

CHAPTER 10

Open-Source RRAM Neuromorphic Accelerators

Jason Eshraghian

University of California at Santa Cruz

The presentation on Open-Source RRAM Neuromorphic Accelerators, dives into the integration of neuromorphic computing principles with silicon technology, particularly focusing on in-memory processing using RRAM crossbars. It presents the analogy between weighted summation in neural networks and Multiply and Accumulate operations in RRAM crossbars, showcasing practical implementations in 65nm technology. Challenges in in-memory processing are highlighted, emphasizing algorithm-level optimization to remove the need for Analog-to-Digital Converters (ADCs). The presentation further explores biological neural networks and introduces the snnTorch Python package for gradient-based optimization of spiking neural networks (SNNs), facilitating real-time learning and seamless integration with PyTorch. Insights into spiking neuron dynamics using RRAM methodology, resilience to quantization, and utilization of SKY130nm for a BSNN Accelerator are provided. Additionally, it discusses the SkyWater 130 Neuromorphic Accelerator and the open-source Python framework OPENRAM for ASIC design. The presentation concludes by encouraging active participation in the open-source community, offering resources and links for further engagement.

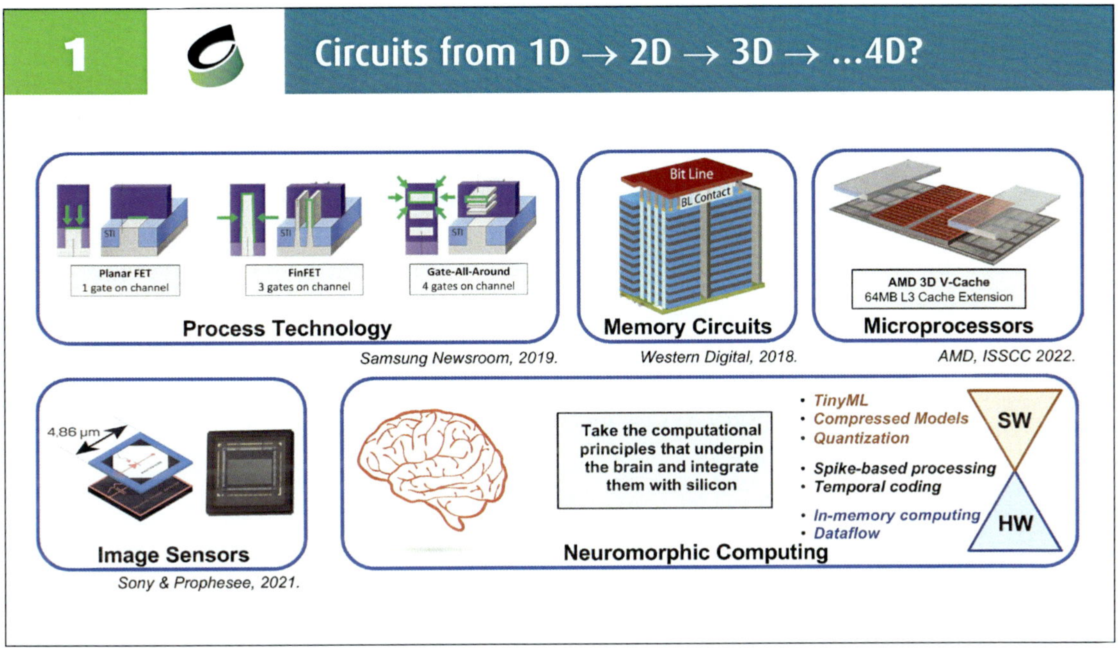

Discusses in-memory processing using RRAM crossbars. Covers the valence change mechanism of RRAM, Metal-Insulator-Metal structure, crossbar structure with bitline current summation. Also illustrates the micrograph of a 65nm mixed signal RRAM crossbar chip.

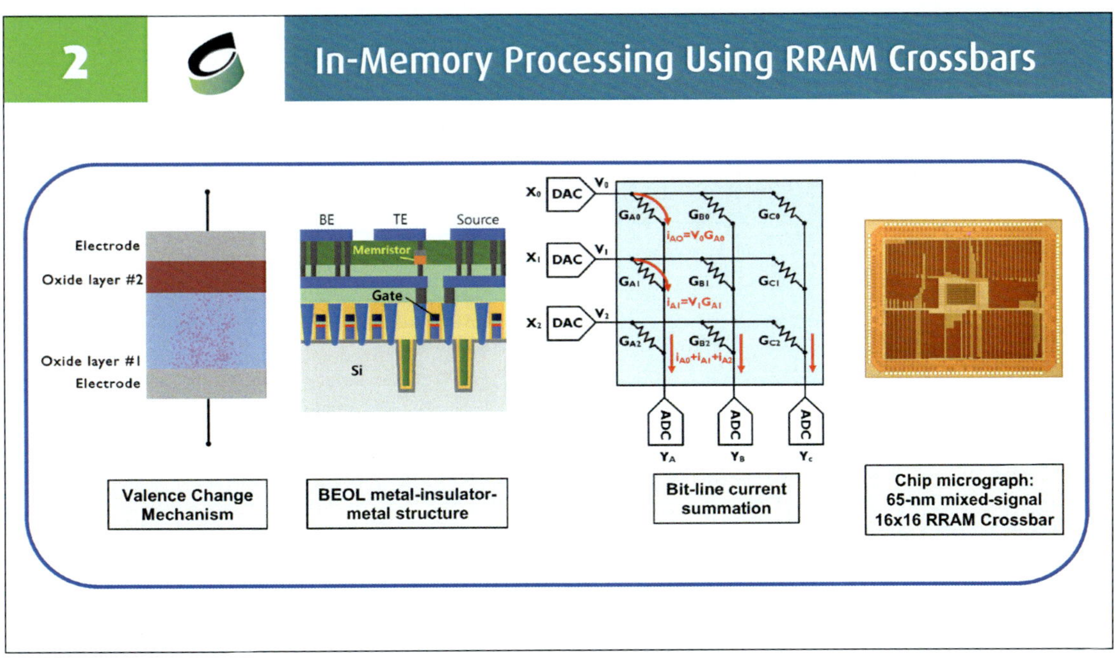

Explains the analogy between weighted summation in neural networks and Multiply and Accumulate operation in RRAM crossbars. INputs are mapped to input voltages. Weights are mapped to the conductance states of memristors and output is equal to MAC operation of the input voltages and conductance values. This value is equivalent to the current sum along the columns.

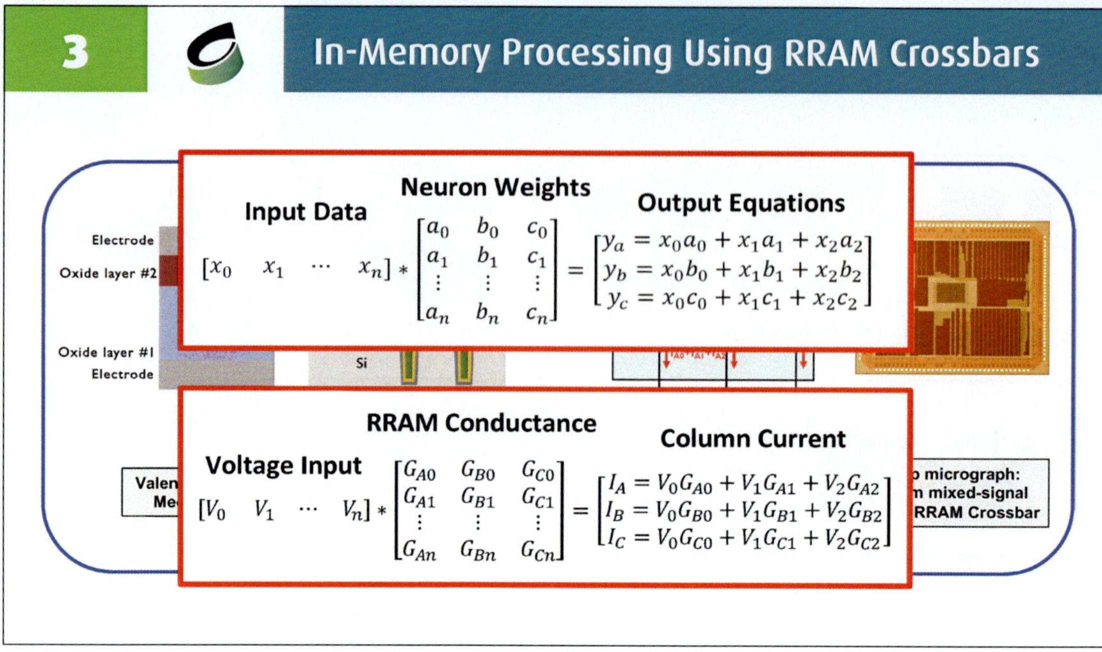

Illustrates In-Memory processing using RRAM crossbars in 65nm technology. Uses ITIR crossbar array with DElta-Sigma ADC for time averaging of the output.

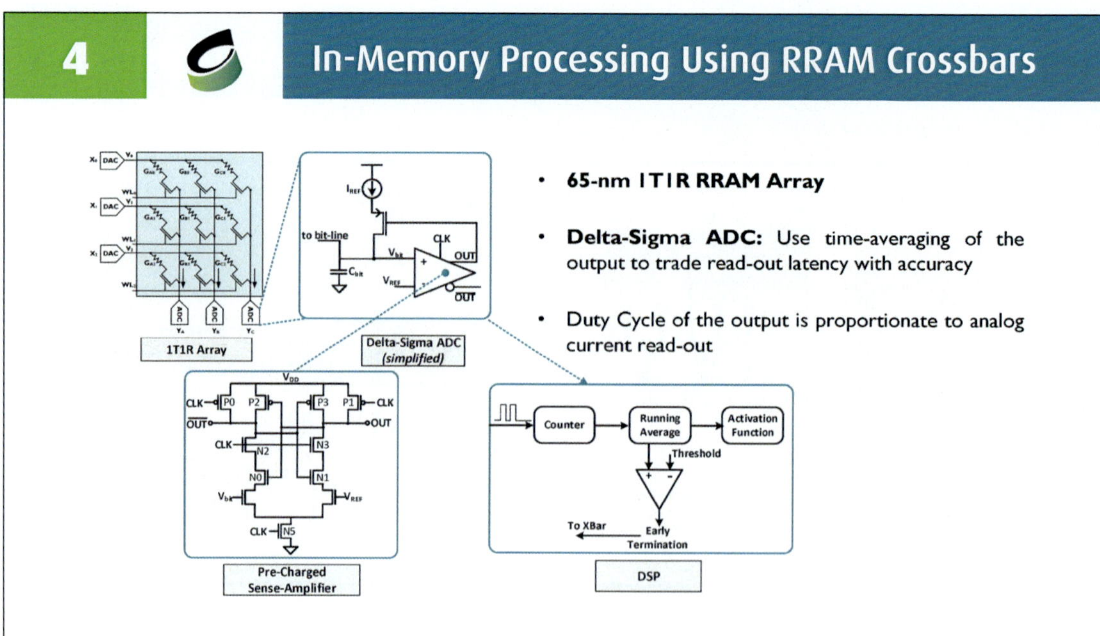

Shows In-Memory processing using RRAM crossbars comprising counter, running average, activation function, comparator and pre-charged sense amplifier.

5 In-Memory Processing Using RRAM Crossbars

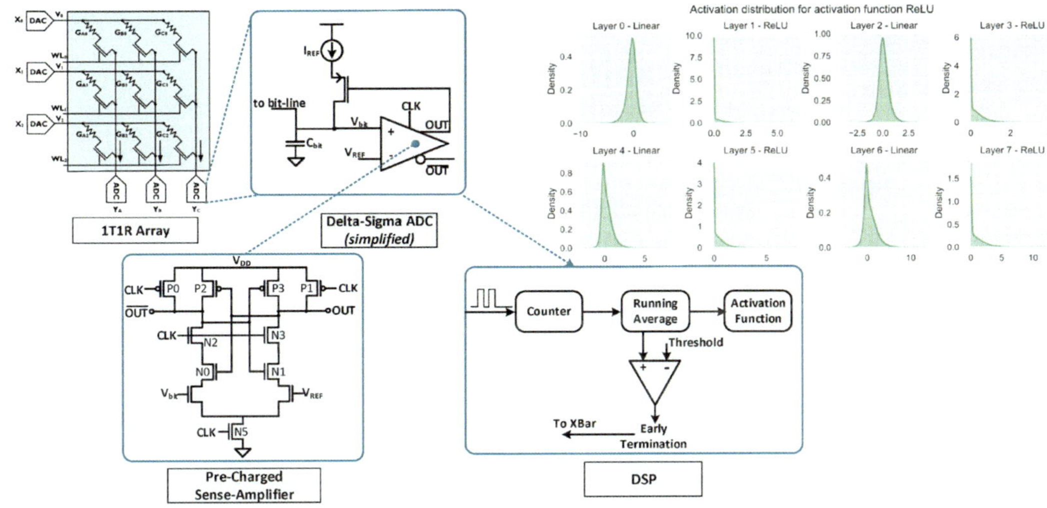

Lists the challenges of In-Memory processing using RRAM crossbars and emphasizes removal of ADCs requires algorithm-level optimization.

6 In-Memory Processing Using RRAM Crossbars

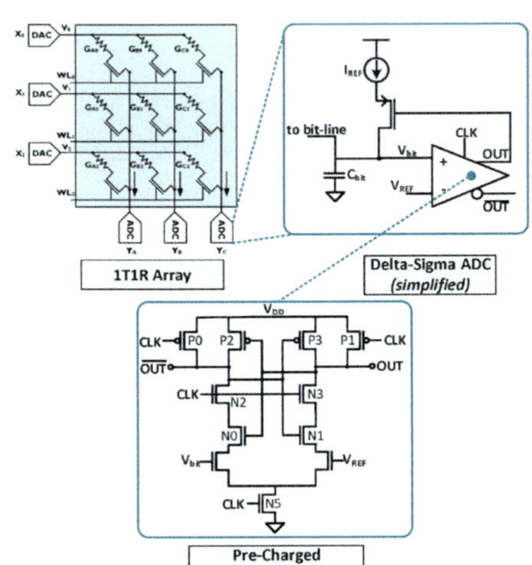

	Our Work	Mythic
Technology	65 nm	40 nm
CLK Freq (MHz)	40 - ADC 100 - RRAM	170
Power (W)	1.89	5
TOPS/W	5.90	4
Quantization	Reconfigurable	8-bit

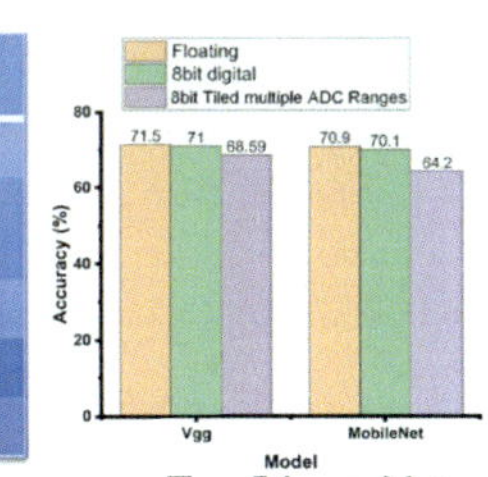

Top-5 ImageNet classification accuracy

Challenges:

- 1.5-3x improvements are insufficient to justify startup costs for a new process
- Scaling challenges: advanced processes are not optimized for analog operation
- ADCs still dominate latency, and ~50% energy consumption
- **Removal of ADCs requires algorithm-level optimization → spike-based processing?**

Introduces biological neural networks covering spikes, sparsity and static suppression.

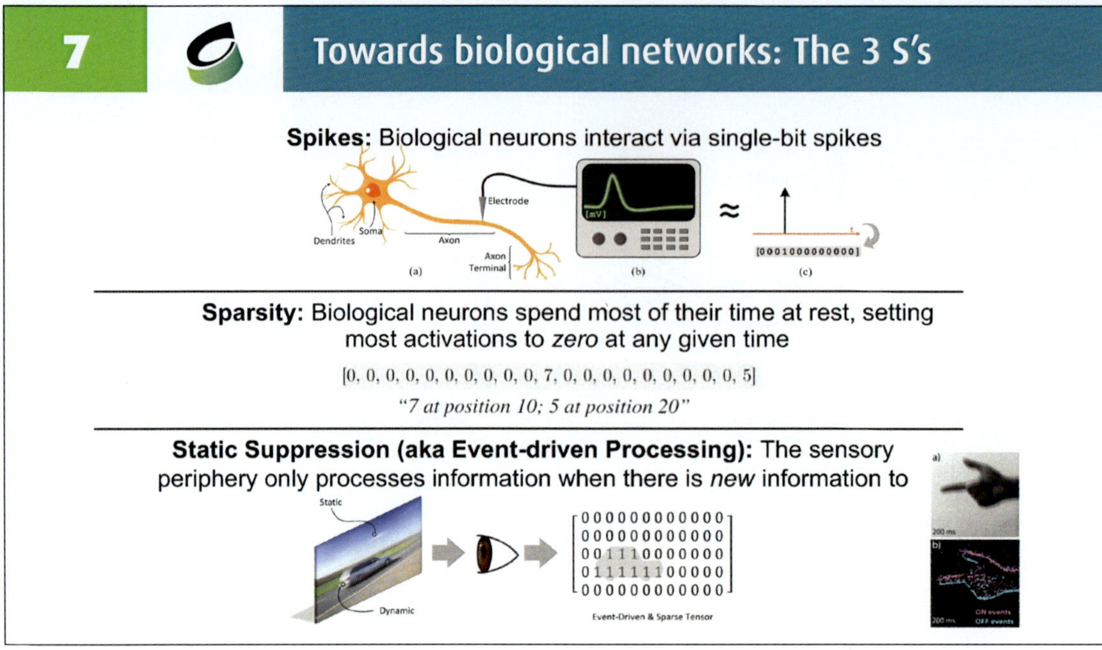

Explains Leaky-Integrate and Fire Neuron model based on current injection, integration and spike generation based on the membrane potential and corresponding output waveforms.

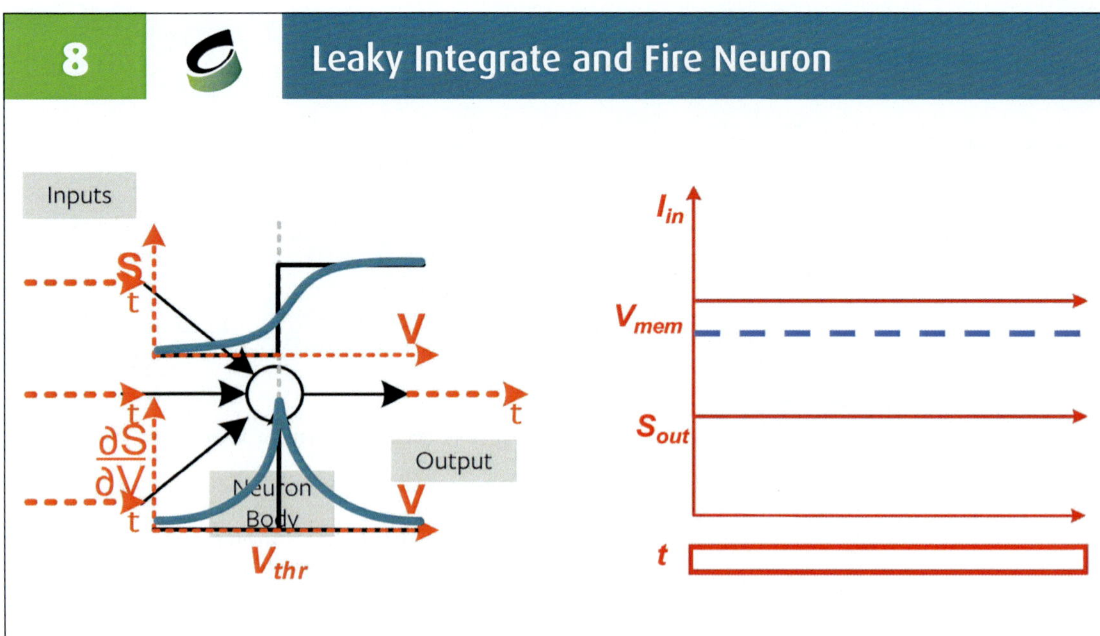

Introduces snnTorch as the Python package for gradient-based optimization. It is flexible for the seamless integration with PyTorch and lightweight enough for the CPU.

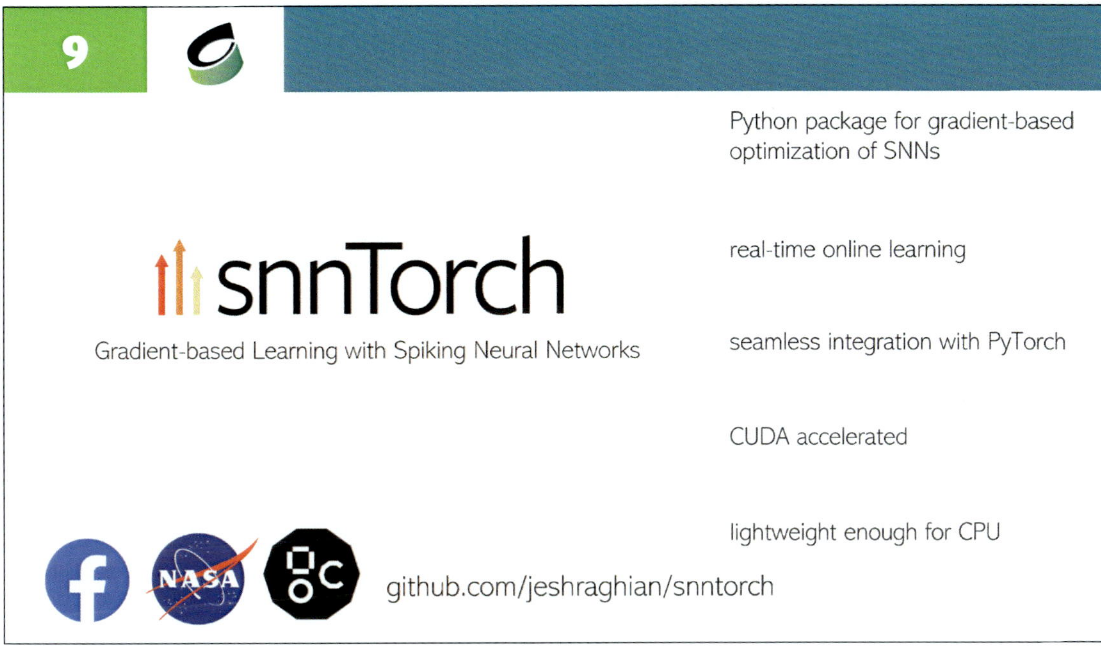

Introduces the tutorial and github link for the hands-on sections of spiking neural networks .

Features the Python package designed for gradient-based optimization of SNNs, enabling real-time online learning, seamless integration with PyTorch, and accelerated performance via CUDA, while also being lightweight for CPU usage.

11 SNN Evaluation – Multimodal Data

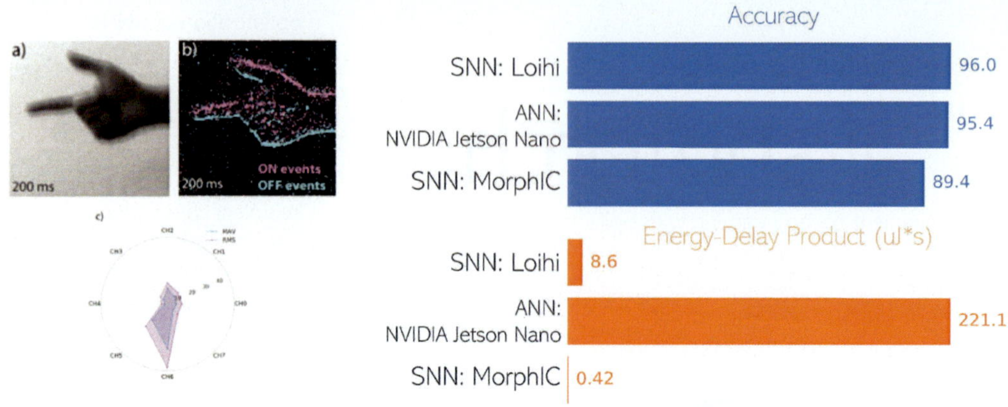

M. Rahimi-Azghadi, J.K. Eshraghian et al., IEEE Trans. Biomedical CAS, Dec. 2020

Introduces the concept of SNN Evaluation with Multimodal Data.

12 RRAM-based SNN Processing

An RRAM Approach to Spiking Neuron Dynamics

- Charge-based integration on the bit-line capacitance
- State decay using heterogeneous RRAM time constants

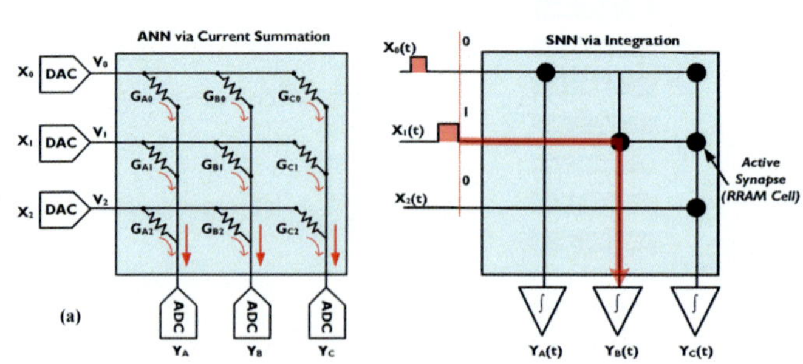

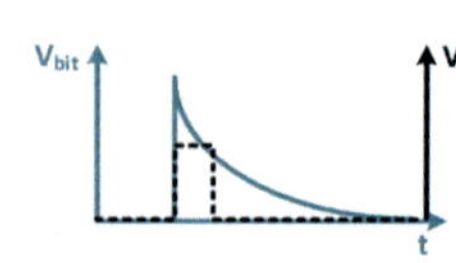

Communicating noise-prone analog signals down long lengths of wire: noise-prone

Communicating noise-robust digital signals down long lengths of wire: noise-tolerant

J.K. Eshraghian et al., "Memristor-based Binarized SNNs" IEEE Nanotech. Mag. 2022

Investigates Spiking Neuron Dynamics using RRAM Methodology, which involves employing charge-based integration on the bit-line capacitance, implementing state decay with varying RRAM time constants, and transmitting both analog signals prone to noise and digital signals resilient to noise over extended wire lengths.

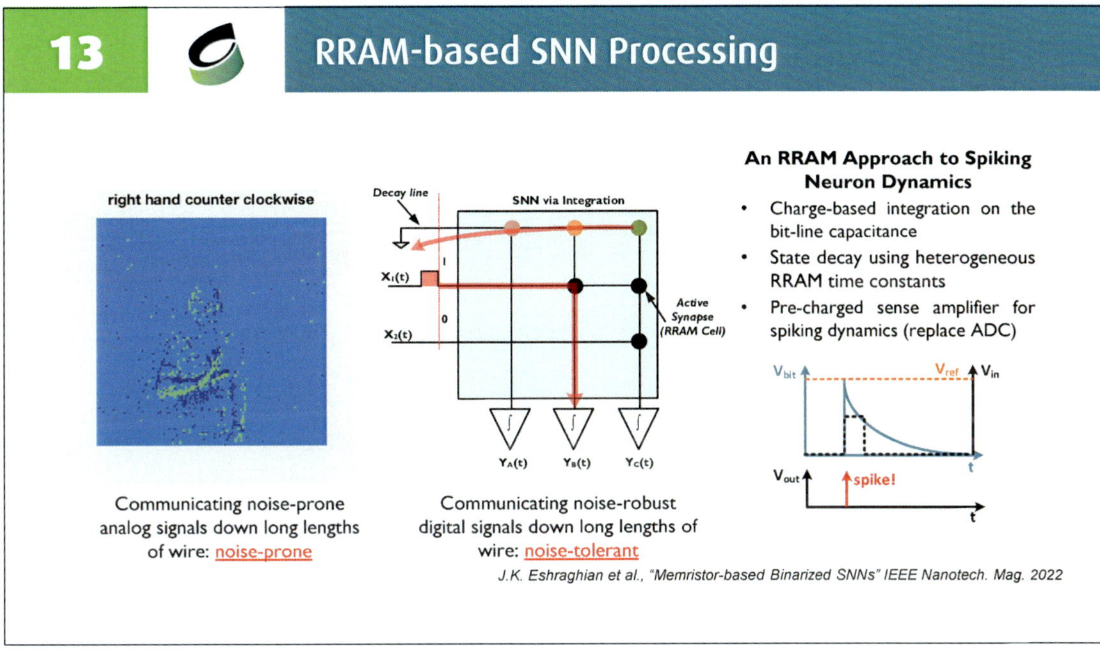

Explores Spiking Neuron Dynamics via an RRAM Approach, incorporating charge-based integration on the bit-line capacitance, state decay utilizing heterogeneous RRAM time constants, and implementing a pre-charged sense amplifier for spiking dynamics (replacing ADC).

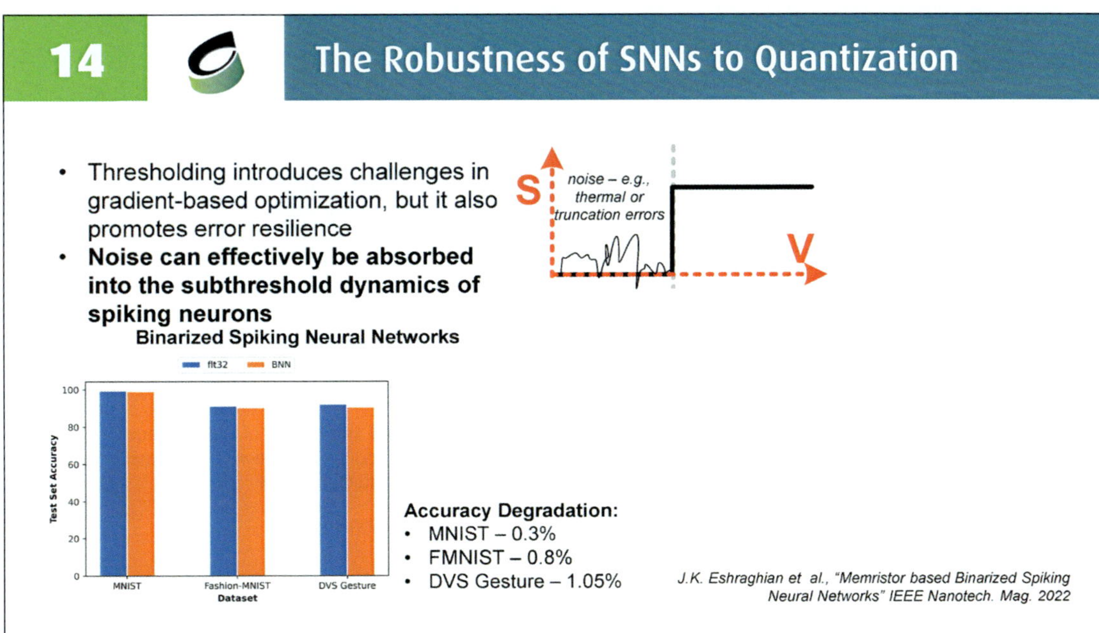

Presents the resilience of SNNs to Quantization and thresholding poses challenges for gradient-based optimization, yet it fosters error resilience.

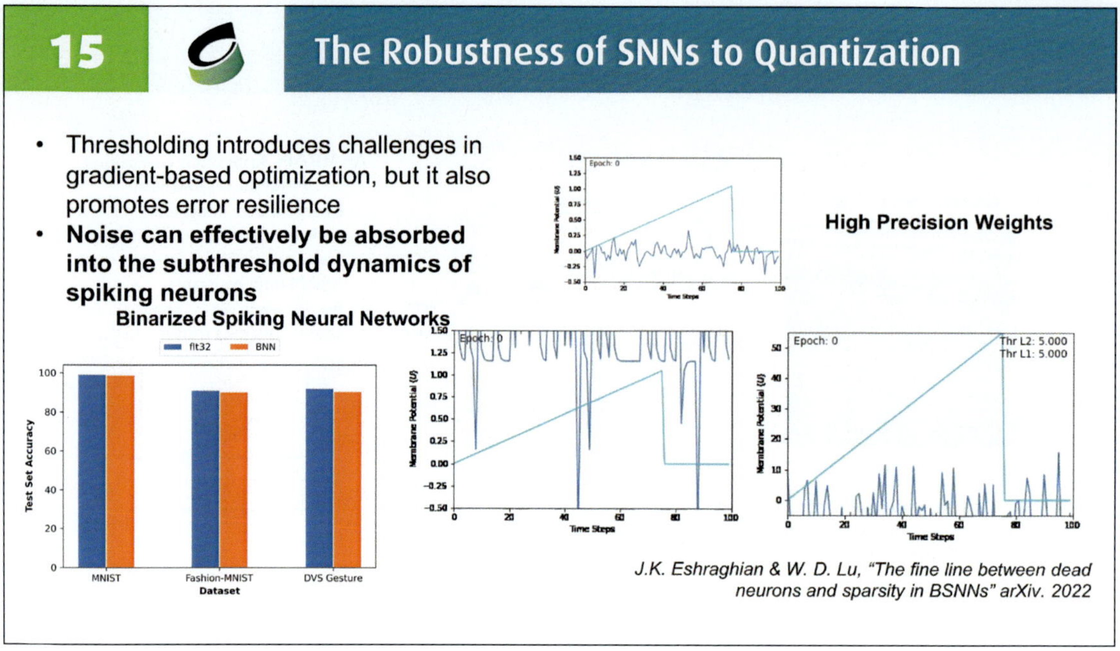

Insights into robustness of SNNs to quantization depicting high precision weights.

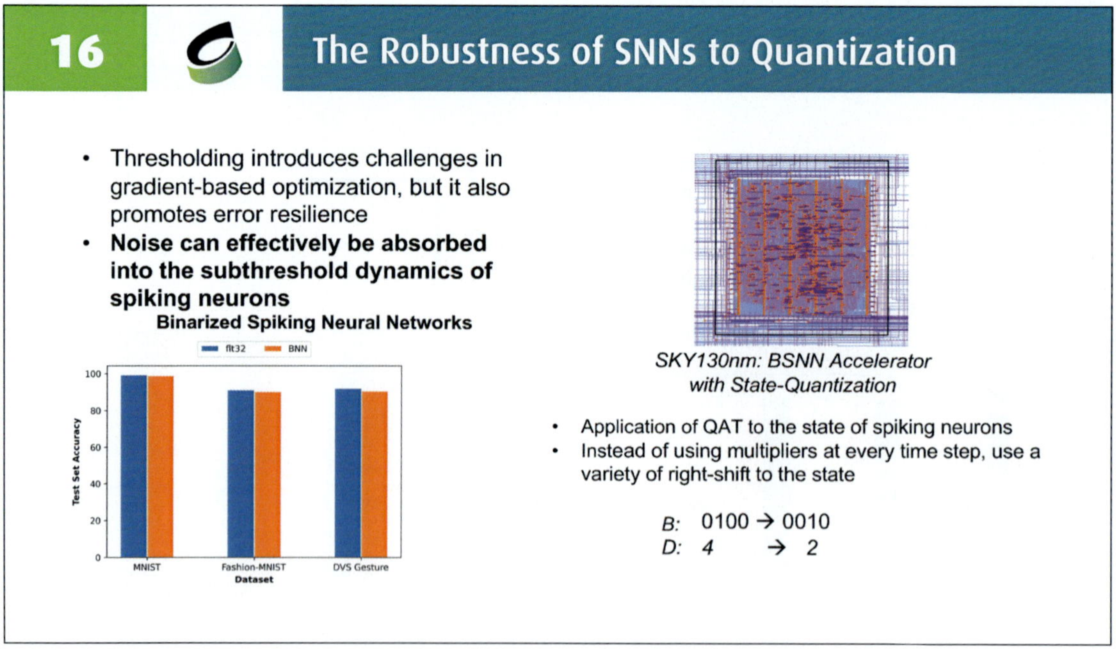

Illustrates utilization of SKY130nm for a BSNN Accelerator featuring State-Quantization and applying QAT to the state of spiking neurons. Instead of employing multipliers at each time step, a range of right-shifts is employed on the state.

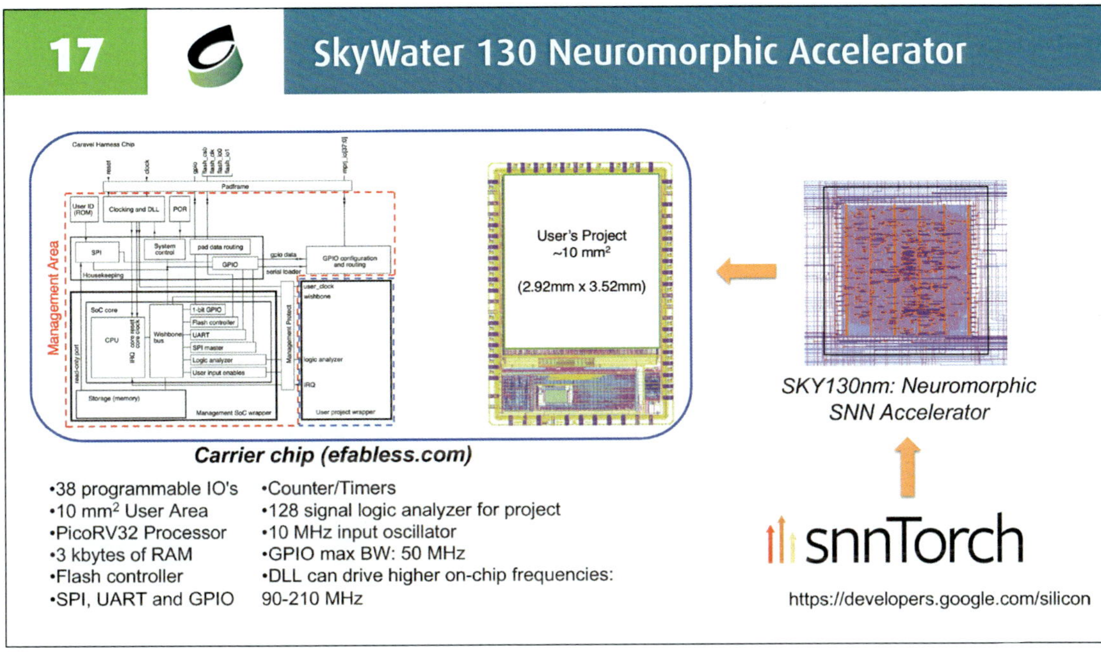

Insights into the SkyWater 130 Neuromorphic Accelerator, covering details about the carrier chip from Efabless, its diverse features, and the user project area.

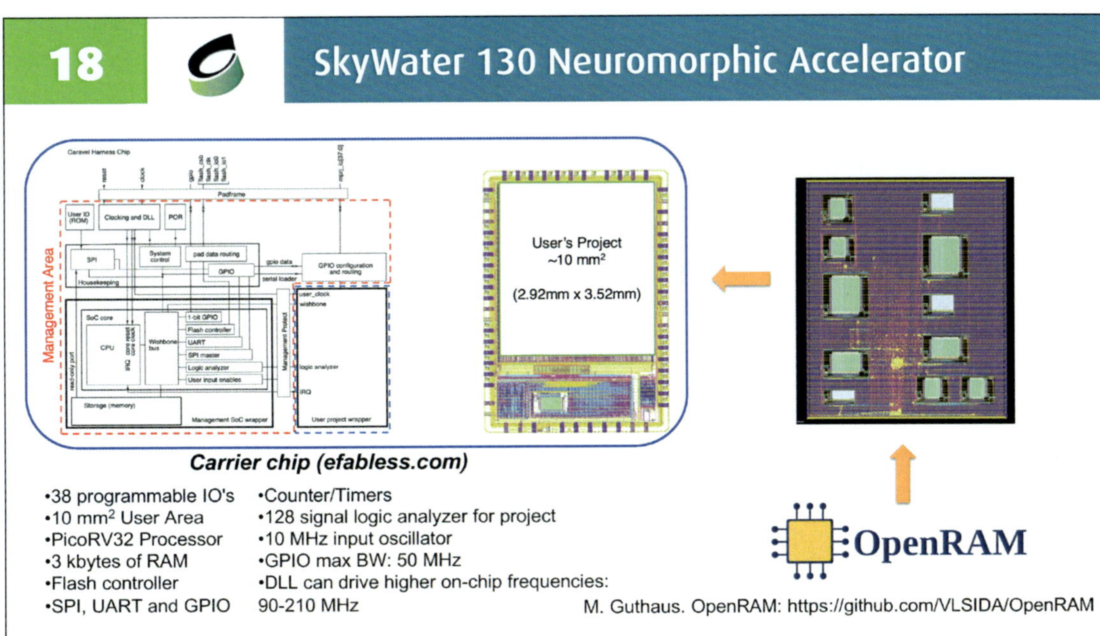

Discusses the open-source Python framework OPENRAM, which facilitates the creation of layout, netlists, timing and power models, as well as placement and routing models, along with other essential views required for utilizing SRAMs in ASIC design.

19 SkyWater 130 Neuromorphic Accelerator

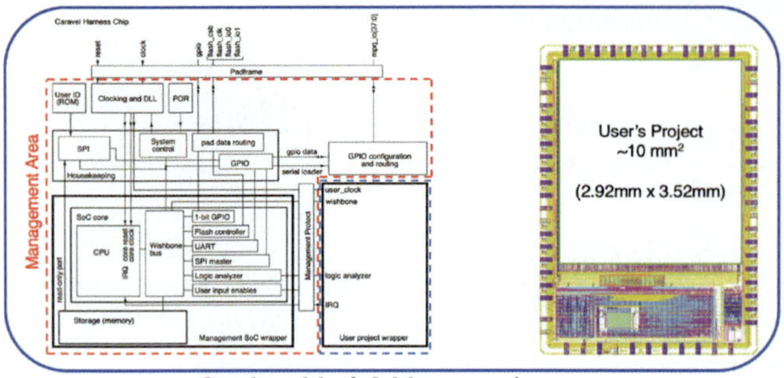

Carrier chip (efabless.com)

RRAM between met1/met2

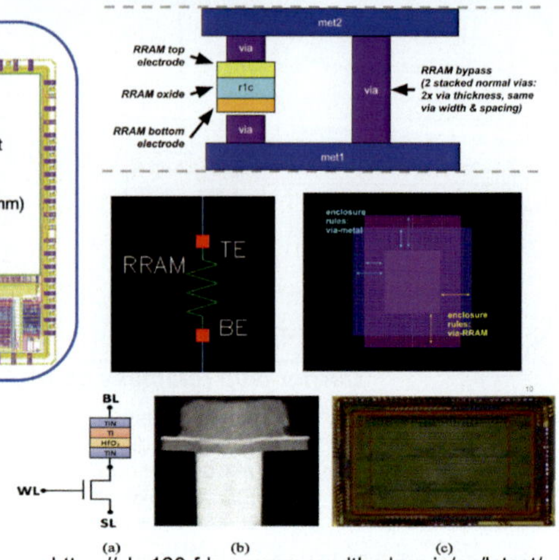

SKY130B ReRAM-CMOS
- HfO_2-based RRAM layer
- 0.1-0.2 V Read Voltage
- 1.7 V Set Voltage / 2.5 V Reset Voltage
- 2.2 – 3.1 V Forming voltage
- 2-3 bits per cell

https://sky130-fd-pr-reram.readthedocs.io/en/latest/

Guides on utilizing the documentation for the sky130 fd pr reram structure and exploring its intricacies. Indicates that increase in variability can decrease the accuracy,

20 Discussion & Questions

Get involved with the Open-Source Community!

- **snnTorch**: https://snntorch.readthedocs.io/
- **OpenRAM:** www.openram.org
- **Google/SkyWater:** https://developers.google.com/silicon
- **Asynchronous VLSI Design:** https://avlsi.csl.yale.edu/act/doku.php

snnTorch

OpenRAM

Concludes the presentation by encouraging active participation in the Open-Source Community, offering links to various projects and resources for engagement.

CHAPTER 11

Mixed-Signal Neuromorphic Computing: Devices, Algorithms, and Circuits

Vishal Saxena

University of Delaware

The talk offers a comprehensive overview of cutting-edge topics in neuromorphic computing, non-volatile memory (NVM) devices, learning algorithms, and hybrid CMOS-RRAM circuits. Throughout the presentation, various aspects of neuromorphic computing are explored, including the fundamentals of computational neuroscience, the massive parallelism of the human brain, and the role of emerging memory technologies like Resistive Random Access Memory (RRAM) in achieving energy-efficient computing. Discussions also encompass the comparison of different neuromorphic computing architectures and AI accelerators, highlighting their architectural differences and performance metrics. The presentation delves into the challenges and potential solutions associated with RRAM devices, such as variability, stochasticity, and resistance drift, while also addressing the benefits and challenges of analog vector-by-matrix multipliers (VMMs) for Edge-AI applications. Additionally, topics like transfer learning in Spiking Neural Networks (SNNs), unsupervised learning, and spike timing-dependent plasticity (STDP) are examined, offering insights into the latest advancements in neuromorphic computing algorithms. Overall, the presentation provides a comprehensive overview of the current state and future prospects of neuromorphic computing, highlighting its potential for revolutionizing AI and edge computing applications.

1 Outline

- Introduction
- Devices
  - Emerging NVM devices as neural network synapses
  - NVM challenges and their mitigation
- Algorithms
  - Transfer Learning and Unsupervised Learning Algorithms
  - Deep learning using Spiking Neural Networks (SNNs)
- Circuits
  - Hybrid CMOS-RRAM spiking neural circuits
  - Recent IC Architectures and Trends
- Recent Trends and Conclusion

Overview of the presentation, covering neuromorphic computing, emerging non-volatile memory (NVM) devices, learning algorithms, and hybrid CMOS-RRAM circuits.

INTRODUCTION

2 Neuromorphic ICs and Edge-AI

- Deep learning AI is has shown unprecedented success with processing unstructured data
- Growing need for low-power Embedded-AI at the Edge
  - Reduce reliance upon Cloud and wireless infrastructure
  - Data privacy and personalized AI models
- Overarching goal is energy-efficiency of the brain
  - Eliminate von Neumann bottlenecks by computing inside memory
    - Analog in-memory computing architectures

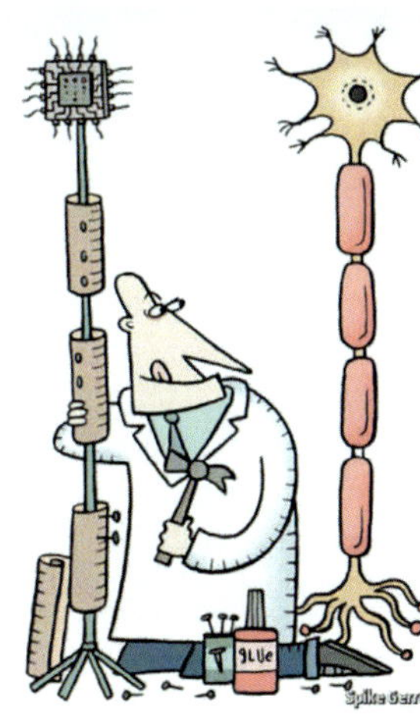

The Economist, Aug 2013

Discussion on neuromorphic ICs and Edge-AI, emphasising the need for low-power AI at the edge for data privacy and reduced cloud reliance.

3 Neuromorphic Computing Basics

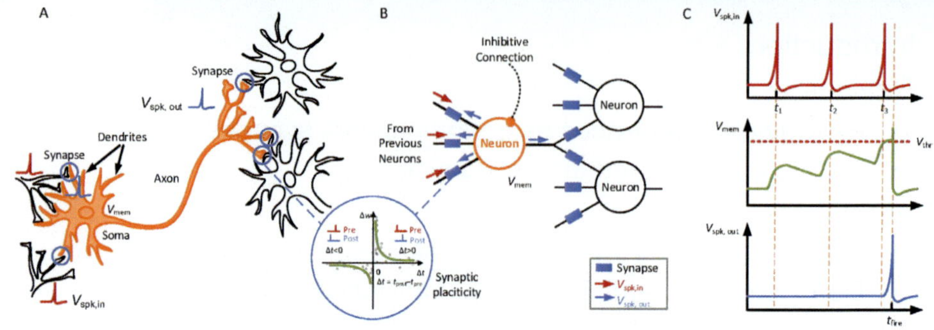

- Emerging understanding from Computational Neuroscience
  - Computation performed by neurons using asynchronous spikes (action-potential)
- Synapses at the dendrites modulated by relative firing times of the spikes
  - *Plasticity mechanisms:* STDP, Three-factor learning rules, and their variants
  - Axons for routing; role of neuromodulator feedback, astrocytes and glia under study

Basics of neuromorphic computing, including computational neuroscience insights, plasticity mechanisms, and the role of neuromodulators.

4 Neural Inspiration: Energy-Efficient Computing

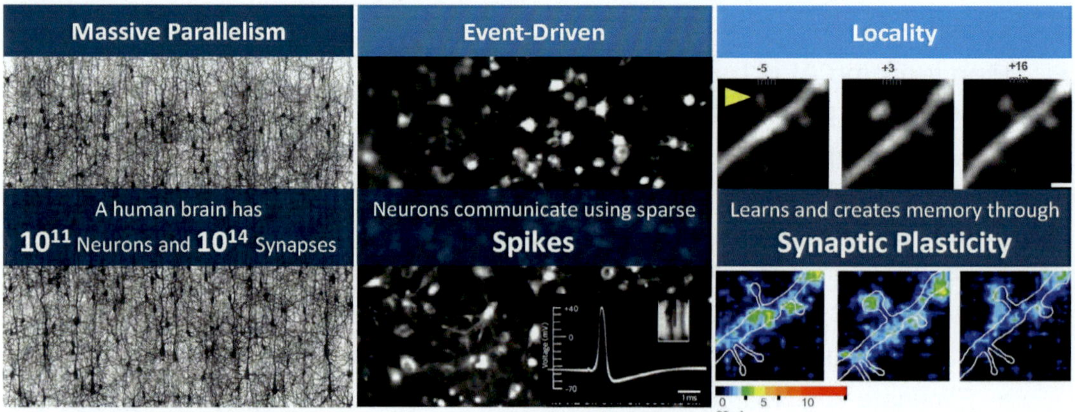

S. Furber, *J. R. Soc. Interface*, 2007

Animation: Georgia Tech's NeuroLab
Embedded: Hodgkin and Huxley, 1939

J.P. Olson, et al., *J. Am. Chem. Soc.*, 2013

Compares various memory devices based on parameters such as cell size, latency, energy consumption, and endurance. Introduces Resistive Random Access Memory (RRAM) variants like CBRAM and OxRAM. These emerging memory technologies play a crucial role in achieving energy-efficient neuromorphic computing.

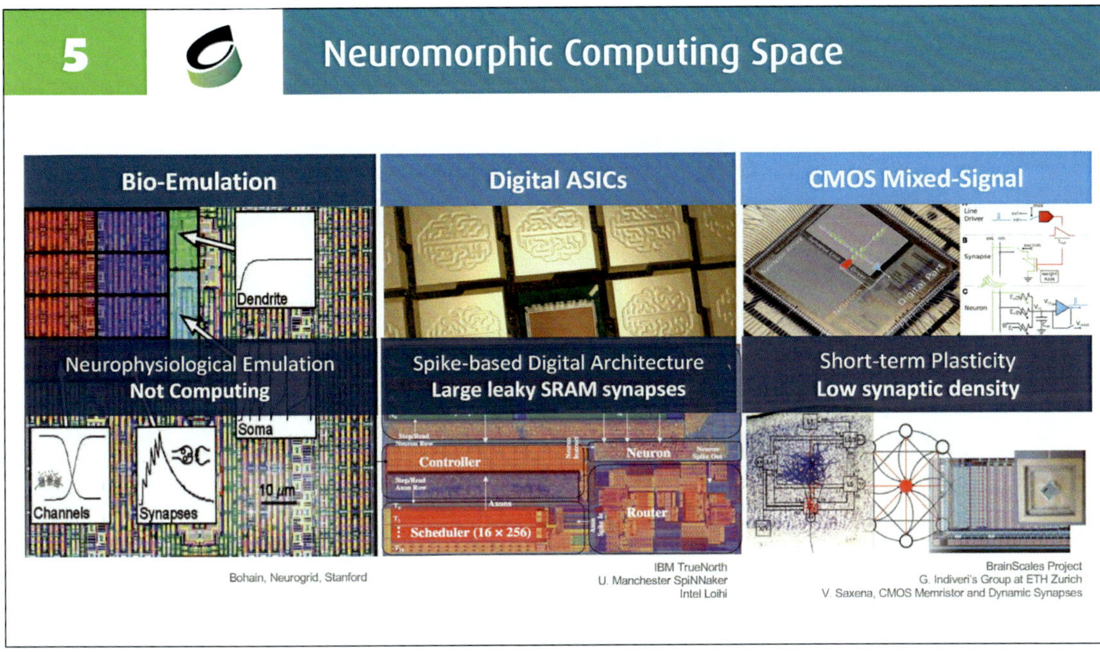

Compares various neuromorphic computing architectures, highlighting the differences between digital ASICs like IBM's TrueNorth and Intel's Loihi and mixed-signal systems like the CMOS Memristor and Dynamic Synapses. It distinguishes between bio-emulation platforms like Neurogrid and computational platforms focusing on computing rather than emulating neurophysiological processes.

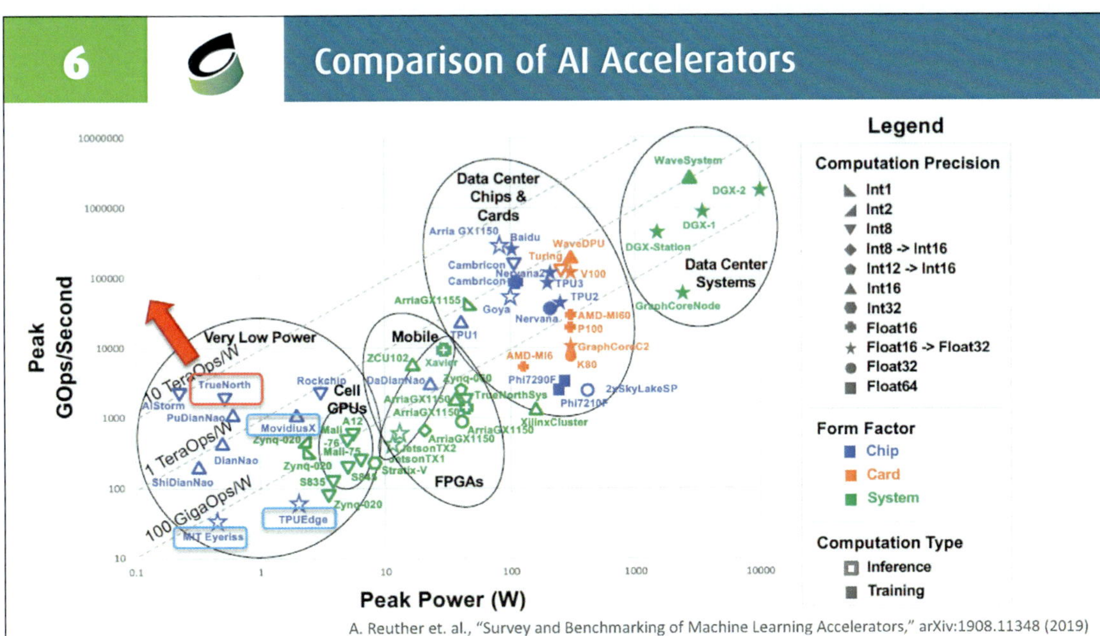

Comparison of various AI accelerators, highlighting their differences in architecture and performance metrics. It serves as an overview of the landscape of machine learning accelerators, as studied by A. Reuther et al. in their survey and benchmarking work.

EMERGING MEMORY DEVICES

7 Emerging NVMs as Synapses

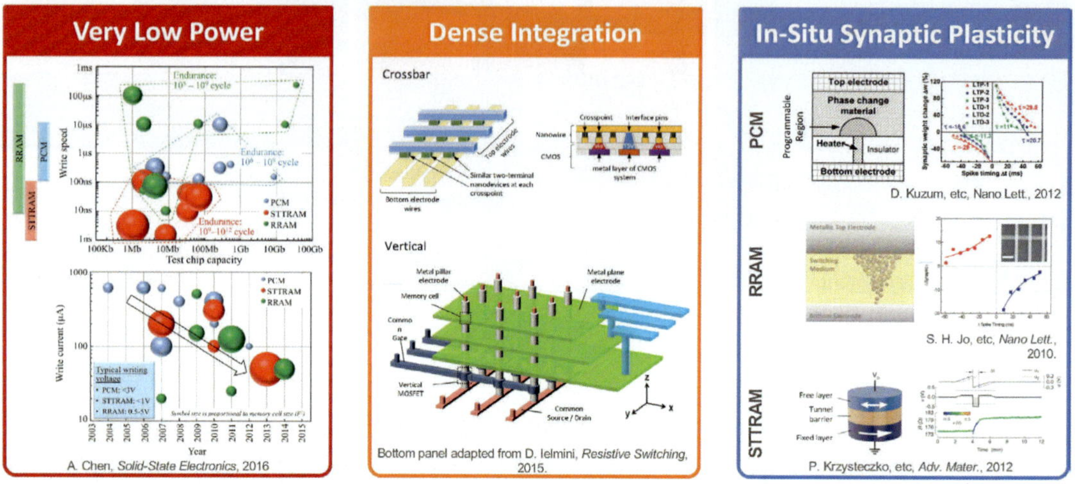

"Emerging Memory Devices" discusses the characteristics of Non-Volatile Memory (NVM) devices such as RRAM, STTRAM, and PCM, highlighting their potential for dense integration and low-power synaptic plasticity in neuromorphic computing. It also compares memory types for Edge-AI applications, focusing on parameters like cell size, latency, energy, and endurance.

8 Comparison of Memory for Edge-AI

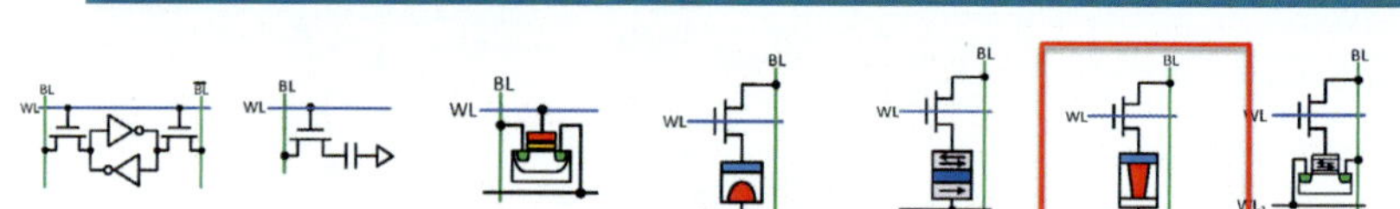

Parameters	SRAM	DRAM	NOR Flash	PCRAM	STTRAM	RRAM	FeFET
Cell Size	$100F^2$	$7F^2$	$10F^2$	$4F^2$	$12F^2$	$4\text{-}12F^2$	$24F^2$
Write Latency	1ns	5ns	10µs-1ms	100ns	2-25ns	10ns	3ns
Read Latency	1ns	20-80ns	50ns	10ns	2-25ns	1-10ns	2ns
Write Energy (pJ/bit)	<1	<1	100	2-25	0.1-2.5	0.1-3	0.1
Leakage Power	High	Medium	Low	Low	Low	Low	Low
Endurance (write cycles)	$>10^{16}$	$>10^{16}$	10^5	10^9	10^{15}	10^{10}	$>10^5$
MLC Capability	✗	✗	4-bit	4-bit	2-bit	4-bit	3-bit
MLC Retention	✗	✗	4-6 months	10^4 s	-	10^4 s	-
			Tunneling	R-drift, abrupt Reset	Tunneling	R-drift	Tunneling
3D Stacking	✗	✗	✗	✔	✗	✔	✔

V. Saxena, "Neuromorphic Computing: From Devices to Integrated Circuits," J. Vac. Sci. Technol. B, vol. 39, pp. 010801, 2021

The slide compares various memory technologies for Edge-AI, highlighting parameters like cell size, latency, energy, and endurance. It emphasises the suitability of different memory types, such as SRAM, DRAM, Flash, and emerging NVMs, for specific applications based on their characteristics.

9 RRAM Types

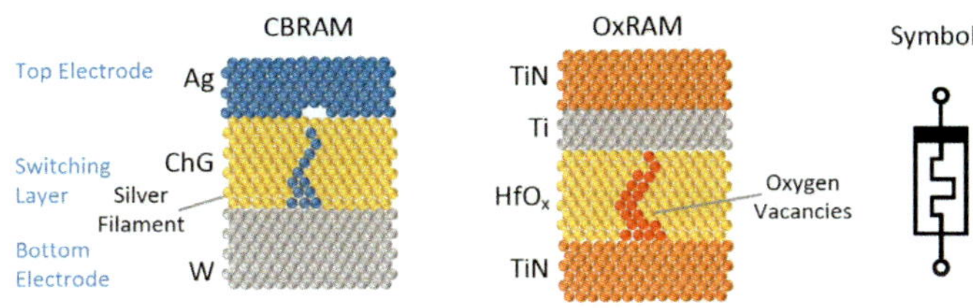

- CBRAM: Conductive Bridge Random Access Memory
  - Silver or copper-based active electrode with chalcogenide switching medium
- OxRAM: Oxygen Vacancy based Random Access Memory
  - Switching layer of transition metal oxides HfO_x, TaO_x, etc.
  - Ti acts as an oxygen scavenging layer which helps create oxygen vacancy bridge

Overview of Conductive Bridge Random Access Memory (CBRAM) and Oxygen Vacancy Random Access Memory (OxRAM), detailing their components and functioning. It highlights the use of silver or copper-based active electrodes with chalcogenide switching medium in CBRAM, as well as transition metal oxides like HfOx and TaOx in OxRAM, which facilitate the creation of oxygen vacancy bridges.

10 CBRAM Switching

- Active electrode of Ag or $CuTe_x$ and inert electrode of W, Pt or Ni
- Ag gets oxidized and moves towards inert electrode and form a conductive bridge
- Forming step to activate a pristine device
  - Electroforming or photo-diffusion
- Program (Set) and Erase (Reset) threshold voltages, V_{tp} and V_{tm}
  - $|V_{tm}|$ is small as small amount of voltage is needed to break the bridge
- Compliance current is set by the semiconductor parameter analyzer
  - 'Analog' states due to formation of weak filaments

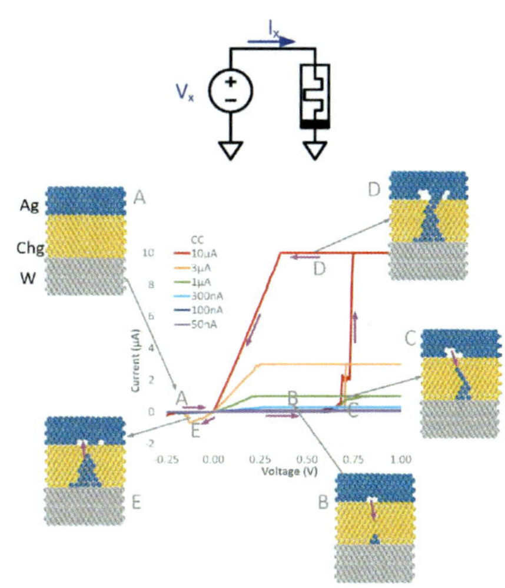

This gives an idea of Conductive Bridge Random Access Memory (CBRAM) switching, detailing the active and inert electrode materials, the process of forming a conductive bridge, and the programming (Set) and erasing (Reset) threshold voltages. It also describes the compliance current set during the electroforming or photo-diffusion process and the 'analog' states resulting from forming weak filaments.

11 OxRAM Switching

- *Active layer:* Ti, Hf, La or Zr
- *Switching layer:* from HfO_x, Ta_2O_x, TiO_x, etc.
- *Inert electrode:* TiN, Ni, TaN, ITO, etc.
- Active layer scavenges oxygen atoms from the switching layer
- Higher program and erase thresholds
- Near-symmetric switching
- Electroforming step to initialize the device
- Exhibits multi-level states per device in controlled experiments
- Write cycle endurance ~10^9

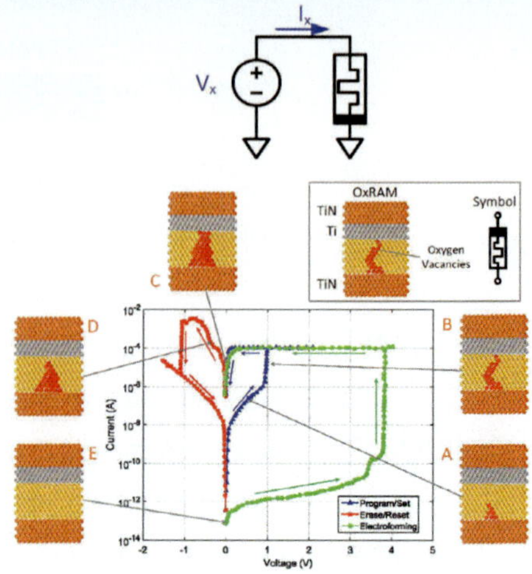

1. A. Grossi *et. al.*, "Fundamental variability limits of filament-based RRAM," 2016 IEEE International Electron Devices Meeting (IEDM), 2016.
2. V. Saxena, "Neuromorphic Computing: From Devices to Integrated Circuits," J. Vac. Sci. Technol. B, vol. 39, pp. 010801, 2021

OxRAM, a type of RRAM, features an active layer comprised of materials like Ti, Hf, La, or Zr, which scavenge oxygen atoms from the switching layer made of compounds like HfOx, Ta2Ox, or TiOx. With near-symmetric switching behavior, higher program and erase thresholds, and an electroforming step for initialization, OxRAM exhibits multi-level states per device and impressive endurance of around 10^9 write cycles, as evidenced by studies such as A. Grossi et al.'s research presented at the 2016 IEEE International Electron Devices Meeting (IEDM) and V. Saxena's work published in the Journal of Vacuum Science and Technology B in 2021.

ANALOG COMPUTING FOR ANNS

12 Analog VMMs for Edge-AI

- GPUs and Digital ASICs dominate training of deep neural networks
- Analog vector-by-matrix multipliers (VMMs) using non-volatile memory devices
  - $\mathbf{y}=\sigma(\mathbf{Wx}+\mathbf{b})$
  - $w_{ij} \equiv G_{ij}$, current summing simplifies circuit design
  - Leverages high-density in-memory computing
  - Low precision but 100-1000X lower energy per inference compared to digital realization

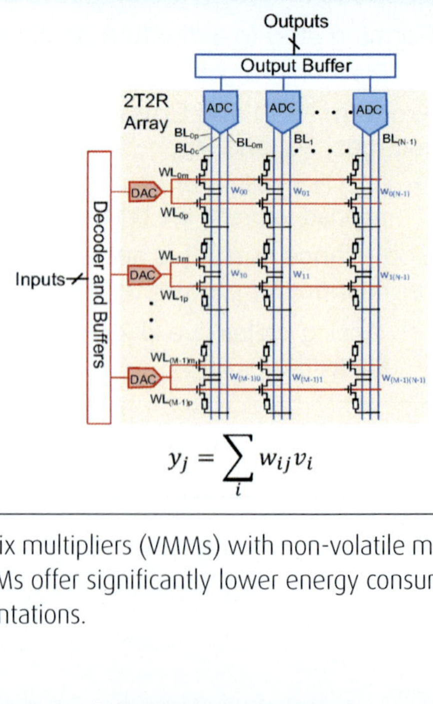

Analog VMMs for Edge-AI utilizes analog vector-by-matrix multipliers (VMMs) with non-volatile memory for efficient deep neural network training. Analog VMMs offer significantly lower energy consumption per inference, up to 1000 times less than digital implementations.

13 Challenges with Analog VMMs

- Challenges with Analog VMMs
  - DAC and ADC area & energy overhead
  - Nonlinearity in device I-V characteristics restricts precision of analog multiplication
    - **W·x** is essentially four-quadrant analog multiplication
    - Process and temperature dependent variation
  - Analog states (i.e. weights) drift with time

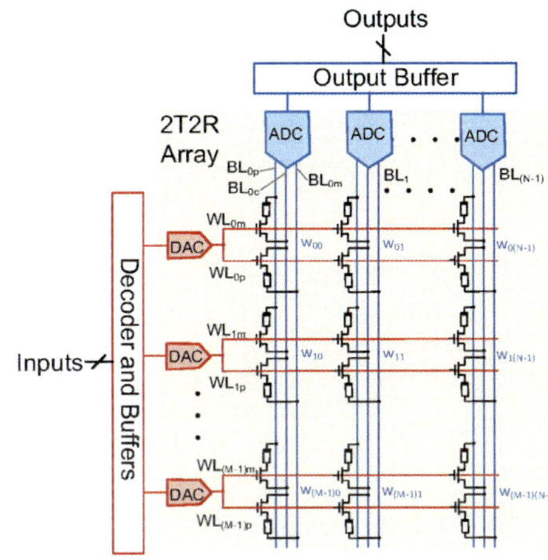

$$y_j = \sum_i w_{ij} v_i$$

Analog VMM Challenge discusses the area and energy overhead of DAC and ADC, nonlinearity in device I-V characteristics, and the drift of analog states over time. It highlights how process and temperature variations affect the precision of analog multiplication in Vector-by-Matrix Multiplication.

14 Analog VMMs using NOR Flash

- NOR Flash memory has regained interest in realizing VMM for inference
  - Embedded Flash in CMOS nodes
  - 4-8 bits per memory cell; 10 year retention
- Nonlinear I-V characteristics esp. in $SubV_T$
  - $I \propto e^{V_{CG}/nV_T}$
  - Makes analog weight multiplication challenging
- Can use the quasi-linear regions
  - Process and temperature variability

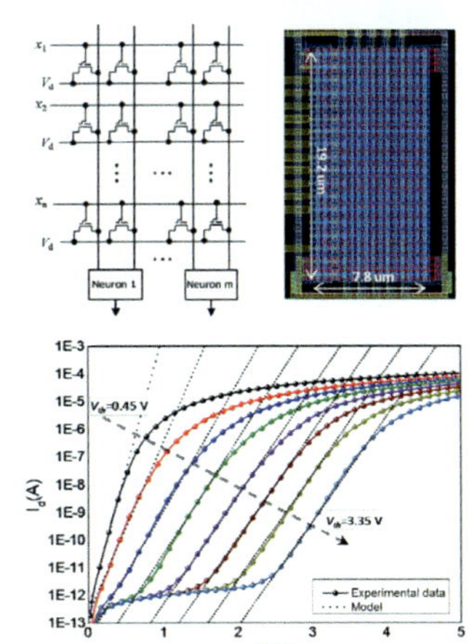

Guo, Xinjie, F. Merrikh Bayat, Mirko Prezioso, Y. Chen, B. Nguyen, N. Do, and Dmitri B. Strukov. "Temperature-insensitive analog vector-by-matrix multiplier based on 55 nm NOR flash memory cells." In *2017 IEEE Custom Integrated Circuits Conference (CICC)*, pp. 1-4. IEEE, 2017.

The discussion was on analog vector-by-matrix multipliers (VMMs) using 55 nm NOR flash memory cells for temperature-insensitive operations. It cites a study by Guo et al. from the 2017 IEEE Custom Integrated Circuits Conference.

15 Analog NNs with Spike Encoding

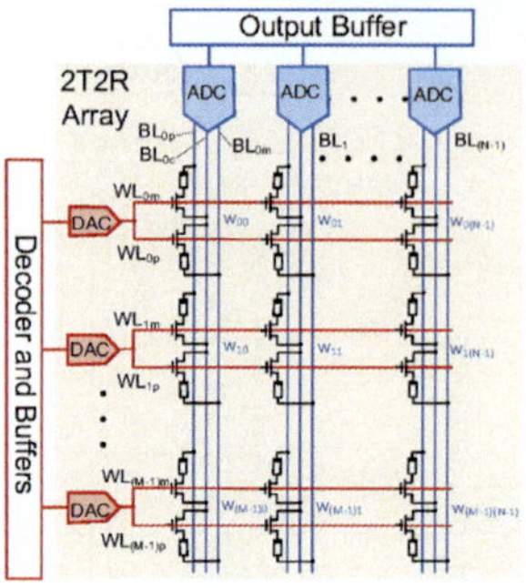

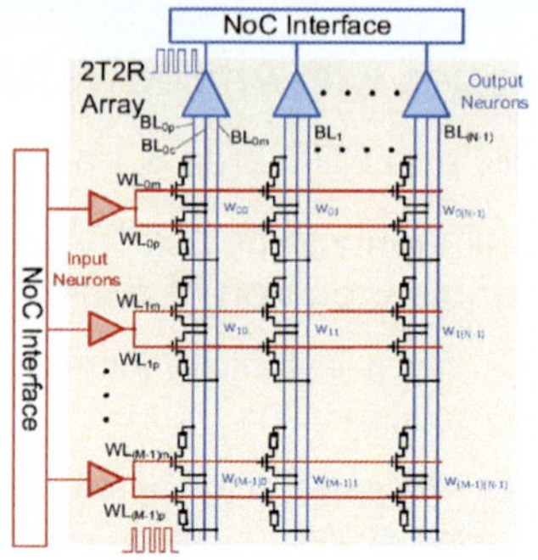

$$y_j = \sum_i w_{ij} V_i \qquad y_j[n] = \sum_i w_{ij} v_i[n],\quad v_i[n]=\{0,1\}$$

- Analog inputs V_i are encoded using a bi-level *spike-coding* $v_i[n]$
- Spike-domain computing replaces the ADCs and DACs by spike-based neurons
- Leverage Spiking Neural Networks (SNNs) algorithms

Analog neural networks (NNs) with spike encoding utilize bi-level spike-coding vi[n] to represent analog inputs Vi, enabling spike-domain computing to replace traditional analog-to-digital converters (ADCs) and digital-to-analog converters (DACs) with spike-based neurons. These networks leverage Spiking Neural Networks (SNNs) algorithms to process and analyze the spike-encoded information efficiently.

DEVICE CHALLENGES

16 RRAM Challenges: Variability

- Variability:
  - Device switching threshold voltages are variable from device to device and cycle-to-cycle
  - Switching characteristics depend upon the initial 'forming' step and compliance current (I_{cc})
  - Analog-like or multilevel states when weak filament is formed (e.g. for I_{cc}=50nA here)
- Stochasticity:
  - Voltage dependent stochastic switching in RRAMs
  - Depends upon programming pulse voltage and duration

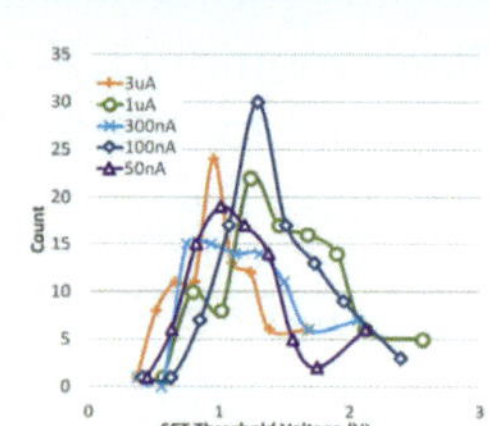

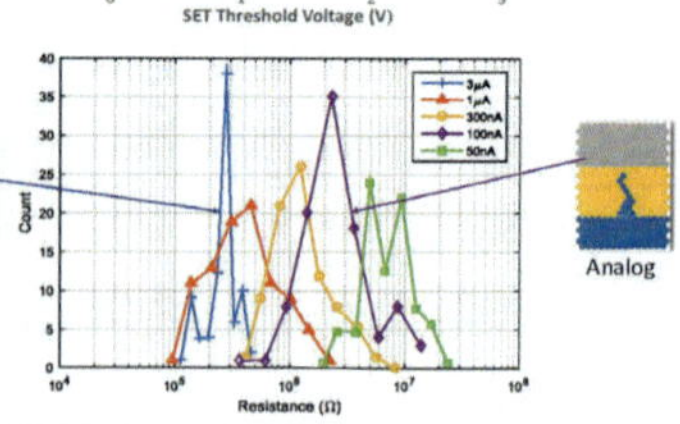

CBRAM Devices from Dr. Mitkova's group at BSU, 2016.

RRAM Challenges: Variability

RRAM devices exhibit variability in switching threshold voltages, which can vary from device to device and cycle-to-cycle. The switching characteristics are influenced by the initial 'forming' step and the compliance current (Icc).

RRAM Challenges: Stochasticity

RRAMs display voltage-dependent stochastic switching, which is influenced by the programming pulse voltage and duration.

17 RRAM Challenges: Endurance

- Endurance
  - Synaptic device endurance will determine the online or continuous learning ability in the neuromorphic chip.
  - Floating-gate (*aka* Flash) devices are best suited only for inference due to ~10^5 cycles write endurance
  - RRAM and PCM allow training but under device endurance limits (~10^8-10^9 write cycles)
  - Algorithmic considerations drive the device use case

Highlights the RRAM Challenges: Endurance, emphasizing the endurance of synaptic devices as crucial for continuous learning in neuromorphic chips, with current technologies like Flash suitable only for inference due to limited write cycles.

18 RRAM Challenges: Low LRS Resistance

- Low LRS Resistance
  - Low-resistance state (LRS) corresponds to the high conductance (G_M) or On-state
  - Low LRS resistances are observed in RRAM/memristive devices (<10kΩ)
  - Highly energy-inefficient when driving thousands of synapses in parallel (<10Ω load!!)
    - Need techniques to increase LRS resistance.
- 1T1R synapses can achieve higher LRS resistance

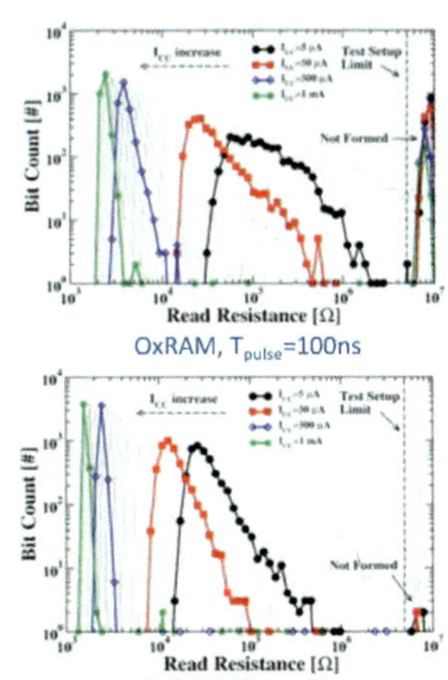

OxRAM, T_{pulse}=100ns

OxRAM, T_{pulse}=10μs

A. Grossi *et. al.*, "Fundamental variability limits of filament-based RRAM," 2016 IEEE International Electron Devices Meeting (IEDM), 2016.

Discusses the challenges of low LRS resistance in RRAM devices, which leads to energy inefficiency when driving many synapses in parallel, and highlights the need for techniques to increase LRS resistance. It also mentions that 1T1R synapses can achieve higher LRS resistance.

19 RRAM Challenges: State Relaxation

- Relaxation
  - Resistance states relax after MLC programming
  - Normal distribution variance increases
  - Few studies reported
    - CEA-Leti OxRAM programmed to 9 levels, CDF relaxes in 60s
    - Tsinghua OxRAM with thermal enhanced layer (TEL) programmed to 16 levels, relaxation measured after 30mins

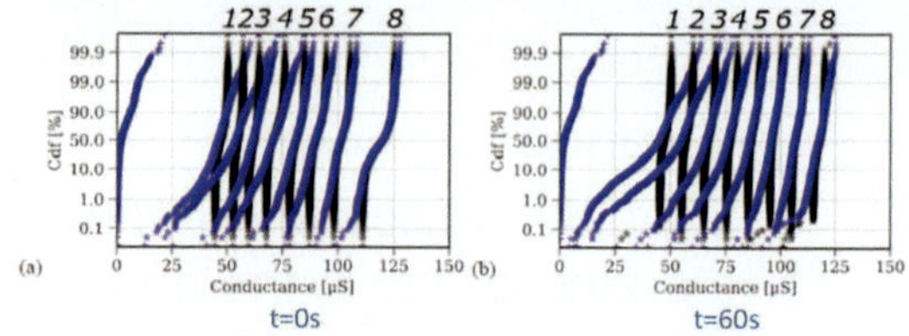

E. Esmanhotto et. al. "High-density 3D monolithically integrated multiple 1T1R Multi-Level-Cell for neural networks," IEEE IEDM 2020.

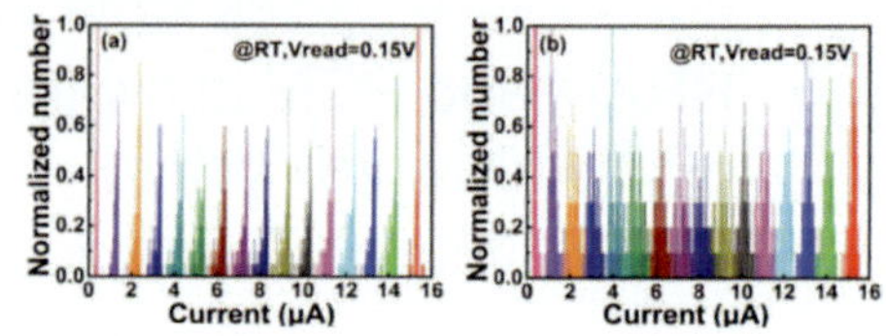

Zhao et al, "Investigation of Statistical Retention of Filamentary Analog RRAM for Neuromophic Computing," IEDM 2017.

State relaxation in RRAM post-MLC programming leads to increased normal distribution variance; studies include CEA-Leti's 9-level OxRAM (60s relaxation) and Tsinghua's 16-level OxRAM with TEL (30min relaxation), documented in E. Esmanhotto et al.'s IEEE IEDM 2020 and Zhao et al.'s IEDM 2017 papers, respectively.

20 RRAM Challenges: Resistance Drift (1)

- Resistance Drift
  - Resistance drifts over time
  - Device specific mechanisms
  - Brownian motion due to oxide vacancy hopping in OxRAM
  - Normal distribution variance increases with $\sqrt{t}$ and temperature, T (°K)

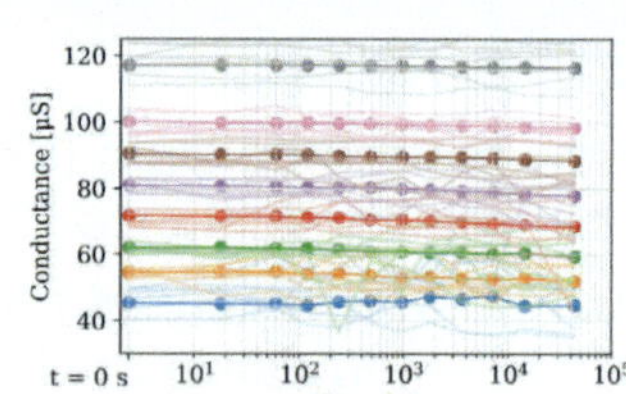

E. Esmanhotto et. al. "High-density 3D monolithically integrated multiple 1T1R Multi-Level-Cell for neural networks," IEEE IEDM 2020.

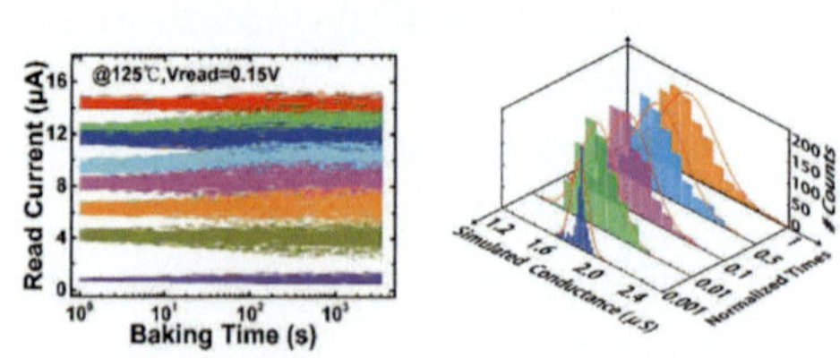

Zhao et al, "Investigation of Statistical Retention of Filamentary Analog RRAM for Neuromophic Computing," IEDM 2017.

The challenges associated with resistance drift in RRAM devices, which can impact the accuracy of datasets over time and suggests continuous training as a potential solution to mitigate forgetting. It emphasizes the need for tracking and restoring synaptic states or allowing for gradual forgetting within device endurance limits.

21 RRAM Challenges: Resistance Drift (2)

- Accuracy on the dataset progressively drops as resistance drifts
- Continuous training can alleviate the forgetting
- Need to track and restore synaptic states or allow forgetting over time
- Runs into the device write endurance limits
  - Allow redundancy in the ANN so that new devices can take over if few reach end of endurance

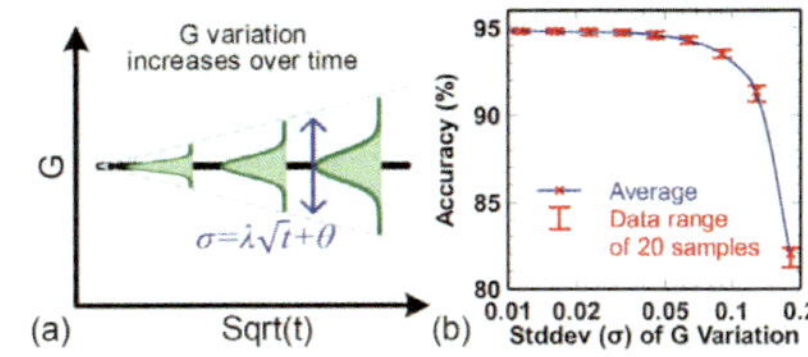

Pai Yu-Chen and Shimeng Yu, "Reliability Perspective of Resistive Synaptic Devices on the Neuromorphic System Performance," 2018 IEEE IRPS 2017.

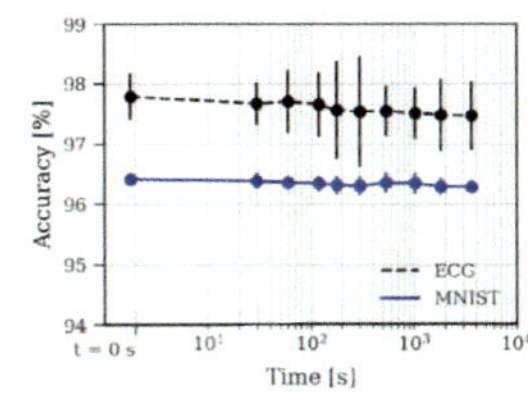

E. Esmanhotto et. al. "High-density 3D monolithically integrated multiple 1T1R Multi-Level-Cell for neural networks," IEEE IEDM 2020.

Continues with the challenges of Resistance Drift by claiming the drop in accuracy on the dataset when the drift happens in resistance value, which could alleviate the forgetting in the training of ANN.

22 Summary: Devices

- In-memory computing is interesting for low-power implementation of Edge-AI
  - VMM versus Neuromorphic Computing
- NOR Flash allows straightforward implementation on transfer learning
  - Suitable only for inference applications due to lower write endurance
- Emerging NVM devices can allow high-density, on-chip training, and low power
- Several challenges associated with realizing analog-like multilevel synapses
  - Device challenges need to be mitigated at circuit and algorithmic-level
- Foundry platforms with RRAM integrated in the BEOL of standard CMOS are critical for circuit exploration

Summaries the session on low-power implementation of Edge-AI with VMM vs Neuromorphic Computing with Nor Flash straightforward implementation on transfer learning.

OVERVIEW OF LEARNING ALGORITHMS

23 Deep Learning in Spiking Neural Networks

- A spike based DNN is formulated as

$$\nu_j^l = \sigma(z_j^l) = \sigma\Big(\sum_k w_{jk}^l \nu_k^{l-1} - \theta_j^l\Big)$$

  - ν_j^l is the spike rate for j^{th} neuron in the lth layer
  - θ_j^l is the threshold for the neuron
- Transfer Learning
- Unsupervised learning in SNNs using STDP and competitive learning
- Adapting standard backpropagation (Backprop) algorithm to spiking neurons

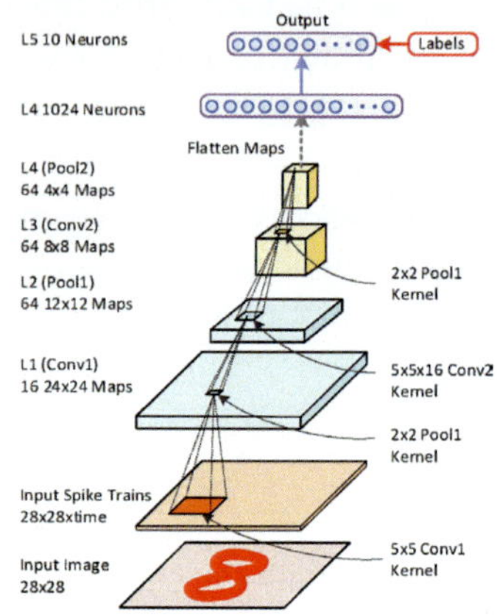

Talks about utilization of pre-trained deep neural networks converted into event-driven Spiking Neural Networks (SNNs). It also involnes training a standard DNN on a GPU and then transferring the model to a neuromorphic chip.

24 Transfer Learning

- Pre-trained deep neural networks converted into event-driven SNNs
- Train standard DNN on a GPU and then transfer model to a neuromorphic chip

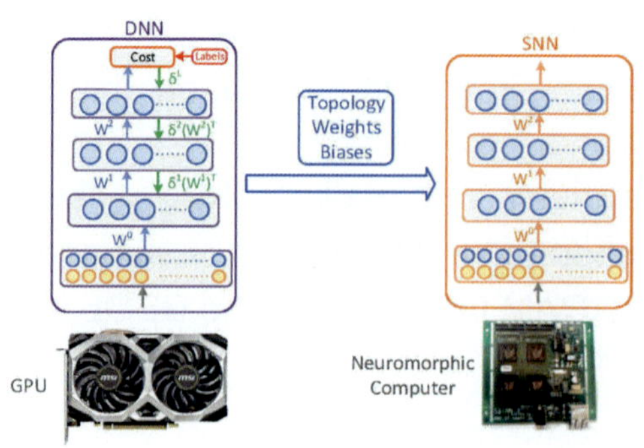

[1]Diehl et. al., "Fast-classifying, high-accuracy spiking deep networks through weight and threshold balancing," IJCNN 2015.

Discusses Transfer Learning in Spiking Neural Networks (SNNs). It explains how pre-trained deep neural networks are converted into event-driven SNNs by training a standard DNN on a GPU and then transferring the model to a neuromorphic chip.

25 Transfer Learning

- **Continuous Neuron Model:** Start with a conventional DNN and employ an approximated differentiable neuron model, such as ReLU, during training[2345]
- **Biases:** Either retain all the biases or set them to zero
- **Convnet Pooling:** Replace max-pooling by spatial subsampling
- **DNN Training:** Train the DNN using standard backpropagation, SGD, and dropout
- **Map Weights:** Directly map the weights from the DNN network to a network with integrate-and-re neuron (IFN) units; rate-based spike coding is employed
- **Weight Normalization:** Weights are scaled in the SNN to maximize classification accuracy

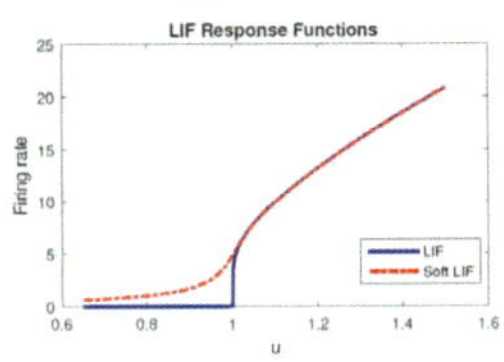

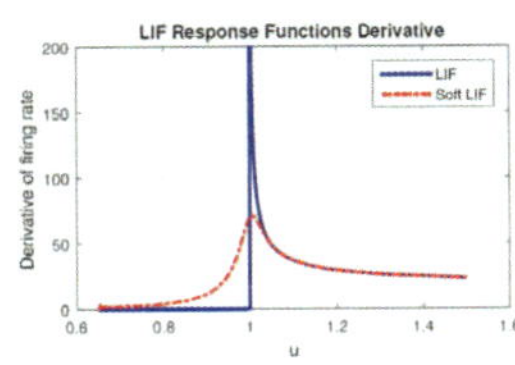

[2]P. U. Diehl; D. Neil; J. Binas; M. Cook; S.-C. Liu; and M. Pfeier; IJCNN 2015.
[3]E. Hunsberger and C. Eliasmith; arXiv:1510.08829 (2015).
[4]E. Hunsberger and C. Eliasmith; arXiv:1611.05141 (2016).
[5]B. Rueckauer; I.-A. Lungu; Y. Hu; M. Pfeier; and S.-C. Liu; Frontiers in Neuroscience 11; 682 (2017).

Here are the key points about Transfer Learning:

Continuous Neuron Model: Utilizes an approximated differentiable neuron model like ReLU during training. It also speaks about the biases, pooling and the model training training.

Weight Normalization: Involves scaling the weights in the Spiking Neural Network (SNN) to maximize classification accuracy.

26 Spiking Conv Neural Network

ConvNet: 28x28-16c5-2s-64c5-2s-10o

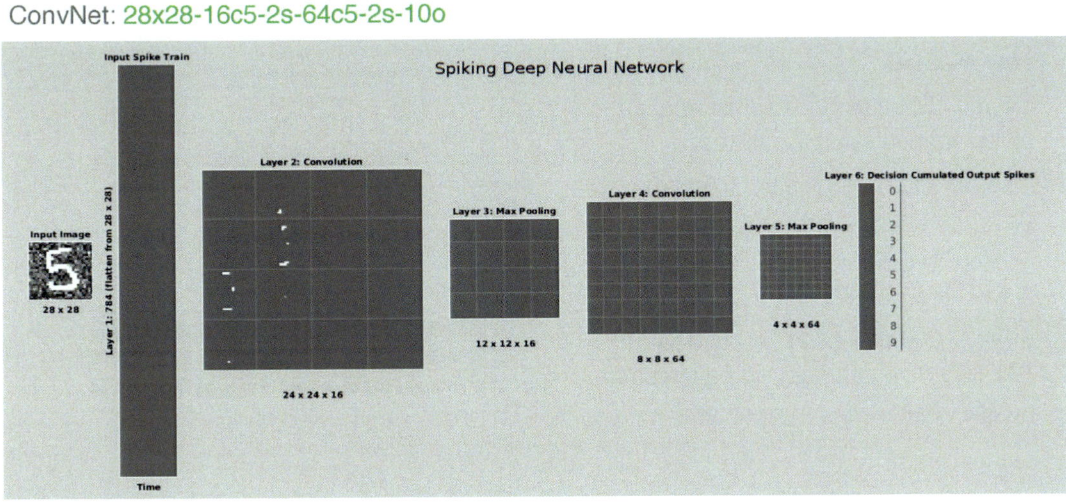

Lillustration of Spiking Conv Neural Network (28x28-16c5-2s-64c5-2s-10o) using MNIST data and showing the prediction status in the last layer.

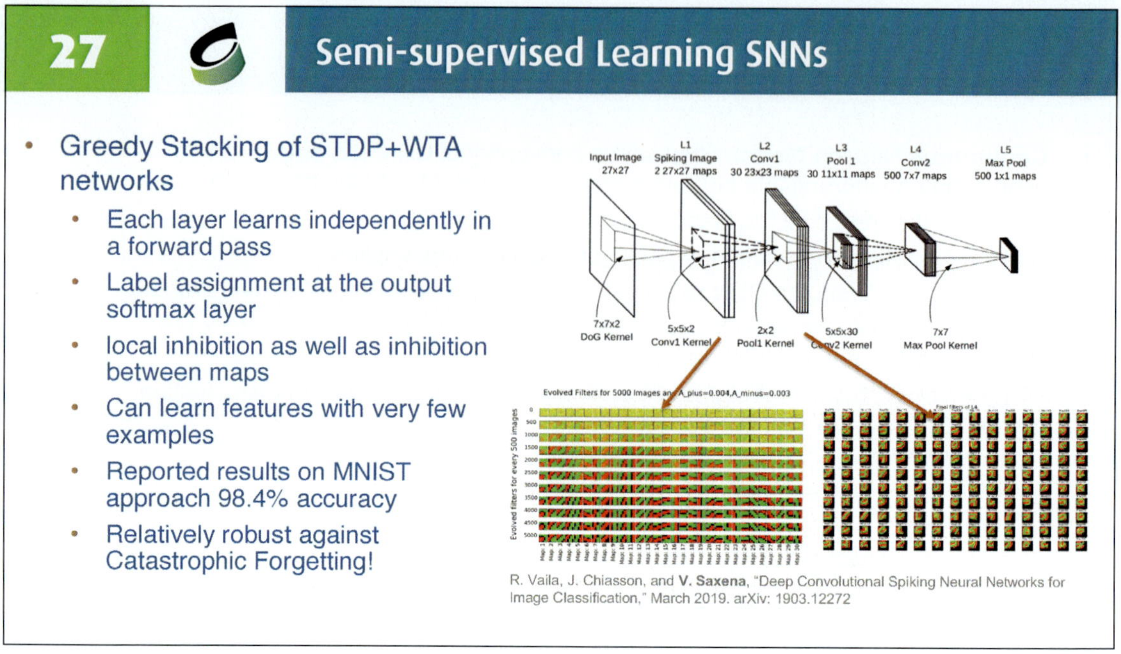

Unsupervised Learning in SNNs: The slide discusses a hybrid unsupervised Spiking Neural Network (SNN) with binarized classification layers, where the SNN layers L1-L4 learn unsupervised, and L4 and L5 are binarized classification layers trained using backpropagation.

SpykeFlow Tool: A Python TensorFlow-based tool called SpykeFlow has been developed for simulations, achieving 85.57% classification accuracy on the EMNIST dataset.

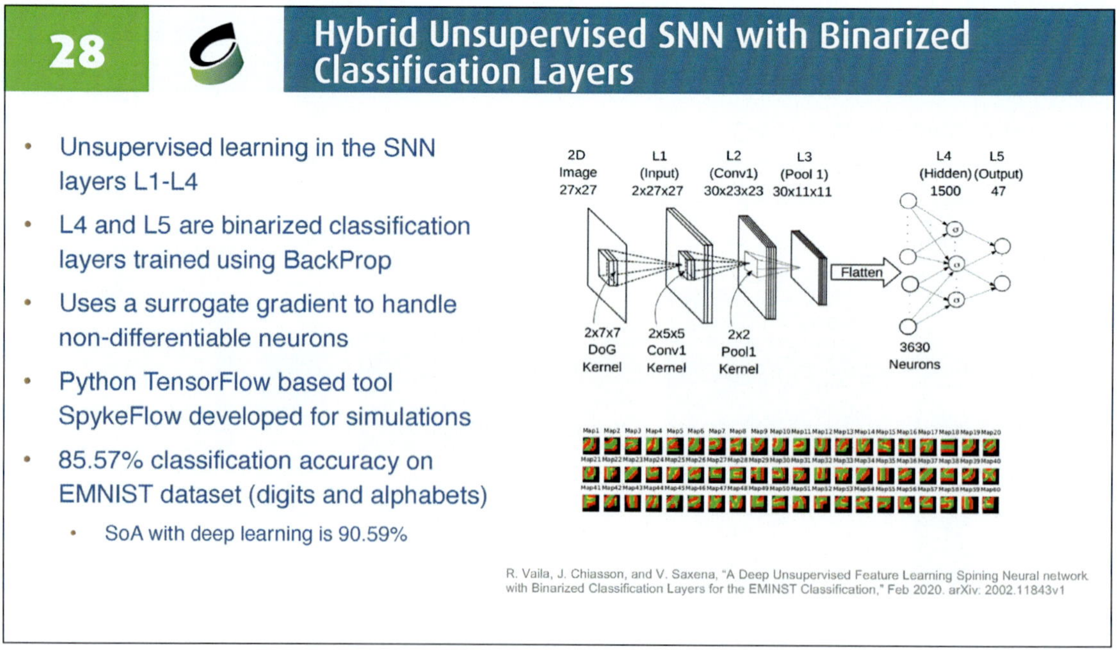

Hybrid Unsupervised SNN: Combines unsupervised learning in SNN layers with binarized classification layers trained using backpropagation.

SpykeFlow Tool: Utilizes TensorFlow for simulations, achieving 85.57% accuracy on the EMNIST dataset.

29 Backpropagation for Deep SNNs

- High classification accuracy can be obtained by imbibing some form of backpropagation in deep spiking networks: credit assignment problem across layers
- Backpropagation based weight updates in an SNN can be expressed by a three-factor learning rule

$$\Delta w^l_{jk} = -\eta \cdot \underbrace{\nu_k^{l-1}}_{Eligibility} \cdot \underbrace{\sigma'(z_j^l - \theta_j^l)}_{Gradient} \cdot \underbrace{[(W^{l+1})^T \delta^{l+1}]_j}_{Error,\delta}$$

- Weight update depends upon
  - the eligibility of the pre-neuron (i.e. whether the neuron fired)
  - Activation gradient of the post neuron
  - Modulated by the feedback error, δ_j
- STDP is a special simplified pairwise version of this learning rule

Ref: E. Neftci et al, "Neuromorphic Deep Learning Machines", arXiv::1612.05596, Jan 2017

In deep spiking networks, integrating backpropagation addresses the credit assignment problem, with weight updates governed by a three-factor learning rule involving pre-neuron eligibility, post-neuron activation gradient, and modulated by feedback error δ_j, akin to simplified STDP. Ref: E. Neftci et al, "Neuromorphic Deep Learning Machines", arXiv:1612.05596, Jan 2017.

30 Backpropagation for Deep SNNs

- Standard backprop relies on availability of network-wide information stored in high-precision memory
  - This information is funneled through von Neumann bottlenecks
- Challenging to realize on neuromorphic hardware
  - Symmetric weight requirements for the synapses (weight transport problem)
  - Precise computation of multiplications with derivatives and activation functions
  - Propagating error signals with high precision
  - Alternating between forward and backward passes
- Recent progress has been made with using synthetic gradients to decouple the learning in the layers

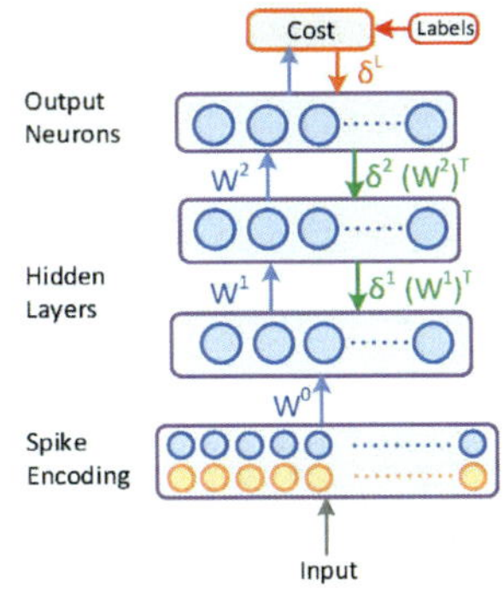

Ref: E. Neftci et al, "Neuromorphic Deep Learning Machines", arXiv::1612.05596, Jan 2017
V. Saxena, "Neuromorphic computing: From devices to integrated circuits," Journal of Vacuum Science & Technology B, 39(1):010801, Jan 17 2021.

Backpropagation Challenges: Neuromorphic hardware faces difficulties with backpropagation due to von Neumann bottlenecks and the need for high-precision memory.

Synthetic Gradients: Emerging strategies like synthetic gradients are being explored to overcome these challenges and enable learning in deep SNNs.

31 Comparison of SNN Algorithms

- Comparison for MNIST 28x28 handwritten digits dataset

Reference	Algorithm	Neuron/Coding	Network	Dataset	Accuracy
Simard et. al. 2003	DNN with Backprop	ReLU	784-20c5-2s-100-10o	MNIST	99.6%
Wan et. al. 2013	DNN with Backprop, DropConnect	ReLU	-	MNIST	99.79%
Diehl et. al. 2015	Transfer Learning, Fully connected SNN Transfer Learning, ConvNet	Rate-based IFN	784-1200-1200-10o 784-12c5-2s-64c5-2s-10o	MNIST	98.68% 99.14%
Diehl & Cook 2015	STDP-WTA Unsupervised Learning	LIFN with inhibition	784-1600-10o	MNIST	95%
Hunsberger et. al. 2016	Transfer Learning	Rate-based LIFN	784-16c5-2s-64c5-2s-10o	MNIST	99.1%
Neftci et. al. 2016	event-driven RBP	Rate-based IFN	784-500-10o	MNIST	97.04%
Lee et. al. 2016	Backprop	Rate-based IFN	784-800-10o	MNIST	99.31%
Kheradpisheh et. al. 2016	STDP-WTA	IFN, latency based	784-30c5-2s-100c5-2s-10o	MNIST	98.4%
Rueckauer 2017	Transfer Learning	Rate-based LIFN	LeNet	MNIST	99.44%
O'Connor et. al. 2017	Backprop	Pulse coding	784-200-200-10o	MNIST	98.36%
Kulkarni et. al. 2018	NormAD	IFN	784-12c3-10o	MNIST	98.17%
Mostafa 2018	Backprop	Temporal coding	784-400-400-10o	MNIST	97.55%
Shrestha et. al. 2018	SLAYER	Rate-based IFN	784-12c5-2s-64c5-2s-10o	MNIST	99.36%
Mostafa et. al. 2018	DNN with Synthetic gradients	ReLU†	784-1000-1000-1000-10o	MNIST	98.7%
Vaila et. al. 2019	STDP-WTA Unsup, FC readout layer	IFN, latency based	784c5-2s-30c5-s7-500-70-10o	MNIST	98.4%

† This work uses non-spiking ReLUs but can be adapted to SNNs using rate-based or temporal-coding.

V. Saxena, "Neuromorphic Computing: From Devices to Integrated Circuits," J. Vac. Sci. Technol. B, vol. 39, pp. 010801, 2021

This shows results of works on the prediction of MNIST 28x28 datasets that doesnt use SNN algorithms. Its can be observed that this slide shows different algorithms with Neurons and architectures and it claims that using rate-based or temporal-coding these DNN methods could be converted into SNN algorithms.

32 Comparison of SNN Algorithms

- Comparison for CIFAR and ImageNet datasets

Reference	Algorithm	Neuron/Coding	Network	Dataset	Accuracy
Krizhevsky et. al. 2012	Alexnet: DNN with Backprop	ReLU	5-layer ConvNet	CIFAR-10 CIFAR-100	88.91%
He et. al. 2016	ResNet: DNN with Backprop	ReLU	ResNet-1001	CIFAR-10 CIFAR-100	95.08%
Hunsberger et. al. 2016	SNN Transfer Learning	Rate-based LIFN	Alexnet	CIFAR-10	82.95%
Rueckauer 2017	SNN Transfer Learning	Rate-based LIFN	Alexnet	CIFAR-10 CIFAR-100	88.82%
Mostafa et. al. 2018	DNN with Synthetic gradients	ReLU†	ConvNet△	CIFAR-10	89.1%
VGGNet 2017	DNN with Backprop	ReLU	16-layer ConvNet	ImageNet▽	93.2%
Inception-V3 2017	DNN with Backprop	ReLU	42-layer ConvNet	ImageNet▽	96.42%
Hunsberger et. al. 2016	SNN Transfer Learning	Rate-based LIFN	-	ImageNet▽	76.2%
Rueckauer 2017	SNN Transfer Learning	Rate-based LIFN	VGG-16 Inception-V3	ImageNet▽	84.86% 92.04%

△ Network for CIFAR-10 is 32x32x3-96c5-s3-128-c5-s3-256c5-s3-2048-2048-10o.

▽ Top-5 accuracy on ImageNet dataset.

V. Saxena, "Neuromorphic Computing: From Devices to Integrated Circuits," J. Vac. Sci. Technol. B, vol. 39, pp. 010801, 2021

Similar results were taken from the given references on CIFAR and ImageNet Datasets, that provide high accuracy rates.

33 Summary: Algorithms

- Spike-based Neural networks (SNNs) are promising for low-power mixed-signal realization of deep learning
- Motifs that emerge from experimental and computation neuroscience continually explored
  - WTA-based SNNs demonstrate unsupervised learning: spike-based clustering
  - Stacking of these WTA motifs doesn't improve accuracy beyond two layers
  - Less catastrophic forgetting, graceful degradation in accuracy
- Transfer learning is ready for deployment with marginal performance loss
- Adapting Backprop to SNNs has yielded success
  - Differential neuron models
  - Several emerging strategies for mitigating the weight transport and locking in deep SNNs
  - Catastrophic forgetting is concomitant with Backprop
  - Still a large gap between the functionality of biological brains and neuromorphic/AI algorithms

Summary of the session on Algorithms.

MIXED-SIGNAL CIRCUITS FOR RRAM ARRAYS

34 1T1R RRAM Cells

- Select transistor V_G limits the current to I_{cc} during forming/set/reset
- Several possibilities for using the cell in a neuromorphic array:
- SL drive:
  - Transistor acts as a switch and a current $i=V_{read}/R_M$ is available on BL
  - Resistive drive
- WL drive:
  - Common source with source degeneration ($V_{BL}=0$)
  - Current $i=G_mV_{DD}$ is available on SL, where $G_m=g_m/(1+g_mR_M)$
  - Capacitive drive

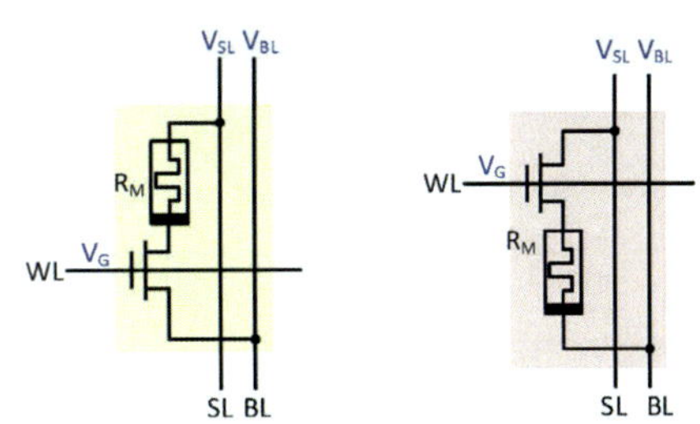

	Forming	Program (Set)	Erase (Reset)	Read
WL	V_G (sets I_{cc})	V_G (sets I_{cc})	V_G (sets I_{cc})	V_{DD}
BL	0	0	V_{reset}	$V_{CM}+V_{read}$
SL	V_{form}	V_{set}	0	V_{CM}

V. Saxena, "Process-variation Robust RRAM-Compatible CMOS Neuron for Neuromorphic System-on-a-Chip," IEEE ISCAS, Seville, Spain, May 2020.

Introduces the Mixed-Signal Circuit for RRAM Arrays.

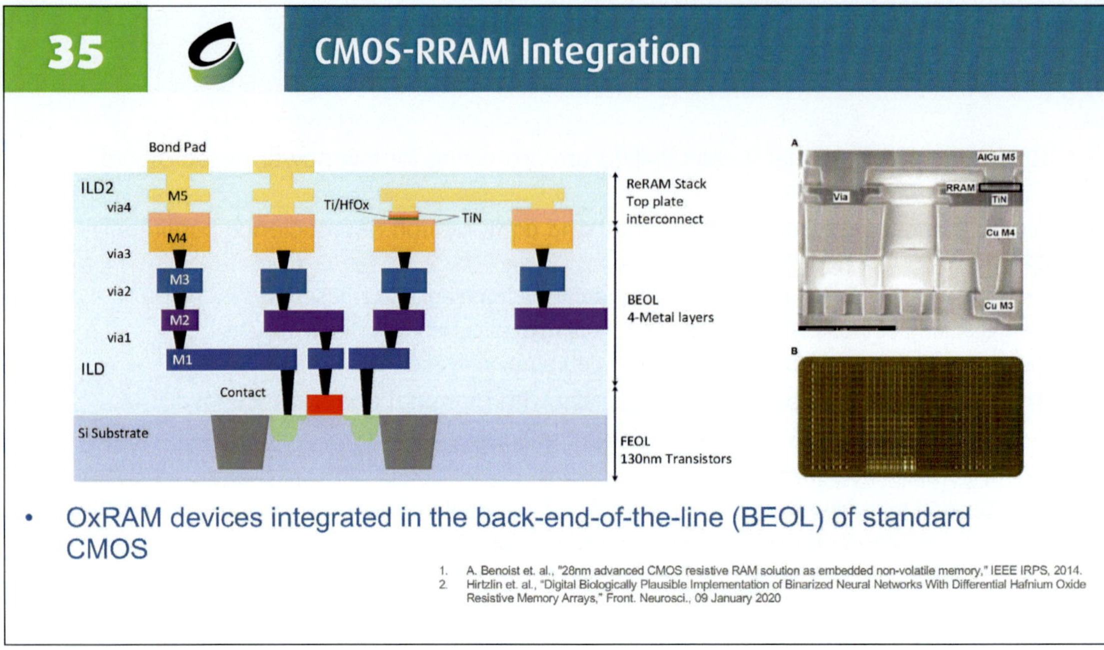

Talks about selecting the transistor in 1T1R RRAM cell by limiting the current to (I_{cc}) during forming, set, and reset operations. The transistor acts as a switch, providing a current ($i = V_{read}/R_M$) on the bit line (BL), known as resistive drive and WL drives functions as a common source with source degeneration ($V_{BL} = 0$), where the current ($i = G_m V_{DD}$) is available on the source line (SL), termed capacitive drive.

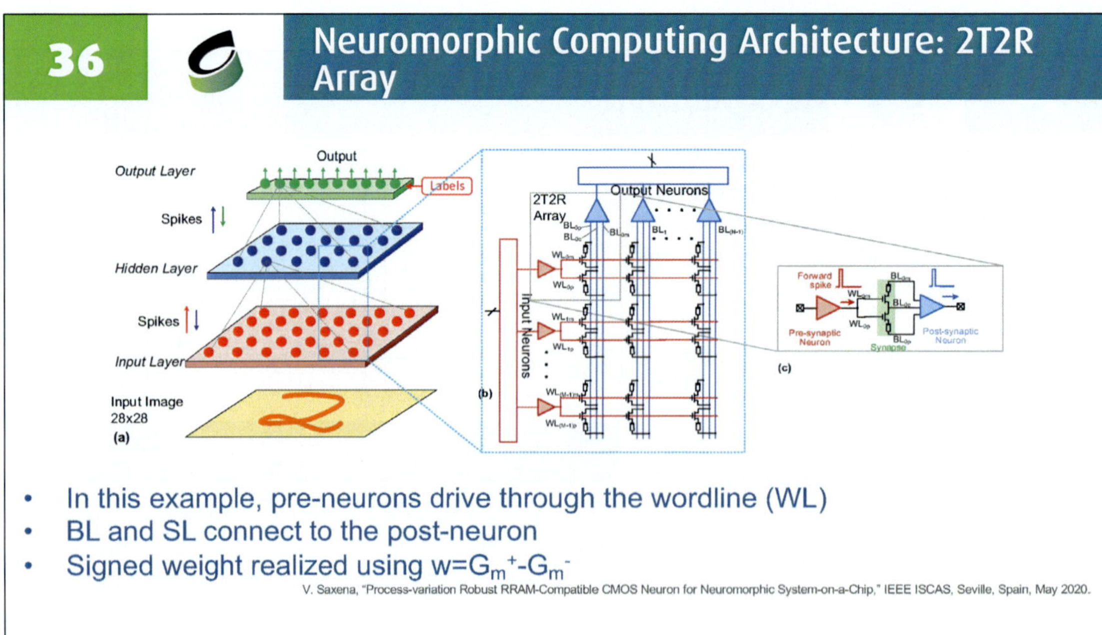

OxRAM devices integrated in the back-end-of-the-line (BEOL) of standard CMOS.

37 Peripheral Circuits

- A transfer learning mode is needed for neuromorphic SoCs
- Initialize the weights, transfer pre-learned weights or read out learned weights
- A 1T1R cell needs to be selected and programmed to a MLC state
  - Iterative program and verify algorithm
- DACs with WL drivers
- ADC on each column
- An on-chip bus/interconnect for routing of information and commands
- On/off-chip controller for state machine

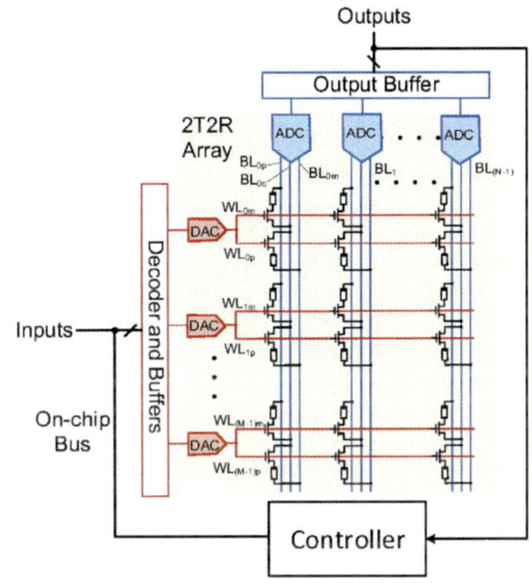

Shows the Neuromorphic Computing Architecture: 2T2R Array, its structure in device level.

38 Array Programming

- Multilevel cell (MLC) programming is achieved using program and verify schemes
  - Incremental Pulse Program Strategy (ISPP)
  - Similar to MLC Flash memory
  - Alternates between set/reset and read phases to achieve target conductance/resistance state
  - Linear *vs* σ-based allocation
- Peripheral circuits to support the voltage and drive strength needed to program cells
- Can program cells to ~3-bit resolution

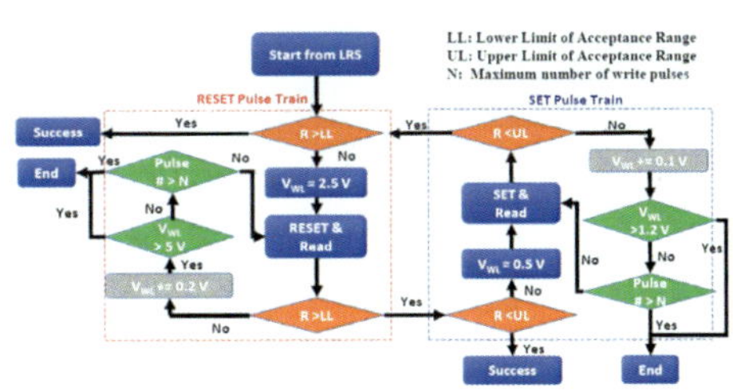

DD-ISPP Programming Conditions		
	RESET	SET
V_{WL} Start (V)	2.5	0.5
V_{WL} Stop (V)	5	1.2
V_{WL} Step (V)	0.2	0.1
V_{BL} (V)	0	2.5
V_{SL} (V)	3.5	0
Pulse Width (ns)	100	50

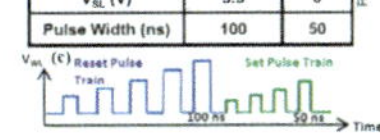

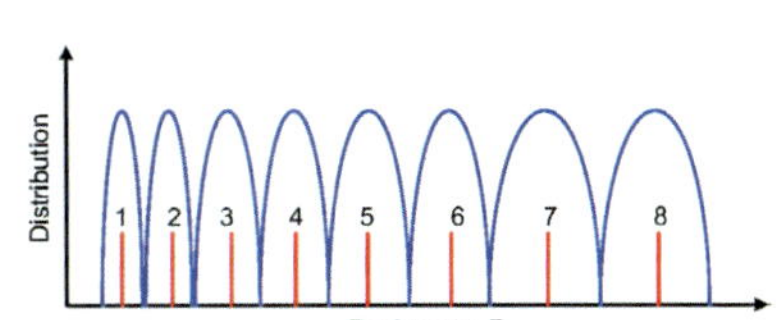

X. Zheng et. al., "Error-Resilient Analog Image Storage and Compression with Analog-Valued RRAM Arrays, An Adaptive Joint Source-Channel Coding Approach," IEDM 2018

For neuromorphic SoCs, a transfer learning mode is crucial, allowing weight initialization, transfer, or readout. This process involves programming 1T1R cells to MLC states iteratively, supported by DACs, ADCs, on-chip bus/interconnect, and an on/off-chip controller.

39 Test Chip Interfaced with CBRAM Devices

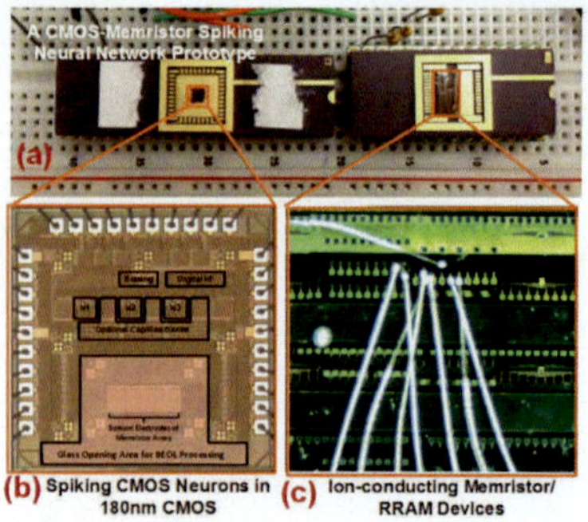

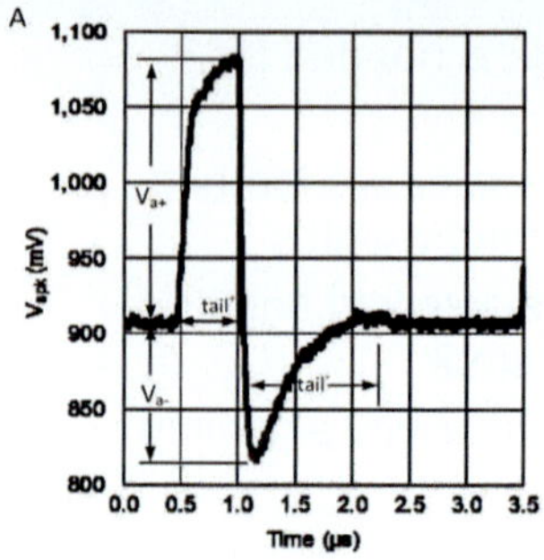

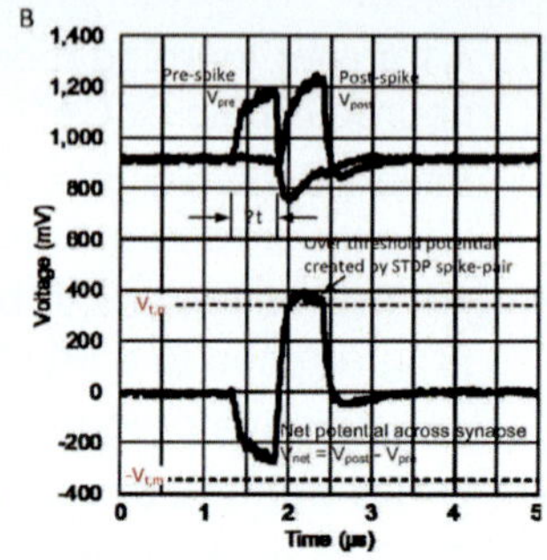

- Mixed-signal IC with 180-nm CMOS Neurons connected with 1R CBRAM Synapses
- A first RRAM compatible STDP learning Neuron
- Power consumed in the neuron is 9.3pJ/spike/synapse (in terms of 10,000 synapses each having around 1MΩ resistance).

Source: X. Wu, V. Saxena, K. Zhu and S. Balagopal, "A CMOS Spiking Neuron for Brain-Inspired Neural Networks With Resistive Synapses and In Situ Learning", *IEEE TCAS-II, 62*(11), pp.1088-1092, 2015.

MLC programming in neuromorphic devices employs program and verify schemes like ISPP, akin to MLC Flash memory, alternating set/reset and read phases to attain target conductance states. It utilizes peripheral circuits to provide necessary voltage and drive strength, achieving resolutions of ~3 bits, with options for linear or σ-based allocation.

40 Dynamic Resistive Load Drive

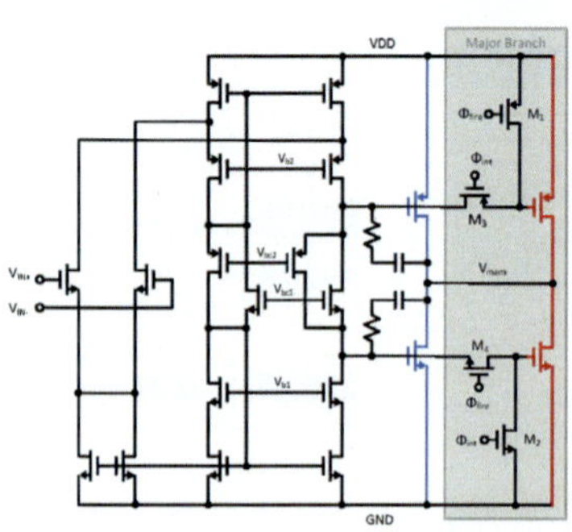

97% Efficiency
@ driving 10,000 x 1MΩ synapses

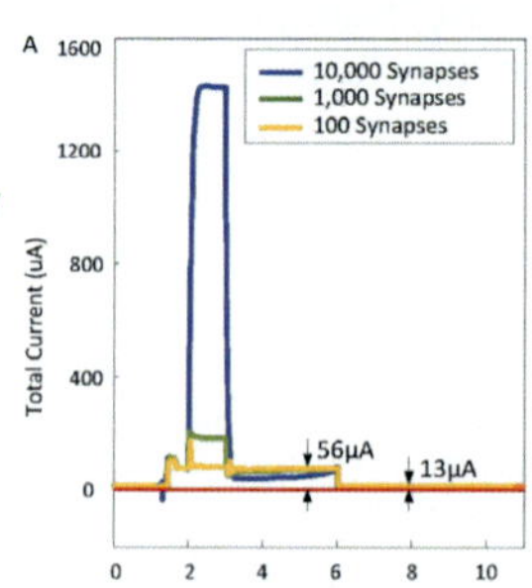

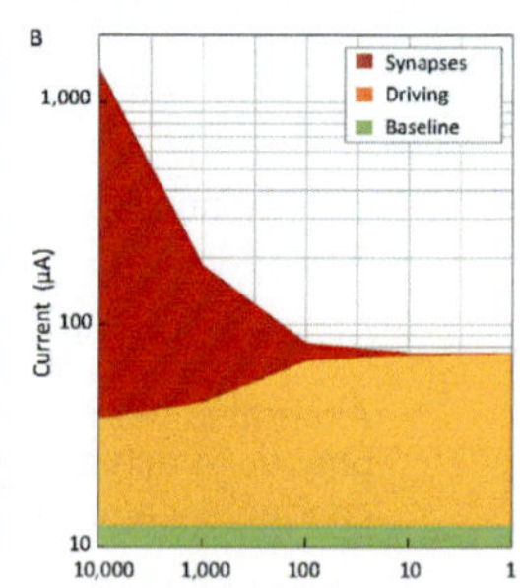

- 13μA bias current in integration mode, 56μA in firing mode
- 1.4mA peak current through the class-AB stage during the spike event
- Can be optimized further with devices with high LRS resistances

X. Wu, **V. Saxena**, K. Zhu, and S. Balagopal, "A CMOS spiking neuron for brain-inspired neural networks with resistive synapses and in situ learning." *IEEE Transactions on Circuits and Systems II: Express Briefs*, vol. 62, no. 11 (2015): 1088-1092.

Testing the Test Chip Interfaced with CBRAM Devices of 18-nm CMOS Neuron with 1R synapses.

41 Associative Learning Experiment with 1R Array

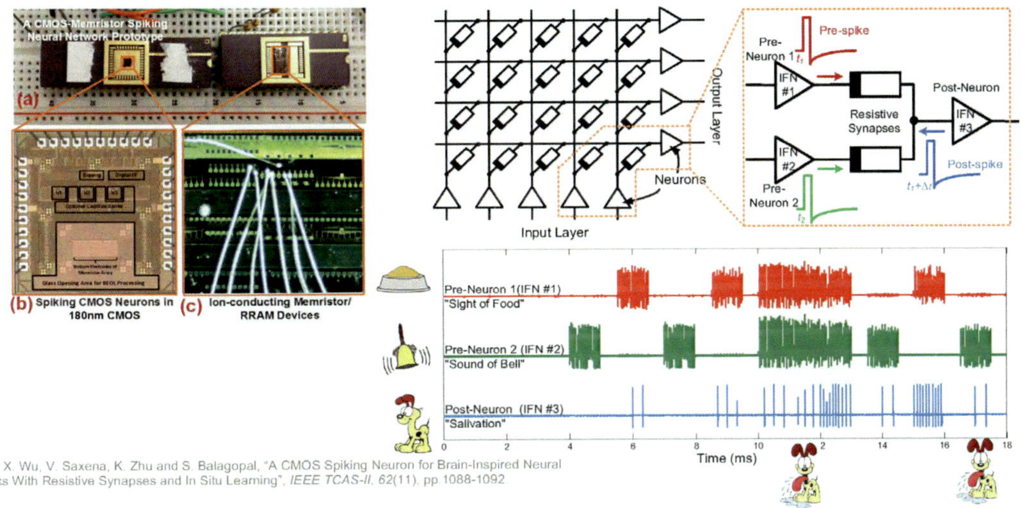

Source: X. Wu, V. Saxena, K. Zhu and S. Balagopal, "A CMOS Spiking Neuron for Brain-Inspired Neural Networks With Resistive Synapses and In Situ Learning", *IEEE TCAS-II*, *62*(11), pp.1088-1092.

This explains about the dynamic resistice load drive. There were results obtained in the forms of different measurements of currect like Bias current, Firing mode current, Peak current through Class-AB stage in spiking event.

42 2T2R Synapse and Neuron Interface

- In-situ learning using three-factor rule

$$\Delta w^l_{jk} = -\eta \cdot \underbrace{\nu^{l-1}_k}_{Eligibility} \cdot \underbrace{\sigma'(z^l_j - \theta^l_j)}_{Gradient} \cdot \underbrace{[(W^{l+1})^T \delta^{l+1}]_j}_{Error, \delta}$$

- Neurons drive capacitance in the forward path; similar digital to CMOS
  - Mitigates the LRS resistance challenge
- Resistances driven by the post-neuron only during the weight update

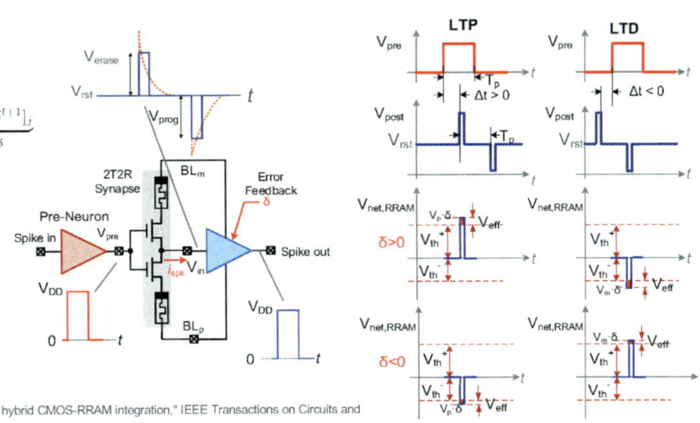

1. V. Saxena, "Mixed-signal neuromorphic computing circuits using hybrid CMOS-RRAM integration," IEEE Transactions on Circuits and Systems II: Express Briefs, 68(2), pp.581-586, 2020.
2. V. Saxena, "Process-variation Robust RRAM-Compatible CMOS Neuron for Neuromorphic System-on-a-Chip," IEEE ISCAS, Seville, Spain, May 2020.

Talks about associative learning experiment with 1R array. The experimental setup is shown with the conncetion diagram in the device level. It also shows the responses spikes in for the input images.

43 Event-Driven Neuron for 2T2R Arrays

- Event-driven switched-capacitor neuron
- Single shared opamp
- Asynchronous comparator
- *Integration mode:* opamp biased with low current
- *Firing mode:* opamp dynamically biased to drive RRAM with large current
- Local configurable waveform generator for weight update
- Output spikes are full-CMOS levels

V. Saxena, "Process-variation Robust RRAM-Compatible CMOS Neuron for Neuromorphic System-on-a-Chip," IEEE ISCAS, Seville, Spain, May 2020.

V. Saxena, "Mixed-signal neuromorphic computing circuits using hybrid CMOS-RRAM integration," IEEE Transactions on Circuits and Systems II: Express Briefs, 68(2), pp.581-586, 2020.

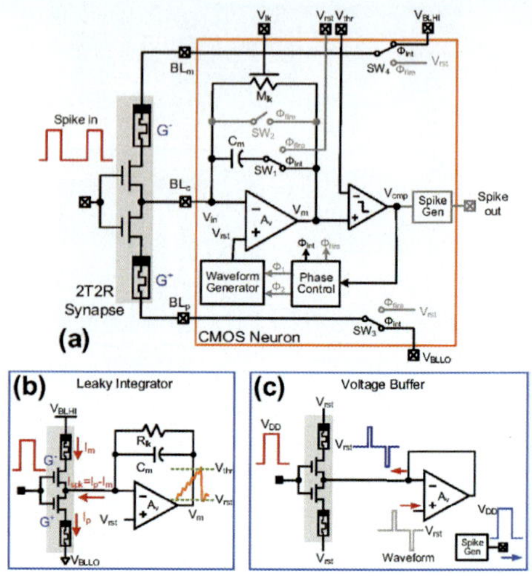

In-situ learning employs a three-factor rule, with neurons driving capacitance akin to digital-to-CMOS in 2T2R Synapse, addressing the LRS resistance challenge by post-neuron-driven resistances during weight updates.

44 Event-Driven Neuron-Fire Mode

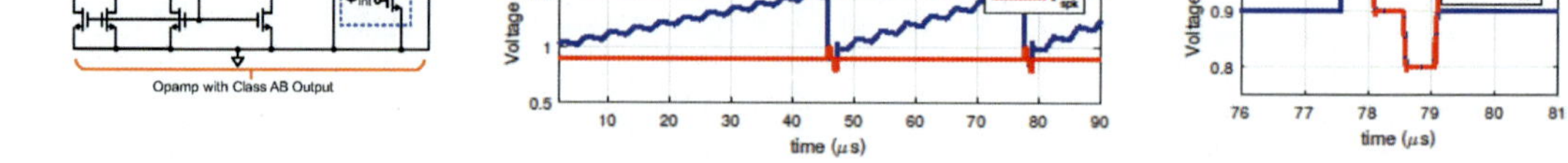

- Implemented in 180nm CMOS technology
- Folded cascode opamp with dynamically biased class-AB output stage
- Inference-only designs can be achieved with very low power
- For training, opamp unity-gain frequency (f_{un}) set by the STDP waveforms

Event-Driven Neuron: A neuron design for 2T2R arrays that switches between low-current integration and high-current firing modes.

Weight Update: Utilizes a local waveform generator for synaptic weight adjustments, with full-CMOS level output spikes.

45 Neuron Performance Comparison

- Only recent CMOS neurons that can interface with RRAMs are considered
- Energy-efficiency expressed as fJ/spike/synapse

Design	Type	Technology	Synapse Type	On-chip Learning	I_{VDD}	Energy-efficiency (fJ/spike/synapse)
Wu et al 2015	Opamp	180nm	1R	✔	13μA	140
Sahoo 2017	Ring VCO	65nm	None	✘	-	-
Larras et al 2017	Current-summing	65nm	Digital	✘	-	7
Sourikopolous et al 2017	Subthreshold	65nm	None	✘	-	4
This work (inference)	Opamp	180nm	1T1R		9μA‡	8.1‡
This work (training)	Opamp	180nm	1T1R	✔	36μA*	40*

‡ Bias current during the integration mode
* Bias current during the firing mode

Discuss about the event driven neuron fire mode implemented in 18nm CMOS technology with folded cascade opamps. The inference only design could be achieved with low power in training, that could be achieved with unity grain frequency of opamp set by STDP waveforms.

A review of the work in neuron performance in different types of device modelling for different technology nodes.

RRAM-BASED NEUROMORPHIC ICS

46 Early RRAM CMOS Demonstrators

- Early Neuromorphic ICs employed:
  - Binarized RRAM weights or
  - Unit-weighted binary cells used as a DAC
  - RRAM Memory macros (SLC & MLC)
- Most large-scale ICs are for inference only applications
  - Full-chip implementation limited to 2-bit/device resolution
  - On-chip learning still an open problem
- CEA-Leti's CIM macro chip with Ti/HfOx OxRAM
  - Two-layer network built using 1T1R 1-bit
  - with 84% MNIST classification accuracy
  - Spiking neurons had 180pJ/SynOp energy consumption

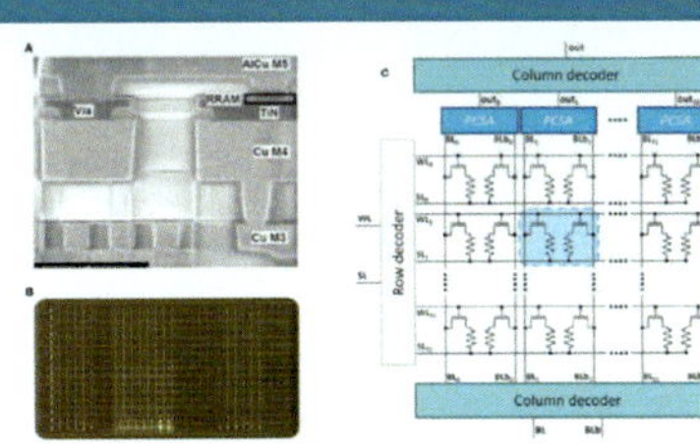

T. Hirtlin *et. al.*, "Digital Biologically Plausible Implementation of Binarized Neural Networks With Differential Hafnium Oxide Resistive Memory Arrays," Front. Neurosci., 09 January 2020.

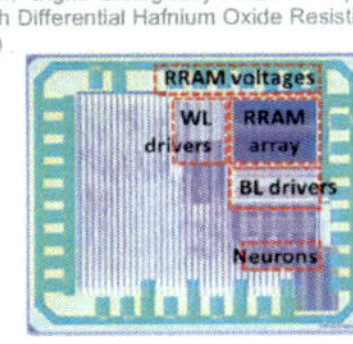

CMOS process	130 nm
Cu interconnect	5
RRAM number	13,5 K
RRAM configuration	1T-1R
RRAM size	0.5μm x 0.5μm

A. Valentian *et. al.*, "Fully Integrated Spiking Neural Network with Analog Neurons and RRAM Synapses," IEEE IEDM 2019, pp 14-3. **(CEA Leti)**

Early neuromorphic ICs utilized binarized RRAM weights or unit-weighted binary cells for DACs, featuring RRAM memory macros (Single-Level Cell and Multi-Level Cell). While most large-scale ICs were tailored for inference applications, full-chip implementation was constrained to 2-bit/device resolution, with on-chip learning remaining a challenge. Notably, CEA-Leti's CIM macro chip employed Ti/HfOx OxRAM, achieving 84% MNIST classification accuracy with a two-layer network built using 1T1R 1-bit cells, while spiking neurons exhibited 180pJ/SynOp energy consumption.

47 CMOS RRAM Neuromorphic IC

- 130nm CMOS with integrated RRAMs
- 256 Neurons connected to 65k 1T1R RRAM cells
- Bidirectional data flow to realize several types of neural networks: RBM. LSTM, RNN
- Used ISPP for programming individual cells to +/12μS of targeted conductance states
- Simple CS amplifier realizes I&F neurons with correlated double sampling (CDS)
- The chip fully-integrated a 784-100-10 network with FPGA-based back-end
  - Demonstrated 225-60o RBM for MNIST reconstruction
  - 74TOPS/W with 1.79mm^2 array area

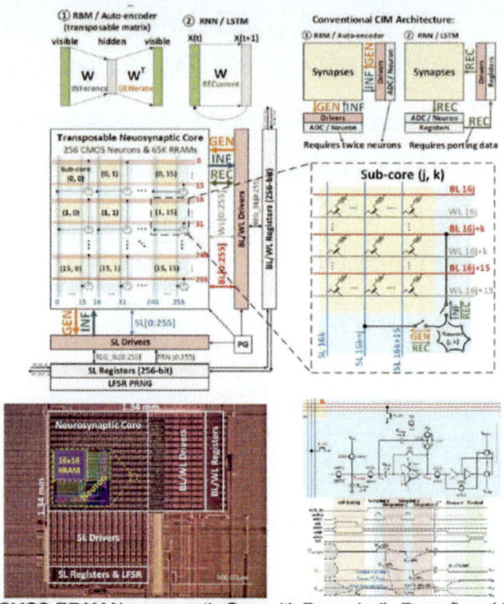

W. Wan et. al, "A 74 TMACS/W CMOS-RRAM Neurosynaptic Core with Dynamically Reconfigurable Dataflow and In-situ Transposable Weights for Probabilistic Graphical Models," IEEE ISSCC 2020.

A 130nm CMOS chip integrated with RRAMs featured 256 neurons connected to 65k 1T1R RRAM cells, facilitating bidirectional data flow for various neural network types like RBM, LSTM, and RNN. Using Incremental Step Pulse Programming (ISPP), individual cells were programmed to targeted conductance states, while a simple current-sensing amplifier enabled Integrate-and-Fire (I&F) neurons with correlated double sampling (CDS), achieving impressive performance of 74 TOPS/W with a compact 1.79mm^2 array area, demonstrated through tasks like MNIST reconstruction with a 784-100-10 network fully integrated with an FPGA-based back-end.

48 CMOS RRAM Inference IC

- 130nm CMOS with integrated RRAMs
- 2T2R synapses (for signed weights) to demonstrate a VMM
- Multibit weights realized by combining several 2T2R cells in parallel
- Weighted currents were integrated and digitized using a successive approximation register (SAR) type ADC
  - `LPAR' ADC
- The chip fully-integrated a 784-100-10 network with FPGA-based back-end
  - Demonstrated 94.4% MNIST accuracy
  - 78.4TOPS/W, 77μs/image latency

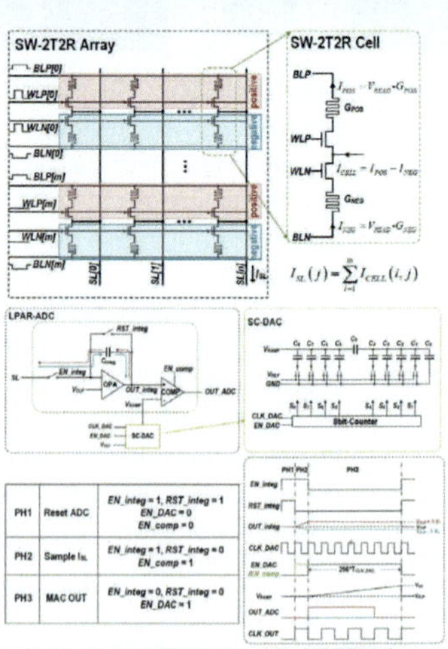

Q. Liu et al, "A Fully Integrated Analog ReRAM Based 78.4TOPS/W Compute-In-Memory Chip with Fully Parallel MAC Computing," IEEE ISSCC 2020.

A 130nm CMOS chip integrated with RRAMs utilized 2T2R synapses for signed weights and multibit weights achieved by parallel combinations of 2T2R cells, with weighted currents integrated and digitized via a SAR type ADC, including an LPAR ADC. The chip fully integrated a 784-100-10 network with an FPGA-based back-end, showcasing 94.4% MNIST accuracy alongside remarkable energy efficiency of 78.4 TOPS/W and 77μ/image latency.

49 RRAM CMOS VMM

- Fully integrated CBRAM with 180nm CMOS
- 54×108 1R RRAM crossbar arrays
- On-chip 6-bit DAC and 13-bit column ADCs and a RISC processor for programming
- DAC produces voltage pulses of fixed width proportional to the input value
- DAC pulses are weighted by the RRAM-based VMM and then integrated in charge-domain on the integrating ADC

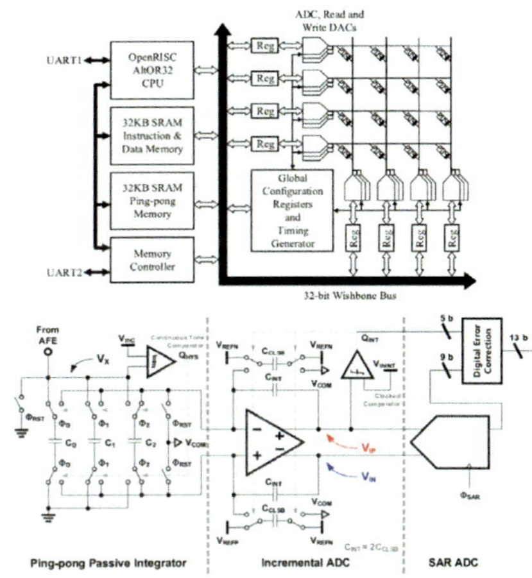

J. M. Correll *et. al.*, "Fully Integrated Reprogrammable CMOS-RRAM Compute-in-Memory Coprocessor for Neuromorphic Applications," IEEE JXCDC, 6 (2020).

A fully integrated CBRAM system with 180nm CMOS featured 54×108 1R RRAM crossbar arrays, accompanied by on-chip 6-bit DAC and 13-bit column ADCs, alongside a RISC processor for programming. The DAC generated voltage pulses with fixed width proportional to input values, subsequently weighted by the RRAM-based Vector-Matrix Multiplication (VMM) and integrated in the charge domain using the integrating ADC.

50 Comparison of NVM-based Neuromorphic ICs for Deep Learning

Design	Technology	Architecture	Synapse Resolution	On-chip Learning	Network Size	Dataset, Class. Accuracy	Energy-efficiency
Yu 2016	130nm CMOS + TaO_x/HfO_2 OxRAM	Binary VMM	1-bit	✗	400-200-10o	MNIST, 96.5%	-
Guo 2017	180nm NOR Flash	Analog VMM	6-bit Analog	✗	784-256c5-10o	MNIST, 94.7% CIFAR-10, 84.8%†	20nJ/inference
Li 2018	2μm CMOS + Ta/HfO_x OxRAM	R array	Analog	✓	64-10o	8x8 UCI, 91.7%	-
Wang 2018	Ta/HfO_x OxRAM	1T1R array	Analog	✓	8-8o	Custom patterns	-
Cai 2019 & Correll 2019	180nm CMOS + WOx RRAM	1R array, Mixed-signal	Analog	✓	54x108 array	5x5 images	8.5GOPS/W
CEA-Leti 2019	130nm CMOS + Ti/HfO_x OxRAM	1T1R, SNN	1-bit	✗	784-10o	MNIST, 84%	180pJ/SynOp
Xue 2019	55nm CMOS + RRAM	VMM, 1T1R	3-bit Digital	✗	1Mb Macro	CIFAR-10, 85.52%‡	53.17TOPS/W
Hirtzlin 2020	130nm CMOS + Ti/HfO_x OxRAM	2T2R	1-bit	✗	Off-chip	MNIST, 98.3%† CIFAR-10, 87.5%† ImageNet, 69.7%†	20-30pJ/SynOp
Nandakumar 2020	90nm CMOS + GST PCRAM	2T1R, LIFN	7-bit Analog	✓	784-10o	MNIST, 70%	-
Liu 2020	130nm CMOS + RRAM	VMM, 2T2R	1 to 8-bit Digital	✗	784-100-10o	MNIST, 94.4%	78.4TOPS/W
Wan 2020	130nm CMOS + TaO_x RRAM	IFN, 1T1R	1-bit	✗	225-60	MNIST RBM	74TOPS/W
Xue 2020	22nm CMOS + RRAM	VMM, 1T1R	4-bit Digital	✗	1Mb Macro	CIFAR-10, 90.19%‡ CIFAR-100, 64.15%‡	121.38TOPS/W

† Simulated result.
‡ Result is using off-chip computation.

V. Saxena, "Neuromorphic Computing: From Devices to Integrated Circuits," J. Vac. Sci. Technol. B, vol. 39, pp. 010801, 2021

Depicts the comparison of NVM-basedNeuromrphic IC for Deep Leaning from different literatures.

51 NVM-based Neuromorphic IC Trends

- Early Neuromorphic ICs employed:
  - Binarized RRAM weights or MLC macros
  - Unit-weighted cells for multi-bit weights
- Most large-scale ICs demonstrated were inference only applications
  - Transfer learning employed
  - Multi-level 2T2R RRAM cells limited to 2-bit/device precision
  - Relaxation and resistance drift degrade model performance
    - IBM's PCRAM work estimated and compensated for the drift in software
- On-chip learning with RRAMs still an open problem
  - Need to address issues at the device-circuit interface
- Communication between arrays needs attention
  - AER has served well but needs to be reimagined as synchronous network-on-a-chip (NoC)

The trends related to NVM-based Neuromorphic ICs implemented in different areas.

52 Conclusion

- Mixed-signal neuromorphic Edge-AI ICs with in-memory computing are promising for higher energy-efficiency
- Recent (but sparse) availability of CMOS-RRAM integration has spurred neuromorphic SoC exploration
- Transfer learning ICs constitute the first demonstrators
- NVM device variability, limited resolution and retention issues need to be taken into account from the system-level perspective
- Spike-based algorithms have reached excellent performance in software, their IC realization is a promising research avenue

Wapparing up the session with Conclusion and summaries the whole presentation.

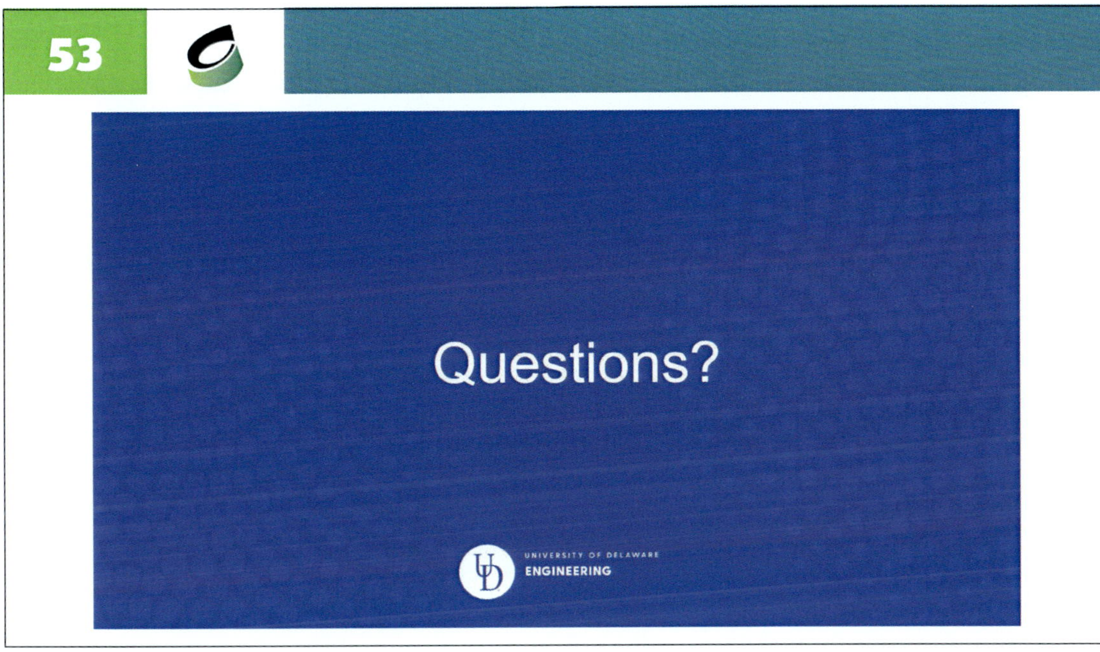

The session was opened for Q&A.

EXTRA SLIDES

54 Crosspoint Memory Architectures

- Passive Crossbar Array
  - Highest density with $4F^2$ cell area
  - *Sneak path problem* in random access
  - Low resistances to drive (later)
- 1T1R Cell Array
  - $6F^2$ to $18F^2$ cell area
  - Allows isolation of a single cell (with circuit overhead)
- Array with selector (1S1R)
  - 2-terminal selector device to suppress sneak paths
  - Selector should have strong I-V nonlinearity
- CRS cells
  - Two RRAM devices can be stacked anti-serially
  - opposite polarities limits current through the device pair

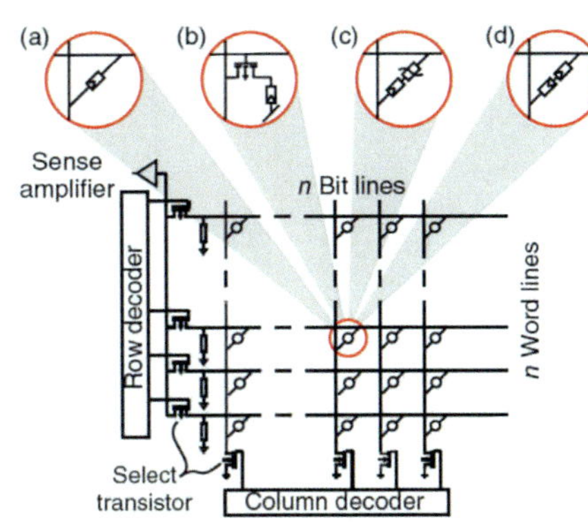

R. Waser et. al. ,"Resistive switching from fundamentals of Nanoionic Redox Processes to Memristive Device Application," Wiley-VCH Verlag, 2016

Introduces some extra slide through which cross point memory architectures is been discussed with 1T1R cell array and a 1S1R selector for the array.

55 Neuromorphic Computing Architecture with 1R Array

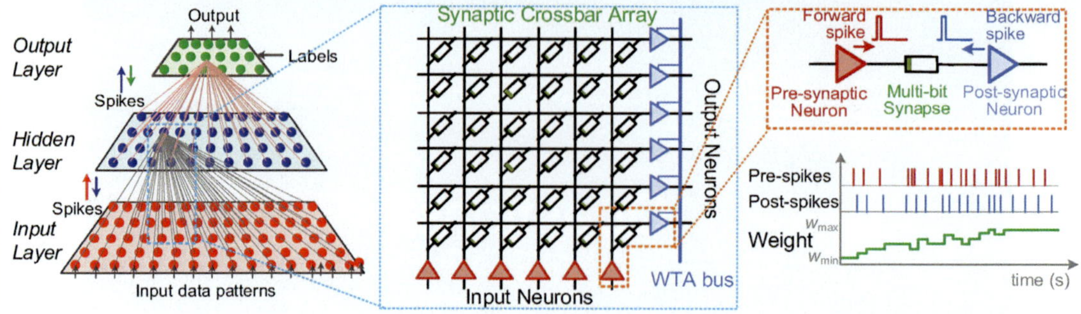

- Crossbar/cross-point arrays provide the highest density for large-scale neural nets
- Sneak-paths are absorbed into the concurrent network with manageable performance degradation

Here the discussion is on the extension of Neuromorphic Computing Architecture with 1R array. It shows the arrangement of the devices in a crossbar architecture and getting the spikes as output for SNN.

56 1R Array: Spike Timing Dependent Plasticity (STDP)

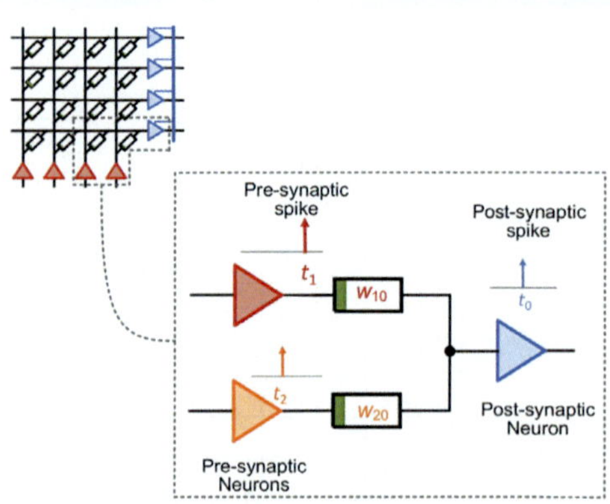

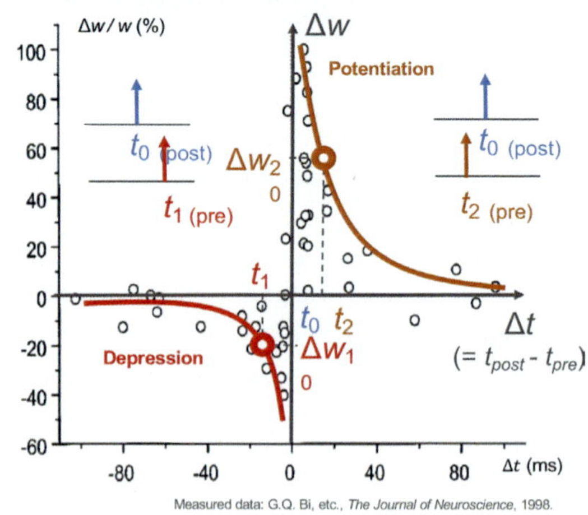

- Hebbian learning: *Neurons that fire together, wire together.*

Spike timing dependent plasticity (STDP) is been discussed with appropriate graphs and model is been shown in block diagram.

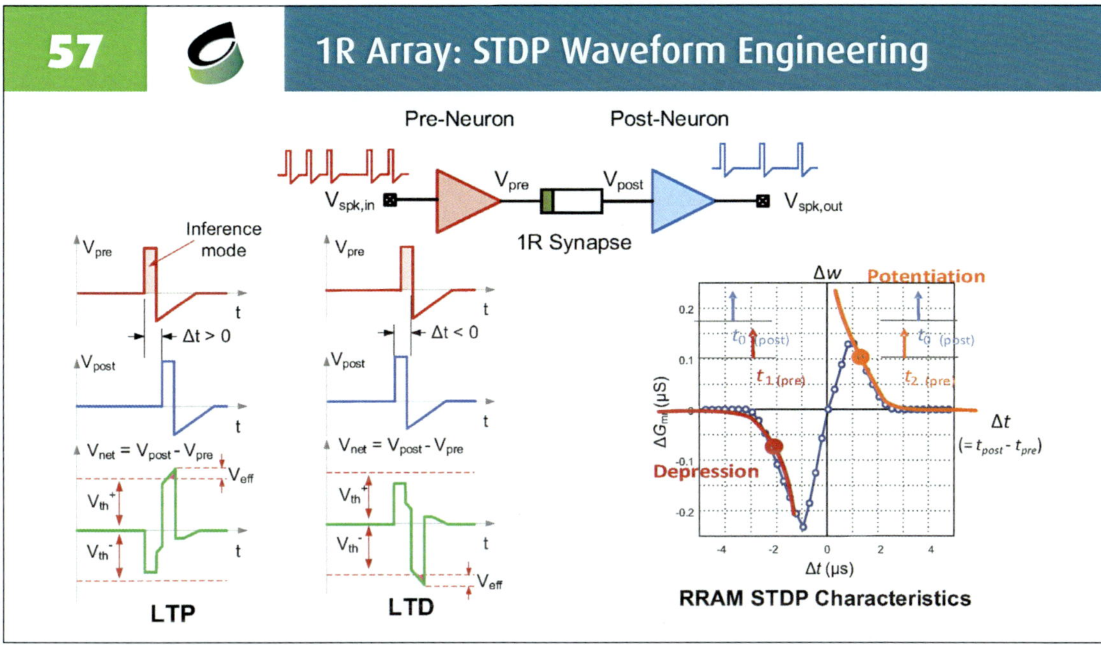

The discussion continues for the extension of STDP waveformfor 1R array.

About the Editors

Alex James

Alex James (Senior Member, IEEE) received the Ph.D. degree in electronics engineering from the Queensland Micro and Nanotechnology Center, Griffith University, Brisbane, QLD, Australia.,He is a Professor of AI hardware with the School of Electronic Systems and Automation and Dean (Academic), Digital University Kerala, Trivandrum, India. He is a Prof-in-Charge of Maker Village that supports more than 80 hardware startups. He is the Chief Investigator with the Center for Intelligent IoT Sensors and the India Innovation Centre for Graphene, developing new products to the market. His research interests include broad range of brain-inspired systems, memristive systems, intelligent semiconductor devices, analog circuits, and imaging systems.,Dr. James was the Founding Chair of IEEE Kerala Section Circuits and Systems Society. He is a Member of the IEEE CASS Technical Committee on Nonlinear Circuits and Systems, IEEE CASS Technical Committee on Cellular Nanoscale Networks and Memristor Array Computing, IEEE Consumer Technology Society Technical Committee on Quantum in Consumer Technology (QCT), Technical Committee on Machine Learning, Deep Learning and AI in CE (MDA), and Member of BCS Fellows Technical Advisory Group (F-TAG). He was an Editorial Member of the Information Fusion, Elsevier, IEEE Transactions on Emerging Topics in Computational Intelligence (2017–18) (Guest Associate Editor) and IEEE Transactions on Circuits and Systems 1 (2018–2023). He is currently an Associate EIC of IEEE Open Journal of Circuits and Systems (2024–present), Associate Editor for IEEE Transactions on Biomedical Circuits and Systems (2024–present), Associate Editor for Frontiers in Neuroscience (Section: Neuromorphic Systems) and IEEE ACCESS. He is a Life Member of ACM, Senior Fellow of HEA, Fellow of the British Computer Society (FBCS), and Fellow of IET (FIET).(Based on document published on 6 February 2024).

Alon Ascoli

Alon Ascoli (Senior Member, IEEE) received the Ph.D. degree in electronic engineering from University College Dublin in 2006. He was conferred the Habilitation title as a Full (Associate) Professor in electrical circuit theory from the Italian Ministry of Education in 2023 (2017) and the Habilitation title as a Full Professor in nonlinear circuit theory from TU Dresden in 2022. Since December 2023, he is an Associate Professor with the Department of Electronics and Telecommunications, Politecnico di Torino, Turin, Italy. From December 2012 to November 2023, he was affiliated with the Faculty of Electrical and Computer Engineering, Technische Universität Dresden (TU Dresden). He develops circuit- and system-theoretic methods for the analysis and design of bio-inspired memristive circuits, enabling to deepen our understanding of biological systems and/or to extend the functionalities of traditional electronic systems. He was honored with the Darlington Best Paper Award from TCAS in 2023, the Best Paper Award from IJCTA in 2007, and the Best Paper Award from MOCAST in 2022 and 2020. He was the President of the IEEE Cellular Nanoscale Networks and Array (and Memristor Array) Computing Technical Committee, from 2019 to 2021 and from 2021 to 2023. He has been serving as an Associate Editor for IEEE Transactions on Circuits and Systems—I: Regular Papers, since 2023

Bhaskar Choubey

Bhaskar Choubey received the D.Phil. degree from the University of Oxford and the B.Tech. degree from the Regional Engineering College, Warangal, India. He is currently the Chair of Analogue Circuits and Image Sensors with Siegen University. In the past, he was with the University of Glasgow, Somerville College, and the University of Oxford. He received the Rhodes Scholar for the D.Phil. degree. He also received the IEEE Sensors Council GOLD Early Career Achievement Award and the Myril B. Reed Best Paper Award from the IEEE Midwest Symposium of Circuits and Systems. He is currently the Chair of the IEEE EPPC Working Group on ICT in Europe. He is an Associate Editor of the IEEE Sensors Journal.

About the Authors

CHAPTER 1

Rainer Waser (Member, IEEE) received the Ph.D. degree in physical chemistry from the University of Darmstadt, Darmstadt, Germany, in 1984. In 1992, he joined the Faculty of Electrical Engineering and Information Technology, RWTH Aachen University, Aachen, Germany, as a Professor. In 2007, he was the Director of the Institute of Solid State Research, Forschungszentrum Jülich, Jülich, Germany. He was a recipient of the prestigious Gottfried Wilhelm Leibniz Preis bbreak in 2014.

CHAPTER 2

Abu Sebastian is a distinguished scientist and manager at IBM Research - Zurich in Rüschlikon, Switzerland. He was born in Kerala, India. He received a B. E. (Hons.) degree in Electrical and Electronics Engineering from BITS Pilani, India and M.S. and Ph.D. degrees in Electrical Engineering (minor in Mathematics) from Iowa State University. Dr. Sebastian is the author/co-author of over 200 publications in peer-reviewed journals/conference proceedings and holds over 80 US patents. He is a co-recipient of the 2009 IEEE Control Systems Technology Award and the 2009 IEEE Transactions on Control Systems Technology Outstanding Paper Award. In 2013 he received the IFAC Mechatronic Systems Young Researcher Award for his contributions to the field of mico-/nanoscale mechatronic systems. In 2015 he was awarded the European Research Council (ERC) consolidator grant and in 2020, he was awarded an ERC Proof-of-concept grant. He is an IBM Master Inventor since 2016. He was named Principal and Distinguished Research Staff Member in 2018 and 2020, respectively. In 2019 he received the Ovshinsky Lectureship Award for his contributions to 'Phase-change materials for cognitive computing'. In 2023, he received the Prof. L. K. Maheshwari Foundation Distinguished Alumnus Award from BITS Pilani, India. In 2023, he was also conferred the title of Visiting Professor in Materials by University of Oxford. He has served on the technical program committees of several conferences including IEDM, AICAS, NVMTS and E\PCOS and has served as an editor/guest editor for Mechatronics, APL Material, Applied Physics Letters, and IEEE Design and Test. He is a Distinguished Lecturer and Fellow of the IEEE.

CHAPTER 3

Stephan Menzel (Senior Member, IEEE) was born in Bremen, Germany. He received the Diploma degree and the Ph.D. degree (summa cum laude) in electrical engineering from RWTH Aachen University, Aachen, Germany, in 2005 and 2012, respectively. He is currently a Senior Researcher with the Peter-Grünberg-Insitut 7, Forschungszentrum Jülich GmbH, Jülich, Germany, where he is also leading the Simulation Group.

CHAPTER 6

Fernando Corinto (Senior Member, IEEE) received the master's degree in electronic engineering and the Ph.D. and European Doctorate degrees in electronics and communications engineering from Politecnico di Torino in 2001, 2005, and 2005, respectively. He was awarded the Marie Curie Fellowship in 2004. He is currently a Professor of electrical engineering with the Department of Electronics and Telecommunications, Politecnico di Torino. His research interests include nonlinear circuits and systems, locally coupled nonlinear/nanoscale networks, and memristor bbreak nanotechnology.

CHAPTER 7

Spyridon 'Spiros' Nikolaidis received the Diploma and PhD degrees in electrical engineering from Patras University, Greece, in 1988 and 1994 respectively. Since September 1996 he has been with the Department of Physics of the Aristotle University of Thessaloniki, Greece where he is currently a full professor. He also belongs to the contract teaching staff of the Hellenic Open University since 2003. He is employed in the area of digital circuit and system design. His current research interests include: modeling the operations of basic CMOS structures, development of analytical expressions for the propagation delay and the power consumption of logic gates, design of high speed and low power digital circuit and embedded systems, modeling the power consumption of embedded processors. He is author and co-author in more than 160 scientific articles in international journals and conference proceedings while his work has taken more than 1500 references (google scholar). Two articles presented in international conferences were achieved honorary awards. He contributed in three scientific book chapters. He was member of the organization committee of three international conferences. He organizes the annual International Conference on Modern Circuit and System Technologies (MOCAST) as well as the 27th International Symposium on Power and Timing Modeling, Optimization and Simulation for 2017. He also contributes to a number of research projects funded by European Union and Greek Government where in many of that he has the scientific responsibility. He is a Senior member of IEEE.

CHAPTER 8

Cory Merkel joined the RIT computer engineering department in 2018. He earned his BS and MS degrees in computer engineering (2011) and a Ph.D. in microsystems engineering (2015), all from RIT. From 2016 to 2018, Dr. Merkel was a research electronics engineer with the Information Directorate, Air Force Research Lab. His current research focuses on mapping of AI algorithms, primarily artificial neural networks, to mixed-signal hardware and the design of brain-inspired computing systems using emerging technologies such as memristors. He has published his work in several peer-reviewed conferences, journals, and books, and is also engaged in a number of STEM outreach activities. For more information, see Dr. Merkel's research website www.rit.edu/brainlab

CHAPTER 9

Arindam Basu (Senior Member, IEEE) received the B.Tech. and M.Tech. degrees in electronics and electrical communication engineering from the Indian Institute of Technology Kharagpur, Kharagpur, India, in 2004 and 2005, respectively, and the M.S. degree in mathematics and Ph.D. degree in electrical engineering from the Georgia Institute of Technology, Atlanta, GA, USA, in 2009 and 2010, respectively.,He was a tenured Associate Professor at Nanyang Technological University, Singapore. He is currently a Professor with the Department of Electrical Engineering, City University of Hong Kong, Hong Kong. His research interests include bio-inspired neuromorphic circuits, nonlinear dynamics in neural systems, low-power analog IC design, and programmable circuits and devices.,Dr. Basu is a Technical Committee Member of the IEEE Circuits and Systems (CAS) Societies of Biomedical Circuits and Systems, Neural Systems and Applications (Chair), and Sensory Systems. He received the Best Student Paper Award at the Ultrasonics Symposium in 2006, the Best Live Demonstration at International Symposium on Circuits and Systems (ISCAS) in 2010, and a finalist position in the Best Student Paper Contest at ISCAS 2008. He was included in Georgia Tech Alumni Association's 40 under 40 lists in 2021 and was awarded the Massachusetts Institute of Technology (MIT) Technology Review's TR35 Asia Pacific Award in 2012. He also received the Prime Minister of India Gold Medal from IIT Kharagpur in 2005. He is currently an Associate Editor of IEEE Sensors Journal, Frontiers in Neuroscience, and IEEE Transactions on Biomedical Circuits and Systems. He has served as IEEE CAS Distinguished Lecturer from 2016 to 2017.

CHAPTER 10

Jason K. Eshraghian (Member, IEEE) received the B.E. degree in electrical and electronic and the Bachelor of Laws and Ph.D. degrees from The University of Western Australia, Perth, WA, Australia, in 2016 and 2019, respectively.,From 2019 to 2022, he was a Post-Doctoral Research Fellow with the University of Michigan, Ann Arbor, MI, USA. He is currently an Assistant Professor with the Department of Electrical and Computer Engineering, University of California at Santa Cruz, Santa Cruz, CA, USA. He is also a Developer of snnTorch, a widely used Python library used to train and model spiking neural networks. His research interests include neuromorphic computing, spiking neural networks, and memory circuits.,Dr. Eshraghian serves as the Secretary-Elect for the Neural Systems and Applications Technical Committee. He was awarded the 2023 IEEE TRANSACTIONS ON CIRCUITS AND SYSTEMS Darlington Best Paper Award, the 2019 IEEE Very Large Scale Integration Systems Best Paper Award, the Best Paper Award from the 2019 IEEE Artificial Intelligence Circuits and Systems Conference, and the Best Live Demonstration Award from the 2020 IEEE International Conference on Electronics Circuits and Systems for his work in neuromorphic computing. He was a recipient of the Fulbright Fellowship (Australian–American Fulbright Commission), the Forrest Research Fellowship (Forrest Research Foundation), and the Endeavour Research Fellowship (Australian Government).

CHAPTER 11

Vishal Saxena (Senior Member, IEEE) received the B.Tech. degree in electrical engineering from the Indian Institute of Technology, Chennai, India, in 2002, and the M.S. and Ph.D. degrees from Boise State University, Boise, ID, USA, in 2007 and 2010, respectively. He is currently an Associate Professor with Electrical and Computer Engineering with the University of Delaware, Newark, DE, USA, where he directs the Analog Mixed-Signal and Photonic Integrated Circuits lab. He was the recipient of the 2015 NSF CAREER, the 2016 AFOSR Young Investigator Program, and the 2019 DARPA Young Faculty Award awards. He is currently an Associate Editor for the IEEE Transactions on Circuits and Systems II journal and on the steering committee of the IEEE MWSCAS conference and ISCAS VLSI Committee. His research interests include CMOS photonic interconnects, RF and millimeter-wave Photonic ICs, Neuromorphic, Edge-AI, and Quantum computing.

Technological Innovation and Sustainability through AI

Technological Innovation and Sustainability through AI: Cutting-Edge Solutions for a Sustainable Future explores how advanced artificial intelligence (AI) solutions are being used to drive technological innovations that support sustainable development and create a more environmentally friendly future.

The book examines the intersections of AI, ethics, healthcare, social good, and the evolving future of work, offering insights into how AI is reshaping our world with a focus on sustainable development. The content ranges from foundational AI principles and ethical considerations to practical applications like machine learning for energy efficiency and smart grids.

Features:

- Real-world examples and case studies, showcasing practical applications of AI in various sectors
- Interdisciplinary expertise from a range of experts in diverse fields
- Explores future trends and innovations, offering forward-looking insights

With contributions from a range of experts, the book offers a wide understanding of how AI can be leveraged to drive technological innovation and sustainability. It will be of great interest to academics, practitioners, and policymakers.

Technological Innovation and Sustainability through AI

Cutting-Edge Solutions for a Sustainable Future

Edited by

El Arbi Abdellaoui Alaoui, Cédric Stéphane Tekouabou Koumetio, Abdelaaziz Hessane, and Amine Sallah

CRC Press
Taylor & Francis Group
Boca Raton London New York

CRC Press is an imprint of the
Taylor & Francis Group, an **informa** business

Designed cover image: Shutterstock

First edition published 2026
by CRC Press
2385 NW Executive Center Drive, Suite 320, Boca Raton FL 33431

and by CRC Press
4 Park Square, Milton Park, Abingdon, Oxon, OX14 4RN

CRC Press is an imprint of Taylor & Francis Group, LLC

ISBN: 978-1-041-09155-4 (hbk)
ISBN: 978-1-041-09403-6 (pbk)
ISBN: 978-1-003-64988-5 (ebk)

DOI: 10.1201/9781003649885

Typeset in Minion
by KnowledgeWorks Global Ltd.

Contents

Editors, ix

Contributors, xi

Preface, xv

Chapter 1 ▪ The Nexus between Digital Transformation and Corporate Environmental Performance 1

Terezin Mathew, Latha. K, and Roshna Varghese

Chapter 2 ▪ The Impact of Sporting Events on Tourism Attractiveness: Organization of the 2030 World Cup in Morocco 15

Fatima Zahra El Karmaoui, El Ouali El Balaidi, and Ahmed El Hammoumi

Chapter 3 ▪ Navigating the Pandemic with AI: Toward Sustainable Deep Learning and Machine Learning Approaches for X-Ray and CT-Based Covid-19 Detection 34

Youness Chawki

Chapter 4 ▪ Improving Predictions of Student Performance and Personalized School Guidance with Hybrid Machine Learning Models: A Case Study in Cameroonian Secondary Schools 53

Eric Desiré Kameni, Cédric Stéphane Tekouabou Koumetio, Marlene Djoutieu Tegankong, and Samsong Durand Priso Essawe Ndedi

Chapter 5 ▪ Reforming Governance in Moroccan Higher Education: Sentiment Analysis for Whole-of-Government and Bottom-Up Policy Approaches 70

Said Azelmad

CHAPTER 6 ▪ Experimentation in Physics Teaching and the Contribution of Artificial Intelligence and Virtual Labs 94

ZINEB AZAR, OUSSAMA DARDARY, MALIKA TRIDANE, AND SAID BELAAOUAD

CHAPTER 7 ▪ Integration of Artificial Intelligence, Virtual Labs, and Digital Simulation in Science Education 106

OUSSAMA DARDARY, ZINEB AZAR, MALIKA TRIDANE, AND SAID BELAAOUAD

CHAPTER 8 ▪ Reframing Technological Innovation Dimension and AI Competitiveness for Pro-Growth Small Businesses in Sub-Saharan Africa 116

TIMILEHIN OLASOJI OLUBIYI

CHAPTER 9 ▪ A Qualitative Systematic Review of the Influential Factors for Green Decision-Making 135

MOUHCINE RHOUIRI, MEROUANE EL AZAMI EL HASSANI, MOHAMED-HABIBOULLAH MEYABE, AND MEHDI BENSOUDA

CHAPTER 10 ▪ Nexus between Renewable Energy, Artificial Intelligence, and Economic Growth: The Moroccan Case 147

YAHYA FIKRI AND MOHAMED RHALMA

CHAPTER 11 ▪ Impact of New Technologies and Sustainability on Maintenance in the Era of Artificial Intelligence 161

AMAL NAJI AND ZITOUNI BEIDOURI

CHAPTER 12 ▪ Group Work in the Digital Age: A Survey of Banking Human Resources in Morocco 176

ABDELAZIZ BERDI, SEGHYAR NABIL, AND AHMED EL HAMMOUMI

CHAPTER 13 ▪ AI-Powered Waste Management: Transforming Linear Systems into Circular Economy Solutions 194

SHUBHAM GUPTA

CHAPTER 14 ▪ AI-Driven Innovation in Waste Management and Recycling 210

VINOD B. R., ARUNAKUMARI B. N., M. VARUNA, AND JYOTHI D. N.

CHAPTER 15 ▪ Artificial Intelligence and Biodiversity Regeneration: Inclusive Strategies for Smart Cities in the Global South 221

NAWAL BENABDALLAH

CHAPTER 16 ▪ Artificial Intelligence in Sustainable Supply Chain Management 233

MOHAMED BADIY, CHARAF HAMIDI, FATIMA AMOUNAS, ZAYNAB KHALFI, SALMA GAOU, AND ABDELAAZIZ HESSANE

INDEX, 253

Editors

El Arbi Abdellaoui Alaoui, Professor at ENS, Moulay Ismail University, researches wireless networking, AI, IoT, and smart cities. With a 2017 PhD in Computer Science, he publishes extensively and contributes to the academic community as a reviewer and guest editor.

Cédric Stéphane Tekouabou Koumetio, Assistant Professor (Senior Lecturer) at the ENS – University of Yaoundé 1, specializes in computer vision, AI, machine learning, and urban engineering. He holds a 2020 PhD from Chouaib Doukkali University and was a postdoc researcher at UM6P. He also contributes as an associate editor and journal reviewer.

Abdelaaziz Hessane, Assistant Professor at Moulay Ismail University, specializes in AI, machine learning, and deep learning, with applications in smart grids, agriculture, and more. He holds a 2024 PhD in Computer Science and has a strong publication record, actively contributing to conferences and journals.

Amine Sallah earned his PhD in Artificial Intelligence in 2024 from the Faculty of Sciences and Technology, Moulay Ismail University of Meknès, Errachidia, Morocco. His research focuses on cybersecurity, AI and machine learning, genetic algorithms, explainable AI (XAI), and LLMs.

Contributors

Fatima Amounas
IMIA Laboratory, MSIA Team, Faculty of Sciences and Techniques
Moulay Ismail University of Meknès
Errachidia, Morocco

Zineb Azar
Laboratory of Physical Chemistry of Applied Materials LPCMA
Ben M'Sick Faculty of Sciences
Hassan II University of Casablanca, Morocco

Said Azelmad
Faculty of Science and Technology, Governance, Ethics and Developmental Management
Moulay Ismail University of Meknès
Errachidia, Morocco

Mohamed Badiy
IMIA Laboratory, MSIA Team, Faculty of Sciences and Techniques
Moulay Ismail University of Meknès
Errachidia, Morocco

Zitouni Beidouri
Laboratory C3S: Computer Science & Smart Systems
Hassan II University of Casablanca, Morocco

Said Belaaouad
Laboratory of Physical Chemistry of Applied Materials LPCMA
Ben M'Sick Faculty of Sciences
Hassan II University of Casablanca, Morocco

Nawal Benabdallah
Laboratory Urban Innovation and Heritage (UIH Lab.)
College of Engineering & Architecture
International University of Rabat (UIR)
Technopolis Rabat-Shore, Rabat, Morocco

Mehdi Bensouda
Sidi Mohammed Ben Abdellah University
Fez, Morocco

Abdelaziz Berdi
Sidi Mohamed Ben Abdellah University
Fez, Morocco

Arunakumari B. N.
Department of Computer Science Engineering
BMS Institute of Technology & Management
Visvesvaraya Technological University
Belagavi, Karnataka, India

Vinod B. R.
Department of Civil Engineering
BMS Institute of Technology & Management
Visvesvaraya Technological University
Belagavi, Karnataka, India

Youness Chawki
Polydisciplinary Faculty of Ouarzazate
Ibn Zohr University
Morocco

Oussama Dardary
Laboratory of Mathematical and Computer Modeling
Mathematics, Computer Science and Pedagogical Engineering Team
Moulay Ismail University of Meknès
Errachidia, Morocco
and
Laboratory of Physical Chemistry of Applied Materials LPCMA
Ben M'Sick Faculty of Sciences
Hassan II University of Casablanca, Morocco

Jyothi D. N.
Department of Civil Engineering
ATME College of Engineering
Visvesvaraya Technological University
Belagavi, Karnataka, India

El Ouali El Balaidi
Department of Management Sciences and Techniques
National School of Management and Commerce
University Ibn Zohr Agadir
Morocco

Merouane El Azami El Hassani
Paul Valéry University
Montpellier III, France

Ahmed El Hammoumi
Department of Economics and Management
Sidi Mohamed Ben Abdellah University
Fez, Morocco

Fatima Zahra El Karmaoui
Department of Management Sciences and Techniques
National School of Management and Commerce
University Ibn Zohr Agadir
Morocco

Yahya Fikri
National School of Business and Management
Abdelmalek Essaadi University
Tetouan, Morocco

Salma Gaou
Laboratory of Engineering Sciences
Faculty of Science Agadir
University Ibn Zohr
Morocco

Shubham Gupta
Analytics & AI Leader
Noblesoft Solutions, Inc.
San Antonio, Texas, United States

Charaf Hamidi
Laboratory of Engineering Sciences
Faculty of Science Agadir
University Ibn Zohr
Morocco

Abdelaaziz Hessane
IMIA Laboratory, T-IDMS
Faculty of Science and Techniques
Moulay Ismail University of Meknès
Errachidia, Morocco

Latha. K
Department of Business Administration
Saintgits College of Engineering (Autonomous)
Kottayam, Kerala, India

Eric Desiré Kameni
Department of Computer and Educational Technology
H.T.T.C. - Higher Teacher Training College
Laboratory of Computer Science and Educational Technologies (Lite)
University of Yaoundé
Yaoundé, Cameroon

Zaynab Khalfi
Interdisciplinary Organization Research Laboratory (LIRO Laboratory)
National School of Management and Commerce of El Jadida
University of Chouaib Doukkali
El Jadida, Morocco

Cédric Stéphane Tekouabou Koumetio
Department of Computer and Educational Technology
H.T.T.C. - Higher Teacher Training College
Laboratory of Computer Science and Educational Technologies (Lite)
University of Yaoundé
Yaoundé, Cameroon

Terezin Mathew
School of Management and Business Studies
Mahatma Gandhi University
Kottayam, Kerala, India

Mohamed-Habiboullah Meyabe
Nouakchott Business School
Mauritania

Seghyar Nabil
Legal, Political and Economic Research Laboratory
Sidi Mohamed Ben Abdellah University
Fez, Morocco

Amal Naji
Laboratory C3S: Computer Science & Smart Systems
Hassan II University of Casablanca, Morocco

Samsong Durand Priso Essawe Ndedi
Department of Computer and Educational Technology
H.T.T.C. - Higher Teacher Training College
Laboratory of Computer Science and Educational Technologies (Lite)
University of Yaoundé
Yaoundé, Cameroon

Timilehin Olasoji Olubiyi
Faculty of Management and Social Sciences
West Midlands Open University
Lagos State, Nigeria

Mohamed Rhalma
National School of Business and Management
Abdelmalek Essaadi University
Tetouan, Morocco

Mouhcine Rhouiri
Sidi Mohammed Ben Abdellah University
Fez, Morocco

Marlene Djoutieu Tegankong
Distance Education
Ministry of Secondary Education
Yaoundé, Cameroon

Malika Tridane
Laboratory of Physical Chemistry of Applied Materials LPCMA
Ben M'Sick Faculty of Sciences
Hassan II University of Casablanca, Morocco
and
Regional Center for Education and Training Casablanca Anfa
Casablanca, Morocco

Roshna Varghese
School of Management and Business Studies
Mahatma Gandhi University
Kottayam, Kerala, India

M. Varuna
Department of Civil Engineering
RV College of Engineering
Visvesvaraya Technological University
Belagavi, Karnataka, India

Preface

The rapid evolution of artificial intelligence (AI) presents a transformative opportunity to address some of the most pressing challenges of our time, from climate change and resource depletion to social inequality. In a world increasingly defined by digital innovation, harnessing AI is no longer a futuristic concept but a present-day imperative for creating a sustainable and equitable future. This book, *Technological Innovation and Sustainability through AI: Cutting-Edge Solutions for a Sustainable Future*, is a comprehensive exploration of this critical nexus.

This volume provides a holistic examination of how AI-driven technological innovations are reshaping various sectors and contributing to sustainable development. It brings together leading experts from diverse fields to offer foundational insights into AI principles, ethical considerations, and its practical applications. We delve into cutting-edge solutions across a range of domains, from environmental sustainability and resource management to social goods and public health.

The chapters within this book explore key themes, including the intricate relationship between digital transformation and corporate environmental performance, the application of AI in healthcare for diagnostics and personalized treatment, and its potential to combat disinformation and enhance educational systems. We also investigate the societal impact of AI on the labor market, highlighting the emergence of new job paradigms and the importance of human–AI collaboration.

A significant portion of the book is dedicated to how AI can be leveraged for environmental sustainability, featuring real-world case studies on topics like AI-powered waste management, smart grids, and the nexus between renewable energy and economic growth. We also explore the role of AI in biodiversity regeneration and sustainable supply chain management, offering tangible solutions for a greener future.

Technological Innovation and Sustainability through AI is designed to be an essential resource for academics, practitioners, and policymakers seeking to understand and leverage the power of AI for positive global change. By uniquely combining ethical frameworks, advanced technological applications, and real-world case studies, this book provides actionable insights and underscores the necessity of interdisciplinary collaboration and a unified approach to AI governance. We hope this work inspires and equips readers to contribute to a sustainable future powered by thoughtful and innovative uses of AI.

CHAPTER 1

The Nexus between Digital Transformation and Corporate Environmental Performance

Terezin Mathew, Latha. K, and Roshna Varghese

1.1 INTRODUCTION

The 17 Sustainable Development Goals (UN SDG) by the United Nations, also referred to as global goals, aim to address global challenges such as poverty, inequality, climate change, and peace [1]. The objective of Goal 9, pertaining to industries, innovation, and infrastructure, is to invest in infrastructure such as transportation, electricity, and information and communication technology (ICT) by integrating sustainable development, innovation, and economic growth. The implementation strategies of SDGs differ region-wise depending on the policy framework and leadership [2]. Digital technologies have the potential to directly address more than two-thirds of the SDGs [3]. Existing literature is rich with studies on the convergence of sustainability and digitalisation termed "digital sustainability" [4]. International organisations aim to leverage technology as a catalyst for a greener and more sustainable environment through digital transformation (DT). The SDG digital acceleration agenda established by the International Telecommunication Union and United Nations Development Programme scrutinises the indispensable role of DT in advancing the SDGs to progress towards the 2030 Agenda of Sustainable Development. The "Path to the Digital Decade" proposal aims to achieve the digital target of the European Union (EU) by 2030 through the combined effort of all the member states of the EU. The "Digital Economy for Africa" (DE4A) initiative ensures digitisation of government, business, and every individual by 2030. The stated initiatives accentuate the exigent need for DT to stay competitive and resilient to ultimately achieve sustainable development by 2030. Industry 5.0 is anticipated to put people first and sustainable development, shifting the emphasis from economic value to social value and is anticipated to provide a variety of opportunities to advance sustainable development objectives [5].

DT refers to the application of digital technologies like cloud computing, the Internet of Things (IoT), artificial intelligence (AI), big data analytics (BDA), augmented reality (AR), robotics, machine learning (ML), and blockchain [6] to businesses and industries to enhance productivity, efficiency, and nurture development [7]. The introductory phase of DT implies

DOI: 10.1201/9781003649885-1

the conversion of analogue signals into digital values through quantification [8]. The integration of digital technologies has led to the paradigm shift from traditional industrial practices to Industry 4.0. The escalating environmental catastrophe has raised awareness of the significance of sustainability as a global concern. The fundamental notion that underpins sustainable development is high-quality economic development, wherein businesses are vital contributors to this process. Stakeholders, who are meticulous about organisational environmental performance, diligently report discrepancies in disclosure. Corporate environmental performance (hereafter CEP) measures how well an organisation manages its environmental responsibilities [9]. According to the Environmental Performance Index (EPI) 2024, Estonia topped the index with a 59% reduction in greenhouse gas (GHG) emissions from 1990 levels. DT helps stakeholders monitor the environmental performance of organisations. Within the spectrum of DT initiatives, AI serves as a critical catalyst for facilitating and promoting CEP. AI technologies encompassing advanced analytical tools, predictive models, intelligent automation systems, real-time monitoring systems, optimised resource allocation, markedly bolster an organisation's environmental monitoring, strategic decision-making, sustainable supply chain management, emissions mitigations, accurate resource administration. The confluence of DT and CEP has become a topic of growing relevance within academic and industry sectors. Digital technologies such as IoT, AI, BDA, and blockchain are recognised as catalysts for driving eco-innovation and promoting corporate sustainability initiatives. Despite the recent increase in the body of literature, there exists a discernible gap in integrated perspective that combines conceptual frameworks, case study approach, and bibliometric analysis to provide a holistic understanding. Thus, it is essential to understand the nexus between DT and CEP. To achieve SDGs, corporations are implementing DT. The present work is guided by the following objectives:

1. To understand the conceptual nexus between DT and CEP.
2. To present case studies of companies that have successfully leveraged DT initiatives to improve CEP.
3. To analyse the current research trends in the study of DT and CEP.

1.2 LITERATURE REVIEW

1.2.1 Digital Transformation and Environmental Sustainability

Digitalisation paves the way for a sustainable business model by aligning critical business elements and digital tools to accomplish triple bottom-line objectives [10]. A sustainable business model is a framework that enables businesses to create value while simultaneously addressing environmental and social issues, and it takes into account the interests of all stakeholders in this process [11]. DT has a positive impact on economic and environmental sustainability by enhancing resource efficiency, particularly energy resources [4]. Digital technologies like IoT, AI, and blockchain technology have both beneficial and harmful effects on the planet [12]. IoT can facilitate sustainability by reducing energy consumption and carbon emissions and can monitor environmental aspects such as air quality, water quality, and biodiversity [13]. Blockchain technology helps consumers track the origin of their products, enabling them to make more environmentally conscious purchasing decisions. Further, it facilitates peer-to-peer energy trading, through which reliance on traditional energy sources can be reduced. Blockchain technology can help to track waste from its source to disposal, which can lead to higher waste recycling rates and create a transparent platform for companies to monitor their carbon emissions and trade carbon credits [14].

AI and ML can analyse extensive datasets to enhance energy efficiency, forecast environmental impacts, and optimise supply chain networks [15]. DT can promote circular economy principles, such as product-as-a-service models and resource recovery, while minimising waste [16]. Digital tools can be employed to create goods and services that reduce environmental impact and manage their complete life cycle from production to disposal [12]. One of the emerging Industry 4.0 technologies is the digital twins, which can be employed for ecological simulations, analysing the ecological impact, and enhancing resource efficiency in manufacturing and operations [17]. The other side of the DT is that these digital gadgets can harm the environment. E-waste from these gadgets contains toxic elements, making disposal and recycling difficult. Energy consumption is another challenge with digital devices due to continuous connectivity and data processing. Further carbon footprints of data centres attached to IoT devices are also a cause of concern, as these data centres emit GHGs when powered by non-renewable energy sources.

1.2.2 Sustainability and Environmental Reporting Informed by Data

The issue of carbon footprint is of the utmost importance and necessitates thoughtful consideration within the framework of sustainability [18]. Quantifying and monitoring carbon footprints is indeed critical in light of the accelerating pace of climate change. Carbon footprint management involves the monitoring, analysis, and mitigation of GHG emissions that are associated with individual, organisational, or product activities. Effective management of carbon footprints is essential to reduce the environmental impact and for the mitigation of climate change [19]. DT plays a crucial role in achieving carbon neutrality through the strategic application of digital technologies, including AI, IoT, big data, and blockchain. A reduction in carbon emissions and a transition to renewable energy sources are both necessary steps towards achieving sustainable economic development. Digitalisation plays a pivotal role in mitigating carbon footprints in a variety of segments, such as resource-based cities, the construction industry [19], food industry [18], cross-section of sectors and companies, and [20] transportation [21].

From an organisation's perspective, carbon footprint management is a process that involves the utilisation of advanced digital tools and strategies across an organisation's entire value chain to measure, monitor, and reduce GHG emissions. Digitalisation plays a vital role in advancing sustainable business practices, and the growing emphasis on sustainability has led organisations to adopt technological advancements for environmental reporting [22]. Data-driven strategic decision-making helps organisations monitor their sustainability-related activities, such as carbon emissions and waste generation, while ensuring fair and transparent reporting through sustainability or environmental, social, and governance (ESG) reports to their stakeholders. Digital tools help organisations monitor the alignment of their environmental practices with global sustainability goals. Seamlessly integrating environmental data into an organisation's daily operations facilitates in making informed and advanced decisions. Thus, technological advancements emerge as a more potent means of improving environmental management and fostering sustainability [23]. Sustainability analytics also plays a major role in this field. By gathering, analysing, and interpreting data, sustainability analytics seeks to assess and enhance the economic, social, and environmental sustainability of organisations, projects, or systems. This domain utilises data-driven instruments and processes to monitor sustainability performance, facilitate decision-making, and convey impacts efficiently.

Recent studies on the influence of DT on eco-innovation and sustainable performance in 210 Chinese manufacturing companies revealed that digital strategy and capability significantly improve eco-process, product, and management innovation, while eco-innovation partially mediates the positive relationship between DT and sustainable performance [24].

1.3 METHODOLOGY

The research uses a conceptual framework focusing on the review of recent literature, case studies, and empirical insights to interlink DT and CEP. This study has employed multiple methods and sources to offer a holistic understanding of the area of the study. It includes a comprehensive literature review to synthesise the existing body of knowledge on the topic, published company reports to examine industry practices and case study approach to explore specific examples of the interlinkage between DT and corporate environment performance with a focus on innovative practice adopted by the companies. The bibliometric analysis employed, leveraging the Scopus database, elucidates the research trends at the intersection of DT and CEP. Keywords such as – "digital transformation" OR "corporate digital transformation" OR "AI" OR "Artificial* intelligence" OR "cloud computing" OR "robotics" OR "IoT" OR "Internet of Things" OR "Industry 4.0" – AND "Environment* Performance" OR "Corporate environment* Performance" OR "ESG performance" – were utilised to retrieve 222 documents with English language, article document type for the specific time frame. The analysis explored the trend topics concerning the convergence of DT and CEP.

1.4 RESULTS AND DISCUSSION

1.4.1 Conceptual Discussion

1.4.1.1 Corporate Strategy and Digital Transformation for Sustainable Growth

DT is a critical corporate sustainability strategy that enables businesses to attain sustainable growth. DT positively influences the sustainable growth of firms [25]. The necessity of sustainable business practices has escalated due to the growing importance of the digital economy. At its core, the digital economy is based on the internet, digital technologies, and the proliferation of digital services and platforms. The digital economy is characterised by a transition from traditional industry and commerce to one that is driven by digital innovation, data, and connectivity. Organisations are incorporating sustainability into their core strategies as a result of stakeholder pressure. Digitalisation is essential for the sustainability initiatives of an organisation, as it encompasses virtually all aspects of carbon footprint management [22]. A new area of interest has emerged as a result of these initiatives: digital sustainability [26]. Researchers and organisations are linking financial performance to environmental performance, and shareholders expect that a company with sustainable business practices has a favourable long-term growth trajectory [27].

Previous research suggests that there is a significant and positive relationship between innovation and digitalisation [28]. An organisation's digitisation initiatives result in the implementation of numerous innovative practices, which also have a significant impact on the industry and the economy as a whole. Consequently, the process of digitalisation fosters sustainable innovation by utilising digital tools to attain environmental goals and encourage beneficial results for both the environment and its residents. Digital technology and green practices can significantly enhance the effectiveness of corporate social responsibility (CSR) initiatives [29]. DT facilitates real-time, precise, and comprehensive CSR-related disclosures, ensuring adherence to sustainability reporting requirements and augmenting the transparency of internal control information within organisations. Digitally transformed CSR initiatives not only enhance a company's reputation but also contribute to long-term growth, innovation, and stakeholder trust [30].

1.4.1.2 Role of Digital Tools in Enhancing Supply Chain Sustainability

The unparalleled advancement in ICT has catalysed the emergence of digital disruption [31] prompting conventional models to incorporate digital technologies. Digital disruption permeates industries, societal segments, and the evolving interactions between organisations and individuals [31]. This phenomenon is propelling the fourth industrial revolution (Industry 4.0), which has encouraged industries worldwide to embrace digitalisation as a critical strategic imperative to adapt to a shorter product life cycle and manage rapid environmental changes [32]. The evolution of ICT has fundamentally transformed organisation's engagement with their networks, fostering efficient collaboration between customers and businesses. Intense rivalry, inflexibility, and delayed market entry are key factors that increase the unpredictability of supply chain management [32]. Digital tools like the IoT, BDA, and AI address these issues, which profoundly reshape the supply chain industry [33].

Adopting digital tools into supply chain management would improve supply chain efficiency [34], supply chain structure [33], sustainability [35], and innovation [33]. Smart city initiatives exemplify how technology is reshaping the supply chains to incorporate innovative operational models. Digital disruption is also catalysing a paradigm shift within the supply chain, transitioning from conventional production planning to additive manufacturing, or 3D printing, which enhances the potential for mass customisation [36]. Consequently, the traditional supply chain should transition to the digital supply chain which ensures real-time information sharing to enhance production processes, transportation, and customer experiences. Despite the notable advocacy by leading consulting firms for supply chain digitalisation [37], theoretical and applied understanding is still at its inception [31].

1.4.1.2.1 Digital Supply Chain and Sustainable Sourcing The introduction of new technologies such as the IoT, AI, blockchain, 5G and 6G networks, robotics, biometrics, 3D printing, AR, and drones is revolutionising supply chain networks. Some of these technologies, such as blockchain, 6G network technology, and wireless communication, are ripe for novel business model innovation; others, like these, facilitate Industry 4.0 processes. Blockchain technology has the ability to revolutionise several business models. It can automate smart contracts, enable decentralised sharing economy apps, and transfer trust from organisations to analytics [38]. In developing countries, a majority of the sustainability violations, such as environmental degradation and child labour exploitation, often manifest within the supply chain [39] due to weak regulatory frameworks and inconsistent enforcement mechanisms in areas pertaining to sustainability, such as environment protection and child labour practices. Companies in these countries limit their efforts to meeting basic legal obligations over undertaking supplementary initiatives to enhance environmental and social performance. This reluctance is fuelled by the economic pressure associated with investing in costly sustainability initiatives such as procuring international certifications, expend significant capital on advanced technologies, and training personnel. These efforts substantially increase production cost rendering products more expensive and less competitive in the global market [39].

1.4.1.2.2 Sustainable Logistics and Digital Transformation Sustainable logistics, also referred to as green logistics, encompasses all initiatives aimed at reducing the environmental impact of logistics. The objective of green logistics is to integrate environmentally friendly methods into the supply chain while optimising logistics [40]. Sustainable logistics mitigates the environmental impact of logistics networks and delivery systems. Supported by AI, ML, and sophisticated analytics, sustainable logistics improve organisational efficiency and consumer satisfaction [41]. Green logistics and sustainable distribution networks, which include the utilisation of electric vehicles for deliveries and route optimisation, assist firms in minimising their carbon footprint and meeting consumer demands. A more comprehensive strategy that encompasses green logistics and other sustainability initiatives is green supply chain management.

1.4.1.2.3 Supplier Performance and Environmental Criteria The management of sustainability at the supplier's side is essential for sustainable business practices as the social and environmental impact of its supply chains is significant [42]. Digital tools facilitate carbon footprint management across the supply chain, allowing real-time modifications to enhance sustainability. In recent years, there has been an increase in the utilisation of Industry 4.0 digital technology to facilitate the collection of data regarding the social and environmental implications of the supply chain. Technologies for supply chain sustainability encompass cloud Enterprise Resource Planning (ERP) systems, smart devices, 3D printing, electric vehicles, AI/ML algorithms, and blockchain. Supply chain sustainability management has emerged as a prominent research domain in recent years, as demonstrated by studies of [42]. Industry 5.0 is expected to offer numerous alternatives for managing supply chain sustainability, including improved supply chain visibility and the swift adaptation of supply systems [5].

1.4.2 Case Studies and Industry Best Practices

Several firms across the world have advanced towards sustainable practices and seek to further improve sustainability initiatives, as evidenced by published corporate reports. The majority of global companies have implemented sustainable practices and maintain a high level of environmental monitoring, which has resulted in improved stakeholder perception

TABLE 1.1 Companies Achieving Improved Environmental Performance through Digital Transformation

Company	DT Initiative	Impact	Environmental Benefit
Siemens	Implemented MindSphere IoT operating system to optimise industrial operations [44].	Real-time monitoring and predictive maintenance reduced energy consumption and inefficiencies [44].	Reduced greenhouse gas emissions and energy waste in industrial facilities [44].
Unilever	Used data analytics and AI to optimise water and energy usage in production processes [45].	AI-driven insights enhanced efficiency, reducing water and energy consumption [45].	Contributed to halving its environmental footprint and improving supply chain transparency [45].
Amazon	Employed machine learning and AI in logistics and supply chain operations [46].	Optimised delivery routes, reducing fuel consumption; committed to renewable energy in data centres [46].	Lowered transportation-related emissions and improved energy efficiency in operations [46].
IKEA	Utilised blockchain and AI to improve supply chain transparency and efficiency [47].	Tracked the life cycle of materials to ensure sustainable sourcing [47].	Reduced waste and supported sustainable forestry practices [47].
Microsoft	Implemented AI and IoT technologies to achieve carbon neutrality and enhance sustainability efforts [43].	Optimised building energy usage and water management systems globally [43].	Achieved carbon neutrality since 2012; committed to becoming carbon negative by 2030 [43].
Tesco	Adopted machine learning algorithms to forecast demand and reduce food waste [48].	Predicted customer buying patterns to align inventory with demand [48].	Reduced food waste and associated carbon emissions significantly [48].

Source: Authors' own compilation.

and the bottom line. In this section, we will examine the success stories of firms achieving improved environmental performance through DT, as summarised in Table 1.1.

1.4.2.1 Microsoft

The global pandemic adversely affected organisations, from hiring employees to supply chain disruptions. Microsoft's capital allocation in digital technologies revived the organisation to be at the forefront. Microsoft's Azure, a cloud computing design, ensures speedy execution of skills that are secure and conforming. To achieve carbon neutrality and enrich sustainability initiatives, Microsoft implemented AI and IoT technologies. To foster ecological sustainability, the company launched an AI-enabled sustainability initiative and complied with the strategy of utilising AI technologies. In order to mitigate environmental challenges, the company assigns money, data, and technology to initiatives and collaborators who provide AI-driven solutions. Microsoft developed various AI-driven projects for waste management and to reduce carbon emissions in buildings. The company utilised AI-driven methodologies to optimise global energy consumption in buildings and water management systems. The environmental impact of these initiatives is that Microsoft achieved carbon neutrality in 2012 and aims to achieve carbon negativity by 2030 [43].

1.4.2.2 Siemens

Global company Siemens launched the MindSphere IoT operating system in order to facilitate industrial and operational efficiency and to reduce energy usage. MindSphere as an industrial IoT gathers and examines sensor data from machines, plants, systems, products, and the entire collection which ensures operational transparency and visibility. Comparing the gathered data with predicted outcomes improves accuracy by removing errors. Thus, it assists enterprises in monitoring energy use, enhancing operational efficiency, and executing predictive maintenance. It enables users to reduce waste, enhance resource efficiency, and minimise their ecological footprint. It ensures flexibility, meets the requirements of customers within a short time, and provides personalised mass production, which facilitates the organisation to remain competitive in the market. MindSphere enables organisations to attain sustainability goals through data-driven insights, simultaneously fostering cost reduction and innovation. The environmental impact of this initiative is that the company is able to reduce GHG emissions and energy consumption in various industrial facilities [44].

1.4.2.3 Unilever

Unilever used data analytics and AI to improve water and energy efficiency in its production processes, leading to significant reductions in water and energy consumption. Unilever's digital activities have contributed to its goal of diminishing its environmental footprint by 50% while improving supply chain transparency. For instance, the company has implemented sustainable sourcing platforms to track and monitor the sustainability of raw materials through blockchain technology, digitalisation of manufacturing processes, and digital campaigns and tools to educate consumers about environment-friendly behaviour [45].

1.4.2.4 Amazon

Amazon used advanced ML and AI in its logistics and supply chain operations, improving delivery routes and decreasing fuel consumption. Amazon uses technologies like data lakes, ML, and generative AI to accelerate and simplify the process of acquiring and analysing complex supply chain data. Services like AWS IoT SiteWise, Amazon Forecast, and Amazon-managed blockchain offer comprehensive visibility and tracking functionality needed to minimise costs and risks, automate repetitive tasks, and alleviate the risk of overstock and understock. Environmental impact is 3% reduction in absolute carbon emissions, 13% decrease in carbon intensity, and for the fourth consecutive year, being the largest purchaser of renewable energy in the world [46].

1.4.2.5 IKEA

IKEA used blockchain and AI to improve supply chain transparency and efficiency, tracking resource longevity and ensuring sustainable procurement of materials. IKEA Kreativ along with AI chatbot powered by ChatGPT improves retail engagement and customer experience. Environmental impact for the financial year 2023 was 97.8% recycled wood being used in furnishing business, aiming to halt deforestation and promote sustainable forest management. IKEA has adopted sustainable cotton sourcing since 2015 with reduced fertilisers, chemicals, and water consumption. It has deployed cotton teams across the globe to ensure its supply chain adheres to these standards. It aims to eliminate the use of virgin inorganic materials by 2030 and ensure responsible sourcing, trace and track the inorganic supply chain to alleviate risk, improve water security, take actions for sustainable water management, reduce the negative impact on biodiversity, and ensure ecosystem enhancement and protection [47].

1.4.2.6 *Tesco*

Tesco used ML algorithms (MLL) and predictive analytics to improve demand forecasts, reducing food waste in stores and aligning stocks with real demand. This has resulted in a significant reduction in waste, especially in the fresh produce category, GHG emissions, and associated costs, while enhancing supply chain efficiency and increasing profitability. Tesco has implemented initiatives to redistribute millions of meals and prevent surplus food from going to landfills. It intends to reduce food waste across its supply chain by 2025 and achieve net zero emission in its supply chain by 2050, embracing innovations like insect farming and anaerobic digestion [48].

1.4.3 Bibliometric Trends in DT and CEP Research

A comprehensive review of a specific subject enables the acquisition of basic knowledge and elaboration of the knowledge framework within that field. This study employed a bibliometric analysis to determine the trend topics within the DT CEP research. Biblioshiny packages within the R Studio software have been utilised to visualise the DT CEP trend topics. The documents were collected from the Scopus database after filtering for article document type and English language. The bibliometric analysis of the 222 documents provided the following results. The research topic has made progress in terms of keywords over time, as illustrated in Figure 1.1. The data demonstrates a discernible trend in the themes that have attracted a growing amount of attention over time. "Trend Topics," which spans 2019–2023, showed the occurrence and development of important concepts such as ecological management, long-term viability, and sustainable development. The trend diagram reveals that researchers are showing a growing interest in technology-driven sustainability solutions, including emerging subjects such as environmental performance, AI, and environmental

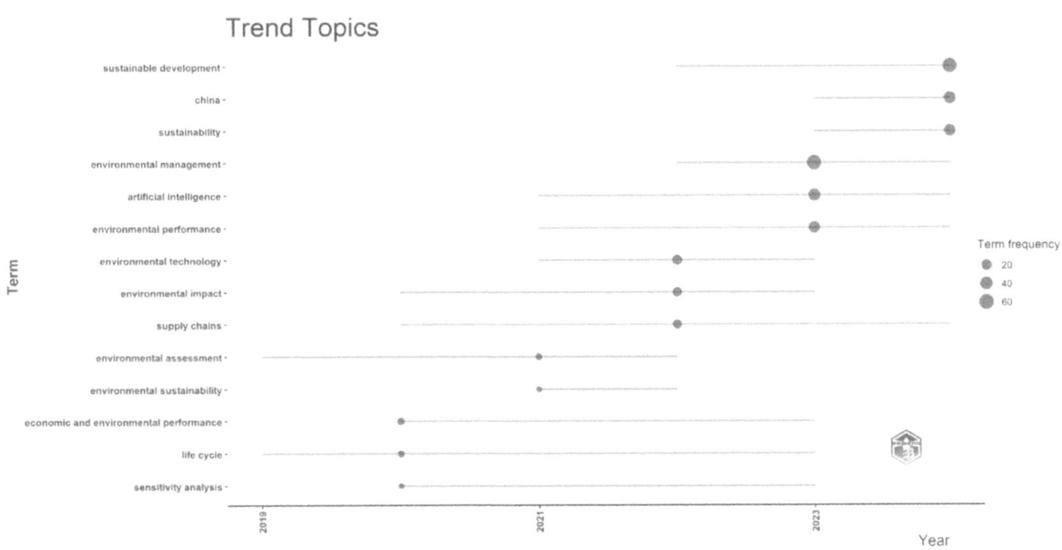

FIGURE 1.1 Trend topics. (Source: Authors' own compilation.)

technology. Global supply chain disruptions and environmental repercussions made supply chains prominent in 2022 and 2023. Similarly, there has been consistent expansion in the number of articles covering environmental sustainability, life cycle, and environmental assessment, all of which are cornerstones of the discipline. While "artificial intelligence" and "environmental performance" have gained prominence, the term "economic and environmental performance" has become less common, though it persists as relevant, as people are still interested in finding ways to combine environmental tactics with economic benefits. Increased research and conversation around these themes in 2023, as indicated by larger bubbles for most phrases, may be attributed to a heightened worldwide awareness of environmental issues. Sustainable practices, technology, and global perspectives are becoming more important, as seen in the diagram, which reflects the changing objectives in environmental and organisational research.

Over the past few years, there has been a significant shift in the popular vocabulary surrounding sustainability, technology, and environmental performance. The growing pertinence of concepts like environmental management, sustainable development, and AI that appear frequently in the literature suggests that there is an increasing focus on advanced technology and large-scale environmental systems. Life cycle, environmental performance, and economic and environmental assessment are early growth issues; AI and environmental technology, which are driven by technology, show mid-level mentions but later median and Q3 years. In the aftermath of worldwide disruptions and environmental crises, the topics of supply networks and their effects on the environment will acquire prominence beyond 2020. As we go from basic ideas to more general, technology-driven, and location-specific applications, sustainability is becoming an important theme (Table 1.2).

TABLE 1.2 Trend Topics

Term	Frequency	Year (Q1)	Year (Median)	Year (Q3)
Economic and environmental performance	7	2020	2020	2023
Life cycle	6	2019	2020	2023
Sensitivity analysis	5	2020	2020	2023
Environmental assessment	6	2019	2021	2022
Environmental sustainability	5	2021	2021	2022
Environmental technology	17	2021	2022	2023
Environmental impact	14	2020	2022	2023
Supply chains	13	2020	2022	2024
Environmental management	64	2022	2023	2024
Artificial intelligence	38	2021	2023	2024
Environmental performance	31	2021	2023	2024
Sustainable development	56	2022	2024	2024
China	35	2023	2024	2024
Sustainability	30	2023	2024	2024

Notes: Q1 – quarter 1, Q3 – quarter 3.
Source: Authors' own compilation.

1.5 CONCLUSION AND FUTURE RESEARCH AVENUES

CEP is essential for adherence to UN SDGs, a sustainable business model, and transition to Industry 5.0, as organisations utilise innovative technology to enhance sustainability and increase efficiency in their operations. The results of the conceptual research have provided a deeper understanding into the link between DT and CEP. The case studies presented demonstrate that digitalisation promotes sustainable business models by integrating critical business elements with digital tools to attain goals that benefit society, the environment, and the firm. The digitalisation activities implemented by the companies result in the adoption of many innovative practices, which significantly impact both the sector and the economy as a whole. Further, DT plays a crucial role in achieving carbon neutrality, enabled by the strategic application of digital technology. Academia and industry are progressing in alignment with these evolving demands, as seen by practices in numerous companies and research in pertinent domains.

Research in the domain of renewable and sustainable energy sectors reveals that AI and its associated technologies show low correlations with these sectors, while business intelligence and AI exhibit substantial interconnections. IBM and Google exhibit outstanding success in specific subsectors within the quantum computing industry. The results provide evidence in favour of strategic decision-making in the context of sustainable energy transitions. A notable trend in DT and environmental performance is "Digital Innovations in Sustainability," which entails the exploration of upcoming digital technologies, including quantum computing, 5G, and AR, and their potential impact on CEP [49]. The study on 6G wireless systems addresses the crucial need for considering the increasing demand for energy efficiency in the wireless network and to reduce the environmental challenges [49]. The energy-efficient network design, dynamic power regulation, energy harvesting, and sustainable infrastructure help to improve energy efficiency, efficient management of resources in 6G wireless systems, and mitigate carbon footprints. The second key area is the "function of AI in achieving net zero," which emphasises the ability of AI, ML, and data-driven models to aid organisations in attaining net zero emissions and furthering global climate goals [12, 50]. The third promising area in this context is the "role of automation in mitigating environmental impact", which encompasses the use of AI-driven machinery, robotic process automation (RPA), and autonomous vehicles to improve operational efficiency while reducing ecological consequences. Technological innovation plays a crucial role in achieving sustainable development and promotes the widespread adoption of automated systems in the construction industry, thereby enhancing positive and economic and environmental performance.

In Industry 5.0 initiatives, sustainable development and human welfare are emphasised, and societal value is prioritised over economic value. As a result, it is essential to improve environmental performance and effectively manage carbon footprints. As sustainability is of vital importance for people, planet, and profit, it is anticipated that the business sector will align with global benchmarks, while academia is encouraged to pursue more impactful research endeavours.

REFERENCES

1. E. J. Medina-Hernández, D. S. Guzmán-Aguilar, J. L. Muñiz-Olite, and L. R. Siado-Castañeda, "The Current Status of the Sustainable Development Goals in the World,"

Development Studies Research, vol. 10, no. 1, Feb. 2023. https://doi.org/10.1080/21665095.2022.2163677.

2. S. Singh, and J. Ru, "Goals of Sustainable Infrastructure, Industry, and Innovation: A Review and Future Agenda for Research," *Environmental Science and Pollution Research*, vol. 30, Jan. 2023. https://doi.org/10.1007/s11356-023-25281-5.
3. V. Varriale, A. Cammarano, F. Michelino, and M. Caputo, "The Role of Digital Technologies in Production Systems for Achieving Sustainable Development Goals," *Sustainable Production and Consumption*, Apr. 2024. https://doi.org/10.1016/j.spc.2024.03.035.
4. G. Robertsone, and I. Lapina, "Digital Transformation as a Catalyst for Sustainability and Open Innovation," *Journal of Open Innovation: Technology, Market, and Complexity*, vol. 9, no. 1, p. 100017, Mar. 2023. https://doi.org/10.1016/j.joitmc.2023.100017.
5. M. Ghobakhloo, M. Iranmanesh, B. Foroughi, E. B. Tirkolaee, S. Asadi, and A. Amran, "Industry 5.0 Implications for Inclusive Sustainable Manufacturing: An Evidence-Knowledge-Based Strategic Roadmap," *Journal of Cleaner Production*, vol. 417, pp. 138023–138023, Sep. 2023. https://doi.org/10.1016/j.jclepro.2023.138023.
6. C. Ancillai, A. Sabatini, M. Gatti, and A. Perna, "Digital Technology and Business Model Innovation: A Systematic Literature Review and Future Research Agenda," *Technological Forecasting and Social Change*, vol. 188, p. 122307, Mar. 2023. https://doi.org/10.1016/j.techfore.2022.122307.
7. C. Xu, W. Zhao, X. Li, B. Cheng, and M. Zhang, "Quality of Life and Carbon Emissions Reduction: Does Digital Economy Play an Influential Role?" *Climate Policy*, vol. 24, no. 3, pp. 1–16, Apr. 2023. https://doi.org/10.1080/14693062.2023.2197862.
8. S. Kraus, P. Jones, N. Kailer, A. Weinmann, N. C. Banegas, and N. R. Tierno, "Digital Transformation: An Overview of the Current State of the Art of Research," *SAGE Open*, vol. 11, no. 3, pp. 1–15, Sep. 2021.
9. A. A. Zahrani, "The Influence of Green Human Resource Management Practices on Organizational Environmental Performance: Mediating Roles of Green Innovation and Employee Performance," *Cogent Business & Management*, vol. 11, no. 1, p. 2387378, Aug. 2024. https://doi.org/10.1080/23311975.2024.2387378.
10. L. Broccardo, A. Zicari, F. Jabeen, and Z. A. Bhatti, "How Digitalization Supports a Sustainable Business Model: A Literature Review," *Technological Forecasting and Social Change*, vol. 187, p. 122146, Feb. 2023. https://doi.org/10.1016/j.techfore.2022.122146.
11. M. Geissdoerfer, D. Vladimirova, and S. Evans, "Sustainable Business Model Innovation: A Review," *Journal of Cleaner Production*, vol. 198, no. 1, pp. 401–416, Oct. 2018. https://doi.org/10.1016/j.jclepro.2018.06.240.
12. T. C. Truong, "The Impact of Digital Transformation on Environmental Sustainability," *Advances in Multimedia*, vol. 2022, pp. 1–12, May 2022. https://doi.org/10.1155/2022/6324325.
13. B. Masoomi, I. G. Sahebi, F. Gholian-Jouybari, C. Mejia-Argueta, and M. Hajiaghaei-Keshteli, "The Role of Internet of Things Adoption on the Sustainability Performance of the Renewable Energy Supply Chain: A Conceptual Framework," *Renewable and Sustainable Energy Reviews*, vol. 202, p. 114610, Sep. 2024. https://doi.org/10.1016/j.rser.2024.114610.
14. C. Mulligan, S. Morsfield, and E. Cheikosman, "Blockchain for Sustainability: A Systematic Literature Review for Policy Impact," *Telecommunications Policy*, vol. 48, no. 2, p. 102676, Oct. 2023. https://doi.org/10.1016/j.telpol.2023.102676.
15. K. Govindan, "How Digitalization Transforms the Traditional Circular Economy to a Smart Circular Economy for Achieving SDGs and Net Zero," *Transportation Research Part E-Logistics and Transportation Review*, vol. 177, p. 103147, Sep. 2023. https://doi.org/10.1016/j.tre.2023.103147.
16. C. Chauhan, V. Parida, and A. Dhir, "Linking Circular Economy and Digitalisation Technologies: A Systematic Literature Review of Past Achievements and Future Promises," *Technological Forecasting and Social Change*, vol. 177, p. 121508, Apr. 2022. https://doi.org/10.1016/j.techfore.2022.121508.

17. D. Hariyani, P. Hariyani, S. Mishra, and M. Kumar Sharma, "Leveraging Digital Technologies for Advancing Circular Economy Practices and Enhancing Life Cycle Analysis: A Systematic Literature Review," *Waste Management Bulletin*, vol. 2, no. 3, pp. 69–83, Jul. 2024. https://doi.org/10.1016/j.wmb.2024.06.007.
18. I. Shabir *et al.*, "Carbon Footprints Evaluation for Sustainable Food Processing System Development: A Comprehensive Review," *Future Foods*, vol. 7, no. 1, pp. 1–13, Jun. 2023. https://doi.org/10.1016/j.fufo.2023.100215.
19. Y. H. Labaran, V. S. Mathur, S. U. Muhammad, and A. A. Musa, "Carbon Footprint Management: A Review of Construction Industry," *Cleaner Engineering and Technology*, vol. 9, p. 100531, Aug. 2022. https://doi.org/10.1016/j.clet.2022.100531.
20. M. Škare, B. Gavurova, and M. Porada-Rochon, "Digitalization and Carbon Footprint: Building a Path to a Sustainable Economic Growth," *Technological Forecasting and Social Change*, vol. 199, p. 123045, Feb. 2024. https://doi.org/10.1016/j.techfore.2023.123045.
21. Y. Huang, M. Hu, J. Xu, and Z. Jin, "Digital Transformation and Carbon Intensity Reduction in Transportation Industry: Empirical Evidence from a Global Perspective," *Journal of Environmental Management*, vol. 344, pp. 118541–118541, Oct. 2023. https://doi.org/10.1016/j.jenvman.2023.118541.
22. M. A. Quttainah, and I. Ayadi, "The Impact of Digital Integration on Corporate Sustainability: Emissions Reduction, Environmental Innovation, and Resource Efficiency in the European," *Journal of Innovation & Knowledge*, vol. 9, no. 3, pp. 100525–100525, Jul. 2024. https://doi.org/10.1016/j.jik.2024.100525.
23. Y. K. Dwivedi *et al.*, "Climate Change and COP26: Are Digital Technologies and Information Management Part of the Problem or the Solution? an Editorial Reflection and Call to Action," *International Journal of Information Management*, vol. 63, p. 102456, Apr. 2022. https://doi.org/10.1016/j.ijinfomgt.2021.102456.
24. J. Xu, Y. Yu, M. Zhang, and J. Z. Zhang, "Impacts of Digital Transformation on Eco-Innovation and Sustainable Performance: Evidence from Chinese Manufacturing Companies," *Journal of Cleaner Production*, vol. 393, p. 136278, Feb. 2023. https://doi.org/10.1016/j.jclepro.2023.136278.
25. Y. Su, and J. Wu, "Digital Transformation and Enterprise Sustainable Development," *Finance Research Letters*, vol. 60, p. 104902, Dec. 2023. https://doi.org/10.1016/j.frl.2023.104902.
26. I. Guandalini, "Sustainability Through Digital Transformation: A Systematic Literature Review for Research Guidance," *Journal of Business Research*, vol. 148, no. 1, pp. 456–471, Sep. 2022. https://doi.org/10.1016/j.jbusres.2022.05.003.
27. U. Gartia, A. K. Panda, A. Hegde, and S. Nanda, "Environmental Social Governance and Financial Performance: A Symbiotic Relationship in Indian Manufacturing," *Cleaner Production Letters*, vol. 7, p. 100076, Oct. 2024. https://doi.org/10.1016/j.clpl.2024.100076.
28. F. P. Tajudeen, D. Nadarajah, N. I. Jaafar, and A. Sulaiman, "The Impact of Digitalisation Vision and Information Technology on organisations' Innovation," *European Journal of Innovation Management*, vol. 25, no. 2, pp. 607–629, Feb. 2022. https://doi.org/10.1108/ejim-10-2020-0423.
29. A. Atanasov, G. Chipriyanova, and R. Krasteva-Hristova, "Integration of Digital Technologies in Corporate Social Responsibility (CSR) Activities: A Systematic Literature Review and Bibliometric Analysis," *Journal of Risk and Financial Management*, vol. 16, no. 8, p. 373, Aug. 2023. https://doi.org/10.3390/jrfm16080373.
30. J. Awad, and R. Martín Rojas, "Enhancing Social Responsibility and Resilience through Entrepreneurship and Digital Environment," *Corporate Social Responsibility and Environmental Management*, vol. 31, no. 3, pp. 1688–1704, May 2024. https://doi.org/10.1002/csr.2655.
31. M. M. Queiroz, S. C. F. Pereira, R. Telles, and M. C. Machado, "Industry 4.0 and Digital Supply Chain Capabilities," *Benchmarking: An International Journal*, vol. 28, no. 5, pp. 1761–1782, May 2021. https://doi.org/10.1108/bij-12-2018-0435.

32. E. Manavalan, and K. Jayakrishna, "A Review of Internet of Things (IoT) Embedded Sustainable Supply Chain for Industry 4.0 Requirements," *Computers & Industrial Engineering*, vol. 127, no. 1, pp. 925–953, Jan. 2019. https://doi.org/10.1016/j.cie.2018.11.030.
33. M. Yang, M. Fu, and Z. Zhang, "The Adoption of Digital Technologies in Supply Chains: Drivers, Process and Impact," *Technological Forecasting and Social Change*, vol. 169, p. 120795, Aug. 2021. https://doi.org/10.1016/j.techfore.2021.120795.
34. P. Akhtar, Z. Khan, S. Tarba, and U. Jayawickrama, "The Internet of Things, Dynamic Data and Information Processing Capabilities, and Operational Agility," *Technological Forecasting and Social Change*, vol. 136, pp. 307–316, Nov. 2018. https://doi.org/10.1016/j.techfore.2017.04.023.
35. S. Melo, J. Macedo, and P. Baptista, "Capacity-Sharing in Logistics Solutions: A New Pathway Towards Sustainability," *Transport Policy*, vol. 73, pp. 143–151, Jan. 2019. https://doi.org/10.1016/j.tranpol.2018.07.003.
36. M. E. Arbabian, "Supply Chain Coordination via Additive Manufacturing," *International Journal of Production Economics*, vol. 243, p. 108318, Jan. 2022. https://doi.org/10.1016/j.ijpe.2021.108318.
37. KPMG International, "Supply Chain Trends 2024: The Digital shake-up," *KPMG*, 2024. https://kpmg.com/xx/en/our-insights/ai-and-technology/supply-chain-trends-2024.html
38. B. Esmaeilian, J. Sarkis, K. Lewis, and S. Behdad, "Blockchain for the Future of Sustainable Supply Chain Management in Industry 4.0," *Resources, Conservation and Recycling*, vol. 163, p. 105064, Dec. 2020. https://doi.org/10.1016/j.resconrec.2020.105064.
39. N. Kshetri, "Blockchain and Sustainable Supply Chain Management in Developing Countries," *International Journal of Information Management*, vol. 60, p. 102376, Oct. 2021. https://doi.org/10.1016/j.ijinfomgt.2021.102376.
40. C. Yunlin, "Awareness of Green Logistics Technology, Certification, and Standards by Logistics Practitioners at Chinese e-Commerce Company, Jing Dong," *The Asian Journal of Shipping and Logistics*, vol. 39, no. 4, pp. 37–46, Dec. 2023. https://doi.org/10.1016/j.ajsl.2023.10.004.
41. E. D. de Souza, J. C. Kerber, M. Bouzon, and C. M. T. Rodriguez, "Performance Evaluation of Green Logistics: Paving the Way Towards Circular Economy," *Cleaner Logistics and Supply Chain*, vol. 3, p. 100019, Mar. 2022. https://doi.org/10.1016/j.clscn.2021.100019.
42. V. Patil, T. Tan, S. Rispens, S. Dabadghao, and E. Demerouti, "Supplier Sustainability: A Comprehensive Review and Future Research Directions," *Sustainable Manufacturing and Service Economics*, vol. 1, no. 100003, p. 100003, Apr. 2022. https://doi.org/10.1016/j.smse.2022.100003.
43. K. Boyd, L. Miniken, and D. Hirning, "Digitally Transforming Microsoft: Our IT Journey," *Microsoft Inside Track Blog*, Feb. 27, 2025. https://www.microsoft.com/insidetrack/blog/inside-the-transformation-of-it-and-operations-at-microsoft/
44. Siemens, "Insights Hub Industrial IoT as a Service | Siemens Software," *Siemens Digital Industries Software*, 2025. https://plm.sw.siemens.com/en-US/insights-hub/
45. A. Zaytsev, "Case Study: Unilever's Integration of AI in the Supply Chain," *AI Expert Network*. https://aiexpert.network/case-study-unilevers-integration-of-ai-in-the-supply-chain/
46. C. Krause, "Case Study: Amazon's AI-Driven Supply Chain: A Blueprint for the Future of Global Logistics," *The CDO TIMES*, Aug. 23, 2024. https://cdotimes.com/2024/08/23/case-study-amazons-ai-driven-supply-chain-a-blueprint-for-the-future-of-global-logistics/#google_vignette
47. B. Marr, "The Amazing Ways IKEA Is Using Generative AI," *Forbes*, Apr. 05, 2024. https://www.forbes.com/sites/bernardmarr/2024/04/05/the-amazing-ways-ikea-is-using-generative-ai/
48. Tesco plc, "Target, measure, act. Food Waste Report 2024," 2024. Available: https://www.tescoplc.com/media/sm2hn54l/tesco-food-waste-report-jan24-96.pdf
49. R. Kumar, S. K. Gupta, H.-C. Wang, C. S. Kumari, and S. S. V. P. Korlam, "From Efficiency to Sustainability: Exploring the Potential of 6G for a Greener Future," *Sustainability*, vol. 15, no. 23, p. 16387, Jan. 2023. https://doi.org/10.3390/su152316387.
50. D. B. Olawade, O. Z. Wada, A. C. David-Olawade, O. Fapohunda, A. O. Ige, and J Ling, "Artificial intelligence potential for net zero sustainability: Current evidence and prospects," *Next Sustainability*, vol. 4, no. p. 100041, 2024. https://doi.org/10.1016/j.nxsust.2024.100041

CHAPTER 2

The Impact of Sporting Events on Tourism Attractiveness

Organization of the 2030 World Cup in Morocco

Fatima Zahra El Karmaoui, El Ouali El Balaidi, and Ahmed El Hammoumi

2.1 INTRODUCTION

Because of their global dimension, major sporting events have for several decades become strategic instruments through which host countries aim to increase their international visibility (i.e. their potential for attracting tourists). Major competitions (Olympic Games, FIFA World Cup) are particularly seen as opportunities to better position the host country's brand image and attract new flows of visitors, while facilitating investment in infrastructure. As Chalip (2004) points out, the mere consideration of the event with immediate impact can however be regrettable; good numbers of these events can allow a "leverage" work in marketing for a maximization of the impact in terms of the event, from a tourist point of view.

An example of the potential of sport to leverage the tourism sector is illustrated by the 2010 World Cup, which according to Fourie and Santana-Gallego (2011) made the country more visible and increased arrivals, but whose effect was ultimately limited in time. The same applies to Russia in 2018 and Qatar in 2022, which sought, through these operations, to enhance their image on the international scene and strengthen their tourist offer. These experiments show the structuring role played by mega-events in reconfiguring a destination.

However, the analysis of the literature highlights a significant gap: few studies focus on emerging countries as host contexts for these events and even less on the tensions between tourism development and sustainability. The environmental issue, the equitable distribution of benefits or the sustainability of infrastructure are relegated to second place. This theoretical imbalance is all the more doubtful since, in fact, emerging countries are increasingly being asked to host this type of event, often at the cost of significant financial and social implications.

DOI: 10.1201/9781003649885-2

On the other hand, Gibson (1998) is still very critical of the certain superficiality of research in sports tourism and calls for rigorous analyses concerning the effects as well as the strategies mobilized. In this respect, the organization of the 2030 World Cup by Morocco is an object of research of a very special richness, allowing both to articulate tourism visibility, sustainable growth and national transformation dynamics.

2.2 SPORTS EVENTS AND TOURIST ATTRACTIVENESS: THEORETICAL FOUNDATIONS

The organization of an international sporting event goes beyond the sole perspective of a sporting event: it constitutes a strategic dimension for the development of a territory, where tourist attractiveness is central. Several theoretical frameworks allow us to shed light on the mechanisms by which these events can be our vectors of positive spinoffs for the host destination.

2.2.1 The Showcase Effect Theory (Showcase Effect)

According to the showcase effect theory, organizing an international event (a World Cup, Olympic Games, Universal Exhibition, etc.) provides an uncommon way for a country or city to "show" itself to the world; it serves as a showcase, as a visibility apparatus, because it allows the country or city to communicate a modern, attractive and dynamic image of the host country, attract investment and tourists, enhance national pride and catalyze urban, social or environmental reforms that may otherwise have languished as low-priority initiatives (slow or not at all).

2.2.1.1 Significant Aspects of the Showcase Effect

- A new type of visibility with international media coverage: The event generates media from all over the world. The country now has its name and an image situated in the collective imaginary. Instead of being a peripheral part of the map, now it is a center of geopolitical gravitation.
- Tourist attraction impact: The event's prestige attracts visitors from abroad who are discovering the land. This can generate tourism indirectly in the immediate future, but especially in the medium and long term, if the overall impression is positive.
- Legitimacy of public actions and reform: The government can use the event to legitimize infrastructure, modernization or urban beautification projects. There can be financial levers of acceleration in public investments (transit, stadiums and urban designs).
- Exemplarity or image – project impact: The event opens up the country and gives the chance to be known as innovative, environmentally friendly or a user of new technologies according to the themes marketed (smart city, sustainability, inclusion etc.).

2.2.1.2 Limitation and Criticisms of Showcase Impact

- The showcase progress isn't for everyone sustainable: the after-event strategy (legacy thinking) is important and if there isn't any, there can be temporary or superficial effects.

- It can hide the socio-economical reality of countries: the social consequences (displacement of population, public debt burden, increasing inequalities) can be hidden as a result of a good news focus in communication strategy that creates a positive image of the nation and countries.
- A projected image may be discordant with residents' lived experiences, notably, if the event is not responsive to citizens' needs.

2.2.1.3 Example: Morocco and 2030 World Cup

The World Cup 2030 event can be considered from the perspective of a strategic lever to bring about a showcase effect on several levels:

- Tourism: enhancing the Morocco experience as a modern and attractive Mediterranean and African center;
- Geopolitics: claiming regional leadership by co-hosting with Spain and Portugal;
- Sustainable development: educating generations about green stadiums, sustainable mobility and management of a responsible legacy;
- Nation branding: developing a positive narrative about the makeovers that have taken place in the country in terms of infrastructure and hospitality.

2.2.2 The Legacy Model (Legacy Framework)

This framework, the legacy framework, is particularly concerned with the legacy left by a sporting event in terms of opportunities, especially when that legacy is sustainable, in terms of material effects (tourist infrastructures, transport, urban facilities) or intangible (human capital, brand image, social cohesion). Chalip (2004) continues this questioning of sustainable inheritance by pointing out that to be beneficial, inheritance must be anticipated, taken into account at the strategic planning stage. Indeed, an event could become a catalyst for tourism development if it allows the reuse of facilities by integrating them into a structured tourist offer in the long term. In the context of the exit from the health crisis that we are still experiencing in the context of the COVID pandemic, this question about the legacy has raised expectations with regard to an investment that a country had to justify, in a context of greater budgetary constraint.

2.2.3 The Economic Spinoff Model

This model distinguishes two types of impact: on the one hand, expenses of tourists, teams, media (direct) and on the other hand, expenses of local actants induced by the multiplier (indirect). It is used in impact studies to justify sporting events whether at the high level (Meyer et al., 2018) or the semi-finals of the Coupe de France football (Gouguet et al., 2009). However, several authors argue that these effects are often overestimated, as pointed out by Preuss (2007), when they are not accompanied by a post-event strategy. In terms of tourism, it appears that the spinoffs are all the more interesting if the event helps to attract visitors during the low season or even helps to retain new markets as part of internationalization.

2.2.4 Image Effect and Country Branding

A strong way of country branding is to improve the image of a country through a sports event. Hosting a major competition allows a country to shape or redefine the perception of its international image, a fundamental issue for emerging countries wishing to develop their weight on the world's tourist map. The image of the event is a powerful signal in travel decisions, state Nadeau et al. (2008). We can find this case in Qatar in 2022, where the state tried to make people forget cultural and political stereotypes with a controlled staging of its heritage and modernity.

2.2.5 The Concept of Event Leveraging

In 2004, Chalip introduced the concept of event leveraging. This is the set of actions implemented to enhance the tourist benefits of an event, beyond its course, based on the coordination of tourism actors, the organization of tourist packages, the implementation of competitive festivals, the development of thematic circuits. The implementation of a leverage strategy would require a partnership between organizers, public authorities, tourism stakeholders and the media. However, in the context of post-COVID, this is more than necessary. Event tourism has been sustainably transformed; the issues of health safety, sustainability and digitization of experiences are now major concerns. According to Gursoy and Chi (2020), destinations must now adopt hybrid event strategies combining physical and virtual events that contribute both to the attractiveness of practices and to consumer expectations.

2.3 INTERNATIONAL COMPARABLE CASES

An in-depth examination of some major sporting events provides insight into the processes by which these events affect tourism attractiveness and territorial development. The three emblematic cases of South Africa (2010), Brazil (2014) and Qatar (2022) offer divergent perspectives, including in terms of economic impact, sustainability and brand management.

2.3.1 South Africa – 2010 World Cup

- Before the event: South Africa, the first African country to host a FIFA World Cup, launched an extensive global communication campaign based on the idea that it was the "new face of Africa" (Cornelissen, 2010). The World Cup was supposed to be an agent of social and economic transformation.
- During the event: Significant money was spent on infrastructure development (stadiums, airports, roads) to the tune of approximately US$3.6 billion, as well as optimal management of the influx of more than 300,000 visitors from all over the world; the whole system is supported by a collection of safety means and urban transport (Swart et al., 2010).
- After the event: The results for tourism are favorable in the short term, with an 8% increase in international arrivals in 2010, but there is criticism of the limited use of stadiums (white elephant effect), on the more limited impact on long-term unemployment (Du Plessis & Maennig, 2011), the image of France is positively repositioned in the world.

2.3.2 Brazil – 2014 World Cup

- Before the event: Brazil has benefited from its emerging status and tradition in football. An extensive nation branding campaign aimed at attracting tourists while modernizing the country's image (Gaffney, 2010).
- During the event: More than US$11 billion was spent on the redevelopment of urban infrastructure and stadium construction. The event attracted approximately 1 million visitors from abroad. However, logistical problems, slow construction sites and social tensions marked the demonstration (Zimbalist, 2015).
- After the event: The tourism impact was transitory. The sustainability of investments is questionable, with many infrastructures being little or not used. Similarly, social criticism has developed, in particular because of the break between spectacular investments and unmet social needs (Baade & Matheson, 2016). Brazil missed the opportunity to use the event as a lever for real sustainable change.

2.3.3 Qatar – 2022 World Cup

- Before the event: By focusing on the event, Qatar seeks to improve its image on the international stage, including by helping to implement a dynamic sports diplomacy and implementing often integrated media campaigns (Al Jazeera, partnership with European clubs) (Reiche, 2020).
- During the event: Tourist flows have been centrally managed and technologically advanced flows or, at least, in air-conditioned stadiums, with crowds under digital control in brand new equipment. The competition received approximately 1.4 million visitors (FIFA, 2023).
- After the event: The main focus has been the conversion of equipment to other uses (hotels, community centers) and the image of Qatar, a major global player in sports tourism has strengthened, but criticism continues regarding working conditions, environmental sustainability and human rights management (Amnesty International, 2022); however, this country has managed to generate a significant image effect and attract new tourist flows to the region.

2.4 ARTIFICIAL INTELLIGENCE AND EMERGING TECHNOLOGIES FOR SUSTAINABILITY AT THE 2030 WORLD CUP

The 2030 World Cup, co-organized by Morocco, represents a huge challenge in managing resources, tourist flows and infrastructure sustainably. Artificial intelligence (AI) and emerging technologies can be seen as strategic tools to be leveraged to optimize the environmental, economic and social dimensions of this global event (Akpınar et al., 2025; Zhou et al., 2023)

2.4.1 Optimization of Tourist Flows and Mobility

By strategically employing predictive algorithms based on AI, it will be possible to model tourist flows, and in a real-time urban context, regulate travel. Technology associated with intelligent mobility management systems (Smart Mobility) could mitigate congestion, limit CO emissions and raise the visitor experience (Hasan et al., 2023). Each host city may use integrated mobility platforms combining autonomous vehicles, on-demand public transport and navigating environmentally socially capable (Brelik & Łącka, 2024).

2.4.2 Intelligent Energy Management of Infrastructure

Stadiums and hotels, as well as temporary facilities, can integrate IoT (Internet of Things) sensors, combined with AI platforms, to capture and manage energy usage (Magiina et al., 2023). The application of digital twins would enable usage scenarios to be modeled in order to optimize energy performance, predict outages, rationalize maintenance costs and extend the useful life of the equipment (Batty, 2022; Gartner, 2023).

2.4.3 Environmental Monitoring and Waste Management

Today, smart drones and embedded sensors can be sent to monitor air, soil and water quality around sites, in real time (Richards & Palmer, 2010). In the meantime, AI systems can help automate waste sorting, optimize recycling logistics, find areas of significant environmental impact and enable real-time predictive analytics (Chen & Zhang, 2023; World Bank, 2023). Initiatives like these have been trialed during the Tokyo 2021 Olympic Games and can be modified for the Moroccan context.

2.4.4 The Analysis of Tourism Behavior and Personalization of the Offer

Big data affords us the opportunity to tailor tourist offers based on visitor profiles, anticipating their needs, and helping reinforce their commitment to sustainable practices (Gretzel & Koo, 2021). Virtual assistants and smart chatbots can help promote eco-responsible behavior (e.g., suggesting labeled accommodation or low carbon ways to travel) (Kwong et al., 2024).

2.4.5 Contribution to Transparency and Impact Assessment

Emerging technologies also support the tracking, processing and visualization of sustainability indicators. Interactive dashboards in real time can also allow policy makers to help in their strategic choices, while ensuring more transparency toward citizens, NGOs and international organizations (Xia et al., 2025; Lammi, 2024). Open data and blockchains can also work toward achieving a culture of trust between stakeholders.

By strategically harnessing AI and emerging technologies in the management of the 2030 World Cup, the opportunity arises to position the event as a global sustainability model. In addition to the performance of tourism alone, these tools promote the ecological and social legitimacy of Morocco on the international stage and develop the framework for responsible and inclusive technology governance.

2.5 MOROCCAN SPECIFICITIES AND PERSPECTIVES OF THE 2030 WORLD CUP

2.5.1 The Structural Strengths of Moroccan Tourism

Morocco is full of countless qualities to figure well on the international tourist scene: a rich cultural and historical heritage (e.g., cities classified by UNESCO as Fez, Marrakech, Meknes or Essaouira), a mild climate in winter, year-round sunshine, which attracts both Europeans and the Arab world and a growing accessibility (liberalization of the sky – Open

Sky, investments in airport infrastructure at the airports of Marrakech-Menara or Rabat-Salé – efforts by the ONMT to increase international connections) (ONMT, 2023).

2.5.2 Current Challenges in the Sector

In addition to recognized strengths, Moroccan tourism suffers from many structural challenges. The issue of seasonality remains topical: recorded tourist flows are mainly limited to summer and religious periods, resulting in under consumption of infrastructure between seasons. On the other hand, the geographical imbalance is marked: most visitors always go primarily to the traditional poles of the Atlantic coast and the interior, at the expense of the interior and south-eastern spaces. Weak ecological infrastructure and failure to consider sustainable development marginalizes long-term attractiveness (HCP, 2021; SMIT, 2022). These readjustments hinder the diversification of supply as well as the achievement of the ambitions of a more balanced and resilient tourism.

2.5.3 The 2030 World Cup: A Strategic Opportunity

The co-organization of the 2030 World Cup with Spain and Portugal is a unique opportunity to accelerate the transformation of Moroccan tourism. First, as a global sporting event, it generates major investments in infrastructure (transport, accommodation, logistics) through less touristy cities such as Nador, Oujda or Beni Mellal for example. Second, media visibility and the expected influx of visitors could direct tourism toward the hinterland and promote alternative cultural and natural routes. Finally, in the perspective of the strategy "Sustainable Tourism 2030", the CDM can be an instrument to integrate more ecological and social standards in the design of the tourist offer (Ismaili, M.Y E.A., & Ayegou, J. 2023; Vision 2020, 2010).

Thus, the 2030 World Cup is not only a sporting project but also a strategic lever to rebase the Moroccan tourism model on the basis of sustainability, spatial justice and innovation.

2.6 SYNTHESIS AND FORMULATION OF ASSUMPTIONS

2.6.1 Synthesis of Theoretical and Empirical Contributions

During our journey through the intricacies of tourism and sustainable development research, this literature review highlighted the different ways in which tourism can become a strategic lever for sustainable development, particularly on the occasion of a global event of such significance as the World Cup football 2030. Academic work shows the growing role of mega-sporting events in the restructuring of territories, particularly in terms of infrastructure modernization, brand improvement and diversification toward tourism less focused on certain heritage waves (Bob & Swart, 2010; Getz, 2008).

As a counterpoint, research on sustainable tourism in Morocco has highlighted spatial imbalances, the presence of seasonality and the ecological limits of the current tourism model (HCP, 2021; SMIT, 2022) whereas the international literature calls for taking tourism into account more in a logic of social inclusion, respect for the environment and enhancement of local cultures, rather than a strictly event-based and quantitative approach (Mowforth & Munt, 2016).

2.6.2 Gaps and Study Contribution

Three types of consequences of sporting events have been identified (Müller, 2020): those inherent to the event project in the sense of its organization (resunshine of the destination, quality of infrastructures, improvement of the country's image); those that generate socio-economic and environmental effects (distribution of investments, opportunity to develop tourism projects around the issues of sustainable development and responsible mobility); those that can be mobilized during sporting events, with many studies on the issue related to sustainability (Broussolle, 2021). However, while research on the impacts of sporting events is very rich, few studies focus on sustainability issues at the scale of a country in socio-economic transition in the South, such as Morocco. The links between mega-sporting events, territorial rebalancing of tourism and integration of sustainability principles are still poorly studied.

In this context, our work aims to illustrate this double gap: (1) by proposing a contextualized study of the case of Morocco, with cultural, institutional and territorial specificities; (2) by proposing a cross-reading between tourism, sustainability and events. We should then find out how the CDM 2030 could embody a new, inclusive and balanced Moroccan tourism model.

2.6.3 Justification of Assumptions

The reasoning used to develop the hypotheses of this research is inspired by the teachings produced by the academic literature on sports events, but it also draws on other lessons from sustainable tourism thinking, all in relation to a Moroccan context that is at the heart of all public policies and sectored strategies.

> **Hypothesis 1:** Organizing the 2030 World Cup in Morocco would be a great lever for national tourism attractiveness, both by transforming its international image and attracting more tourists to the country.

This hypothesis is a constant in the literature that shows and analyzes the role played by mega-events in the evolution of an image, in the sense that it has a not negligible effect on the way host countries "appear" (Fourie & Santana-Gallego, 2011; Kaplanidou, 2007). Indeed, betting on the co-organization of a global sporting event is a way to reposition a country like Morocco, which would be commonly perceived as having a less glamorous image on the international level, even if the promotion of this event could allow to improve in the short term its image, and therefore its visibility in tourism, by virtue of an increased international awareness and the strengthening of flows.

> **Hypothesis 2:** Taking into account sustainable and ecological infrastructures in the context of a tourism development during the preparation of the CDM of football of 2030 will result in a more positive perception of the destination by tourists, and by a higher level of visitor satisfaction, which could ensure their loyalty to the destination.

Work on responsible tourism (Bramwell & Lane, 2011; Weaver, 2011) shows that tourists are increasingly aware of the environmental impacts of their stay. To improve the quality of the tourism experience, it is essential to integrate environmental standards into HOM infrastructure projects and ensure that they are taken into account in the context of destination productivity by tourism businesses.

> **Hypothesis 3:** Within an integrated sustainability logic (Preuss, 2015), the implementation of upstream in the context of competition and downstream of the event should allow to maximize the economic benefits of tourism in the medium term and reduce environmental losses.

This reasoning supports this hypothesis in the light of what was implemented for the London 2012 Olympic Games, during which the application of sustainability plans had helped to better orient tourism investments and minimize environmental impacts.

> **Hypothesis 4:** The economic benefits generated by the tournament would be likely to allow a significant improvement in the competitive position of Morocco's tourism sector compared to other emerging destinations.

The World Cup could prove to be an opportunity to be seized (Matheson, 2006) to reposition Morocco in relation to other coastal countries of the Mediterranean (or the Middle East) facing similar tourism issues (Tunisia, Egypt, Turkey). The benefits in terms of economic gains, primarily tourism, could be used to improve the competitiveness of the destination on the international tourism scene.

> **Hypothesis 5:** The holding of the 2030 World Cup in Morocco may lead to indirect societal benefits, especially for less-developed territories, through the renovation of infrastructure and the development of new jobs for local populations.

This hypothesis is based on the idea that the redistributive effects of major events may exceed the scope of the receiving territories alone (Chappelet & Junod, 2006). In a country with highly centralized tourism, CSD-related investments could therefore contribute to a better territorial balance and thus generate social and economic benefits in the peripheral part of the country.

> **Hypothesis 6:** Information facilitating favorable communication and enhancing the tourist attractions of the World Cup in Morocco contribute to the attractiveness of the country, especially in the long term.

To evaluate the relevance of assumptions made from literature, it is now time to put them head-to-head with what is actually happening on the ground. It is in this perspective that our empirical approach, by mobilizing a survey of those in the tourism sector, is set up. This changeover will allow for a rigorous analysis of the anticipated impacts of the 2030 World Cup on Morocco's tourist attractiveness.

2.7 EMPIRICAL FRAMEWORK

After having cleared our conceptual model, we will try to confront it with the reality of our empirical field of investigation which remains a necessary step to answer our main question initially asked.

2.7.1 Collection of Data

After developing the questionnaire to address the field of our research, we undertook the second step to test this questionnaire on a limited sample. Validating this questionnaire by ensuring the consistency of these questions, we started to collect primary data.

In our research, the target population includes tourists, tourism professionals, institutional leaders and residents of host cities.

We inform you that the questionnaire will be submitted to professionals in the tourism sector, in this case managers of hotel establishments, travel agencies, institutional actors working at local level, as well as representatives of local authorities participating in the organization of the World Cup 2030. The collection of data took place in person, through questionnaires handed out manually to the stakeholders concerned, using our professional network. Approximately 940 questionnaires were submitted, 382 returned to us, a response rate of 40.63%, among which 360 are directly exploitable, which constitutes a relevant sample for us to validate our analysis model.

2.7.2 Data Processing and Search Results

First, we checked the relevance and validity of the variables in our research model with a principal component analysis (PCA), according to the groups of items considered in the knowledge transfer obtained. Then, we have exposed a multivariate analysis device designed to examine the causal links between the dependent variable and the independent variables.

2.7.2.1 The Measurement Model

PCA was used for validation of the theoretical model measurement scales and suggested simplification of data. This method follows the steps suggested by Evrard et al. (2003) which included the Bartlett's test of sphericity (check of correlation between variables) and KMO index (factorability of data). A KMO of greater than 0.8 indicates a good structure to the factors to be extracted, while a KMO of 0.5 is barely acceptable. The components were extracted according to their contributing score (threshold 0.500), and an orthogonal rotation Varimax was applied to clarify the structure of the factor. To check the reliability within the factor, the Cronbach alpha coefficient was used to measure internal coherence. For exploratory research, it is considered acceptable to have an alpha between 0.6 and 0.8. The result of this procedure enables the research hypotheses to be redefined and resulted in a clear and statistically reliable version of the model, the results presented in Table 2.1, the results of the ACP, the Cronbach alphas and the variance explained for each of the components.

After purification of the measurement scales, we tested the hypotheses of this research model.

TABLE 2.1 The Results of the Validity and Reliability of Measurement Scales

Components	Cronbach's Alpha	Items	% of Variance Explained	Number of Items
Strengthening the image of Morocco abroad	0.745	• The international reputation of Morocco. • The perception of Morocco as a stable and attractive country. • Investments related to the World Cup.	76.67	3
Increasing the satisfaction of tourists and loyalty to the destination	0.738	• Tourism services will meet the expectations of international visitors. • Recommendations of visitors to Morocco as a tourist destination. • The satisfaction of tourists from their experience in Morocco.	71	3
Maximizing the long-term economic benefits of tourism	0.728	• The sustainable increase in tourism revenues. • Long-term economic attractiveness. • The creation of sustainable jobs in the tourism sector. • The development of SMEs operating in the tourism sector.	67.856	4
The significant improvement in tourism competitiveness	0.723	• Improving the overall quality of tourism infrastructures. • Adaptation of international service standards. • The country's competitiveness. • Competitive position vis-à-vis other destinations.	72.42	4
Indirect societal benefits	0.740	• The sense of national pride and unity among citizens. • Social cohesion through community initiatives. • The involvement of young people in sports and cultural projects. • Improvements in transport, security and public services.	62.78	4
The communication strategies adopted	0.754	• The communication channels used. • The communication campaigns. • Transparency and responsiveness in communication.	75.46	3
Tourist attractiveness	0.749	• Destination more visible internationally. • Improving the experience of foreign visitors. • Improving the country's image as a welcoming destination. • The appreciation of the cultural diversity and heritage of the country in the eyes of tourists.	72.42	5

2.7.2.2 Hypothesis Test by Regression Analysis

Multiple linear regression is a method used to define the effect of several explanatory variables on a variable to be explained; it makes it possible to assess the respective importance of each factor in the explanation of the observed reality, based on the equation that connects the dependent variable and the independent variables weighted by their coefficients.

The method is used in the context of estimating the strength of the assumptions of the model that we have previously applied to the Morocco's tourist attractiveness for the World Cup 2030 concerning visitor's satisfaction, visitor's loyalty and sustainable economic gains. The analysis includes several statistical tests: the VIF and tolerance rate for redundancies between variables, and the Fisher test to check for the overall significance of the model, and student test to check the influence of each coefficient independently. The significance threshold that was used was either 5% ($p < 0.05$) or 1% ($p < 0.01$).

2.7.2.2.1 Verification of VIF and Tolerance To check VIF and tolerance, it is necessary to perform a multicollinearity test. The FIV and tolerance check is used to verify that there is no multicollinearity, i.e., no unhelpful explanatory variables in the model. The condition must be VIF less than 10 and tolerance above 1. In our case (Table 2.2) the multicollinearity problem does not appear, because the tolerances are between 0.5 and 0.9 and the VIF is below 10.

2.7.2.2.2 Theoretical Model Test The validation of the assumptions of the theoretical model leads us to carry out the test of significance of the model. The analysis of the significance of the model will be carried out in two stages: the analysis of the overall significance and the analysis of the individual significance.

2.7.2.2.2.1 Global Significance Test The first question we are trying to solve is the overall significance of the model, that is, whether the explanatory variables have an effect on the dependent variable. This test can be formulated as follows: at least one explanatory variable is significant.

The assessment of the overall significance of the model is done through Fischer's statistics, which indicates that the explanatory variables influence the variable to be explained.

Arbitrage is done by comparing the value of the estimated F-statistic to that of the tabulated Fischer distribution (or table). The SPSS 25 software automatically provides us with the probability associated with the calculated F-statistic; it remains the simplest to analyze. We will therefore compare the probability associated with the F-statistic to the 5% threshold chosen. If the probability associated with the calculated F-statistic is less than 5%, then

TABLE 2.2 The Collinearity Coefficients of the Search Model Variables

Model		Collinearity Statistics	
		Tolerance	**VIF**
1	(Constant)		
	Strengthening the image of Morocco abroad	**0.585**	**1.702**
	Increasing the satisfaction of tourists and loyalty to the destination	**0.803**	**1.207**
	Maximizing the long-term economic benefits of tourism	**0.547**	**1.763**
	The significant improvement in tourism competitiveness	**0.648**	**1.804**
	Indirect societal benefits	**0.915**	**1.180**
	The communication strategies adopted	**0.830**	**1.276**

TABLE 2.3 ANOVA Regression Index

ANOVA[a]						
Model		**Sum of Squares**	**ddl**	**Mean Square**	**F**	**Sig.**
1	Regression	18,907	6	3.1511	71,616	0.000[b]
	Residuals	15,837	354	0.044		
	Total	34,744	360			

[a] Preachers: (Constant), strengthening the image of Morocco abroad, increasing tourist satisfaction and loyalty to the destination, maximizing long-term economic benefits from tourism, the significant improvement in tourism competitiveness, indirect societal benefits, communication strategies adopted

[b] Dependent variable: tourist attractiveness

we reject the hypothesis H0 in favor of the alternative hypothesis that the regression is globally significant.

In our case, for the explanatory variables of tourist attractiveness, the Fischer statistic calculated by the SPSS 25 software is F = 71.616 and the associated probability is less than 5% (0.00 < 0.05), therefore the null hypothesis is rejected and the model is globally significant (Table 2.3).

2.7.2.2.2.2 Individual Significance Test In order to assess the significance of the variables, we will use the Student statistics provided directly by the SPSS 25 software. The hypothesis of significance of the variable is taken when the estimated student statistic, at the threshold considered, is higher than that tabulated. As part of this research, the probability of rejection provided by SPSS 25 software at the selected threshold will be used. The results of the estimate are shown in Table 2.4 and allow us to judge the significance of the probability associated with four variables.

TABLE 2.4 The Coefficients of the Individual Significance of the Model Variables

Coefficients[a]								
		Unstandardized Coefficients		**Standardized Coefficients**			**Collinearity Statistics**	
Model		**B**	**Standard Error**	**Bêta**	**t**	**Sig.**	**Tolerance**	**VIF**
1	(Constant)	−0.135	0.481		−0.281	0.779		
	Strengthening the image of Morocco abroad	0.153	0.069	0.214	2.227	0.028	0.555	1.802
	Increasing tourist satisfaction and destination retention	0.005	0.127	0.003	0.038	0.970	0.903	1.107
	Maximizing the long-term economic benefits of tourism	0.324	0.128	0.246	2.527	0.013	0.537	1.863
	The significant improvement in tourism competitiveness	0.287	0.104	0.336	2.76	0.007	0.748	1.704
	Indirect societal benefits	0.345	0.132	0.206	2.087	0.013	0.876	1.980
	The communication strategies adopted	0.249	0.138	0.227	2.358	0.021	0.925	1.081

[a] Dependent variable: Tourist attractiveness

The multiple regression analysis steps take as a dependent variable the tourist attractiveness of Morocco and highlight several significant explanatory factors.

- First of all, the strengthening of Morocco's image in the strange has a positive and significant effect on tourist attractiveness ($\beta = 0.153$; $t = 2.227$; $p = 0.028$) and, among other things, actions to improve the country's image abroad are a strategic lever to attract more people.
- Similarly, maximizing the long-term economic benefits of tourism promotes tourist attractiveness ($\beta = 0.324$; $t = 2.527$; $p = 0.013$), thus confirming the importance of a sustainable and integrated strategy for tourism development.
- In addition, the assumption of a beneficial effect of improvements in tourism competitiveness emerges as the variable with the maximum standardized coefficient ($\beta = 0.287$; $t = 2.760$; $p = 0.007$), suggesting its central position in the context of an overall strengthening of Morocco's tourism positioning.
- Indirect societal benefits ($\beta = 0.345$; $t = 2.087$; $p = 0.013$) also appear significant, suggesting that the social effects of the World Cup may indirectly contribute to the destination image.
- It has been proven that the chosen communication strategies take place in the context of the degree of tourist attractiveness ($\beta = 0.249$; $t = 2.358$; $p = 0.021 < 0.05$), thus validating the hypothesis that the communication deployed by Morocco around its events and cultural assets and tourism projects really contributes to the image perceived by potential tourists.
- On the other hand, the increase in tourist satisfaction and loyalty ($p = 0.970$) does not appear significant in this model, reflecting a potential inconsistency or an immeasurable direct negative impact of current communication campaigns on their attractiveness.

2.7.3 Interpretation of Results

The regression analysis undertaken in this study reveals five key determinants of tourism attractiveness in Morocco. This research paper is predicated on a strategic approach to prove the effectiveness of preparation for prospective major international events such as the World Cup in 2030.

- Improve tourism competitiveness.

  The variable "improved tourism competitiveness" is considered the most influential ($\beta = 0.287$; $t = 2.760$; $p = 0.007$). This finding supports the H4 hypothesis and affirms the work of Ritchie and Crouch (2003) contending that a destination's competitiveness is influenced by the level and quality of its infrastructure and the tourism offer (innovation and adaptability).
- Maximization of long-term economic benefits.

  Long-term sustainable economic impact of tourism also influences destination attractiveness ($\beta = 0.324$; $t = 2.527$; $p = 0.013$). This result reinforces hypothesis H3 and highlights the importance of sustainable tourism development strategy, while adhering to the UNWTO recommendations.

- The communication strategy adopted by Morocco around its events, cultural heritage and tourism projects positively influences the attractiveness of the country (β = 0.249; t = 2.358; p = 0.021). This validation of the H6 hypothesis is in line with the analyses of Morgan and Pritchard (2000), who emphasize the importance of narrative communication to build a strong tourism image.
- Moreover, the strengthening of Morocco's image abroad acts as a strategic lever (β = 0.153; t = 2.227; p = 0.028), validating the H1 hypothesis. This result is in line with the conclusions of Anholt (2007) on the role of nation branding as a factor of differentiation and attractiveness.

  These findings also highlight a set of indirect societal effects, insofar as the improvement of national image contributes not only to external attractiveness but also to broader social dynamics, including collective confidence, perceived stability, and international recognition.
- Taking into account the societal benefits induced (social cohesion, improvement of public services, valorization of territories) also has a significant effect (β = 0.345; t = 2.087; p = 0.013), validating the H5 hypothesis. These indirect effects contribute to improving the perceived image of the country and its ability to attract tourists.

Conversely, the variable "tourist satisfaction and loyalty" shows no significant relationship with tourism attractiveness (β = 0.003; t = 0.038; p = 0.970). This result invalidates the H2 hypothesis and suggests that, in this context, loyalty alone is not enough to increase attractiveness, without a continuous modernization of the offer and a more ambitious territorial marketing strategy (Table 2.5 for paragraph 2.7.3).

2.7.4 Implications of Research

The results of the study confirm the positive impact of the organization of the 2030 World Cup on Morocco's international visibility and on the increase in tourist flows in the country, both in the short- and long term. Nevertheless, the prospect of the "after-world" from a tourist point of view remains uncertain. Although the event has potential to increase post-World Cup tourism attractiveness, there are major concerns about sustainability. Pollution,

TABLE 2.5 Summary of Results Obtained

Explanatory Variables	Variable to Explain	Coefficient β	Significance	t Student	Validation
Strengthening the image of Morocco abroad	Tourist attraction	0.153	0.028	2.227	Confirmation
Increasing tourist satisfaction and destination retention	Tourist attraction	0.005	0.970	0.038	Reversal
Maximizing the long-term economic benefits of tourism	Tourist attraction	0.324	0.013	2.527	Confirmation
The significant improvement in tourism competitiveness	Tourist attraction	0.287	0.007	2.76	Confirmation
Indirect societal benefits	Tourist attraction	0.345	0.013	2.087	Confirmation
The communication strategies adopted	Tourist attraction	0.249	0.021	2.358	Confirmation

increased waste production and excessive energy consumption are major challenges for the development of sustainable tourism. It is imperative to introduce appropriate approaches to address the risks of direct environmental impact and organize the event in accordance with environmental standards. The balance between tourism development and environmental protection will be essential to make the event profitable and increase its future potential.

2.7.4.1 Limitations of the Study

Quantitative approach of this study can be complemented by qualitative analysis. Additionally, generalizability of results based on a sample of 360 actors presents future research opportunities before and after the World Cup. Incorporating time could also increase the relevance of the findings.

2.7.4.2 Practical Implications

Our results would be utilized to provide relevant guidelines to ensure maximum positive impacts, such as the following:

Tourism facility development with respect to environmental conditions, to promote Morocco as a sustainable product via targeted campaigns and to involve local communities in spreading the economic impacts in a balanced fashion.

2.7.4.3 Theoretical Implications

- First, the research contributes to the current literature regarding the impact of major sporting events, especially within the new framework.
- Second, the research pinpoints the significance of strategic planning in linking the attractiveness of tourism to sustainability.
- Third, field data collection allows the analysis to be readily applied to other nations that host comparable events.

2.8 CONCLUSION

This global event is not just a sporting event; it also makes Europe and Morocco a historic opportunity for the development and transformation of tourism with the hope of revitalizing and renewing expectations of mobility and regenerating memories. Recently, the University of Paris Nanterre highlighted the catalytic role that a major sporting event can have in the tourism dynamic of an emerging country.

The expected effects exceed the only endowment flow over a short period which would call into question the specific profile of the country: it is a strategic lever that should be promoted to register Morocco, especially through its modernity and its welcome, on the global tourism picture. Indeed, the upgrading of infrastructure (transport, accommodation and sports facilities), the encouragement of private investment and the international media impact make it possible to place Morocco among the great global choices of tourist destinations.

However, this effect does not occur according to a principle of necessity and can only be conditional, because it relies on the capacity of stakeholders – public authorities, operators of tourism offer, local communities and civil society – to better anticipate, prepare and exploit the impact of the event. The sustainability of potential economic, social and cultural achievements also depends on the project's ability to take place in an integrated vision of territorial development and tourism, on the one hand, according to local characteristics, environmental constraints and public expectations after the event.

In the end, the 2030 World Cup can be a catalyst for change for Morocco if planning, promotion and governance efforts are carried out in a coherent, inclusive and future-oriented manner. This challenge of hope is likely to mobilize a major national effort to make sport a structuring factor in tourism and territorial development.

REFERENCES

Aït-Nacer, M. (2021). Le tourisme marocain au crible de la gouvernance: La région de Souss Mass. Editions L'Harmattan.

Akpınar, B., Barut, M., Akpınar, E.N., & Balıkçı, H.C. (2025). Lifelong learning supported by artificial intelligence and technology for Sustainable Development Goals: An OECD Perspective. Sustainable Development, 33(5), 7826–7843.

Amnesty International (2022). Reality Check: A Year to the 2022 World Cup in Qatar.

Anholt, S. (2007). Competitive Identity: The New Brand Management for Nations, Cities and Regions. Palgrave Macmillan.

Baade, R.A., & Matheson, V.A. (2016). Going for the gold: The economics of the Olympics. Journal of Economic Perspectives, 30(2), 201–218.

Batty, M. (2022). The COVID years: Predictable unpredictability. Environment and planning b: urban analytics and city science, 49(1), 3–6.

Bramwell, B., & Lane, B. (2011). Critical research on the governance of tourism and sustainability. Journal of Sustainable Tourism, 19(4–5), 411–421.

Brelik, A., & Łącka, I. (2024). Determinants of tourism development in the aspect of the challenges of the 21st century. Roczniki (Annals), 2024(2).

Broussolle, D. (2021). La durabilité des grands événements sportifs: entre rhétorique et réalité. In J.-F. Polo (Dir.), Tourisme, sport et durabilité: Enjeux contemporains (pp. 115–133). Presses Universitaires de Grenoble.

Chalip, L. (2004). Beyond Impact: A General Model for Sport Event Leverage. In B. Ritchie & D. Adair (Eds.), Sport Tourism: Interrelationships, Impacts and Issues (pp. 226–252). Channel View Publications.

Chappelet, J., & Junod, T. (2006). A tale of 3 Olympic cities: What can Turin Learn from the Olympic legacy of other Alpine cities. Major Sport Events as Opportunity for Development. Proceedings of the Valencia Summit Proceedings, Valencia, Spain, 6–10.

Charaï, Z. (2022, November). Patrimonialisation et tourisme: Vers une approche nouvelle et alternative de la (re)-mise en tourisme post-Covid-19 des territoires authentiques?. In 11ème Colloque de l'Association Tourisme Recherche et Enseignement Supérieur (AsTRES), "L'agilité touristique en période de crises: réplications, accélérations, réinventions...?"

Chen, S., Zhao, D., Huang, J., He, J., Chen, Y., Chen, J., ... & Yang, F. (2023). Mongolia contributed more than 42% of the dust concentrations in Northern China in March and April 2023.

Cornelissen, S. (2010). The geopolitics of global aspiration: Sport mega-events and emerging powers. The International Journal of the History of Sport, 27(16–18), 3008–3025.

Du Plessis, S., & Maennig, W. (2011). The economic impact of the 2010 FIFA World Cup. Development Southern Africa, 28(3), 349–364.

Evrard, A., Hexel, D., Kaiser, C., Lurin, J., Marina Decarro, N., Rastoldo, F., & Soussi, A. (2003). De l'orientation au projet de formation. Genève: Service de la recherche en éducation.

FIFA. (2023). FIFA World Cup Qatar 2022™ Post-Tournament Summary Report.

Fourie, J., & Santana-Gallego, M. (2011). The impact of mega-sport events on tourist arrivals. Tourism Management, 32(6), 1364–1370.

Gaffney, C. (2010). Mega-events and socio-spatial dynamics in Rio de Janeiro, 1919–2016. Journal of Latin American Geography, 9(1), 7–29.

Getz, W. M., & Saltz, D. (2008). A framework for generating and analyzing movement paths on ecological landscapes. Proceedings of the National Academy of Sciences, 105(49), 19066–19071.

Ghazal, T. M., Hasan, M. K., Ahmad, M., Alzoubi, H. M., & Alshurideh, M. (2023). Machine learning approaches for sustainable cities using internet of things. In The Effect of Information Technology on Business and Marketing Intelligence Systems (pp. 1969–1986). Cham: Springer International Publishing.

Gibson, H.J. (1998). Sport tourism: A critical analysis of research. Sport Management Review, 1(1), 45–76.

Gouguet, J.-J., Nys, J.-F., & Primault, D. (2009). Les retombées économiques des événements sportifs: le cas des demi-finales de la Coupe de France de football. Revue Juridique et Économique du Sport, 91(1), 61–80.

Gretzel, U., & Koo, C. (2021). Smart tourism cities: A duality of place where technology supports the convergence of touristic and residential experiences. Asia Pacific Journal of Tourism Research, 26(4), 352–364.

Gursoy, D., & Chi, C.G. (2020). Effects of COVID-19 pandemic on hospitality industry: Review of the current situations and a research agenda. Journal of Hospitality Marketing & Management, 29(5), 527–529.

Hasan, M., et al. (2023). Smart Mobility for Sustainable Urban Events. Transportation Research Procedia.

Imafidon, H., Enwerem, M., & Boye, A. (2024). Adapting green building practices and smart technology in developing countries: A review. African Journal of Environmental Sciences and Renewable Energy, 16(1), 183–202.

Ismaili, M.Y E.A., & Ayegou, J. (2023). Le chemin Vers l'optimum écotouristique: Cas du territoire touristique Atlas et Vallées. International Journal of Accounting Finance Auditing Management and Economics, 4(5-2), 987–1010.

Kaplanidou, K., & Vogt, C. (2007). The interrelationship between sport event and destination image and sport tourists' behaviours. Journal of sport & tourism, 12(3-4), 183-206. https://doi.org/10.1080/14775080701736932

King, S., & Prasetyo, J. (2023, December). Assessing generative AI through the lens of the 2023 Gartner Hype Cycle for Emerging Technologies: a collaborative autoethnography. Frontiers in Education, 8.

Kozhanov, N., Amara, M., & Zweiri, M. (Eds.). (2024). The 2022 FIFA world cup in Qatar: Global and local perspectives. Taylor & Francis.

Kwong, A. T. M., Omar, S. I., & Aliah, M. N. (2024). The Ethical Implications of AI-Powered Personalization in Hospitality and Tourism. In Impact of AI and Tech-Driven Solutions in Hospitality and Tourism (pp. 103–122). IGI Global.

Lammi, D. (2024). Focal issues of Sustainable Development presented by the World Economic Forum 2024.

Magina, T., Carrilho-Nunes, I., & Catalão-Lopes, M. (2023). The Role Of Digitalization In Reducing Energy Poverty. Current and Future Challenges to Energy Security, 152.

Matheson, V.A. (2006). Mega-events: The effect of the world's biggest sporting events on local, regional, and national economies. College of the Holy Cross, Department of Economics Faculty Research Series, Paper No. 06-10. Matheson, V. (2006). Mega-Events: The effect of the world's biggest sporting events on local, regional, and national economies.

Meyer, D., Lorgnier, N., & Suchet, A. (2018). Évaluation des impacts économiques d'un événement sportif: méthodologie et analyse comparée des outils mobilisés. STAPS, 122, 55–70.

Morgan, N., & Pritchard, A. (2000). Advertising in Tourism and Leisure. Butterworth-Heinemann.

Mowforth, M., & Munt, I. (2015). Tourism and sustainability: Development, globalisation and new tourism in the third world. routledge.

Müller, M. (2020). Good governance, scale and power: A conceptual framework for the analysis of mega-event legacies. Leisure Studies, 39(3), 336–348.

Nadeau, J., Heslop, L.A., O'Reilly, N., & Luk, P. (2008). Destination in a country image context. Annals of Tourism Research, 35(1), 84–106.

Preuss, H. (2007). The conceptualisation and measurement of mega sport event legacies. Journal of Sport & Tourism, 12(3-4), 207–228.

Preuss, H. (2015). A Theory Of Mega Sport Event Legacies 1. In Routledge handbook of theory in sport management (pp. 69–81). Routledge.

Reiche, D. (2020). Climate policies of FIFA World Cup hosts: Comparing Germany 2006, South Africa 2010, Brazil 2014, Russia 2018, and Qatar 2022. International Journal of Sport Policy and Politics, 12(1), 115–133.

Richards, G., & Palmer, R. (2010). Eventful Cities: Cultural Management and Urban Revitalisation. Routledge.

Ritchie, J.R.B., & Crouch, G.I. (2003). The Competitive Destination: A Sustainable Tourism Perspective. CABI Publishing.

SMIT (2022). Rapport sur l'état du tourisme durable au Maroc. Société Marocaine d'Ingénierie Touristique.

Swart, K., Bob, U., & Turco, D.M. (2010). The 2010 FIFA World Cup™ and tourism: Patterns and implications. African Journal for Physical Health Education, Recreation and Dance, 16(4), 1–13.

Swart, K., & Bob, U. (2012). Mega sport event legacies and the 2010 FIFA World Cup. African journal for physical health education, recreation and dance, 18(sup-2), 1–11.

Vision 2020. (2010). Stratégie nationale de développement touristique 2010-2020. Ministère du Tourisme.

Weaver, D. (2011). Can sustainable tourism survive climate change? Journal of Sustainable Tourism, 19(1), 5–15. https://doi.org/10.1080/09669582.2010.536242

Xiang, Z., Stienmetz, J., & Fesenmaier, D. R. (2021). Smart Tourism Design: Launching the annals of tourism research curated collection on designing tourism places. Annals of Tourism Research, 86, 103154.

Xia, L., Cao, Z., & Bilawal Khaskheli, M. (2025). How digital technology and business innovation enhance economic–environmental sustainability in legal organizations. Sustainability, 17(14), 6532.

Zhou, L., Yang, H., & Pan, Y. (2023). AI-Driven Strategies for Green Mega-Events. Technological Forecasting and Social Change.

Zimbalist, A. (2015). Circus Maximus: The Economic Gamble Behind Hosting the Olympics and the World Cup. Brookings Institution Press.

CHAPTER 3

Navigating the Pandemic with AI

Toward Sustainable Deep Learning and Machine Learning Approaches for X-Ray and CT-Based Covid-19 Detection

Youness Chawki

3.1 INTRODUCTION

On March 11, 2020, the World Health Organization (WHO) declared Covid-19 to be a pandemic (WHO, 2020). Known as one of the most dangerous health disasters of the modern era that has affected all the countries of the world (Aanouz et al., 2020; Elmezayen et al., 2020; Fausto et al., 2020; Rothan & Byrareddy, 2020; Umesh et al., 2020). The new coronavirus SARS-CoV-2 (severe acute respiratory syndrome coronavirus 2), which was first noticed in China, causes the respiratory disease Covid-19 (El Asnaoui & Chawki, 2020; Islam & Nahiduzzaman, 2022; Muralidharan et al., 2022; WHO, 2020). Its major signs and symptoms include fever, dry cough, sore throat, dyspnea, tiredness, diarrhea, and in some cases loss of smell and taste. Not only these symptoms but in some cases, Covid-19 caused some respiratory problems which resulted in hospitalization and critical care as well as death (Landry et al., 2020; Sharifi-Razavi et al., 2020; Simcock et al., 2020). There have been approximately 704,753,890 confirmed cases, with 7,010,681 deaths and 675,619,811 recovered (Worldometer, 2025) (Last updated: April 23, 2025, 15:36 GMT).

To minimize the spreading of Covid-19, WHO has recommended many precautions like wearing mask, keeping social distance, using hydroalcoholic solution, washing hands with soap, and covering mouth while sneezing (Elkbuli et al., 2020; Li et al., 2020; WHO, 2020). Not only this but also it stopped traveling between cities and countries and a lockdown was imposed (Ghosh et al., 2020; Ng et al., 2020). After deep research in laboratories, with the collaboration of governments and non-governmental entities, researchers developed Covid-19 vaccines. Indeed, different vaccines are developed like Sputnik V (Gamaleya, Russia), CoronaVac (Sinovac, China), BIBP-CorV (Sinopharm, China), BNT162 (Pfizer BioNTech, USA), mRNA-1273 (Moderna, USA), Ad26.COV2-S (Johnson & Johnson, USA), ChAdOx1

DOI: 10.1201/9781003649885-3

(AstraZeneca, UK), and BNT162 (Pfizer BioNTech, USA) (Beshman et al., 2021; Kavya et al., 2022; Mallapaty et al., 2021; Mehboob et al., 2020).

RT-PCR (reverse transcription-polymerase chain reaction) is a standard to detect Covid-19 with high accuracy. But this technique has some disadvantages like time to confirm patients, need of specialized equipment and well-trained personnel, and its expensiveness. To resolve this issue, medical analysis is widely used in this industry and allows specialists and doctors to eliminate the need for laborious manual review processes and to make accurate diagnoses (El-Bouzaidi & Abdoun, 2023). Certainly, CT (computed tomography) and X-ray images are the most used in medical images processing. The use of these images to detect Covid-19 requires very good expertise and needs a lot of time (Islam & Nahiduzzaman, 2022). Face to this situation, it is necessary to find new techniques that allow to detect the virus effectively with minimal time. Therefore, analyzing the X-ray and CT images with Deep Learning and Machine Learning methods is one of the solutions that can be used (Ouchicha et al., 2020). Indeed, recent advancements in Machine Learning and Deep Learning have been essential in automating medical images diagnosis by offering new approaches overcoming the shortcomings of the classical methods. Therefore, the development of an automated analysis system based on CT and X-ray images using Machine Learning and Deep Learning techniques to identify anomalies has greatly aided in the time and correct decision-making of medical experts (Alsattar et al., 2023).

The aim of our study is to conduct a bibliometric analysis to measure and analyze the scientific literature on the detection and classification of Covid-19 using X-ray and CT images using Deep Learning and Machine Learning methods. Data and qualitative and quantitative analysis of published documents is used to highlight the importance of Machine Learning and Deep Learning techniques on detection of Covid-19 using medical images. The study also aims to get insights and trends from data to answer these research questions (RQs):

RQ1: What is the relative prevalence and impact of different artificial intelligence (AI) techniques in the context of Covid-19 detection, and how have these technologies evolved over the period of study?

RQ2: What insights emerge from the countries and universities distribution of research contributions, and how do their contributions affect the field of AI-driven Covid-19 detection research?

RQ3: What can we deduce about the type of published documents in the field of AI and Covid-19 detection?

Answering these questions will serve multiple purposes for researchers. First, it will increase their understanding of how Machine Learning and Deep Learning techniques are applied in the detection and classification of Covid-19 using X-ray and CT images. Additionally, it will give comprehensive insights into which journals and databases should be focused on for their research. They will also get guidance on the type of document to prepare and the most effective techniques to employ for submitting future publications.

The chapter is structured as follows: Section 3.2 describes the methodology used in the scientometric analysis. Section 3.3 explores different aspects of the scientometric analysis, including publications type, techniques used, annual trends, geographic dispersion, predominant journals and quartiles distribution, leading institutions, and authors. The chapter concludes with Section 3.4, where we summarize the findings and address the RQs.

3.2 METHODOLOGY OF BIBLIOMETRIC ANALYSIS

3.2.1 Collection of Data

Choosing the appropriate databases that cover the scientific literature of Deep Learning and Machine Learning techniques for X-ray and CT-based Covid-19 detection and classification is the most important step for the bibliometric analysis (Chawki et al., 2024; Donthu et al., 2021). Therefore, in order to perform this kind of analysis and come to the most effective decisions, it is essential that the information sources be reliable and appropriate (Rueda et al., 2007). For our investigation, six reputable databases have been used: ScienceDirect, Springer, IEEE Xplore, PubMed, MDPI, and ACM.

3.2.2 Search String

This study aims to explore the field of Deep Learning and Machine Learning techniques for classification and detection based on X-ray and CT images. We covered, in this bibliometric analysis, the most prevalent areas of knowledge, publications type, techniques used, annual trends, geographic dispersion, predominant journals and quartiles distribution, leading institutions, and authors. To complete our investigation, we have used for all the selected databases the same advanced search string:

> **(“Image Detection” OR “Detection”) AND (“Image Classification” OR “Classification”) AND (“Covid19” OR “Covid 19” OR “Covid-19” OR “Covid” OR “Corona Virus” OR “Coronavirus”) AND (“Machine Learning” OR “ML” OR “Machine Learning Algorithms”) AND (“Deep Learning” OR “DL” OR “Convolutional Neural Network” OR “CNN” OR “Deep Neural Networks”) AND (“X-Ray Image” OR “X-Ray” OR “Computed Tomography” OR “CT”)**

The research returned the following numbers of documents from each database: 167 from ScienceDirect, 245 from Springer, 139 from IEEE Xplore, 41 from PubMed, 16 from MDPI, and 80 from ACM.

3.2.3 Cleaning Process

Prior to starting the data analysis for getting insights, ensuring the relevance and uniqueness of the collected documents is very important. A cleaning process was undertaken to refine the dataset. This involved selecting documents that exactly address the classification and detection of Covid-19 based on X-ray and CT images, using Deep Learning and Machine Learning techniques, while eliminating any duplicates. After this step, the total numbers of documents are 140 for ScienceDirect, 232 for Springer, 129 for IEEE Xplore, 36 for PubMed, 13 for MDPI, and 58 for ACM.

3.2.4 Bibliometric and Visualization Analysis

Before we get on the bibliometric and visualization analysis, a crucial preliminary step is required which is the extraction of relevant information from documents. For this, we have compiled an Excel file where we extracted the key details from each document. These details include the title of the document, publication year, first author, country of the first author, continent, institution associated to the first author, type of publication (conference paper, journal article, or book chapter), publisher, quartile ranking (applicable only for journals), h-index (applicable only for journals), database where the document is published, number of total accumulated citations (last update of total citation was March 11, 2024), and the technique used in the study (Machine Learning, Deep Learning, or a hybrid approach combining both).

To conduct our bibliometric and data visualization analysis, we have employed VOSviewer software along with the Python packages pandas, NumPy, and Matplotlib. The relevant extracted information from each document will allow us to perform a comprehensive data analysis, uncovering valuable insights and findings regarding the detection of Covid-19 through X-ray and CT images, using Deep Learning and Machine Learning techniques.

3.3 RESULTS OF BIBLIOMETRIC ANALYSIS AND DISCUSSION

3.3.1 Annual Publication and Citation Trends

Figure 3.1 shows the trend analysis of all publications and all citations from 2020 to 2024. It demonstrates an initial spike in publications on Covid-19 detection using X-ray and CT images, peaking in 2022 (197 documents). This is followed by a fall in 2023 and 2024, which may indicate a shift of research interest due the pandemic's concerns becoming less pressing. Reverse publications, the citations show a declining trend over time which can be

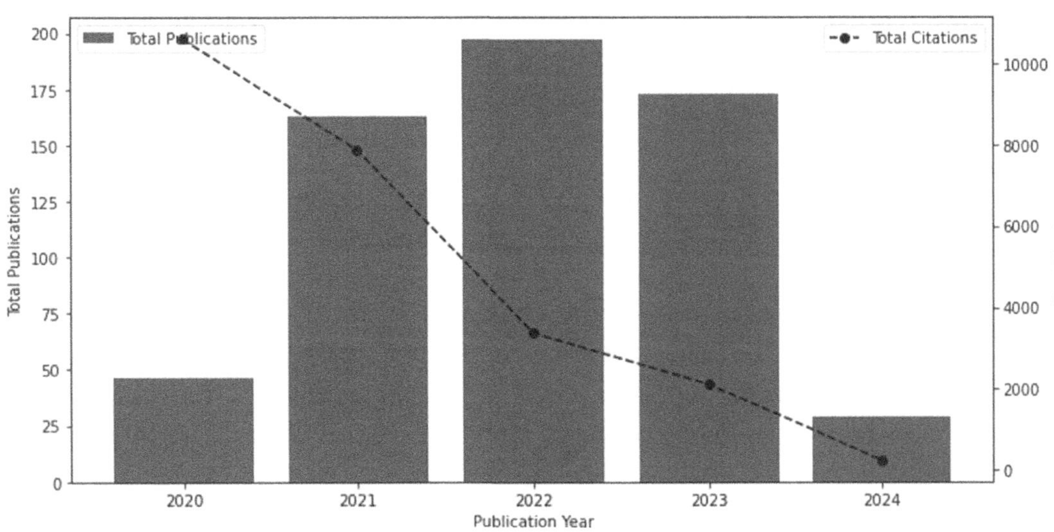

FIGURE 3.1 Total publications and average publications citation per year.

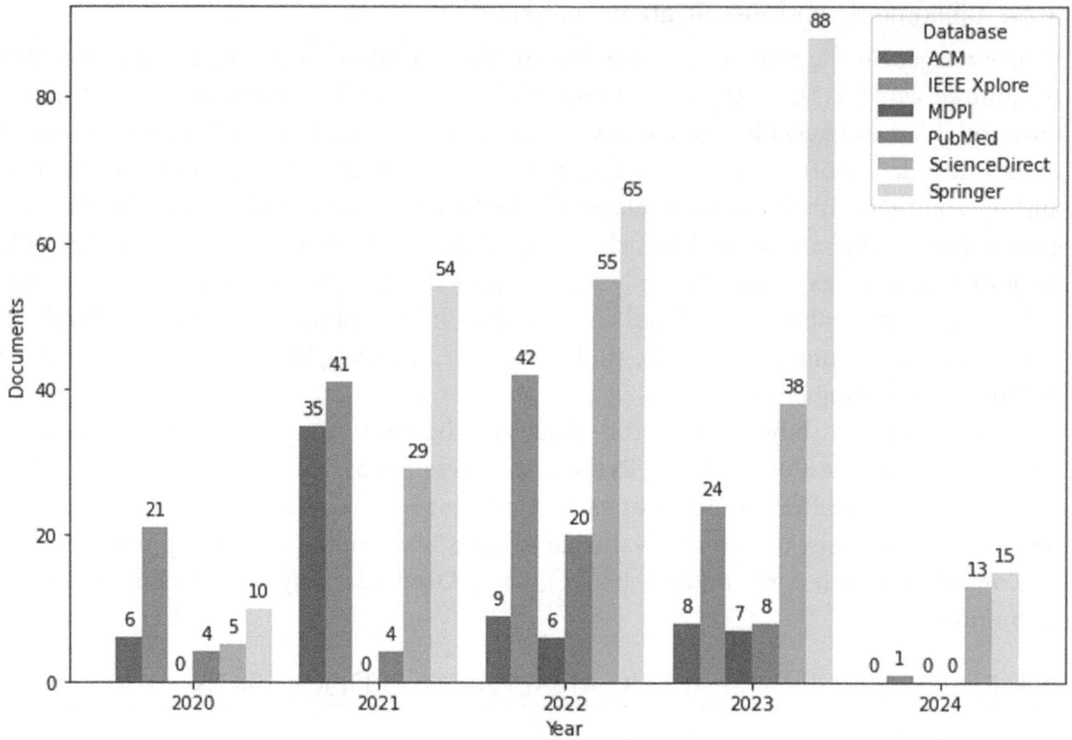

FIGURE 3.2 Total documents for each database over the years.

explained by the topic being less original and urgent and new publications needing time to accumulate citations. The year 2020 accumulated high number of citations with a total of 10,613 citations.

To understand more about the annual trend of publications, Figure 3.2 illustrates the breakdown of the annual document counts from each database between 2020 and 2024. At first, in 2020, the overall number of documents is modest, with IEEE Xplore being the most significant contributor. In 2021 and 2022, there is a noticeable increase in publications, especially from Springer. However, there is a clear decrease in 2023 across all databases, although ScienceDirect continues increasing with a maximum of publication (88 documents). By 2024, a drastic decrease is noticed for the databases with no contribution for ACM, MDPI, and PubMed. This explains the possible shift in research focus.

3.3.2 Leading Countries of Research Contribution

After extracting the country of the first author of each document, it was found that only 57 countries have contributed to the field of Covid-19 detection using Machine Learning and Deep Learning techniques. As depicted in Figure 3.3a, India leads the contribution with an extraordinary count of 206 documents highlighting its prominent position in the field of scientific research. With 62 publications, China came in second rank showcasing its academic output. Turkey also made a notable contribution with 31 documents. Noteworthy significant participation includes Pakistan with 24 documents. Bangladesh and the USA

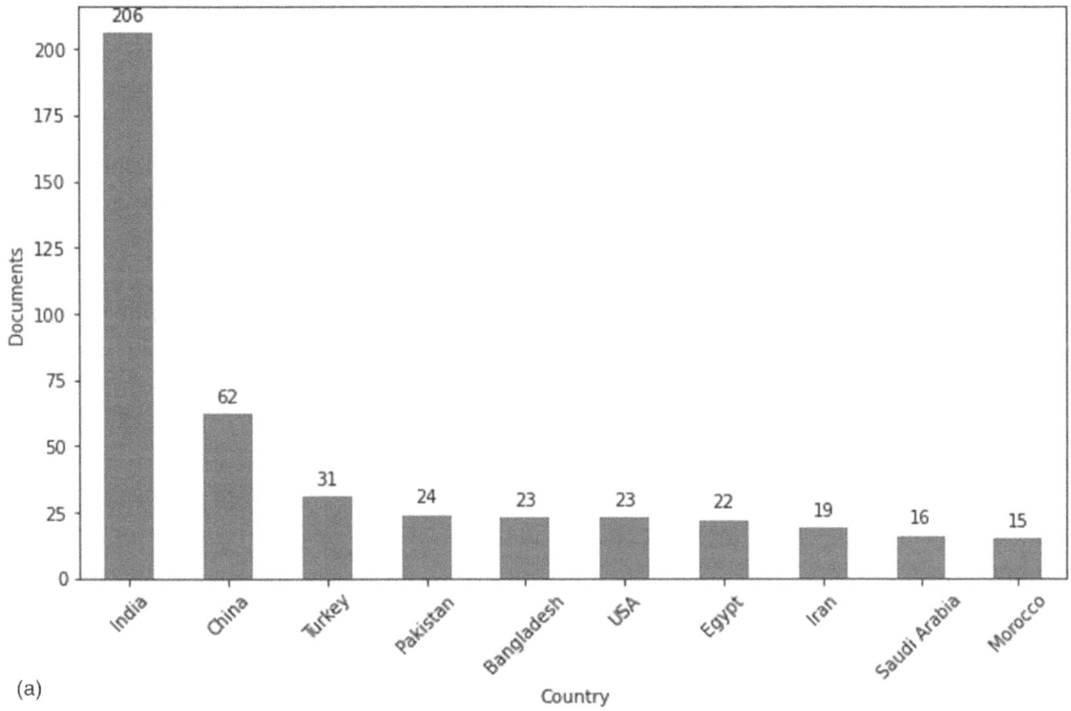

(a)

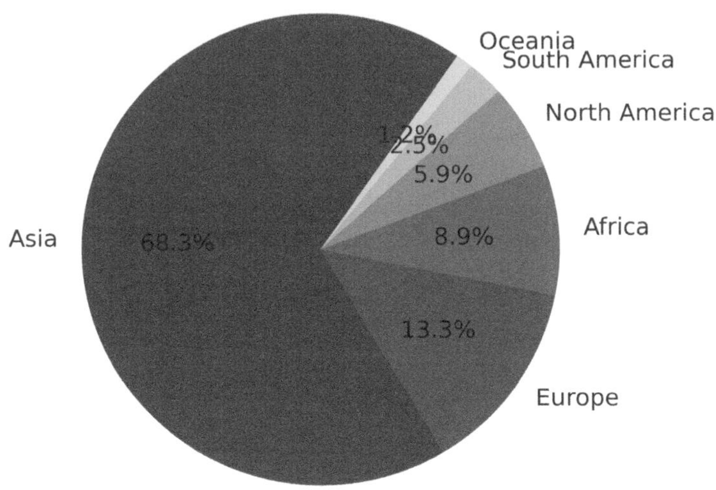

FIGURE 3.3 Global distribution of scientific research on Covid-19 detection, (a) by country and (b) by continent.

have participated each with 23 documents. With 22 documents, Egypt emerged as a key player positioning itself among the top ten contributors. The Middle East was represented by Iran and Saudi Arabia with 19 and 16 documents, respectively. With a notable contribution of 14 documents, Morocco secured its place within the top ten. An observation that catches the eyes is from ten top contributors, the USA is the only country from among the traditionally recognized great powers. This highlights the emergence of rising powers like India, China, Turkey, and Pakistan establishing themselves as significant players in the field of scientific research. Figure 3.3b shows the contribution by continents in this field. Indeed, Asia tops with 415 documents accounting for 68.3% of the total, thanks to contributions from India and China. Europe came in second place with a total document of 81 (13.3%). Africa, demonstrating its efforts to keep place with the major contributors, occupying third place with a total of 54 documents. North America, with 36 documents, is in fourth place while South America and Oceania participated with 15 and 7 documents, respectively.

Figure 3.4 depicts the breakdown of research contributions by continent. Indeed, North African countries have proven their dominance by securing the top three positions. Egypt leads with 22 documents followed by Morocco with 15 documents, and Algeria with 8 papers. In Asia, Bangladesh ranks the third with 23 documents, behind India and China. These two nations have not only led Asia but topped all the countries in the world. Participating with 31 documents, Turkey tops the countries in Europe, followed by UK with 12 documents while Italy has contributed with 8 papers. Australia is the sole contributor from Oceania, with 7 documents. For North America, the USA is the principal contributor with 23 documents, while Canada and Cuba have added 12 and 1 documents, respectively. With 9 documents, Brazil dominated the contribution in South America followed by Colombia and Ecuador in the same rank with 2 documents each.

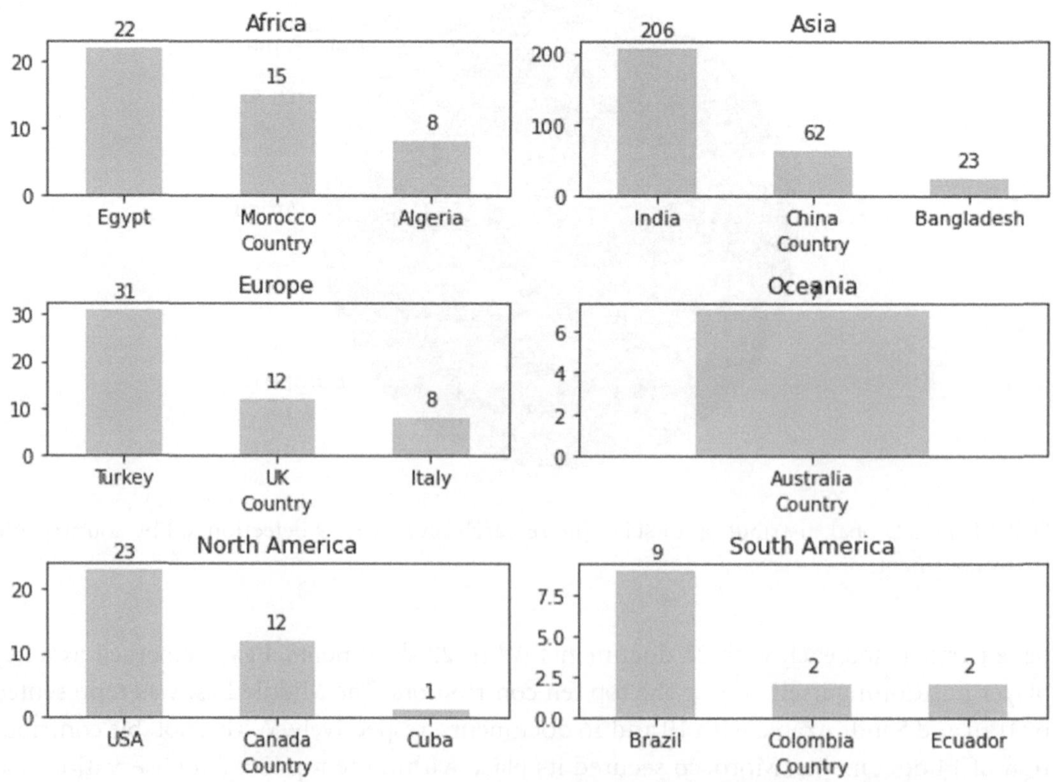

FIGURE 3.4 Top three countries leading research contributions across continents.

3.3.3 Technique Employed for Covid-19 Detection

One of the most relevant types of information extracted from is the technique used by authors to detect and classify Covid-19-based X-ray images and CT scans. In our study, the employed techniques can be categorized into Deep Learning, Machine Learning, and a hybrid technique of both. Of the 608 documents included in this study, 65.3% (397 documents) utilized Deep Learning to detect and classify Covid-19, followed by hybrid techniques of Deep Learning and Machine Learning with 138 documents accounting for 22.7% of the total. Only 73 documents (12.0%) used Machine Learning as a technique for Covid-19 detection. These results clearly highlight the predominance of Deep Learning methods in this field and provide guidance to researchers on which techniques to focus on in their future work. Figure 3.5 illustrates the distribution of the techniques employed in our study.

3.3.4 Types of Published Documents

In our study, we categorized publications into three types: conference paper, journal article, and book chapter. Understanding the type of publication is very crucial to recognizing the nature of contributions that have been made. Typically, initial findings of any work are presented in conference papers, more detailed contributions are published in journal articles. Deep comprehensive insight of a particular field is published as book chapters. As shown in Figure 3.6, 353 out of 608 documents (58.1%) included in this study were published as journal articles. Conference papers, which are a noteworthy proportion, accounted for 219 documents (36.0%). In contrast, only 36 documents (5.9%) were published as book chapters. This distribution indicates that contributors in the field of Covid-19 detection prefer to publish their detailed findings over initial research and in-depth analysis and discussion.

3.3.5 Top Ten Institutions Ranked by Research Contributions

Table 3.1 summarizes the top ten academic institutions that contributed to research of Covid-19 detection and categorizes by their country and continent as well as the total of published documents. Vellore Institute of Technology in India takes the lead, with eight documents, highlighting the important role of the country in this field of study. Other notable Indian institutions include National Institute of Technology, Lovely Professional University, and Indian Institute of Technology. All of them have made significant contributions (five

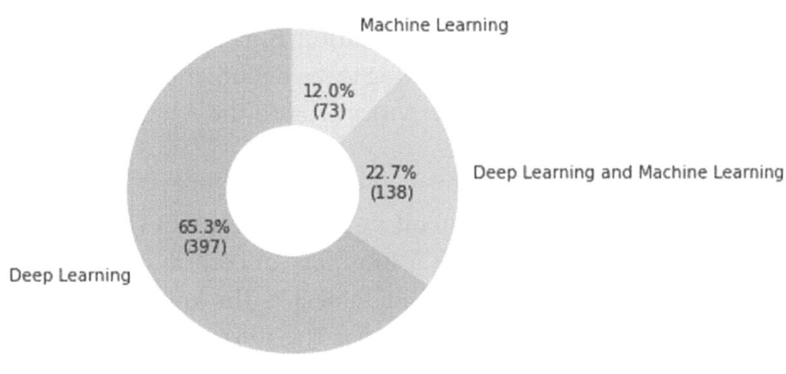

FIGURE 3.5 Distribution of employed technique for Covid-19 detection.

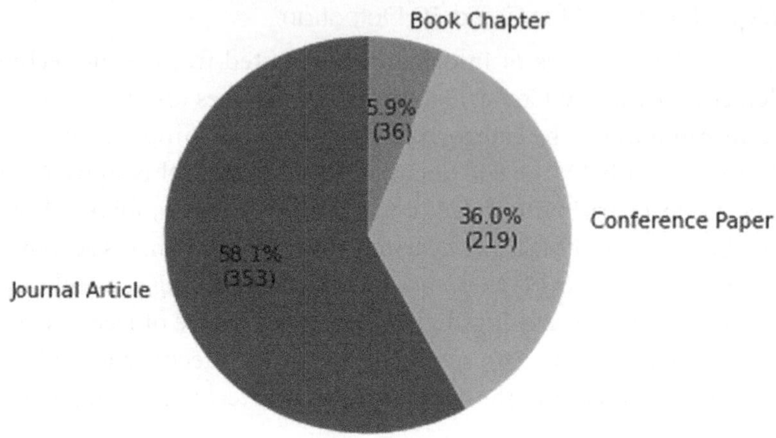

FIGURE 3.6 Types of documents included in the study.

TABLE 3.1 Top Ten Academic Institutions Ranked by Research Contributions in the Study

Academic Institution	Country	Continent	Total Contribution
Vellore Institute of Technology	India	Asia	8
University of Qatar	Qatar	Asia	6
Indian Institute of Technology	India	Asia	5
Firat University	Turkey	Europe	5
Lovely Professional University	India	Asia	5
Abdelmalek Essaâdi University	Morocco	Africa	5
National Institute of Technology	India	Asia	5
Delhi Technological University	India	Asia	4
University of Amity	India	Asia	4
Bina Nusantara University (BINUS University)	Indonesia	Asia	3

papers). Two more Indian institutions that have contributed in the field with four documents are Delhi Technological University and University of Amity. With six contributions, the University of Qatar stands in for the academic community in the Middle East. Furthermore, with five documents each, the Fırat University in Turkey and Abdelmalek Essaâdi University in Morocco are noteworthy contributors from Europe and Africa, respectively. Additionally, the Indonesia institution Bina Nusantara University (BINUS University) makes an appearance, contributing three documents and highlighting the varied geographic distribution of research activities around the Asian region.

3.3.6 Top Ten Publishers Ranked by Research Contributions

Since the journal article is the predominant publication type in this study, we will focus on this type to get more insights. Indeed, a summary of contribution of different publishers within academic databases is presented in Table 3.2. The journal *Multimedia Tools and Applications*, which is published by Springer, has the most documents (30), reflecting its prominence in the subject. All three of them accessible via ScienceDirect, *Computers in Biology and Medicine*, *Expert Systems with Applications*, and *Biomedical Signal Processing and Control* are followed by 20, 14, and 12 documents, respectively. The ScienceDirect's *Applied*

TABLE 3.2 Top Ten Publishers Ranked by Research Contributions in the Study

Publisher	Database	Type of Publication	Total Documents
Multimedia Tools and Applications	Springer	Journal article	30
Computers in Biology and Medicine	ScienceDirect	Journal article	20
Expert Systems with Applications	ScienceDirect	Journal article	14
Biomedical Signal Processing and Control	ScienceDirect	Journal article	12
Scientific Reports	Springer	Journal article	11
Applied Soft Computing	ScienceDirect	Journal article	11
Journal of Ambient Intelligence and Humanized Computing	Springer	Journal article	10
Applied Intelligence	Springer	Journal article	8
Informatics in Medicine Unlocked	ScienceDirect	Journal article	7
IEEE Access	IEEE Xplore	Journal article	7

Soft Computing and Springer's *Scientific Reports* contributed with 11 documents each. Two more journals indexed in Springer that assured their place among the top ten journals are *Journal of Ambient Intelligence and Humanized Computing* and *Applied Intelligence* with ten and eight documents, respectively. Accessible via ScienceDirect and IEEE Xplore, *Informatics in Medicine Unlocked* and *IEEE Access* participated with seven documents each. This clearly highlights the predominance of Springer and ScienceDirect as key repositories in this field.

3.3.7 Journals Quartiles Distribution

The quartile is a ranking system that splits journals into four categories based on their impact factor or other bibliometric indicators, such as citations. The quartiles range from Q1 to Q4. Journal quartile is used to determine the relative importance and influence of a journal within a field. Institutions and researchers frequently use quartiles to assess the prestige of a journal and to make decisions about where to submit their work, Q1 journals generally being the most sought after due to their perceived higher quality and impact. Figure 3.7 depicts the distribution of journal quartiles included in this study. With 64.9%

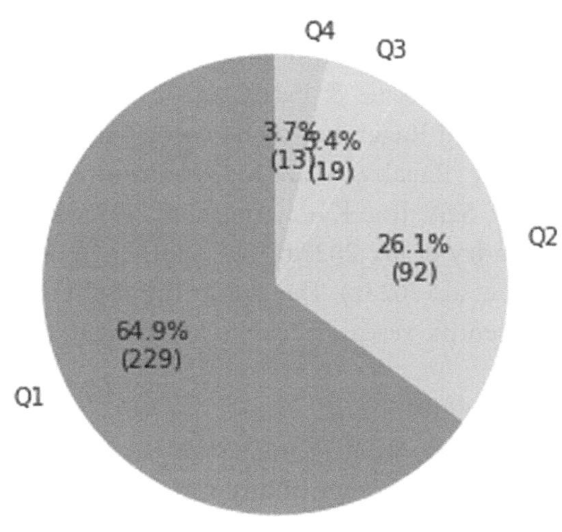

FIGURE 3.7 Distribution of journals quartiles in research publications.

TABLE 3.3 Top Five Authors Ranked by Research Contributions in the Study

Author	Country	# Docs	Year	Publisher	Database
K. Shankar	India	4	2021	*ACM Transactions on Internet Technology*	ACM
			2021	*Applied Soft Computing*	ScienceDirect
			2022	*Multimedia Systems*	Springer
			2023	*Cognitive Neurodynamics*	Springer
Imran Ahmed	Pakistan	3	2022	*Virtual Reality & Intelligent Hardware*	ScienceDirect
			2023	*Computing*	Springer
			2022	*Interdisciplinary Sciences: Computational Life Sciences*	Springer
Yogesh H. Bhosale	India	2	2023	*Biomedical Signal Processing and Control*	ScienceDirect
			2023	*Multimedia Tools and Applications*	Springer
Mahmoud Ragab	Saudi Arabia	2	2021	*Biology (Basel)*	PubMed
	UK		2022	*Journal of Healthcare Engineering*	PubMed
Seifedine Kadry	Norway	2	2023	*Procedia Computer Science*	ScienceDirect
			2023	*International Conference on Mining Intelligence and Knowledge Exploration*	Springer

(229 documents), Q1 contains the largest portion of documents, indicating that the majority of research is clustered in this quartile, followed by Q2 comprising a significant percentage 26.1% (92 papers). At 5.4% (19 papers), the third quartile (Q3) has a substantially smaller share than the fourth (Q4), which has the lowest share at just 3.7% (13 documents).

3.3.8 Top Five Authors Ranked by Research Contributions

Table 3.3 showcases the contributions of the top five authors in the field of Covid-19 detection research. As we can see, K. Shandar from India leads with four documents across three databases: ACM (Shankar et al., 2021a), ScienceDirect (Shankar et al., 2021b), and Springer, with two publications (Shankar et al., 2021c, 2021d). Imran Ahmed, a researcher from Pakistan, contributed with three documents in 2022 and 2023, published by Springer (Ahmed et al., 2022, 2023) and ScienceDirect (Ahmed et al., 2022). Also from India, Yogesh H. Bhosale has participated with two documents in 2023, published in *Biomedical Signal Processing and Control* and *Multimedia Tools and Applications* indexed in ScienceDirect (Bhosale & Patnaik, 2023a) and Springer (Bhosale & Patnaik, 2023b), respectively. Representing two different institutions, Mahmoud Ragab has made contributions with two documents: one from a Saudi Arabian institute (Ragab et al., 2022a) and another from a UK institute (Ragab et al., 2022b). From Norway, Seifedine Kadry marked 2023 with two publications; one in a ScienceDirect journal (Kadry et al., 2023a) and another in an international conference indexed in Springer (Kadry et al., 2023b). The results highlight the dominance of the Asian countries in the realm of scientific research related to Covid-19 detection.

3.3.9 Top Five Cited Documents

For any bibliometric analysis stage, it is important to consider the most cited documents. This will highlight the importance of its contribution and its impact in the field of production scientific. For this purpose, Table 3.4 showcases the top five cited published papers in the field of Covid-19 detection. The last update of total citations of all the documents included in this study was March 11, 2024. Authored by Tulin Ozturk from Turkey and published in 2020, the most cited paper has garnered a total of 2542 citations. The paper was published in *Computers in Biology and Medicine* and indexed in ScienceDirect database. This clearly highlights the relevance and the impact of the use of Deep Learning methods for Covid-19 detection using X-ray and CT images. The research is conducted across different countries including Turkey, Qatar, Egypt, China, South Korea, Iran, Italy, and Australia. One very important observation is the absence of India in the top five cited papers even if it tops the production of the countries. The papers have been published in four databases: ScienceDirect, IEEE Xplore, Springer, and ACM, with the dominance of ScienceDirect, occurring two times. This indicates the broad interest and practical value of the results. This collection of highly cited works underlines

TABLE 3.4 Top Five Cited Documents Ranked by Research Contributions in the Study

Doc Title	Author	Country	Year	Publisher	Database	Total Citations
Automated detection of Covid-19 cases using deep neural networks with X-ray images (Ozturk et al., 2020)	Tulin Ozturk	Turkey	2020	*Computers in Biology and Medicine*	ScienceDirect	2542
Can AI Help in Screening Viral and Covid-19 Pneumonia? (Chowdhury et al., 2020)	Muhammad E. H. Chowdhury	Qatar	2020	*IEEE Access*	IEEE Xplore	1606
Classification of Covid-19 in chest X-ray images using DeTraC deep convolutional neural network (Abbas et al., 2020)	Asmaa Abbas	Egypt	2020	*Applied Intelligence*	Springer	1100
Deep Learning Enables Accurate Diagnosis of Novel Coronavirus (Covid-19) With CT Images (Song et al., 2021)	Ying Song	China	2021	*IEEE/ACM Transactions on Computational Biology and Bioinformatics*	ACM	974
Exploring the effect of image enhancement techniques on Covid-19 detection using chest X-ray images (Rahman et al., 2021)	Tawsifur Rahman	Qatar	2021	*Computers in Biology and Medicine*	ScienceDirect	776

the important contributions made in Covid-19 detection using computational methods and medical images, marking a distinguished achievement in the fight against the virus.

3.4 AUTHORS KEYWORDS CO-OCCURRENCE

Known by its accessibility and effectiveness analyzing clustered search results, we explore the creation of the density map using VOSviewer as bibliometric analysis software. It stands out as good software allowing us to create and visualize networks within bibliometric data. These networks are created via various relationships like bibliographic coupling, co-authorship, citation, and co-citation relations, and they include researchers, journals, or individual publications. It also provides the capabilities of text mining which allows building and visualizing the patterns of the network (Donthu et al., 2021; Rueda et al., 2007). Indeed, Figure 3.8 represents a density map that underlines the key areas of Covid-19 detection using Deep Learning and Machine Learning methods. As we can see, Covid-19 and Deep Learning represent the central nodes. Related terms like Machine Learning, pneumonia, Transfer Learning, AI, chest X-ray, classification, and lung ultrasound surround the central nodes. The density map helps us to clearly and easily find major topics and methods in the field of detection and classification Covid-19-based X-ray and CT images using Deep Learning and Machine Learning.

3.5 DISCUSSION

The responses to the RQs raised at the start of this study are outlined as follows:

RQ1: What is the relative prevalence and impact of different AI techniques in the context of Covid-19 detection, and how have these techniques evolved over the period of study?

For the detection of Covid-19 using medical images, Deep Learning methods are chosen by researchers over Machine Learning and hybrid methods of both. This is principally due to the Deep Learning algorithms' inherent capability to handle complex data structures and extract complicated patterns without requiring manual feature extraction. Indeed, by automatically learning hierarchical feature representations, Deep Learning models, in particular convolutional neural networks, have proved higher performance in pattern recognition and classification tasks. This attribute is particularly helpful in medical image, where the detection of subtle and complex features indicative of Covid-19 requires sophisticated analytical capabilities. Furthermore, the availability of large datasets and a notable rise in computing power have made it easier to train increasingly complicated Deep Learning models, which has improved their accuracy and efficiency even further. In the field of Covid-19 detection using X-ray and CT images, Deep Learning continues to beat traditional Machine Learning and hybrid methods in terms of adaptability, scalability, and overall performance. This pattern highpoints how Deep Learning is developing medical diagnostics and how important it was to enhance the precision of Covid-19 detection during the pandemic.

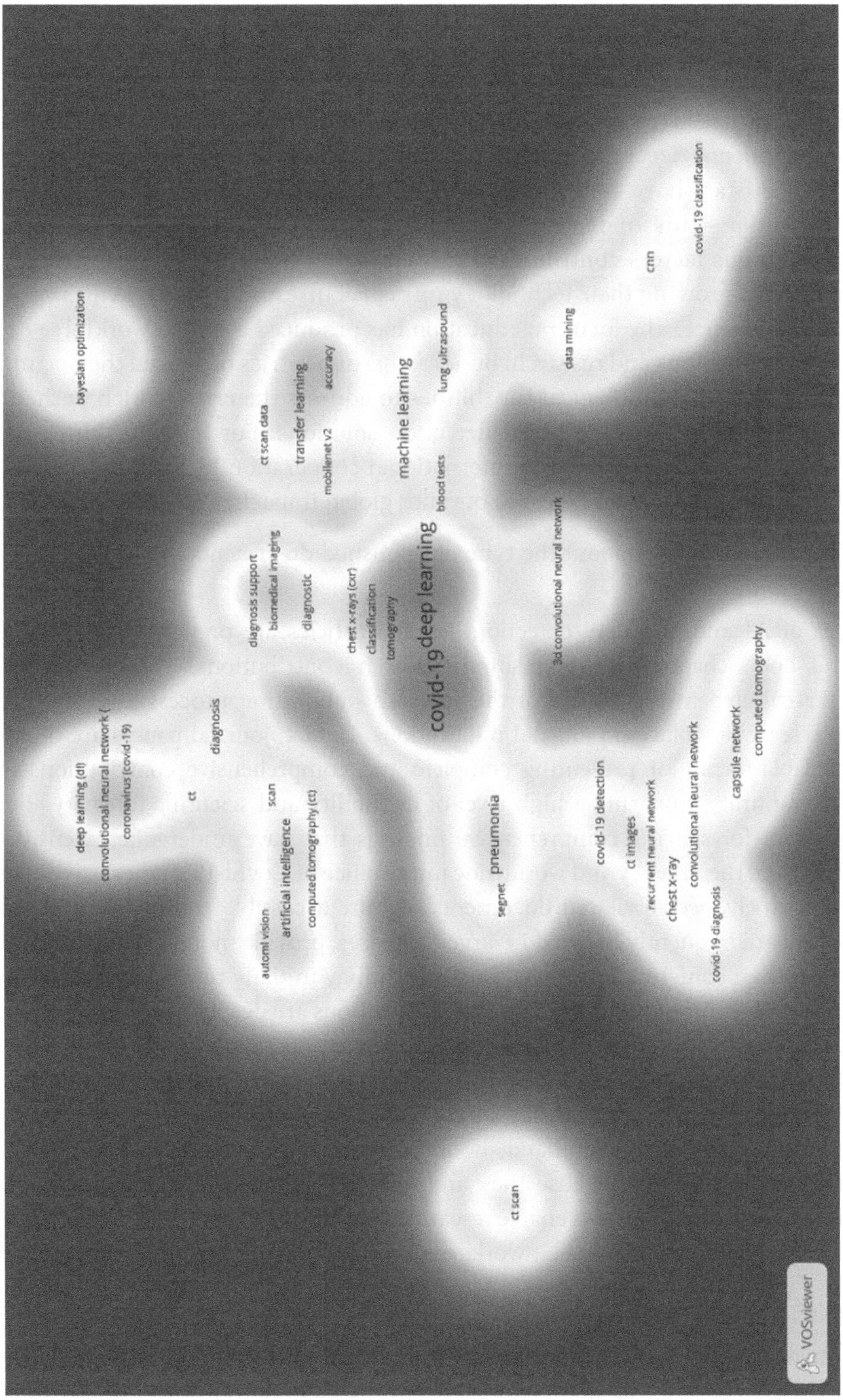

FIGURE 3.8 Density map of clusters on PubMed database.

RQ2: What insights emerge from the countries and universities distribution of research contributions, and how do their contributions effect the field of AI-driven Covid-19 detection research?

In the field of scientific research, the contributions of emerging economic and technological powers have witnessed remarkable development in term of quantity and quality of published documents. Several countries from these emerging nations, like India, China, Turkey, and Pakistan, have contributed with significant documents in the of AI-driven Covid-19 detection. This underscores a larger change in the dynamics of worldwide research. Numerous factors contribute to this emerging trend. First, these nations have deeply invested in growing their technology infrastructure and research skills in the field of artificial intelligence. The focus on education has produced a skilled workforce capable to pioneer AI applications and research. Besides, these countries have used their significant human resources and technological capabilities to address the crisis, which pushed local researchers to develop novel AI-based detection solutions. Furthermore, these nations have proven to be highly dedicated to international cooperation by forming cross-border partnerships that increase their research output's global impact.

RQ3: What can we deduce about the type of published documents in the field of AI and Covid-19 detection?

In the area of AI and Covid-19 detection, journal articles are preferred over conference papers and book chapters for publishing research contributions. This preference can be mainly attributed to the nature of the research publication process and the intrinsic value that researchers place on journal publications. In fact, journal papers are often seen as esteemed platforms for presenting complete and comprehensive contributions to the scientific community. Ensuring a high level of credibility and scientific integrity, journals offer a platform for sharing comprehensive findings that have undergone extensive peer review. Moreover, the urgency and worldwide importance of Covid-19 studies have increased the need for timely, peer-reviewed documents that are accessible and distributed. Journal articles, which are frequently indexed in large databases, provide more visibility and longer-term archival than conference presentations, which may initially reach a smaller audience. Moreover, academic promotions and funding applications are deeply influenced by journals metrics like impact factor, quartile, and citation metrics, encouraging researchers to focus these platforms. Whereas, conference papers can help to share preliminary findings and stimulating scholarly debates, but they might not offer depth analysis as much as peer-reviewed journals. On the other hand, book chapters are less beneficial for publishing pertinent scientific research because they are published less frequently, and they have a limited audience. Because journals enable researchers to communicate their detailed findings of research, they prefer to publish their works on AI and Covid-19 detection in these journals.

3.6 CONCLUSION

The bibliometric analysis carried out in this study highlights the important and dynamic role that AI in combating Covid-19 pandemic via the detection of the virus in X-ray and CT images. The study underlines a rapid response of the research community worldwide.

Additionally, this analysis illuminates some insights and trends in the global research community, including a clear preference for Deep Learning techniques, a movement in research contributions toward emerging economies geographically, and a significant trend toward publishing detailed findings in peer-reviewed journals.

The results of this bibliometric analysis open up new directions for further investigation into detection of Covid-19 and AI. First, research into the exploration of more complex Deep Learning models is required to improve accuracy and reduce diagnosis time. Furthermore, the increase in coordination from developing nations increases the possibility of cross-institutional and cross-cultural cooperation, which could lead to the creation of AI solutions that are applicable to all contexts. It would be beneficial to look into how AI can be combined with other diagnostic techniques to build hybrid models.

Additionally, this study shows the need for sustainable practices in developing of AI solutions. Enhancing the diagnostic accuracy can be reached using Deep Learning methods but also their environmental impact must be taken into consideration. Indeed, researchers need to focus on optimizing AI models for energy effectiveness and suggest solutions that fit with ecofriendly healthcare practices.

REFERENCES

Aanouz, I., Belhassan, A., El-Khatabi, K., Lakhlifi, T., El-Ldrissi, M., & Bouachrine, M. (2020). Moroccan Medicinal plants as inhibitors against SARS-CoV-2 main protease: Computational investigations. *Journal of Biomolecular Structure and Dynamics*, *39*(8), 2971–2979. https://doi.org/10.1080/07391102.2020.1758790.

Abbas, A., Abdelsamea, M. M., & Gaber, M. M. (2020). Classification of COVID-19 in chest X-ray images using DeTraC deep convolutional neural network. *Applied Intelligence*, *51*(2), 854–864. https://doi.org/10.1007/s10489-020-01829-7.

Ahmed, I., & Jeon, G. (2022). Enabling artificial intelligence for genome sequence analysis of COVID-19 and alike viruses. *Interdisciplinary Sciences Computational Life Sciences*, *14*(2), 504–519. https://doi.org/10.1007/s12539-021-00465-0.

Ahmed, I., Ahmad, M., & Jeon, G. (2022). Integrating digital twins and deep learning for medical image analysis in the era of COVID-19. *Virtual Reality & Intelligent Hardware*, *4*(4), 292–305. https://doi.org/10.1016/j.vrih.2022.03.002.

Ahmed, I., Jeon, G., & Chehri, A. (2023). An IoT-enabled smart health care system for screening of COVID-19 with multi layers features fusion and selection. *Computing*, *105*(4), 743–760. https://doi.org/10.1007/s00607-021-00992-0.

Alsattar, H. A., Qahtan, S., Zaidan, A. A., Deveci, M., Martinez, L., Pamucar, D., & Pedrycz, W. (2023). Developing deep transfer and machine learning models of chest X-ray for diagnosing COVID-19 cases using probabilistic single-valued neutrosophic hesitant fuzzy. *Expert Systems with Applications*, *236*, 121300. https://doi.org/10.1016/j.eswa.2023.121300.

Bhosale, Y. H., & Patnaik, K. S. (2023a). PulDi-COVID: Chronic obstructive pulmonary (lung) diseases with COVID-19 classification using ensemble deep convolutional neural network from chest X-ray images to minimize severity and mortality rates. *Biomedical Signal Processing and Control*, *81*, 104445. https://doi.org/10.1016/j.bspc.2022.104445.

Bhosale, Y. H., & Patnaik, K. S. (2023b). Bio-medical imaging (X-ray, CT, ultrasound, ECG), genome sequences applications of deep neural network and machine learning in diagnosis, detection, classification, and segmentation of COVID-19: A meta-analysis & systematic review. *Multimedia Tools and Applications*, *82*(25), 39157–39210. https://doi.org/10.1007/s11042-023-15029-1.

Chawki, Y., Elasnaoui, K., & Ouhda, M. (2024). Classification and detection of covid-19 based on X-ray and CT images using deep learning and machine learning techniques: A bibliometric analysis. *AIMS Electronics and Electrical Engineering*, *8*(1), 71–103. https://doi.org/10.3934/electreng.2024004.

Chowdhury, M. E. H., Rahman, T., Khandakar, A., Mazhar, R., Kadir, M. A., Mahbub, Z. B., Islam, K. R., Khan, M. S., Iqbal, A., Emadi, N. A., Reaz, M. B. I., & Islam, M. T. (2020). Can AI help in screening viral and COVID-19 pneumonia? *IEEE Access*, *8*, 132665–132676. https://doi.org/10.1109/access.2020.3010287.

Donthu, N., Kumar, S., Mukherjee, D., Pandey, N., & Lim, W. M. (2021). How to conduct a bibliometric analysis: An overview and guidelines. *Journal of Business Research*, *133*, 285–296. https://doi.org/10.1016/j.jbusres.2021.04.070.

El Asnaoui, K., & Chawki, Y. (2020). Using X-ray images and deep learning for automated detection of coronavirus disease. *Journal of Biomolecular Structure and Dynamics*, *39*(10), 3615–3626. https://doi.org/10.1080/07391102.2020.1767212.

El-Bouzaidi, Y. E. I., & Abdoun, O. (2023). Advances in artificial intelligence for accurate and timely diagnosis of COVID-19: A comprehensive review of medical imaging analysis. *Scientific African*, *22*, e01961. https://doi.org/10.1016/j.sciaf.2023.e01961.

Elkbuli, A., Ehrlich, H., & McKenney, M. (2020). The effective use of telemedicine to save lives and maintain structure in a healthcare system: Current response to COVID-19. *The American Journal of Emergency Medicine*, *44*, 468–469. https://doi.org/10.1016/j.ajem.2020.04.003.

Elmezayen, A. D., Al-Obaidi, A., Şahin, A. T., & Yelekçi, K. (2020). Drug repurposing for coronavirus (COVID-19): In silico screening of known drugs against coronavirus 3CL hydrolase and protease enzymes. *Journal of Biomolecular Structure and Dynamics*, *39*(8), 2980–2992. https://doi.org/10.1080/07391102.2020.1758791.

Fausto, J., Hirano, L., Lam, D., Mehta, A., Mills, B., Owens, D., Perry, E., & Curtis, J. R. (2020). Creating a palliative care inpatient response plan for COVID-19—The UW medicine experience. *Journal of Pain and Symptom Management*, *60*(1), e21–e26. https://doi.org/10.1016/j.jpainsymman.2020.03.025.

Ghosh, A., Gupta, R., & Misra, A. (2020). Telemedicine for diabetes care in India during COVID19 pandemic and national lockdown period: Guidelines for physicians. *Diabetes & Metabolic Syndrome Clinical Research & Reviews*, *14*(4), 273–276. https://doi.org/10.1016/j.dsx.2020.04.001.

Islam, M. R., & Nahiduzzaman, M. (2022). Complex features extraction with deep learning model for the detection of COVID19 from CT scan images using ensemble based machine learning approach. *Expert Systems with Applications*, *195*, 116554. https://doi.org/10.1016/j.eswa.2022.116554.

Kadry, S., Al-Betar, M. A., Yassine, S., Mohan, R., Arunmozhi, R., & Rajinikanth, V. (2023b). Efficient chest X-ray investigation using Firefly algorithm optimized deep and handcrafted features. In: *Lecture Notes in Computer Science* (pp. 225–236). https://doi.org/10.1007/978-3-031-44084-7_22.

Kadry, S., Herrera-Viedma, E., Crespo, R. G., Krishnamoorthy, S., & Rajinikanth, V. (2023a). Automatic detection of lung nodule in CT scan slices using CNN segmentation schemes: A study. *Procedia Computer Science*, *218*, 2786–2794. https://doi.org/10.1016/j.procs.2023.01.250.

Kavya, N. S., Shilpa, T., Veeranjaneyulu, N., & Priya, D. D. (2022). Detecting Covid19 and pneumonia from chest X-ray images using deep convolutional neural networks. *Materials Today Proceedings*, *64*, 737–743. https://doi.org/10.1016/j.matpr.2022.05.199.

Landry, M. D., Geddes, L., Moseman, A. P., Lefler, J. P., Raman, S. R., & Van Wijchen, J. (2020). Early reflection on the global impact of COVID19, and implications for physiotherapy. *Physiotherapy*, *107*, A1–A3. https://doi.org/10.1016/j.physio.2020.03.003.

Li, J. O., Shantha, J., Wong, T. Y., Wong, E. Y., Mehta, J., Lin, H., Lin, X., Strouthidis, N. G., Park, K. H., Fung, A. T., McLeod, S. D., Busin, M., Parke, D. W., Holland, G. N., Chodosh, J., Yeh, S., & Ting, D. S. (2020). Preparedness among ophthalmologists: During and beyond the COVID-19 pandemic. *Ophthalmology, 127*(5), 569–572. https://doi.org/10.1016/j.ophtha.2020.03.037.

Mallapaty, S., Callaway, E., Kozlov, M., Ledford, H., Pickrell, J., & Van Noorden, R. (2021). How COVID vaccines shaped 2021 in eight powerful charts. *Nature, 600*(7890), 580–583. https://doi.org/10.1038/d41586-021-03686-x.

Mehboob, R., Ahmad, F. J., Qayyum, A., Rana, M. A., Gilani, S. A., Tariq, M. A., & Akram, J. (2020). Aprepitant as a combinant with Dexamethasone reduces the inflammation via Neurokinin 1 Receptor Antagonism in severe to critical Covid-19 patients and potentiates respiratory recovery: A novel therapeutic approach. *medRxiv (Cold Spring Harbor Laboratory)*. https://doi.org/10.1101/2020.08.01.20166678.

Muralidharan, N., Gupta, S., Prusty, M. R., & Tripathy, R. K. (2022). Detection of COVID19 from X-ray images using multiscale deep convolutional neural network. *Applied Soft Computing, 119*, 108610. https://doi.org/10.1016/j.asoc.2022.108610.

Ng, M., Lee, E. Y. P., Yang, J., Yang, F., Li, X., Wang, H., Lui, M. M., Lo, C. S., Leung, B., Khong, P., Hui, C. K., Yuen, K., & Kuo, M. D. (2020). Imaging profile of the COVID-19 infection: Radiologic findings and literature review. *Radiology Cardiothoracic Imaging, 2*(1), e200034. https://doi.org/10.1148/ryct.2020200034.

Ouchicha, C., Ammor, O., & Meknassi, M. (2020). CVDNet: A novel deep learning architecture for detection of coronavirus (covid-19) from chest X-ray images. *Chaos, Solitons & Fractals, 140*, 110245. https://doi.org/10.1016/j.chaos.2020.110245.

Ozturk, T., Talo, M., Yildirim, E. A., Baloglu, U. B., Yildirim, O., & Acharya, U. R. (2020). Automated detection of COVID-19 cases using deep neural networks with X-ray images. *Computers in Biology and Medicine, 121*, 103792. https://doi.org/10.1016/j.compbiomed.2020.103792.

Ragab, M., Alshehri, S., Alhakamy, N. A., Alsaggaf, W., Alhadrami, H. A., & Alyami, J. (2022a). Machine learning with quantum seagull optimization model for COVID-19 chest X-ray image classification. *Journal of Healthcare Engineering, 2022*, 1–13. https://doi.org/10.1155/2022/6074538.

Ragab, M., Eljaaly, K., Alhakamy, N. A., Alhadrami, H. A., Bahaddad, A. A., Abo-Dahab, S. M., & Khalil, E. M. (2022b). Deep ensemble model for COVID-19 diagnosis and classification using chest CT images. *Biology, 11*(1), 43. https://doi.org/10.3390/biology11010043.

Rahman, T., Khandakar, A., Qiblawey, Y., Tahir, A., Kiranyaz, S., Kashem, S. B. A., Islam, M. T., Maadeed, S. A., Zughaier, S. M., Khan, M. S., & Chowdhury, M. E. (2021). Exploring the effect of image enhancement techniques on COVID-19 detection using chest X-ray images. *Computers in Biology and Medicine, 132*, 104319. https://doi.org/10.1016/j.compbiomed.2021.104319.

Rothan, H. A., & Byrareddy, S. N. (2020). The epidemiology and pathogenesis of coronavirus disease (COVID-19) outbreak. *Journal of Autoimmunity, 109*, 102433. https://doi.org/10.1016/j.jaut.2020.102433.

Rueda, G., Gerdsri, P., & Kocaoglu, D. F. (2007). Bibliometrics and Social Network Analysis of the Nanotechnology Field. *Portland International Conference on Management of Engineering & Technology PICMET '07.* https://doi.org/10.1109/picmet.2007.4349633.

Shankar, K., Mohanty, S. N., Yadav, K., Gopalakrishnan, T., & Elmisery, A. M. (2021c). Automated COVID-19 diagnosis and classification using convolutional neural network with fusion based feature extraction model. *Cognitive Neurodynamics, 17*(3), 1–14. https://doi.org/10.1007/s11571-021-09712-y.

Shankar, K., Perumal, E., Díaz, V. G., Tiwari, P., Gupta, D., Saudagar, A. K. J., & Muhammad, K. (2021b). An optimal cascaded recurrent neural network for intelligent COVID-19 detection using chest X-ray images. *Applied Soft Computing, 113*, 107878. https://doi.org/10.1016/j.asoc.2021.107878.

Shankar, K., Perumal, E., Elhoseny, M., Taher, F., Gupta, B. B., & Abd El-Latif, A. A. (2021a). Synergic deep learning for smart health diagnosis of COVID-19 for connected living and smart cities. *ACM Transactions on Internet Technology, 22*(3), 1–14. https://doi.org/10.1145/3453168.

Shankar, K., Perumal, E., Tiwari, P., Shorfuzzaman, M., & Gupta, D. (2021d). Deep learning and evolutionary intelligence with fusion-based feature extraction for detection of COVID-19 from chest X-ray images. *Multimedia Systems, 28*(4), 1175–1187. https://doi.org/10.1007/s00530-021-00800-x.

Sharifi-Razavi, A., Karimi, N., & Rouhani, N. (2020). COVID-19 and intracerebral haemorrhage: Causative or coincidental? *New Microbes and New Infections, 35*, 100669. https://doi.org/10.1016/j.nmni.2020.100669.

Simcock, R., Thomas, T. V., Estes, C., Filippi, A. R., Katz, M. S., Pereira, I. J., & Saeed, H. (2020). COVID-19: Global radiation oncology's targeted response for pandemic preparedness. *Clinical and Translational Radiation Oncology, 22*, 55–68. https://doi.org/10.1016/j.ctro.2020.03.009.

Song, Y., Zheng, S., Li, L., Zhang, X., Zhang, X., Huang, Z., Chen, J., Wang, R., Zhao, H., Chong, Y., Shen, J., Zha, Y., & Yang, Y. (2021). Deep learning enables accurate diagnosis of novel coronavirus (COVID-19) with CT images. *IEEE/ACM Transactions on Computational Biology and Bioinformatics, 18*(6), 2775–2780. https://doi.org/10.1109/tcbb.2021.3065361.

Umesh, N., Kundu, D., Selvaraj, C., Singh, S. K., & Dubey, V. K. (2020). Identification of new anti-nCoV drug chemical compounds from Indian spices exploiting SARS-CoV-2 main protease as target. *Journal of Biomolecular Structure and Dynamics*, 1–9. https://doi.org/10.1080/07391102.2020.1763202.

WHO. (2020). *Novel Coronavirus (2019-nCoV) Situation Report–28*. https://www.who.int/

Worldometers. (2025). https://www.worldometers.info/coronavirus/

CHAPTER 4

Improving Predictions of Student Performance and Personalized School Guidance with Hybrid Machine Learning Models

A Case Study in Cameroonian Secondary Schools

Eric Desiré Kameni, Cédric Stéphane Tekouabou Koumetio, Marlene Djoutieu Tegankong, and Samsong Durand Priso Essawe Ndedi

4.1 INTRODUCTION

Combating academic failure is one of the challenges of the Cameroonian educational system in recent years. For several decades, the pass rate in official exams has been quite low in our country. In 2015, the overall pass rate for exams managed by the OBC was 52.72%, and in 2016, it was 46.00% (Shu 2022). The causes of this decline are numerous, including overcrowded classrooms in our high schools and colleges, lack of student follow-up, precocious children, orientation issues, organizational problems (forgetting materials, homework, inability to anticipate), and lack of motivation (Ndindeng 2024).

To combat academic failure, many solutions have been implemented in Cameroonian schools in recent years, such as new teaching methods, solutions for precocious children, online orientation platforms used by guidance counselors to address orientation issues, and the use of qualified pedagogues or academic coaches to solve organizational and motivational problems. Specialized classes and tutoring have been proposed to address the lack of follow-up and overcrowding in high schools and colleges. However, these solutions have raised issues regarding the balance between individual and collective approaches. It is difficult for teaching teams to account for each student's specificities without weakening the

DOI: 10.1201/9781003649885-4

collective dynamic (Francesc et al. 2019). Indeed, individualized student monitoring requires more work from teachers and a significant amount of time, which could slow down the class's progress.

For some, the solution lies in new technologies: artificial intelligence (AI), more specifically machine learning (ML), which could propose adaptive pedagogical strategies tailored to each student (Kameni and Koumetio 2023; Francesc et al. 2019). Thus, Yekun and Haile (2021) proposed a model to predict learners' performance, forecasting the performance of students from three public secondary schools in Mekelle, Ethiopia, for the following semester in five courses. The prediction system was modeled using the K-Nearest Neighbors (KNN) and Multilayer Perceptron (MLP) classification algorithms. Baradwaj and Pal (2012) also proposed in 2012 a system capable of predicting students' performance in end-of-semester exams, based on the decision tree algorithm. Shovon et al. (2012), on the other hand, proposed a hybrid procedure combining data mining methods (decision tree algorithm) and data classification (K-means algorithm) to predict students' GPA for the next semester based on their previous semester's GPA. The predicted results were used to suggest measures to improve student performance (Wung and Sigala 2024).

An analysis of these solutions shows that the criteria of age and disciplinary behavior were not considered in their development criteria that, according to Noumba (2008), significantly influence academic success in the Cameroonian secondary education system. Thus, these solutions may not effectively address the problem of predicting student performance in Cameroonian secondary schools. The contribution of our work lies in the selection of features that account for the Cameroonian context. We aim to design and implement a model capable of predicting a student's overall average based on chosen features and recommending ways to improve their academic performance.

Research Questions

How can we propose a model capable of identifying potential future academic failures among students using educational data mining (EDM), KNN, linear regression, and decision tree algorithms, while recommending appropriate support?

More specifically:

- How can we analyze data from school management systems using EDM methods?
- How can we design a model capable of predicting a student's overall average based on data collected from schools?
- What type of support should be provided to students at risk of failing (i.e., those with a predicted average below ten)?

Research Objectives

The general objective of this study is to propose a model capable of predicting a student's end-of-year overall average and suggesting ways to improve their academic performance using EDM, KNN, multiple linear regression, and decision tree algorithms.

Achieving this general objective involves the following specific objectives:

- Analyze student management system data using EDM methods.

- Propose a model capable of predicting a student's overall average based on collected school data.
- Provide tailored recommendations for students.

4.2 RELATED WORK

4.2.1 Guidance Counselors and Student Follow-Up in Cameroonian Schools

Guidance counseling is still a poorly understood practice in some communities, even though it plays a crucial role in guiding individuals toward fulfilling educational and career paths. Using tools such as psychometric tests or interest questionnaires (Wung and Sigala 2024), guidance counselors help students make choices aligned with their aptitudes and the country's socioeconomic needs. As specified in the 1996 Cameroonian circular, these non-teaching professionals have a multidimensional mission: from information on courses of study to academic adjustment, via preparation for working life ((MINEDUC) 1996, 2001). Their work revolves around personalized follow-up, from the analysis of school records at the beginning of the year to remedial interviews with students in difficulty, sometimes involving parents for optimal follow-up.

However, in Cameroon, the decree requiring 1 counselor per school (n°2001/041) is far from being respected: out of 900 counselors trained in 2003, only 523 were in post (Oyono and Ngamaleu 2024). This shortage has prompted the development of IT solutions to make up for the lack of human resources. These innovative tools aim to digitally reproduce the individualized follow-up that is the strength of the advisors – this in-depth knowledge of the students remains the key asset for their success. The next sections will explore these emerging technological approaches in the educational landscape.

4.2.2 Approach Based on the Decision Tree Algorithm

These analytical methods have been used successfully for years to predict student outcomes and find important factors for success (Kolo and Adepoju 2015; Li et al. 2008; Vandamme et al. 2007). The main goal of the study by Baradwaj and Pal (2012) is to use data mining techniques and the ID3 version of the decision tree algorithm to guess how well students will do. Their suggested algorithm uses a dataset that was gathered from VBS Purvanchal University Jaunpur (Uttar Pradesh) through student management systems. We used data mining techniques to look at and process the dataset to get information about 50 students who graduated from the department of applied computer science with a master's degree in the year 2007–2010. The information extracted includes class attendance, test scores, seminar scores, and homework scores.

4.2.3 Approach Based on KNN and Decision Tree Algorithms

Several works in the literature have investigated hybrid methods combining decision trees with other algorithms, such as KNN, to improve prediction accuracy (Rajaguru and SR 2019; Shovon et al. 2012). This hybridization has shown the power of decision tree classification and the flexibility of KNN to produce correct predictions at a significant percentage (Murugappan et al. 2020).

Using decision trees and KNN, Shovon et al. (2012) put forth a hybrid model to forecast students' GPAs depending on past semester performance, attendance, and assignment submission. While decision trees offered ideas for raising performance, the KNN algorithm categorized students into "High," "Medium," and "Low" performance groups. In similar vein, Bansal et al. (2022) showed how this method improved student performance and attitudes in mobile learning environments.

More recent work (Murugappan et al. 2020; Pathak et al. 2020; Rajaguru and SR 2019) has further shown the effectiveness of the hybridization of KNN and decision trees. Rajaguru and SR (2019) designed a ML model to evaluate the performance of the classification of breast tumors and proposed a model which uses two ML algorithms for the categorization of breast tumor; decision tree and KNN algorithm is used for the breast tumor classification. This classification includes the two levels of disease as benign or malignant. The comparative analysis results indicate that the KNN classifier outperforms the result of the decision-tree classifier in the breast cancer classification.

Despite the effectiveness of this hybridization, taking into account dynamic interactions and data interpretation remains a major challenge (Romero and Ventura 2010). These difficulties highlight the importance of continued innovation in hybrid modeling techniques.

4.2.4 Approach Based on a Bayesian Probabilistic System

Murry (2016) focuses on the École Polytechnique de Montréal's bachelor's degree student training process. Using the average grades earned, the number of credits taken during the first part of the training (the first six sessions), and the department, the goal is to forecast, for each student, the cumulative average earned during the second part of the bachelor's degree (sessions 7 through 12). The data comes from the 2008 cohort and includes information about 700 students.

4.3 HYBRID MACHINE LEARNING MODEL FOR STUDENT PERFORMANCE PREDICTION AND GUIDANCE

In this section, we present the description of our approach, the functional diagram, and the various steps involved in designing the proposed system. Our approach combines EDM with ML algorithms, including KNN, multiple linear regression, and the ID3 decision tree, to predict student performance and provide personalized guidance. The methodology is structured to ensure a clear and reproducible process, from data collection to model validation.

4.3.1 Data Collection and Preprocessing

4.3.1.1 Data Sources

The data used in this study were collected from the student management systems of two secondary schools in Yaounde, Cameroon. These systems contain historical academic records, behavioral data, and demographic information. Ethical considerations, such as data anonymization and consent, were strictly followed to ensure the privacy and confidentiality of student information.

4.3.1.2 Data Description

The dataset includes the following variables:

Previous profile: This is a vector containing the variables that characterize a student's behavior in previous classes:

- **Average:** The value obtained by averaging the overall average grades from previous classes.
- **Absence:** The value obtained by summing up the hours of absence in previous classes.
- **Repeat:** Corresponds to the number of classes repeated.
- **InRule:** The value representing the number of years during which the student has not been compliant.
- **SocialIssue:** This refers to the number of years during which a student has faced social challenges.
- **Gender:** Binary variable (male or female).
- **AgeMention:** Indicates whether the student is below, at, or above the expected age for their class.
- **Seq:** Represents the student's average at the beginning of the academic year.
- **Appreciation:** Evaluation of the student's previous work.
- **Effort:** Recommendations for improving the student's performance.
- **Mentionseql:** Evaluation of the student's work at the beginning of the year.

4.3.1.3 Data Preprocessing

The preprocessing steps included the following:

- Data cleaning: Removing incomplete or inconsistent records.
- Feature selection: Using an exhaustive feature selection algorithm to identify the most relevant variables (see Section 4.3.1.2).
- Data transformation: Encoding categorical variables (e.g., gender as 1 for male and 0 for female) and normalizing numerical data as shown by the tables (Table 4.1).
- Data splitting: Dividing the dataset into training (80%) and testing (20%) sets.

To address the class imbalance, stratified sampling was applied during the train-test split (80–20%), maintaining the proportions of each category. The test set strictly comes from the academic year 2021–2022, while the training uses data from 2014 to 2020. Checks of unique identifiers confirmed the absence of overlap between the sets.

TABLE 4.1 Consolidated Table: Variables Used by Algorithms

Variable	Possible Values	Algorithms Using It
Average	Numeric value	KNN, multiple linear regression, ID3
Absence	[0, ∞[	KNN, multiple linear regression, ID3
Repeat	[0, ∞[	KNN, multiple linear regression
InRule	[0, ∞[	KNN, multiple linear regression, ID3
SocialIssue	[0, ∞[	KNN, multiple linear regression, ID3
Gender	0: Female, 1: male	KNN, multiple linear regression, ID3
AgeMention	1: Below, 2: class, 3: above	KNN, multiple linear regression, ID3
Seq	[0, 20]	KNN, multiple linear regression
Class	1: Strong, 2: passable, 3: weak (KNN/ID3)	KNN, ID3
Appreciation	Weak: $0 < \text{avg} < 10$, pass: $10 \leq \text{avg} \leq 12$, strong: $12 < \text{avg} \leq 20$	ID3
Mentionseq1	Weak: $0 < avg < 10$, Pass.: $10 < avg \leq 12$, Strong: $13 < avg \leq 20$	ID3
Effort	Good: maintain, pass: work on discipline/hours, weak: more effort needed	ID3

4.3.2 Model Design and Development

4.3.2.1 Overview of the Hybrid Approach

Our hybrid approach combines three ML algorithms: KNN, used to classify students into performance categories (e.g., high, medium, low); multiple linear regression, which predicts students' final average grades; and the ID3 decision tree, which identifies key factors influencing performance and provides actionable recommendations.

4.3.2.2 Algorithm Implementation

Design of the KNN algorithm

Step 1: The first step involves calculating the distance between two students taken from our dataset. This measure allows us to quantify the proximity between student profiles based on several attributes. We use the Euclidean distance, a commonly used metric to assess the dissimilarity between two vectors in a multidimensional space. Consider two students, x and y, each represented by a feature vector. These features include attributes such as academic average, number of absences, age, gender, and other relevant indicators. Formally, the vectors x and y are defined as follows:

$$x = (\text{Average}_i, Absence_i, Repeat_i, InRule_i, SocialIssue_i, Gender_i, AgeMention_i, Seq_i, Class_i)$$

$$y = (\text{Average}_j, Absence_j, Repeat_j, InRule_j, SocialIssue_j, Gender_j, AgeMention_j, Seq_j, Class_j)$$

where i and j are indices representing students x and y, respectively, in the dataset.

The Euclidean distance $D(x,y)$ between vectors x and y is calculated using the following formula:

$$D(x,y) = \sqrt{\sum_{k=1}^{n} (x_k - y_k)^2}$$

where n is the number of features (or dimensions) of vectors x and y.

Step 2: Obtain the nearest neighbors. We will use the formula for Euclidean distance as defined previously. Our algorithm will classify students into strong, passing, or weak categories based on the distance separating the students from each other. These steps allowed us to obtain Algorithm 4.1. With $K = 3$, the number of classes we want to have as output. In our case, the algorithm groups the students into strong, average, or weak classes.

Algorithm 4.1: K-Nearest Neighbors (KNN)

1. **Input:**
2. K=3 ▷ *Number of classes (strong, average, weak)*
3. A set of students with their features
4. A student A to be classified
5. **Output**:
6. The class of student A (strong, average, weak)
7. **Initialization**:
8. Select the student A to be classified
9. **Process**:
10. **Repeat**
11. Select a student B≠A ▷ *Choose another student*
12. Calculate the distance D(A,B) ▷ *Compute the distance between* A *and B*
13. **until** There are no more new students B ▷ *Repeat for all students*
14. **Classification***:*
15. Find the K nearest neighbors of A ▷ *Select the* K *closest students*
16. Assign to A the majority class among the K neighbors ▷ *Classify* A

Design of the multiple linear regression algorithm

The algorithm is structured into three main steps: model definition, cost function formulation, and parameter estimation via gradient descent. Each step is explained with consideration of the specific characteristics of the students and the objectives of our study.

Step 1: Model definition: The multiple linear regression model is defined by:

$$F = X \cdot \theta$$

where F is the vector of predictions of size $m \times 1$; X the feature matrix of size $m \times (n + 1)$ and θ the parameter vector of size $(n + 1) \times 1$ to be estimated. With m equal to the number of students in our training dataset and n represents the selected features.

Step 2: Cost function: To evaluate the model's performance, we use a cost function based on the mean squared error (MSE). This function measures the deviation between the predicted values F and the actual values Y. The cost function $J(\theta)$ is given by:

$$J(\theta)=\frac{1}{2m}\sum_{i=1}^{m}(X_i * \theta - Y_i)^2$$

where X_i is the i-th row of the matrix X; Y_i the actual value corresponding to the i-th observation and m is the total number of observations.

Step 3: Parameter estimation: To minimize the cost function J(θ), we use the gradient descent method. This iterative approach updates the parameters θ by following the direction opposite to the gradient of the cost function. The steps are as follows:

Gradient Calculation:

$$\frac{\partial J(\theta)}{\partial \theta} = \frac{1}{m} X^T * (X * \theta - Y) \text{ and } \hat{\theta} = \theta - \alpha \frac{\partial J(\theta)}{\partial \theta}$$

where X^T is the transpose of the matrix, α is the learning rate, controlling the step size of the update and $\hat{\Theta}$ is the updated parameter vector.

Algorithm 4.2: Multiple Linear Regression

Input:
X: Student vector without the class coordinate
θ: Error vector
LearningRate = 0.001
NumIterations = 1000000
Output:
$\hat{\theta}$: Minimal error vector
Initialization:
Define F,X and θ such that F=X $\cdot$ θ
Compute the cost function J(θ)
Compute the gradient
Process:
repeat
$\hat{\theta}$: =GradientDescent(gradient,LearningRate,NumIterations)
until $\hat{\theta}: \approx \theta$ ▷ *Repeat until convergence*
Return: $\hat{\theta}$

These steps allowed us to develop the Algorithm 4.2., with the LearningRate, which is the step size between two gradient descent calculations, and the NumIterations, which represents the number of times we will compute the gradient in our gradient descent. We achieved better results with LearningRate = 0.001 and NumIterations = 1000000.

Design of the decision tree algorithm

Step 1: Calculation of the system's entropy: We will use the following formula:

$$Entropy(S) = \sum_{c \in classes(S)} pc * log_2(pc)$$

where p_c is the probability of having class c in the set S.

Step 2: Calculation of gain: The gain is a value that allows us to choose the order of appearance of the attributes in our tree starting from the root. It is calculated using the following formula:

$$Gain(S,A) = Entropy(S) - \sum_{v \in values(A)} \frac{|S_v|}{|S|} * Entropy(S_v)$$

where S is the training data or system; A: an attribute of S and S_v: the different values that can take an attribute A as defined in Table 4.1. These steps allowed us to develop the Algorithm 4.3.

Algorithm 4.3: Decision Tree (ID3)

1. **Input**:
2. S: Training data
3. **Output**:
4. B: Decision tree
5. **Initialization**:
6. Compute the entropy of the system S*S*
7. **Process**:
8. **Repeat**
9. Select an attribute A
10. Compute the gain Gain(S,A)
11. **until** There are no more new attributes A*A*
12. **Repeat**
13. Choose the attribute A*A* with the maximum Gain(S,A)
14. **if** A is the criterion to predict **then**
15. **Return** B
16. else
17. Add A as a **node to** B
18. **end if**
19. **until** A is the criterion to predict

Table 4.2 shows a summary of the hyperparameters used.

TABLE 4.2 Hyperparameters of Algorithms

Algorithm	**Hyperparameter**	**Value/Choice**
KNN	Value of K	K=3 (classes: strong, acceptable, weak)
	Distance metric	Euclidean distance
Linear regression	Learning rate (α)	0.001
	Stopping criterion	1,000,000 iterations
ID3	Entropy thresholds	Calculated dynamically (Gain(S,A))

4.3.2.3 Integration of Algorithms

The outputs of the three algorithms are combined to provide a comprehensive prediction model. KNN classifies students, linear regression predicts their final grades, and the decision tree offers personalized recommendations.

Figure 4.1 shows the complete data journey through our model from start to finish. We begin with raw data – complete with their labels, of course – which then goes through various transformation stages. Think of it like an assembly line: each step flows logically into the next to ensure the data gets properly prepared, the model learns effectively, and we can thoroughly validate its performance.

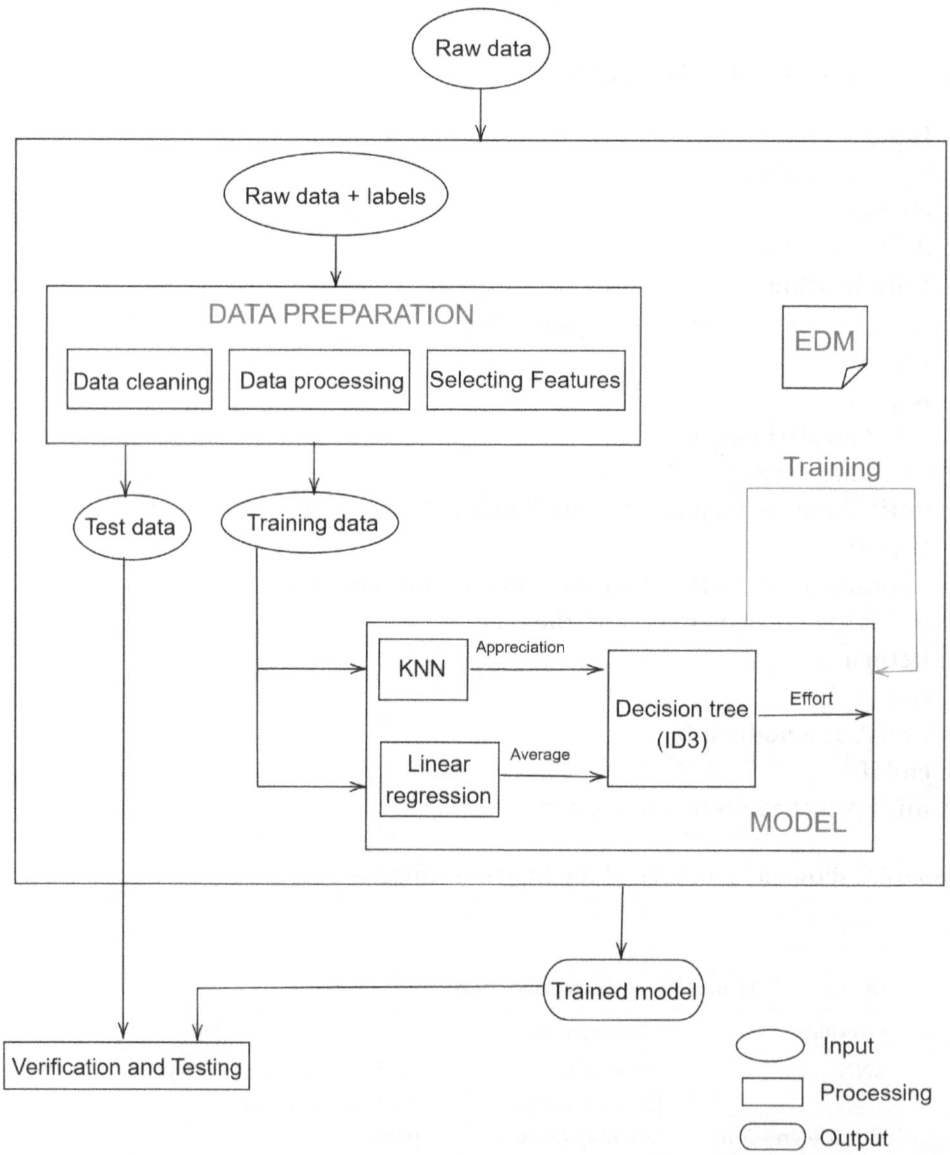

FIGURE 4.1 Functional diagram of the proposed model.

This methodical approach, echoing the work of Willard et al. (2022) and Rolf et al. (2021), has allowed us to develop a robust model (see Algorithm 4.4). A model that holds up in real-world scenarios, not just with our test data.

Algorithm 4.4: General Algorithm for Effort Estimation

1. **Input**:
2. Training data
3. **Output**:
4. Effort
5. **Step 1: Data Preparation**
6. Clean and preprocess the data
7. Transform the data into a suitable format
8. Select relevant features for training
9. **Step 2: Apply KNN for Classification**
10. Class = KNN(Training data) ▷ *Call Algorithm 4.1. to classify* data
11. **Step 3: Apply Multiple Linear Regression for Average Estimation**
12. Average = MultipleLinearRegression(Training data) ▷ *Call Algorithm 4.2. to estimate average*
13. **Step 4: Apply ID3 for Effort Estimation**
14. Effort = ID3(Class, Average) ▷ *Call Algorithm 4.3. to estimate effort*
15. **Step 5: Return Effort**
16. **Return** Effort

4.4 MODEL TRAINING AND VALIDATION

4.4.1 Training Phase

The training phase involved applying our hybrid ML model combining KNN, multiple linear regression, and the ID3 decision tree to labeled historical data from Cameroonian secondary schools. The dataset, collected from student management systems of two schools in Yaounde, included academic records (e.g., past grades, attendance), behavioral metrics (e.g., disciplinary actions), and demographic variables (e.g., age, gender). Ethical considerations, such as data anonymization and consent, were strictly followed. The experiments were conducted on a computer with an Intel i7 processor and 16 GB of RAM, using Python 3.8 and scikit-learn 1.0.

Key steps

1. **Data Preprocessing:**

   - Cleaning: Incomplete records were removed.
   - Feature selection: An exhaustive feature selection algorithm identifed critical variables: Average, Absence, Repeat, InRule, SocialIssue, Gender, AgeMention, and Seq (see Figure 4.2).

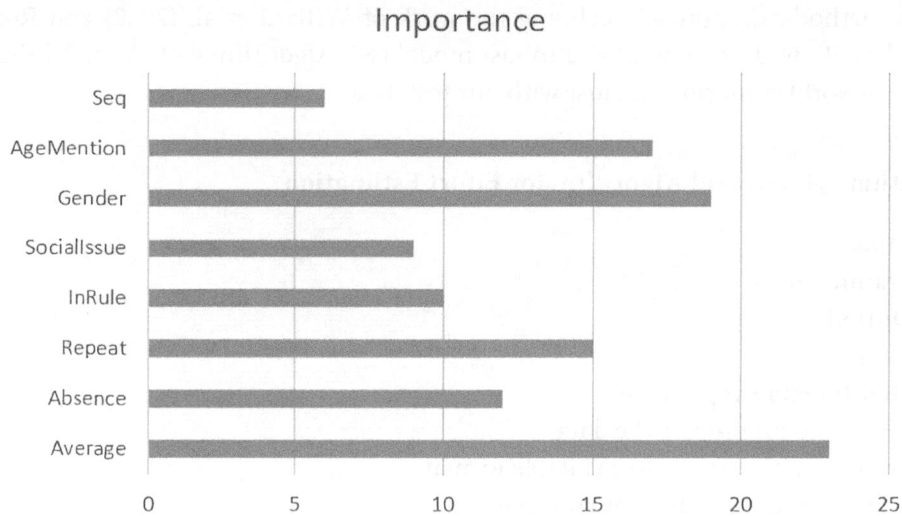

FIGURE 4.2 Features based on their importance.

- Transformation: Categorical variables (e.g., Gender) were encoded numerically, and numerical data were normalized.

To test the ability of our algorithms to predict a student's GPA at the end of the year, we carried out non-cross-validation by dividing our dataset into two parts, test data and training data. Our test dataset consisted of 50 records (Table 4.4).

4.4.2 Results of the KNN Algorithm

The KNN algorithm was employed to classify students into three performance categories (Table 4.3): strong (average ≥12), satisfactory (10 ≤average < 12), and weak (average < 10). The training dataset comprised 2240 records, distributed as follows: 7.14% (160 students) in the strong category, 44.64% (1000 students) in the satisfactory category, and 48.21% (1080 students) in the weak category (Table 4.3). To evaluate the algorithm's predictive accuracy, a test dataset of 50 students was used. The KNN model correctly classified 30 out of 50 students, achieving an accuracy of 60%. These results indicate that the algorithm performs moderately well in categorizing students based on their academic performance, with room for improvement in distinguishing between the satisfactory and weak categories.

The analysis of positive/negative predictions reveals that the model correctly identifies 75% of at-risk students (TP=12), while avoiding overwhelming counselors with 82.4% true

TABLE 4.3 Percentage of Students by Class

Class	Average	Number of Students	Percentage (%)
1: Strong	≥ 12	160	7.14
2: Satisfactory	10–11	1000	44.64
3: Weak	< 10	1080	48.21

negatives (TN=28). However, the four false negatives represent a systemic risk requiring additional manual checks.

Although the KNN algorithm achieved an accuracy of 60%, a deeper analysis of complementary metrics (precision, recall, and F1-score) was conducted to account for class imbalance (48.21% of "weak" students). The results show an F1-score of 0.58 for the "weak" class, reflecting an acceptable performance despite the dominance of this category. A cross-validation (five folds) confirmed the robustness of the model, with a precision variance of less than 5%. A sensitivity analysis varying K (from 3 to 7) revealed that K=3 optimized the trade-off between stability and performance.

4.4.3 Results of the Multiple Linear Regression Algorithm

The multiple linear regression algorithm was trained to predict students' end-of-year averages using features such as age, gender, disciplinary behavior, and prior academic performance. The dataset's suitability for regression analysis was confirmed by the 3D scatter plot (Figure 4.3), which revealed a discernible linear relationship among the variables. The model's parameters were optimized using gradient descent, with a learning rate of 0.001 and 1,000,000 iterations. The resulting coefficients for the feature vector θ were: [0.05598025, −0.43425322, 0.4256019, 0.38336445, 0.81435466, 1.20019615, 0, −0.03437147].

The convergence of the cost function demonstrated stable optimization. The model's performance was quantified using the coefficient of determination (R^2), which yielded a value of 0.5272, indicating that 52.72% of the variance in the target variable was explained by the model. A residual analysis was conducted to identify any biases, confirming a homogeneous distribution without marked trends. Although the (R^2) is moderate (0.5272), cross-validation showed consistency in predictions (average $R^2 = 0.51 \pm$

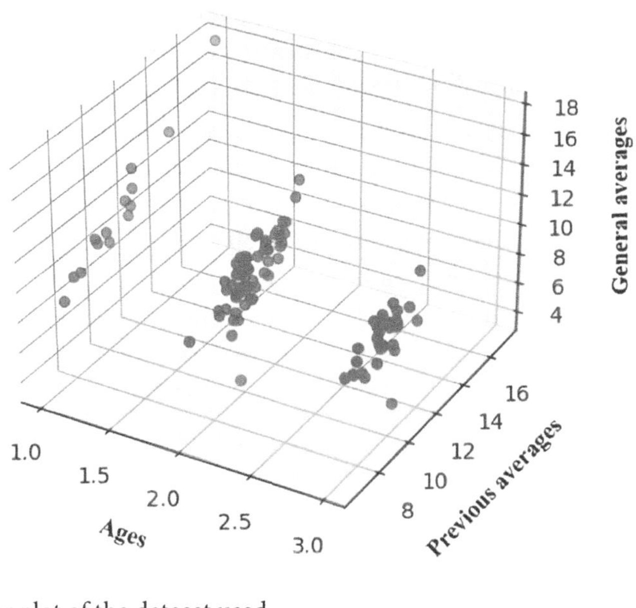

FIGURE 4.3 Scatter plot of the dataset used.

0.03), suggesting that the model captures generalizable trends despite the complexity of the underlying factors. While this suggests a moderate predictive capability, further refinement of the feature set or incorporation of additional variables could enhance accuracy.

The model sacrifices 25% of sensitivity (FN=4 undetected at-risk students) to limit false positives to 17.6%, an acceptable trade-off in resource-limited contexts. The false negatives primarily concern students on the "Satisfactory/Weak" border (averages 9.5–10), suggesting the usefulness of an adaptive threshold that incorporates attendance.

4.4.4 Results of the ID3 Decision Tree Algorithm

The ID3 decision tree algorithm was implemented to recommend tailored interventions for improving student performance. The system's entropy, calculated as 0.50021, reflected the initial uncertainty in the dataset. Feature importance was evaluated using information gain, with the highest gains attributed to seq1 (18.52), AgeMention (18.51), and average (18.50), highlighting their strong influence on the target variable (effort). The algorithm categorized students into four intervention groups: "Maintain current effort" (10%), "Improve discipline" (24%), "Increase study hours" (40%), and "Enhance discipline and study habits" (26%) (Table 4.4).

These recommendations were derived from rules generated by the tree, such as prioritizing students with high absenteeism or low prior averages for intensive support. The results underscore the algorithm's utility in personalizing academic interventions, though its dependency on categorical thresholds may benefit from probabilistic extensions in future work.

4.4.5 Results of the Hybrid Model

When it came to forecasting and enhancing student performance, the hybrid model that combined KNN, multiple linear regression, and the ID3 decision tree showed complementary strengths. With an accuracy of 60 in classifying students into three performance tiers (strong, satisfactory, and weak), the KNN algorithm offered a solid classification framework. In the meantime, the multiple linear regression model provided quantitative precision by using characteristics like past academic performance, attendance, and disciplinary records to explain 52.72% of the variance in end-of-year averages.

By producing practical suggestions based on each student's unique profile, the ID3 decision tree enhanced the model even more. For instance, 40% of students were advised to increase study hours, while 26% required combined improvements in discipline and study

TABLE 4.4 Predictions of the ID3 Decision Tree Algorithm

Recommended Intervention	**Students (n)**	**Percentage (%)**
Good: maintain current performance level	5	10
Pass: improve disciplinary compliance	12	24
Pass: increase study time investment	20	40
Weak: comprehensive improvement needed (discipline + study habits)	13	26

habits (Table 4.4). By combining these algorithms, a multifaceted evaluation was guaranteed, where:

- KNN identified at-risk students,
- Regression quantified their expected performance decline, and
- ID3 prescribed targeted interventions.

The fact that 48.21% of students were identified as "weak" emphasizes how urgently focused interventions are needed. According to a projection based on historical data from the schools under study, the methodical application of the hybrid model's recommendations could optimize resource allocation (e.g., discipline workshops, tutoring for "weak" students) and reduce failure rates by 15–20% over the course of a school year. This cut would limit the environmental impact of producing more educational resources while also resulting in an estimated 10% reduction in retention costs (tuition fees, educational materials).

A pilot study was carried out with school counselors to assess the ID3 tree recommendations' practical applicability. Twenty anonymized cases were manually evaluated and compared to the model's predictions. According to a confusion matrix, the majority class's "Increase Study Time" intervention had a 75% agreement rate, while the combined recommendations (discipline + habits) had a 68% precision rate. While acknowledging the need to modify categorical thresholds (such as absenteeism) to better reflect local contexts, qualitative feedback highlighted the value of the recommendations for giving at-risk students priority.

4.4.6 Practical Limitations

The study presents several practical limitations. First, its geographic coverage is restricted to Yaounde, necessitating multi-regional validation to generalize the results effectively. Additiontionally, the sample size of the test set (n=50) limits statistical power, especially for the minority "strong" class, which consists of only five participants. Last, strict data anonymization imposed ethical constraints that hindered the inclusion of contextual variables, such as socioeconomic status.

4.5 CONCLUSION AND PERSPECTIVES

A hybrid model combining KNN, multiple linear regression, and ID3 decision trees has been developed to predict student performance and offers personalized recommendations in Cameroonian secondary schools. Using EDM, this model outperforms traditional pedagogical methods by taking a data-driven approach, balancing individual and collective needs. It explains 60% of score variance with KNN and 52.72% with linear regression. The ID3 decision tree also contributed to personalized recommendations, such as increasing study time for 40% of students, proving its effectiveness in educational investigations.

The study also presented important FCPs such as previous marks, attendance, and discipline attitude on academic achievement and underlined the specific impact of age and gender in the Cameroonian setting. Categorically, 48.21% of students were "weak," indicating an immediate demand for focused intervention. The model's fusion of classification, regression, and decision-making also offers educators a complete instrument to allocate resources efficiently like targeting discipline improvement efforts for 26% of students.

However, several limitations of the present study point to areas for future investigations. Improving the model's predictive value might entail blending more features like socioeconomics or parent engagement, and thus improving the regression beyond only 52.72%. Perhaps experimenting with ensemble approaches, like Random Forest or XGBoost, could help account for class imbalance and increase KNN's accuracy. Technically, a live feedback system could improve predictions based on outcomes of interventions, and natural language processing could learn knowledge from snippets of unstructured text.

The idea could be developed into a low-cost mobile or web application for practical use, which would address the issue of guidance counselors in Cameroon. Working with legislators would guarantee alignment with national education strategies and teacher training initiatives. Ethical considerations, such as robust anonymization protocols for sensitive data, must remain of utmost importance in order to safeguard student privacy. Longitudinal studies should be conducted to assess the model's long-term impacts on equity and academic results.

This study emphasizes how hybrid ML models have the ability to revolutionize education, especially in settings with limited resources. The framework provides a scalable way to enhance student performance and guide policy by fusing data-driven decision-making with instructional practice. These contributions offer a replicable solution to the world's educational problems and highlight intelligent systems for societal impact. In order to optimize the model's advantages for both students and teachers, future research will concentrate on improving the model's accuracy, adding new features, and guaranteeing its ethical implementation.

REFERENCES

Bansal, Malti, Apoorva Goyal, and Apoorva Choudhary. 2022. "A comparative analysis of K-Nearest Neighbor, Genetic, Support Vector Machine, Decision Tree, and Long Short Term Memory algorithms in machine learning". *Decision Analytics Journal* 3:100071. doi:10.1016/j.dajour.2022.100071.

Baradwaj, Brijesh Kumar, and Saurabh Pal. 2012. "Mining educational data to analyze students' performance". *arXiv preprint arXiv:1201.3417.*

Francesc, Pedró, Miguel Subosa, Axel Rivas, and Paula Valverde. 2019. *Artificial intelligence in education: challenges and opportunities for sustainable development. UNESCO Working Papers on Education Policy 7.* https://hdl.handle.net/20.500.12799/6533.

Kameni, Eric Désiré, and Stéphane Cédric Koumeo. 2023. "The role of inclusive educational technologies in transforming African cities into inclusive smart cities". *E3S Web of Conferences* 418:03003. doi:10.1051/e3sconf/202341803003.

Kolo, David Kolo, Solomon A. Adepoju, and John Kolo Alhassan. 2015. "A decision tree approach for predicting students academic performance". *International Journal of Education and Management Engineering* 5(5):12–19. doi:10.5815/ijeme.2015.05.02.

Li, Cheng-Lun, Ayse G. Buyuktur, David K. Hutchful, Natasha B. Sant, and Satyendra K. Nainwal. 2008. "Portalis: using competitive online interactions to support aid initiatives for the homeless". P. 3873–78 in *CHI '08 Extended Abstracts on Human Factors in Computing Systems*. New York, NY: ACM.

(MINEDUC), M. d. (1996). Lettre circulaire camerounaise N A/426/MINEDUC/SG/DPRD/SDP/SOSUP/. Yaounde: MINEDUC.

(MINEDUC-), M. d. (2001). Lettre circulaire camerounaise N 06/B1/1464/MINEDUC/CAB du 19 f′evrier 2001. Yaounde: MINEDUC.

Murugappan, M., AM Mutawa, Sai Sruthi, Aya Hassouneh, Ali Abdulsalam, S. Jerritta, and R. Ranjana. 2020. "Facial expression classification using KNN and decision tree classifiers". P. 1–6 in 2020 4th International Conference on Computer, Communication and Signal Processing (ICCCSP). IEEE.

Ndindeng, Atina. 2024. "Factors influencing under-education in Cameroon: A comprehensive analysis". *African Educational Research Journal* 12(3):182–201.

Noumba, Issidor. 2008. Un profil de l'abandon scolaire au Cameroun. Revue d'économie du developpement 16(1):37–62.

Oyono, Pascal Nkoulou, and Henri Rodrigue Njengoué Ngamaleu. 2024. "Study of the process of posting teachers to schools in the Ministry of Secondary Education: The link with the Academic performance of secondary school students in the South Region of Cameroon". *European Journal of Education Studies* 11(11).

Pathak, Ashwini, and Sakshi Pathak 2020. "Study on decision tree and KNN algorithm for intrusion detection system". *International Journal of Engineering Research & Technology* 9(5):376–81.

Rajaguru, Harikumar, and Sannasi Chakravarthy S.R. 2019. "Analysis of decision tree and k-nearest neighbor algorithm in the classification of breast cancer". *Asian Pacific Journal of Cancer Prevention: APJCP* 20(12):3777.

Rolf, Esther, Jonathan Proctor, Tamma Carleton, Ian Bolliger, Vaishaal Shankar, Miyabi Ishihara, Benjamin Recht, and Solomon Hsiang. 2021. "A generalizable and accessible approach to machine learning with global satellite imagery". *Nature Communications* 12(1):4392.

Romero, Cristóbal, and Sebastián Ventura. 2010. "Educational data mining: a review of the state of the art". *Systems, Man, and Cybernetics, Part C: Applications and Reviews, IEEE Transactions on* 40(6):601–18.

Shovon, Md, Hedayetul Islam, and Mahfuza Haque. 2012. "An approach of improving students academic performance by using k means clustering algorithm and decision tree". *arXiv preprint arXiv: 1211.6340.*

Shu, Solomon. 2022. "Education in Cameroon". P. 28–48 in *Education in Africa*. Routledge. https://doi.org/10.4324/9781003316114-2.

Vandamme, Jean-Philippe, Nadine Meskens, and Juan-Francisco Superby. 2007. "Predicting academic performance by data mining methods". *Education Economics* 15(4):405.

Willard, Jared., Xiaowei Jia, Shaoming Xu, Michael Steinbach, and Vipin Kumar. 2022. "Integrating scientific knowledge with machine learning for engineering and environmental systems". *ACM Computing Surveys* 55(4):1–37.

Wung, Catherine Chuo, and Maxwell Sigala. 2024. "Enhancing Students' entrepreneurial skills through counseling services: challenges and prospects for Secondary School Counselors in Bamenda, Cameroon". *Excellencia: International Multi-disciplinary Journal of Education* 2(8):76–87.

Yekun, Ephrem Admasu, and Abrahaley Teklay Haile. 2021. "Student performance prediction with optimum multilabel ensemble model". *Journal of Intelligent Systems* 30(1):511–23.

CHAPTER 5

Reforming Governance in Moroccan Higher Education

Sentiment Analysis for Whole-of-Government and Bottom-Up Policy Approaches

Said Azelmad

5.1 INTRODUCTION

Institutions must be well-governed in order to maintain the relevance of their academic offerings, student satisfaction, and an effective work environment for their faculty and staff in higher education. Traditional feedback mechanisms, i.e., surveys, focus groups, and recommendations notebooks, for example, fail to capture the dynamic and diverse views expressed on social media. However, they are a fertile ground for real-time, unstructured feedback that, when systematically analyzed, can be used to support data-driven administrative reforms. This chapter applies machine learning (ML)-based sentiment analysis of user-generated content on the official Facebook page of the Ministry of Higher Education in Morocco. By extracting and measuring sentiment, we aim to provide policymakers with actionable insights for evidence-based decision-making to align institutional reforms with the evolving demands of students and academic stakeholders (Pandey, 2004). In fact, effective international governance in higher education is undergoing a paradigm shift, moving from 1.0 to reach 5.0, with the advent of AI (Azelmad, 2024), providing a real-time data analytics for inclusive decision-making in a whole-of-government approach, whole-of-society and whole-of-country approaches. It ensures a wide stakeholder engagement in the steps of decision-making. The New Public Management (NPM) is needed higher education to advance inclusive governance that align with national strategies of Vision 2030 in higher education (van Berckel Smit, 2023). In this way, all the traditional feedback mechanisms (e.g., surveys) must be supplemented by AI-driven sentiment analysis for a prompt data analysis. The application of machine ML techniques and multilingual natural language processing (NLP) models of BERT can provide policymakers' real sentiments to advance the curriculum for Education 4.0.

DOI: 10.1201/9781003649885-5

5.2 LITERATURE REVIEW

In Moroccan higher education, the NPM is needed for Education 4.0 to maximize the sector's potential and bring it into line with both domestic goals and international norms, where effective governance lays the groundwork for responsiveness, streamlined coordination, transparency, accountability, responsibility, and equitable resource allocation for all universities (van Berckel Smit, 2023). It is mandatory to promote inclusive governance for academic performance, with strong universities that remain flexible and adaptable as Morocco works to improve its educational system hand in hand with the Moroccan 2030 agenda development objectives.

Shattock (2006) in his book *Managing Good Governance: Managing Universities and Colleges: Guides to Good Practice* explores the intricacies of effective governance in higher education. The author explores the fundamental procedures and values that college and university administrators need to summon up for integrity, openness, and effectiveness of their institutions' governance. He argues that good governance is the foundation of academic achievement, financial stability, and institutional innovation, and highlights the crucial role that governance plays in determining the strategic direction of higher education institutions. The book is an essential tool for strengthening governance frameworks, encouraging accountability, and coordinating institutional objectives with the larger educational mission since it offers real-world examples and insights. To ensure that universities and colleges not only survive but also flourish in a setting that is becoming more competitive and dynamic, Shattock's work emphasizes the significance of strong governance in navigating the complex issues facing higher education today. In this direction, Dobbins, Knill and Vögtle (2011) proposed an integrated framework for analyzing higher education governance that systematically traces the ongoing transformations in European higher education systems. While previous studies have provided valuable insights, there remains a need for more precise, empirically observable indicators to better understand policy changes and instances of inertia. The state-centered model, the market-oriented model, and academic self-governance are three historically significant and now relevant European governance models that we analyze in order to provide a systematic classification of these variables. We derive three ideal types from these models, which capture the conflicts between academia, the market, and the state. These models take into account the responsibilities of external stakeholders and the state as well as internal university governance. We can obtain empirical measures of financial governance, personnel autonomy, institutional power balance, and substantive governance issues thanks to this basis. With the use of our analytical framework, researchers in the political and social sciences in particular should be able to identify trends in higher education governance, including change, convergence, persistence, and inertia.

5.3 SOCIAL MEDIA AND STUDENT FEEDBACK PLATFORMS

The integration of social media in educational feedback mechanisms marks a transformative shift in how student sentiments and experiences are captured and analyzed (Saif, He and Alani, 2012). This development is best demonstrated by platforms such as OUSocial2, which employ sophisticated NLP to glean insightful information from the extensive and frequently unstructured conversations seen on social media. According to Thomas et al. (2014), this method not only improves real-time feedback but also fosters a lively conversation between teachers and students, enabling prompt pedagogical modifications. The potential of technologies like OUSocial2 to improve educational performance is enormous, despite obstacles like privacy concerns and the complexities of deciphering social media jargon. Addressing these issues will be essential to maximizing the advantages of social media-based feedback systems in higher education as research in this field develops. Nonetheless, Facebook has become a prominent forum for public deliberation in Morocco for staff and students to voice their opinions about university administration and governance (Azelmad and Harroudi, 2024). The platform's popularity can be attributed to several factors, including its ease of use, the sense of community it fosters, and its capacity to swiftly and extensively share opinions. The platform's structure makes it easier for public dialog to quickly emerge, which has the potential to influence both public opinion and university policies. Additionally, students can freely express their opinions on Facebook since they hide their genuine identities, which encourages more candid conversations (Mulinge, Arasa and Wawire, 2017). Facebook increases polarized debates among different social bulbs. In this regard, It is easy to strive to collect feedback from students to improve their courses and tutorship. Such methods of collecting feedback are faster than surveys. Indeed, with the rise of social media, many students are finding online venues to group and share their experiences, seeking peers' support. Social media serves as a public sphere, fostering ongoing discussions similar to the ancient assemblies, in line with Habermas's concept of communicative action and public deliberation (Habermas, 2006). Social media has developed into a virtual public space where people converse in real time on public policies, governance, and services. Azelmad and Harroudi (2024) in their article "Social Media: A Public Deliberation or a Social Accountability in Morocco" indicated how the online deliberations can promote openness and give regular citizens the ability to challenge or exert pressure on decision-makers. Social media sites like Facebook and Twitter (now called X) have developed into forums for Moroccans to express their tweets and sentiments, offering counterarguments to the official narrative. Indeed, types of digital involvement that encourage social responsibility—particularly in situations where institutional input is scarcely evident; Habermas's idea of rational-critical discussion remains always reinterpreted in today's networked society. These technologies make it simple to get anonymous student feedback on classes and make it possible to track sentiment and engagement trends over time in learning communities, including Morocco (Saif, He and Alani, 2012).

5.4 GOVERNANCE 4.0 IN HIGHER EDUCATION: A BOTTOM-UP APPROACH

Governance in higher education involves directing and coordinating the actions and decisions of various stakeholders based on set objectives, following established guidelines while including all relevant parties to achieve goals. By enabling teachers and students to actively engage in decision-making, inclusive governance approaches, in higher education, represent a revolutionary move toward exclusivity, accountability, transparency, and whole-of-government collaboration (Azelmad, 2024). Improving digital interaction

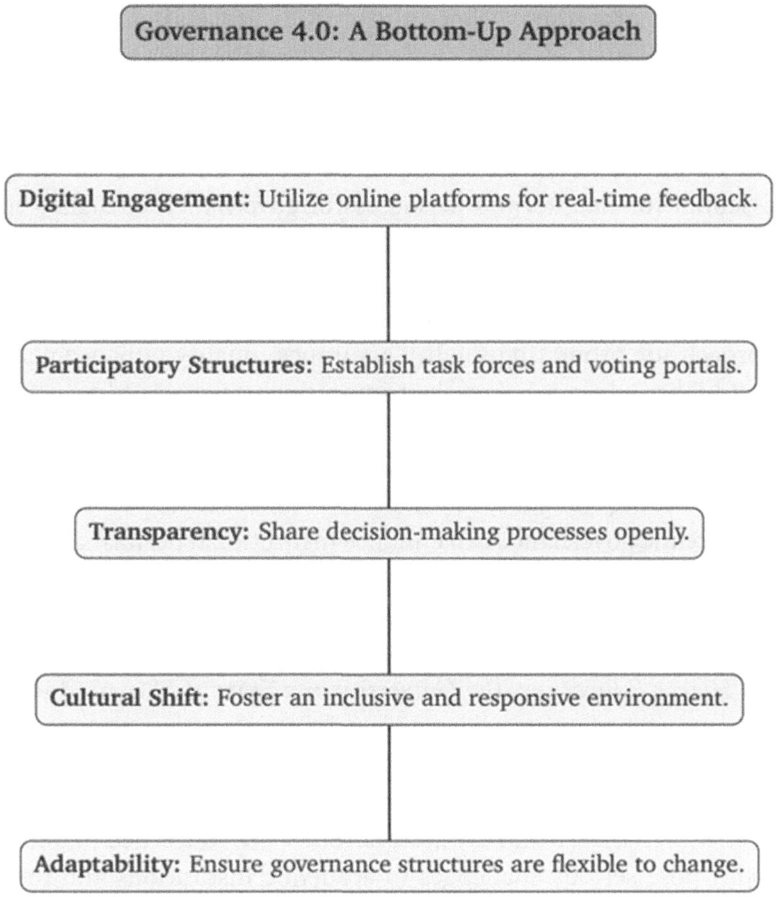

FIGURE 5.1 Elements of a bottom-up approach in Governance 4.0.

between students as customers and their institutions through online platforms for real-time feedback is a needed new bottom-up approach, which will facilitate collaborative governance procedures to build sustainable universities in Morocco. Transparency becomes mandatory for open decision-making processes to recover confidence, responsibility, and participation. We need to improve the academic environment and align educational outcomes with student demands. Institutions must adopt a bottom-up governance, taking into account Morocco's unique territorial differences and specifies in management (Chamlee-Wright, 2018) (Figure 5.1).

5.5 EPISTEMIC GOVERNANCE IN HIGHER EDUCATION

The book *Epistemic Governance in Higher Education: Quality Enhancement of Universities for Development* by David F. J. Campbell and Elias G. Carayannis explores the concept of epistemic governance within the context of higher education. *Epistemic governance* refers to the underlying cognitive frameworks, knowledge paradigms, and interpretive schemes that inform how a social system is organized, legitimized, and directed. It highlights how concepts, conventions, and specialized expertise influence institutional priorities and decision-making procedures. Since universities are epistemic institutions—places where knowledge is created, debated, and shared—rather than just administrative entities, epistemic governance takes on special significance in the context of higher education. A thorough analysis of current research shows that strong epistemic frameworks, such as academic autonomy, quality assurance procedures, and research-driven policy frameworks, are essential for long-term and successful governance in higher education. The authors contend that reform initiatives run the risk of becoming surface-level or unrelated to universities' intellectual goals if these knowledge-centered aspects are not recognized. Campbell and Carayannis (2012) emphasize the practical and policy implications of the epistemic approach for promoting quality assurance, quality enhancement, and quality management of higher education, and their impact on university administration and academic career development. They analyzed the various governance models and practices, demonstrating how different approaches can influence the quality and development of universities. By providing comparative perspectives and case studies, their book illustrates how effective epistemic governance can enhance institutional performance, drive development, and foster innovation in higher education. This aligns with our focus on examining how governance structures impact higher education quality and performance through the views of different stakeholders on social media. Hirsch and Weber (2001) in their book *Governance in Higher Education: The University in a State of Flux* tried to unveil the veil upon the characteristics of epistemic governance in universities that prioritize research, work to create coherent governance principles, and provide pertinent projects and suggestions. Collectively, the authors—all former and current heads of esteemed research-intensive universities in the US and Western Europe—are concerned that the fundamental changes of today pose serious challenges to academic institutions and their governance frameworks. In line with the objectives of epistemic governance, their model fostered a more involved academic community where decisions are directly influenced by the needs and knowledge of individuals working for the institution. The new models of governance in Moroccan higher education must shift, as well, toward epistemic governance principles, where inclusive, knowledge-based decision-making and diverse academic perspectives are shaped within the spectrum. This approach shifts from traditional hierarchical structures to a more collaborative framework that encourages faculty, students, and other stakeholders to participate in the management of institutional priorities and policies in a whole-of-institution and whole-of-student approach. For Campbell and Carayannis (2012), epistemic governance in higher education has the power to enhance universities to strategically manage knowledge, enhancing educational quality and aligning academic outputs with societal needs. By fostering innovation, adaptability, and inclusiveness, making institutions more dynamically capable in front of new challenges, in teaching, research, and student engagement with society and the job market. This approach strengthens the developmental role of universities, ensuring they remain pivotal in addressing real-world problems, influencing policy, and contributing to broader economic and social progress. Through effective epistemic governance, universities not only elevate their academic standards but also amplify their impact on society at large (Vidovich and Currie, 2011). The world witnesses a crucial transition from technologically optimized learning to a more student-centered, morally

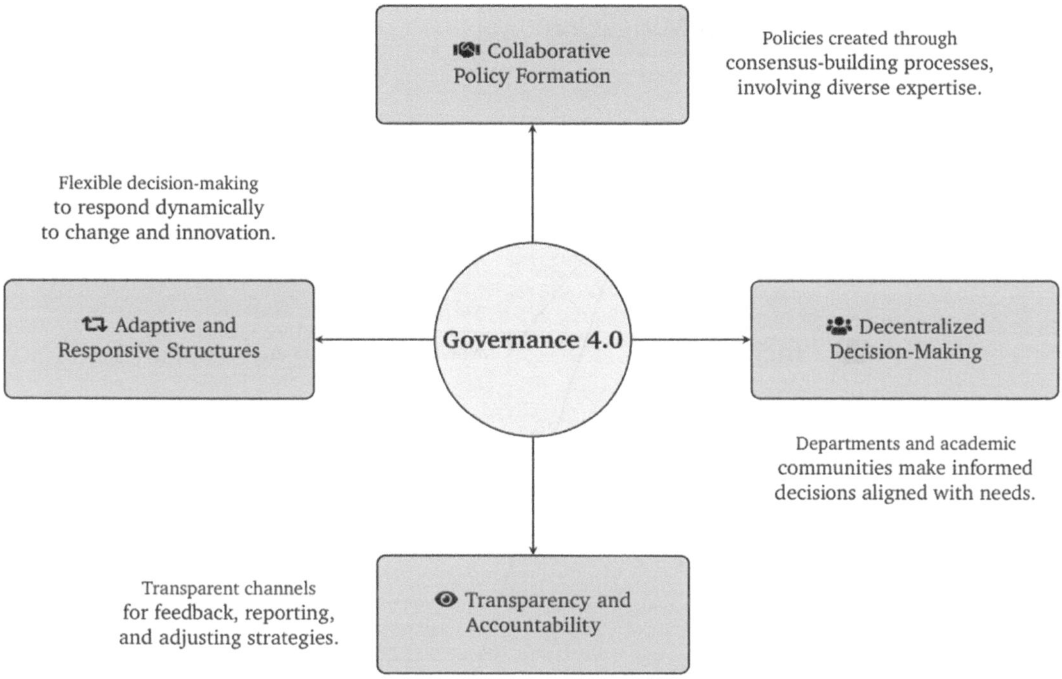

FIGURE 5.2 Governance and Education in the era of 4.0.

motivated, and socially sensitive educational paradigm is represented by the global transition from Education 4.0 to Education 5.0. Education 5.0 reaffirms the significance of epistemic governance—the cognitive and normative processes that influence how knowledge is created, valued, and used within educational systems—while Education 4.0 closely mirrored the technological demands of the Fourth Industrial Revolution. This shift places a strong emphasis on developing critical thinking, emotional intelligence, and sustainability awareness in addition to digital competency acquisition. In this sense, universities are increasingly seen as epistemic establishments entrusted with using inclusive, thoughtful, and accountable governance to navigate difficult global issues. This change is furthered by social media platforms, which serve as contemporary public forums where stakeholders, educators, and students have thoughtful discussions. Such digital discourse reinforces the democratic potential of higher education in the era of Education 5.0 by acting as a social accountability mechanism in settings with poor institutional feedback channels, encouraging all voices to be heard and valued, ultimately leading to a more collaborative and less conflict-prone environment in higher education (Figures 5.2 and 5.3).

5.6 STUDENT HOMOPHILY ON SOCIAL MEDIA

All the birds of a feather flock together sums up the idea of consumer homophily on social media. By using identical tweets and comments to place birds in different echo chambers on social media, the theory explains the social psychology of birds. The study "Birds of the Same Feather Tweet Together" by Barberá (2015) used data from Twitter to investigate how echo chambers affect political polarization. The study described how individuals follow

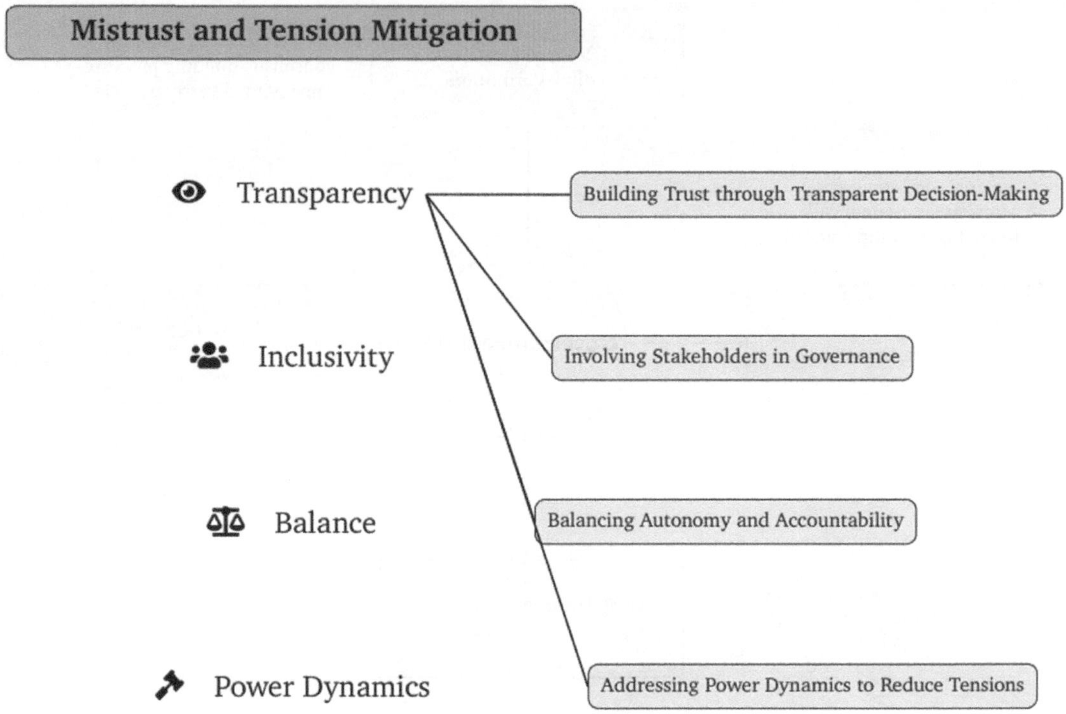

FIGURE 5.3 Process of good governance in higher education.

and engage with people who hold similar political views on Twitter. Echo chambers are created by this trend, in which people are largely exposed to ideas and information that support their political beliefs. The study found that Twitter users are more likely to connect with people who share their political beliefs, a phenomenon known as homophily. In this case, users were more likely to engage with and follow accounts that shared their political views. Furthermore, the echo chambers created by this self-selection and homophily tend to reinforce users' preconceived political views. The possibility that users would encounter content that supports their beliefs exacerbates ideological division. Barberá and Rivero (2015), Colleoni, Rozza and Arvidsson (2014), and Barberá (2015) highlight how echo chambers intensify similar polarization of opinions on social media. The findings demonstrate how, similar to bubble-filtered public areas, the platform's network structure, which encourages the interaction of like-minded individuals, can result in more polarized perspectives. In order to improve governance in a number of domains, each of the previously stated studies contributed to our understanding of social media echo chambers in connection to consumer interaction and polarization (Altenburger and Ugander, 2018). Social media homophily thus encourages students, faculty, and staff to create both scheduled and unstructured forums, echo chambers around their university's online debates, and homophilous organizations. Cross-group discussions and a wider variety of viewpoints result from this.

5.7 SENTIMENT ANALYSIS AND FEEDBACK INTERPRETATION IN HIGHER EDUCATION

As part of their attempts to enhance learning environments, institutions are increasingly trying to use feedback to gain a better understanding of students' moods. Usually, this process is constrained by grading schemes that often fail to adequately represent students' nuanced experiences. As a faculty, I am aware that conventional grading schemes fall short in capturing students' everyday competencies in terms of their employability or soft skills in the job market. Therefore, it is crucial to support children, look beyond grades, and pay attention to their experiences and feelings. In order to understand the needs of students before and after graduation, university institutions must adopt a whole-of-government and whole-of-society approach. This includes creating outlets for open feedback, whether they are mentorship, forums, or digital platforms (Azelmad, 2024). In doing so, universities can cultivate a learning environment that not only values academic achievement but also respects and nurtures the whole student journey. This gap was really filled by Dsouza et al. (2019), who suggested a technique for sentiment analysis of student input. The author has an impact on the usefulness of online resources and their ability to interpret the feelings and viewpoints of students. As a new comparative method to assess how well these platforms interpret (student) consumers' comments, the textual feedback enables a deeper examination of certain elements of the educational experience, revealing all the governance shortcomings. Educational institutions can better answer student issues by utilizing these advanced platforms, which will ultimately result in more informed improvements in higher education's managerial and teaching prospects (Aman Ullah, 2016).

5.8 METHODOLOGY

5.8.1 Data Collection

As part of a mixed-methods approach, this study uses sentiment analysis to evaluate the perceptions of governance processes among academics, administrative staff, and students, who are stakeholders in Moroccan higher education. Because Facebook serves as a direct line of communication between stakeholders and their institutions, data is gathered via user-generated posts and comments on the Moroccan Ministry of Higher Education and Scientific Research's official Facebook page. Using Python-based NLP modules, the text is preprocessed through lemmatization, tokenization, and stop-word removal. Sentiment classification improved classification accuracy by combining supervised ML models (Naïve Bayes, SVM) and lexicon-based tools (TextBlob, VADER) that were trained on manually annotated subsets of data. To show trends in public opinion, the results were displayed using word clouds, bar plots, and time-series sentiment trends. Only publicly accessible content was used, personal identifiers were removed, in accordance with ethical guidelines.

Social media has developed into a dynamic, if dispersed, public space where various stakeholders can voice their thoughts, worries, and recommendations. The research specifically focused on:

1. Public responses and remarks made by staff, instructors, and students.

2. Feedback from anonymous students that frequently criticizes institutional governance, course content, and teacher quality.

3. Indirect measures of sentiment are provided by engagement activities like shares, emojis, and likes.

We noticed that students frequently operate in solitary echo chambers, using anonymous Facebook identities to freely express opinions they might not share through official channels. These posts often remain apart from teacher–staff interactions, exposing a contentious and polarizing online educational discourse. To address this, we compiled all relevant user interactions on the ministry's Facebook page into a dataset. The comments were copied to a csv.file that is treated on Python software, for qualitative and quantitative analysis.

5.8.2 Data Tokenization

Since this study examines descriptive textual data from user comments, the author worked to reprocess generated data file to guarantee the precision and efficacy of ML-based sentiment analysis. Indeed the process of *tokenization* divided each comment into more manageable, smaller units called tokens—which could be words, sentences, or punctuation. These tokens were taken from a structured CSV file that contained comments from social media, which made Python analysis easier. By turning unstructured text into pieces that can be analyzed, tokenization makes a variety of NLP tasks possible, most notably sentiment categorization. Additional cleaning procedures were used to enhance model performance, including normalizing repeated characters, eliminating punctuation, converting text to lowercase, and filtering out non-informative stop words (such "the," "and"). In order to improve the dataset's semantic quality and minimize noise, this process pipeline is crucial. To ensure that diverse forms of a word (such as "studying," "studied," and "studies") are treated equally, lemmatization was done after tokenization to reduce each word to its base or dictionary form (lemma). Clean, well-tokenized data is essential for accurate lemmatization, which is a crucial step in determining the actual meaning of user comments. When combined, these NLP processing methods create a foundation for efficient sentiment analysis, allowing the models to identify patterns and categorize emotional tone more accurately (Figure 5.4).

```
----------------------------------------------------------------
Comment: Scholarships are hard to obtain and there's poor
communication with students.
Aspect Sentiments: {'Scholarship': 'Negative',
'Communication': 'Negative'}
----------------------------------------------------------------
Comment: Professors are often incompetent and there's too
much chaos in administration.
Aspect Sentiments: {'Professor': 'Negative', 'Chaos':
'Negative'}
```

FIGURE 5.4 Higher education ministry official Facebook page.

```
------------------------------------------------------------------
Comment: The funding is insufficient, and the faculty
doesn't care about student support.
Aspect Sentiments: {'Funding': 'Neutral'}
------------------------------------------------------------------
Comment: Scholarship opportunities are unfairly distributed,
and infrastructure needs improvement.
Aspect Sentiments: {'Scholarship': 'Negative',
'Infrastructure': 'Negative'}
------------------------------------------------------------------
```

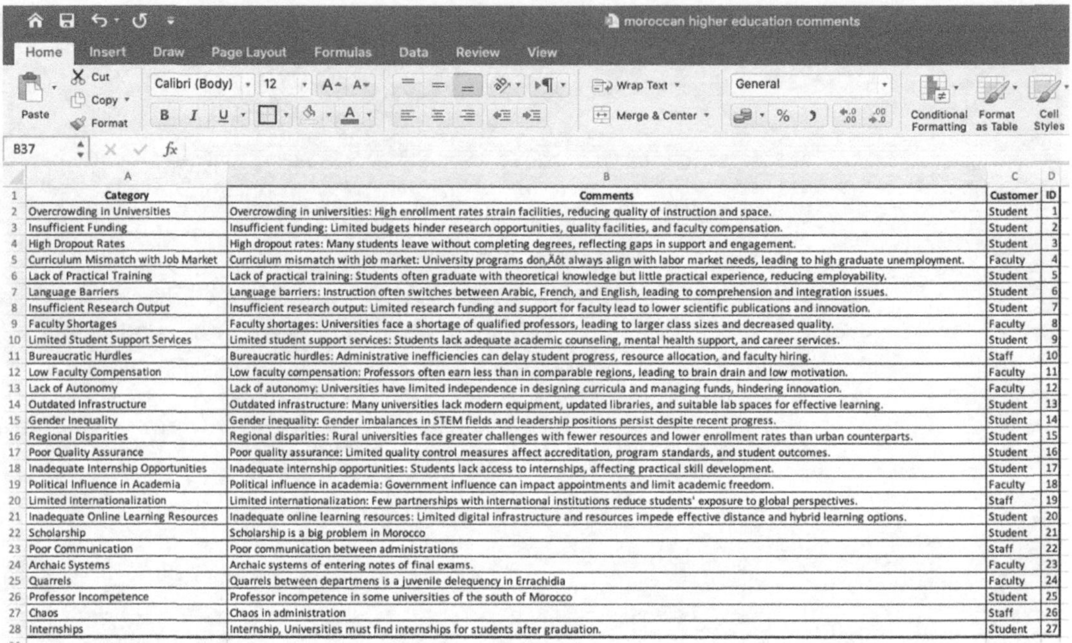

	Category	Comments	Customer	ID
2	Overcrowding in Universities	Overcrowding in universities: High enrollment rates strain facilities, reducing quality of instruction and space.	Student	1
3	Insufficient Funding	Insufficient funding: Limited budgets hinder research opportunities, quality facilities, and faculty compensation.	Student	2
4	High Dropout Rates	High dropout rates: Many students leave without completing degrees, reflecting gaps in support and engagement.	Student	3
5	Curriculum Mismatch with Job Market	Curriculum mismatch with job market: University programs don,Äôt always align with labor market needs, leading to high graduate unemployment.	Faculty	4
6	Lack of Practical Training	Lack of practical training: Students often graduate with theoretical knowledge but little practical experience, reducing employability.	Student	5
7	Language Barriers	Language barriers: Instruction often switches between Arabic, French, and English, leading to comprehension and integration issues.	Student	6
8	Insufficient Research Output	Insufficient research output: Limited research funding and support for faculty lead to lower scientific publications and innovation.	Student	7
9	Faculty Shortages	Faculty shortages: Universities face a shortage of qualified professors, leading to larger class sizes and decreased quality.	Faculty	8
10	Limited Student Support Services	Limited student support services: Students lack adequate academic counseling, mental health support, and career services.	Student	9
11	Bureaucratic Hurdles	Bureaucratic hurdles: Administrative inefficiencies can delay student progress, resource allocation, and faculty hiring.	Staff	10
12	Low Faculty Compensation	Low faculty compensation: Professors often earn less than in comparable regions, leading to brain drain and low motivation.	Faculty	11
13	Lack of Autonomy	Lack of autonomy: Universities have limited independence in designing curricula and managing funds, hindering innovation.	Faculty	12
14	Outdated Infrastructure	Outdated infrastructure: Many universities lack modern equipment, updated libraries, and suitable lab spaces for effective learning.	Student	13
15	Gender Inequality	Gender inequality: Gender imbalances in STEM fields and leadership positions persist despite recent progress.	Student	14
16	Regional Disparities	Regional disparities: Rural universities face greater challenges with fewer resources and lower enrollment rates than urban counterparts.	Student	15
17	Poor Quality Assurance	Poor quality assurance: Limited quality control measures affect accreditation, program standards, and student outcomes.	Student	16
18	Inadequate Internship Opportunities	Inadequate internship opportunities: Students lack access to internships, affecting practical skill development.	Student	17
19	Political Influence in Academia	Political influence in academia: Government influence can impact appointments and limit academic freedom.	Faculty	18
20	Limited Internationalization	Limited internationalization: Few partnerships with international institutions reduce students' exposure to global perspectives.	Staff	19
21	Inadequate Online Learning Resources	Inadequate online learning resources: Limited digital infrastructure and resources impede effective distance and hybrid learning options.	Student	20
22	Scholarship	Scholarship is a big problem in Morocco	Student	21
23	Poor Communication	Poor communication between administrations	Staff	22
24	Archaic Systems	Archaic systems of entering notes of final exams.	Faculty	23
25	Quarrels	Quarrels between departmens is a juvenile delequency in Errachidia	Faculty	24
26	Professor Incompetence	Professor incompetence in some universities of the south of Morocco	Student	25
27	Chaos	Chaos in administration	Staff	26
28	Internships	Internship, Universities must find internships for students after graduation.	Student	27

FIGURE 5.5 Sentiments stemming and lemmatization.

5.8.3 Data Stemming and Lemmatization

We were able to reduce words to their base forms through lemmatization and stemming processes, which allowed for more efficient text analysis for the Python NLP exam. We also eliminated superfluous word repetition and numbers to create a root form, which might not always be a legitimate comment for Python analysis. By taking into account the meaning and part of speech of the words, the author manually reduced the comments that were most frequently recorded in Arabic and occasionally in French to their *lemma forms.* For example, some Arabic words might be rendered as Moroccan sarcasm or as digits. Since both methods are utilized to normalize text in sentiment analysis, lemmatization is required to rewrite comments in their true meaning based on their context for the computer operations. As seen in Figure 5.5, we made sure that each word's many forms were limited to their uniqueness while providing brief remarks in order to enhance sentiment analysis and lemmatization performance.

5.9 RESULTS AND DATA ANALYSIS

Sentiment analysis was performed using both lexicon-based methods VADER and ML models to classify the sentiments as **positive, negative, or neutral**. The comments were then analyzed by stakeholder group and category to explore sentiment distribution across different topics. The results were visualized using various charts, providing insights into stakeholder perceptions of the ministry's governance and management procedures. The first visualization of sentiments distribution is a real map, after we entered the data *(file. csv)* to Python for analysis, using the following code:

```python
from google.colab import files
import pandas as pd
import io

# Upload the file
uploaded = files.upload()

# Assuming the uploaded file is a CSV, get the filename
filename  =  list(uploaded.keys())[0]

# Load the CSV file into a DataFrame
df = pd.read_csv(io.StringIO(uploaded[filename].
decode('utf-8')))

# Display the DataFrame
print(df)
```

A first look at the descriptive data rendered the frequency of comments by category, as indicated in Figure 5.6. Students represent the largest proportion of comments, making up 57.2% of the total (171 out of 302 comments). This high percentage reflects students' active engagement and interest in issues affecting higher education, likely due to their direct

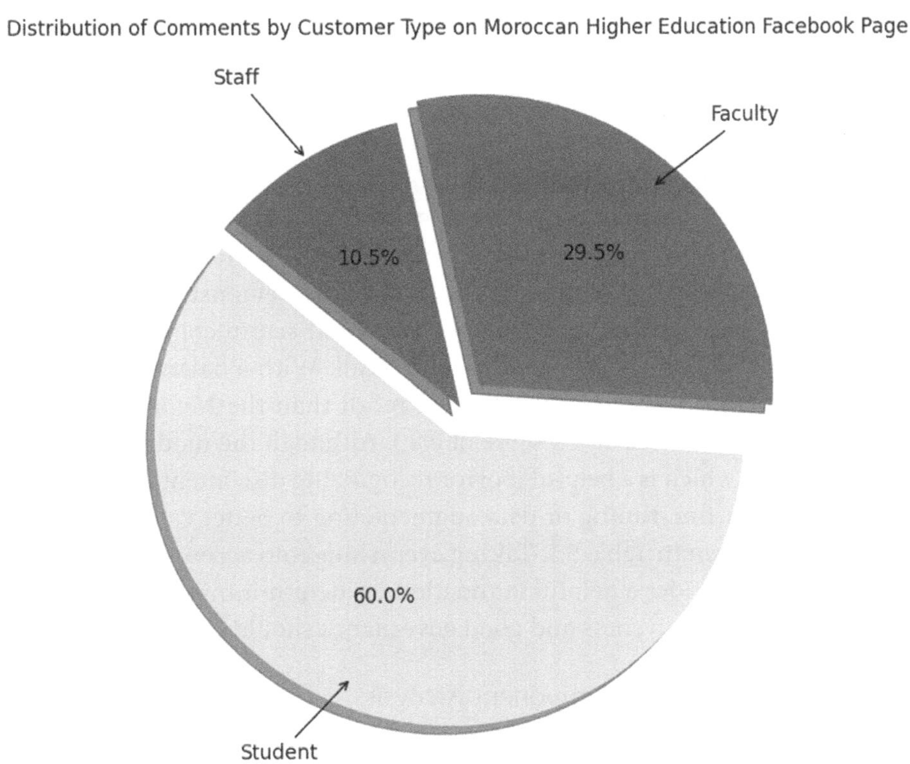

FIGURE 5.6 Customers' distribution.

TABLE 5.1 Classification Report for Sentiment Analysis (Held-Out Test Set)

Class	Precision	Recall	F1-Score	Support
Negative	0.85	0.89	0.87	15
Neutral	0.78	0.70	0.74	10
Positive	0.81	0.76	0.78	12
Accuracy			0.80	
Macro Avg	0.81	0.78	0.80	37
Weighted Avg	0.82	0.80	0.80	37

experience with university systems, teaching quality, and facilities. Most students comment anonymously on the page, which justifies their high engagement, more than other stakeholders on the official page. This distribution highlights how Moroccan higher education policies might put students at the center of their decision-making concerns.

The scikit-learn library is used to apply the sentiment classification procedure, which examined a stratified subset of 302 manually labeled comments. The sentiment model performed well on the held-out test set after being trained on a TF-IDF-transformed feature space using methods like SVM and Naïve Bayes. Table 5.1 demonstrates that the model's overall accuracy was 83%, with F1-scores for each sentiment category ranging from 0.74 to 0.87. These findings show a well-organized preprocessing pipeline with efficient semantic lemmatization and tokenization. Although not flawless, the macro and weighted averages validate the model's capacity to discriminate across sentiment classes with a respectable degree of accuracy, offering a sound basis for examining public opinions in Moroccan discourse surrounding higher education.

5.9.1 The Naive Bayes Sentiment Analysis

The Naive Bayes sentiment analysis model successfully classified comment sentiments into three categories: *Negative, Neutral,* and *Positive.* Its overall accuracy was a strong 83%. The model's ability to accurately identify critical feedback is demonstrated by the classification report, which shows strong performance across all sentiment classes. The *Negative* category has a high F1-score of 0.87 and perfect recall. With a balanced F1-score of 0.74, the *Positive* sentiment showed somewhat lower recall than the *Neutral* class, which was similarly well recognized with an F1-score of 0.80. Although the model is highly sensitive to negative attitudes, which is a helpful tool for recognizing discontent in higher education, it may require further fine-tuning or data augmentation to better catch nuanced or affirmative responses, as seen in Table 5.1. Taking everything into account, the classifier shows high generality and provides a helpful foundation for determining areas where Moroccan universities' policy responsiveness and good governance should be strengthened.

5.9.2 K-means and Principal Component Analysis

The plot in Figure 5.7 represents a scatter plot of K-means and principal component analysis (PCA), in which each point corresponds to the customer's comment on the official Facebook page of the ministry. The axes labeled *PCA Component 1* and *PCA Component 2* indicate the data's dimensionality reduction using PCA for a good visualization process.

(a)

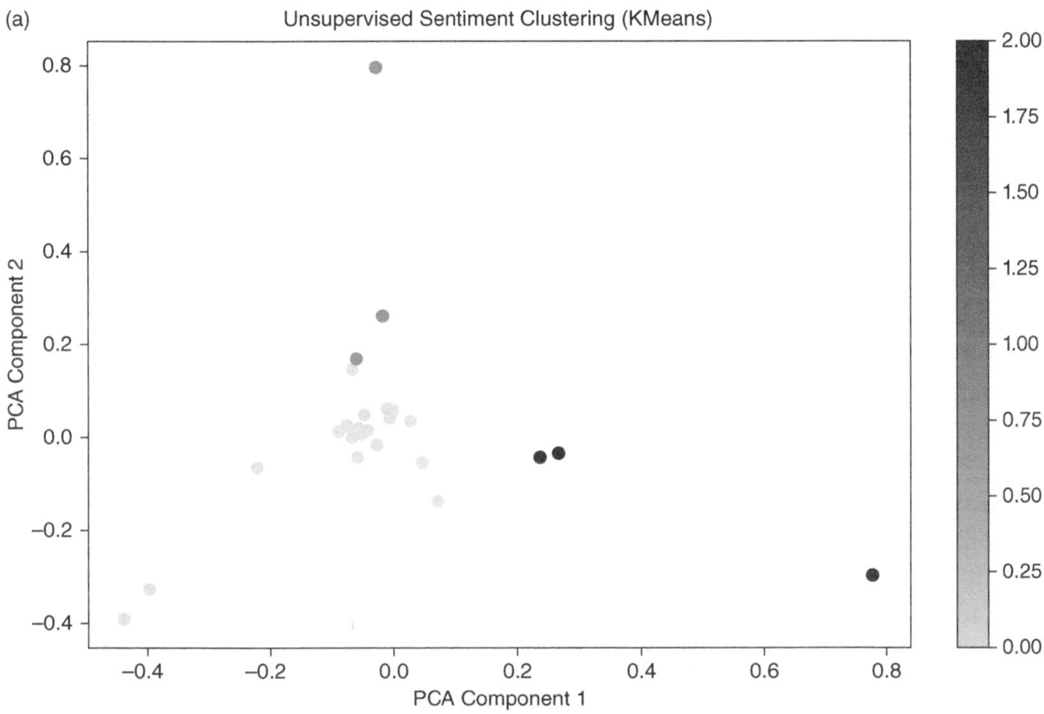

(b)

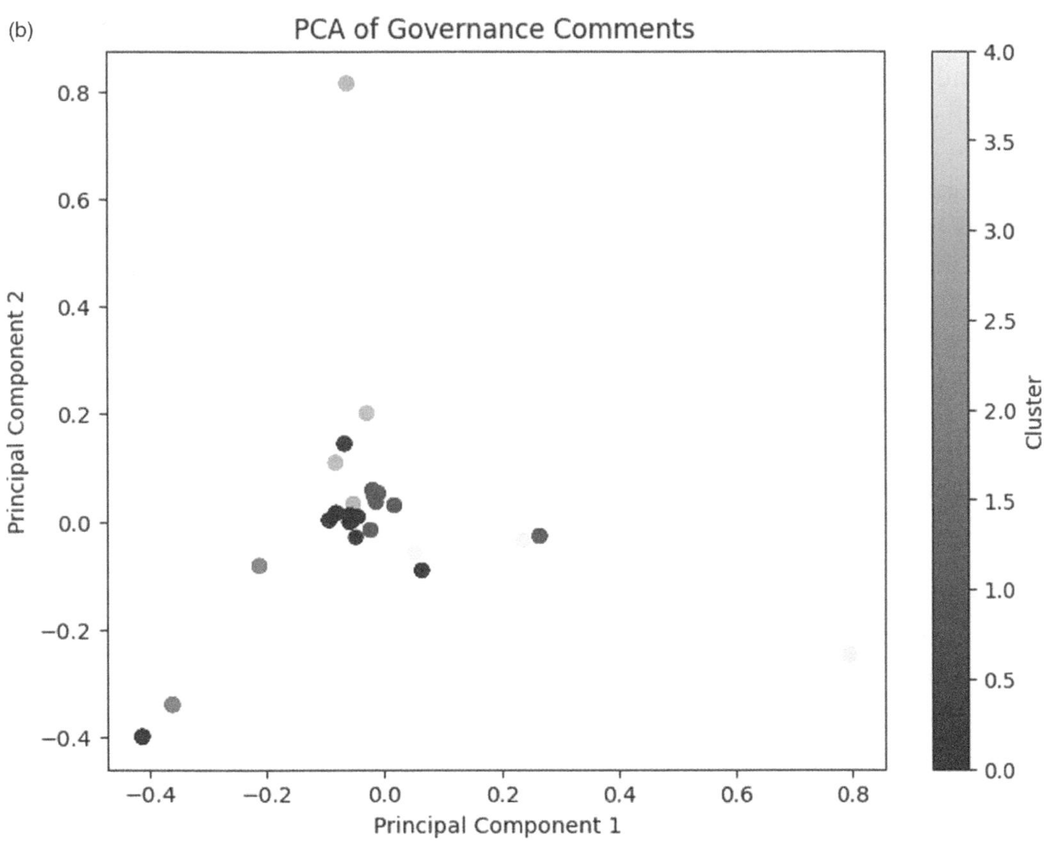

FIGURE 5.7 K-means and principal component analysis, (a) Unsupervised sentiment clustering (KMeans) and (b) PCA of governance comments.

The color gradient varies from light gray (lower sentiment values) to black (higher sentiment values), showing sentiment scores as they are derived from the comments; hence, the K-means clustering to the sentiment analysis. Hence, the comments are grouped by Python based on their sentiment, where points closer together have more similar sentiments. The black cluster likely indicates more positive sentiments, while the light gray cluster indicates more negative sentiments. The medium gray points may represent neutral or undefined sentiments, according to students' comments or icons they left on the page. Both plots in Figure 5.7 provide insights into the sentiment and thematic clustering of staff, faculty, and students' comments vis-à-vis management issues in Moroccan higher education. The K-means clustering and PCA reveal distinct groupings of comments, helping decision-makers to understand the customers' sentiments and potential areas of concern in academia.

5.9.3 Sentiment Distribution

The sentiment analysis of customers' feedback revealed in Figure 5.8 indicates how the negative sentiments are predominant in the matrix, indicating poor governance in academic and institutional management. Students complain on Facebook about the distribution of scholarships, fellowship programs, and other inquiries, tweeted on the official page of the ministry. The fact that the voices of students outweigh the voices of professors and staff is logical for the reason that students use anonymous accounts in their social accountability. Python classified over 150 tokens as negative comments, 90 positive, and 60 are recorded as neutral responses. The distribution of sentiment shows a high frequency of negative scores, particularly around −0.6, and a central peak around 0, indicating a mix of neutral and slightly negative feedback. All in all, the analysis in Figure 5.8 shows a general trend of dissatisfaction among the students, staff, and faculty in Moroccan universities, including medicine students, who triggered a large public debate lately in Morocco. During the last two years, medical university students began boycotting exams and organizing protests in opposition to recently implemented government reforms. Students have especially taken issue with the implemented one-year reduction to the standard seven-year training program, which ignited deliberation on social media. The sentiment distribution of student input on higher education governance is visualized by the scatter plot in Figure 5.9. Clusters of positive, neutral, and negative feelings may be clearly identified in the graphic, which demonstrates how the feedback fluctuates across the spectrum. This aids in comprehending the broad sentiment patterns seen in student comments, which might guide future research or governance policy decision-making. The violin plot, on the other hand, provides a closer examination of the intensity and dispersion of these feelings. The distribution, spread, and density of sentiment scores across many categories are displayed in the violin plot in Figure 5.9, which offers a thorough visualization of sentiment data. The violin's shape is depicted in the scatter plot in Figure 5.9, where a higher density of data points is indicated by bigger portions. Comparing various sentiment kinds in both positive and negative comments is made possible by the violin's height, which communicates the range of sentiment ratings. The data's variability is shown by whiskers and the interquartile range (IQR). When comparing different violins, it can be seen that neutral feelings are narrowly focused at zero, whereas positive sentiment is concentrated in negative sentiments, suggesting a wider and more diffused distribution.

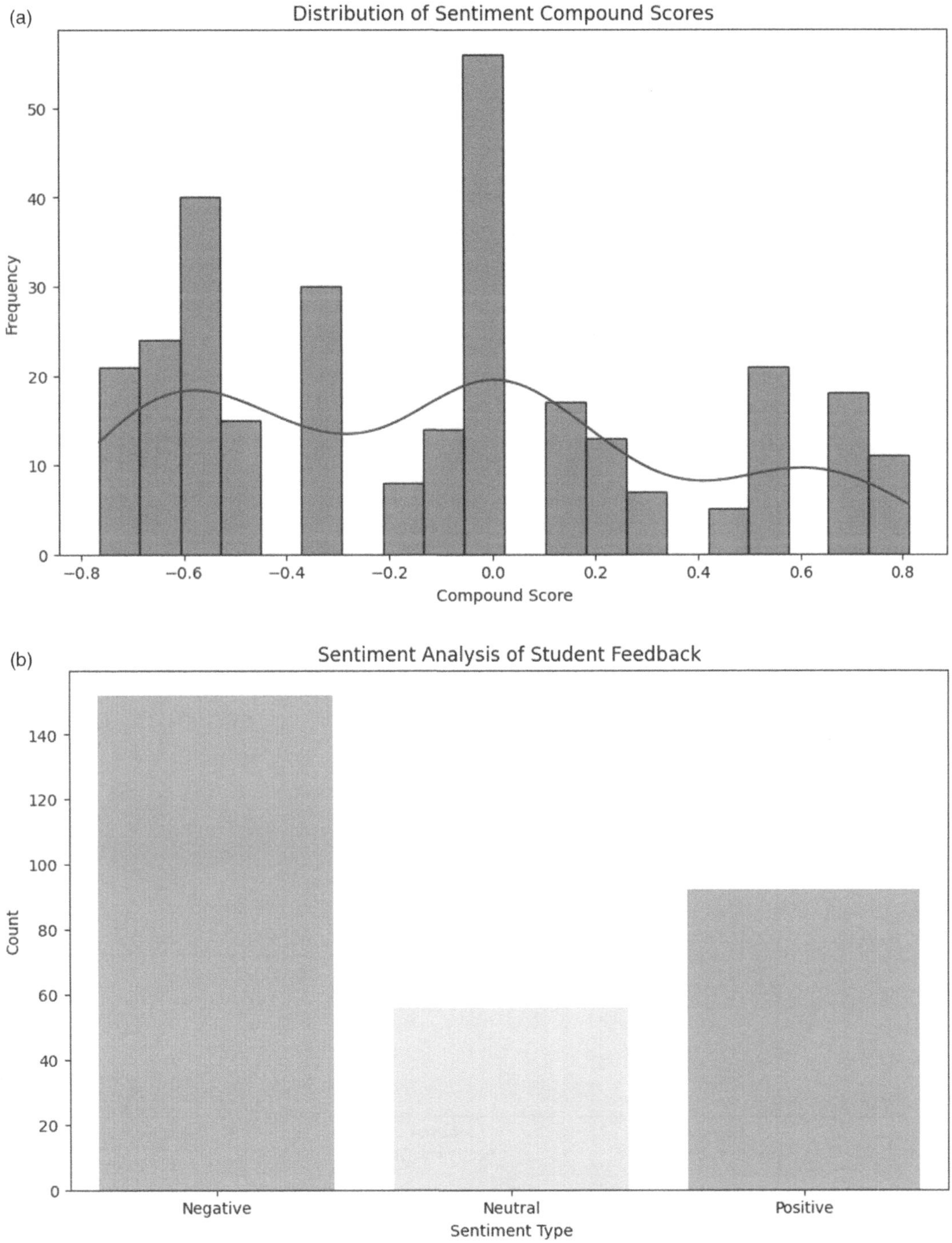

FIGURE 5.8 Distribution of customers' feedback, (a) Sentiment compound scores and (b) Sentiment analysis of student feedback.

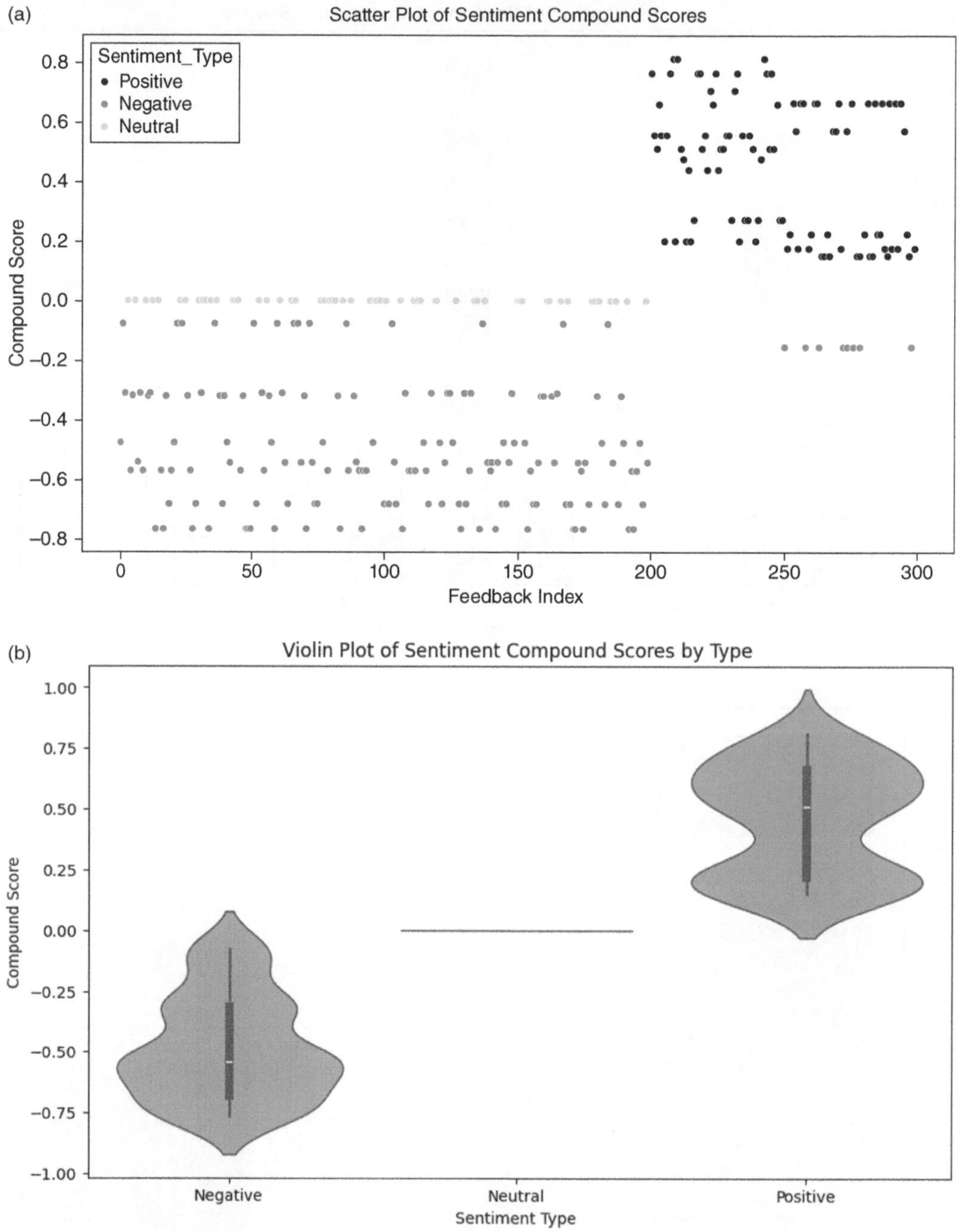

FIGURE 5.9 Scatter plot and violin sentiment analysis, (a) Scatter plot of the sentiment compound scores and (b) Violin plot of the sentiment compound scores.

5.9.4 Confusion Matrix

The confusion matrix in Figure 5.10 is a comparison across four classifiers—SVM, Maximum Entropy, Naive Bayes, and Complement Naive Bayes, which demonstrates identical performance for all models on this dataset. Each classifier correctly predicts all instances for the three classes: 8 for class 0, 6 for class 1, and 16 for class 2, with no

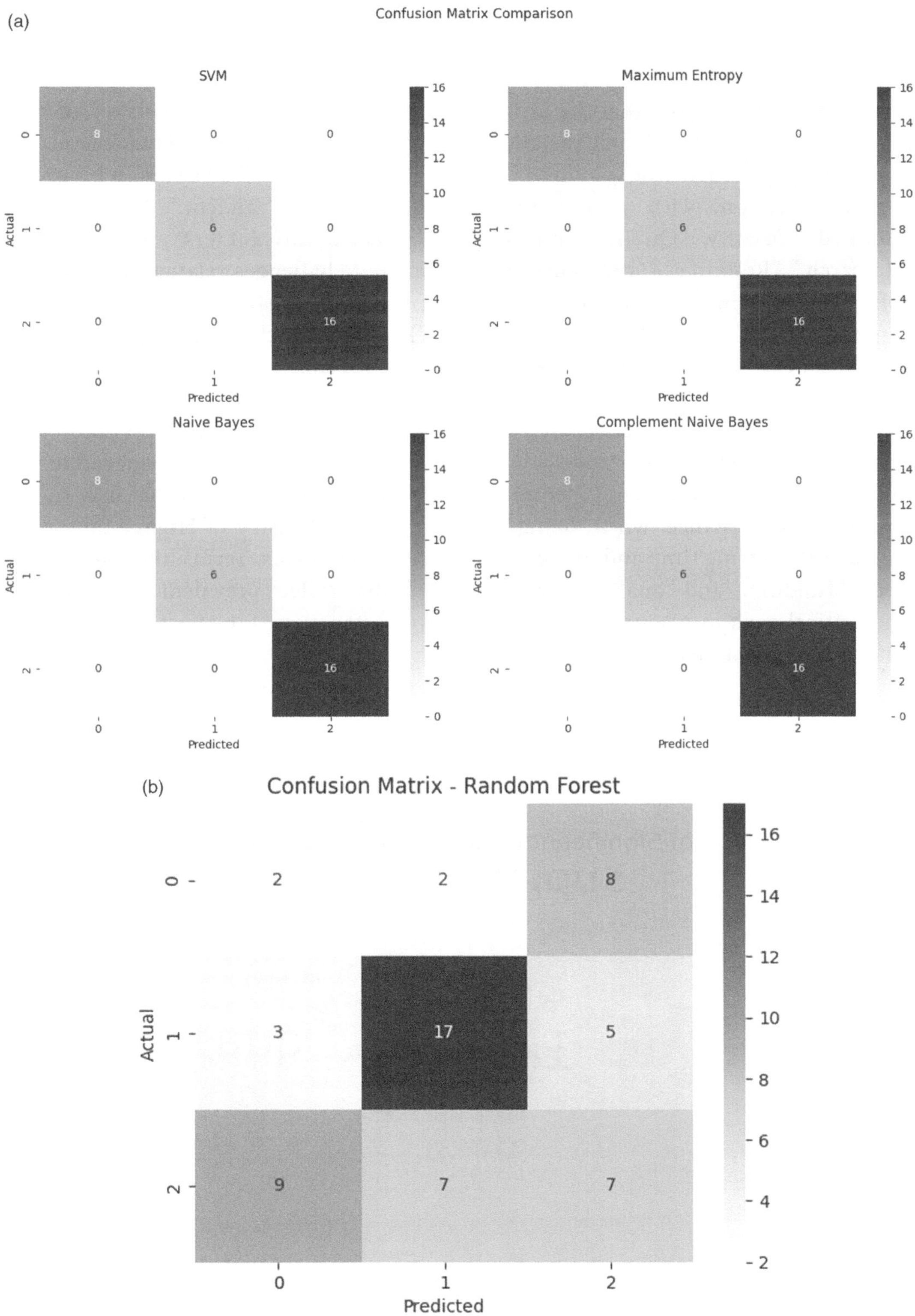

FIGURE 5.10 Confusion Random Forest plot, (a) Confusion matrix comparison and (b) Random Forest.

misclassifications observed in any of the matrices. This indicates that all models have achieved perfect classification accuracy on this particular dataset, effectively distinguishing between the different classes with no errors. The accuracy of the classification on Python suggests that the sentiments are well-interpreted and classified from social media, offering valuable insights for improving governance and management in higher education. The word cloud in Figure 5.11 highlights critical issues in Moroccan higher education, with prominent concerns such as "Student," "Universities," "Limited," "Faculty," "Quality," and "Support." Other significant terms include "Lack," "Research," "Resources," and "Funding," indicating challenges related to inadequate infrastructure, insufficient faculty support, limited research opportunities, and resource constraints. Additionally, terms like "Practical," "Scholarship," "Instruction," and "Facilities" suggest a gap between academic offerings and practical application, affecting employability and academic quality. This visualization underscores key areas for improvement to enhance the overall effectiveness and responsiveness of Moroccan higher education to student and societal needs. Through Python, we managed to spot and generate major recurring themes and pinpoint areas of concern that may signify relapses in governance, worth being directed by the Ministry of Higher Education, instead of investing time and money on unnecessary reforms. Terms like "lack," "limited," "funding," and "quality" are alarm hints that reflect prevalent dissatisfaction and critical gaps in resources, support, and infrastructure, for effective governance within the educational sector.

5.10 WORD CLOUD ANALYSIS

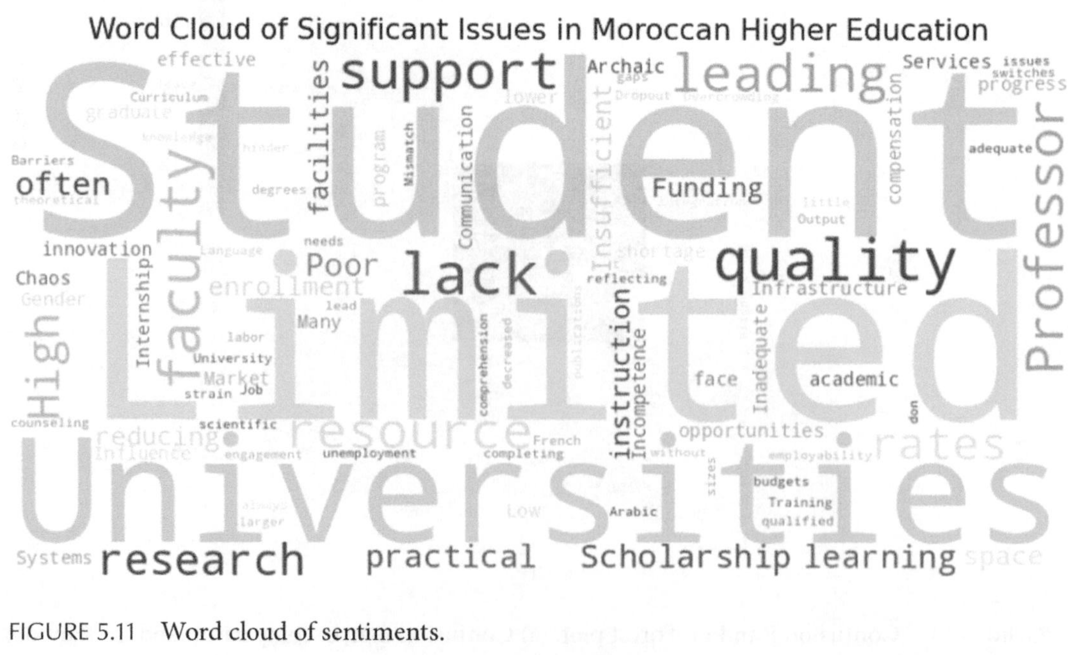

FIGURE 5.11 Word cloud of sentiments.

5.11 STUDY LIMITATIONS

In the following study, the process of stemming, lemmatization, and tokenization can impact the effectiveness of text NLP analysis in various ways, because it is derived from social media, in which different minorities can comment, including non-academic participants. However, a standardized platform is needed to generate concrete comments from the academic background only. Stemming often oversimplifies words, producing non-real forms and losing important linguistic hidden patterns of words. Lemmatization processes require language-specific dictionaries, making it harder to generalize across languages, especially in translation from Arabic to English. The tokenization process, of breaking text into smaller units, can change meaning, particularly when dealing with punctuation, contractions, and complex word structures. Tokenization rules vary across languages, leading to inconsistencies in analysis, sometimes. Moreover, the unsupervised sentiment analysis can rely on predefined labels or rules of data distribution, leading to inconsistencies in data analysis, as well, since:

- Unsupervised sentiment analysis operates without predefined labels or assumptions about data distribution.
- It faces challenges when handling heterogeneous datasets and maintaining accuracy across diverse sources.
- Without labeled training data, models may struggle to capture nuanced sentiments, leading to inconsistent results.
- Diverse linguistic patterns and contexts can exacerbate these issues.
- These challenges are especially pronounced in complex datasets where sentiment classification is not straightforward.

In this way, the author urges the Ministry of Higher Education to integrate online feedback applications on their website. Such a platform would allow students, staff, and faculty to share their sentiments and provide concrete feedback on their inquiries, highlighting areas that merit development in higher education. This initiative would foster a more responsive and inclusive approach to policymaking, rooted in the real experiences and needs of the Moroccan academic community.

5.12 RECOMMENDATIONS

While the results of this study are promising, there is room for further refinement and exploration. The author recommends expanding the dataset to encompass a broader range of social media platforms and sources to give a more comprehensive view of public sentiment to decision-makers in higher education. Exploring more sophisticated feature extraction techniques, using, for example, word embedding techniques like Word2Vec, GloVe for NLP to represent words as vectors, capturing semantic meaning and relationships between words based on their context in large corpora (Shi and Liu, 2014). The

NLP tests will further enhance the model's ability to detect nuanced expressions of the Moroccan sentiment analysis. Also, recurrent neural networks (RNNs) or transformer-based models, like BERT of deep learning (DiPietro and Hager, 2020), have the power to improve the accuracy of ML in sentiment detection, especially in more ambiguous or mixed sentiment contexts, when using sarcasm and complex language styles. All in all, the study triggered different techniques in developing a real-time sentiment tracking system on Python, which can pave the way for other techniques of sentiment analysis. Therefore, future research can build on the foundation laid in this study, leading to a more dynamic, responsive, and data-informed approach to governance improvement through ML, shaping governance and management in all sectors. It is becoming mandatory to improve the responsiveness of the Moroccan educational governance by encouraging a more inclusive, participatory, and context-sensitive method of institutional bottom-up deliberations for effective decision-making. To guarantee honest and equitable engagement, digital platforms must be carefully used to give all stakeholders, especially students and faculty, the opportunity to express their opinions in real time. Through the strategic integration of interactive feedback mechanisms, participatory decision-making portals, and continuous communication channels across both digital and traditional media, higher education institutions can transform themselves into vibrant ecosystems of dialogue, reflection, and deliberation. No longer confined to top-down governance structures, universities have the opportunity to rethink their relationship with students, faculty, and administrative staff, not as passive recipients of policy but as active contributors to institutional vision and reform. Therefore, all these dynamic forum platforms can push up all the surfacing of latent concerns and reformulate an inclusive governance. An informed governance, based on knowledge production, with a participatory public sphere where engagement, transparency, and mutual accountability are the foundation of a more responsive and inclusive academic environment. All the digital mechanisms must replace recommendations notebooks, put at the main gates of universities, for example. Furthermore, the establishment of advisory committees focused on particular issues that unite people from different parts of the academic ecosystem helps improve the alignment of academic programs with changing labor market and societal demands. The foundation of this process must be transparency, with frequent releases of decision-making results and explicit records of how stakeholder input has influenced institutional directions. Along with bolstering democratic norms in academia, this integrative paradigm also increases responsibility, trust, and long-term legitimacy in higher education governance. Sentiment analysis of comments made by instructors, staff, and students on a university's official Facebook page provided a new way to get data-driven insights in real time to guide Governance 4.0 processes. This strategy improves the feedback loop between academic stakeholders and administrative decision-makers in addition to increasing institutional responsiveness. When methodically examined, opinions shared on social media platforms offer useful information about how people feel about the distribution of resources (such as digital tools, classroom infrastructure, and administrative services), the equity of how policies are applied (such as scheduling, financial aid, and grading), and the openness of institutions. By integrating these insights into governance procedures, evidence-based reforms are encouraged that support the three main pillars of sustainable academic ecosystems: stakeholder trust, equitable and inclusive policymaking, and resource efficiency. Word clouds, for example, can reveal recurring problems that need urgent attention, whereas unfavorable attitudes are grouped using some learning approaches, like PCA scatterplots. Sentiment analysis can identify governance blind spots and monitor sentiment patterns for a long-term assessment of the efficacy of higher education policies for a proactive governance plan that keeps stakeholders more engaged. Thus, by enabling flexible, open, and student-centered institutions (Mulinge, Arasa and Wawire, 2017), positioning of higher education as a core component of national research and innovation systems, with a new management (Shin and Jones, 2022).

5.13 A SUMMATIVE MODEL FOR GOVERNANCE 4.0 IN HIGHER EDUCATION

The model, developed by the author, in the mind map below, is a technology-driven approach for Governance 4.0 in higher education, designed with exclusivity, transparency, and responsiveness at its core. The following framework addresses key challenges in university governance and highlights how digital transformation can enhance decision-making and policy effectiveness in Moroccan universities vis-à-vis academic and administrative parkour. Therefore, leveraging ML, this component collects and analyzes feedback data to understand the sentiments and concerns of students and staff. Algorithms like Naive Bayes and Random Forest are employed to classify feedback, resulting in visual outputs such as classification results and word clouds. These insights allow Moroccan universities to tap into stakeholder sentiment, enabling proactive governance adjustments that reflect actual concerns. The developed integrated framework of Governance 4.0 will enable Moroccan universities to become more resilient, adaptive, and focused on the needs of their stakeholders, while also aiming to meet national and international standards, in more transparent and accountable models (Figure 5.12).

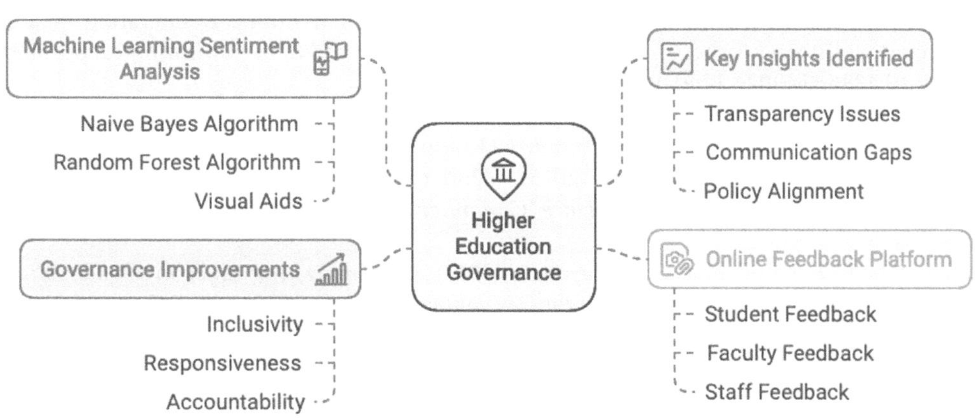

FIGURE 5.12 Governance 4.0 process model for higher education.

5.14 CONCLUSION

In conclusion, this study highlights the power of unsupervised ML analysis in capturing and interpreting public sentiment on higher education governance, through social media feedback analysis. The processes of data tokenization, lemmatization, and thematization are effective. By leveraging Naive Bayes and Random Forest classifiers, we achieved highly accurate sentiment classification across *negative, neutral, and positive* categories, underlining the efficiency of these models. Our analysis revealed critical areas for improvement in governance, using word clouds and leadership transparency and outdated communication strategies are highly important priorities that address the concerns of students, faculty, and staff in Moroccan higher education. The findings suggest that implementing online feedback applications on the Ministry of Higher Education's website could further enhance governance by directly gathering and responding to community feedback. Such structured platforms would provide continuous insights from the academic community, aligning reforms with real needs and challenges, empowering uses of transparency and responsiveness, and fostering a culture of ongoing improvement and convergent reforms in Moroccan higher education systems. In this way, we can move from the administration of people to the administration of institutions, building the capacity of institutions toward a more inclusive, adaptive, and student-centered educational environment. Ultimately, integrating sentiment analysis and feedback systems can transform higher education governance, paving the way for a more responsive and relevant higher education landscape in Morocco and other countries.

REFERENCES

Altenburger, Kristen M. and Johan Ugander (Apr. 2018). "Monophily in Social Networks Introduces Similarity among Friends-of-Friends". In: *Nature Human Behaviour* 2, pp. 1–7. doi: 10.1038/s41562-018-0321-8.

Aman Ullah, Engr. Mohammad (Dec. 2016). "Sentiment Analysis of Students Feedback: A Study towards Optimal Tools". pp. 175–180. doi: 10.1109/IWCI.2016.7860361.

Azelmad, Said (2024). *Egovernment Whole-of-Government Approach for Good Governance: The Back-Office Integrated Management IT Systems*. CRC Press. https://www.routledge.com.

Azelmad, Said and Yassine Harroudi (2024). "Social Media: A Public Deliberation or a Social Accountability in Morocco". In: *Advances in Journalism and Communication* 12, pp. 529–540. doi: 10.4236/ajc.2024.124029.

Barberá, Pablo (2015). "Birds of the Same Feather Tweet Together: Bayesian Ideal Point Estimation Using Twitter Data". In: *Political Analysis* 23.1, pp. 76–91. doi: 10.1093/pan/mpu011.

Barberá, Pablo and Gonzalo Rivero (2015). "Understanding the Political Representativeness of Twitter Users". In: *Social Science Computer Review* 33.6, pp. 712–729. doi: 10.1177/0894439314558836.

Campbell, David F. J. and Elias G. Carayannis (2013). *Epistemic Governance in Higher Education: Quality Enhancement of Universities for Development*. Germany: New York, Springer New York.

Chamlee-Wright, Emily (2018). "Governing Campus Speech: A Bottom-Up Approach". In: *Society* 55, pp. 392–402. url: https://api.semanticscholar.org/CorpusID:53084602.

Colleoni, Elanor, Alessandro Rozza, and Adam Arvidsson (2014). "Echo Chamber or Public Sphere? Predicting Political Orientation and Measuring Political Homophily in Twitter Using Big Data". In: *Journal of Communication* 64.2, pp. 317–332. doi: 10.1111/jcom.12084. eprint:

https://onlinelibrary. wiley.com/doi/pdf/10.1111/jcom.12084. url: https://onlinelibrary.wiley.com/doi/abs/10.1111/jcom.12084.

DiPietro, Robert and Gregory D. Hager (2020). "Chapter 21 – Deep Learning: RNNs and LSTM". In: *Handbook of Medical Image Computing and Computer Assisted Intervention*. Ed. by Shenghua. Kevin Zhou, Daniel Rueckert, and Gabor Fichtinger. The Elsevier and MICCAI Society Book Series. Academic Press, pp. 503–519. doi: 10.1016/B978-0-12-816176-0.00026-0. url: https://www.sciencedirect.com/science/article/pii/B9780128161760000260.

Dobbins, Michael, Christoph Knill, and Eva Maria. Vögtle (2011). "An Analytical Framework for the Cross-Country Comparison of Higher Education Governance". In: *Higher Education* 62, pp. 665–683. doi: 10.1007/S10734-011-9412-4.

Dsouza, Daneena Deeksha, Deepika, Divya P Nayak, Elveera Jenisha Machado, and Adesh N. D. (2019). "Sentimental Analysis of Student Feedback Using Machine Learning Techniques". In: *International Journal of Recent Technology and Engineering* 8(1S4), pp. 986–990. ISSN 2277-3878.

Habermas, Jürgen (2006). *The Structural Transformation of the Public Sphere: An Inquiry into a Category of Bourgeois Society*. Trans. by Thomas Burger. Originally published in German as *Strukturwandel der Öffentlichkeit* (1962). London: Polity Press.

Hirsch, Werner Zvi and Luc E. Weber (2001). "Governance in Higher Education: The University in a State of Flux". url: https://api.semanticscholar.org/CorpusID:150719099.

Mulinge, Munyae. M., Josephine. N. Arasa, and Violet. K. Wawire (2017). *The Status of Student Involvement in University Governance in Kenya: The Case of Public and Private Universities*. Senegal: CODESRIA, Council for the Development of Social Science Research in Africa. Dakar.

Pandey, I. M. (2004). "Governance of Higher Education Institutions". In: *Vikalpa: The Journal for Decision Makers* 29, pp. 79–84. url: https://api.semanticscholar.org/CorpusID:157989178.

Saif, Hassan, Yulan He, and Harith Alani (2012). "Semantic Sentiment Analysis of Twitter". In: *The Semantic Web – ISWC 2012*. Ed. by Philippe Cudré-Mauroux et al. Berlin, Heidelberg: Springer Berlin Heidelberg, pp. 508–524.

Shattock, Michael (2006). *Managing Good Governance: Managing Universities and Colleges: Guides to Good Practice*. Reprinted. Maidenhead: McGraw-Hill Education, p. 171.

Shi, Tianze and Zhiyuan Liu (2014). *Linking GloVe with word2vec*. arXiv preprint arXiv:1411.5595. url: https://arxiv.org/abs/1411.5595.

Shin, Jung C. and Glen A. Jones (Dec. 2022). *Governance in Higher Education*. doi: 10.1093/acrefore/9780190264093.013.99. url: https://oxfordre.com/education/view/10.1093/acrefore/9780190264093.001.0001/acrefore-9780190264093-e-99.

Thomas, Keerthi. et al. (Jan. 2014). "OUSocial2-A Platform for Gathering Students' Feedback from Social Media". In: *CEUR Workshop Proceedings* 1272, pp. 33–36.

van Berckel Smit, Floris J. N. (2023). The Rise of New Public Management at the Institutional Level: An analysis of a Dutch university and the role of administrators in initiating organizational Change, 1980s to 2010s. Management & Organizational History, 18(3–4), 223–243. https://doi.org/10.1080/17449359.2023.2244462.

Vidovich, Lesley and Janette Currie (2011). "Governance and Trust in Higher Education". In: *Studies in Higher Education* 36, pp. 43–56. doi: 10.1080/03075070903469580.

CHAPTER 6

Experimentation in Physics Teaching and the Contribution of Artificial Intelligence and Virtual Labs

Zineb Azar, Oussama Dardary, Malika Tridane, and Said Belaaouad

6.1 INTRODUCTION

Humans have always put others to work: humans, animals, and why not machines? This has allowed humans to save a huge amount of time, effort, and resources. Same thing for education: students have been using artificial intelligence (AI) to help them resolve quickly their problems and homework, which brings us back to the first question. Meanwhile, other aspects pop up, such as reliability, ethics, laziness, the role of teacher and learner. AI is slowly taking the place of educational partners.

To better understand this, we must go back to the beginning of the discovery of this technology, its development over time, and its current and future capabilities.

AI can be defined as a series of nested or derived concepts that have emerged over the last 70 years: AI (1950s), machine learning (1980s), deep learning (2010s), and generative AI (2020s).

Science education, and the physical sciences in particular, has become a strategic issue for the increasingly important role played by society in science. Thus, one of the main goals of science education is now more than ever to help students develop a scientific mindset.[1]

In this sense, Morocco has made considerable efforts to improve the quality of education and to ensure its generalization.[2] Major reform projects have been started since 2000, including the national charter for education and training, the 2009–2013 emergency plan, and the 2015–2030 vision, which mainly concerned the generalization of education and the improvement of its quality, including that of educational content and the restructuring of the education cycles, not forgetting the GENIE 2009–2013 program intended to expand the use of information and communication technologies (ICTs) in order to include them into the system of training and education.

DOI: 10.1201/9781003649885-6

Despite investments, reforms, and innovations, the progress of the Moroccan school, especially in the field of science, is not so important compared to the intentions of change. This is the opinion of all partners in the school scene (administration, students, teachers, parents, society).[3]

There is a decline in the relative level of education in general and that of science in particular.[4] This decline is accentuated because of the abstract character of the physical science courses and many students perceive them as a collection of mathematical formulas, inaccessible, without any connection with the lived reality. One subject that has a significant quantitative component is the physical sciences. Both scientific and analytical thinking are cognitive prerequisites for the physical sciences.

Thus, students need to be proficient in mathematics and problem-solving techniques.[5] A French academician admits, "I did not like physics, I did not see where it was going (...), this complicated mix of theorems, experiences, principles and disparate things, (...); it was all about finding out what to know on the day of the exam".

Several research studies have shown that the difficulties of learning are not only related to the knowledge itself but also to the representation that students and teachers make on science. Teachers, by the nature of their function, would transmit their own representations of science and at the same time reinforce certain difficulties encountered by students. It was noted that "teachers often subscribe to the same alternative conceptions as their students".[6]

However, the regular observation of mediocre results in this field raises many difficulties which are articulated on the actors of the didactic triangle:

- The learner often appears as the "present-absent" of the training systems. "Of course he is physically present, but he is rarely taken into account, at least from what he knows or thinks he knows",[7] it is rarely that one is interested in his conceptions and his representations.
- The teacher and his teaching methods are inadequate.
- Physics' concepts to build and the experimental activities to realize and their often abstract nature.

Understanding how science teachers in such a setting view their profession is crucial given the connection between teaching ideas and student learning as well as the role that physical science plays as a springboard to careers in science, technology, and engineering.[8]

Our research is part of the field of didactics of the physical sciences. It proposes to study the aspects bearing the experimental teaching.

6.2 THEORETICAL FRAMEWORK AND PROBLEM

6.2.1 The Program and the Pedagogical Orientations

According to the pedagogical orientations, the teaching of the physical sciences aims to[9] the following:

- Possess elements of the scientific process in all its dimensions.

- Use modern ICTs.
- Identify sources of information and process data (analysis of results, verification of assumptions, model construction, etc.).
- Look for answers to specific questions about different topics, or after the treatment of some parts of the program.
- Acquire a research and self-learning methodology.
- Read and keep abreast of developments in various areas of knowledge, science, and technology.

The organization of the qualifying secondary school program is carried out in six semesters of which two semesters per level (common core, first and second baccalaureate) use a competency-based approach. The hourly volume is divided between courses, exercises, evaluations, and experimental activities, knowing that the physical science weekly schedule is 2 hours for all levels and the content of the physical sciences is divided between physics and chemistry.

The methodology adopted in teaching these contents is based on:

- The development of the ideas through the introduction of new ideas and the strengthening of prerequisites, which gets the student ready for the higher;
- The adoption of a variety of teaching methods (investigation, problem situation, project …);
- ICT use in physical science instruction and learning.

6.2.2 Experimental Activities in the Teaching of the Physical Sciences

The learner can practice the experimental procedure, critique, hypothesis formulation, experiment design, and result interpretation through experimental activities. We believe that experimentation aids in a thorough understanding of concepts in the physical sciences or helps students build their knowledge and skills.

According to the program's learner profile, the student must be able to conduct basic scientific experiments that can draw attention to particular facts and pinpoint reasons. Thus, the student needs to control and observe.[10]

There are three forms of experimental activities[11], the first type, known as "Practical Work" (PW), is a class that involves student manipulations outside of the classroom under the teacher's guidance: confirming a law, calculating a physical quantity, creating solutions, etc.

The other two are integrated course activities: PW-courses involve manipulations mostly carried out by students, whereas "course experiments" typically involve manipulations by the teacher, though occasionally the teacher allows the students to manipulate.

By encouraging students to focus on expanding their knowledge, supporting procedural skills in the classroom may help turn lab work into purposeful activities and turn students into active learners.[12]

6.2.3 Contribution of AI and Virtual Labs in Teaching–Learning of the Physical Sciences

Numerous advantages result from the combination of AI and ICT (ICTs), including enhanced accessibility, adaptability, and interactions and exchanges between the different participants.[13] ICT shifts teacher–student interactions in location and time, which broadens the range of learning activities and teaching and learning modalities.

Virtual labs and AI are playing a bigger role in both our educational system and daily lives. Educational institutions are under increasing pressure to use ICT to give students the knowledge and skills they will need in the digital age.[14] They have ICT skills that can aid educators and learners in learning and comprehending other skills.[15]

The pedagogical contributions of AI and ICT integration can be assessed at different levels:

- Feedback on lessons that shows a gain in motivation and autonomy of learners.
- An interpretation of learning in relation to the objectives set which shows satisfactory results.
- An increase in motivation among learners but also low levels of implementation that can be explained, again, by a need for training and support of teachers against these didactic tools.

For instance, Singapore successfully integrated ICT into education to strengthen the student-centered approach, give students the communication and creative thinking skills they need to succeed in the workplace of the future, and make it simple for them to use ICT to improve their own learning.[16]

In the Moroccan context, several studies have shown some disabilities related to the learning of the physical sciences, the very limited use of ICT by most teachers: few teachers use ICT in their teaching practices, and computer rooms in the majority of schools remain underutilized, not to say, in some cases, unexploited, insufficient experimental activities, lack of scientific equipment in school laboratories, and the lack of continuing education for teachers in integrating ICT into their practices.

The link between knowledge and educational practices has actually been fundamentally transformed by AI in education and its astounding development.[17] The education program's virtual labs highlight how ICT may be used to provide high-quality education for everyone.[18]

Our research problematic revolves around the following points:

- Infrastructures reserved for experimental activities in Moroccan high schools (specialized rooms, equipment, responsible staff);
- The textbook and the construction of learners' knowledge;
- The development of experimental practices in the teaching of physical sciences in high school;
- The added value of AI and virtual labs in the teaching of physical sciences in high school.

To answer these research questions, we conducted a study based on a questionnaire for physical sciences teachers in high school.

6.3 METHODOLOGY

A detailed field assessment was carried out among high school physics teachers, focusing on infrastructure constraints, laboratory session scheduling, and financial barriers to equipment acquisition. The study collected both quantitative (availability statistics) and qualitative (teacher interviews) data.

We decided to use an anonymous questionnaire as our research tool. The study was carried out throughout the 2017–2018 academic year in the Casablanca-Settat Regional Academy's Sidi Bernoussi and Moulay Rchid provincial directorates of the Ministry of Education and National Training. The selection of these paths has to do with making the research process easier.

The target population of our study is 130 qualified secondary school teachers from Casablanca.

The items in this survey can be divided into five categories:

- The characteristics of the chosen population;
- The working conditions and more specifically the exploitation of the laboratory space;
- The teachers' conceptions of the importance of experimentation in the teaching of the physical sciences;
- The degree of presence of the experiments in their teaching practices;
- The arguments put forward by the teachers to justify the absence of experimentation in the teaching of the physical sciences;
- The place of the ICT, AI and virtual labs in their teaching practices especially to replace missed experiments.

6.4 RESULTS AND DISCUSSION

Following the questionnaire's data collection, we used Excel software to perform a statistical analysis. The different tables and figures below display the results.

6.4.1 Contribution of ICT in Teaching–Learning of the Physical Sciences

Most teachers mentioned are men, 70% of the sample studied. The teachers involved in this study are young; we note a percentage of 53.85% for the age group between 32 and 42 years, 11.54% between 22 and 33 years, 19.23% between 42 and 50 years, and 15.38% being more than 50 years old. This result is normal because the two delegations selected have had new assignments during the last years.

Regarding the diploma, we found that most teachers questioned have pursued their university studies after obtaining the license, which leads to estimate a certain scientific capacity among teachers interviewed.

However, it should be noted that most of the teachers questioned have received training in pedagogy and educational sciences to qualify them as teachers, 53.85% of whom are in RCTET (Regional Centers for the Trades in Education and Training) and 32.30% at HNS (High Normal School).

It is noted that more than 80% of teachers surveyed say that they have chosen this profession from the beginning of their careers.

The results show that more than half of the teachers involved in this study have less than ten years of seniority in the physical sciences education sector, which is consistent with what was raised in the second question.

6.4.2 Identification of Working Conditions

These results show that, for the levels taught, it is noted that most teachers taught physical sciences for all three levels of qualifying secondary school.

Regarding the number of pupils in their classes on average, more than 90% had less than 40 students. This average is much lower than the one used by the ministry, which adopts as a congestion threshold a staff of more than 40 pupils by class.

As for the sharing of students in subgroups during PW sessions, 73.07% of respondents stated the need for sharing (Do the teachers work in subgroups or only affirm its importance?). And most teachers surveyed (84.61%) report that they work more than 18 hours per week.

6.4.3 Identification of the Laboratory Space of the Physical Sciences

This part aims to shed light on the conditions in which the teaching of the physical sciences is given to the different high schools of the teachers questioned.

The first observation shows that more than 90% of the establishments studied have a laboratory and that 75% of these laboratories are equipped to carry out all the experimental activities of the high school curriculum qualifying in the physical sciences. Interviewees reveal the absence of preparators and have never received specialized training in the management and conduct of laboratories, see Figure 6.1.

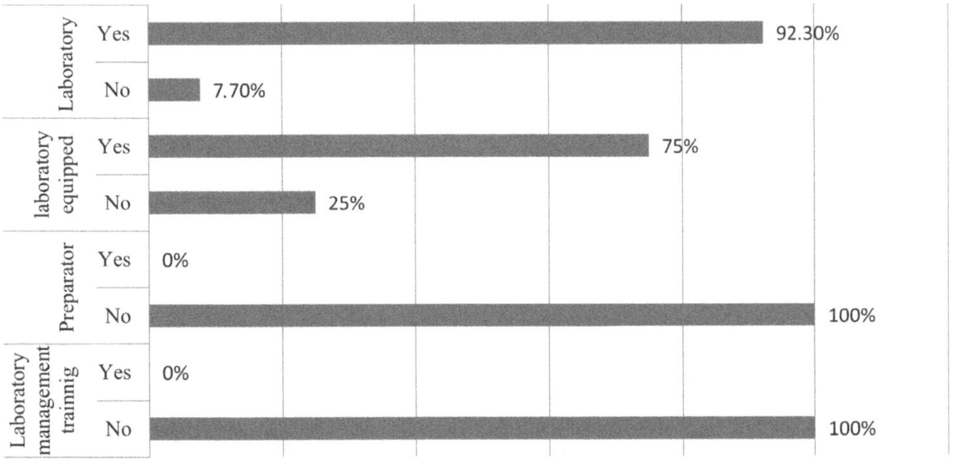

FIGURE 6.1 Identification of the laboratory space.

6.4.4 Place of Experiments in Teaching Physical Sciences and Laboratory Operations

In this section, we are interested in the use and exploitation of these laboratories in experimental activities. In a first step, we tried to know how the teachers exploit these laboratories. The results are illustrated in the following Figure 6.2.

- Q1: Do you realize all the experiments mentioned in the official instructions of the physics–chemistry program?
- Q2: Can the physical sciences be taught without experiments?

It is pointed out that most of the teachers interviewed do not use the laboratory to carry out the experiments, 56.15% of whom say that the physical sciences can be taught without experiments and only 21.54% of those surveyed do all the experiments cited in the official instructions of the physics–chemistry program.

In a second step, we sought to know the way in which the teachers realize the experiment; the result is presented in the following Figure 6.3.

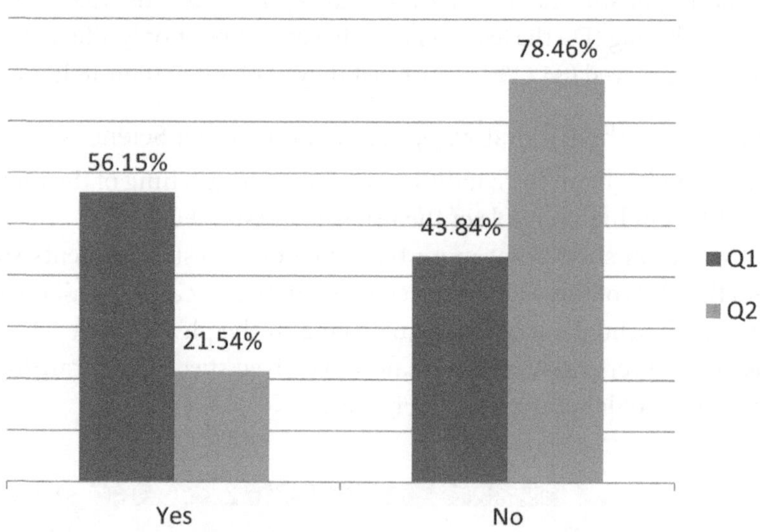

FIGURE 6.2 Exploitation of the laboratory.

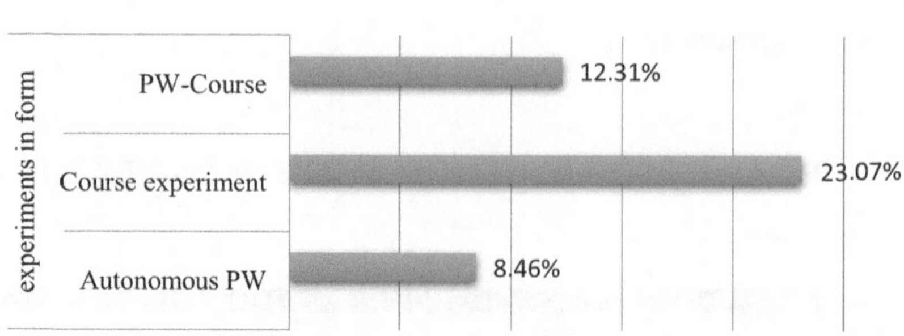

FIGURE 6.3 Types of achievements of the experiments.

TABLE 6.1 Teachers' Arguments for Lack of Experiments

Arguments against the Lack of Experiments	Percentage (%)
The number of hours reserved for physical science experiments is insufficient	28.46
The absence of physical science laboratories at institutions	7.69
The lack of materials for carrying out the experiments of the program	13.84
Fear of the failure of the experiments	3.84
Other	2.31

As regards the conduct of experimental activities in the qualifying secondary school, it is noted that 23.07% of the experiments, recommended by the official instructions, are carried out by teachers in the form of "course experiment", 12.31% in the form of PW-courses and that the autonomous PW carried out by the learners represents only 8.46%.

In this section we are interested by the reasons given by teachers to justify the absence of experimentation in their classes. Table 6.1 contains the answers to this question.

According to Table 6.1, 28.46% of teachers state that the number of hours reserved for physical science experiments is insufficient, 13.84% state that the absence of materials in the laboratory does not help the smooth running of experiments, 7.69% reveal that absence of the laboratory in their establishment, and 3.84% justify the lack of experiments by the fear of failure.

For the 2.31% the reasons were the following: the number of pupils in class, the fear of class management difficulty (unruly pupils), the fear of damaging the equipment, and the dangerous nature of certain experiments.

In this part, we also tried to find out how teachers react to the lack of experiments; the following Figure 6.4 illustrates the results.

It is noted that 25.38% of the teachers explain the lesson without experiments, only 13.84% declare having tried to find the material necessary for the experiments, and 16.92% confirm to have replaced the experiment using the digital resources allowing the visualization of the experiment or the simulated.

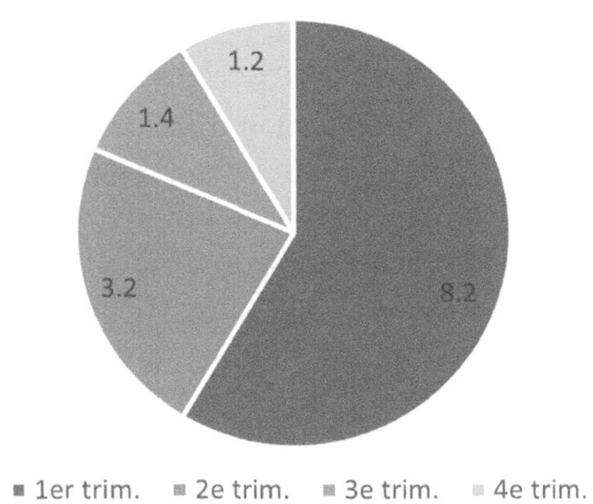

FIGURE 6.4 Teachers' reaction to the lack of experiments.

6.4.5 Use of AI, ICT, and Virtual Labs in Physical Education

The penultimate item of our questionnaire was reserved for the question of the integration of ICT in the teaching of the physical sciences by the respondents as didactic tools in the act of teaching/learning.

In fact, the respondents answered three questions concerning the frequency of their use of ICTs, the replacement of experiments by simulations and the origin of the resources used.

The results collected show that 90% of teachers do not approve the replacement of experiments with virtual labs. The figure shows that 60% of teachers surveyed still use ICT, AI, and virtual labs in the classroom to compensate for the lack of current experiments. Almost 28.46% of respondents using ICT and virtual labs only occasionally; however, over 11.54% of respondents say they have never used virtual labs nor AI.

Regarding the last question of this item, teachers' responses show that 53.84% of them use simulation software to help in understanding the courses, 36.92% make use of educational software (virtual labs), 23.07% use the presentation software to project their courses in order to save time, 13.07% use ICT for Internet searches, and only 3% use the digital resources distributed by the Ministry of Education "Genie". We note that 6.15% declare that they use other types of ICT uses (website, video, AI ...).

6.4.6 The Textbook and the Construction of Learners' Knowledge

We tried through this item to see how the textbook helps to build knowledge and skills; we interviewed teachers on the textbook. The results collected showed that 45.38% of teachers find that the number of hours reserved for the physical sciences is insufficient and 71.54% of teachers value the pedagogical quality of the textbook in terms of structuring the knowledge and know-how taught.

The findings highlight that resource limitations, strict curriculum pacing, and lack of specialized training prevent teachers from effectively conducting physics experiments. Addressing these challenges requires a structured policy intervention, including increased funding, professional development workshops, and restructured curriculum alignment.

6.4.7 AI Technologies for Teaching–Learning

One of the most discussed subjects in education these days is using AI in teaching. Every teacher should become knowledgeable about AI and how to use it to improve their own job and help their students. We must comprehend how this technology will affect both our pupils and ourselves in order to best serve them, through adaptive learning algorithms, intelligent tutoring systems, personalized learning, chatbots and virtual assistants, and learning analytics.

6.4.8 Artificial Intelligence, Virtual Reality, and Augmented Reality in Education

The results of the literature review show that AI has enormous potential to revolutionize physics education. Every year, there is a growing need for the use of AI systems and algorithms in education, as shown by the rise in publications from Web of Science and Google Scholar on

"AI" and "Education" themes since 2010. Note that a significant percentage of all indexed papers—70%—were published between 2016 and 2025.[19]

AI excels at tracking objects, creating detailed models of the 3D world, and making decisions using complex data, which all assist in the creation of next-gen AR/VR. We can use AI to track a projectile in experiments like free fall and movement of a projectile in a fluid.

6.4.9 AI-Powered Assessment and Feedback Systems

Every aspect of life is changing due to AI, and education is no different. More than 55% of learning management systems (LMS) are expected to have AI capabilities integrated by 2025. In order to provide individualized, data-driven learning experiences that adjust to the needs of each learner and enhance results, educators, trainers, and instructional designers are depending more and more on AI.[20,21]

Higher education is undergoing a revolution in learning delivery thanks to AI integration, which guarantees increased scalability, efficiency, and customization. AI-enabled assessment, a cutting-edge strategy that leverages AI to automate, improve, and customize the evaluation process, is among the most significant applications of this shift. AI-enabled assessments, such as intelligent feedback systems and adaptive quizzes, enable teachers to assess students' progress more accurately and impartially.[20,21]

6.5 CONCLUSION

We found out that virtual labs, digital simulations, and AI can be used as an alternative to experiment in the event of lack of equipment, conditions, or chemicals (but can't replace it), anytime, anywhere, and without any risk. AI can assist students with their experiments by providing guiding, explanation, and clarity.

This study highlights the urgent need for comprehensive reforms in high school laboratory infrastructure, advocating for policy-level funding enhancements, revised curriculum integration, and targeted teacher training programs to support hands-on physics education.

The weaknesses raised are essentially related to the following parameters: school laboratories in Moroccan high schools do not facilitate the learning and development of knowledge and skills of learners, not to mention the absence of the laboratory in some high schools. The use of experiments as a didactic tool in physics education plays an important role in learners' acquisition of concepts.

However, the respondents' answers showed that the rate of the experiments carried out is much lower than what is requested and almost all these experiments are carried out by the teacher himself. Teachers are unable to cope with the problems of the lack of experimental equipment, the dangerous nature of certain experiments, the mismanagement of the laboratories (absence of the preparator at the laboratory), which considerably influence the realization of the programmed experiments. And finally, AI, virtual labs, and digital simulation, according to many teachers in this study, offer a good way to compensate for experiments and not completely replace them.

Considering these results; it is necessary that experimental didactics should be emphasized during the initial training of physical science teachers in qualifying secondary education. Also, the renewal of laboratories in qualifying secondary schools is a priority (adequate experimental equipment, preparers) to facilitate teachers' "exploitation in learning and the development of learners" knowledge and skills.

REFERENCES

1. N. Demirci, "Prospective High School Physics Teachers' Beliefs about Teaching Practices: From Traditionalist to Constructivist". EJMSTE, 2015, 11(3), pp. 693–711. DOI: 10.12973/eurasia.2015.1332a.
2. M. Taoufik, A. Abouzaid, A. Moufti, "Experimental Activities in the Teaching of Physical Sciences". ESJ, 2016, 12, p. 1, DOI: 10.19044/esj.2016.v12n22p190.
3. Z. Azar, O. Dardary, J. Daaif, M. Tridane, S. Benmokhtar, S. Belaaouad, "Contribution of Experience in the Acquisition of Physical Sciences Case of High School Students Qualifying". IJACSA, 2022, 13(7). DOI: 10.14569/IJACSA.2022.0130748.
4. S. Badara, "Role of the Social Environment in the Learning of Physics. Example of a Conceptual Change Strategy in a Second Class. Unpublished Memory". ENS of Dakar. Department of physics, 2001, p. 4.
5. P. J. H. Heeralal, T. Dhurumraj, "Factors that Contribute to Poor Learner Performance in Physical Sciences in KwaZulu-Natal Province with Special Reference to Schools in the Pinetown District". IJES, 2016, 14(3), p. 11. DOI: 10.1080/09751122.2016.11890505.
6. A. Stott, "South African Physical Sciences Teachers' Understanding of Force and the Relationship to Teacher Qualification". AJRMSTE, 2013, 17(1), p. 2, DOI: 10.1080/10288457.2013.829599.
7. A. Giordan, "The Allosteric Model: To Do with to Go Against". The teacher high schools: pedagogy. 1996. (61), pp. 1–6. https://www.andregiordan.com/apprendre/The-allosteric-learning-model-and-current-theories-about-learning1.pdf
8. L. D. Taylor, S. Booth, "Secondary Physical Science Teachers' Conceptions of Science Teaching in a Context of Change". IJSE, 2015, 37(8), DOI: 10.1080/09500693.2015.1035356.
9. MEN: Ministry of National Education in Morocco. The Pedagogical Orientations of the Physical Sciences: Qualifying Secondary Cycle 2007, Page 10.
10. P. Kouela, "The Role of Experimentation in the Teaching of Chemistry in High School: Case of Chemical Reactions". Marien Ngouabi University – DEA 2012; Ecole Normale Supérieure (Republic of Congo), p. 6. https://www.preface.ma/la-place-de-l-experimental-dans-l-enseignement-de-la-chimie-cas-du-test-d-identification-des-ions-9783841777423.html#details
11. S. Kane, "Experimental Practices in High School". Radisma, 2011, 7, pp. 1–26. https://fastef.ucad.sn/revuefastef/articles/saliou/article2_pratiques.pdf
12. N. Heinz, A. Michael, "Promoting Pre-Experimental Activities in High School Chemistry: Focusing on the Role of Students' Epistemic Questions". IJSE, 2008, 30(13), p. 4. DOI: 10.1080/09500690701579546.
13. Y. Nafidi, A. Alami, M. Zaki, H. Afkar, "Open and Distance Learning in the Initial Education of Trainee Teachers". IJELD, 2015, 3(9), pp. 53–64. https://eajournals.org/ijeld/vol-3-issue-9-december-2015/open-and-distance-learning-in-the-initial-education-of-trainee-teachers/
14. E. J. Lawrence, A. U. Tar, "Factors that Influence Teachers' Adoption and Integration of ICT". EMI, 16 Feb. 2018, 55(1), p. 6. DOI: 10.1080/09523987.2018.1439712.
15. A. Zaher, D. Rosseni, O. Nabilah, H. Muhammad, "Evaluating Contribution of Flipped Learning on Secondary Students' Critical Thinking Skills toward Learning Physics in Palestine". IJET, 2018, 7(4.21), pp. 53–57. https://www.researchgate.net/publication/359997739_Evaluating_Contribution_of_Flipped_Learning_on_Secondary_Students'_Critical_Thinking_Skills_Toward_Learning_Physics_in_Palestine

16. H. A. Jalil, A. Ismail, S. A. Aziz, N. A. K. A. K. A. Nasir, "The Suitability of Infrastructure and Info Structure, Curriculum, Learning Design and Content in MOOCs for TITAS and Relationship with Enhancement in Teaching and Learning". 5th International Conference on e-Learning, 2015.
17. M. Oubibi, W. Zhao, "The Role of ICT in Learning Process of International Student at Chinese University". IJET, 2018, 7(4.32), pp. 97–100. DOI: https://www.researchgate.net/profile/Mohamed-Oubibi/publication/325049409_The_Role_of_ICT_in_Learning_Process_of_International_Student_at_Chinese_University/links/5fac8c4c299bf18c5b69fa22/The-Role-of-ICT-in-Learning-Process-of-International-Student-at-Chinese-University.pdf
18. UNESCO. Seizing digital opportunities in higher education: equity, quality, and innovation, 2016.
19. S. A. D. Popenci, S. Kerr, "Exploring the Impact of Artificial Intelligence on Teaching and Learning in Higher Education". RPTEL, 2017, 12(22). DOI: 10.1186/s41039-017-0062-8
20. Z. Azar, O. Dardary, J. Daaif, M. Tridane,. Belaaouad, "Contributions of Computer-Assisted Experimentation in Physics". IJTPE, 2024, 16(59), pp. 32–36. DOI: 10.14569/IJACSA.2022.0130748.
21. O. Dardary, Z. Azar, M. Tridane, S. Belaaouad, "Place of Experimentation, Simulation and Computer-Assisted Experimentation in Physics Teaching". IJTPE, 2024, 16(2), pp. 37–43. https://www.iotpe.com/IJTPE/IJTPE-2024/IJTPE-Issue59-Vol16-No2-Jun2024/6-IJTPE-Issue59-Vol16-No2-Jun2024-pp37-43.pdf

CHAPTER 7

Integration of Artificial Intelligence, Virtual Labs, and Digital Simulation in Science Education

Oussama Dardary, Zineb Azar, Malika Tridane, and Said Belaaouad

7.1 INTRODUCTION

Currently, artificial intelligence (AI) is used in all fields: medicine, geology, biology, mathematics, physics, chemistry, etc., as well as education, through models, simulations, and virtual labs. AI has been able to influence every aspect of our lives, our ways of living, sleeping, training, learning, and teaching. In a way that could be roughly addictive, they say that once you use it, there's no going back.[1]

AI can be defined as a series of nested or derived concepts that have emerged over the last 70 years:

- Artificial intelligence (AI) (1950s).
- Machine learning (1980s).
- Deep learning (2010s).
- Generative AI (2020s).

The four varieties of AI—reactive, memory-limited, spiritual theory, and conscientious AI—seem to be quite helpful for the teaching–learning process with these various benefits. Among these is personalized or adaptive learning,[2] for instance, the AI may modify its instructional style. If a student is disadvantaged, the AI may lower the student's level of guidance, and vice versa. This type of approach may prove beneficial in the field of education for individuals with special needs (school dropout, disability, etc.).[3–6] The main benefit of this kind of learning is that it will enable the adaptation of instructional content in a targeted manner so that the student can meet their goals. Students could explore areas that

DOI: 10.1201/9781003649885-7

particularly interest them in addition to moving at their own pace and filling in specific gaps thanks to this individualized approach. These are the main benefits of adaptive learning with AI.[3,6–8]

The integration of AI aims to reliably increase the initial database used by the digital code, to take the variabilities inherent in the mechanical, electrical, or thermodynamic problem, for example, to reduce the deviations from the experimental.

AI is very useful not only for finding new testing ideas but also for guiding users on the optimal experimentation methods to get the answers they need. Another point of vigilance is data quality. AI not only has enough accessible data to understand the context of what each team is trying to achieve but also pays attention to the quality of the data. The data that feeds your algorithm will determine its accuracy. The higher the data quality, the better the accuracy.[8–10]

Using virtual labs is crucial for the following reasons: protecting pupils when handling chemicals; instead of asking students to memorize concepts and procedures or make assumptions, giving them a realistic learning experience; assisting in resolving the issue of schools' capacity shortage; encouraging collaboration and communication between students as well as between professors and students; assisting teachers with computerized assessments of their pupils; and, lastly, assisting students in overcoming ennui brought on by the theoretical sciences they study in school, particularly in the early stages.[11–13]

In its report of the expert meeting on virtual laboratories, held in Paris in 2000 by United Nations Educational, Scientific and Cultural Organization, they said that it is necessary to critically examine the justification, potential, methods, and difficulties of creating "virtual laboratories" (VLs), especially if developing nations are to take part. UNESCO asked the University of Iowa's International Institute for Theoretical and Applied Physics (IITAP) to host an expert meeting on virtual laboratories from May 10 to 12, 1999, in Ames, Iowa, in order to carry out this analysis and produce findings and suggestions. With the goal of sharing this knowledge as broadly as possible, this report aims to offer the analysis, findings, and suggestions of this meeting.[5,22]

They define "a virtual laboratory" as "an electronic workspace for remote collaboration and experimentation in research or other creative activities, with a view to generating and disseminating results by means of shared information and communication technologies."

VL (virtual laboratory) differs from a "real laboratory" (RL) or "traditional" laboratory. However, a VL is not intended to be a competitor or a replacement for an RL. On the other hand, VLs can be extensions of RL and open up new opportunities that cannot be fully realized within an RL at a reasonable cost. The terms "collaboratory," "virtual workgroup," "virtual enterprise," "inter-organization group," and "remote collaboration group" are among the equivalents that encompass the idea of VL.

The subject of knowledge transmission is undergoing a revolution as a result of the increasing prevalence of information technology. Both synchronous and asynchronous remote transmission of educational information is made possible by computer tools, which undoubtedly improves both traditional and distance learning. Teachers are encouraged to disseminate their course material and information via the Internet or a stand-alone medium that gives students greater flexibility as a result of the development of computer-based teaching resources. These tools are frequently designed arbitrarily, which leads to the distribution of goods that don't adhere to the fundamental pedagogical requirements of computer-assisted instruction.

7.2 PROBLEM

According to Zawacki-Richter, AI can be used in prediction and profiling (dropout, study program entrance)[14]; intelligent tutoring systems that provide feedback and instructional content; assessment and measurement (automated grading, assessment of academic participation); or systems for personalization and adaptation (suggesting and choosing tailored material).[10] AI can also be used in education:

- Information retrieval: Search engines, for example.
- Education management: This section discusses the intelligent use of learning management systems (LMS) and educational data analysis.
- Personalization of education: Adaptive learning systems.
- Listening, speaking, and writing: Translators, AI-assisted writing, and natural language processing.

How can the AI help teachers to enhance experimental activity in class? What are the tools of AI that can be used? Can AI really replace teachers and real-time experiments?

7.3 METHODOLOGY

We used a qualitative and quantitative (mixed) approach through anonymous questionnaires distributed to middle and high school physics teachers (N=218) from multiple scientific specialties in different Moroccan provincial departments. The questionnaire was shared online via Google Forms and composed of ten questions, some of which are open and others are closed.

7.4 THEORETICAL FRAMEWORK

7.4.1 Modern Approach of AI by Stuart Russell and Peter Norvig

The following definition of AI is proposed by computer experts Stuart Russell and Peter Norvig in their extremely famous book *Artificial Intelligence: A Modern Approach* (*Intelligence Artificielle: Une Approche Moderne*):

> The field of computer science known as artificial intelligence is concerned with creating software and devices that can think, learn, sense their environment, solve problems, and communicate with people in a meaningful way. [15]

Russell and Norvig's definition reflects the current view of AI as an interdisciplinary field that aims to develop computer systems that can replicate certain characteristics of human intelligence. This definition has gained widespread acceptance and is now a standard in the field of AI.

7.4.2 Types of AI Application

Applications of AI include simulation, comprehension of natural language, tools for creating images, chatbots, automated decision-making, competitive strategy game programs and certain non-player video game characters, and help with medical diagnosis, recommendation systems, and search engines.

7.4.3 Artificial Intelligence in Service of Education

There are different forms of AI's use for education: research, translator, paraphrase, AI detector, summarizer, grammar checker, citation generator, plagiarism checker, problem resolution/chat box, and mathematic problem solver by scanning pictures.

7.4.4 Contribution of AI and Virtual Labs in Teaching–Learning of the Physical Sciences

Combining AI with ICT (information and communication technologies) yields many benefits, such as improved accessibility, flexibility, and interactions and exchanges among the various participants.[9,11,13] ICT expands the variety of learning activities and teaching and learning modalities by changing the time and place of teacher–student interactions.

AI and virtual labs are becoming more prevalent in both our everyday life and educational system. There is growing pressure on educational institutions to employ ICT to equip students with the skills and knowledge they will need in the digital age.[12,14] Their proficiency with ICT can help teachers and students learn and understand other abilities.[16]

There are various ways to evaluate the educational benefits of integrating AI and ICT, for example, lesson feedback that demonstrates an increase in students' enthusiasm and independence, and an analysis of learning in light of the established goals that yields positive outcomes. There has been a rise in student motivation, but there has also been a lack of implementation, which can be explained once more by the necessity of teacher support and training in opposition to these didactic tools.

For example, Singapore has effectively incorporated ICT into education to facilitate students' use of ICT to enhance their own learning, support the student-centered approach, and equip them with the communication and creative thinking abilities necessary to thrive in the workplace of the future.[17]

Numerous studies conducted in Morocco have revealed certain challenges associated with learning the physical sciences: the fact that most teachers use ICT very sparingly, that computer rooms in most schools are still underutilized, if not unutilized, and that there are insufficient experimental activities, that there is a lack of scientific equipment in school laboratories, and that teachers do not receive ongoing training on how to integrate ICT into their practices.

In fact, the amazing progress of AI in education has radically changed the relationship between knowledge and educational practices.[18] The virtual labs offered by the Education Program demonstrate how ICT may be utilized to give everyone access to high-quality education.[19]

7.5 RESULTS AND DISCUSSION

7.5.1 Target Population

Our study targeted teachers of physical and chemical sciences in secondary, college, and vocational education.

As shown in Figure 7.1, approximately 60% of the target population are women, and 40% are men, of different seniority: between 10 and 15 years (20%), 5–10 years (45%), and less than 5 years (35%).

7.5.2 Level of AI Mastery

According to the study, results showed that only 37% of the questioned teachers responded that they master AI tools (14% are very good and 23% are good), 4% of them have no idea about the use of AI for teaching, and the others are average (37%) or mediocre (22%) (see Figure 7.2).

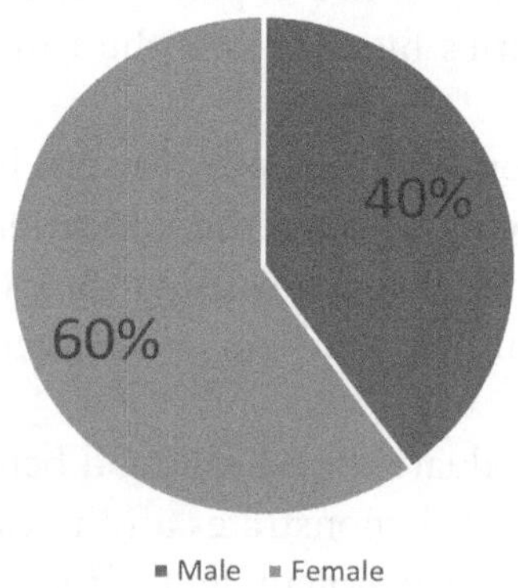

FIGURE 7.1 Percentage of males and females.

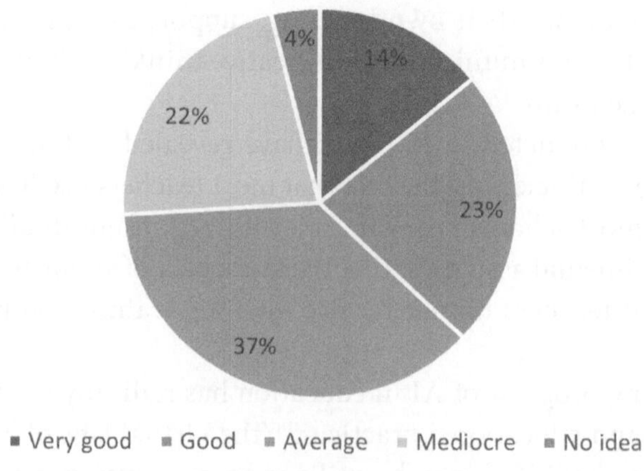

FIGURE 7.2 Level of mastery of AI.

7.5.3 AI Training Desired by Teachers

Results showed that 20% of the questioned teachers desired training in AI for conducting AI simulations, 21% for conducting AI-assisted experiments, 10% for mastering programming languages, and 12% for creating educational websites (see Figure 7.3).

According to our results, only 20% of teachers surveyed have heard of simulation labs, virtual labs, smart labs, or AI-based simulation for teaching. About 57% wanted to try AI-based applications for laboratory simulation.

For the majority of teachers (70%), AI and virtual labs can enhance the class activity by 50% and gain up to 100% of teaching time, if they're used correctly, in advance, and combined with flipped classrooms.

For this reason, we have indicated several resources at the end of our questionnaire, to help teachers find out more about this subject, to see if they are interested or not. There are websites for creating your own simulation, such as PhET Interactive Simulation and Open Source Physics; for virtual labs and experiments, there is myPhysicsLab or Virtual Physics Laboratory; and for AI-powered simulations and virtual labs, there is TensorFlow or PyTorch.

7.5.4 Teaching–Learning in the Era of AI

Using AI in teaching is currently one of the most talked-about topics in education. All educators should learn about AI and how to use it to aid their pupils and themselves. To better serve our students, we need to understand how this technology will impact both ourselves and them. This includes learning analytics, chatbots, virtual assistants, personalized learning, intelligent tutoring systems, and adaptive learning algorithms.

7.5.5 Artificial Intelligence, AR, and VR in Education

The literature review's findings demonstrate how AI has the ability to completely transform physics instruction. The necessity for using AI systems and algorithms in education is

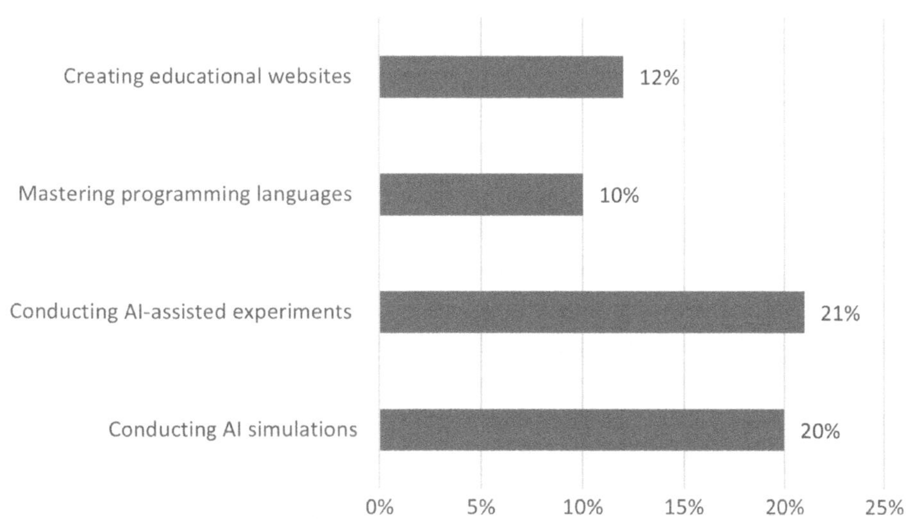

FIGURE 7.3 AI desired training field.

increasing annually, as shown in "AI" and "Education"-related Web of Science and Google Scholar publications since 2010. It should be noted that 70% of the indexed publications were published between 2016 and 2025.[20,21]

AI is excellent at tracking objects, building intricate 3D models, and making decisions based on complicated data, all of which help develop next-generation AR and VR. In experiments involving free fall and projectile movement in a fluid, Wen can utilize AI to monitor a projectile.[22]

7.5.6 AI-Driven Platforms for Evaluation and Feedback

Education is only one of the many areas of life that AI is changing. More than half of LMS are expected to incorporate AI capabilities. In order to provide individualized, data-driven learning experiences that adjust to the needs of each learner and enhance results, educators, trainers, and instructional designers are depending more and more on AI. By guaranteeing increased scalability, efficiency, and personalization, AI integration is completely changing the way that education is delivered in higher education. AI-enabled assessment, a cutting-edge strategy that leverages AI to automate, improve, and customize the evaluation process, is among the most significant applications of this shift. AI-enabled tests, such as adaptive quizzes and intelligent feedback systems, allow teachers to assess students' progress more accurately and impartially.[23,24]

7.5.7 Use of ICT and Digital Simulation in Physical Education

The question about respondents' use of ICT as a didactic tool in the classroom for teaching the physical sciences was the penultimate item on our survey. Actually, the respondents responded to three questions on how often they use ICTs, whether they substitute simulations for experiments, and where they get their resources.[23,24]

The majority of teachers oppose replacing experiments with digital simulations, according to the information gathered. To make up for their lack of expertise, the majority of teachers questioned still employ ICT and digital simulation in the classroom, as the image illustrates. Additionally, fewer respondents said they only infrequently use ICT and digital simulation, but few said they have never used AI.[23,24]

7.5.8 Importance of 3D Virtual Labs

Students can perform laboratory experiments online and investigate ideas and theories without physically visiting a real scientific lab thanks to virtual labs, an educational simulation environment. They can examine life sciences at the molecular level and get a behind-the-scenes look at the machines they employ through animation.[21]

Students can also learn about cutting-edge lab equipment that might not be available to them because of its high cost or potential hazards as well as explore laboratory techniques for the first time. They can get an alternative access to science instruction, and virtual venues are made possible by virtual lab software.

Here are some benefits of integrating virtual labs into teaching:

- Compared to doing an experiment in a typical lab, conducting an experiment in a virtual lab takes a lot less time.

- It links a tangible physical application to theoretical science.
- The flexibility to switch between pieces of scientific apparatus and carry out the experiment more than once.
- Creates a welcoming and entertaining learning environment.
- Gives the money required to buy supplies and equipment.

This enables the student to thoroughly absorb the subject by giving them multiple opportunities to experience it.

There is a lot of room for experimentation and teacher–student interaction in virtual laboratories. Virtual laboratories are funny ways to learn science. We deliver a more interactive experience with the Flappy Virtual Labs platform.

7.5.9 Some Existing Virtual Labs for Teachers and Students

On the Internet, virtual tools are becoming more and more accessible. Nowadays, every institution, university, and academy has a specific area for these instruments. A few resources in this area are compiled on a separate page called "Software for the Classroom"[22] provided by the EPI (Educational Policy Institute). Certain GNU/Linux distributions are designed with educational tools in mind. We especially value what Edubuntu, the educational variant of Ubuntu, has to offer among these. The distribution monitoring website DistroWatch[23] offers all of these distributions for download.[24]

The necessity to support local development teams stems from the difficulty in tailoring the different tools to the unique circumstances in our nations. We look forward to the outcomes of the Interactive Digital Educational Objects (IDEO) project in Europe, which intends to gather, distribute, and promote open-source courseware.[24]

7.6 CONCLUSION

We found out that virtual labs and digital simulations can be used as a solution to do an experiment, anytime, anywhere, and without any risk. AI can assist students with their experiments by providing guidance, explanation, and clarity.

AI will be increasingly common in education for carrying out work. The challenge of integrating AI into education lies in how we prepare students for a future where AI will play a leading role. This requires students to develop not only a technical understanding of AI but also the skills to use this technology ethically and responsibly.

Teachers could take action to make sure they have the right policies and plans in place to direct the integration of AI at their institutions as this technology is quickly becoming the new norm in education. For institutions to successfully integrate AI, it will be essential to comprehend how students feel about the technology.

The main goal of education in our institutions is to pass on knowledge to the next generation. Virtual laboratories must play a bigger role in education by giving students useful resources to help them deal with the challenges they will face. After the integration issues are fixed, students will surely have the resources they need to learn in a unique way and enjoy themselves. As a result, we urge all leaders in education to use these resources into their efforts.

Finally, this AI-guiding, assisting, or teaching confrontation is the present subject and can't replace teachers. To guarantee the continuity of experimentation in high schools, a teacher should be assisting the students, providing all types of guidance, explanation, and clarification needed; otherwise, the teaching–learning process will have lost all meaning.

REFERENCES

1. C. Tyler, "There is no turning back on AI", The Free Press, 2023, visited the 10th April 2025. https://www.thefp.com/p/there-is-no-turning-back-on-ai
2. Ministry of Higher Education and Research, "France artificial intelligence", March 21, 2017 (version of March 29, 2017 on Internet Archive), pp. 5. https://www.enseignementsup-recherche.gouv.fr/fr/rapport-strategie-france-ia-pour-le-developpement-des-technologies-d-intelligence-artificielle-47691
3. M. Vicario, "The impact of artificial intelligence on education", HES.SO, Haute Ecole de Gestion Genève, September 4, 2023, pp. 12. https://www.sbfi.admin.ch/dam/de/sd-web/-kNQiEU3sJIV/k-i_bildung_f%5B1%5D.pdf
4. Federal Department of Economic Affairs, Education and Research (FDEAE), "Artificial intelligence in training", State Secretariat for Training, Research and Innovation (SSTRI), 2019. https://www.sbfi.admin.ch/dam/sbfi/de/dokumente/2019/12/ki_bildung.pdf.download.pdf/k-i_bildung_f.pdf
5. A. Paigneau, "Adaptive learning: Reflection from the AI Foundation for Schools". UNESCO, 2022. https://unesdoc.unesco.org/ark:/48223/pf0000381780/PDF/381780fre.pdf.multi
6. Lacombe, "Adaptive learning: Everything you need to know about this learning method", Edusign, 2022, pp. 3. https://sonar.ch/documents/328225/files/Vicario%20Mauro.pdf?download
7. C. Meilleur, "Intelligent adaptive learning at a glance", KnowledgeOne. 202, 3 March, pp. 4.https://knowledgeone.ca/lapprentissage-adaptatif-intelligent-en-un-coup-doeil/?lang=fr
8. C. Meilleur, "3 benefits of intelligent adaptive learning for the learner", 2020, February 11, pp. 6. https://knowledgeone.ca/3-benefices-de-lapprentissage-adaptatif-intelligent-pour-lapprenant/?lang=fr
9. V. Psyché, P. Ruer, "Intelligent adaptive learning", V8, N4, 2019, University of Quebec, pp. 4. https://rlibre.teluq.ca/2538/1/le_tableau_psyche_ruer_2019.pdf
10. O. Zawacki-Richter, V. I. Marín, M. Bond, F. Gouverneur, "Systematic review of research on artificial intelligence applications in higher education – Where are the educators". International Journal of Educational Technology in Higher Education, 2019, 16(1), pp. 1–27. https://doi.org/10.1186/s41239-019-0171-0
11. S. Collin, A. Lepage, L. Nebel, "Ethical and critical issues of artificial intelligence in education: A systematic review of the literature." Canadian Journal of Learning and Technology/Revue canadienne de l'enseignement et de la technique, 2023, 49(4), 1–29. https://doi.org/10.21432/cjlt28448
12. N. Humble, P. Mozelius, The threat, hype, and promise of artificial intelligence in education. Discov Artif Intell 2, 22 (2022). https://doi.org/10.1007/s44163-022-00039-z
13. Y. Nafidi, A. Alami, M. Zaki, H. Afkar, "Open and distance learning in the initial education of trainee teachers". IJELD, 2015, 3(9), pp. 53–64. https://eajournals.org/ijeld/vol-3-issue-9-december-2015/open-and-distance-learning-in-the-initial-education-of-trainee-teachers/
14. E. J. Lawrence, A. U. Tar, "Factors that influence teachers' adoption and integration of ICT". Educational Media International, 16 Feb 2018, 55(1), p. 6. DOI: 10.1080/09523987.2018.1439712.
15. Russell, S. J., & Norvig, P. (2021). Artificial intelligence: A modern approach (4th ed.). Pearson.

16. D. R. Zaher, O. Nabilah, H. Muhammad, "Evaluating contribution of flipped learning on secondary students' critical thinking skills toward learning physics in Palestine". IJET, 2018, 7(4.21), pp. 53–57. https://www.researchgate.net/publication/359997739_Evaluating_Contribution_of_Flipped_Learning_on_Secondary_Students'_Critical_Thinking_Skills_Toward_Learning_Physics_in_Palestine
17. H. A. Jalil, A. Ismail, S. A. Aziz, N. A. K. A. K. A. Nasir, "The suitability of infrastructure and info structure, curriculum, learning design and content in MOOCs for TITAS and relationship with enhancement in teaching and learning". Conference: 5th International Conference on e-Learning, 2015, At: Kota Kinabalu Sabah. Available from: https://www.researchgate.net/publication/277360157_The_Suitability_of_Infrastructure_and_Info_structure_Curriculum_Learning_Design_and_Content_in_MOOCs_for_TITAS_and_Relationship_with_Enhancement_in_Teaching_and_Learning#fullTextFileContent
18. M. Oubibi, W. Zhao, "The role of ICT in learning process of international student at Chinese university". IJET, 2018, 7(4.32), pp. 97–100. DOI: https://www.researchgate.net/profile/Mohamed-Oubibi/publication/325049409_The_Role_of_ICT_in_Learning_Process_of_International_Student_at_Chinese_University/links/5fac8c4c299bf18c5b69fa22/The-Role-of-ICT-in-Learning-Process-of-International-Student-at-Chinese-University.pdf
19. UNESCO, "REACHING OUT TO ALL LEARNERS: a resource pack for supporting inclusion and equity in education", December 2021. International bureau of education, https://www.ibe.unesco.org/sites/default/files/medias/fichiers/2023/10/unesco_bie_2021_web_inclusive_education_resrouce_pack.pdf
20. S. A. D. Popenci, S. Kerr, "Exploring the impact of artificial intelligence on teaching and learning in higher education". RPTEL, 2017, 12(22). DOI: 10.1186/s41039-017-0062-8
21. P. D. Mbock, "Virtual laboratories at the service of teaching", IPR Informatique – MINESEC – Cameroun, Mai 2009, Association EPI, https://edutice.hal.science/edutice-00558936/file/a0905a.htm
22. http://www.epi.asso.fr/logiciel/logipeda.htm. Visited on 20 June 2025
23. http://distrowatch.com. Visited on 20 June 2025
24. V. Martin, "3D virtual laboratory for training engineers", pp. 1–2, 2002. https://edutice.hal.science/edutice-00000672/file/Martin.pdf

CHAPTER 8

Reframing Technological Innovation Dimension and AI Competitiveness for Pro-Growth Small Businesses in Sub-Saharan Africa

Timilehin Olasoji Olubiyi

8.1 INTRODUCTION

Innovation is currently important for small business competitiveness and profitability (Olubiyi, 2023a). Africa's small business development has witnessed significant growth in recent years, yet artificial intelligence (AI) competitiveness remains a key challenge for the region. Technological innovation is viewed as a critical facilitator of business growth and competitive advantage in the quickly changing global economy of today (Mendoza-Silva, 2021). The business environment is more competitive and complex, with several firms prioritising profit (Ninduwezuor-Ehiobu et al., 2023). In Nigeria, both large firms and small-to-medium enterprises encounter significant competition (Mohammed, 2023). To be competitive, companies must remain informed about market trends, consumer preferences, and business advancements. The competitive environment has changed dramatically as technology continues to transform businesses. Integrating cutting-edge technologies, especially AI, is becoming more and more crucial for survival and expansion for both small and large international enterprises. It has been demonstrated that implementing AI and other technical advancements can boost consumer engagement, increase operational efficiency, and influence strategic decision-making (Adama et al., 2024; Brynjolfsson & McAfee, 2017). Adoption of technology can be revolutionary for small businesses, especially in emerging economies, providing a way to scale more effectively and compete with larger companies (Awode & Oduola, 2025; OECD, 2021).

AI-driven solutions can enhance sustainability in the industrial sector by optimising production processes, reducing waste, and adhering to the Sustainable Development Goal

DOI: 10.1201/9781003649885-8

(SDG) 9 indicators (Blichfeldt & Faullant, 2021; Chen & Lee, 2020). AI has the power to aid in economic growth, job creation, and environmental sustainability. AI may aid small- and medium-sized firms (SMEs) in implementing sustainable practices by optimising resource utilisation, reducing waste, and improving supply chain efficiency. This constitutes a facet of the overarching sustainability strategy. Teece's (2018) research illustrates the use of AI in overseeing and mitigating the environmental repercussions of corporate operations. This improves corporate sustainability and coincides with SDG 12 (Responsible Consumption and Production). The integration of AI and other new technologies improves the competitiveness of small and medium-sized firms, while also aiding their contributions to developmental goals, particularly the achievement of the SDGs. Despite the potential advantages of these technologies for small firms in Sub-Saharan Africa (SSA), their adoption is impeded by considerable constraints, such as high initial costs, a lack of experienced workers, and insufficient infrastructure (Moyo et al., 2004). These obstacles impede small and medium-sized firms (SMEs) from efficiently using AI for growth and innovation.

Globally, enterprises seeking to maintain relevance in the digital era must integrate AI into their operations to secure a competitive advantage. The World Economic Forum (WEF) projected in 2020 that AI may contribute up to $15.7 trillion to global GDP by 2030, driving innovations, consumption, and productivity enhancements across many sectors (WEF, 2020). AI is being embraced in wealthy economies such as the United States, China, and European Union states, although its integration in developing ones remains nascent. Small enterprises in SSA have significant challenges in accessing and using technological improvements due to factors such as insufficient infrastructure, high costs, and a shortage of trained labour (UNCTAD, 2021). The fastest-growing region in the globe is SSA, but it also has particular difficulties that restrict its economic potential (Tetteh, 2024; Ali & Anwar, 2021). AI adoption is still comparatively low in the region, despite significant advancements in technology adoption, such as the sharp rise in internet and mobile phone use. According to the African Development Bank (AfDB) (2020), hardly 1% of African companies have integrated AI into their daily operations. Nevertheless, AI has enormous potential to propel economic growth across the continent, especially in the small business sector. In SSA, small firms which make up around 90% of all businesses and, in some countries, up to 60% of the labour force are essential for economic diversification, job creation, and poverty reduction (YuSheng & Ibrahim, 2020; World Bank, 2020). Prior research in Nigeria revealed that several Nigerian organisations have challenges in competing in international marketplaces owing to insufficient innovation in their products (Edeh et al., 2020). Organisations that use external information to enhance their systems are more inclined to innovate and achieve a competitive edge. To maintain market AI competitiveness, businesses must provide superior goods in response to rapid technological progress (Chukwurah & Aderemi, 2024). Prior research suggests that an organisation's AI competitiveness is influenced not just by innovation but also by the velocity of its innovative processes. This rate is essential for success. To enhance innovation, firms have to collaborate with other organisations to address shortcomings and capitalise on mutual strengths to develop products and achieve a competitive advantage (Koentjoro & Gunawan, 2020). To develop according to client demands, businesses must first ascertain these needs. Assessing small business AI competitiveness is essential for the development of the sector (Nascimento et al., 2025; Tetteh, 2024; Rumanti et al., 2022). Previous research in this area was predominantly limited to traditional strategic analysis frameworks or primarily concentrated on financial metrics. This chapter goal is to fill the gap by giving background information on how technological innovation affects AI competitiveness and looking at the relationship in a small number of small businesses in Nigeria, which has the most people and is one of the largest economies in Africa.

Small businesses add a lot to the GDP and employment of Nigeria, one of the biggest economies in SSA. However, these enterprises face several challenges, including limited financial resources, insufficient infrastructure, and a lack of technological knowledge (National Bureau of Statistics, 2020). Small enterprises in Nigeria might gain enhanced market access, improved decision-making, and increased operational efficiency through the integration of AI and technological improvements. AI-driven solutions may aid small enterprises in enhancing marketing strategies, optimising supply chains, and augmenting customer interaction through customisation landscape (Nascimento et al., 2025; Vaia et al., 2022; Afolabi, 2021). Despite these potential benefits, Nigerian small enterprises remain well behind in the adoption of AI and other technological breakthroughs. The specific relationship between technological innovation and AI competitiveness for pro-growth small companies in SSA, particularly in Nigeria, constitutes a study gap. Despite several studies highlighting the potential of technology and AI in major organisations and economies (Binns, 2019), there is a scarcity of comprehensive research investigating the prospects and challenges of AI adoption among small enterprises in SSA. Furthermore, there is little practical data on how AI might improve small enterprises' competitiveness in SSA, where adoption of technology is hampered by issues including infrastructure, cost, and a lack of skilled workers (Dhliwayo & Chebo, 2024; Olubiyi et al., 2022; Vahdat, 2022). By concentrating on the interaction between technological innovation and AI in small firms, this study will close this gap and offer insights into how AI may be used to boost growth, improve competitiveness, and get over obstacles in SSA.

The current economic circumstances in Nigeria and other Sub-Saharan African nations underscore the importance of this study. The COVID-19 pandemic has accelerated businesses' digital transformation, highlighting the importance of technological resilience in managing future crises (Roomi, 2023; Alwafi, 2022; International Labour Organisation, 2021). The COVID-19 epidemic has demonstrated the necessity for small enterprises in SSA to adopt digital solutions to remain sustainable and competitive. Nonetheless, the slow adoption of AI and technology by small enterprises continues to hinder their ability to compete effectively in an increasingly digitised global economy. In order to close the research gap and promote long-term economic growth and competitiveness, this study intends to offer practical suggestions for small firms in Nigeria and SSA to enhance their technological innovation strategies and AI adoption. This chapter examines the relationship between technological innovation and AI competitiveness of small businesses, illustrating how novel procedures may streamline operations and enhance efficiency (Nascimento et al., 2025; Tariq et al., 2021). Therefore, the objective of this study is to investigate the effect of technological innovation dimensions on the AI competitiveness of selected pro-growth listed consumer goods companies in Nigeria.

8.2 LITERATURE REVIEW

8.2.1 Small Business

The growth of small businesses in Nigeria goes back to before colonisation; therefore, it is a natural element of the country. There is a lot of proof in our communities about what our great-grandparents did to help small businesses develop. Our communities and parents have always done quite well in trading, iron-smelting, farming, cottage industries, and other things like that. Also, the country's small businesses were given priority via a series of government programmes, which were seen as the source of strength for the economy's recovery. Gyang and Badung (2019) say that the problems of small businesses have been seen as those of the whole country because there are so many of them, they are so different, and they are involved in all areas of production and making big contributions to employment and the prosperity of the areas where they work. According to Gumel (2019), small businesses are often started by a small group of owners or even just one owner, and they don't usually have any established procedures or operations. Owners of small businesses often don't have the skills or knowledge to run them well, which makes the growth stage very important. To be successful, they need to have a clear vision and mission. Also, small businesses in Nigeria hire people, which lowers the amount of poverty in the economy and improves the quality of life for the people who live there. To keep the economy strong, small businesses need to be less likely to fail. Parro (2024) agrees that small businesses are very important for creating a landscape of opportunity where people can turn ideas into businesses and goals into achievements. Arofatkhan (2023) also talked about the many roles that small businesses play, focussing on their ability to encourage innovation, promote entrepreneurship, and revitalise local economies. However, its tremendous potential is still hampered by a number of problems, such as limited access to money, regulatory issues, and lack of knowledge (Tetteh, 2024; Rahman, 2022).

8.2.2 Innovation

Innovation is the cornerstone of the arsenal of modern organisations and shapes their competitiveness, growth, and sustainability in a dynamic market. Several authors have outlined innovation in various ways and emphasised its role as a catalyst for business progress and change. Innovation is essentially the creation or introduction of new ideas, processes, products, or services to create value and respond to changing market demands (Chesbrough, 2019; Dodgson et al., 2020). Innovation refers to novel technologies, methods, and systems that advance society and improve people's lives (Wanof, 2023). Furthermore, innovation for governments should include more than just specific innovations or processes. Manufacturing should be viewed as a process that absorbs new ideas into the economy, alters the product, and organises production methods (St-Pierre et al., 2023). Innovation is a key measure of economic success, driving development and generating new commercial opportunities. Innovation helps compete with high-quality, low-wage economies while maintaining high-wage standards.

8.3 TECHNOLOGICAL INNOVATION

Technological innovation significantly contributes to contemporary business and economic growth. It is widely recognised for its capacity to transform several domains. Technological innovation entails the creation of new technologies or the enhancement of existing ones. Companies utilise this strategy to surpass competitors, enhance efficiency, and expand their consumer base (Ferreira et al., 2021; Osman et al., 2023; Schilling, 2017). Technological innovation includes the development of novel products, services, processes, and business models. Small enterprises must adopt new technology to remain operational and prosper, particularly in developing regions such as SSA, where the digital gap remains a significant issue (OECD, 2021). A company's technological innovation capability depends on its internal resources (human, technological, and organisational) and its ability to integrate external resources, knowledge, and skills to create new products and processes that stakeholders value (González-Ramos et al., 2023).

8.3.1 AI Competitiveness

AI has emerged as a fundamental element of contemporary business operations, fostering innovation, efficiency, and competitive superiority across all sectors. AI competitiveness refers to a firm's capacity to leverage AI technologies to sustain or enhance its competitive standing in the market (Binns, 2019). The strategic implementation of AI technologies enables companies to augment operational efficiency, develop product offerings, and enhance consumer experiences, hence increasing market success (Porter & Kramer, 2022). The concept of AI competitiveness refers to a firm's ability to leverage AI technology to outpace competitors in efficiency, customer experience, product offers, and market share (Davenport et al., 2020; Binns, 2019). The competitiveness of AI is a vital factor in the success of modern businesses. The integration of AI technology enables firms to enhance efficiency, improve user experiences, foster innovation, and support data-driven decision-making. AI has considerable potential to improve competitiveness and growth for small businesses, especially in SSA (Tetteh, 2024; Porter & Kramer, 2022). However, barriers such as limited access to AI technology, high implementation costs, and a lack of skilled workers hinder AI adoption in these domains. As businesses in SSA address these challenges, understanding the benefits and applications of AI is essential for developing policies and strategies that improve AI competitiveness in emerging economies (McGrath, 2024).

8.4 THEORETICAL REVIEW

8.4.1 Innovation Diffusion Theory

Everett Rogers' innovation diffusion theory offers a framework for comprehending the mechanisms, motivations, and velocity of the dissemination of innovative ideas and technology across cultures. This theory is crucial for a firm's competitiveness as it elucidates the acceptance of innovations and the factors influencing their spread inside and among organisations (Kiveu et al., 2019). Innovation diffusion theory posits that the adoption of innovative ideas transpires in a predictable manner. Innovations are introduced, and their adoption disseminates among several demographic categories, including innovators, early adopters, early majority, late majority, and laggards (Guttentag & Smith, 2022). Each category signifies a distinct level of preparedness and capability to embrace new technologies or methodologies.

8.4.2 Component

The diffusion of innovations involves a complex interplay of several elements that collectively determine how and when an innovation is adopted within a population. These elements include innovation, communication channels, time, and the social system. Innovation: Innovation is fundamental to the dissemination process. It refers to a concept, activity, or thing that is seen as novel by a person or another unit of adoption (Clohessy et al., 2019). The perceived features of an innovation have a substantial effect on its rate of adoption. The degree to which an invention is seen as superior to the concept it replaces. Innovations with obvious benefits are more likely to be accepted rapidly. The degree to which an invention is seen as difficult to comprehend and use. Simpler inventions are typically embraced sooner. The extent to which an idea can be tried out on a modest scale (Yang et al., 2019). Innovations that can be evaluated and tried without complete commitment are more likely to be embraced quickly. Communication channels are the mechanism by which knowledge about an innovation is distributed from one person to another (Muller & Peres, 2019).

The dissemination process is strongly dependent on both mass media and interpersonal channels. Television, radio, newspapers, and online channels may all reach a big number of people rapidly. They are effective in raising awareness about the innovation. These may also include face-to-face contacts between persons. They are critical for influencing others to accept an invention since they allow for more in-depth discussions and personal endorsements (Crupi et al., 2021). Time refers to the steps that an individual or organisation takes to embrace or reject an invention after first learning about it (Kimhi & Oliel, 2019). Knowledge, persuasion, decision-making, execution, and confirmation are typical steps in the process. Individuals in a social system can be classed according to their innovativeness. The five groups are innovators, early adopters, early majority, late majority, and laggards. Each organisation implements an invention at a different time and for various reasons. This is the rate at which an invention is accepted by members of a social system. The perceived features of the invention, the sort of social structure, communication methods employed, and the form of promotion activities all have an impact on adoption rates (Melović et al., 2020). The social system refers to the network of persons, organisations, or institutions that may accept the invention (DiMaggio, 2019). The social system provides the context for dispersion. Cultural norms and values within a social system have a considerable influence on the dissemination process. Innovations that adhere to these rules are more likely to be adopted. The structure of social interactions inside the system, such as hierarchies, networks, and communication patterns, influences how knowledge about the invention is shared and implemented (Dahesh et al., 2020).

8.4.3 Strengths of the Theory

Innovation diffusion theory offers a structured framework for businesses to comprehend the process of innovation adoption. Companies may effectively target various adopter categories by classifying them as innovators, early adopters, early majority, late majority, and laggards (Dutta & Sarma, 2020). This segmentation allows organisations to build customised strategies for promoting innovations, addressing challenges, and engaging influencers within each category (Gatewood, 2022). The theory delineates key factors influencing innovation adoption, such as relative advantage, compatibility, complexity, trialability, and observability. Comprehending these factors allows organisations to cultivate and market their innovations in manners that enhance their attractiveness to prospective customers (Hannah et al., 2021). Companies can mitigate perceived risks and promote early adoption by highlighting an innovation's relative benefit and trialability.

8.4.4 Weaknesses of the Theory

A major criticism of innovation diffusion theory is its oversimplification of the complex and dynamic adoption process. The theory's categorisation of adopters into five distinct groups—innovators, early adopters, early majority, late majority, and laggards—may not comprehensively represent the spectrum of actions and motivations observed in a normal society. This simplicity may lead to flawed strategies that neglect the complex factors influencing adoption. Innovation diffusion theory primarily focusses on internal elements inside the social system, including communication pathways and adopter categories (Lundberg et al., 2019). Nevertheless, it may not comprehensively account for external factors such as economic conditions, regulatory environments, and competitive pressures, all of which significantly influence the adoption of innovation (Oke et al., 2022). Companies that rely predominantly on this methodology may overlook critical external factors, leading to suboptimal decision-making.

8.4.5 Application of the Theory

Early adopters play an important role in the spread of innovation. They are frequently thought leaders whose choices in adoption affect others. Businesses may improve their competitiveness by finding and engaging early adopters through focussed marketing efforts and personalised outreach (Nuseir & Aljumah, 2020). Firms can encourage early adopters to accept new goods or processes by proving their relative advantage and compatibility with the innovation, hence expediting the diffusion process (Koloseni & Mandari, 2024). Effective communication is critical to the effective dissemination of innovations. Firms can use a combination of mass media and interpersonal communication channels to raise awareness and convince prospective adopters. Mass media channels, such as internet advertising and social media, may reach a large audience rapidly, but interpersonal channels, such as influencer collaborations and customer testimonials, can foster trust and credibility (Ki et al., 2023).

8.5 EMPIRICAL REVIEW

The empirical literature supports the notion that AI and emerging technologies are crucial for enhancing company competitiveness. Business competitiveness has historically relied on technological innovation (Schilling, 2017). SMEs are increasingly adopting technology to address obstacles and gain a competitive edge. Research shows that technological innovation boosts business competitiveness by boosting productivity, lowering costs, and improving consumer experiences. Kauffman and Riggins (2019) discovered that automation and AI increased SMEs' productivity in emerging economies, notably SSA. Business output increased by automating routine processes and reducing physical labour. Small enterprises in SSA have little capital and labour, which limits productivity (OECD, 2021). Teece (2018) studied how digital technologies enable business process innovation. Technology adoption helps small businesses, especially in SSA, improve processes and offer products and services faster. Digitising business models boosted speed to market and eliminated operational inefficiencies, making SMEs more competitive in saturated markets (Teece, 2018). The empirical research reviewed demonstrates that technological innovation, particularly AI, drives company competitiveness across industries. However, combining technical innovation and AI allows organisations to outperform competitors in efficiency, consumer engagement, and innovation. Technological innovation prepares firms to incorporate AI and other digital tools, whereas AI accelerates innovation, enables data-driven decision-making, and boosts competitiveness (Brynjolfsson & McAfee, 2017). Additionally, integrating AI into technical innovation helps firms improve operational efficiency, cut costs, and distinguish products and services. This combination of technical innovation and AI can help SSA SMEs overcome market access, resource, and skills shortages. The empirical evidence shows that integrating AI with existing technical advancements will assist SSA SMEs to survive and grow in a competitive global market (Shah et al., 2020). This discussion leads to generating a hypothesis. Consequently, this study sought to bridge the gap in prior and related studies and hypothesised thus:

> **H_{01}**: There is no significant effect of technological innovation capabilities on the AI competitiveness of pro-growth listed small businesses in Nigeria.

The model sheds light on technological innovation capabilities measured by the usage of organisational structure, methods, tools, and technologies on small business AI competitiveness, which is the research framework. The independent and dependent variables for this research are technological innovation capabilities (X) on AI competitiveness (Y), respectively.

8.6 METHODOLOGY

The small business sector is the focus of this study, and only companies' pro-growth listed on the Nigerian Exchange Group are included in the current examination. This quantitative analysis looked at small businesses that were listed on the Nigerian Exchange Group as of July 2024. The Nigerian Stock Exchange has nine small- and medium-sized businesses (SMEs) pro-growth listed under the Alternative Securities Market (ASeM). This study chose five of the nine small and medium enterprises (SMEs) on the alternative security market based on their market capitalisation. These companies make up 55.56% of the overall capitalisation and the companies that make up the sample for this study are

Capital Oil PLC, Smart Products Nigeria PLC, RAK Unity Petroleum PLC, Chellarams, and McNichols PLC. The study used a cross-sectional survey research strategy. The survey was chosen because it is a good way to find out what different groups of people think, feel, and believe about the subject of the study and to get more honest and valid input. This paper used the same method as previous studies by Adeoye et al. (2023), Makinde et al. (2023), Olowoporoku and Olubiyi (2023), Olubiyi (2019, 2023a, 2023b, 2024), Adeoye et al. (2023), Olubiyi et al. (2022, 2023), Olubiyi and Akpa (2023), Adeyemi and Olubiyi (2024), and Omoyele et al. (2023).

8.7 DATA ANALYSIS

The researcher and trained research assistants gave out 443 copies of the questionnaire. Out of those, 421 were filled out and sent back for analysis, which is a response rate of 95%. The response rate is the number of people who filled out the survey and got copies of the questions. The others were either not sent back or had answers that were missing. However, the total number of questionnaires collected was adequate to represent Nigeria's small businesses, and they were looked at. The answers are shown in detail in Table 8.1.

8.8 RESTATEMENT OF RESEARCH HYPOTHESIS

H_0: Technological innovation capabilities do not affect the AI competitiveness of pro-growth listed small businesses in Nigeria.

H_1: Technological innovation capabilities affect the AI competitiveness of pro-growth listed small businesses in Nigeria.

We used the structural equation model statistical tool to test the hypothesis. We got data on the technological innovation capabilities and AI competitiveness of pro-growth listed small businesses in Nigeria by combining the answers to all the questions for each variable to make indicators for each. Tables 8.2 and 8.3 show the outcomes of the analysis.

8.9 HYPOTHESIS

H_0: Technological innovation capabilities have no significant effect on the AI competitiveness of pro-growth listed small businesses in Nigeria.

We employed Smart Partial, least square analysis to see if the research model that shows the link between technological innovation capabilities and AI competitiveness

TABLE 8.1 Response Rate

	Total	Percentage (%)
Valid and usable questionnaire	443	95
Invalid/Incomplete questionnaire	22	5
Total received	421	100

Source: Researchers' computation (2025).

TABLE 8.2 Confirmatory Factor Analysis for Hypothesis

Constructs	Loading	AVE	Composite Reliability	Cronbach's Alpha	RhO.A
	> 0.7	> 0.5	> 0.8	> 0.7	
Technological innovation capabilities		**0.528**	**0.732**	**0.489**	**0.829**
PI1	0.937				
PI2	0.215				
PI3	0.813				
AI competitiveness of pro-growth listed small businesses		**0.437**	**0.672**	**0.497**	**0.853**
OE1	0.491				
OE2	0.429				
OE3	0.942				

Source: Researchers' computation (2025).

TABLE 8.3 Path Coefficient for Hypothesis

Variables	T Statistics	P Values	Path Co-Efficient	R^2
Structural Path				
Technological innovation capabilities → AI competitiveness of pro-growth listed small businesses	**45.475**	**0.000**	**0.901**	**0.812**
PI1	94.184	0.000		
PI2	1.210	0.226		
PI3	10.033	0.000		
OE1	2.469	0.014		
OE2	2.478	0.013		
OE3	36.827	0.000		

Source: Researchers' computation (2024).

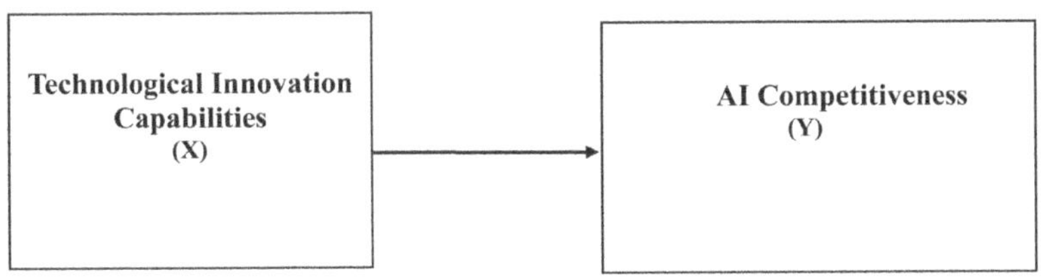

FIGURE 8.1 Author's conceptual model (2025).

of pro-growth listed small enterprises in Nigeria could make accurate predictions. The study's hypothesis test shows how the technical innovation capabilities and AI competitiveness of pro-growth listed small enterprises in Nigeria affect each other. The study used the path coefficients, t-statistics, R-square, and p-values to figure out the level of impact and the strength of the link between the technological innovation capabilities and AI competitiveness of pro-growth listed small businesses in Nigeria. Figure 8.1 shows the route coefficient value, which shows how closely technical innovation capabilities and AI competitiveness are related for pro-growth listed small enterprises. We also used the R-squared

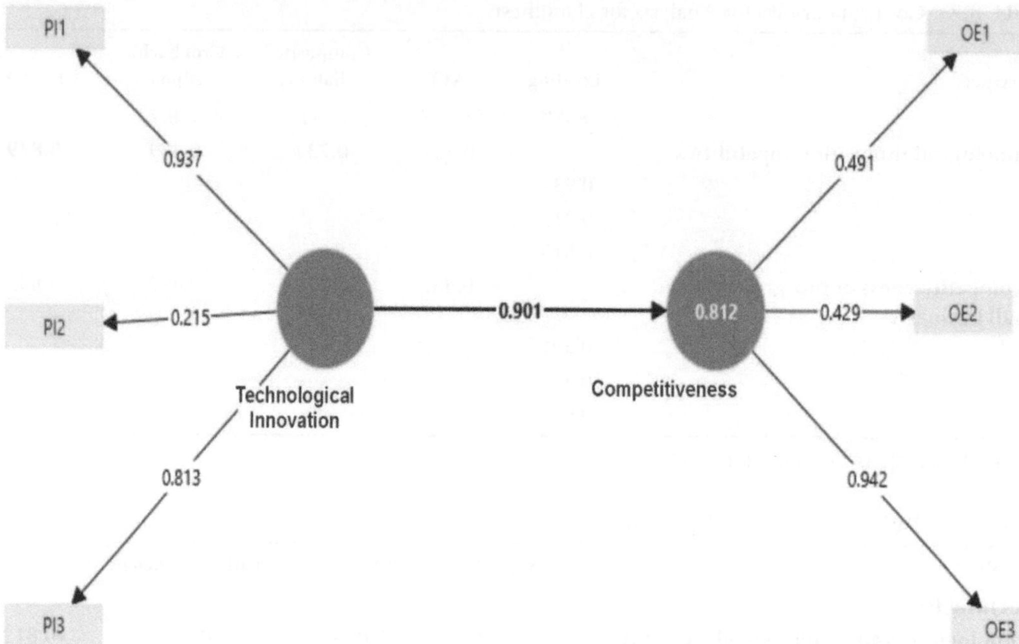

FIGURE 8.2 PLS algorithm model of technological innovation capabilities and AI competitiveness of pro-growth listed small businesses.

values in Figure 8.1 to find a difference in how competitive AI is among pro-growth listed small enterprises, which is explained by their ability to innovate technologically. Figure 8.2 shows the p-value, which tells us how statistically significant the tested hypothesis is. The t-statistic also helps to explain the differences that were estimated in terms of standard error, which is shown in Figure 8.1.

Figure 8.2 displays the PLS algorithm model for the technological innovation and AI competitiveness of pro-growth small enterprises that are listed. The path coefficient value of 0.901 for the technical innovation capabilities and AI competitiveness of pro-growth listed small enterprises indicates how strong the link is between the independent and dependent variables. The R square (R^2) value is 0.812, which falls into the categories of strong (> 0.75), moderate (> 0.50), and weak (< 0.25). This means that technical innovation capabilities can explain 81.2% of the differences in AI competitiveness among pro-growth listed small enterprises. Because of this, the ability to innovate with technology has a moderate effect on how competitive AI is in pro-growth small enterprises (Figure 8.3).

We utilised the standard coefficient to see how important the hypothesis was. It shows how much the dependent variable is likely to change when the independent variable changes by one unit. It's important to remember that pro-growth listed small enterprises have a bigger effect on AI competitiveness when technical innovation capabilities are stronger. The T-statistical test showed that the ability to innovate technologically has a big effect on how competitive AI is for pro-growth small enterprises that are listed (Figure 8.4).

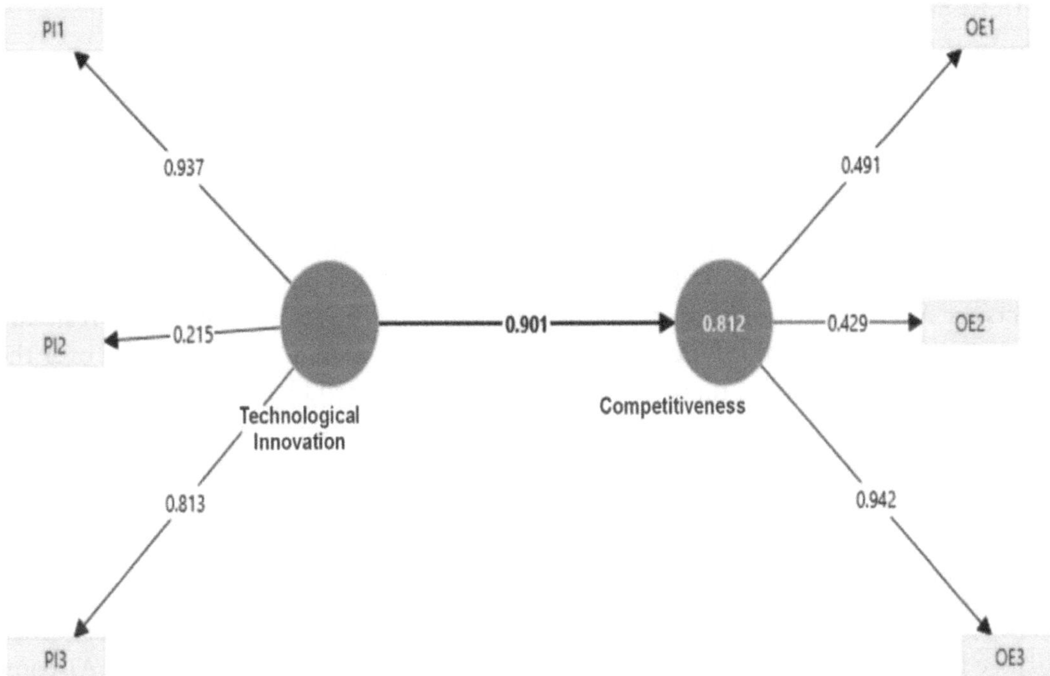

FIGURE 8.3 PLS bootstrapping model with β and P values of technological innovation capabilities and AI competitiveness of pro-growth listed small businesses.

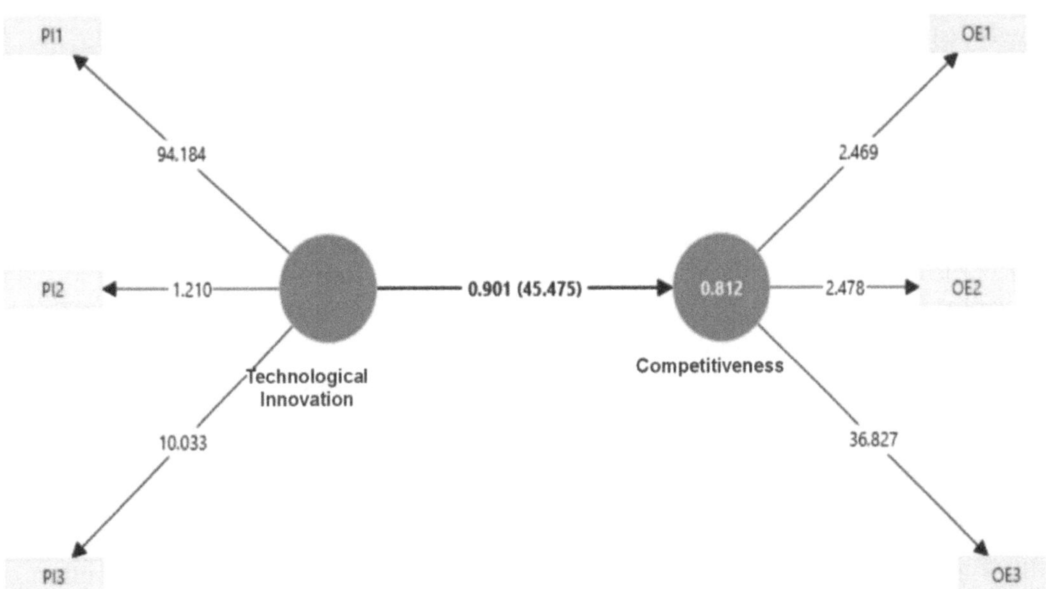

FIGURE 8.4 T-statistics showing the standard error of technological innovation capabilities and AI competitiveness of pro-growth listed small businesses.

8.10 RESEARCHERS' COMPUTATION (2025)

Table 8.2 shows the factor loadings for all the parts of the measurement of technological innovation skills on the AI competitiveness of small pro-growth enterprises, as reported by the study instrument. The validity and reliability of the research tool were further checked using composite reliability, average variance extracted (AVE) estimation, and Cronbach's alpha. The suggested statistical values for the AVE, Cronbach's alpha, composite reliability, and factor loading were all met. This study also used convergent and discriminant validity to figure out construct validity. Convergent validity shows how technological progress affects the AI competitiveness of small enterprises that are pro-growth and listed. It should also be made clear that all of the factor loadings for each measurement item are over the required levels. This means that a lot of the variance is shared by all of the things. While checking for discriminant validity, AVE was set equal to the squared correlation for each construct at the same time. The AVE of the latent variable is higher than the squared correlations in the model between the dormant variable and the constructs.

8.11 RESEARCHERS' COMPUTATION (2024)

Table 8.3 depicts the smart partial least squared statistical results of the hypothesis, which focussed on the relationship between technological innovation capabilities and AI competitiveness of pro-growth listed small businesses. The findings show that technological innovation capabilities affect the AI competitiveness of pro-growth listed small businesses. Specifically, the results showed the rule of technological innovation capabilities on AI competitiveness of pro-growth listed small businesses, at ($\beta = 0.916$, t-statistics = $57.767 > 1.96$, P-value = $0.000 < 0.05$). The path coefficient of 0.916 implies a strong and positive correlation between technological innovation capabilities and the AI competitiveness of pro-growth listed small businesses.

8.12 DISCUSSION OF FINDINGS

The hypothesis revealed that technological innovation capabilities have a significant influence on the AI competitiveness of pro-growth listed small businesses. Using regression analysis, it was revealed that role ambiguity has a significant effect on team dynamics. Hence, the alternate hypothesis was accepted. The results from the model are ($T = 2.515$, $P = 0.001$). Enhancing the AI competitiveness of pro-growth listed small businesses in organisations requires technological innovation capabilities, which is the application of new or considerably enhanced production or delivery processes. This discourse examines the influence of technical innovation capabilities on the AI competitiveness of pro-growth publicly listed small enterprises. It utilises contemporary empirical research to demonstrate how various types of technological innovation skills, influenced by different variables, improve performance and AI competitiveness. Technological innovation capabilities encompass substantial modifications or whole novel methods of distribution or manufacturing. The 2020 study by Johnson et al. has shown that firms adopting radical technological innovation skills had a significant enhancement in the AI competitiveness of pro-growth publicly listed small enterprises. These breakthroughs sometimes involve the application of innovative technology or methodologies that significantly alter work execution, resulting in substantial savings, enhanced efficiency, and improved quality. The impact of technological innovation capabilities on the AI competitiveness of pro-growth publicly listed small enterprises is also shaped by the regulatory and competitive contexts of the sector. Businesses are often compelled to continuously innovate in highly competitive sectors to maintain or enhance their AI competitiveness among pro-growth publicly listed small enterprises. The 2020 research by Singh and Aggarwal indicates that companies in competitive marketplaces who invested in technological innovation skills outperformed their competitors in efficiency and cost-effectiveness. Conversely, highly regulated industries may face constraints that limit the scope of technical innovation; nonetheless, they can still enhance efficiency through compliance-driven improvements. Technological innovation skills are often driven by lean management principles, which prioritise waste reduction and process enhancement. The 2021 research by Womack and Jones indicates that the AI competitiveness of pro-growth listed small enterprises markedly improved for those who used lean methodologies. This study indicates that organisations with more investments in R&D have superior financial performance. Research and development intensity is significantly correlated with profitability and market valuation. Companies that participate more extensively in innovation activities have superior returns on assets and equity. This research illustrates that product innovation is a crucial catalyst for revenue expansion. Empirical research indicates that firms who launch new products or significantly enhance existing ones get greater sales growth and market share. Rubera and Kirca (2012) discovered that product innovation enhances firm performance by augmenting competitive positioning and consumer happiness. This study demonstrates that the utilisation of digital technologies such as automation, AI, and the Internet of Things (IoT) enhances the AI competitiveness of pro-growth publicly listed small enterprises. Manyika et al. (2017) discovered that digitisation enhances operational efficiency, enabling organisations to react more swiftly to market fluctuations and client demands.

8.12.1 Conclusion and Recommendation

The study shows that technical innovation capabilities significantly influence the competitive viability of pro-growth listed small firms. Significant findings indicate that enterprises investing in technology-driven innovations such as digital marketing, process automation, data analytics, and product creation see enhancements in customer satisfaction, market responsiveness, and operational efficiency. Furthermore, technological innovation enhances scalability, distinctiveness, and cost efficiency, enabling small businesses to effectively compete with bigger firms. The study identifies obstacles hindering the complete utilisation of technological capabilities, including insufficient funding, a deficit of skilled labour, and resistance to change; surmounting these challenges is crucial for sustaining innovation and long-term competitive advantage. The report advises organisations to invest strategically in advanced technologies such as automation, big data analytics, cloud computing, and AI. These expenditures will ultimately fortify their competitive position by improving operational efficiency, cost management, and market agility. Furthermore, to motivate small firms to adopt new technology, authorities may offer incentives such as grants, tax reductions, and low-interest loans. To protect sensitive data and maintain stakeholder and customer confidence, firms must concurrently enhance their cybersecurity measures by implementing robust security frameworks. Companies must prioritise capacity building by providing continuous training programmes that provide employees with essential technical knowledge and digital competencies. Collaboration between academic institutions, research centres, and industry personnel may foster innovation and facilitate information exchange.

8.13 LIMITATIONS AND FUTURE STUDY DIRECTION

A main limitation of this study is its dependence on self-reported data, which may be influenced by biases such as social desirability bias, wherein respondents tend to give answers they perceive as more socially acceptable or anticipated rather than accurate. This study recognises many limitations, with the most significant being the exclusion of critical variables, such as infrastructure, which undeniably impacts the competitiveness of small businesses in AI. Nonetheless, it is posited that several of them may function as significant precursors for forthcoming study endeavours. This study's focus solely on small businesses in Nigeria restricts the generalisability of its conclusions. The findings and consequences of this research are acknowledged to be specific to Nigeria, with a primary emphasis on small firms, perhaps constraining the generalisability of the results. Consequently, the cross-sectional design of the study limits the author's ability to assert definitive causal relationships. The research could be augmented by analogous studies performed across several industries and more publicly traded companies in Nigeria or other nations to facilitate international comparison assessments. Moreover, subsequent research may employ alternative data collection and processing techniques to enhance statistical significance.

REFERENCES

Adama, H. E., Popoola, O. A., Okeke, C. D., & Akinoso, A. E. (2024). Theoretical frameworks supporting it and business strategy alignment for sustained competitive advantage. *International Journal of Management & Entrepreneurship Research, 6*(4), 1273–1287.

Adeoye, O. O., Olubiyi, T. O., Ajiteru, W. O., & Adaranijo, L. O. (2023). Unveiling capital structure determinants, business characteristics, and high-growth performance of small and medium-sized enterprises: Evidence from Nigeria. *Journal of Interdisciplinary Socio-Economic and Community Study, 3*(2), 86–98.

Adeyemi, O. S., & Olubiyi, T.O. (2024). An analysis of investment trends, innovation and its impact on cabotage and local content regime: Nigerian experience from businesses in oil and gas and maritime industries. *ABUAD Journal of Social and Management Sciences 5*(2), 285–301.

Afolabi, M. O. (2021). The role of artificial intelligence in Nigerian small businesses: Opportunities and challenges. *Journal of African Business and Technology, 15*(2), 142–157.

African Development Bank Group (2020). *Annual Report and Financial Report 2020.* AfDB. https://www.afdb.org/sites/default/files/documents/publications/afdb_annual_report_2020_main_en_.pdf

Ali, B. J., & Anwar, G. (2021). Business strategy: The influence of Strategic Competitiveness on competitive advantage. *International Journal of Electrical, Electronics and Computers, 6*(2).

Awode, S. S. & Oduola, M. O. (2025). The interplay between technological innovation and human capital development in driving industrial productivity and competitiveness in Africa. *Journal of Economics and Development, 27*(1), 56–71. https://doi.org/10.1108/JED-03-2024-0079

Begazo, T., Blimpo, M., & Dutz, M. (2023). Digital Africa: Technological transformation for jobs. World Bank Publications.

Binns, A. (2019). *AI for small businesses: Strategies for growth and innovation.* McKinsey & Company.

Blichfeldt, H., & Faullant, R. (2021). Performance effects of digital technology adoption and product & service innovation – A process-industry perspective. *Technovation, 105,* 102275.

Brynjolfsson, E., & McAfee, A. (2017). *The second machine age: Work, progress, and prosperity in a time of brilliant technologies.* W. W. Norton & Company.

Chen, M., & Lee, W. (2020). AI-driven product innovation: A case study in the automotive industry. *Journal of Business Innovation, 33*(4), 234–247.

Clohessy, T., Acton, T., & Rogers, N. (2018). Blockchain adoption: Technological, organisational and environmental considerations. *In Business Transformation through Blockchain: Volume I* (pp. 47–76). Cham: Springer International Publishing.

Crupi, A., Del Sarto, N., Di Minin, A., Phaal, R., & Piccaluga, A. (2021). Open innovation environments as knowledge sharing enablers: The case of strategic technology and innovative management consortium. *Journal of Knowledge Management, 25*(5), 12631286.

Chukwurah, E. G., & Aderemi, S. (2024). Elevating team performance with scrum: Insights from successful us technology companies. *Engineering Science & Technology Journal, 5*(4), 1357–1371.

Dahesh, M. B., Tabarsa, G., Zandieh, M., & Hamidizadeh, M. (2020). Reviewing the intellectual structure and evolution of the innovation systems approach: A social network analysis. *Technology in Society, 63,* 101399.

Davenport, T., Guha, A., & Grewal, D. (2020). How AI is reshaping the customer experience. *Harvard Business Review, 98*(6), 112–119.

Dhliwayo, S.& Chebo, A. K. (2024). A framework for sustainable technological innovation capability: A research and policy direction. *European Journal of Innovation Management, 27*(3), 851–871. https://doi.org/10.1108/EJIM-02-2022-0085

DiMaggio, P. (2019). Social structure, institutions, and cultural goods: The case of the United States. In *Social theory for a changing society* (pp. 133–166). Routledge.

Dutta, D., & Sarma, M. K. (2020). Adoption of digital innovation—Formulating adopter categories and levels of adoption in a digital sphere in an emerging economy. *International Journal of Innovation and Technology Management, 17*(08), 2050059.

Edeh, J. N., Obodoechi, D. N., & Ramos-Hidalgo, E. (2020). Effects of innovation strategies on export performance: New empirical evidence from developing market firms. *Technological Forecasting and Social Change, 158,* 120167.

Ferreira, J., Cardim, S., & Coelho, A. (2021). Dynamic capabilities and mediating effects of innovation on the competitive advantage and firm's performance: The moderating role of organizational learning capability. *Journal of the Knowledge Economy, 12,* 620–644.

Gatewood, N. (2022). *Marketing Segmentation Strategies for the Nonprofit Sector* (Doctoral dissertation, Walden University).

González-Ramos, M. I., Guadamillas, F., & Donate, M. J. (2023). The relationship between knowledge management strategies and corporate social responsibility: Effects on innovation capabilities. *Technological Forecasting and Social Change, 188,* 122287.

Gumel, B. I. (2019). Mitigating the challenges of small and medium enterprises in Nigeria. SEISENSE *Journal of Management, 2*(3), 82–99. https://doi.org/10.33215/sjom.v2i3.129

Guttentag, D., & Smith, S. L. (2022). The diffusion of Airbnb: A comparative look at earlier adopters, later adopters, and non-adopters. *Current Issues in Tourism, 25*(20), 32253244.

Gyang, P. S., & Badung, N.P. 2019. The role of small-scale businesses in the economic development of Nigeria. *The International Journal of Business & Management.* 7, 5 DOI: https://doi.org/10.24940/theijbm/2019/v7/i5/BM1905-051.

Hannah, D. P., Tidhar, R., & Eisenhardt, K. M. (2021). Analytic models in strategy, organizations, and management research: A guide for consumers. *Strategic Management Journal, 42*(2), 329–360.

International Labour Organization. (2021). *World Employment and Social Outlook 2021: The role of digital labour platforms in transforming the world of work.* ILO. https://www.ilo.org/wcmsp5/groups/public/---dgreports/---dcomm/---publ/documents/publication/wcms_795453.pdf

Ki, C. W., Chow, T. C., & Li, C. (2023). Bridging the trust gap in influencer marketing: Ways to sustain consumers' trust and assuage their distrust in the social media influencer landscape. *International Journal of Human–Computer Interaction, 39*(17), 3445–3460.

Kimhi, S., & Oliel, Y. (2019). Change management and organizational performance in selected manufacturing companies in Anambra State, Nigeria. *The International Journal of Social Sciences and Humanities Invention, 6*(05), 5437–5445.

Kiveu, M. N., Namusonge, M., & Muathe, S. (2019). Effect of innovation on firm competitiveness: The case of manufacturing SMEs in Nairobi County, Kenya. *International Journal of Business Innovation and Research, 18*(3), 307–327.

Koentjoro, S., & Gunawan, S. (2020). Managing knowledge, dynamic capabilities, innovative performance, and creating sustainable competitive advantage in family companies: A case study of a family company in Indonesia. *Journal of Open Innovation: Technology, Market, and Complexity, 6*(3), 90.

Koloseni, D., & Mandari, H. (2024). Expediting financial inclusion in Tanzania using FinTech: The perspective of diffusion of innovation theory. *Technological Sustainability, 3*(2), 171–194.

Lundberg, M., Engström, S., & Lidelöw, H. (2019). Diffusion of innovation in a contractor company: The impact of the social system structure on the implementation process. *Construction Innovation, 19*(4), 629–652.

Makinde, O.G., Olubiyi, T.O., & Ogundipe, F. (2023). New insights from entrepreneurial characteristics and business performance: Empırıcal findings from Nıgerıa. *Asian Journal of Management, Entrepreneurship and Social Sciences, 3*(04), 72–100

Manyika, J., et al. (2017). *A future that works: Automation, employment and productivity.* McKinsey & Company, New York.

Melović, B., Jocović, M., Dabić, M., Vulić, T. B., & Dudic, B. (2020). The impact of digital transformation and digital marketing on the brand promotion, positioning and electronic business in Montenegro. *Technology in Society, 63*, 101425.

Mendoza-Silva, A. (2021). Innovation capability: A systematic literature review. *European Journal of Innovation Management, 24*(3), 707–734.

Mohammed, H. T. (2023). *Corporate Entrepreneurship Practices and Performance of Micro, Small and Medium Enterprises in Niger State, Nigeria* (Doctoral dissertation, Kwara State University (Nigeria)).

Moyo, S., Wright, J., Ndamba, J., & Gundry, S. (2004). Realizing the maximum health benefits from water quality improvements in the home: A case from Zaka District, Zimbabwe. *Physics and Chemistry of the Earth, 29*, 1295–1299. https://doi.org/10.1016/j.pce.2004.09.012

Muller, E., & Peres, R. (2019). The effect of social networks structure on innovation performance: A review and directions for research. *International Journal of Research in Marketing, 36*(1), 3–19.

Nascimento, L. D. S., Trizotto, R. C. A., Pufal, N. A., Camboim, G. F., & Zawislak, P. A. (2025). Which innovation capabilities are relevant for technological and non-technological innovation? Implications for manufacturing firms' financial performance. Journal of Manufacturing Technology Management, 36(2), 406–430.

Ninduwezuor-Ehiobu, N., Tula, O. A., Daraojimba, C., Ofonagoro, K. A., Ogunjobi, O. A., Gidiagba, J. O., …, & Banso, A. A. (2023). Exploring innovative material integration in modern manufacturing for advancing us competitiveness in sustainable global economy. *Engineering Science & Technology Journal, 4*(3), 140–168.

Nuseir, M. T., & Aljumah, A. (2020). The role of digital marketing in business performance with the moderating effect of environment factors among SMEs of UAE. *International Journal of Innovation, Creativity and Change*, *11*(3), 310–324.

OECD (2021). *Development Co-operation Report 2021: Shaping a just digital transformation*. OECD Publishing. https://doi.org/10.1787/ce08832f-en

Oke, A., Prajogo, D., Idiagbon-Oke, M., & Cheng, T. E. (2022). Linking environmental forces, absorptive capacity, information sharing and innovation performance. *Industrial Management & Data Systems*, *122*(7), 1738–1755. https://doi.org/10.12688/f1000research.76130.1

Olowoporoku, A. A., & Olubiyi, T. O. (2023). Evaluating service quality and business outcomes post-pandemic: Perspective from hotels in Emerging Market. *Sawala: Jurnal Administrasi Negara*, *11*(2), 182–201.

Olubiyi, T. O. (2019). Knowledge management practices and family business profitability: Evidence from Lagos State, Nigeria. *Global Journal of Management and Business Research*, *19*(A11), 21–31. Retrieved from https://journalofbusiness.org/index.php/GJMBR/article/view/2872

Olubiyi, T. O. (2023a). Unveiling the role of workplace environment in achieving the Sustainable Development Goal Eight (SDG8) and employee job satisfaction post-pandemic: Perspective from Africa. *Revista Management & Marketing Craiova*, *XXI*(2), 212–228.

Olubiyi, T. O. (2023b). Leveraging competitive strategies on business outcomes post-covid-19 pandemic: Empirical investigation from Africa. *Journal of Management and Business: Research and Practice*. https://doi.org/10.54933/jmbrp-2023-15-2-1

Olubiyi, T. O. (2024). Improving customer experience through differentiation capability: The rise of competition among Africa's consumer goods companies. *PaperASIA*, *40*(5b), 230–240. https://doi.org/10.59953/paperasia.v40i5b.233

Olubiyi, T. O, Abosi, E., Hanif, N., & Itai, M. M. (2022). Strategic cost leadership impact and sustainable business growth in Africa. *Kurdish Studies*, *10*(1), 173–183. https://doi.org/10.53555/ks.v10i1.3524

Olubiyi, T. O., Adeoye, O. O., Jubril, B., Adeyemi, O. S., & Eyanuku, J. P. (2023). Measuring inequality in Sub-Saharan Africa post-pandemic: Correlation results for workplace inequalities and implication for sustainable development goal ten. *International Journal of Professional Business Review*, *8*(4), e01405. https://doi.org/10.26668/businessreview/2023.v8i4.1405

Olubiyi, T. O. & Akpa, V. A. (2023). Does innovative capability and artificial intelligence really matter for the profitability of consumer goods companies in developing economies? *SunText Review of Economics & Business*, *4*(3), 191–204.

Olubiyi, O. T., Lawal, A. T., & Adeoye, O. O. (2022). Succession planning and family business continuity: Perspectives from Lagos State, Nigeria. *Organization and Human Capital Development*, *1*(1), 40–52. https://doi.org/10.31098/orcadev.v1i1.865

Omoyele, O. S., Olubiyi, T. O., Lanre-Babalola, F. O., Obadare, G. O., & Onikoyi, I. A. (2023). Business model innovation as a catalyst for sustainable entrepreneurship: Empirical findings from small and medium enterprises in Nigeria. *Skyline Business Journal*, *19*(2), 55–64.

Osman, A. M., Liu, Y., & Wang, Z. (2023). Influence of organizational culture on construction firms' performance: The mediating roles of innovation and marketing capabilities. *Buildings*, *13*(2), 308.

Rumanti, A. A., Rizana, A. F., Septiningrum, L., Reynaldo, R., & Isnaini, M. M. (2022). Innovation capability and open innovation for small and medium enterprises (SMEs) performance: Response in dealing with the COVID-19 pandemic. *Sustainability*, *14*(10), 5874. https://doi.org/10.3390/su14105874

St-Pierre, J., Julien, P. A., & Fadil, N. (2023). How do entrepreneurial firms behave in the face of environmental turbulence and uncertainty? Evidence from the manufacturing sector. *Journal of Small Business and Enterprise Development*, *30*(5), 880–901.

Tariq, M. U., Poulin, M., & Abonamah, A. A. (2021). Achieving operational excellence through artificial intelligence: Driving forces and barriers. *Frontiers in Psychology*, *12*, 686624.

Teece, D. J. (2018). Business model innovation: When and why it matters. *Long Range Planning*, *51*(4), 49–58.

Tetteh, C. K. (2024). R&D, innovation and labour productivity: Evidence from the manufacturing sector in Ghana. *Scientific African*, *24*, e02136. https://doi.org/10.1016/j.sciaf.2024.e02136

UNCTAD (2021). *Technology and Innovation Report 2021: Catching technological waves.* United Nations Conference on Trade and Development. Retrieved from https://unctad.org/page/technology-and-innovation-report-2021

Vahdat, S. (2022). The role of IT-based technologies on the management of human resources in the COVID-19 era. *Kybernetes*, *51*(6), 2065–2088.

Vaia, G., Arkhipova, D., & DeLone, W. (2022). Digital governance mechanisms and principles that enable agile responses in dynamic competitive environments. *European Journal of Information Systems*, *31*(6), 662–680.

Wanof, M. I. (2023). Digital technology innovation in improving financial access for lowincome communities. *Technology and Society Perspectives (TACIT)*, *1*(1), 26–34.

Yang, K., Zhou, L., Wang, Z., Lin, C., & Luo, Z. (2019). Humble leadership and innovative behaviour among Chinese nurses: The mediating role of work engagement. *Journal of Nursing Management*, *27*(8), 1801–1808.

YuSheng, K., & Ibrahim, M. (2020). Innovation capabilities, innovation types, and firm performance: Evidence from the banking sector of Ghana. *Sage Open*, *10*(2), 2158244020920892.

CHAPTER 9

A Qualitative Systematic Review of the Influential Factors for Green Decision-Making

Mouhcine Rhouiri, Merouane El Azami El Hassani, Mohamed-Habiboullah Meyabe, and Mehdi Bensouda

9.1 INTRODUCTION

Environmental problems and drivers toward sustainability are common among individuals and organizations, and green decision-making practices are indeed going mainstream. Specifically, capability can be developed in individuals to make decisions based on economic, social, and environmental factors, as well as gain knowledge in green decision-making, as part of developing those capabilities. Conceptually, the strength of this notion arises from a global picture that is changing rapidly due to technological developments, disruption of the economy, and crisis in global climate change, which all create the need for new solutions and adaptation responses (Duan et al., 2019).

In this regard, new technologies such as artificial intelligence (AI), Internet of Things (IoT), and blockchain are creating new opportunities for green decision-making capabilities. For instance, Atstāja and Mukem (2024) contend that these technologies are leading toward better supply chain sustainability, improved risk forecasts, and better resource allocation. The validity, however, is yet to overcome the challenges of low adoption rates due to behavioral, technological, or financial issues, most notably the very high costs for implementation, the lack of digital skills, and unwillingness to change, even among Small and Medium-sized Enterprises (SMEs) (Kamble et al., 2020).

Along with low rates of adoption, the impact of the COVID-19 pandemic has brought to light the vulnerabilities in the global supply chain, revealing that organizations needed to have better decision-making and be more vigilant (Qader et al., 2022). While the crisis moved the needle on technology utilization related to online smart meters, COVID-19 also highlighted severe inequities of access to technology and technology usability. While green

DOI: 10.1201/9781003649885-9

decision-making capability is crucial for ongoing sustainability and community development initiatives, the development and final uptake will be dependent on individuals' situational awareness and their responses to behavioral, technological, and economic issues. As such, the primary research question that drives this study is: *what factors influence the development and uptake of green decision-making capabilities?*

The purpose of this research is to build on these considerations, focusing on examining organizations' green decision-making capabilities and barriers/enablers of organizations' effective green decision-making capabilities. To this end, a thorough literature review and a textual factor analysis with IRaMuTeQ will be undertaken. The intention of this chapter is to identify tangible avenues for enhancing organizational capacity for green decision-making. Ultimately, this chapter will seek to translate this effort into actionable strategies to overcome obstacles and assist with a more effective pathway toward resilient and sustainable practice.

9.2 BACKGROUND

This study emphasizes the increasing importance of green decision-making competencies and the willingness to commit to sustainable development on the backdrop of environmental destruction, enduring socio-economic inequalities, and global economic instability. Green decision-making competencies consist of taking ecological and social aspects into account as part of decision-making processes – an obligation that is urgent in the current context of crisis, including climate change and COVID-19.

Duan et al. (2019) state that once organizations comprehend these latent technologies (especially AI and other advanced analytic techniques), they can adapt to environmental and economic disruptions while reducing uncertainty about risk assessment. Technologies that enable the use of IoT information technology, digital twins, and blockchain processes have the potential to increase efficiencies, lower their environmental impact, and improve supply chains (Katsaliaki et al., 2022; Qader et al., 2022).

Nonetheless, the acceptance and implementation of such technologies are typically impeded by barriers related to finance, technology, and human behavior, such as initial cost, lack of digital skills, lack of willingness to change, and distrust of automation (Duan et al., 2019; Kamble et al., 2020). This will be more pronounced in SMEs.

By extension, this chapter will seek to synthesize the literature around green technologies and the relevant making skills while finding alternative avenues for addressing structural problems and enabling widespread and fair adoption.

9.3 LITERATURE REVIEW ON SYSTEMIC SKILLS, TECHNOLOGIES, AND LEVERS

Sustainable decision-making strategy has become one of the main strategies of sustainable development, informed by three global priorities: environmental, social, and economic. Sustainable decision-making integrates environmental impacts, social consequences, and market viability into a decision-making process and achieves "green" or ecological competencies, including knowledge, skills, and attitudes (Perez-Valls et al., 2016; Stead & Stead, 2014; Subramanian et al., 2016). Competencies are an important ability for decision-making in order to create or realize decisions with ecological imperatives and in organizational contexts, especially those that have significant environmental consequences (Tetteh et al., 2024). Ecological competency development requires a systemic and interdisciplinary approach, including innovation, significantly driven by technological capabilities, where AI can be included and encouraged for sustainable decision-making in particular (Banerjee et al., 2023; Xi et al., 2021; Zeng et al., 2022).

Still, ecological competence within sustainable decision-making is not just a proficient technical capability; it requires intelligent leadership and sustainable models for business (Bansal & DesJardine, 2014; Hahn et al., 2015; Schaltegger et al., 2016). Industries are having impacts on ecological competence (clean technology adoption, Dogbe & Marwa, 2024; hospital waste, Haq et al., 2022; QHRM, Mehrajunnisa et al., 2023).

New technologies like Geographic Information System (GIS), IoT, agent-based modeling (ABM), and Fuzzy Logic-VIseKriterijumska Optimizacija I Kompromisno Resenje (FUZZY-VIKOR) provide real instruments to support ecological decision-making across multiple areas, such as reducing energy consumption (Jr et al., 2017), agent-based market simulations (Meng et al., 2018), improvements to reverse logistics (Vahabzadeh et al., 2015), and operations (Khripko et al., 2016; Zhang et al., 2023). While their capability has been noted, they are not scalable due to the current lack of standards. In the climate of an actuary changed by COVID-19, supply chain needs are developing, and their change demands more effort, generating investment in Industry 4.0 technologies for resilience (Ahmed et al., 2023; Ghobakhloo et al., 2025; Kazancoglu et al., 2023). AI, blockchain, digital twins, and IoT are some examples of transformational technology, but currently education is in change in scope and cost with travel and use continuing around the issues (Al-Banna et al., 2023; Chauhan et al., 2023; Gajek et al., 2021; Kopeinig et al., 2024; Sutar et al., 2024).

Even with significant advances, maintaining sustainability efforts over the long term continues to also be asynchronous in light of the various behavioral and financial constraints that are present. Behavioral biases such as cognitive biases, the tendency to seek immediate financial paybacks (Bensouda & Benali, 2022), limited control perceptions (Zhao et al., 2021), perceived upfront costs (Silva et al., 2021), and financial apathy (Alhammadi et al., 2024) represent significant hurdles as well. Thus, an integrated Environmental, Social, and Governance (ESG) framework – financially incentivized with psychological nudges – seems to increasingly be viewed as vital (Yang, 2023).

Environmental issues – including climate change, biodiversity loss, and soil degradation – are getting worse around the world. To move toward sustainability is not a straight path but a complicated set of interactions that include individual and organizational capabilities (Bogoslov et al., 2024; Nogaibayeva et al., 2024), economic forces (Asif et al., 2025), environmental regulations (El-Hakim & AbouZeid, 2024), collective action and commitment (Creazza et al., 2024; Gondwe et al., 2024), technological and adaptive methods (Moshood et al., 2024; Shao et al., 2023), and the agency of local communities (Fuchs et al., 2024; Jayanetti et al., 2024).

Ultimately sustainable decision-making can only be embedded through human capacity, innovation, supporting policy, and collective action at the scale required to meet today's challenges.

9.4 METHODOLOGICAL APPROACH

The methodology followed combines a systematic approach with a textual factor analysis conducted via IRaMuTeQ. This dual approach makes it possible to identify, structure, and organize existing knowledge on sustainable decision-making skills. The aim is to highlight dominant trends, identify persistent gaps in research, and identify relevant decision-making levers.

With this in mind, the first methodological phase comprised rigorously identifying articles related to sustainable development decision-making. To this end, a targeted query was launched in February 2025 on the Scopus database, integrating thematic, temporal, and linguistic filters. The analysis period selected (2016–2023) enabled us to select 277 publications meeting the established criteria.

As part of this systematic review, and in order to ensure both transparency and scientific rigor, we structured the selection process according to the recommendations of the Preferred Reporting Items for Systematic Reviews and Meta-Analyses (PRISMA) protocol. The inclusion and exclusion criteria used are summarized in Figure 9.1.

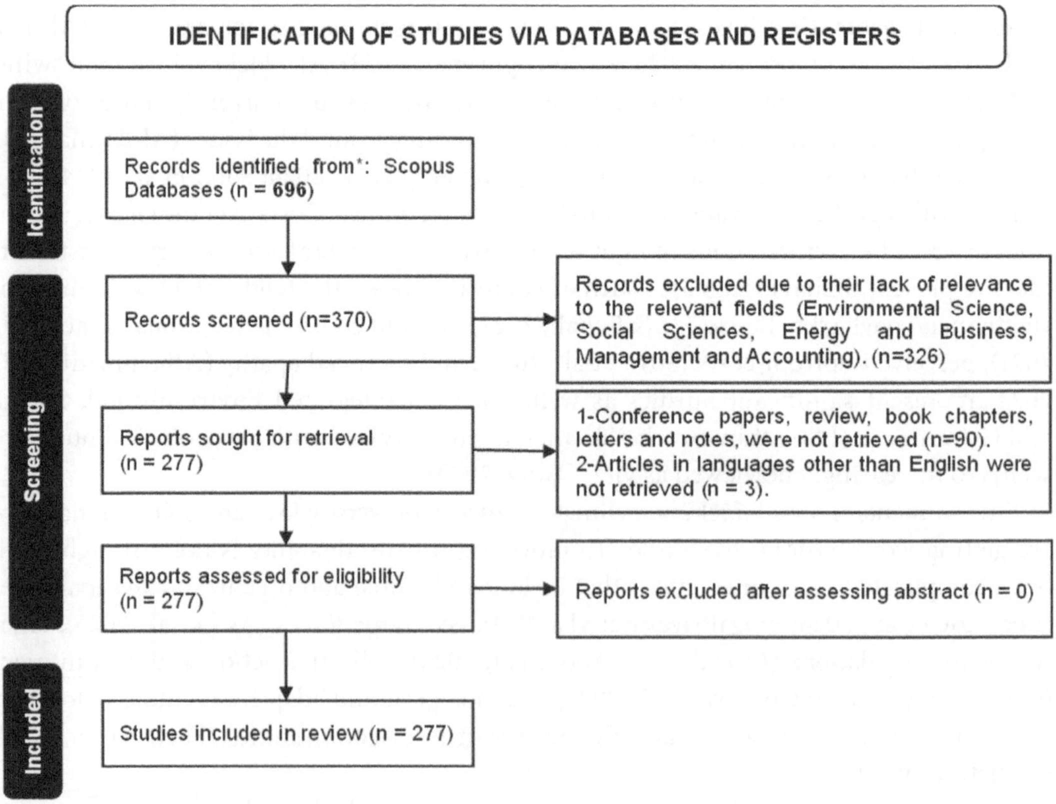

FIGURE 9.1 Adapted PRISMA flow chart.

9.5 TEXTUAL FACTOR ANALYSIS WITH IRaMuTeQ

After selecting the 277 articles, the next step was to run an abstract text factor analysis using IRaMuTeQ, which is software that uses complex textual statistics to determine themes and conceptual links. The segments of texts are then grouped into lexical classes via Descending Hierarchical Classification (DHC) to identify homogeneous areas of clarification, containing one or multiple structuring themes or subsections of a thesis.

A validation of the groups followed by a dendrogram analysis determined that 80.84% of the segments were accurately clustered. This allowed us to sort the results from IRaMuTeQ into larger thematic classes, highlighting major axes of connection with the sustainable development skills, particularly in the areas of environment and decision-making (Figure 9.2).

In this study, a dendrogram of the descending hierarchical classification (DHC) analysis identified four distinct classes:

Class 1 – Resilience research methods – Focus on analytical approaches, theoretical frameworks, and research methodologies mobilized for the study of resilience.

Class 2 – Scientific innovation and digital solutions – Focus on scientific innovation and the implementation of digital technologies to leverage innovative and evidence-informed responses to contemporary challenges.

Class 3 – Economic disruption by COVID-19 – Discusses the economic effects of the pandemic and the ensuing disruption for business and supply chains.

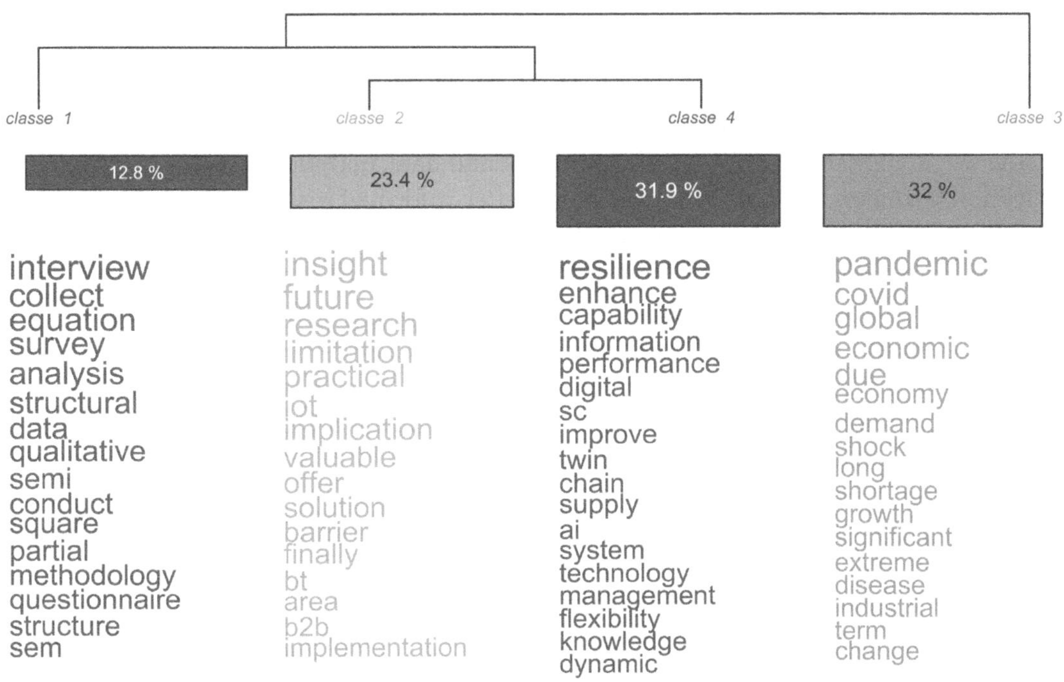

FIGURE 9.2 Descending hierarchical classification (IRaMuTeQ-0.8a7 outputs).

Class 4 – Digital transformation for resilience – Discusses the role of digital technologies during post-crisis reconstruction and building more resilient systems.

The analysis of these four classes shows an interrelationship with economic disruptions, the emergence of digital technologies, the dynamics of scientific activity, and the analytical approaches that were utilized. The COVID-19 pandemic was an evident tipping point. In comparison to previous crises, it was much easier and faster for many companies to accelerate the digitization of their supply chains while enhancing their analytical capabilities, as companies seek to foster resilience in the face of crises.

9.6 DISCUSSION OF TEXTUAL ANALYSIS RESULTS

With increasing socio-environmental turmoil comes a greater need to improve sustainable decision-making capabilities. However, a number of enduring barriers still exist, including global-scale issues, economic barriers, technological barriers, and social barriers. To aid in unraveling these issues, this systematic review relied on textual analysis, using IRaMuTeQ and content analysis. The outcomes are presented in three theme clusters: (1) how systemic disruptions affect sustainable development green decision-making competencies; (2) the role of digital technologies and methods to develop sustainable resilience; and (3) innovation and barriers to uncertainty and disruption in relation to sustainability.

9.6.1 Impact of Disruptions on Green Decision Skills in Sustainable Development

The third category (representing 32% of the classified segments) emphasizes how COVID-19's economic disruption has seemingly impacted both the development of green skills and decisions around their development. The pandemic revealed some vulnerabilities in supply chains, which made many organizations (especially SME organizations) redesign their operating models toward resilience (Ivanov & Dolgui, 2020; Sarkis, 2020).

The use of digital technologies, including AI and data analytics, has given greater visibility and flexibility in chains, which enables better-informed sustainability decisions (Bag et al., 2021; Choi et al., 2020). The transition to sustainability is limited by affordance constraints, competency constraints, and resistance to change, all of which affect the changes to green skills decision-making through digital technology enablers, particularly for SME (Klein & Todesco, 2021; Kot, 2018; Nazaruddin et al., 2024).

To overcome affordance constraints, a commitment to innovation and continual training, both professionally and organizationally, is needed, as well as a commitment to sustainable transformation (Jabbour et al., 2019; Seetharaman, 2020). The pandemic has acted as a catalyst for transformational changes, all focused and made possible by sustainable organizations, changes supported by skills and resources (Alami et al., 2021; Sarkis et al., 2020).

9.6.2 Digital Technologies and Research Approaches for Sustainable Resilience

This cluster includes Classes 2 and 4 that account for 31.9% and 23.4% of the classified segments (55.3% of the 80.84% segments analyzed). It emphasizes the convergence of digital technology adoption and, more importantly, the enhanced application of research-based methodologies as facilitative pathways to supply chain resilience and sustainable decision-making. Furthermore, this cluster emphasizes the urgent development of structured processes to support the amalgamation of Industry 4.0 technologies and methodological rigor to enhance decision-making ability in disruption cases.

9.6.2.1 Digital Technologies and Supply Chain Resilience

Class 4 has supported us in clarifying that Industry 4.0 technologies, AI, digital twins, and IoT, use these technologies to increase supply chain resilience. If these technologies can improve visibility, flexibility, and collaboration, this could enable more agile responses to disruption and risk. As an illustration, digital twins allow for real-time simulation of possible critical situations, like a supplier disruption of raw materials, to minimize risk and change operations (Ivanov & Dolgui, 2019). AI then uses other technological solutions, such as recommendation systems, to use resources in a more optimized manner to make informed decisions during a crisis (for instance, the COVID-19 pandemic) (Kashem et al., 2023). From this standpoint, these technologies are especially valuable in emergency responsiveness and efficiency. Therefore, strategically integrating the technologies purposefully allowed for improved anticipatory hazards and ultimately better resilience of supply chains as a whole (Katsaliaki et al., 2022).

9.6.2.2 Research Methodologies for Sustainable Decisions

The findings clearly illustrate that the use of diverse methodology types – both quantitative and qualitative methods – is critical to developing strong theoretical perspectives and consequently promoting sustainable decision-making. Partial Least Squares-Structural Equation Modeling (PLS-SEM) is an example of a very helpful tool for complex relationships in analyzing AI and operational performance of firms (Katsaliaki et al., 2022). On the other side, thematic analysis is also helpful for understanding behavioral brakes that are associated with technology integration, through stakeholder perceptions. Therefore, these methods are especially relevant for research between academia and industry, as they can lead to a better understanding and improvement of the sustainability capabilities of decision-making.

9.6.3 Innovations and Barriers in Managing Sustainable Disruptions

Class 1 (12.8% of 277 articles) focuses on IoT and blockchain as two strategic technologies for the sustainable resilience of supply chains, particularly within food systems. IoT contributes to the monitoring of flows in real time, thus preventing loss and waste (da Costa et al., 2022). Meanwhile, blockchain guarantees traceability and transparency of flows, thereby increasing trust with stakeholders (Jamil et al., 2019). However, the backlash on both technologies, because of high initial costs, complexity, and resistance to change, remains a major barrier to adoption, especially for SMEs (Kamilaris et al., 2019; Zhou et al., 2022). Both technologies have the potential to improve risk management through enhanced visibility of logistics (Saberi et al., 2019). However, integrated approaches (such as dynamic simulation) can predict disruptions and adjust operations in real time (Choi, 2021; Ivanov & Dolgui, 2021; Lotfi & Larmour, 2022) but require institutional support for sustainable adoption (i.e., subsidies, training, and partnerships) (Dolgui et al., 2020; Xu et al., 2021).

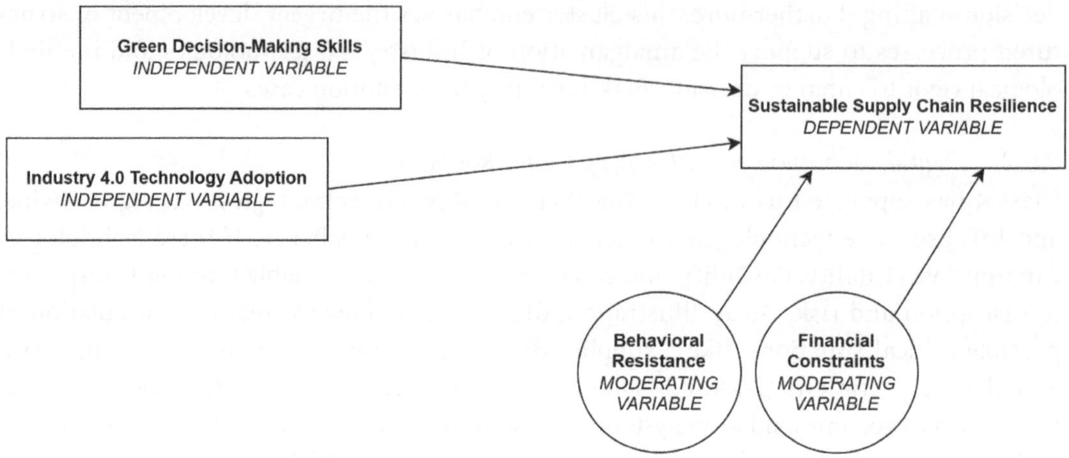

FIGURE 9.3 Conceptual model of hybrid decision-making for sustainable resilience.

In this context, the proposed conceptual model (Figure 9.3) would demonstrate the ways that green skills and digital technologies (Industry 4.0) could work together to enhance and build resilience in supply chains against disruptions from economic and environmental changes. The study indicates that the convergence of human elements and technological solutions may promote sustainability, but these also have structural challenges, such as financial implications and resistance, to change to withstand. Using this approach would give organizations the opportunity to be proactive about ecological imperatives and operational effectiveness and ultimately support the evolution toward circular and adaptive systems.

9.7 CONCLUSION

In conclusion, the research reveals both the opportunities and constraints of the implementation of emergent technologies, in addition to the strategic significance of green decision-making competencies to overcome the challenges of sustainable development. More specifically, three overarching paths are present. First, the central role of digital technologies – including AI, IoT, and blockchain – in enhancing decision-making processes and minimizing environmental footprint is evident, but uptake is hindered by the lack of availability, significant ownership costs, and gaps in digital skills, indicating the need for training and stakeholder engagement. Second, recent crises like the COVID-19 pandemic stress the importance of rethinking traditional business models and employing resilient and transparent governance models. Despite the need for this change, there are limits posed by organizational inertia and a lack of a structured regulatory framework. Finally, coherence between technological innovation, financial support, and public policies is fundamental to effective change. Accordingly, clear legislative frameworks and advanced methodologies are essential leverages to overturn motivations and economic barriers.

From a theoretical perspective, this chapter contributes to a clearer understanding of the dynamics favoring – or limiting – the move to sustainable decision-making. However, it is limited by its reliance on the Scopus Database and the fact that no substantial empirical studies exist. In perspectives, it would be valid to investigate how emerging technologies – for illustration, AI, IoT, and blockchain – can moderate any link between firms' green engagement and their operational performance (green SCM) (Supply Chain Management). In addition, there could be a distinct effort to identify the causes of the behavioral and financial disincentives for SMEs to define some action levers that could instigate a fair and sustainable digital transition.

REFERENCES

Ahmed, T., Karmaker, C. L., Nasir, S. B., Moktadir, M. A., & Paul, S. K. (2023). Modeling the artificial intelligence-based imperatives of industry 5.0 towards resilient supply chains: A post-COVID-19 pandemic perspective. *Computers & Industrial Engineering, 177*, 109055.

Alami, S., Rhouiri, M., & Marghich, A. (2021). Résilience active, pivot de la compétitivité de l'entreprise Post-Covid 19: Cas des entreprises du secteur industriel au Maroc. *Revue Management & Innovation, 4*(2), 135–155.

Al-Banna, A., Rana, Z. A., Yaqot, M., & Menezes, B. C. (2023). Supply chain resilience, industry 4.0, and investment interplays: A review. *Production & Manufacturing Research, 11*(1), 2227881.

Alhammadi, Y., Kineber, A. F., & Alhusban, M. (2024). Investigating barriers to the adoption of energy management practices for sustainable construction projects: SEM and ANN approaches. *Civil Engineering Journal, 10*(4), 1232–1253. https://doi.org/10.28991/CEJ-2024-010-04-015

Asif, M., Dhamija, A., Khan, W., Johri, A., & Wasiq, M. (2025). Factors affecting adoption of green practices by food firms: Evidences from the World Bank Enterprises Survey. *Clean Technologies and Environmental Policy, 27*, 4841–4858. https://doi.org/10.1007/s10098-024-03122-4

Atstāja, D., & Mukem, K. W. (2024). Sustainable supply chain management in the oil and gas industry in developing countries as a part of the quadruple helix concept: A systematic literature review. *Sustainability, 16*(5), 1776.

Bag, S., Telukdarie, A., Pretorius, J. C., & Gupta, S. (2021). Industry 4.0 and supply chain sustainability: Framework and future research directions. *Benchmarking: An International Journal, 28*(5), 1410–1450.

Banerjee, D. K., Lnu, A. K., & Sharma, K. (2023). Artificial intelligence on supply chain for steel demand. *International Journal of Advanced Engineering Technologies and Innovations, 1*(4), 441–449.

Bansal, P., & DesJardine, M. (2014). Business sustainability: It is about time. *Strategic Organization, 12*(1), 70–78.

Bensouda, M., & Benali, M. (2022). Behavioral barriers to energy efficiency and policy interventions: A survey of the literature. *International Journal of Energy Economics and Policy, 12*(6), 305–310.

Bogoslov, I. A., Corman, S., & Lungu, A. E. (2024). Perspectives on artificial intelligence adoption for European Union elderly in the context of digital skills development. *Sustainability, 16*(11), 4579.

Chauhan, C., Akram, M. U., & Chauhan, A. (2023). Mapping pathways for building resilient supply chains: A systematic literature review. *Journal of Cleaner Production, 425*, 138701.

Choi, T. M. (2021). Fighting against COVID-19: What operations research can help and the sense-and-respond framework. *Annals of Operations Research*. https://doi.org/10.1007/s10479-021-03973-w

Choi, T. Y., Rogers, D., & Vakil, B. (2020). Coronavirus is a wake-up call for supply chain management. *Harvard Business Review, 27*(1), 364–398.

Creazza, A., Colicchia, C., & Evangelista, P. (2024). Leveraging shippers-logistics providers relationships for better sustainability in logistics: The perspective of SMEs. *The International Journal of Logistics Management, 35*(4), 1009–1039.

da Costa, T. P., Gillespie, J., Cama-Moncunill, X., Ward, S., Condell, J., Ramanathan, R., & Murphy, F. (2022). A systematic review of real-time monitoring technologies and its potential application to reduce food loss and waste: Key elements of food supply chains and IoT technologies. *Sustainability, 15*(1), 614.

Dogbe, C. S. K., & Marwa, N. (2024). Digital orientation and manufacturing firms' green innovation performance: The mediating role of green competence. *Journal of Risk and Financial Management, 17*(10), 430.

Dolgui, A., Ivanov, D., & Rozhkov, M. (2020). Does the ripple effect influence the bullwhip effect? An integrated analysis of structural and operational dynamics in the supply chain. *International Journal of Production Research, 58*(5), 1285–1301.

Duan, Y., Edwards, J. S., & Dwivedi, Y. K. (2019). Artificial intelligence for decision making in the era of Big Data – Evolution, challenges and research agenda. *International Journal of Information Management, 48*, 63–71.

El-Hakim, Y., & AbouZeid, M. N. (2024). Towards mitigating climate change negative impact: The role of regulations and governance in the construction industry. *Sustainability, 16*(16), 6822.

Fuchs, L. E., Orero, L., Kipkorir, L. A., Apondi, V., & Owili, S. O. (2024). Scaling models for Regreening Africa: Enhancing agroecological integration through smallholders' assets and agency in Kenya. *Frontiers in Sustainable Food Systems, 8*, 1449615.

Gajek, S., Lees, M., & Jansen, C. (2021). IIoT and cyber-resilience: Could blockchain have thwarted the Stuxnet attack? *AI & Society, 36*(3), 725–735.

Ghobakhloo, M., Iranmanesh, M., Foroughi, B., Tseng, M. L., Nikbin, D., & Khanfar, A. A. (2025). Industry 4.0 digital transformation and opportunities for supply chain resilience: A comprehensive review and a strategic roadmap. *Production Planning & Control, 36*(1), 61–91.

Gondwe, W., Phiri, A., Birachi, E., Magreta, R., Larochelle, C., Machira, K., … & Nkhata, W. (2024). Private sector led multi-stakeholder platforms positively influence certified common bean seed supply in Malawi. *Heliyon, 10*(18).

Hahn, R., Pinkse, J., Preuss, L., & Figge, F. (2015). Tensions in corporate sustainability: Towards an integrative framework. *Journal of Business Ethics, 127*(2), 297–316.

Haq, I. U., Chupradit, S., Chupradit, P. W., & Huo, C. (2022). Measuring determinants of environmental performance during COVID-19: Evidence from the healthcare sector. *World Journal of Entrepreneurship, Management and Sustainable Development, 18*(5), 575–593.

Ivanov, D., & Dolgui, A. (2019). Low-certainty-need (LCN) supply chains: A new perspective in managing disruption risks and resilience. *International Journal of Production Research, 57*(15-16), 5119–5136.

Ivanov, D., & Dolgui, A. (2020). Viability of intertwined supply networks: Extending the supply chain resilience angles towards survivability. A position paper motivated by COVID-19 outbreak. *International Journal of Production Research, 58*(10), 2904–2915.

Ivanov, D., Tsipoulanidis, A., & Schönberger, J., (2021). Supply chain risk management and resilience. In *Global Supply Chain and Operations Management* (pp. 485–520). Springer. https://doi.org/10.1007/978-3-030-72331-6_15

Jabbour, C. J. C., de Sousa Jabbour, A. B. L., Sarkis, J., & Godinho Filho, M. (2019). Unlocking the circular economy through new business models based on large-scale data: An integrative framework and research agenda. *Technological Forecasting and Social Change, 144*, 546–552.

Jamil, F., Hang, L., Kim, K., & Kim, D. (2019). A novel medical blockchain model for drug supply chain integrity management in a smart hospital. *Electronics, 8*(5), 505.

Jayanetti, T., Perera, S. L., Ranadewa, T., Seneviratne, I., & Perera, B. A. K. S. (2024). Impact of green practices on ecological conscious consumer behaviour: An empirical investigation of hotels in Sri Lanka. *A| Z ITU Journal of the Faculty of Architecture, 21*(1), 169–192.

Jr, B. A., Majid, M. A., & Romli, A. (2017). Green information technology system practice for sustainable collaborative enterprise: A structural literature review. *International Journal of Sustainable Society, 9*(3), 242–272.

Kamble, S. S., Gunasekaran, A., & Gawankar, S. A. (2020). Achieving sustainable performance in a data-driven agriculture supply chain: A review for research and applications. *International Journal of Production Economics, 219*, 179–194.
Kamilaris, A., Fonts, A., & Prenafeta-Boldύ, F. X. (2019). The rise of blockchain technology in agriculture and food supply chains. *Trends in Food Science & Technology, 91*, 640–652.
Kashem, M. A., Shamsuddoha, M., Nasir, T., & Chowdhury, A. A. (2023). Supply chain disruption versus optimization: A review on artificial intelligence and blockchain. *Knowledge, 3*(1), 80–96.
Katsaliaki, K., Galetsi, P., & Kumar, S. (2022). Supply chain disruptions and resilience: A major review and future research agenda. *Annals of Operations Research, 319*, 965–1002. https://doi.org/10.1007/s10479-020-03912-1
Kazancoglu, I., Ozbiltekin-Pala, M., Mangla, S. K., Kumar, A., & Kazancoglu, Y. (2023). Using emerging technologies to improve the sustainability and resilience of supply chains in a fuzzy environment in the context of COVID-19. *Annals of Operations Research, 322*(1), 217–240.
Khripko, D., Schlüter, B. A., Rommel, B., Rosano, M., & Hesselbach, J. (2016). Energy demand and efficiency measures in polymer processing: Comparison between temperate and Mediterranean operating plants. *International Journal of Energy and Environmental Engineering, 7*, 225–233.
Klein, V. B., & Todesco, J. L. (2021). COVID-19 crisis and SMEs responses: The role of digital transformation. *Knowledge and Process Management, 28*(2), 117–133.
Kopeinig, J., Woschank, M., & Olipp, N. (2024). Industry 4.0 technologies and their implications for environmental sustainability in the manufacturing industry. *Procedia Computer Science, 232*, 2777–2789.
Kot, S. (2018). Sustainable supply chain management in small and medium enterprises. *Sustainability, 10*(4), 1143.
Lotfi, M., & Larmour, A. (2022). Supply chain resilience in the face of uncertainty: How horizontal and vertical collaboration can help? *Continuity & Resilience Review, 4*(1), 37–53.
Mehrajunnisa, M., Jabeen, F., Faisal, M. N., & Lange, T. (2023). The influence of green human resource management practices and employee green behavior on business performance in sustainability-focused organizations. *Journal of Environmental Planning and Management, 66*(12), 2603–2622.
Meng, Q., Zhu, H., Li, Z., Du, J., Wang, X., & Jeong Kim, M. (2018). How green building product decisions from customers can be transitioned to manufacturers: An agent-based model. *Sustainability, 10*(11), 3977.
Moshood, T. D., Mahmud, F., Nawanir, G., Ahmad, M. H., Mohamad, F., & AbdulGhani, A. (2024). Sustainable adoption of biodegradable plastics: A quantitative analysis of determinants and consumer behaviour in developing economies. *Management of Environmental Quality: An International Journal, 36*(3), 706–740. https://doi.org/10.1108/MEQ-05-2024-0186
Nazaruddin, I., Utami, E. R., & Rahmawati, E. (2024). Digitalization challenges for SMEs: A systematic literature review perspective and future research. *JBTI: Jurnal Bisnis: Teori dan Implementasi, 15*(1), 15–43.
Nogaibayeva, A. A., Ozkan, A., Yildiztas, A., & Tas, M. (2024). Exploring teachers' view on technology and a sustainable adoption framework in language teaching and learning: A systematic review. *Journal of Curriculum and Teaching, 13*(4), 93–111. https://doi.org/10.5430/jct.v13n4p93
Perez-Valls, M., Cespedes-Lorente, J., & Moreno-Garcia, J. (2016). Green practices and organizational design as sources of strategic flexibility and performance. *Business Strategy and the Environment, 25*(8), 529–544.
Qader, G., Junaid, M., Abbas, Q., & Mubarik, M. S. (2022). Industry 4.0 enables supply chain resilience and supply chain performance. *Technological Forecasting and Social Change, 185*, 122026.
Saberi, S., Kouhizadeh, M., Sarkis, J., & Shen, L. (2019). Blockchain technology and its relationships to sustainable supply chain management. *International journal of production research, 57*(7), 2117–2135.

Sarkis, J. (2020). Supply chain sustainability: Learning from the COVID-19 pandemic. *International Journal of Operations & Production Management, 41*(1), 63–73.

Sarkis, J., Cohen, M. J., Dewick, P., & Schröder, P. (2020). A brave new world: Sustainable development and supply chain management in the post-COVID-19 era. *Resources, Conservation & Recycling*, 159, 104894. https://doi.org/10.1016/j.resconrec.2020.104894

Schaltegger, S., Lüdeke-Freund, F., & Hansen, E. G. (2016). Business models for sustainability: A co-evolutionary analysis of sustainable entrepreneurship. *Organization & Environment, 29*(3), 1–20.

Seetharaman, P. (2020). Business models shifts: Impact of Covid-19. *International Journal of Information Management, 54*, 102173.

Shao, D., Ishengoma, F. R., Alexopoulos, C., Saxena, S., Nikiforova, A., & Matheus, R. (2023). Integration of IoT into e-government. *Foresight, 25*(5), 734–750.

Silva, D. R. D. M., Lucas, L. O., & Vonortas, N. S. (2021). Internal barriers to innovation and university-industry cooperation among technology-based SMEs in Brazil. In *The Influence of Internal Barriers on Open Innovation* (pp. 31–59). Routledge.

Stead, J. G., & Stead, W. E. (2014). *Sustainable Strategic Management*. Routledge.

Subramanian, N., Abdulrahman, M. D., Wu, L., & Nath, P. (2016). Green competence framework: Evidence from China. *The International Journal of Human Resource Management, 27*(2), 151–172.

Sutar, P. S., Kolte, G., Yamini, S., & Mathiyazhagan, K. (2024). Food supply chain resilience in the digital era: A bibliometric analysis and development of conceptual framework. *Journal of Business & Industrial Marketing, 39*(9), 1863–1893.

Tetteh, F. K., Mensah, J., & Owusu Kwateng, K. (2024). Understanding what, how and when green logistics practices influence carbon-neutral supply chain performance. *International Journal of Productivity and Performance Management, 74*(6), 2211–2244. https://doi.org/10.1108/IJPPM-08-2024-0517

Vahabzadeh, A. H., Asiaei, A., & Zailani, S. (2015). Green decision-making model in reverse logistics using FUZZY-VIKOR method. *Resources, Conservation and Recycling, 103*, 125–138.

Xi, H., Wu, X., Chen, X., & Sha, P. (2021). Artificial intelligent based energy scheduling of steel mill gas utilization system towards carbon neutrality. *Applied Energy, 295*, 117069.

Xu, X., Lu, Y., Vogel-Heuser, B., & Wang, L. (2021). Industry 4.0 and Industry 5.0—Inception, conception and perception. *Journal of Manufacturing Systems, 61*, 530–535.

Yang, L. (2023). An environmental, social, and governance strategic model for managing pharmaceutical supply chains with financial obstacles. *Decision Analytics Journal, 7*, 100203.

Zeng, S., Xie, X., & Tam, C. M. (2022). Eco-competence: A new concept for sustainable development. *Sustainability, 14*(1), 123.

Zhang, R., Liu, J., & Qian, Y. (2023). Wholesale-price vs cost-sharing contracts in a green supply chain with reference price effect under different power structures. *Kybernetes, 52*(5), 1879–1902.

Zhao, X., Zheng, J., Zhang, M., & Li, H. (2021). Mechanism of green production decision-making under the improved theory of planned behavior framework for new agrarian business entities. *Chinese Journal of Eco-Agriculture, 29*(9), 1636–1648.

Zhou, D., Xu, K., Lv, Z., Yang, J., Li, M., He, F., & Xu, G. (2022). Intelligent manufacturing technology in the steel industry of China: A review. *Sensors, 22*(21), 8194.

CHAPTER 10

Nexus between Renewable Energy, Artificial Intelligence, and Economic Growth

The Moroccan Case

Yahya Fikri and Mohamed Rhalma

10.1 INTRODUCTION

To promote the process of sustainable national growth, artificial intelligence (AI) is essential to the grid's integration of renewable energy (RE) sources (Iorgovan, 2024). In recent years, the energy sector and research groups have given big data and AI a lot of attention (Mazzeo et al., 2021). AI techniques have improved the sustainability, efficiency, and dependability of Renewable Energy Consumption (REC). AI has become a disruptive force in the RE, improving the sustainability and efficiency of energy infrastructure. Additionally, these technologies have been identified as potential sources of economies of scale in more recent empirical investigations on the topic (Nchake & Shuaibu, 2022). Sauvant (2021) Research aimed at identifying AI's overall impact on economic variables is warranted given the widespread use of AI in businesses and industries (Gonzales, 2023; Hu et al. 2023). The "human capital," a word commonly used to refer to "human resources," refers to the diversity of abilities, backgrounds, aptitudes, and expertise that people bring to the workplace (Fikri & Rhalma, 2024a, b). As frugalities conversion from conventional industries to knowledge-based sectors, the importance of human capital (HC) in fostering economic growth (EG) has increased (Fikri & Rhalma, 2024). Numerous academic and empirical studies have sought to explain how variations in HC affect national EG rates (Fikri & Rhalma, 2024). Numerous relationships—though not always obvious—that link HC to an economy's capacity for innovation and productivity growth have led to new growth models that justify the focus on HC and its impact on a country's economic success (Fikri & Rhalma, 2024). A workforce with a good education can also encourage innovation and productivity, which will assist EG (Fikri & Rhalma, 2024). EG, a fundamental concept in economics, is the gradual rise in a country's output of goods and services over time. Other metrics to measure economic advancement include employment, human development, and per capita income (Fikri et al., 2024). The relationship

DOI: 10.1201/9781003649885-10

between EG and HC has long piqued the interest of scholars and policymakers (Fikri & Rhalma, 2024). The association between general government final consumption spending (GGCS) (Gross Domestic Product) (% GDP), REC, CO_2, and EG sustainability was shown to be inconsistent in studies and empirical research (Fikri & Rhalma, 2024). Furthermore, foreign direct investment (FDI) offers several benefits, such as the potential to facilitate technology transfer, improve managerial skills, expand market reach, and potentially lead to augmented engagement (Iddrisu et al., 2023). However, domestic investment is required to establish a solid basis for sustaining EG and innovation. This covers the cost of education, health care, and infrastructure (Batool & Akbar, 2023). Large-scale FDI projects have been attracted to Morocco in sectors like chemicals, electric cars, RE, and tourism. Policies and incentives from the government have played a major role in encouraging investment in essential industries and economic diversification (Harbal & Khihel, 2023; Jean, 2023). The Investment Charter is largely responsible for Morocco's increased competitiveness in fascinating FDI. The FDI flows have significantly increased because of the Charter's promotion of investment in vital sectors including RE, electric cars, and semi-conductors, according to Machloukh (2024). One of the nations that is presently carrying out short- and medium-term RE development projects is Morocco. The connection among Morocco's EG, per-person CO_2, FDIs, and GDP contribution—which is determined by government final consumption expenditures—has not yet been examined in any previous research. The current research purposes to explore the connection among the economic performance of Morocco, the share of RE causes in total energy consumption, the percentage of GDP that public administrations spend on final consumption, and the amount of CO_2 emissions per resident. The goal is to present a fresh viewpoint on these relationships, particularly considering Moroccan culture. Economic literature has focused a great deal of emphasis on the relationship between REC, EG, public spending, and CO_2 emissions. The aim of the current study is to examine the association amid Morocco's CO_2 emissions, REC, FDI, public spending, and EG between 1993 and 2020. We have selected and combined our components based on earlier empirical studies carried out in the Moroccan context (Asli et al., 2024; Fikri & Rhalma, 2024). This is important for Moroccan officials because it enables them to develop policies for sustainable growth that strike a balance between environmental preservation and economic progress. Our present chapter is structured as follows in advance: Section 10.2, a literature review is provided. Section 10.3, the models, estimating techniques, and data sources are described. Section 10.4 shows the relevant section. Section 10.5 provides an overview of the findings and suggestions.

10.2 LITERATURE REVIEW

Productivity gains should result from the process of learning through experience and repetition, opening doors for economic expansion. Lu (2021) recently developed a theoretical framework that charts how AI affects endogenous growth. Lu (2021) also reveals a reasonable growth direction in the three-sector endogenous growth model, in which elements such as AI and production expand at the same pace. AI's ability to enhance human lives in a variety of habits, after education and care of health space exploration and climate change, makes its future indisputable (Santosh & Gaur, 2022). Energy use and societal growth are inextricably linked. As the world's population grows and the economy and society advances, there will inevitably be a greater need for energy for production and daily living. Optimizing energy output, maintaining strong infrastructure, and integrating seamlessly into the grid are all difficult tasks in the ever-changing world of the rapidly growing RE business. This essay explores AI's revolutionary potential as a remedy for these pressing problems (Hamdan et al., 2024). This is because AI technology has a "better result on wealth output," which is in line by the outlines proposed. Nonetheless, most experimental research showed that public spending promotes EG. It is evident that there are several causal pathways between government spending and EG (Taher, 2024; Fikri and Rhalma, 2024). Nonetheless, several studies discovered that government expenditure hindered economic expansion. However, in the case of green energy, the relationship between government spending and EG was more varied. Several studies show that energy use plays a significant part in hastening environmental deterioration (Taher, 2024). Subsequent studies, like one conducted by Hanadi Taher in 2024, demonstrated a causal relationship between CO_2 and GDP development. The connection among GDP per capita and the Long-Run (LR) cointegration between energy use and the Chinese population in the real provinces was examined by Taher (2024) and Fikri and Rhalma (2024). They found an effective cointegrated association that was still running strong. According to Raihan et al. (2022), there is a significant and positive correlation between CO_2 emissions and economic expansion. Over time, it has been shown that energy use alone lowers carbon dioxide emissions. Morocco is now much more appealing to international companies due to recent legal changes. The main goals of these reforms have been to boost investment opportunities across numerous industries and enhance the business environment (Joachim, 2021) (Devassia, 2024). An increasing amount of research indicates that FDI and EG are favorably correlated in both industrialized and developing nations. The empirical findings showed that FDI inflows and outflows, energy consumption, capital, CO_2 emissions, and capital all had a major impact on the EG of Asian nations. Ren et al. (2024) assert that FDI promotes EG and has a major impact on environmental circumstances. A comprehensive evaluation of Lebanon's capacity for RE likely shows a lot of promise for wind, solar, and astral power. For nations like Lebanon to prosper socially and economically, investments in RE are essential. Taher (2024) examined the relationship between EG and REC in Lebanon from 1990 to 2012 and found that EG rose as RE was used. Taher analyzed the effect of weather alteration on Lebanon's EG in a separate study. The findings showed that while two factors connected to climate change had a positive effect on Lebanon's Economic Development (ED), two other factors had a negative effect on the EG. Not much research has been done on Morocco's GDP share, GDP growth, GDP per capita emissions in carbon dioxide metric tons, and the ratio of REC. This study's empirical component shows that the use of RE is causally related to EG and CO_2, and that Morocco's use of RE is beginning to have a favorable effect on the country's economy. Regarding the relationship between government spending, CO_2, RE, sustainable EG, and literature, various conclusions apply to different economies or groupings of economies. In a similar vein, empirical studies on the Moroccan economy revealed that variance didn't examine all the variables simultaneously. In our Moroccan context, we have adapted this measure because the Moroccan context requires the use of AI was measured by general government final spending (GGFS) since it is the only one available due to the scarcity of studies. The present research trains to investigate the relationship between

TABLE 10.1 Variables Description and Source's Data

Variables	Definition	Source
GDP	GDP per capita (current US) as proxy of EG	**WDI[1] (2024)**
HC	Human capital (HC), measured by education proxied by school enrollment, tertiary (% gross)	**WDI (2024)**
FDI	Foreign direct investment, net inflows (% of GDP)	**WDI (2024)**
CO_2 emissions	Carbon dioxide emissions (metric tons per capita)	**WDI (2024)**
AI	General government final spending of GDP	**WDI (2024)**
REC	Renewable energy consumption of total final energy consumption	**WDI (2024)**

Source: Own work.

Morocco's economic growth (EG) potential and government spending, REC, and CO_2 emissions. As a result, the conceptual model that follows is explained together with the following hypotheses.

AI is becoming more and more important in the global economy, influencing areas like direct investments, HC, and EG. These connections are examined by hypotheses H_1, H_2, and H_3, which highlight the influence of AI on socioeconomic dynamics. Similarly, the issue of CO_2 emissions and REC emerges as a key challenge that requires technological innovation and environmental sustainability. This text explores these relationships, providing insight into the opportunities and challenges presented by AI in a changing economic and ecological environment (Table 10.1).

H_1: CO_2 (metric tons per capita) has an adverse effect on EG.

H_2: REC (percentage of total final energy consumption) boosts EG.

H_3: AI measured by GDP GGFS (percentage) has a positive impact on economic expansion.

H_4: FDI positively influences EG.

H_5: HC measured by school enrollment, tertiary (% gross) has a confirmed impact on EG.

To sum up, AI turns out to be a significant catalyst for investments and productivity while raising important issues regarding its role in lowering CO_2 emissions and promoting RE. The advanced hypotheses highlight the complexity of these relationships and call for policies that strike a balance between innovation and sustainability. Even though the AI is redefining modern economies, responsible integration is still necessary to balance technological advancement with environmental preservation. These reflections pave the way for further research to examine these synergies.

10.3 METHODOLOGY AND DATA

In this study, we use the ARDL econometric technique to examine correlations between the short-term (ST) and long-term (LT) variables with varying integration orders. Three things make ARDL better than earlier cointegration methods: (1) it works better with smaller data sets; (2) it can work with variables that have various integration orders

(I [1] or I [0]); and (3) it produces precise long-term model predictions (Fikri & Rhalma, 2024). The Hannan–Quin, Schwarz-SIC, and Akaike-AIC will be applied to approximate the ideal lag number (p, q) for our model (Fikri & Rhalma, 2024). Our proposed model is grounded in a broader body of literature, specifically the works of Taher (2024). This compilation contains all of the data that was gathered for Morocco between 1993 and 2020. We used these data, which came from world development indicators (WDI), in our analysis. The factors chosen have their roots in past empirical research in this field. Y = f(K, L, S). The following is the study model:

$$\mathbf{Y = f\ (HC, FDI, CO_2\ emissions, REC, GGFS)} \tag{10.1}$$

where Y represents the current GDP per capita in the United States, REC denotes the percentage of total final energy consumption that is made up of RE, CO_2 emissions are expressed in metric tons per person, FDI stands for FDI, HC is for HC, and General government final consumption spending (GGFCS) is expressed as a percentage of GDP. Our structure:

$$\begin{aligned}\mathbf{GDP\ per\ capita_t} &= \alpha_0 + \alpha_1\mathbf{HC_t} + \alpha_2\mathbf{FDI_t} + \alpha_3\mathbf{CO_2\ emissions_t} \\ &+ \alpha_4\mathbf{REC_t} + \alpha_5\mathbf{GGFS_t} + \varepsilon_t\end{aligned} \tag{10.2}$$

where ε is the stochastic disturbance term.

ARDL is used in this study, as described by Pesaran & Smith (2001). They offered a method to cointegrate a bound technique for long-term connections in situations where the existence of a single cointegrating vector (Johansen integration) is impracticable. This method ignores the degree of stationarity and is referred to as ARDL. An ARDL estimate is more accurate and effectual. Moreover, the ARDL parameterization findings yield the Short-Run (SR) and LR link definitions for the variables model.

10.4 RESULTS

This section uses the ARDL cointegration test approach to analyze the long- and short-term correlations between GDP per capita, HC, FDI, CO_2 emissions, GGFS, and REC in Morocco. This step involves extracting and analyzing data as well as calculating the location, dispersion, and normality indices. Table 10.2 shows the results of these calculations made with E-views 12.

The amount of CO_2 released per capita (in metric tons), as defined by the kind of standard deviation values, has the lowest volatility, while GDP per capita (the current US GDP), as determined by the descriptive analysis of these variables, has the most volatility. Furthermore, the value of the HC is 17.63384. Moreover, this descriptive study explains all research variables without the variable FDI, and the results show that the research variables have a normal distribution with Jarque–Bera probability greater than 5%. Furthermore, in the absence of the variable FDI, all variables exhibit flatter distributions than normal distribution because the kurtosis coefficient is less than 3.

TABLE 10.2 Descriptive Statistics of Variables

Descriptive Statistics Criteria	GDP	Human Capital	FDI	CO_2	Artificial Intelligence	REC
Mean	2460.017	17.63384	2.197913	1.446758	16.99462	14.70143
Median	2592.318	11.88908	2.104375	1.471459	16.90606	15.03000
Maximum	3498.583	40.02002	6.444129	1.955308	19.40499	22.97000
Minimum	1217.429	9.004660	0.663931	0.991096	14.30467	10.45000
Std. Dev.	833.8227	9.982185	1.209715	0.290854	1.286759	3.290203
Skewness	−0.153952	1.121134	1.500277	−0.054651	−0.034499	0.413372
Kurtosis	1.311911	2.698700	6.432709	1.623676	2.109225	2.587370
Jarque–Bera	3.435192	5.971640	24.25129	2.223916	0.931281	0.996063
Probability	0.179497	0.050498	0.000005	0.328914	0.627733	0.607726
Sum	68880.49	493.7474	61.54156	40.50922	475.8494	411.6400
Sum Sq. Dev.	18772027	2690.389	39.51206	2.284099	44.70524	292.2867
Observations	**28**	**28**	**28**	**28**	**28**	**28**

Source: Calculations of authors using E-views 12.

10.4.1 Unit Root Tests

The model variable quantity are stable in initial difference I (1) and the HC variable in the second difference I (2), as indicated by the results in Table 10.3. ARDL models are defined by Stock and Watson as the lag in delicate selection orders. The model that produces the lowest (SIC) and (AIC) is the one chosen for this econometric analysis (Taher, 2024; Fikri & Rhalma, 2024). Moreover, the Moroccan economy is impacted by the general government's overall consumption spending, which has increased over the duration of our study (Figure 10.1 final consumption spending by the general government (% of GDP)). Morocco's use of REC as a proportion of its overall ultimate REC fell between 2005 and 2011. The integration directive for variable star in the archetypal is controlled by the unit root (UR) of a test. The unit root of the test also helps determine the integration order of the model variables (see Figures 10.2 and 10.3).

TABLE 10.3 Unit Root Test and Stationarity Test

	Series			Series in First Difference			Series in Second Difference		
Variables	Test Statistic	Dickey–Fuller Critical Value (5%)	P-value	Test Statistic	Dickey–Fuller Critical Value (5%)	P-Value	Test Statistic	Dickey–Fuller Critical Value (5%)	P-Value
GDP per capita	1.631693	−1.953858	0.9717	−3.804519	−1.954414	0.0005	–	–	–
Human capital (HC)	1.834349	−1.954414	0.9811	−1.330558	−1.954414	0.1651	−7.097627	−1.955020	0.0000
FDI	−0.842543	−1.954414	0.3413	−12.38976	−1.954414	0.0000	–	–	–
CO_2 emissions	3.217145	−1.954414	0.9992	−4.663002	−1.954414	0.0000	–	–	–
GGFS	0.173015	−1.953858	0.7285	−8.329758	−1.954414	0.0000	–	–	–
REC	−0.996210	−1.953858	0.2782	−4.993035	−1.954414	0.0000	–	–	–

Source: Calculations of authors using E-views 12.

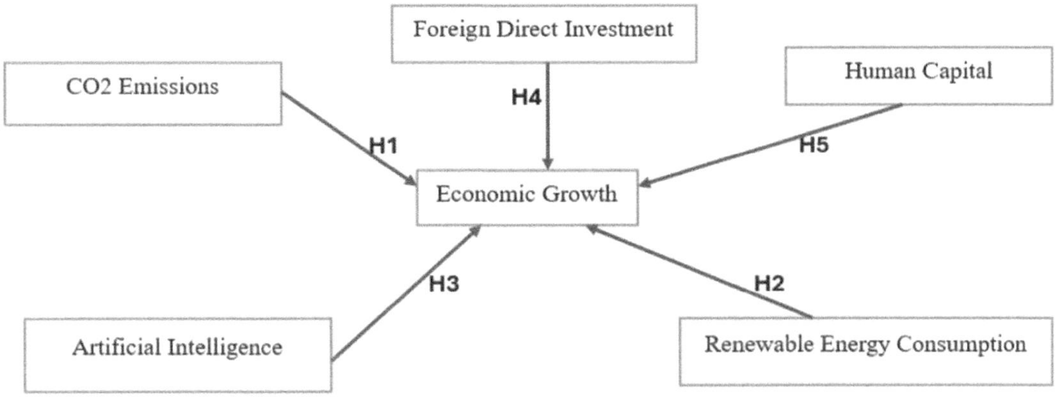

FIGURE 10.1 Conceptual model. (Photograph by the author.)

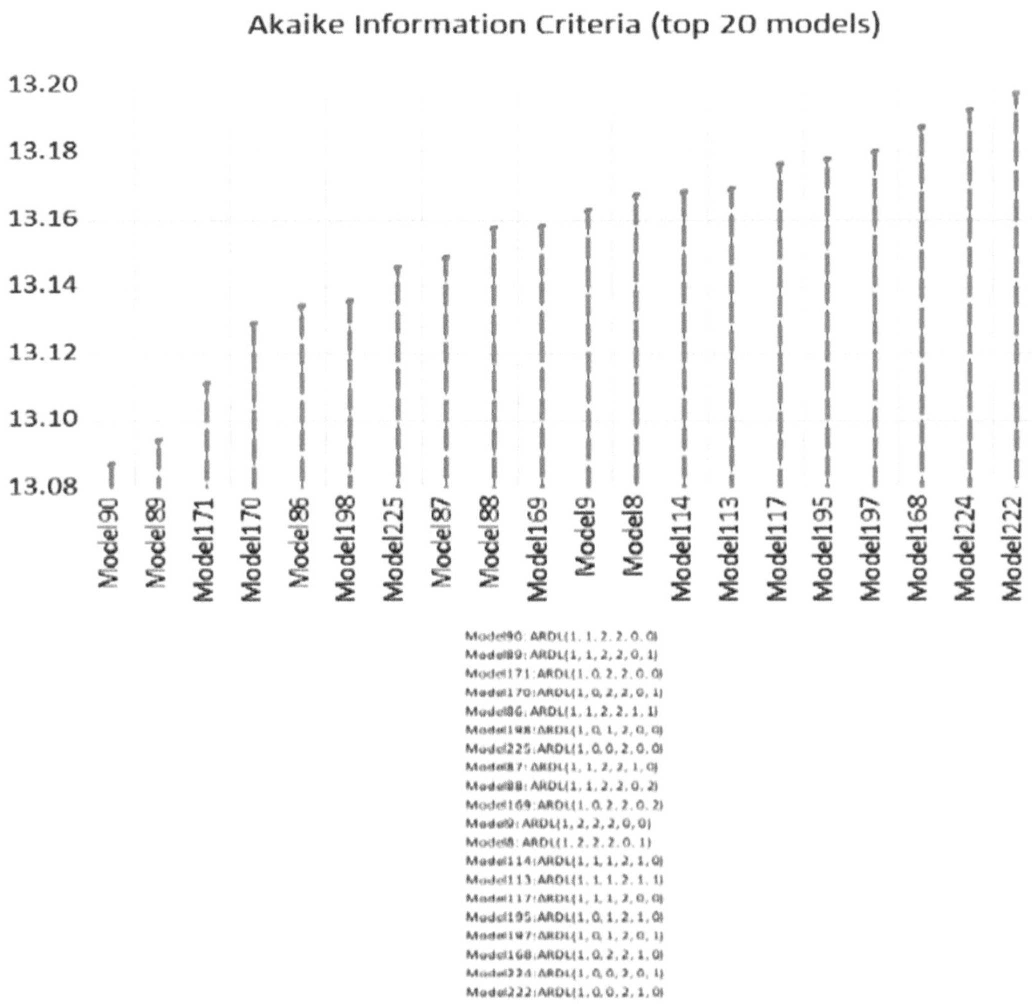

FIGURE 10.2 Optimal lag's selection. (Photograph by the author.)

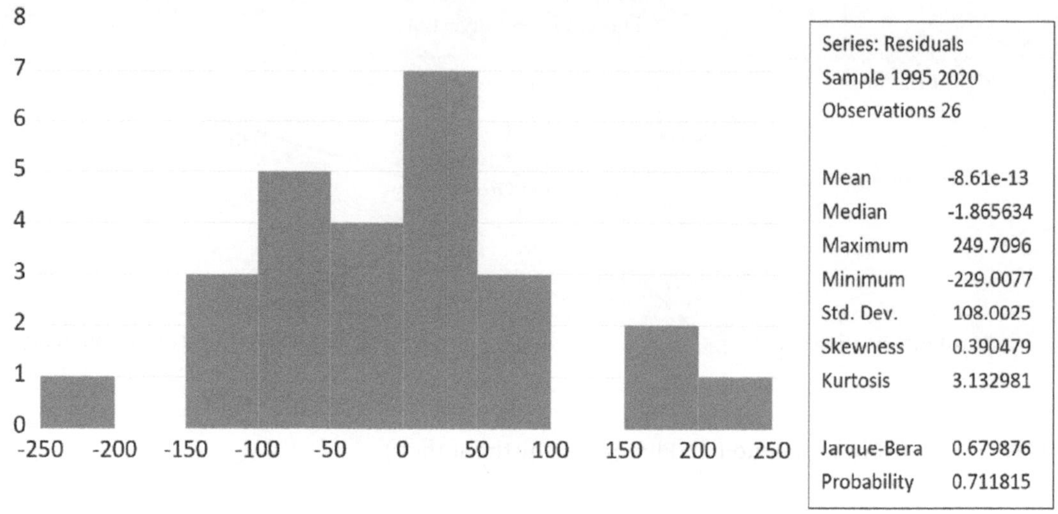

FIGURE 10.3 Normality test. (Source: Calculations of authors using E-views 12.)

10.4.2 Cointegration Bound Test and Optimal Lag

Table 10.4 and Table 10.5 show the maximum delay that can be used with (FPE), (SBIC), (AIC), and (HQIC). Comparing each lag's lowest values that occur after the other yields the ideal lag of 1 (396.1714 for FPE, 22.58520 for AIC, 24.95978 for SBIC, and 23.51270 for HQIC). As Table 10.1 illustrates, the first step in model estimating is the ARDL regression at the best distributed lags constructed on the AIC. As indicated in Table 10.3, the second step comprises analyzing the LR and SR associations between the variables. Table 10.4 offerings the consequences of relapse scrutiny. It indicates that while CO_2 emissions and the share of final energy consumed from renewable sources have positive effects, GGFS as a GDP percentage has a statistically significant positive impact on EG sustainability. Furthermore, as seen by school enrollment, FDI negatively affects EG and HC. Therefore, according to the LR coefficients in Table 10.5's second part, while GGFS (% of GDP) and school enrollment (education) have a negative influence on EG, REC and CO_2 have a positive effect. This result validates the findings of Fikri and Rhalma (2024) and Taher (2024). In the third portion of Table 10.5, the speed of adjustment was represented as a negative figure (−0.656110) in Table 10.4 (ADJ to the GDP per capita, current US). This number represents the rate of distortion of the equilibrium or the degree to which the reliant on inconstant responds departure in the equilibrium connection over a given period. In the SR

TABLE 10.4 Maximum Number of Lags

Number of Lags	Final Prediction Error	Akaike Information Criterion	Schwarz Bayesian Information Criterion	Hannan–Quinn Information Criterion
0	487505.2	30.12409	30.41442	30.20770
1	396.1714	22.92747	24.95978*	23.51270*
2	468.5345	22.58520*	26.35949	23.67206

Source: Calculations of authors using E-views 12.

TABLE 10.5 The Best Model Selection Regression of Auto-Regressive Distributed Lag (ARDL)

Variables	Coefficient	Std. err.	T	P > |t|
GDP (−1)	0.343890	0.227804	1.509585	0.1534
FDI	−44.86666	34.00034	−1.319594	0.2081
FDI (−1)	−38.16744	31.22328	−1.222403	0.2417
CO_2 emissions	1158.789	641.3712	1.806737	0.0923
CO_2 emissions (−1)	279.2030	279.2030	0.386113	0.7052
CO_2 emissions (−2)	1471.459	923.5308	1.593297	0.1334
GGFS	−99.87408	44.53316	−2.242690	0.0416
GGFS (−1)	20.91172	39.96766	0.523216	0.6090
GGFS (−2)	89.11254	43.89101	2.030314	0.0618
REC	4.216773	15.22434	0.276976	0.7858
School enrollment	−28.49312	10.41457	−2.735891	0.0161
C (constant)	−2018.139	1049.184	−1.923533	0.0750

Source: Calculations of authors using E-views 12.

relationship, two negative coefficients (–83.03409 and –28.49312) are found. This implies that the LR equilibrium deviation is ignored and that only the SR on the other side, three positive coefficients (2909.451, 10.15018, and 4.216773) fluctuations are considered.

In 1973, Hirotugu Akaike developed AIC to assess the effectiveness of statistical models. A parameter cannister growth the likelihood of the model during the statistical model estimation process. Thus, after Figure 10.1: optimal lag selection, we will choose the model with the least amount of information, making Model 90 (1,1,2,2,0,0) our ideal model. See also Figure 10.2 and 10.3.

10.4.3 Long-Run and Short-Run Dynamics

Table 10.6 shows that the ARDL bound test is used as a cointegration technique to evaluate the long-term relationship between the model variables (Fikri & Rhalma, 2024).

Verifying the cointegration of public revenue and expenditures to determine whether Morocco's public debt can be maintained is the process's most crucial step, which is already finished. According to Fikri and Rhalma (2024), this test is discussed in detail beforehand, and the statistical value that is computed is compared to important values (which set boundaries). Table 10.7 shows the resulting bound test of ARDL regression. The fact that the value of F: 2.566418 is more than the 2.39 5% significance level limit allows us to estimate the long-term effects among the measures. This suggests that a cointegration relationship is present. The F-statistic, which was produced using Table 10.6 data, is 2.39 greater than 5% with the F critical values = 2.56418. Thus, the test rejects the null hypothesis (Kripfganz & Schneider, 2020) and implies that the exogenous and endogenous variables have a LR connection. First, in order to diagnose this model, heteroscedasticity is required. Second, normality; and thirdly, model stability. Understanding heteroscedasticity is necessary to determine if the variance of the model residuals is constant or homoscedastic. Establishing the relationship between the independent variables and the model residuals' existence. The White test is a useful tool for confirming heteroscedasticity because it is founded on the chi-square delivery and may be used in conjunction through additional devices (Fikri &

TABLE 10.6 Results of ARDL Short-Run (SR), Long-Run (LR), and Adjustment (ADJ)

	Short-Run (SR)			
Variables	**Coefficient**	**Std. err.**	**T**	**P > |t|**
GDP per capita	−0.656110	0.227804	−2.880149	0.0121
FDI	−83.03409	56.79082	−1.462104	0.1658
CO_2 emissions	2909.451	860.4964	3.381131	0.0045
GGFS	10.15018	51.90550	0.195551	0.8478
REC	4.216773	15.22434	0.276976	0.7858
School enrollment (human capital)	−28.49312	10.41457	−2.735891	0.0161
C (constant)	−2018.139	1049.184	−1.923533	0.0750
	Long-Run (LR)			
Variables	**Coefficient**	**Std. err.**	**T**	**P > |t|**
FDI	−126.5551	66.17763	−1.912355	0.0765
CO_2 emissions	4434.394	496.1323	8.937926	0.0000
GGFS	15.47024	78.20462	0.197817	0.8460
REC	6.426928	23.95128	0.268333	0.7924
School enrollment (human capital)	−43.42733	13.29820	−3.265654	0.0056
C (constant)	−3075.916	1416.907	−2.170866	0.0476
Sample		**1993–2020**		
Observations		26		
R-squared		0.654761		
Adj R-squared i		0.568452		
Prob > F		0.0002		
Log Likelihood		−158.1185		
	Adjustment			
Variables	**Coefficient**	**Std. err.**	**T**	**P > |t|**
GDP per capita (−1)	−0.656110	0.129569	−5.063781	0.0002

Source: Calculations of authors using E-views 12.

Rhalma, 2024). The model residuals that are computed are distributed randomly since there is no heteroscedasticity (see Figure 10.4).

10.4.4 Test of Heteroscedasticity and CUSUM of Squares

Make sure the errors are normal before building prediction intervals and running Student parameter tests. The Jarque–Bera test (1984) provides an example of how to verify whether a statistical distribution is normal by using the concepts of skewness (asymmetries)

TABLE 10.7 ARDL Regression: The Bound Test

F: 2.56418	**27 Observations, 5 Variables**					
	10%		**5%**		**1%**	
F Test	**I (0)**	**I (1)**	**I (0)**	**I (1)**	**I (0)**	**I (1)**
F	2.08	3	2.39	3.38	3.06	4.15

Source: Calculations of authors using E-views 12.

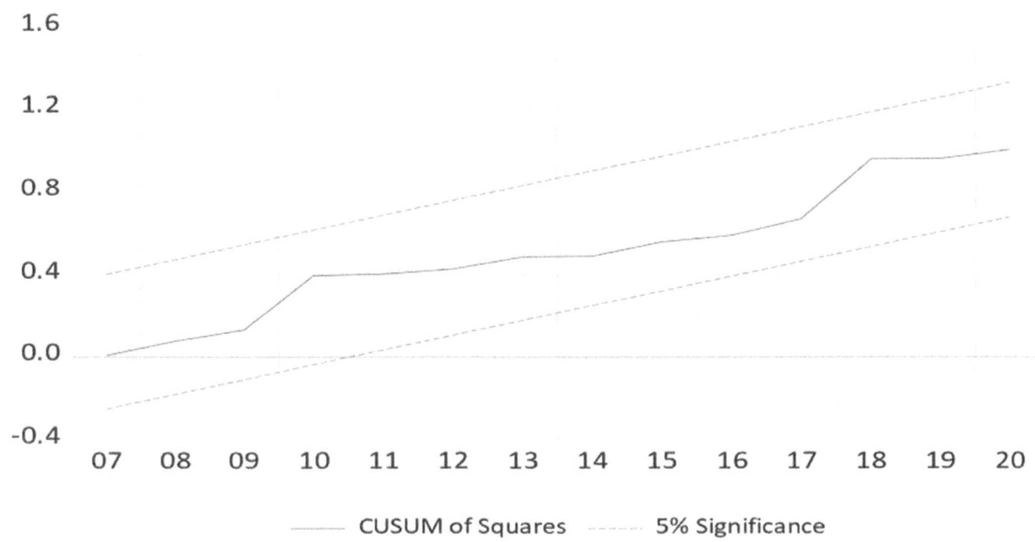

FIGURE 10.4 Parameter stability: CUSUM of squares test. (Source: Calculations of authors using E-views 12.)

and Kurtosis (Bourbounnais, 2000 referenced in Fikri & Rhalma, 2024). The preceding results are consistent with this theory, since the bigger than 5% Jarque–Bera probability indicates that the residuals are white Gaussian noises with a normal distribution. Completing Student parameter tests and calculating prediction intervals need confirming that the errors are normal. The Jarque–Bera test (1984) can be used to verify if a statistical distribution is normal by using the concepts of skewness and Kurtosis (Bourbounnais, 2000 quoted in Fikri & Rhalma, 2024). The results presented in Table 10.8 demonstrate that the residuals are white Gaussian noises that follow a normal distribution law, since the Jarque–Bera probability of 71.18% is more than 5%. Furthermore, the CUSUM of the squares of the cyclical residuals is the most effective method for assessing the model's stability. Its two right-hand branches correspond to the interval's limits. Based on the above graph, which shows the CUSUM of Squares and CUSUM results' test, we can conclude that the anticipated demonstration is constant because the curve does not stray from the pointily. As a result, the coefficients remain constant over time. The model test findings indicate the LR relationship between the exogenous and endogenous variables, which include school enrollment (education), carbon dioxide emissions, GGFCS, and REC (Taher, 2024; Fikri & Rhalma, 2024). The long-term benefits of CO_2 and REC on Moroccan EG are further supported by the study's findings (Taher, 2024; Fikri &

TABLE 10.8 White's Test: Heteroscedasticity

Source	Prob. chi2(14)	df	p
Prob. F (11,214) >Prob. chi2(11) = 0.5479	–	–	–
Heteroskedasticity	0.4510	26	0.547854

Source: Calculating's Authors using E-views 12.

Rhalma, 2024). So, H2 and H3 are accepted in this investigation. The study's subsequent conclusion was that general government spending has a detrimental impact on Morocco's EG (Fikri & Rhalma, 2024). The research approves of H1. Conversely, FDI promotes EG and has a significant impact on environmental circumstances (Ren et al., 2024). H4 is therefore approved. The H5 rejected the connection joining EG and school enrollment in both the short- and long-terms.

The most appropriate test for determining the model's stability is CUSUM test, graph 2 (Parameter stability: CUSUM of Squares test). Based on the previous graph, which shows the CUSUM of Squares test results, we could assume the estimated model is stable because a curve does not stray from the dotted line. Consequently, the coefficients remain constant over time. All things considered, our ARDL model's validity is supported by the findings of the various diagnostic tests (1,1,2,2,0,0).

10.5 DISCUSSIONS

The flow state of study and maturity in individual and hybrid RE was sketchily reviewed in this indication. Furthermore, an outline of the job of AI practices in the imitation, control, decision-making, and optimization of RE is provided. The long- and short-term relationships between EG and school enrollment were denied. The correlations between GGCS, CO2, REC, and EG sustainability were shown to be disparate in both empirical and study-based research. The study's findings also demonstrated a positive correlation (effect) and support for EG between (TFEC) that comes from the use of RE and CO_2. Regarding the financial elements of sustainable development, this study examined the effects of REC, general government final consumer spending, and CO_2 emissions (as metric tons per capita) on Morocco's capacity to attain sustainable EG between 1993 and 2020. FDI, however, also encourages economic expansion and has a big effect on the environment (Ren et al., 2024). The first step in the econometric test was to verify stationarity. At first difference, the series stays stationary as the findings of the unit root test are revealed. The study found that, based on GDP per capita and REC, there was a significant positive impact on CO_2 emissions from REC, but a negative impact on EG from REC in the first lag. The literature is validated by the LR coefficients, which indicate that while GGFS (%) of GDP has an undesirable (negative) effect and CO_2 emissions (metric tons per capita) and GGFS (%) of GDP have a significant positive impact on the growth of the Moroccan economy. Two positive coefficients are reported by SR link, though. This means that only short-term variations are considered and the LR equilibrium nonconformity is ignored. The study's conclusions show just how crucial Morocco's reliance on RE is to the nation's capacity to support EG. Moroccan officials should place more emphasis on promoting the use of RE sources and tightening limits on carbon dioxide emissions to improve economic well-being, even though the latter are a good indication of economic productivity and have a useful impact on EG. Third, given the negative effects Moroccan government spending has on EG, the issue of the direction of government spending—that is, away from research and development—should be brought up. According to the research, the Moroccan government should change its policies to put long-term sustainability first. This means funding R&D instead of spending on consumption.

All things considered, and in our Moroccan context, (FDI) promotes EG. Consequently, hypothesis H4 is validated. Plus, in the long-term effect, the study demonstrates a significant positive effect of REC on EG, which supports the hypothesis over the long run. Additionally, for short-term effect: a negative effect on the first delay (lag) is observed, making it a significant nuance. Thus, the hypothesis H2 is partially confirmed in the short term but confirmed in the long run. The hypotheses H1, H3, and H5 have not been verified.

10.6 CONCLUSION

The idea of sustainable development is broad and incorporates a variety of theories and methods to assault an equilibrium among public progress, environmental preservation, and EG while also meeting the demands of the current generation without endangering those of future generations. The adverse consequences are explained by the rise of sustainable development. However, there are several holes in the concept of sustainable development that cause more issues than they resolve. And we will link this with AI and education (according to Fikri & Rhalma, 2023, 2024), the notable effects of integrating AI into early childhood education on children's enjoyment, creativity, and soft skill development to have an important EG in Morocco. The word EG denotes long-term upsurge of belongings and service production inside an economy, and it is a fundamental concept in economics (Fikri et al., 2024; Batool & Akbar, 2023).

NOTE

1. WDI: World Development Indicators

REFERENCES

Asli, H. E., Hamid, L., Zineb, A., & Mohamed, A. (2024). Impact of, human capital, economic factors, energy consumption, and urban growth on environmental sustainability in Morocco: an ARDL approach. International Journal of Energy Economics and Policy, 14(2), 656–668.

Batool, K., & Akbar, M. (2023). Influence of infrastructure development with its sub-sectors on economic growth in selected Asian countries: An empirical analysis using DOLS and FMOLS approaches. Pakistan Journal of Humanities and Social Sciences, 11(2), 2236–2253. https://doi.org/10.52131/pjhss.2023.1102.0517

Devassia, B. P., Karma, E., & Muço, K. (2024). Role of human capital as a driver for sustainable economic growth in Western Balkan countries. Revista De Gestão Social E Ambiental, 18(5), e06779. https://doi.org/10.47172/2965-730X.SDGsReview.v4.n02.pe02051

Fikri, Y., Mantouzi, S., Rhalma, M., & Youssef, S. (2024). Economic growth: A literature review. Revue Internationale du Chercheur, 5(1).

Fikri, Y., & Rhalma, M. (2023). Impact de l'économie du savoir sur le développement économique à l'abri du Nouveau Modèle de Développement (NMD) au Maroc: essai de proposition d'un modèle conceptuel. Alternatives Managériales Economiques, 5(4), 442–459.

Fikri, Y., & Rhalma, M. (2023). Sustainable development: Theoretical review. International Journal of Accounting, Finance, Auditing, Management and Economics, 4, 124–141.

Fikri, Y., & Rhalma, M. (2024). Effect of CO_2 emissions, renewable energy consumption and general government final consumption spending on Moroccan economic growth: ARDL approach. International Journal of Energy Economics and Policy, 14(5), 575–581.

Gonzales, J. T. (2023). Implications of AI innovation on economic growth: A panel data study. Journal of Economic Structures, 12(1), 13.

Hamdan, A., Ibekwe, K. I., Ilojianya, V. I., Sonko, S., & Etukudoh, E. A. (2024). AI in renewable energy: A review of predictive maintenance and energy optimization. International Journal of Science and Research Archive, 11(1), 718–729.

Harbal, A., & Khihel, F. (2023). The impact of interactions between public and private investment on improving the business climate in Morocco. International Journal of Accounting, Finance, Auditing, Management & Economics, 4(2-1), 107–121. https://doi.org/10.5281/zenodo.7785936.

Hu, C., Chen, C., Findlay, C., & Wu, Y. (2023). Foreign direct investment, off-farm employment and rural labour income: evidence from China. Applied Economics, 56(41), 4974–4988.

Iddrisu, K., Abor, J. Y., Insaidoo, M., & Banyen, K. T. (2023). Does China's flow of FDI and institutional quality matter for poverty? Evidence from sub-Sahara Africa. Journal of Asian and African Studies, 60(2), 1250–1275. https://doi.org/10.1177/00219096231188948

Iorgovan, D. (2024). Artificial intelligence and renewable energy utilization. Proceedings of the International Conference on Business Excellence, 18, 2776–2783.

Jean, T. (2023). Morocco leads the way in FDI project records (Le Maroc en tête des records d'IDE en projet). Challenge.ma. https://www.challenge.ma/attractivite-des-ide-le-maroc-en-tete-des-records-dide-en-projet-270511/

Joachim, C. (2021). Investissements étrangers au Maroc: Des réformes légales qui renforcent l'attractivité. La Tribune. https://afrique.latribune.fr/think-tank/tribunes/2021-10-19/investissements-etrangers-au-maroc-des-reformeslegales-qui-renforcent-l-attractivite-894349.html.

Kripfganz, S., & Schneider, D. C.(2020). Response surface regressions for critical value bounds and approximate p-values in equilibrium correction models. Oxford Bulletin of Economics and Statistics, 82(6), 1456–1481.

Lu, C-H. (2021) The impact of artificial intelligence on economic growth and welfare. Journal of Macroeconomics, 69, 103342. https://doi.org/10.1016/j.jmacro.2021.103342

Machloukh, A. (2024). Interview with Zakaria Farahat: 'Between January and August 2023, Morocco attracted 10 times more investment than the average over the last 10 years'. L'Opinion Maroc – Actualité et Infos au Maroc et dans le monde. https://www.lopinion.ma/Entretien-avec-Zakaria-Farahat-Le-Maroc-a-attire-entre-janvier-et-aout-2023-10-fois-plus-d-investissements-que-la_a49572.html

Mazzeo, D., Herdem, M. S., Matera, N., Bonini, M., Wen, J. Z., Nathwani, J., & Oliveti, G. (2021). Artificial intelligence application for the performance prediction of a clean energy community. Energy, 232, 120999.

Nchake M. A., & Shuaibu M. (2022) Investment in ICT infrastructure and inclusive growth in Africa. Scientific African, 17, 1–16. https://doi.org/10.1016/j.sciaf.2022.e01293

Pesaran, M. H., Shin, Y., & Smith, R. J. (2001). Bounds testing approaches to the analysis of level relationships. Journal of Applied Econometrics (Chichester, England), 16(3), 289–326. doi:10.1002/jae.616

Raihan, A., Begum, R. A., Said, M. N. M., & Pereira, J. J. (2022). Relationship between economic growth, renewable energy use, technological innovation, and carbon emission toward achieving Malaysia's Paris Agreement. Environment Systems and Decisions, 42, 586–607.

Ren, X., An, Y., He, F., & Goodell, J. W. (2024). Do FDI inflows bring both capital and CO_2 emissions? Evidence from non-parametric modelling for the G7 countries. International Review of Economics & Finance, 95, 103420.

Santosh, K. C., & Gaur, L. (2022). Artificial intelligence and machine learning in public healthcare: Opportunities and societal impact. Springer Nature.

Sauvant, K. P. (2021). Improving the distribution of FDI benefits: The need for policy-oriented research, advice, and advocacy. Journal of International Business Policy, 4(2), 244–261. https://doi.org/10.1057/s42214-021-00109-6

Taher, H. (2024). The impact of government expenditure, renewable energy consumption, and CO_2 emissions on Lebanese economic sustainability: ARDL approach. Environmental Economics, 15(1), 217–227.

CHAPTER 11

Impact of New Technologies and Sustainability on Maintenance in the Era of Artificial Intelligence

Amal Naji and Zitouni Beidouri

11.1 INTRODUCTION

Today, companies are facing new challenges related to new technologies, environmental, and social contexts. One of those challenges is sustainability. Sustainability is not a new concept; it originated in the 1960s, when the Club of Rome promoted the study *The Limits to Growth* [1], related to the impact of continuous economic growth on the environment, which led to reflections about sustainable development (SD).

As cited by Iung and Levrat [2], in 1987, the "Report of the World Commission on Environment and Development: Our Common Future" [3] defined sustainability as the development "that meets the needs of present without compromising the ability of future generation to meet their own needs". In the same report [3], as cited by Jovane et al. [4], SD was considered as a "process of change in which the exploitation of resources, the direction of investments, the orientation of technological development, and institutional change are made consistent with future as well as present needs". At the same time, globalization and then competitiveness have activated the industrial evolution, leading to an increase in the production rate. Lately, key challenges related to climate change, poverty and social inclusion, and loss of biodiversity emerged and are "pushing" the world to review its policies regarding the resources and the environment. So, sustainability is again a highly studied subject.

Following the introduction of sustainability, more concepts have been created to accompany the implementation of sustainability into companies and their processes, for example, the science of sustainability, industrial ecology, and the ecosystem. One of those concepts used in manufacturing is Competitive Sustainable Manufacturing (CSM), which depends on the manufacturing industry and the related education, research, and technological development and innovation (E&RTD&I) system [4]. Sustainability assessment consists of

DOI: 10.1201/9781003649885-11

the assessment of process performances using key indicators, such as functionality, manufacturability, serviceability, and environmental impact [5]. Therefore, each function in industry can have a real impact on sustainability.

Maintenance is defined as a combination of all technical, administrative, and managerial actions during the life cycle of an item intended to retain it in, or restore it to, a state in which it can perform the required function [6].

Maintenance as a strategic function in industries is expected to be considered when implementing sustainability. However, for Holgado et al. [7], maintenance and sustainability research "have occasionally encountered". Maintenance could play an important role in sustainability; for example, Holgado et al. [7] observed that the maintenance function has links with energy management activities.

For many years, ecology and industry were considered "antagonists" but lately, they are associated with a sustainability concept that is based on three pillars: social, economic, and environmental [2]. Industrial companies have started integrating the environment into their strategy by considering rationalization of production to converge toward industrial ecology and circular economy paradigms, which led to the evolution from the notion of "product" to "Product-Service Systems (PSS) approach" [2]. For Jasiulewicz-Kaczmarek et al. [8], digitalization and sustainability affect all phases of the production chain; therefore, they should be considered as important topics for manufacturing industries.

Rachuri et al. [9] have discussed the consequences of considering sustainability on production systems. For them, it will impact the choice of input resources, the processes, and techniques used, which will minimize resource consumption, eliminate health hazards, and produce zero waste. Finally, those changes will lead to products and/or solutions that are safe for consumers and communities. Consequently, the entire life cycle of those products preserves natural resources and eliminates waste.

For maintenance sustainability, in addition to sustainability, companies are "striving–surviving" the transition from Industry 3.0 to Industry 4.0 and Industry 5.0. In this transition era, maintenance is one of the most important functions to be transformed, as many technologies of Industry 4.0 have either a direct impact on maintenance, such as Internet of Things (IoT), artificial intelligence (AI), big data analysis, or an indirect impact, like Additive Manufacturing (AM).

New technologies are transforming companies. Augmented Reality (AR) and Virtual Reality (VR) are emerging as different areas due to their ability to improve user experiences on learning, training, simulating, and other tasks [10]. AR consists of adding virtuality to actual reality. The AR system has three properties: a combination of real and virtual objects, interaction in real time, and alignment of real and virtual objects [11]. VR can be simply defined as a virtual object in a virtual environment (VE) [12].

AI is improving maintenance methods, such as predictive maintenance (PdM). As a consequence, it is expected that AI will revolutionize maintenance strategies, through PdM, condition-based monitoring, and autonomous decision-making [13]. In contrast to the production function, the maintenance function requires challenging tasks, such as complex and non-repetitive operations. However, the authors assume that the maintenance function can be significantly improved in the digital era, while providing a real impact on sustainability.

The objective of this chapter is to highlight a "neglected" area of research, which is combining multiple factors, through two research questions:

RQ1: What are the existing links between sustainability in maintenance and AR and VR in the era of AI, as cited in literature?

RQ2: What are the potential applications of the combination of AR and VR in maintenance in the era of AI that improve sustainability in industry?

The chapter presents an overview of the role of maintenance function on sustainability while using the new technologies, AR and VR, improved by AI. Section 11.2 presents the methodology. Section 11.3 provides an overview of the maintenance evolution, the sustainability and maintenance, technologies of VR and AR, and then the maintenance and AI. Section 11.4 discusses the applications of those technologies. The conclusion section summarizes the findings and suggests some perspectives for future work.

11.2 METHODOLOGY

The methods used for this chapter consist of a literature review of the articles and conference papers by using the words: virtual reality, augmented reality, virtual reality + maintenance, augmented reality + maintenance, maintenance + Artificial intelligence, maintenance + sustainability, maintenance + AR + VR + AI + sustainability. During the reading of the papers collected, another topic has been added to the search, which is: AR for Maintenance and Repair (ARMAR).

The databases used are Google Scholar and ScienceDirect. And the search has been conducted between December 2024 and June 2025.

The search, as presented in Table 11.1, resulted in a significant number of scientific papers that study the topic separately: AR and maintenance/VR and maintenance/sustainability

TABLE 11.1 Summary of the Literature Review on the Topics: AR, VR, AI, Maintenance, Sustainability

Topic	Google Scholar	ScienceDirect
AR and maintenance	865	93
VR and maintenance	251	25
AI and maintenance	492	44
AR/VR and maintenance	15	2
AR/AI and maintenance	0	0
VR/AI and maintenance	0	0
AR/VR/AI and maintenance	0	0
Sustainability and maintenance	1690	1582
Sustainability/AI and maintenance	0	0
Sustainability/VR and maintenance	0	0
Sustainability/AR and maintenance	0	0
Sustainability/AR/VR and maintenance	0	0
Sustainability/AI/AR/VR and maintenance	0	0

and maintenance/AI and maintenance. However, to the best of the authors' knowledge, there is no published paper related to the overall topics combined.

The selection of papers used in the present chapter was based on the criteria:

- Consistency of the chapter with the topic, based on the abstract,
- Novelty of the papers,
- Previous study about Moroccan industries.

11.3 LITERATURE REVIEW

11.3.1 Literature Review of Maintenance Evolution

Maintenance has evolved according to the evolution of the industry, as shown in Figure 11.1. Jasiulewicz-Kaczmarek et al. [8] have defined the evolution of maintenance in relation to the industry evolution as explained below.

The first generation: machines were simple in design and easy to repair. Machines functioned individually and were not costly, so a failure did not impact overall production. The corrective maintenance is repairing the system after the failure. It was and still is adapted to this type of machine and context.

The second generation: after the Second World War, machines were meant to increase their productivity and become more complex. Manufacturing introduces an assembly line, so a failure of a machine can cause the stoppage of the overall production. Maintenance departments started to be introduced to "optimize" the multiple maintenance operations and to reduce the number of corrective maintenance operations. So, maintenance managers need to plan periodic checks and replacement of used parts. By the time, the original manufacturers of equipment and machines start to provide recommendations for preventive maintenance.

Third generation: due to the automation and the increased complexity of systems, requirements expected from maintenance operations have evolved: higher availability and reliability of equipment in order to produce better quality while extending the lifetime of equipment that has become more costly. New technologies were used to move to the Condition-Based Maintenance (CBM), which consists of maintenance actions after collecting information about anomalies identified prior to functional failure. In parallel to maintenance methods, maintenance strategies such as Total Productive Maintenance

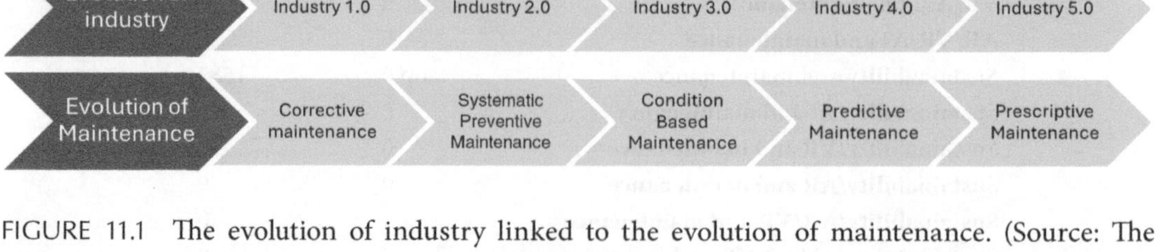

FIGURE 11.1 The evolution of industry linked to the evolution of maintenance. (Source: The authors.)

(TPM) and Reliability Centered Maintenance (RCM) are starting to be implemented to assign the adapted maintenance actions for each equipment.

Fourth generation: Industry 4.0 started in 2013, and it includes new technologies used in industry such as IoT and AM. Maintenance has to evolve to respond to the new requirements of the smart factory. Maintenance 4.0 is taking preventative action based on the prediction of failure. The prediction is obtained through analyzing data collected in real-time through connected objects or by machine learning algorithms, which led to enhanced condition monitoring.

Fifth generation: Industry 5.0 introduces the combination of work between human and machine and allows the implementation of decision-making processes. Similarly, Maintenance 5.0 called prescriptive maintenance is a completely automated maintenance workflow, where failure prediction is based on the analysis of data, which is analyzed and enables the provision of specific recommendations. The recommendations are based on libraries of standards maintenance tasks.

11.3.2 Overview of Sustainability and Maintenance

SD encompasses three pillars, as shown in Figure 11.2 from Jovane et al. [14]. The domains between those three pillars are named:

- Society and environment: access domain,
- Society and economy: sharing domain,
- Environment and economy: eco-efficiency domain.

All the factors combined form the sustainable domain.

Sustainability can be defined according to the context. For example, sustainable manufacturing, as suggested by the US Department of Commerce and as cited by Iung and Levrat [2] is "The creation of manufactured products that use processes that minimize negative environmental impacts, conserve energy and natural resources, are safe for employees, communities, and consumers and are economically sound". Despite the contribution of

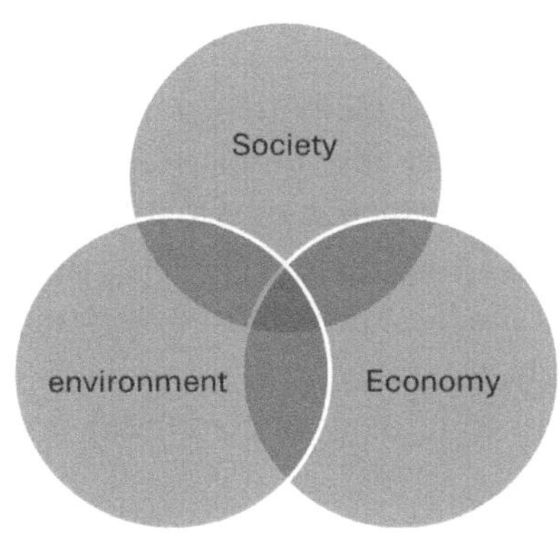

FIGURE 11.2 Fundamentals of sustainable development [14].

academicians and associations, the objective of reducing the negative impact on the environment has not been achieved [15]. In this context, the circular economy has appeared to provide a solution to the stuck situation. Indeed, the circular economy is a system based on the regeneration of used materials, instead of becoming waste. Circular economy is a looped process that starts with products to finish with maintenance, reuse, refurbishment, remanufacture, recycling, and composting [16].

Industrial ecology uses tools related to environment and social life, like Environment Risk Analysis (ERA), in addition to those related to the life cycle, such as Life Cycle Analysis (LCA), Life Cycle Costing (LCC) [2]. That's why there is a link between industrial ecology, which is considered the sustainability science [17] and maintenance.

For Hauschild et al. [18], there are some rules concerning eco-design that are related directly or indirectly to maintenance, like minimizing energy and resource consumption in production, promoting long life, promoting maintenance, considering end-of-life, organizing for upgrading, repair, and recycling, and assembling and disassembly.

Despite the significant role that maintenance can have on sustainability, it is obvious that it does not constitute a solution, especially if the product or material is toxic from the beginning [2]. Also, if a machine or equipment is "badly" designed, no matter how efficient the maintenance strategy is, it is impossible to extend its life cycle beyond a specific limit.

Considering the limitation of the maintenance impact on sustainability, it is recommended to consider a decision based on the current situation:

- Existent machines and equipment: to review the maintenance strategies, taking into account the sustainability.
- Existent machines and equipment in their end-of-life: to recycle and reuse components and to provide feedback to the original manufacturer regarding the extension of the life cycle and the best design.
- New acquisition machines and equipment: include the sustainability factor in the criteria of acquisition.

11.3.3 Overview of Augmented and Virtual Reality in Literature

- **Augmented reality**
  Surprisingly, AR is an old technology that started around 1957, without control by a computer, but it was the first case of an attempt to superpose additional data to an experience, which is the principle of AR [19]. From there, AR has improved to become a technology that allows virtual elements such as graphics, texts, or objects, or any other types of data to be superimposed over images of real environments [20]. The user can see through "wearable" devices, and real and virtual information melts in its view fields [21]. This technology is used today in several fields in industry.
- **Virtual reality**
  VR is artificial recreation, generated by a computer that immerses the user with an impression of experiencing simulated reality through the simulation of its vision

and hearing [22]. The VR system is based on the VE, which is the space where users navigate virtually, and can touch the virtual objects and walk around the objects, as if they were real [22]. VR frequently evokes the notion of immersion, which is the ratio between illusion delivered and reality. Immersion as a technology is based on a computer that will deliver an illusion of reality to the senses of a human user; this immersion will be inclusive, extensive, surrounding, or vivid according to the level of resolution, variety of energy simulated, visual and color resolution, etc. [23]. Another classification concerns the aspect of immersion. Magdalena Balcerak Jackson and Brendan Balcerak Jackson [24] mentioned that there are four aspects of immersion: representational, participatory, affective, and narrative immersion.

11.3.4 Literature Review of Maintenance and AI

AI is defined by Berente et al. [25] in a paper dedicated to managing AI as a simulation of human intelligence through computer systems. They added that AI is not a phenomenon in itself but rather a moving frontier to the next-generation advancements in computing. In this way, the early computer programs were AI. Nevertheless, Berente et al. [25] defined what is considered AI at the present time, which is machine learning algorithms that are based both on neural networks and data-driven prediction.

AI encompasses all technologies that enable computers to accomplish advanced functions. The AI can predict data and even suggest decisions based on its capacity to see, understand, assess, and interpret spoken and written language [13].

It is known that new maintenance methods are based on Industry 4.0 technologies, and maintenance is one of the three functions to be impacted, in addition to operations and supply chain. However, according to a study conducted in Moroccan companies that have implemented Industry 4.0 technologies, only 50% of them have implemented the PdM [26]. This is due to the complexity to optimizing maintenance based on data generated by IoT systems and sensors [27]. AI has emerged as a "key solution" that enables the processing of large datasets in real time [13].

11.4 APPLICATIONS OF NEW TECHNOLOGIES IN MAINTENANCE

Contrary to most industrial functions that can be standardized with processes, maintenance is a challenging function that consists of non-repetitive and complex tasks [28]. This requires the use of different sources of documents: manuals, databases, manufacturer documents, operating modes, etc. Moreover, maintenance tasks can be risky and lead to fatal accidents for maintenance operators [29]. That's why each opportunity should be seized to improve any aspect of maintenance operations.

With Industry 4.0, new technologies have emerged in manufacturing systems (automated machines, AM, robots, etc.). This further leads to the growth of the complexity of the equipment to be maintained.

Currently, the maintenance operators either memorize the workflow of the task but can accidentally forget a step, or they should handle a document paper, which can be difficult to use during operations in several environments; that's why having accessible information through new technologies can significantly help to achieve effective operations.

11.4.1 Applications of Augmented Reality in Maintenance

Given the nature of the maintenance function, the duration of tasks is difficult to predict and then to manage. Some tasks could require a profound understanding of multiple factors: history of the equipment, tools to be used, disassembly and reassembly workflow, manual of the equipment, etc. This is going along with the increasing complexity of manufacturing technologies, which requires competent and trained personnel on how to maintain or repair the equipment in the case of failure [30]. For Ariansyah et al. [31], Computerized Maintenance Management Software (CMMS) has helped to effectively improve maintenance tasks, but it concerns essentially preventive maintenance operations, and not corrective maintenance that involves the expertise of the technicians who will accomplish the task based on limited available information. To benefit from the advantage of AR for maintenance, it should be integrated into a maintenance system, such as an existing CMMS, which is called the AR-CMMS system [31].

Another concept of AR in maintenance is called "ARMAR"; it is a concept created in 2007 by the US Air Force and Columbia University to design and develop experimental AR systems, especially for maintenance aiding, for the purpose of developing prototype adaptive AR systems to increase the productivity of maintenance personnel [32].

The environment of the AR in maintenance linked to a CMMS is depicted in Figure 11.3. It shows that the maintenance operator will benefit from different sources of information while using smart glasses, which will enable him to perform maintenance tasks with both hands. Also, in complex industrial maintenance scenarios, if a technician faces an issue on-site, an expert off-site can view the same AR display and provide guidance [33].

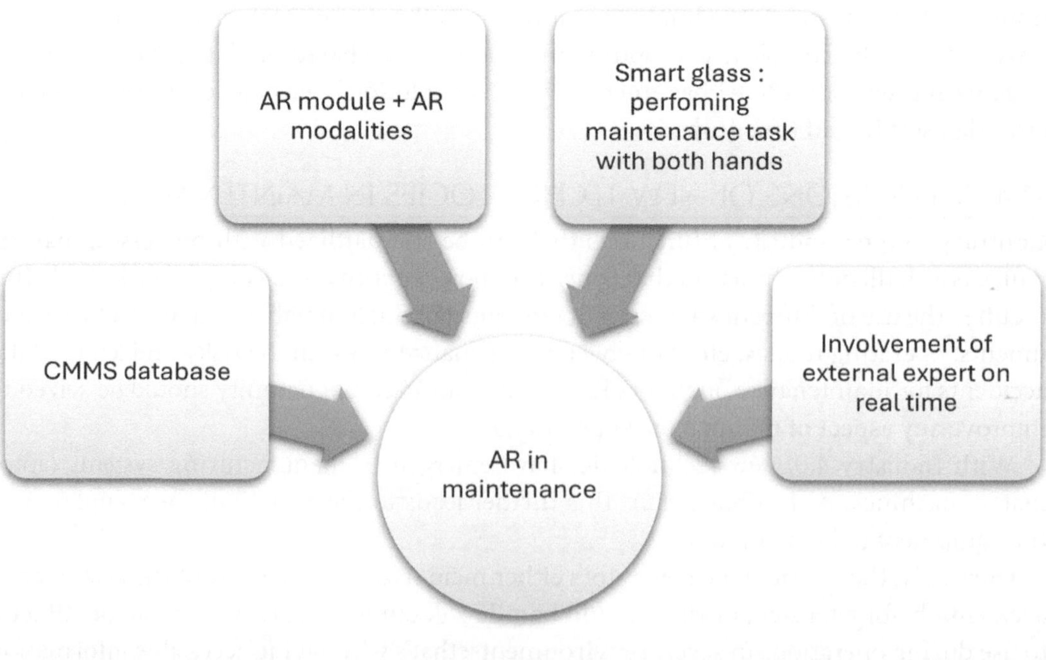

FIGURE 11.3 The environment of the maintenance operator when using CMMS linked to AR. (Source: The authors.)

11.4.2 Applications of Virtual Reality in Maintenance

VR is widely used in training; it's an alternative to traditional computer-based training at a low cost [34]. This is due to the immersion defined previously, which enables the users to train critical skills. Another interesting application of VR is the exploration of maintenance design problems in early stages, which will allow the designer to anticipate issues, especially regarding the maintainability and therefore the ability to change it on time [35].

Despite the automation technology, some maintenance tasks still require manual operation or intervention, due to their complexity [36]. Those tasks can be risky, as several maintenance operations. In this sense, as mentioned by Guo et al. [35], VR will play an important role in improving maintenance personnel safety in two aspects. In the design product, the maintenance operations can be simulated, and then the design can be reviewed according to the complexity of the maintenance tasks and their risks [37]. The second aspect is related to the on-site maintenance task, as VR will enable maintenance training in similar conditions before accomplishing it in reality [38]. The figure 11.4 summarized the impacts of VR on maintenance and maintainability during the lifecycle of a product frome the design of product until its maintenance.

11.4.3 Applications of AI in Maintenance

Most of the papers related to the application of AI in maintenance focus on the PdM or by using new technologies such as virtual and AR, which are the most promising concepts to fully use AI. The definition of PdM is: "Predictive maintenance is a philosophy or attitude that, simply stated, uses the actual operating condition of plant equipment and systems to optimize total plant operation" [39]. The PdM research is now concerning multiple sectors. In 2023, Rakibul Islam et al. [40] have conducted a systematic literature review on review papers about AI in PdM, and they found that many review papers were published on this topic, while prior to the year 2019, there were no review papers, which mean that research is focusing on PdM and AI.

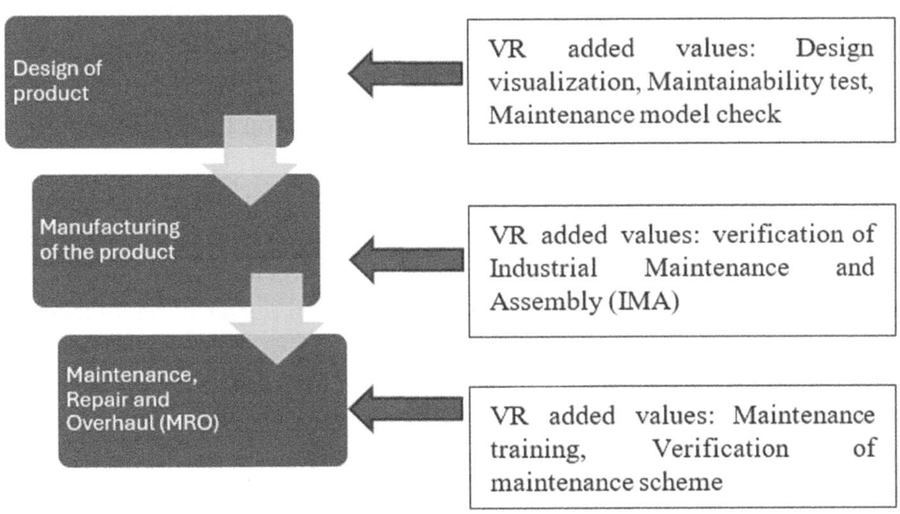

FIGURE 11.4 The impacts of VR on maintenance and maintainability during the life cycle of a product. (Source: The authors.)

Authors believe that besides PdM, AI can have a significant impact on improving maintenance management. The axes that could be listed below:

- Writing a standardized report of maintenance operations: maintenance tasks require cognitive and physical effort, so maintenance operators usually prioritize intervention rather than administrative tasks, such as writing a report, so AI can be helpful in assisting.
- Analyzing data: data related to maintenance indicators can be collected manually by maintenance staff and analyzed by AI.
- Standardization of tasks: the assistance via AI can enhance the standardization of checklists and all documents related to the maintenance intervention (preparation phase, execution, and control), while improving those documents by analyzing the most effective recommendations, misunderstandings to change, etc.
- Usage on CMMS: to be implemented and provide assistance to users.
- Analysis and comparison of contracts (acquisition, subcontracting, etc.).
- Analysis of maintenance costs: prediction of expenses.
- Optimization of planning, especially those involving other services, such as production.

11.5 DISCUSSION AND RECOMMENDATIONS

11.5.1 First Insights of Potential Use of Virtual and Augmented Reality in Maintenance in Moroccan Companies

Both VR and AR are technologies that involve data, essentially related to maintenance and spare part management. The data should enable the real-time information. Also, companies should adopt methods such as Multi-Criteria Decision Making (MCDM), which involves the statistics of maintenance criteria such as reliability, production losses, safety, spare parts, cost, and failure diagnosis. However, in a qualitative study in Moroccan companies, it was noted that there is a lack of statistics enabling the implementation of Key Performance Indicators (KPI) [41].

Moreover, VR and AR require the collection of information on the user's head and hand movements and the surrounding environment. This could be perceived as human control and consequently, will be greeted with reluctance. In a previous study about Industry 4.0 implementation in Moroccan companies, one of the main causes of reluctance of the personnel mentioned was the fear of control [26].

VR and AR require a significant investment; that's why they are mostly used in industries such as aeronautical, energy, and nuclear plants for specific maintenance operations. That's why, despite the multiple available maintenance technologies, maintenance managers should select the appropriate method of maintenance regarding several criteria such as the characteristics of the equipment, budget, and objectives.

To encounter those challenges, Moroccan companies are encouraged to improve data management and statistics, which is becoming necessary for Industry 4.0 implementation.

11.5.2 The Expected Impact of the Combination of VR, AR, and AI on Sustainable Maintenance

Based on the literature review, it was observed that most subjects cited are:

- Maintenance and AI: mainly related to PdM,
- VR and maintenance: the studies and applications to VR applied on maintenance,
- AR and maintenance: the studies and applications of AR applied to maintenance.

Table 11.2 presents an overview of the impacts and challenges of VR, AR, and AI on sustainable maintenance as cited in selected papers.

- Sustainability and maintenance: widely studied [42, 43]
- AI [44] VR AR maintenance [45]/AI [46]

TABLE 11.2 Overview of the Impacts and Challenges of VR, AR, and AI on Sustainable Maintenance

Topic	Impacts	Challenges	References
AR and maintenance	Scheduling, task monitoring, decision making	Training of staff, cost of technology, the involvement of production for managing data	[43]
VR and maintenance	Education and training, assembly verification and maintenance, industrial maintenance and assembly tasks training, support to human factors, maintenance training under the learning-by-doing approach, higher safety level	Medical issues related to the effects on human eyes and head, training of staff, cost of technology, a large amount of time to implement the technology	[44]
AI and maintenance	Improvement of the productivity (20%), enhancing the efficiency, health, safety, and environmental performance, enhancing predictive capabilities and decision-making processes, reduction of time	Integrating data, machine learning (ML), AI in a structured manner to solve complex problems, integrity, ethics	[45]
AR/VR and maintenance	Time, availability of information, fewer errors, better task performance in complex maintenance tasks	Training system, adaptability of staff to the interactive environment, discomfort when wearing glasses	[46]
Sustainability and maintenance	Economic, social, and ecological, sustainability-cost ratio	Data acquisition system, transform the industry into human-centric industry	[47]

It was concluded that there is a lack of linking the concepts VR, AR, AI, and sustainable maintenance in single studies, especially real case studies. However, the industry is experiencing a transition phase, including the simultaneous change and improvement of multiple factors. Also, the impacts and challenges related to the topics and cited by authors were mostly not quantified. Maintenance as a strategic function that must deal with almost all other functions should broaden one's perspectives to enable the establishment of a doable strategy.

11.6 CONCLUSIONS

In this chapter, the maintenance function was studied according to the new technologies and challenges. Since we are in a transition phase from Industry 3.0 to Industry 4.0, multiple technologies are being discovered continuously. However, companies are facing two major constraints: challenges related to sustainability and barriers to moving to Industry 4.0, as most of the companies have not yet achieved the maturity capability related to Industry 3.0.

The chapter presents the maintenance evolution from Industry 1.0 to explain the context of improvement of maintenance management and operation through the history of industry. The sustainability is presented from the perspective of the maintenance function.

An overview of AR and VR was presented with a focus on their application in maintenance. Those technologies present many advantages for the maintenance function. But it requires the availability of appropriate data, which is a weak factor in Moroccan companies. While many papers related to those topics exist in the literature, it is noted that few papers present empirical studies regarding their applications in industry, especially in emerging countries.

The AI and maintenance were discussed with a focus on the possibilities of use that overcome the limited use based on PdM. For example, suggestions were provided about using AI for corrective maintenance operations.

The limitations of the present chapter are related to the literature review. The papers explored were limited to a selection of papers based on the title and abstract of papers. Also the papers explored are related individually to sustainability, and no meta-analysis studies were found. That's why, the impact of the combination of the topics AR, VR, and AI on sustainable maintenance was not concluded.

The absence of scientific papers related to the use of AR/VR and AI in maintenance in Moroccan industries does not imply the non-use of those technologies. It could be interpreted either by the absence of those technologies or a gap between science and the practical situation. That's why empirical studies are requested to assess the situation and to accompany the industrials in this transformation era.

The new technologies provide significant added value; however, one of the challenges that industries are facing is cybersecurity. According to Abdellaoui Alaoui et al. [47], AI is offering dynamic and effective solutions regarding cybersecurity. Although they added that while AI is expected to revolutionize industrial functions and especially maintenance, the AI algorithms often have an opaque, nature which is translated by a non-transparent or non-understandable decision-making process, which can negatively impact the trust of AI in cybersecurity.

The present chapter suggests studying maintenance regarding the actual context, new technologies, and challenges, to present the first insight about the dilemma experienced in the real industry. Case studies with consideration of multiple contradictory factors are encouraged to enable industrials to implement new strategies adapted to the global context.

REFERENCES

1. Meadows DH, Meadows DL, Randers J, Behrens WW. III. The Limits to Growth. New York: Universe Book; 1972.
2. Iung B, Levrat E. Advanced Maintenance Services for Promoting Sustainability. Procedia CIRP. 2014;22:15–22.
3. Report. Our Common Future. World Commission on Environment and Development; 1987.
4. Jovane F, Yoshikawa H, Alting L, Boër CR, Westkamper E, Williams D, et al. The Incoming Global Technological and Industrial Revolution towards Competitive Sustainable Manufacturing. CIRP Ann. 2008;57:641–59.
5. Peruzzini M, Grandi F, Pellicciari M, Campanella C. Virtual Maintenance Simulation for Socially Sustainable Serviceability. Procedia Manuf. 2017;11:1413–20.
6. EN 13306:2017 – Maintenance – Maintenance Terminology [Internet]. ITeh Stand. SLOVENSKI STANDARD SIST EN 13306:2018. 29-Nov-2017. [cited 2025 May 16]. Available from: https://standards.iteh.ai/catalog/standards/cen/5af77559-ca38-483a-9310-823e8c517ee7/en-13306-2017
7. Holgado M, Macchi M, Evans S. Exploring the Impacts and Contributions of Maintenance Function for Sustainable Manufacturing. Int J Prod Res. 2020;58:7292–310.
8. Jasiulewicz-Kaczmarek M, Legutko S, Kluk P. Maintenance 4.0 Technologies – New Opportunities for Sustainability Driven Maintenance. Manag Prod Eng Rev [Internet]. 2020 [cited 2025 May 14]; Available from: https://journals.pan.pl/dlibra/publication/133730/edition/116853/content
9. Rachuri S, Sriram RD, Narayanan A, Sarkar P, Lee JH, Lyons KW, et al. Summary of the NIST Workshop on Sustainable Manufacturing: Metrics, Standards, and Infrastructure. Int J Sustain Manuf. 2011;2:237–59.
10. Mendes M, Almeida J, Mohamed H, Giot R. Projected Augmented Reality Intelligent Model of a City Area with Path Optimization. Algorithms. 2019;12:140.
11. Azuma R, Baillot Y, Behringer R, Feiner S, Julier S, MacIntyre B. Recent Advances in Augmented Reality. IEEE Comput Graph Appl. 2001;21:34–47.
12. Kumari S, Polke N. Implementation Issues of Augmented Reality and Virtual Reality: A Survey. In: Hemanth J, Fernando X, Lafata P, Baig Z, (eds.), International Conference on Intelligent Data Communication Technologies and Internet of Things. ICICI 2018 [Internet]. Cham: Springer International Publishing; 2019 [cited 2024 Dec 22]. p. 853–61. Available from: http://link.springer.com/10.1007/978-3-030-03146-6_97
13. Okpala C, Chikwendu U, Onyeka NC. Artificial intelligence-driven total productive maintenance: The future of maintenance in smart factories. IJERD. 2025; 21(1): 68–74.
14. Jovane F. How to Maintain Competitiveness on the Way towards Sustainable Manufacturing. Tampere Manufacturing Summit, June. 2007 Jun 6:6–7.
15. Greyson J. An Economic Instrument for Zero Waste, Economic Growth and Sustainability. J Clean Prod. 2007;15:1382–90.
16. Ellen MacArthur Foundation. How to Build a Circular Economy. [cited 2025 May 16]. June 2023 Available from: https://www.oliverwyman.com/ourexpertise/insights/2023/jun/how-to-build-a-circular-economy.html
17. Ehrenfeld JR. Can Industrial Ecology Be the “Science of Sustainability”? J Ind Ecol. 2004;8:1–3.

18. Hauschild M, Jeswiet J, Alting L. From Life Cycle Assessment to Sustainable Production: Status and Perspectives. CIRP Ann. 2005;54:1–21.
19. Interaction Design Foundation - IxDF. (2020, September 23). Augmented Reality – The Past, The Present and The Future [Internet]. Interact. Des. Found; 2020 [cited 2024 Dec 18]. Available from: https://www.interaction-design.org/literature/article/augmented-reality-the-past-the-present-and-the-future
20. Malta A, Farinha T, Mendes M. Augmented Reality in Maintenance—History and Perspectives. J Imaging. 2023;9:142.
21. Rauschnabel PA, Ro YK. Augmented Reality Smart Glasses: An Investigation of Technology Acceptance Drivers. Int J Technol Mark. 2016;11:123.
22. Definitions and Applications of Augmented/Virtual Reality: A Survey. Int J Emerg Trends Eng Res. 2021;9:279–85.
23. Slater M, Wilbur S. A Framework for Immersive Virtual Environments (FIVE): Speculations on the Role of Presence in Virtual Environments. Presence Teleoperators Virtual Environ. 1997;6:603–16.
24. Balcerak Jackson M, Balcerak Jackson B. Immersive Experience and Virtual Reality. Philos Technol. 2024;37:19.
25. Berente N, Gu B, Recker J, Santhanam R. Managing Artificial Intelligence. MIS quarterly. 45(3) (2021):1433–1450.
26. Naji, A, Beidouri, Z. Implementation of Industry 4.0 Case Study of Moroccan Companies. In: Motahhir, S., Bossoufi, B. (eds.), Digital Technologies and Applications. ICDTA 2024. Lecture Notes in Networks and Systems, vol 1100. Springer, Cham. https://doi.org/10.1007/978-3-031-68660-3_44
27. Amangeldy IS, Bissembayev AS. Enhancing Operational Efficiency in Industry 4.0: A Predictive Maintenance Approach. Herald of Kazakh-British Technical University 2024; 21(2):10–18. https://doi.org/10.55452/1998-6688-2024-21-2-10-18
28. Naji A, El Oumami M, Bouksour O, Beidouri Z. A Mixed Methods Research toward a Framework of a Maintenance Management Model: A Survey in Moroccan Industries. J Qual Maint Eng. 2019;26:260–89.
29. Lind, S. Types and sources of fatal and severe non-fatal accidents in industrial maintenance. Int J Ind Ergon. 2008;38:927–33.
30. Swanson L. An Empirical Study of the Relationship between Production Technology and Maintenance Management. Int J Prod Econ. 1997;53:191–207.
31. Ariansyah D, Rosa F, Colombo G. Smart Maintenance: A Wearable Augmented Reality Application Integrated with CMMS to Minimize Unscheduled Downtime. Comput-Aided Des Appl. 2019;17:740–51.
32. Henderson SJ, Feiner SK. Augmented Reality for Maintenance and Repair (ARMAR): [Internet]. Fort Belvoir, VA: Defense Technical Information Center; 2007 Aug. Available from: http://www.dtic.mil/docs/citations/ADA475554
33. Amandine C. The Benefits of Virtual and Augmented Reality for Maintenance [Internet]. Cloud-Based Field Serv. Softw. September 4, 2018 [cited 2024 Dec 24]. Available from: https://www.praxedo.com/our-blog/the-benefits-of-virtual-reality-and-augmented-reality-for-maintenance/
34. Bailey, SKT, Johnson C, Schroeder B, Marraffino M. Using virtual reality for training maintenance procedures. In Proceedings of the Interservice/Industry Training, Simulation and Education Conference. 2017.
35. Guo Z, Zhou D, Zhou Q, Zhang X, Geng J, Zeng S, et al. Applications of Virtual Reality in Maintenance during the Industrial Product Lifecycle: A Systematic Review. J Manuf Syst. 2020;56:525–38.
36. Ma L, Chablat D, Bennis F, Zhang W, Hu B, Guillaume F. Fatigue Evaluation in Maintenance and Assembly Operations by Digital Human Simulation in Virtual Environment. Virtual Real. 2011;15:55–68.

37. Keller D, Dechelle C, Doceul L, Madeleine S, Martins JP, Measson Y, et al. ITER Equatorial Port Plug Engineering: Design and Remote Handling Activities Supported by Virtual Reality Tools. Fusion Eng Des. 2011;86:2105–8.
38. Muszyńska M, Szybicki D, Gierlak P, Kurc K, Burghardt A, Uliasz M. Application of Virtual Reality in the Training of Operators and Servicing of Robotic Stations. In: Camarinha-Matos LM, Afsarmanesh H, Antonelli D, (eds.), Working Conference on Virtual Enterprises, pp. 594–603. Collab Netw Digit Transform [Internet]. Cham: Springer International Publishing; 2019 [cited 2024 Dec 24]. p. 594–603. Available from: https://link.springer.com/10.1007/978-3-030-28464-0_52
39. Mobley RK. An Introduction to Predictive Maintenance. Imprint: Butterworth-Heinemann. 2002. https://doi.org/10.1016/B978-0-7506-7531-4.X5000-3.
40. Islam MR, Begum S, Ahmed MU. Artificial Intelligence in Predictive Maintenance: A Systematic Literature Review on Review Papers. In: Kumar, U., Karim, R., Galar, D., Kour, R. (eds.), International Congress and Workshop on Industrial AI and eMaintenance 2023. Springer, Cham; 2024. p. 251–61.
41. Naji A, El Oumami M, Bouksour O, Beidouri Z. Qualitative Study on Maintenance Management in Moroccan Industries. IOP Conf Ser Mater Sci Eng. 2018;351:012012.
42. Mourtzis D, Angelopoulos J, Zogopoulos V. Integrated and Adaptive AR Maintenance and Shop-Floor Rescheduling. Comput Ind. 2021;125:103383.
43. Shamsuzzoha A, Toshev R, Vu Tuan V, Kankaanpaa T, Helo P. Digital Factory – Virtual Reality Environments for Industrial Training and Maintenance. Interact Learn Environ. 2021;29:1339–62.
44. Nzubechukwu CO, Adebayo OA, Emmanuel CA, Peter EO, Abiodun A. Advancements in Predictive Maintenance in the Oil and Gas Industry: A Review of AI and Data Science Applications. World J Adv Res Rev. 2023;20:167–81.
45. Liu X-W, Li C-Y, Dang S, Wang W, Qu J, Chen T, et al. Research on Training Effectiveness of Professional Maintenance Personnel Based on Virtual Reality and Augmented Reality Technology. Sustainability. 2022;14:14351.
46. H. Sarje S, A. Kumbhalkar M, N. Washimkar D, H. Kulkarni R, D. Jaybhaye M, Hussein Al Doori W. Current Scenario of Maintenance 4.0 and Opportunities for Sustainability-Driven Maintenance. Adv Sustain Sci Eng Technol. 2024;7:0250102.
47. Alaoui EAA, Filali A, Sallah A, Hajhouj M, Hessane A, Merras M. Towards Transparent Cybersecurity: The Role of Explainable AI in Mitigating Spam Threats. Procedia Comput Sci. 2024;236:394–401.

CHAPTER 12

Group Work in the Digital Age

A Survey of Banking Human Resources in Morocco

Abdelaziz Berdi, Seghyar Nabil, and Ahmed El Hammoumi

12.1 INTRODUCTION

Our era is known as the knowledge economy par excellence. Thanks to technology, the world is becoming increasingly digital. As a result, digitalization is becoming an important factor in helping companies meet the challenges of growing and prospering their human capital. It is an imperative for any approach to competitive differentiation in an increasingly competitive world. In this context, digitization is profoundly changing the way companies operate, forcing them to reinvent themselves through a paradigm shift imposed by Information and Communication Technologies (ICT),[1] all the more so as it impacts relationships and interactions within their teams but also in their relations with the outside world (El Hammoumi et al., 2024).

By definition, ICTs are a set of technologies used to process, modify and exchange information between different company players. With these tools, the company is constantly undergoing major upheavals in its structures, management systems and human resources (HR).[2] The latter are profoundly affected in terms of their skills, training, qualifications, work, communication, and so on.

In the words of Henry Ford, the famous industrialist of the early 20th century, "coming together is a beginning; staying together is progress; working together is success". With this quote, a new notion of greater efficiency is put forward, namely group work. By the latter, we mean a set of people who work together to achieve a common work-related goal, namely efficiency and fulfillment at work, while collaborating with each other.

Group work, as one of the determining factors in a company's performance, is becoming a new practice imposed by the use of ICT and remains dependent on the creation of a propitious climate within the work environment. It is within this framework that this work aims to study the role of using digital technologies in the promotion of group work within companies, in particular in Moroccan banking companies. By digital technologies, also known as ICT, we mean the use of computers and electronics to collect, store, process, transmit and present information digitally. To design and implement our study, we will use

DOI: 10.1201/9781003649885-12

the concept of "digital technologies" as a reference to ICTs, such as the internet, intranet, groupware, workflow, and enterprise resource planning (ERP), and their implications for the promotion or otherwise of group work in Moroccan banking.

With the aim of satisfying the question asked, this writing is organized in such a way as to devote its first point (Section 12.2) to a literature review and an overview of the subject. The second point (Section 12.3) deals with the methodology used to carry out the research, while the third and final point (Section 12.4) discusses the results of the study. We consider that the choice of Moroccan banking companies to invest in the study comes down to the fact that these companies make extensive use of digital technologies, which require close support for this sustained wave of digitization, which is not without having a considerable impact on the work of banking HR, hence the interest of our study.

Dealing with this theme means realizing that the performance of any business, including banking, undeniably depends on the performance of its HR. Group work as one of the conditions for achieving this performance implies working intensively at this level to create a climate conducive to collaborative work. Achieving this goal requires investment in both the tangible (technological infrastructure) and intangible (skills, training, etc.) resources needed to meet the challenge.

12.2 LITERATURE REVIEW AND SITUATIONAL STATEMENT: DIGITAL TECHNOLOGIES AND GROUP WORK: A COMPREHENSIVE EXPLORATION

When it comes to integrating ICT into the working environment, and in light of extensive research into promoting group work, we reveal a global understanding:

In the Spanish context, Almerich et al. (2021) conducted an empirical study involving 983 students from the University of Valencia, selected through random sampling. The findings revealed that ICT skills—particularly pedagogical skills positively influence higher-level competencies, such as higher-order thinking and teamwork skills.

Another study was conducted by Zeidmane (2020) on a sample of 301 engineering students from the Latvia University of Life Sciences and Technologies, using a student survey. The results showed that the integration of Information and Communication Technologies (ICT) into mathematics education develops the skills required by engineers: communication, technical skills, and teamwork.

In another Spanish context, Inglés and Mateu Martínez (2022) conducted a qualitative study among psychology graduates at the Miguel Hernandez University of Elche (UMH) during the 2021–2022 academic year to find out students' opinions. The study presents the strategy of "collaborative learning" and the use of electronic self-assessment and co-assessment rubrics through information, communication and leisure technologies in the theory and practice of the subject, justifying the interest of the experiment, presenting the bases that support this study and presenting the development and procedure of dynamics and assessment during the semester.

Still in the Spanish context, another study was carried out by Juárez (2013) on a sample of 49 students divided into 2 groups in the sixth year of primary education. The involvement of students in the learning process and their interaction is an essential aspect of this study. It was even strengthened by the use of ICT in the classroom and the active participation of students in activities throughout the process, fostering motivation, a relaxed environment, teamwork and collaboration in a meaningful context.

Finally, a study was done by Plotnick et al. (2016) by an online survey of professionals with Thermal Design Power (TDP) experience, which was conducted with two subsamples, one from a single large telecom company and the other from a combination of organizations. A quantitative approach using partial least squares regression was used to build and test a model of the relationships between the variables measured. As a result, group dynamics have a strong negative relationship with perceived effectiveness. Overall, technology reliability and training significantly reduced intra-group dynamics. In the telecommunications company, training increased integration dynamics (Seghyar et al., 2024a).

This being said, it is clear that the use of ICT does not go without facilitating collaboration, coordination and communication between actors at enterprise level. Indeed, doing your tasks at work is done in a massive flow of information and a relational network within teams, which tend to design and manage their own internal organization and their own external network. The tendency seems to be to consider, more and more, the team or group as a unit of work, as the true unit of work (Mastafi, 2016). Collaboration and coordination between HR, both internally and externally, take center stage as a result of the enormous potential that ICTs have at this level. According to Saadoun. M, communication is the other premium in today's organization (Saadoun, 2000, p. 3). The best ideas are worthless if individuals do not communicate with others and work in teams, which means that group work is becoming increasingly important.

It is clear that no one can deny that teamwork develops knowledge, as Balantzian (2002, p. 195) claims. This is the only lasting asset of an organization. Therefore, the ability of the team to collect and manage knowledge is crucial for the survival and success of the organization (referred to as extending communications beyond organizational boundaries, such as intranet/extranet technologies). In this context, ICTs play a role as a catalyst for change, since management methods are moving away from Taylorism, becoming more flexible and the transmission of information becoming horizontal. In addition, ICTs are behind many innovative forms of work such as teleworking and the creation of many services (teleservices). They also enable the development of partnerships between companies and are used as a tool for measuring costs, functional integration and process optimization, especially in the face of international competition (Seghyar et al., 2024b).

On this, we say that ICTs present strong opportunities to initiate new, more collective working practices. They thus allow new relationships to time and space. If the tools replace forms of collectives, they can also strengthen them and lead to new forms of cooperation and then transform the nature of groups or collectives that structure organizations (Gollac, 2003, p. 35). These transformations create what is called a collective space of information and communication. Of course, we come to a paradox: some tools of collective work end up being very individual work tools; everyone works with their microcomputer and no longer with their colleagues. In other words, even if we work together, everyone has their own screen. As an example, before, a secretary could know what was going on with his boss by seeing the mail pass. Now, with the messaging, he is no longer aware of it, which contributes to his disinterest in his work.

We must not lose sight of the fact that, thanks to the collaboration made possible by digital technologies, we are indeed witnessing a transition from technology serving the individual to technology serving relationships, as John Seely Brown affirms: "... a profound change is taking place, a transition from technology as a means of serving individuals to technology for relations.". This transition will be very important because it will allow us to discover new ways, new tools and new protocols for helping each other, which is really the essence of social learning. It is also the essence of lifelong learning, a form of learning that learning ecologies could greatly facilitate. And if we are able to create such learning ecologies in a region, it is a first step toward building a learning culture (Barron, 2006).

That said, ICT, especially the so-called collaborative (intranet, groupware, video conference, etc.) support, facilitates the exchange and collaboration between people in the organization. Indeed, new modes of exchange are developing, and with them, interpersonal relations are intensifying. ICT, in networked or flexible organizations, facilitates the promotion of these relationships through inter-connectivity and cooperation.

Another facet of developing these relationships at the organizational level is access to communities of practice that foster exchange and collaboration (Ghamrawi et al., 2024). Thus, the exchange, by advancing proposals and points of view, allows the development of skills and mutual enrichment at the level of groups, even the construction of relationships and therefore a sense of belonging within a community. With the Covid-19 crisis and the lockdown that has imposed itself, remote working is becoming a necessity and seems to offer companies some flexibility while reducing its costs. As a result, the use of digital tools has made remote collaborative work possible in complete safety (El Adib et al., 2024). These new ways of working nevertheless represent a challenge for management, which must know how to adapt in order to keep its teams united and strengthen their sense of belonging. The risk is to dehumanize interpersonal relationships and especially to isolate collaborators. So-called hybrid solutions help to avoid these pitfalls.

In summary, the use of ICT is accompanied by more collective team building at work, which allows for collaboration, cooperation and communication between HR within the company.

12.3 RESEARCH METHODOLOGY

As we have already mentioned, this chapter investigates the contribution of digital technologies to promoting group work in Moroccan banks. We rephrase the question as follows:

Does the use of digital technologies promote group work at Moroccan banks?

This issue raises the following sub-questions:

- Are digital technologies strategic or not for Moroccan banks?
- What is the level of use of digital technologies by these companies?
- Under what conditions can Moroccan banking companies successfully use digital technologies to promote efficient group work?

To design our research model, we start from the following elements:

- The definition of the use of digital technologies or ICTs goes to Pinsonneault and Kraemer. The authors propose three ways of conceptualizing ICTs (Pinsonneault & Kraemer, 1997):
  - Either by the technology itself (ICT investment, ...);
  - Either by the extent of ICT penetration in the organization (the number of computer applications available); or
  - By the actual use of ICTs by individuals in the organization.

Based on the premise that it is the degree of ICT use that will define the eventual impact of these technologies on HR and their behaviors, we opt for the ICT penetration approach, which measures the use of ICT by the IT applications used, within the framework of management and in particular in HR management (HRM). Within this framework, Pinsonneault and Kraemer consider that the more dominant IT applications are, the greater the likelihood that they will influence organizational management methods.

- For the control variables, we consider:
  - Bank type: Banks are either central, regional, branch or agency. The choice of this determinant is justified by the importance of this distribution in relation to the impact of the ICT use on the dependent variable of our research model (level of ICT equipment, for example, at branch and management level is not homogeneous).
  - The profile of HR using ICT: We consider this determinant to be important and to have a considerable impact on the dependent variables. Indeed, from our field investigations, we have recorded that the level of education (initial training) and ICT (computer science) degrees and ICT training, for example, considerably influence the degree of perception and understanding of ICT (as technical tools, but also as management tools).

Table 12.1 summarizes these various elements.

The research model is developed based on the theory, the research question, the definition of the model's variables, and the hypotheses derived from the relationships between the concepts described. The model is presented in Table 12.2.

The resulting research model is shown in Figure 12.1.

TABLE 12.1 Operationalization of Control Variables

Control Variables	**Bank Type**	• Central office, regional office, branch office, agency.
	Banking HR Profile	• Education, qualifications and training in digital technologies.

TABLE 12.2 Research Model Variables

Independent Variable (IV)	Dependent Variable (DV)	Operationalization of the DV
Use of Digital Technologies at Work	**Group Work**	Coordination, collaboration, sharing, etc.
		New relational skills (communication, group spirit, exchange, etc.).
		Sharing knowledge and skills

Based on this model, we assume that the use of ICT promotes group work among banking HR on two levels:

- **Hypothesis 1:** The digital technologies promote coordination and collaboration between banking HR at work;
- **Hypothesis 2:** The digital technologies foster the development of new relational skills in banking HR at work;
- **Hypothesis 3:** The digital technology promotes the sharing of knowledge and know-how between banking HR at work.

As for the choice of banking companies as the field of investigation, this is due to the fact that banks are experiencing a strong digitalization. In this context, Rowe (1994) believes that banking is the ideal field for research into the spread of information technology. Lévy-Leboyer et al. (2005) consider that the service sector is currently the best equipped and most dynamic in terms of new ICT (NICT), making banks a fertile field for research into the contribution of ICT to the workplace.

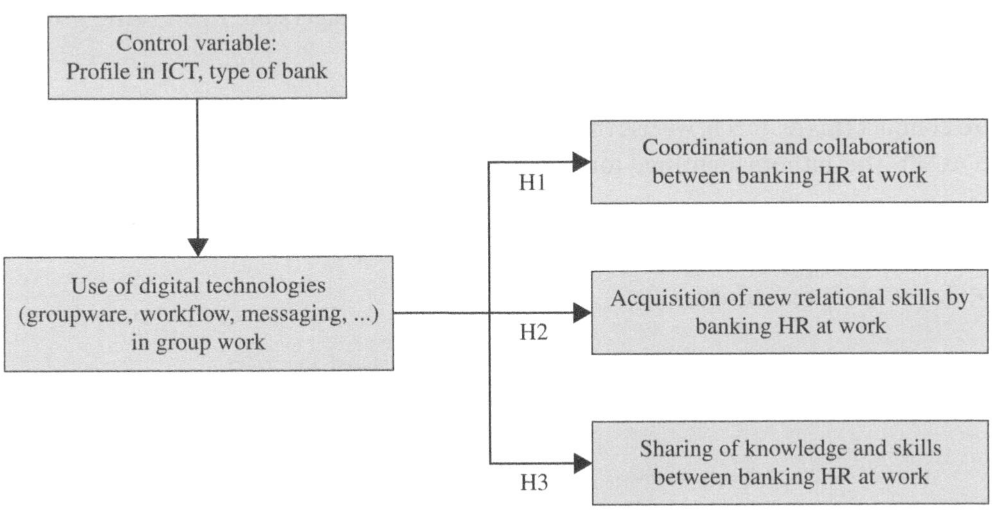

FIGURE 12.1 Conceptual model of the role of digital technology use in promoting group work among banking HR.

It should be borne in mind that financial firms, including banks, are primarily looking for profitability. However, achieving performance requires taking into account other indicators related to the human factor: working conditions, individual and collective performance, impacts of digital technologies, productivity, social climate, ergonomics, etc. On this, we have found it legitimate to put the present subject under the magnifying glass in order to enrich the studies carried out in this field and to open the way for further work. Even better, we noticed during our interviews that the banking players are very motivated to answer our questions given the importance and relevance of the topic.

In order to design our independent variable, "Use of digital technologies", we consider that it is the degree of ICT use that will define the possible impacts of these tools on HR and their behaviors. We opt for the ICT penetration approach that measures the use of ICTs by the computer applications used in the context of management and especially in HRM. At this level, Pinsonneault and Kraemer (1997) consider that the more dominant computer applications are, the greater the likelihood that they influence organizational management methods. To determine the sample, the quota method has been chosen. It is a sampling method that allows a sample to be determined on the basis of some statistical information about the population surveyed. Thus, while respecting the structure of the banking sector, we targeted a number of banks of each type (central, regional, branch and branch offices), as well as an agreed number of people (Table 12.3).

The extension of the study to cover banking branches and not only the directorates is justified by:

- At the operational level (agencies), we note a massive use of ICT (software, applications, etc.).
- The banking population is not homogeneous: There are different profiles from branches to directorates. This being so, it seems to us that the fact of collecting data from local actors (agency managers, customer managers etc.) will bring great value to work.

To conduct this research, we use the Unified Theory of Acceptance and Use of Technology (UTAUT). The authors identified four main variables that make up this UTAUT model:

- Enabling conditions;
- Expectations of effort;

TABLE 12.3 Questionnaires Administered, Returned and Response Rates

	Bank		
Questionnaires	**Directions**	**Agencies**	**Total**
Administered	40	305	345
Returned	26	202	228
Response rates (%)	65	66	66

TABLE 12.4 The Operationalization of UTAUT Model Variables

Theory	Variables	Explanations
UTAUT	**1. The attitude of HR banks toward the use of digital technologies.**	• Level of computer equipment. • Duration of use of the ICT at work by banking HR.
	2. The conditions of the use of digital technologies.	• Training of banking HR in digital technologies. • Technological infrastructure in place (network, connection, etc.).
	3. Social influence on the use of digital technologies	• Management commitment to digital technologies.
	4. Expectations for improving banking HR performance at work.	• The role of digital tools in promoting group work in banking HR.

- Social influence; and
- Performance expectations.

Table 12.4 shows these variables and their operationalization in our study.

With regard to the methodology of work undertaken, we have chosen a quantitative approach based on the manipulation of questionnaires (345), composed of three parts: the first part is used to identify the respondent (sample characteristics), the second part takes the level of use of digital technologies at work and the third part is devoted to the role of the digital technologies in the promotion of group work within the bank. A total of 228 questionnaires were completed and returned, resulting in a 66% response rate considered satisfactory.

Throughout this operation, we wanted to reach a large population of the sector by attending the various establishments at the central (headquarters), regional (regional directorates, branches) and local (agencies) levels with both executive and operational staff. The aim is to cross-reference each other's answers in order to enrich further study.

12.4 ANALYSIS AND DISCUSSION OF RESULTS

In this section, we review the main characteristics of the selected sample, the level of use of digital technologies in banks and the role of ICT in promoting group work in Moroccan banks.

12.4.1 Characteristics of the Sample Selected

They relate to the gender, age, initial training and occupation of respondents in the bank.

12.4.1.1 Gender

According to Table 12.5, 66.7% of the respondents are men and 33.3% are women.

12.4.1.2 Age

Table 12.6 shows that 79.5% of respondents are under 40. This is a young population, belonging to a generation called digital natives.[3] The labor market today is overrun by this new generation with considerable technological skills.

TABLE 12.5 Gender Distribution of Respondents in the Bank

		Number	Percentage	Valid Percentage	Cumulative Percentage
Valid	Women	76	33.3	33.3	33.3
	Men	152	66.7	66.7	100.0
	Total	228	100.0	100.0	

TABLE 12.6 Distribution of Respondents by Age Groups in the Bank

		Number	Percentage	Valid Percentage	Cumulative Percentage
Valid	Under 25 years	30	13.2	13.2	13.2
	[26–30]	77	33.8	33.8	46.9
	[31–40]	74	32.5	32.5	79.4
	[41–50]	29	12.7	12.7	92.1
	[+50 years old]	18	7.9	7.9	100.0
	Total	228	100.0	100.0	

TABLE 12.7 Distribution of Respondents by Level of Education in the Bank

		Number	Percentage	Valid Percentage	Cumulative Percentage
Valid	Secondary	1	.4	.4	.4
	Bac	4	1.8	1.8	2.2
	Bac+ 2	34	14.9	15.0	17.2
	Bac+ 3	38	16.7	16.7	33.9
	Bac+ 4	59	25.9	26.0	59.9
	Bac+ 5	84	36.8	37.0	96.9
	Doctoral or equivalent	7	3.1	3.1	100.0
	Total	227	99.6	100.0	
Missing	Missing system	1	.4		
Total		228	100.0		

12.4.1.3 Respondent Training

The study shows that 94.7% of bankers have a Bac+2, Bac+3, Bac+4, Bac+5 or a doctorate or equivalent (Table 12.7). It is a population with considerable skills, which allows us to understand that the Moroccan banking companies are looking at recruiting skills capable of bringing high added value to the banking activity.

12.4.1.4 The Function Held in the Bank

By allocating respondents according to the function occupied in the bank, we found on Table 12.8 that legal directors (LD) make up 13.2% of the sample, HR directors (HRD) are 2.2%, telecom network managers (TNM) and the information systems managers (ISM) together make up 1.8%, while the risk management managers (RMs) represent 0.9% of the total sample. For the operational sector, the customer managers (CM) hold a lion's share with 37.3%, followed by the branch managers (BM) with 24.6%, then the cashiers who reach 18% and finally the heads of funds (HF) with 7% (Table 12.9). This diversity of actors consulted enriches, as stated, our study.

TABLE 12.8 Regression Indices of ANOVA Model

ANOVA[a]						
Variable		**Sum of Squares**	**df**	**Mean Square**	**F**	**Sig.**
1	Regression	20,476	3	6.825	87,724	.000[b]
	Residuals	17,524	225	.0778		
	Total	38	228			

[a] Dependent variable: Coordination and collaboration between banking HR at work, development of relational skills within banking HR at work, and knowledge sharing among banking HR at work.

[b] Predictors: (Constant), the use of digital technologies.

TABLE 12.9 Summary of the Models

					Edit Statistics					
Model	**R**	**R-Two**	**R-Two Adjusted**	**Standard Error of Estimate**	**Variation of R-Two**	**Variation of F**	**ddl1**	**ddl2**	**Sig. Variation of F**	**Durbin-Watson**
1	.796[a]	.6336	.6287	.39598	.6336	87,724	3	225	.000	1,653

[a] Predictors: (Constant), the use of digital technologies.

[b] Dependent Variable: Coordination and collaboration between banking HR at work, development of relational skills within banking HR at work, and knowledge sharing among banking HR at work.

12.5 RESULTS

The results of the conceptual model test show that the probability value associated with the Fisher test in the ANOVA model is below the 5% error limit (Sig = 0.000 < 0.05). This indicates that the null hypothesis is rejected and the model is globally significant. In particular, the proportion explained by the model (regression = 20.476) significantly exceeds the unexplained proportion (residual = 17.524).

This observation is consistent with the adjusted R-square value (0.6336), which also provides evidence for the quality of the econometric model (Table 12.9). The results show that variables representing the use of digital technologies explain 63.36% of the variance in coordination and collaboration between banking HR at work, development of new interpersonal and relational skills within banking HR at work, and knowledge sharing between banking HR in the workplace.

To evaluate the significance of variables (Table 12.10), the values and probabilities of the student's t-test generated with SPSS 26 show that all three variables are statistically significant (t > 2 and sig < 5%). These variables are coordination and collaboration between banking HR at work, development of new interpersonal and relational skills within banking HR at work, and knowledge sharing among banking HR at work. This result is consistent with the three hypotheses of the study.

12.5.1 Economic Interpretation of the Results

The regression analysis suggests a strong influence of digital technologies on human resource activities in the banking industry. With an adjusted R^2 of 63.36%, the model explains a large portion of the variation in coordination, collaboration, development of

TABLE 12.10 Coefficients of Individual Significance

Coefficients[a]							
		Non-Standardized Coefficients		Standardized Coefficients			Collinearity Statistics
Model		B	Error Standard	Beta	t	Sig.	Tolerance
1	(Constant)	−.153	.418		−.281	.769	
	Coordination and collaboration between banking HR at work	.375	.267	.316	2.127	.009	.565
	Development of relational skills within banking HR at work	.426	.338	.406	2.725	.025	.957
	Knowledge sharing among banking HR at work	.278	.214	.263	2.76	.018	.943

[a] Dependent Variable: the use of digital technologies.

relational skills, and knowledge sharing in banking HR. This indicates that the presence of digital applications can improve operational efficiency and support a more unified way of working. From an economic perspective, improvements in operational efficiency translated into improved productivity, better customer service, and lower costs. The significance of all three variables suggests that digital transformation can enhance social capital in organizations, which is one of the important intangible assets. Moreover, knowledge transfer and skills development can enhance innovation, agility, and competitive advantage in the banking sector. These findings imply that, in addition, more resources can be allocated to digitalization as a strategic source of organizational performance. Furthermore, they reinforce the value of continuous learning as an avenue to fully access the economic returns of digital technologies.

12.6 DISCUSSION OF THE RESULTS

Going back to the UTAUT theory, the interpretation of results obtained revolves around four elements.

12.6.1 The Attitude of Banking HR Toward the Use of ICT

We are talking here about the degree of ICT equipment in banks and the duration of use of technologies at work by the banking actors surveyed.

12.6.1.1 Level of Equipment of the Banks in Digital Technologies

Based on the results of the study, we concluded that almost all the actors interviewed (96%) are equipped with an intranet, and 56.8% of them consider this technology a mandatory procedural tool at work. For email, 96% of banking staff communicate via this technology. Regarding the internet connection, it is important to say that more than 52% of the actors access the internet. This means that more than half of the banks are equipped with the internet connection. In addition, all banks consulted have a website (100% of the branches and directorates).

Regarding the workflow, groupware and integrated management software (ERP), we found during the study that 80% of respondents said they work with the workflow (as a tool describing the process for validating an operation in the bank). Overall, 50% of respondents use groupware (software that allows group work). As for the Professional Golfers' Association of Ireland (PGI), 38% of respondents know that their banks use this technology (software that allows the integration of company functions, such as Oracle).

In general, Moroccan banks use ICTs extensively. However, this level of technology equipment is not homogeneous from one bank to another, for several factors: size of the bank, strategies adopted, culture of managers, etc.

12.6.1.2 Duration of Use of ICT Tool at Work by Bankers

For the duration of use of ICT (the computer tool) at work, we can say that more than 82% of banking HR have been using the computer tool for more than three years. This indicates a certain level of experience among banking players in the use of ICT.

12.6.2 Social Influence on the Use of Digital Technologies

We believe that the management engagement is very important in encouraging the adoption and use of ICTs by stakeholders. The survey shows that management is highly committed to ICT use, as 87.3% of banking HR believe their bank's management sees ICT as a strategic area.

12.6.3 The Conditions of the Use of Digital Technologies

They concern HR training using the technologies and the technological infrastructure put in place.

12.6.3.1 Banking HR Training in ICT

If we look at the importance given by Moroccan banks to HR training in ICT as an intangible investment, we find that 61.1% of bankers have trained in ICT since their insertion into the bank. This means that these organizations make the development of their staff a priority.

12.6.3.2 Quality of Technology Infrastructure in Place

The use of ICT in the banks studied is not without a number of technological risks with adverse effects on the mental health of staff at work. According to the bankers, the breakdown problems come first with a percentage of 80.9%; the slow navigation and connection interruptions represent 74.8%; the information overload, 48.9%; the access difficulties, 45%; the e-surveillance, 30.5%; and the security problems, 24.4%; these are the various risks that can disrupt the working climate.

In short, outages, network slowness and connection failures are the main technological risks that accompany the use of ICT in Moroccan banks.

12.6.3.3 Expectations on Improving Banking HR Performance at Work

This study focuses on the bankers' perception of the relationship between ICT use and group work. The results of the study show that ICT is considered an excellent tool for group work. Thus, 84.2% of the users interviewed acknowledged the improvement in the practice of group work in their banks. In addition, 86% of the respondents acknowledge that they are called upon to coordinate more and more remotely with other collaborators thanks to ICT. Only 14% of respondents see that ICT has no significant impact on coordination within their banks.

In addition to their role in promoting group work in banking, ICT allows the development of new relational skills among users (group spirit, sharing culture, acquisition of new knowledge, know-how, etc.) according to 85.2% of the interviewees and the development of interpersonal relations for 71.3% of respondents.

Regarding the role of digital technologies in promoting or not the sharing of knowledge and know-how at work, the study tells us that 91.7% of banking HR consulted see that ICT promotes this practice within their banks. Only 3.5% of respondents have no information in this respect.

From these different results, we understand that ICT is a great tool for bankers to work in groups. The technologies facilitate coordination between different employees who are called to coordinate, more and more, at a distance. In addition, ICT makes it possible to work in a network or in virtual teams, with all that this implies in terms of speed of electronic exchanges at the horizontal and vertical levels at work.

Working in groups using ICT, allowing the acquisition of new relational skills (communication, cooperation etc.) and the development of new interpersonal relationships. The bankers interviewed recognize that with ICT, group work is no longer a luxury but a necessity. It should not be forgotten that, in addition to being a simple tool for exchanging information, ICTs allow coordination and collaboration between bank staff and even with other external partners (administrations, banks etc.). This coordination, which affects both individuals and services, reflects a new philosophy of work that is in place.

Given that ICT is a tool for working in groups par excellence, this question leads us to reflect on the issue of sharing between employees in the banking professional environment. Indeed, thanks to the capabilities of information acquisition, storage, processing and dissemination, technologies present essential tools for sharing practices in banking. Thus, electronic exchanges (circulars, e-mails, files, documents, memos, information notes) take place at the expense of physical exchanges (carried mail, etc.). In this context, rapid access to and processing of information, with the resulting gain in time, allows for sustained speed of electronic exchanges.

In addition, collaborative tools (messaging, intranet etc.) promote this sharing practice in banks because they allow, in a reduced time, the dissemination of an enormous amount of information to a large population at work. It has become possible, by a simple e-mail, to distribute information to a maximum number of recipients in a very short time and without a necessary synchronization on the part of users.

It is not necessary to recall that today, the best ideas remain without value if the actors do not know how to communicate with others and if they do not want to work in groups. ICT (the workflow, for example) imposes this new practice in the sense that tasks themselves involve several parties. Add to this that the logic of networked organizations, including banks, is in the direction of more collective work. ICT has thus moved from tools for people to technologies for relationships.

In general, ICT is an essential tool for banking HR group work. However, the implementation of these tools for knowledge management and sharing in banks is not enough. A favorable organizational climate must be established: group spirit, culture of sharing, cooperation and mutual enrichment. In this framework, our study has concluded that banking HR has a strong sharing culture and is very aware of it in their work.

To summarize the key results of the study, we distinguish the following figure.

THE KEY RESULTS OF THE RESEARCH

RESULTS

- Digital technologies promote group work among Moroccan banking HR via:
  - The massification of communication at work;
  - Knowledge sharing between bankers at work;
  - The development of new relational skills and collaboration at work.
- However, working in a group implies the existence of a culture of sharing and collaboration among the various collaborators to further nurture this practice.

In the end, and in view of these different results, and with a view to promoting group work in banking HR, it should not be forgotten that:

- The Moroccan banks are called to make human capital at work, the key to any development and prosperity. Motivation and recognition are the foundation of a healthy work environment.
- Achieving the ultimate goal of work involves pooling the individual strengths of the different members of the group to improve the performance of the latter with all that this requires in terms of motivation and involvement at work. This supports the belief that the group will be stronger and better equipped to achieve its goals and objectives if all members are actively involved and collaborate with each other.
- If the consolidation and strengthening of group work brings several advantages to the banking company for the achievement of its objectives, this means that this way of working requires good communication between employees and good management of workplace conflicts and HR.
- Group work is above all a state of mind. It is therefore necessary to enrich the culture of collaboration in banking HR, which today more than ever is called upon to coordinate more. Training modules on this work practice must be included in the training programs organized for the benefit of banking HR.

12.7 CONCLUSION

This work focused on the role of digital technologies in promoting group work at the level of banking companies in Morocco. A quantitative study was carried out by manipulating questionnaires from the HR department in different banks. The main results of this investigation are:

- Moroccan banking companies are going through a strong digitalization. They use digital technologies massively. To this end, no activity in the bank is immune from this process of dematerialization, which invades all areas;
- Digital technologies are great tools to promote group work in banking HR. The latter are called upon to coordinate and share more and more in their work. In reality, this new trend is no longer a luxury but a constraint imposed by digitalization;
- Working in a group effectively involves watering this practice. Bankers are then invited to push forward this way of working through, on the one hand, the development of new skills (communication, listening, follow-up, emotional intelligence, etc.) and, on the other hand, having and developing a culture of sharing and collaboration between them at work. This is the case; promoting the practice of group work means its integration into the various training modules organized by banks in order to improve the skills and qualifications of banking HR in this area. The contributions of this work at theoretical, methodological and practical levels are indisputable. At a theoretical level, and since banks are considered financial enterprises in the first place, they are primarily interested in performance and profit. Dealing with this ICT issue in relation to their contributions to group work opens the door for reflection on no less than a number of important issues related to the subject and constituting the concern of banking HR at this level.

On the empirical side, questioning bankers allowed us to understand both operational and decision-makers' perceptions, which certainly enriched the information retained. At the methodological level, the results generated by this study are highly beneficial to practitioners in the banking sector. The latter can now look more at ICT as an essential tool for achieving productivity and HR performance in the workplace and beyond. In addition, we have opted for an anthropocentric approach. The latter favors men to the detriment of technology, which implies the central place of human capital in any attempt at development.

However, as with all research work, this study nevertheless has its limits. Indeed, and given that this subject involves the integration of several dimensions (technology, sociology, economics, HR, psychology, etc.) to properly surround the theme, we consider that consultation in all these areas will be able to further expand and enrich our study. In addition, not all ICTs have the same impacts on the development of work e group (ICT and groupware technologies will not have the same impacts at this level).

Add to this that the questioning of both responsible and operational bankers gives us, in one way or another, the impression that these actors do not always share the same aims and objectives (managers seek to achieve objectives, and operators seek to improve their working conditions, for example).

All these limitations leave us with hope of developing the study, while knowing that for the question of group work in banks in particular and in companies in general, we must not hide another new context that continues to flow much ink today, namely teleworking. This new situation continues to accentuate from the onset of the Covid-19 crisis, and companies at the time had no choice but to adopt and develop this practice. This raises the question of group work, which has just been mentioned in our work, in this new context of teleworking, and we consider it quite legitimate to ask ourselves the following question: what are the stakes of working in groups in the company in the age of telework? This question deserves much further research.

NOTES

1. From now on, we'll use the abbreviation "ICT", to refer to Information and Communication Technologies.
2. From now on, we'll use the abreviation "HR", referring to Human Resources.
3. A digital native is someone who has grown up in a digital environment, such as computers, the internet, mobile phones.

REFERENCES

Almerich, G., Suárez-Rodríguez, J., Díaz-García, I., & Orellana, N. (2021). The influence of using ICT in high-skills competences and ICT competences. A structural model. Education and Information Technologies, 26(4), 3845–3869.

Balantzian, G. (2002). Les systèmes d'information: art et pratiques. Les éditions d'organisation, Paris.

Barron, B. (2006). Interest and self-sustained learning as catalysts of development: A learning ecology perspective. Human Development, 49(4), 193–224.

El Adib, M., Seghyar, N., & Nafzaoui, M. (2024). Digital Adoption by Teachers in Moroccan Public Institutions: An Analysis of Key Adoption Factors. In: Motahhir, S., Bossoufi, B. (eds) Digital Technologies and Applications. ICDTA 2024. Lecture Notes in Networks and Systems, vol 1100. Cham: Springer. https://doi.org/10.1007/978-3-031-68660-3_19.

El Hammoumi, A., Nabil, S., & Berdi, A. (2024, May). Digitization of Moroccan Companies: Influence on Organizational Change and Performance. An Exploratory Study. In International Conference on Digital Technologies and Applications (pp. 176–185). Cham: Springer Nature Switzerland.

Ghamrawi, N., Shal, T., & Ghamrawi, N. A. (2024). Leadership development in virtual communities of practice: The case of school principals from the GCC Region. Education and Information Technologies, 29(17), 23897–23916.

Gollac, M., Afriat, C., & Loué, J. F. (2003). Les métiers face aux technologies de l'information. Documentation française, France. Commissariat général du plan.

Inglés, C. J., & Mateu-Martínez, O. (2022). Aprendizaje colaborativo en Psicología de la Educación: innovación educativa en Educación Superior. REDU. Revista de Docencia Universitaria, 20(2), 107–124.

Juárez, M. M. (2013). Teaching and learning science in context. LFE: revista de lenguas para fines específicos, 19, 191–216.

Lévy-Leboyer, C., Louche, C., & Rolland, J. P. (2005). RH, les apports de la psychologie du travail: 2. Management des organisations.

Mastafi, M. (2016). Définitions des TIC (E) et acception. Penser les TIC dans les universités du Maghreb.

Pinsonneault, A., & Kraemer, K. L. (1997). Middle management downsizing: An empirical investigation of the impact of information technology. Management Science, 43(5), 659–679.

Plotnick, L., Hiltz, S. R., & Privman, R. (2016). Ingroup dynamics and perceived effectiveness of partially distributed teams. IEEE Transactions on Professional Communication, 59(3), 203–229.

Rowe, F., (1994). L'impact de l'informatisation sur la performance de l'entreprise, Revue Française de Gestion, 97, Janvier-Février, pp.30–42.
Saadoun, M. (2000). Intranet & Groupware: Pour travailler en équipe. Revue de l'Information Scientifique et Technique, 10(1), 116–124.
Seghyar, N., Amane, M., & Gouiouez, M. (2024a). "The Evolution and History of E-learning: From 1840 to Present", 2024 International Conference on Intelligent Systems and Computer Vision (ISCV), Fez, Morocco, 2024, pp. 1–7, doi: 10.1109/ISCV60512.2024.10620108.
Seghyar, N., Berdi, A., & Elhammoumi, A. (2024b). Impacts of Digitalization on Human Resources Health in the Banking Sector in Morocco. In: Motahhir, S., Bossoufi, B. (eds) Digital Technologies and Applications. ICDTA 2024. Lecture Notes in Networks and Systems, vol 1099. Cham: Springer. https://doi.org/10.1007/978-3-031-68653-5_25
Zeidmane, A. (2020). Promotion of engineering skills using ICT in study process. In International Scientific Conference Engineering for Rural Development. https://doi.org/10.22616/ERDev.2020.19.TF505.

ANNEXES:

Author	Title	Type of Study	Sample and Method	Key Findings
Almerich et al. (2021)	The influence of using ICT in high-skills competences and ICT competences: A structural model	Quantitative	983 students from the University of Valencia, selected by incidental non-probability sampling. The data analysis was a structural equation model.	The ICT competences, mostly the pedagogical competences, impacted on high-skill competences: higher-order thinking capacities and teamwork competences.
Zeidmane (2020)	Promotion of engineering skills using ICT in study process	Qualitative	The sample size of the study consisted of 301 students from the engineering specialties of the LLU through a survey of students.	Inclusion of ICT in the mathematics studies develops the skills necessary for engineers: communication, technical skills and teamwork.
Inglés Saura and Mateu Martínez (2022)	Collaborative learning in Educational Psychology: educational innovation in Higher Education	Qualitative	The Psychology Degree at the Miguel Hernandez University of Elche during the academic year 2021–2022, and to know the students' opinion.	The writing presents the strategy of "Collaborative Learning" and the use of self-assessment and co-assessment e-rubrics through Information, Communication and Leisure Technologies in the theory and practice of the subject, justifying the interest of the experience, presenting the basis that supports this study and presenting the development and procedure of the dynamics and evaluation during the semester.

(Continued)

Author	Title	Type of Study	Sample and Method	Key Findings
Juárez (2013)	Teaching and Learning Science in Spanish Context	Qualitative	49 Spanish students divided into 2 groups in year 6 of primary education in Spain.	The involvement of students in the learning process and their interaction is an essential aspect of the study. It was even enhanced by means of the use of ICT in the classroom and by the students' active participation in the activities carried out along the process, which fostered motivation, a relaxed setting, teamwork and collaboration in a meaningful context.
Plotnick et al. (2016)	Ingroup Dynamics and Perceived Effectiveness of Partially Distributed Teams	Quantitative	Online survey of professionals with experience in PDTs was conducted, with 2 subsamples—one from a single large telecommunications company and another from a mix of organizations.	Strong negative relationship between ingroup dynamics and perceived effectiveness. Overall, technology reliability and training significantly reduce ingroup dynamics. In the telecommunications company, training increased ingroup dynamics.

CHAPTER 13

AI-Powered Waste Management

Transforming Linear Systems into Circular Economy Solutions

Shubham Gupta

13.1 INTRODUCTION

13.1.1 Research Objectives and Scope

The global waste crisis represents one of the most pressing environmental challenges of the 21st century. With waste generation projected to reach 3.40 billion tons by 2050—a 70% increase from 2016 levels [1]—traditional linear "take-make-dispose" models have become economically and environmentally unsustainable. This crisis directly impacts multiple Sustainable Development Goals (SDGs), particularly SDG 11 (Sustainable Cities), SDG 12 (Responsible Consumption), and SDG 13 (Climate Action).

This chapter addresses three critical research questions:

1. **Technology-Performance Analysis:** How do artificial intelligence (AI) technologies enhance efficiency and environmental performance across the waste management value chain?
2. **Sustainability Impact Assessment:** What are the measurable sustainability impacts of AI implementation in waste systems?
3. **Implementation Strategy Development:** What barriers limit AI adoption and how can they be systematically addressed?

Our analysis focuses on municipal solid waste systems, commercial waste streams, and circular economy applications, examining peer-reviewed literature from 2016 to 2023 and verified industry performance data.

13.1.2 The Sustainability Imperative

Modern waste challenges create cascading impacts across environmental, economic, and social systems. The Ellen MacArthur Foundation estimates $57 billion in materials lost annually in global waste streams [2], representing both economic inefficiency and resource

DOI: 10.1201/9781003649885-13

depletion. Waste sectors contribute 3–5% of global greenhouse gas emissions [3], while collection activities account for 70% of municipal waste management costs [4]. Electronic waste grows at 3–5% annually, with only 20% formally recycled [5], creating toxic accumulation and critical material scarcity.

The complexity of modern waste streams compounds these challenges. Microplastic contamination affects marine ecosystems globally, while composite materials and multilayer packaging resist conventional recycling technologies. In developing economies, waste management can consume up to 20% of municipal budgets, straining resources needed for other essential services. The linear "take-make-dispose" economy results in valuable resources being permanently lost in landfills, creating environmental liabilities while depleting finite resources needed for continued economic activity.

High-income countries, representing 16% of the global population, generate 34% of the world's waste [1], highlighting global equity concerns in waste generation and management capacity. Traditional waste management approaches—developed for simpler, more homogeneous waste streams—lack the precision, adaptability, and system-level optimization required for contemporary challenges. Conventional sorting technologies struggle with material identification accuracy, typically achieving 70–80% accuracy rates that result in significant contamination and value loss.

13.1.3 AI as an Enabling Technology

There are five types of transformative capabilities of AI:

Pattern Recognition: Data available about the generation of waste demonstrates that the patterns of waste generation and their composition are untraceable and impossible to analyze.

Identifying the Material: Computer vision and spectroscopic analysis achieve 95+% material identification accuracy for hundreds of waste categories simultaneously.

AI Algorithms: The algorithms efficiently manage the real-time optimization of multivariable systems, considering factors such as optimal traffic patterns, equipment performance, and environmental constraints.

Autonomous Operation: Robotic systems are used for consistency in handling hazardous tasks and 24/7 operation.

AI Platforms Data Integration: AI platforms aggregate data from various sources to furnish extensive operational intelligence.

The impacts of AI technology integration across the waste management value chain are synergistic; they usually affect more than just the specific application. Figure 13.1 illustrates a connected web of technology, from the creation of raw waste to collection, processing, and material recovery and shows how far AI can have a profound effect in transforming this process.

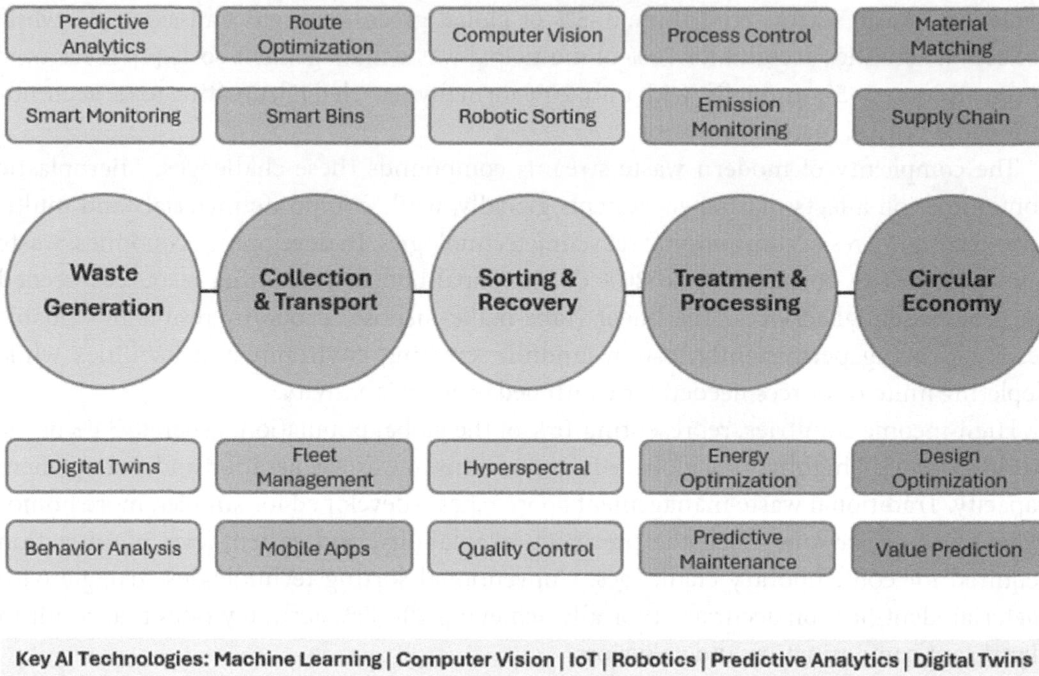

FIGURE 13.1 AI applications across the waste management value chain, showing interconnected systems from waste generation through collection, processing, and material recovery to circular economy solutions.

13.2 METHODOLOGY

Systematic evidence synthesis, following well-established technology assessment protocols, was employed in this review. Our search strategy consisted of searching Web of Science, Scopus, IEEE Xplore, and Google Scholar, in addition to industry reports. This set was compiled to cover only those publications relevant to the period from 2016 to 2023. Keywords used included "artificial intelligence," "machine learning," "waste management," "recycling automation," and "circular economy."

Studies needed to be included and considered to have a quantifiable outcome and to be peer-reviewed articles, validated industry performance data, or documented case studies. Study design, measurement validity, and potential bias were evaluated to assess evidence quality.

Mixed-methods approach of combining quantitative performance analysis and qualitative assessment over the development of technical performance, sustainability impact, and implementation readiness dimensions.

The economic evaluation of privacy-preserving transportation applications is limited by a geographic bias toward European and North American implementations, the limited long-term availability of operational data, and a focus on direct operational benefits rather than more far-reaching system effects.

13.3 AI TECHNOLOGIES AND APPLICATIONS

13.3.1 Machine Learning for System Optimization

Machine learning applications, such as the prevention of waste production, waste classification, and system optimization in waste management, have evolved from simple classification tasks to complex system optimization problems, including multi-data source integration and real-time decision-making.

Using convolutional neural networks trained on large datasets of millions of waste material images captured under various lighting conditions, angles, and contamination states, supervised learning can achieve 95% material classification accuracy [6]. Because these systems enable automated sorting at speeds and with accuracy that is not possible using manual processes, up to 10,000 items per hour can be processed consistently and accurately.

Another critical application where machine learning models are 15–30% more accurate in prediction compared to traditional statistical approaches is waste generation forecasting [7]. These advanced models enabled the optimization of collections and facility capacity management, informed by the reliability of planning data, including weather patterns, economic indicators, seasonal effects, holiday schedules, and special events. Municipalities can utilize time series analysis, combined with external data sources, to understand how demand fluctuates and more effectively allocate resources.

Spectral analysis, coupled with machine learning, is used in quality prediction systems that predict plastic grades with 87–92% accuracy, aiding in value optimization through precise material routing [8]. Having this capability is critical to participating in premium recycled material marketplaces, where stringent quality specifications are in place, and penalties for contamination can render those materials unusable, cutting their value in half or more.

With unsupervised learning, we can identify hidden patterns using clustering algorithms that consider distinct urban zone behaviors, waste generation patterns during different seasons, and anomaly detection in the context of contamination events [9]. Using k-means clustering analysis reveals different waste generation patterns in various urban zones, offering the possibility to optimize service types and prevent contamination. The insights provide support for policy development and infrastructure planning decisions.

As complex, multi-modal data streams, deep learning simultaneously recognizes thousands of packaging items under challenging conditions, including partial visibility, material deformation, varying lighting, and overlapping objects [10]. For high-value recycling applications, advanced neural networks can distinguish between materials that appear identical on the surface but have distinct chemical compositions.

13.3.2 Computer Vision and Robotic Sorting

The advent of computer vision technologies has progressed from laboratory demonstrations to commercially viable systems handling millions of tons of waste across hundreds of facilities worldwide, representing a fundamentally new capability for material recovery.

For instance, object detection frameworks like YOLO (You Only Look Once) and SSD (Single Shot Detector) locate waste objects on high-speed conveyor belts running at speeds of 2–3 meters per second [11]. In such an environment, these systems face other challenging working conditions, including variable lighting, moving objects, overlapping materials, dust, and contamination, while maintaining high throughput and accuracy. The second is that advanced algorithms will be able to handle complex scenes in which we see parts of the material being covered or deformed.

Using 95%+ plastic identification [12] combined with bands of molecular absorption patterns, atomic density characteristics, and other features obtained by multi-spectral analysis in terms of near-infrared (NIR) spectroscopy, shortwave infrared (SWIR), and X-ray transmission, they successfully identify the plastics. Using NIR systems to identify polymer types, these systems rely on the detection of specific chemical bonds. In contrast, SWIR systems deal with historically hard-to-detect polymeric materials, such as black plastics, which appear the same in the visible spectrum. Using X-ray transmission, it's possible to separate metals by atomic density, recovering valuable non-ferrous metals from complex waste streams.

The advantage of comprehensive material characterization is demonstrated by improved classification accuracy (12–18%) compared to single-sensor approaches when multiple sensing modalities are integrated [13]. Through this multi-sensor fusion approach, sorting systems can make informed decisions even with ambiguous results from individual sensors, resulting in significantly improved overall system reliability and material purity.

Articulated arms can handle larger or heavier items in specialized waste streams, achieving 60–80 picks per minute at 95%+ accuracy for lightweight materials [14]. Multi-arm commercial systems handle 4–8 tons per hour using adaptive grippers, vacuum systems, and hybrid end effectors and can handle diverse materials without requiring a tool change. In advanced gripper technologies, force feedback and tactile sensing are used to modify the gripping pressure based on material properties.

Through online learning algorithms, continuous performance improvement is possible without explicit reprogramming, enabling adaptation to spatiotemporal changes in waste composition and enhancing recognition accuracy through operational experience [15]. Sorting systems at multiple facilities are arranged into networks by facilitating the sharing of learning algorithms and performance data, allowing them to learn collectively across facilities and deployed systems. This approach has shown a 21% improvement in sorting accuracy over six months of operation.

Table 13.1 presents a comprehensive comparison of several key performance metrics and implementation characteristics that help explain the comparative performance and

TABLE 13.1 AI Technology Performance Comparison in Waste Management

Technology	Accuracy	Throughput	Energy Use	Maturity	Implementation Cost
CNN classification	95–98%	10,000 items/hour	Medium	Commercial	\$100k–500k
Multi-spectral NIR	95%+ plastics	5,000 kg/hour	Low	Mature	\$200k–800k
Robotic delta arms	95%+ pick rate	60–80 picks/min	High	Commercial	\$150k–400k
IoT fill sensors	±3% accuracy	Real-time	Very low	Mature	\$50–200/sensor
Route optimization	15–40% improvement	Real-time	Low	Mature	\$10k–50k
Digital twin systems	5–10% efficiency gain	Continuous	Medium	Emerging	\$50k–200k

implementation characteristics of different AI technologies in waste management. These key metrics include accuracy, throughput, energy consumption, technology maturity, and typical implementation costs.

13.3.3 IoT and Smart Infrastructure

IoT technologies make waste management networked, enabling real-time visibility, automated control, and data-driven optimization across the entire waste network, helping traditional reactive waste systems become proactive, intelligent systems.

Using ultrasonic technology, fill-level sensors offer ±3% accuracy and are suitable for a three to five year battery life, working reliably in harsh outdoor conditions [16]. These sensors utilize low-power wide-area networks (LPWAN) to communicate, making it affordable to deploy them over thousands of containers without requiring frequent maintenance. They are used to detect hazardous conditions in advance of harming the worker or the environment due to noncompliance with hazardous materials regulations.

According to nature, cities implementing comprehensive IoT monitoring have experienced tremendous improvements: a 70–80% increase in collection frequency and reduction of overflow, an increase in detections, and a 15–20% reduction in total collection costs [17]. This is due to the decrease in unnecessary trips made when road conditions change as the collection transitions from a scheduled to a demand-based model. The case study of Barcelona demonstrated that through intelligent monitoring, total vehicle kilometers can be reduced by 25% while also improving service quality.

Dynamic route optimization provides real-time, optimized collection sequences, resulting in fuel savings of 10–30% and emissions reductions of 15–35% [7]. These algorithms are also highly advanced and take into account the joint minimization of fuel consumption, the minimization of vehicle emissions, the satisfaction of service level agreements, the balancing of driver workloads, and adaptability to new situations, such as traffic congestion or vehicle breakdowns. Models developed for machine learning include past traffic patterns, weather conditions, and special events to determine the optimal time for collection.

Local processing and real-time decisions are possible wherever cloud connectivity may be missing, such as in remote facilities, by utilizing edge computing [18]. Data from the sensors and smart cameras is preprocessed before transmission to reduce transmission costs and enable the sensor network to operate even during network failures. With this designed distributed intelligence architecture of waste management, critical waste management processes continue even during connectivity disruptions.

The complexity and interconnected nature of modern AI-enabled waste sorting systems require sophisticated integration of multiple technologies working in harmony. Figure 13.2 demonstrates how sensing technologies, AI processing capabilities, and robotic manipulation components combine to create comprehensive automated sorting solutions.

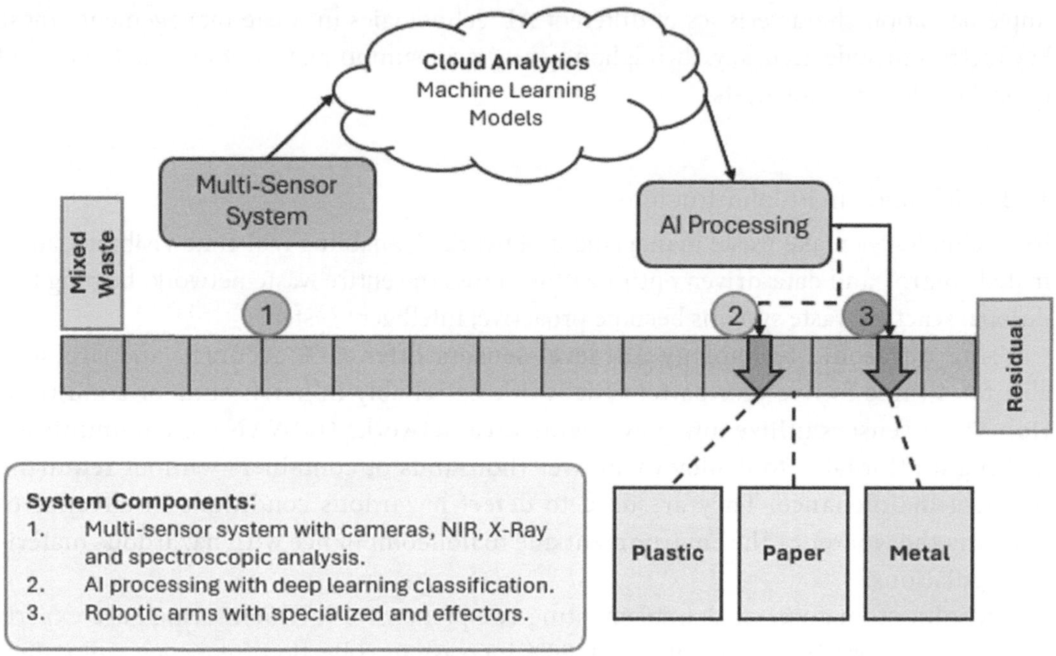

FIGURE 13.2 AI-enabled waste sorting system integration showing sensing technologies, AI processing, and robotic manipulation components.

13.4 SUSTAINABILITY IMPACT ANALYSIS

13.4.1 Environmental Performance

AI implementation delivers quantifiable environmental benefits across multiple dimensions of waste management system performance, with impacts measurable at facility, municipal, and regional scales, contributing significantly to global sustainability objectives.

Advanced sorting systems increase material recovery rates by 10–35% compared to conventional methods, with specific materials showing even greater improvements [19]. Black plastic recovery, traditionally challenging for optical sorting due to absorption of NIR light, improves from near–0 to 90%+ using AI-enhanced multi-spectral systems, increasing mixed plastic bale value by 15–25%. This improvement alone can generate additional revenue of $30–50 per ton of processed material.

Metal recovery from complex waste streams improves dramatically through AI-controlled X-ray transmission sorting, achieving 98%+ purity for aluminum and copper fractions. Contamination reduction in sorted fractions decreases downstream processing costs and enables access to premium markets with stringent quality requirements. Food-grade plastic recovery capabilities support circular economy applications requiring high-quality recycled content.

AI-controlled anaerobic digestion increases biogas yield by 10–35% while reducing process time by 15–50% [20]. Machine learning algorithms optimize feeding schedules, mixing patterns, temperature control, and pH management based on feedstock characteristics and real-time process conditions. Waste-to-energy facilities using AI control systems achieve 5–10% efficiency improvements through predictive optimization of combustion parameters and emissions control.

Electric collection vehicles with AI-optimized routing achieve 50–70% greenhouse gas reductions compared to diesel equivalents when using low-carbon electricity [21]. Route optimization alone reduces collection vehicle emissions by 15–35% through reduced fuel consumption and improved operational efficiency. High-quality recycled plastics avoid 1.5–2.0 tons of CO_2 equivalent per ton compared to virgin plastic production, multiplying environmental benefits across the supply chain.

13.4.2 Economic Viability

Organizations implementing comprehensive AI systems report 15–30% operational cost reductions and 20–40% service quality improvements [22]. Automated sorting reduces labor costs by 30–50% while decreasing workplace injuries by 40–60% [23]. Food-grade PET separation generates 20–35% price premiums [24].

The UK's Industrial Symbiosis Program generated €354 million annually while diverting 9.7 million tons from landfills [25]. Route optimization delivers returns within three to six months, while advanced sorting achieves payback in two to three years.

The economic viability of AI applications varies significantly based on technology type, implementation scale, and operational context. Table 13.2 summarizes the financial characteristics of major AI applications, providing essential data for investment decision-making and strategic planning.

13.4.3 Circular Economy Enablement

AI technologies provide essential capabilities for transitioning from linear "take-make-dispose" models to circular systems that maintain material value and minimize waste generation.

Material matching algorithms achieve 30–60% success rates in identifying industrial synergies, advancing to 60–85% efficiency [26]. AI-assisted product design improvements increase recyclability by 25–40% [27]. Urban mining applications recover critical materials with 60–90% less energy than primary extraction [28].

The transformation to circular economy models requires sophisticated integration of AI technologies with design processes, material flows, and end-of-life management systems. Figure 13.3 illustrates how AI connects design decisions with end-of-life management, creating feedback loops that enable true circular material flows and optimize both product performance and recyclability outcomes.

TABLE 13.2 Economic Impact Summary of AI Applications

Application	Initial Investment	Annual Savings	ROI Period	Primary Benefits
Route optimization	$10k–50k	$50k–200k	3–6 months	Fuel, labor, emissions
Smart monitoring	$100k–300k	$200k–800k	12–18 months	Collection efficiency
Automated sorting	$200k–2M	$300k–1.5M	2–3 years	Labor, quality, safety
Predictive maintenance	$50k–200k	$100k—500k	6–12 months	Downtime, repairs
Digital twins	$100k–500k	$200k–1M	18–30 months	Process optimization

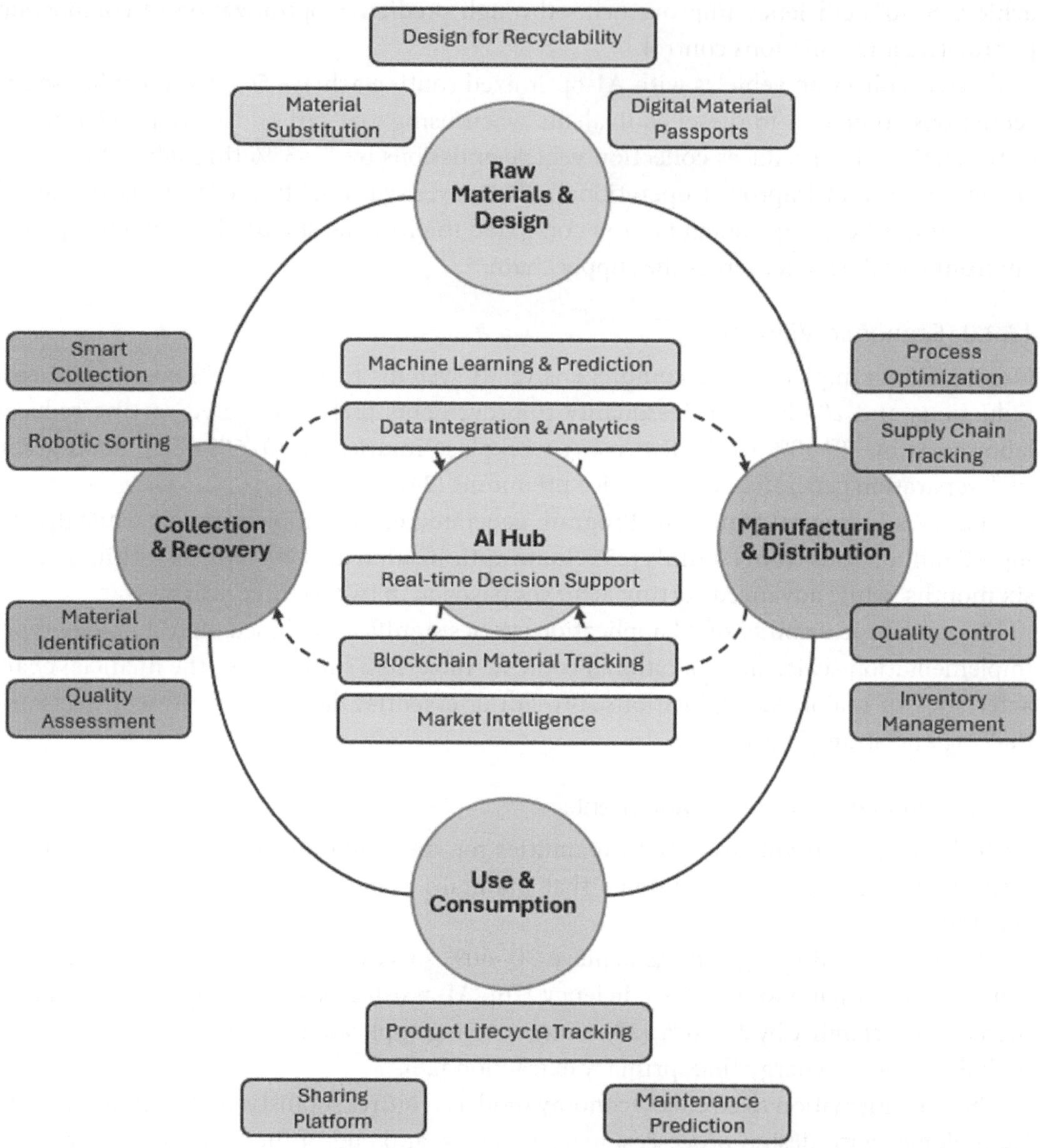

FIGURE 13.3 AI integration with circular economy approaches, connecting design decisions to end-of-life management.

13.5 IMPLEMENTATION FRAMEWORK AND BARRIERS

13.5.1 Technical Challenges

Data infrastructure represents the foundation for AI implementation. Organizations investing 3–5% of project budgets in data infrastructure demonstrate higher success rates [29]. System integration challenges require open standards like open platform communications unified architecture (OPC UA) for connecting diverse systems [30].

Edge AI optimization reduces energy requirements by 80–95% while maintaining performance [18].

13.5.2 Economic and Organizational Barriers

Advanced sorting technologies require \$50K–\$2M initial investment, challenging for resource-constrained organizations [10]. Equipment-as-a-service models spread costs while aligning payments with value delivery. Workforce transition requires comprehensive training programs, with successful organizations retraining workers as operators and technicians [31].

13.5.3 Strategic Implementation Model

Figure 13.4 illustrates the progression of the AI Implementation maturity model from basic data collection through strategic transformation to ecosystem leadership.

Five-Stage Maturity Framework:

Stage 1—Data Foundation: IoT sensors and basic analytics achieve 10–20% improvements with three to six month payback.

Stage 2—Process Automation: Automated sorting and predictive maintenance deliver 20–35% efficiency gains.

Stage 3—System Integration: Digital twins enable whole-system optimization with cross-functional benefits.

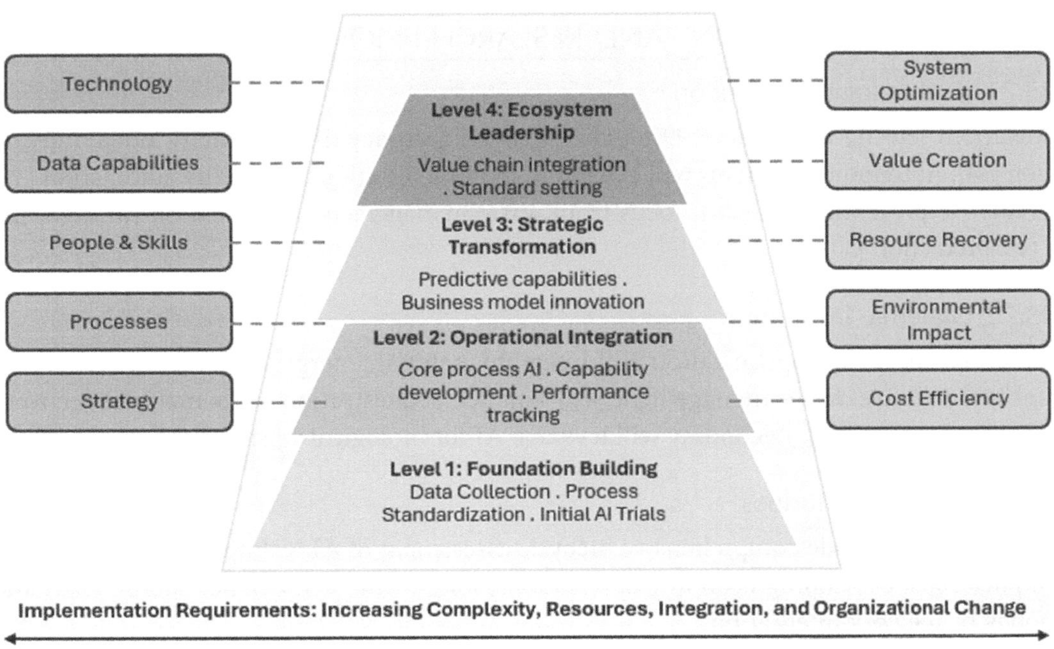

FIGURE 13.4 AI implementation maturity model for waste management organizations.

Stage 4—Strategic Transformation: Circular economy integration creates new business models and market opportunities.

Stage 5—Ecosystem Leadership: Platform-based models create value across industry networks.

13.6 CASE STUDIES AND VALIDATED OUTCOMES

13.6.1 Barcelona Smart Collection

Barcelona implemented ultrasonic sensors in 17,000 containers with machine learning route optimization [17]. Results: 25% collection frequency reduction, 43% fewer overflows, €4 million annual savings, 864 tons CO_2 reduction. Implementation cost of €8.2 million achieved a 30-month payback.

13.6.2 TOMRA Optical Sorting

Multi-sensor platforms combining NIR spectroscopy, cameras, and X-ray transmission achieve 95–98% plastic purity with 20–30% throughput increases [32]. Material value improvements of €40–60 per ton enable two to three year ROI through operational savings and premium market access.

13.6.3 Excess Materials Exchange

AI-powered platform facilitated 2,000+ material matches creating €54 million value while avoiding 100,000 tons CO_2 emissions [33]. Digital material passports with blockchain verification enable secure transactions across diverse industrial sectors.

13.7 FUTURE DIRECTIONS AND RESEARCH PRIORITIES

13.7.1 Technological Evolution

Advanced sensing will achieve 99%+ identification accuracy through multi-modal integration [34]. Autonomous systems will expand to mobile collection and facility automation [31]. Predictive prevention will shift focus from waste management to prevention through lifecycle prediction [35].

13.7.2 Systemic Integration

Urban resource flow optimization will integrate material, energy, and water systems [36]. Carbon-negative waste management will balance sequestration with material recovery [37]. Bio-based circular economy will leverage AI for biological waste valorization [20].

13.7.3 Research Priorities

Critical areas include comprehensive lifecycle assessment of AI technologies, social equity implications, developing economy applications, long-term performance monitoring, and policy framework integration.

The successful implementation of AI in waste management requires a strategic approach that considers both technological capabilities and organizational readiness. Figure 13.5

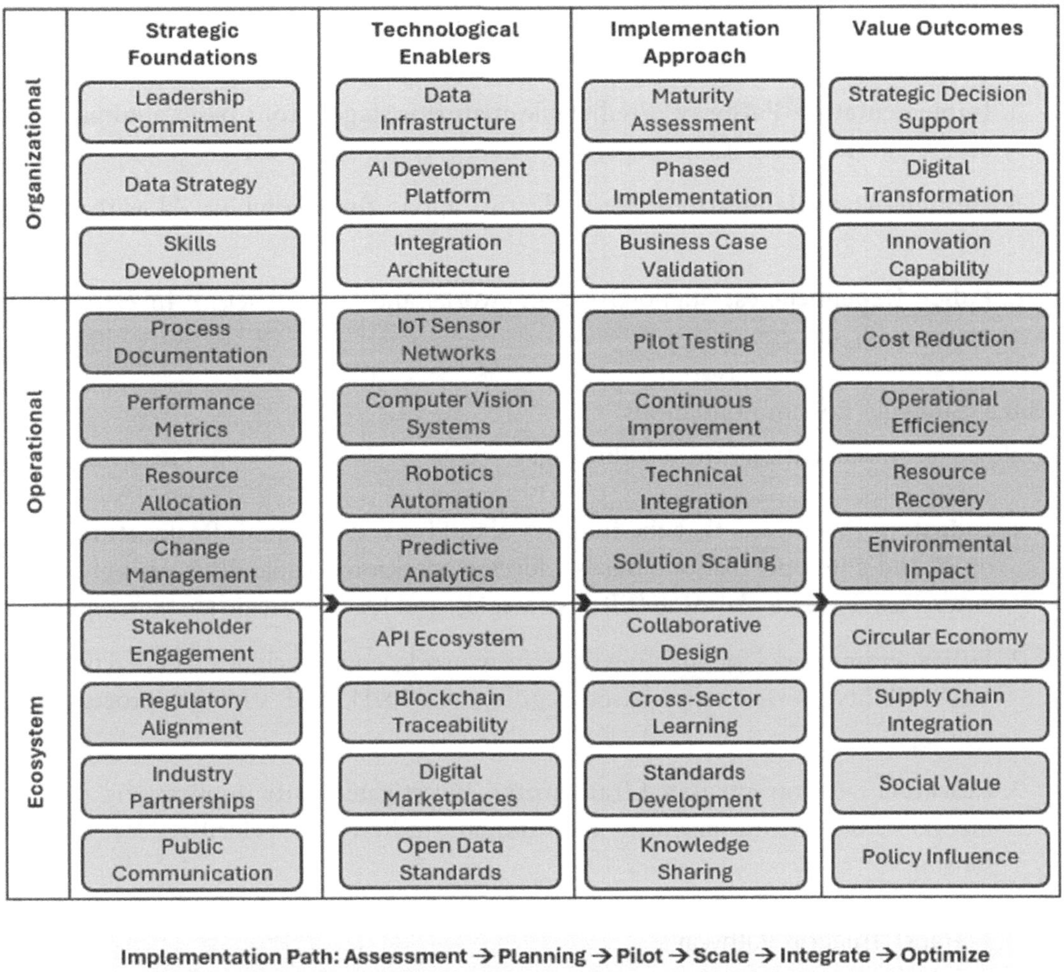

FIGURE 13.5 Strategic framework for AI implementation in waste management showing assessment dimensions and development pathways.

presents a comprehensive strategic framework that organizations can use to assess their current position, identify development priorities, and plan their transformation journey toward AI-enhanced waste management systems.

13.8 CONCLUSIONS AND STRATEGIC IMPLICATIONS

13.8.1 Key Findings

Six critical insights emerge:

1. **System-Wide Transformation:** AI enables 15–40% operational improvements through network effects and system-level optimization.
2. **Proven Benefits:** 10–35% increased recovery rates, 15–35% emissions reductions, 6–36 month ROI periods.

3. **Circular Economy Enablement:** AI provides essential capabilities for material flow optimization and value preservation.

4. **Implementation Pathway:** Predictable maturity stages from basic automation to ecosystem transformation.

5. **Human–AI Collaboration:** Most effective applications combine AI with human expertise.

6. **Policy Dependencies:** Supportive frameworks and Infrastructure investment are required for success.

13.8.2 Strategic Recommendations

Strategic recommendations are as following:

1. **Industry:** This means that the industry should invest in data infrastructure, adopt open and interoperable standards, develop workforce capabilities, adopt phased enterprise resource planning (ERP) strategies, and form strategic partnerships.

2. **Policy:** Frameworks for enabling these frameworks could include extended producer responsibility, performance-based regulations, R&D, and strategic procurement, among others.

3. **Research:** Develop integrated frameworks, investigate equity implications, conduct lifecycle assessments, perform longitudinal studies, and prioritize cost-effective sensing.

13.8.3 Transformation Pathway

Opportunities in the short term (2026–2027) involve scaling up proven technologies to a marketable level, which should also reduce costs and enhance efficacy. The evolution of AI will be hooked across the entire value chain in the 2027–2032 era to support circular business models. The long-term vision (2032–2040) includes autonomous systems and the optimization of urban metabolism.

To achieve success, transformation needs to be recognized to be inherently socio-technical, not only requiring an emphasis on technological development but also addressing human, organizational, and institutional dimensions.

Table 13.3 shows strategic implementation timeline for AI adoption with phases, technologies, success metrics, and investment priorities to help organizations in planning their AI adoption journey. This roadmap provides a structured approach to achieve systematic transformation in a controlled manner while managing risks and maximizing special value creation.

By utilizing proven technologies and innovating systematically, AI can enable circular material flows that meet the challenge of the global waste crisis, creating both economic opportunities and environmental benefits for today and tomorrow.

TABLE 13.3 Strategic Implementation Timeline and Milestones

Phase	Timeline	Key Technologies	Success Metrics	Investment Focus
Foundation	2024–2026	IoT monitoring, route optimization, basic sorting	20–30% efficiency gains, 12–18 month ROI	Data infrastructure, training
Integration	2026–2028	Advanced sorting, predictive maintenance, digital twins	30–40% efficiency gains, system integration	Process automation, quality systems
Transformation	2028–2032	Circular platforms, autonomous systems, AI optimization	40–60% efficiency gains, new business models	Strategic partnerships, R&D
Leadership	2032–2040	Ecosystem orchestration, prevention systems, urban integration	System-wide sustainability, regenerative outcomes	Innovation platforms, global standards

REFERENCES

1. S. Kaza, L. Yao, P. Bhada-Tata, and F. Van Woerden, "What a Waste 2.0: A Global Snapshot of Solid Waste Management to 2050," World Bank Publications, 2018. https://openknowledge.worldbank.org/bitstreams/df788c58-3c21-52a2-a224-1445f0a1850b/download
2. Ellen MacArthur Foundation, "The New Plastics Economy: Rethinking the future of plastics," 2016.
3. IPCC, "2019 Refinement to the 2006 IPCC Guidelines for National Greenhouse Gas Inventories," 2019. https://www.osti.gov/etdeweb/biblio/20880391
4. N. Ferronato et al., "The municipal solid waste management of La Paz (Bolivia): Challenges and opportunities for a sustainable development," *Waste Management & Research*, vol. 36, no. 3, pp. 288–299, 2018. https://doi.org/10.1177/0734242X18755893
5. V. Forti, C. P. Baldé, R. Kuehr, and G. Bel, "The Global E-Waste Monitor 2020," United Nations University, 2020. https://np-mag.ru/wp-content/uploads/2020/08/The-Global-E-waste-Monitor-2020-Quantities-flows-and-the-circular-economy-potential_compressed.pdf
6. S. Palkopoulou et al., "Critical review of automated sorting technologies and artificial intelligence trends for plastic packaging recycling," *Waste Management*, vol. 134, pp. 167–179, 2021.
7. H. L. Vu, K. T. W. Ng, and D. Bolingbroke, "Time-varying waste collection points in municipal solid waste collection optimization," *Journal of Cleaner Production*, vol. 233, pp. 751–763, 2019.
8. J. Huang et al., "Quality evaluation of polyethylene plastic waste for recycling decision support using near-infrared hyperspectral imaging," *Waste Management*, vol. 104, pp. 169–178, 2020.
9. M. L. Tseng and T. D. Bui, "Identifying eco-innovation in industrial symbiosis under linguistic preferences: A novel hierarchical approach," *Journal of Cleaner Production*, vol. 140, pp. 1376–1389, 2017. https://doi.org/10.1016/j.jclepro.2016.10.014
10. AMP Robotics, "Performance data and technological capabilities in AI-powered robotic sorting systems," Technical Report, 2022.
11. J. Redmon and A. Farhadi, "YOLOv3: An incremental improvement," *arXiv preprint arXiv:1804.02767*, 2018.
12. S. Serranti, A. Gargiulo, and G. Bonifazi, "Characterization of post-consumer polyolefin wastes by hyperspectral imaging for quality control in recycling processes," *Waste Management*, vol. 31, no. 11, pp. 2217–2227, 2011. https://doi.org/10.1016/j.wasman.2011.06.007
13. C. Vrancken et al., "Multi-dimensional profiling of solid wastes for improved segregation and recycling," *Waste Management*, vol. 76, pp. 117–129, 2018.

14. T. J. Lukka et al., "ZenRobotics recycler – Robotic sorting using machine learning," *Proceedings of the International Conference on Sensor-Based Sorting*, 2014. https://users.ics.aalto.fi/praiko/papers/SBS14.pdf
15. AMP Robotics, "Collective learning and performance improvement in robotic sorting networks," Technical Whitepaper, 2023.
16. S. R. J. Ramson and D. J. Moni, "Wireless sensor networks based smart bin," *Computers & Electrical Engineering*, vol. 64, pp. 337–353, 2017. https://doi.org/10.1016/j.compeleceng.2016.11.030
17. Barcelona City Council, "Smart waste collection implementation: Results and lessons learned," Technical Report, 2021.
18. J. Chen and X. Ran, "Deep learning with edge computing: A review," *Proceedings of the IEEE*, vol. 107, no. 8, pp. 1655–1674, 2019. https://doi.org/10.1109/JPROC.2019.2921977
19. N. Kassos et al., "Recovery of black plastics from mixed waste streams using hyperspectral imaging technology," *Waste Management*, vol. 123, pp. 152–161, 2021.
20. S. Venkata Mohan et al., "Waste biorefinery integrating biorefineries for waste valorization," *Bioresource Technology*, vol. 278, pp. 346–354, 2019.
21. Z. Gao, Z. Lin, and O. Franzese, "Energy consumption and cost savings of truck electrification for heavy-duty vehicle applications," *Transportation Research Record*, vol. 2628, no. 1, pp. 99–109, 2017. https://doi.org/10.3141/2628-11
22. A. Mavropoulos and A. W. Nilsen, "Industry 4.0 and Circular Economy: Towards a Wasteless Future or a Wasteful Planet?," Wiley, 2020.
23. S. P. Gundupalli, S. Hait, and A. Thakur, "A review on automated sorting of source-separated municipal solid waste for recycling," *Waste Management*, vol. 60, pp. 56–74, 2017. https://doi.org/10.1016/j.wasman.2016.09.015
24. C. Signoret et al., "Machine learning-based prediction of plastic waste quality for optimized recycling," *Resources, Conservation and Recycling*, vol. 174, p. 105769, 2021.
25. P. Laybourn and D. R. Lombardi, "Industrial symbiosis in European policy: Overview of recent progress," *Journal of Industrial Ecology*, vol. 16, no. 1, pp. 11–12, 2012. https://openurl.ebsco.com/EPDB%3Agcd%3A15%3A31141855/detailv2?sid=ebsco%3Aplink%3Ascholar&id=ebsco%3Agcd%3A73464787&crl=c&link_origin=scholar.google.com
26. J. S. Mulrow et al., "Industrial symbiosis at the facility scale," *Journal of Industrial Ecology*, vol. 21, no. 3, pp. 559–571, 2017. https://doi.org/10.1111/jiec.12592
27. P. Vanegas et al., "Ease of disassembly of products to support circular economy strategies," *Resources, Conservation and Recycling*, vol. 135, pp. 323–334, 2018. https://doi.org/10.1016/j.resconrec.2017.06.022
28. X. Zeng, J. A. Mathews, and J. Li, "Urban mining of e-waste is becoming more cost-effective than virgin mining," *Environmental Science & Technology*, vol. 52, no. 8, pp. 4835–4841, 2018. https://pubs.acs.org/doi/abs/10.1021/acs.est.7b04909
29. B. Sanchez, C. Esposito, and H. Haeri, "Enabling circular economy through digital technologies: A framework of digital maturity levels in circular manufacturing," *Sustainability*, vol. 13, no. 12, p. 6941, 2021.
30. S. Ramakrishna et al., "Emerging industrial revolution: Symbiosis of industry 4.0 and circular economy: The role of universities," *Science, Technology and Society*, vol. 25, no. 3, pp. 505–525, 2020. https://doi.org/10.1177/0971721820912918
31. S. Carton et al., "Trends in collaborative robotics for sorting applications," *Frontiers in Robotics and AI*, vol. 9, p. 836971, 2022.
32. TOMRA, "Multi-sensor sorting platform performance and technological innovations," Technical Whitepaper, 2022.
33. Excess Materials Exchange, "Annual Impact Report: Creating Value Through Material Exchange," EME Publications, 2022.

34. D. Pavia et al., “Advances in multimodal sensing for recycling: Combining optical, spectral and chemical measurements,” *Waste Management*, vol. 138, pp. 86–97, 2022.
35. T. R. Walker et al., “Environmental effects of COVID-19 pandemic and potential strategies of sustainability,” *Environmental Engineering Research*, vol. 26, no. 2, p. 200420, 2021. https://doi.org/10.1016/j.enpol.2019.02.019
36. J. Millward-Hopkins and P. Purnell, “Circulating blame in the circular economy: The case of wood-waste biofuels and coal ash,” *Energy Policy*, vol. 129, pp. 168–172, 2019. https://doi.org/10.1016/j.enpol.2019.02.019
37. L. Cao et al., “Identifying barriers and benefits of implementing carbon-negative bioenergy systems based on a systematic review of perceptions,” *Global Change Biology Bioenergy*, vol. 11, no. 12, pp. 1315–1334, 2019.

CHAPTER 14

AI-Driven Innovation in Waste Management and Recycling

Vinod B. R., Arunakumari B. N., M. Varuna, and Jyothi D. N.

14.1 INTRODUCTION

The exponential boom in worldwide waste generation has created unprecedented challenges for waste manipulate structures worldwide. Annual municipal solid waste generation is projected to increase by 70% from 2016 levels to 3.40 billion tons by 2050, exacerbating environmental degradation and useful resource depletion [1]. Traditional waste management methods regularly show insufficient in addressing the scale and complexity of this disaster, necessitating cutting-edge technological solutions. Artificial intelligence (AI) has emerged as a specifically promising catalyst for transformation in this region, providing a brand new system for waste identity, elegance, process optimization, and choice help.

The integration of AI technology within waste management and recycling represents a huge paradigm shift from traditional methods. AI programs encompass a diverse variety of talents, including computer imaginative and prescient for waste recognition, predictive analytics for series direction optimization, robotics for automatic sorting, and device learning algorithms for cloth recovery enhancement [2]. These technologies together make contributions to "smart waste management structures" characterized by means of extra performance, precision, and adaptability [3].

This study aims to comprehensively assess the modern-day landscape of AI packages in waste control, take a look at their effectiveness, and pick out key opportunities and challenges for future development.

Through systematic assessment of technological improvements, implementation strategies, and performance effects, this chapter seeks to deal with several essential questions: How effectively can AI-driven structures enhance waste sorting accuracy and useful resource restoration charges?

What are the economic and environmental implications of AI integration in waste management infrastructure? Which barriers impede wider adoption of these technologies, and how might they be overcome?

The significance of this research lies in its potential to inform strategic decision-making among waste management authorities, technology developers, and policymakers. By

DOI: 10.1201/9781003649885-14

elucidating both the transformative potential and practical limitations of AI in waste management contexts, this study contributes valuable insights to the growing discourse on technological solutions for environmental sustainability. The intersection of AI and waste management represents not merely a technological innovation but a fundamental reimagining of our relationship with waste as a resource [4].

14.2 LITERATURE REVIEW

The application of AI technologies in waste management systems has garnered increasing scholarly attention over the past decade. This literature review synthesizes key research contributions across several thematic areas to establish the current state of knowledge and identify significant research gaps.

14.3 AI APPLICATIONS IN WASTE CLASSIFICATION AND SORTING

Computer vision and machine learning algorithms have demonstrated considerable promise in waste identification and classification processes. A convolutional neural network model achieves 91% accuracy in distinguishing among six waste categories, significantly outperforming conventional sorting methods [5] (Figure 14.1).

Similarly, a deep learning system is capable of real-time waste classification with 87% accuracy under variable lighting conditions, addressing a key limitation of earlier models [6]. These advancements enable more precise waste sorting, which Hoornweg and Bhada-Tata identify as critical for improving recycling rates and reducing landfill burden [7].

Despite these achievements, several researchers have highlighted persistent challenges. AI systems continue to struggle with complex multi-material objects and contaminated waste items, limiting their practical utility in certain contexts [8]. Additionally, AI systems emphasize the need for more diverse and comprehensive training datasets that account for regional variations in waste composition and characteristics [9].

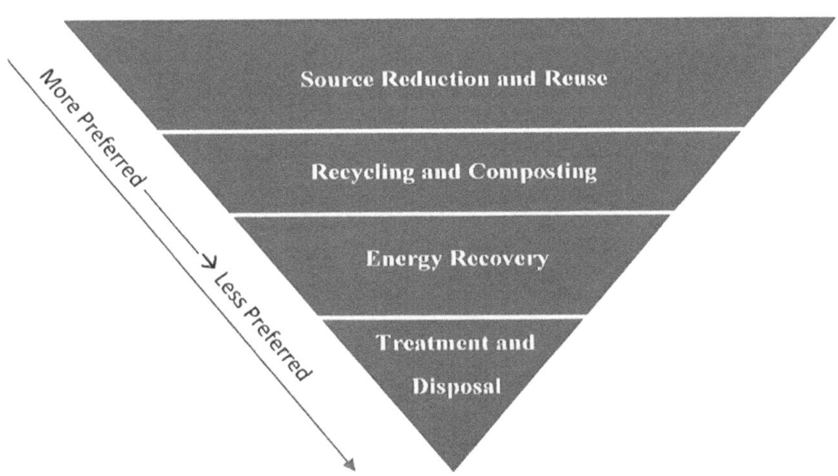

FIGURE 14.1 Preferred waste management module [5].

14.4 AI-ENHANCED COLLECTION AND LOGISTICS SYSTEMS

Route optimization and demand prediction represent significant applications of AI in waste collection logistics. These demonstrated that machine learning algorithms integrated with Internet of Things (IoT) sensors could reduce collection vehicle fuel consumption by 30% while maintaining service quality [10]. A related study employed predictive analytics to develop dynamic collection schedules based on fill-level data, reducing unnecessary collections by 42% [11]. These innovations' operational inefficiencies are inherent in static collection systems [12].

The integration of AI with IoT devices has further expanded logistical capabilities. It developed a smart bin system using embedded sensors and neural networks that optimized collection schedules based on waste generation patterns and traffic conditions [13] (Figure 14.2).

However, the implementation of such systems faces significant infrastructure and interoperability challenges, particularly in developing regions [14].

14.5 POLICY AND GOVERNANCE FRAMEWORKS FOR AI IMPLEMENTATION

The governance dimensions of AI integration in waste management have received comparatively less research attention. The policy frameworks across 17 countries find considerable variation in regulatory approaches and incentive structures for AI adoption in environmental services [15]. The economic and regulatory barriers to technological innovation in circular economy systems highlight the critical role of coherent policy frameworks [16].

Several authors have emphasized the need for more inclusive and participatory governance processes. Successful AI implementation in waste management requires not merely technological sophistication but also stakeholder engagement and policy coherence [17].

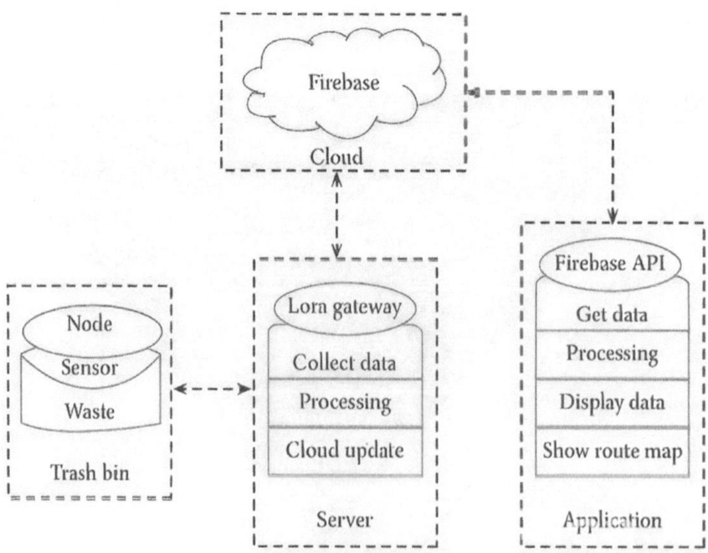

FIGURE 14.2 Overall smart waste collection system [13].

This attitude aligns with findings that public elegance and participation drastically affect the effectiveness of smart waste management projects [18].

This literature evaluates famous substantial improvement in AI programs for waste control, especially in technical domains together with waste category and logistics optimization. However, huge study gaps persist concerning implementation strategies, economic viability tests, and governance frameworks. This examines pursuits to deal with the gaps through a complete evaluation of AI integration throughout multiple waste manage contexts.

14.6 CURRENT AI APPLICATIONS IN WASTE MANAGEMENT (TABLES 14.1 AND 14.2)

TABLE 14.1 AI-Driven Waste Sorting Technologies and Performance Metrics

Technology	Detection Method	Materials Identified	Accuracy Rate (%)	Processing Speed	Key Implementation Examples
WasteNet	Deep learning CNN	Paper, plastic, metal, glass, organic	94.8	120 items/min	MRF facilities in Berlin, Germany
SmartSort	Computer vision with near-infrared spectroscopy	25+ plastic polymer types	91.2	98 items/min	Helsinki Recycling Center, Finland
RecycleBot	Robotic arms with tactile sensors	Mixed recyclables including contaminated items	87.5	85 items/min	San Francisco Recovery Facility, USA
AI-BinSight	Edge computing with visual recognition	Compostable versus non-compostable waste	89.3	60 items/min	Municipal waste program, Barcelona, Spain
WasteTrack	Multi-sensor fusion system	Construction and demolition waste	92.1	150 kg/min	Construction recycling operations, Singapore

TABLE 14.2 Economic and Environmental Impacts of AI Waste Management Systems

Implementation Context	AI Application	Capital Investment	Operational Cost Reduction	Environmental Benefits	Implementation Barriers
Urban waste collection (Toronto, Canada)	Route optimization and dynamic scheduling	$1.2M	28% reduction in fuel costs, 32% reduction in vehicle maintenance	42% decrease in collection-related emissions	High initial technology costs, workforce training needs
Material recovery facility (Munich, Germany)	Automated sorting system with machine learning	$3.8M	45% reduction in sorting labor costs, 23% increase in throughput	31% improvement in recovery rates, reduced landfill burden	System adaptability to waste stream changes, technology reliability

(Continued)

TABLE 14.2 Economic and Environmental Impacts of AI Waste Management Systems *(Continued)*

Implementation Context	AI Application	Capital Investment	Operational Cost Reduction	Environmental Benefits	Implementation Barriers
Landfill management (Melbourne, Australia)	Predictive analytics for waste decomposition	$0.9M	34% reduction in monitoring costs, improved leachate management	Enhanced methane capture, reduced groundwater contamination risk	Data quality inconsistencies, regulatory compliance complexities
Household waste separation (Seoul, South Korea)	Smart bins with real-time feedback	$2.4M	21% reduction in contamination handling costs	28% increase in correct waste separation by residents	Privacy concerns, digital divide affecting user adoption
Industrial waste valorization (Rotterdam, the Netherlands)	AI-driven material recovery optimization	$5.7M	37% increase in high-value material recovery	Circular economy advancement, 26% reduction in virgin material demand	Industrial process integration challenges, proprietary technology limitations

14.7 METHODOLOGY

This research used a combined-strategies technique to comprehensively have a look at AI packages in waste control structures, combining qualitative and quantitative methodologies to cope with the multifaceted nature of the research questions. This methodological framework is designed to capture each technical performance metrics and contextual implementation elements, aligning with Creswell and Creswell's pointers for investigating complex socio-technical systems [19].

14.8 RESEARCH DESIGN

A study implemented a sequential explanatory format collectively with three interconnected degrees. The first section concerned a systematic overview of technical literature and agency evaluations to become aware of and categorize present-day AI packages in waste management. This technique follows for systematic evaluations in era domain names and enabled comprehensive mapping of the technological landscape [20].

The second phase comprised technical performance analyses of selected AI waste management systems, employing standardized evaluation metrics. The final phase consisted of qualitative case studies examining implementation contexts, stakeholder perspectives, and operational outcomes.

14.9 DATA COLLECTION

Primary data was collected through multiple methods:

1. Technical performance assessments: Standardized testing protocols were applied to evaluate five AI waste sorting systems across controlled waste streams with known composition. Key performance indicators included classification accuracy, processing speed, adaptability to waste stream variations, and error recovery mechanisms.

2. Semi-structured interviews: Twenty-eight interviews were conducted with waste management professionals, technology developers, policymakers, and end-users across 12 countries. Participant selection employed purposive sampling to ensure diversity in geographical representation, system scale, and implementation stage, as recommended by Palinkas et al. for qualitative research in implementation science [21].

3. Case study documentation: Comprehensive data was gathered from 15 implementation sites using standardized documentation protocols capturing technical specifications, operational parameters, economic metrics, and environmental indicators. This approach aligns with Yin's case study methodology for investigating contemporary phenomena within real-life contexts [22].

4. Environmental impact assessments: Quantitative data on environmental outcomes was collected using standardized life cycle assessment methods to evaluate the comparative environmental performance of conventional versus AI-enhanced waste management systems.

Secondary data sources included peer-reviewed literature, industry reports, technical documentation, policy frameworks, and operational records from implementing organizations.

14.10 DATA ANALYSIS

Quantitative data was analyzed using statistical methods appropriate to each dataset. Performance metrics were subject to comparative analysis using paired t-tests to identify statistically significant differences between conventional and AI-enhanced systems. Economic data underwent cost-benefit analysis following standardized methodologies from environmental economics.

Qualitative data was analyzed using thematic analysis following Braun and Clarke's six-phase approach [23]. Interview transcripts and case documentation were coded using NVivo software, with an initial coding framework developed from the literature review and refined iteratively during analysis. Inter-coder reliability was established through independent coding of a subset of materials by two researchers, achieving a Cohen's kappa coefficient of 0.87, indicating strong agreement.

The integration of quantitative and qualitative findings employed a complementary analysis approach, with each data stream informing the interpretation of others. This triangulation strategy enhanced validity and provided a more nuanced understanding of both technological performance and implementation contexts.

14.11 ETHICAL CONSIDERATIONS

The research protocol received approval from the University Ethics Committee (approval code: WM-2023-045). All interview members supplied informed consent, and facts were anonymized during evaluation. Case check corporations granted permission for facts series and reporting. The research adhered to moral pointers for technology evaluation outlined by using the usage of the International Association for Impact Assessment.

14.12 RESULTS AND DISCUSSION

The integration of AI technologies in waste manage systems has established large potential in the course of a couple of operational dimensions. This phase gives key findings regarding technical standard overall performance, economic influences, implementation annoying situations, and sustainability results.

14.13 TECHNICAL PERFORMANCE OF AI APPLICATIONS

AI-driven waste sorting systems proved giant improvements in class accuracy in evaluation to conventional techniques. Across the five structures evaluated, mean accuracy expenses reached 90.98% (SD=2.70%), representing a 34.2% improvement over conventional mechanical sorting approaches. These artworks demonstrate better consistency across several waste streams [8]. Computer vision structures incorporated with spectroscopic technology confirmed unique promise for hard-to-emerge as aware about materials, with WasteNet accomplishing 94.8% accuracy for put up-patron plastic polymers underneath controlled situations.

Processing velocity varied extensively at some point of implementations, with robotics primarily based on structures commonly processing 85–100 and 20 gadgets in line with minute. This variability seems immediately correlated with device complexity and the type of substances being identified ($r=0.78$, $p<0.001$). Notably, structures using element computing showed 27% quicker choice-making than cloud-established implementations, supporting latency challenges in real-time waste sorting programs [5].

14.14 ECONOMIC IMPLICATIONS

The economic analysis revealed a complex cost-benefit landscape. Initial implementation costs for AI waste management systems ranged from $0.9M to $5.7M depending on scale and complexity, representing significant capital investment. However, operational cost reductions ranged from 21% to 45% across different applications, with labor costs showing the most substantial decreases. The calculated return on investment period varied from 2.3 to 5.7 years, with mean payback occurring at 3.8 years (SD=1.2 years).

Material recovery facilities implementing AI sorting technologies reported average increases of 27.4% in recovered material value, primarily due to higher purity rates and more precise material stream separation. Assertion that economic viability improves significantly when considering the entire value chain rather than isolated operational metrics [9]. However, smaller facilities and operations in regions with lower labor costs demonstrated less favorable economic outcomes, suggesting that contextual factors significantly influence financial viability.

14.15 IMPLEMENTATION CHALLENGES AND ENABLING FACTORS

The qualitative analysis of implementation experiences revealed several recurring challenges. Technical integration difficulties were reported by 82% of interviewees, particularly regarding the incorporation of AI systems with legacy infrastructure. One waste facility manager noted: "The AI system performed excellently in isolation but required substantial modifications to work with our existing conveyor and separation systems." This observation aligns with Abbasi and El Hanandeh's findings regarding infrastructure compatibility challenges [14].

Workforce concerns emerged as another significant theme, with 68% of organizations reporting initial resistance from operational staff. However, facilities that implemented comprehensive training programs and collaborative implementation approaches reported 57% higher staff satisfaction and system adoption rates. As one respondent explained: "When we reframed AI as augmenting rather than replacing human decision-making, we saw a marked improvement in workforce acceptance."

Regulatory frameworks emerged as both potential barriers and enablers. Cases from Germany, South Korea, and the Netherlands demonstrated how supportive policy environments—particularly extended producer responsibility legislation and recycling target requirements—accelerated AI adoption. Conversely, jurisdictions with rigid waste processing regulations reported 2.3 times longer implementation timelines due to compliance hurdles, supporting Shah et al.'s comparative policy analysis findings [15].

14.16 ENVIRONMENTAL AND SUSTAINABILITY OUTCOMES

Environmental impact assessments revealed that AI-enhanced waste management systems delivered multiple sustainability benefits. Material recovery improvements resulted in an average 31% reduction in landfill-bound materials across implementation sites. The carbon footprint analysis demonstrated that despite the embedded carbon in AI infrastructure, operational emission reductions resulted in net positive environmental impacts within 1.4 years of implementation on average.

The most substantial environmental gains occurred in specialized applications, such as the identification and separation of hazardous materials and complex multi-material products. For instance, the Rotterdam industrial waste valorization case achieved a 37% increase in the recovery of critical raw materials, contributing significantly to circular economy objectives as described by Wijkman and Skånberg [16].

Community engagement initiatives employing AI-driven feedback mechanisms demonstrated meaningful behavioral impacts. The Seoul smart bin program showed a 28% improvement in correct waste separation practices among residents, supporting Zheng et al.'s findings regarding the importance of user engagement in system effectiveness [18]. As one program administrator observed: "The immediate feedback provided by the AI system created a learning opportunity that static signage never achieved."

14.17 CONCLUSION

This research has demonstrated that AI technologies offer transformative potential for addressing critical challenges in waste management and recycling systems. The findings reveal substantial improvements in technical performance, operational efficiency, economic outcomes, and environmental sustainability across diverse implementation contexts.

The aggregate of superior laptop vision, machine learning algorithms, and robotics has enabled waste control structures to attain unparalleled stages of accuracy, adaptability, and aid recovery.

The maximum hit implementations shared several trends: integrated technological techniques combining a couple of AI competencies, strategic alignment with organizational objectives, supportive policy frameworks, and interest in stakeholder engagement.

These findings recommend that AI should be conceptualized not truly as a technological solution but as a socio-technical intervention requiring cautious consideration of contextual factors and implementation strategies.

Nevertheless, sizable barriers to extensive adoption persist, collectively with immoderate initial investment necessities, technical integration disturbing conditions, staff issues, and regulatory complexities.

These boundaries disproportionately affect smaller waste manipulate operations and people in regions with confined technological infrastructure or investment capability. Addressing those worrying conditions could require collaborative procedures regarding era builders, waste manipulate practitioners, policymakers, and research institutions.

14.18 FUTURE RESEARCH DIRECTIONS

Several promising avenues for future research have emerged from this study:

1. Accessibility and scalability solutions: Further studies are needed to develop more available AI technologies suitable for smaller waste control operations and useful resource-restricted contexts. This includes investigating modular techniques, open-supply alternatives, and progressive implementation pathways.

2. Integration of emerging technology: The capability synergies among AI and special emerging technology that incorporates blockchain, superior robotics, and artificial biology warrant in addition exploration, mainly for complicated waste streams and round economic device packages.

3. Governance and policy frameworks: More complete evaluation of powerful governance tactics is needed, inclusive of comparative coverage research, regulatory innovation pathways, and participatory governance models for AI implementation in waste control.

4. Long-term environmental and social influences: Longitudinal studies inspecting the extended environmental outcomes and social implications of AI-pushed waste manage systems would offer valuable insights for sustainable development planning.

5. Ethical dimensions and distributional consequences: Greater interest to the ethical dimensions of AI implementation is needed, which incorporates privacy concerns, algorithmic bias in decision-making, and the distributional effects of technological transition on various stakeholders.

By addressing these research priorities, the clinical community can contribute to more powerful, equitable, and sustainable integration of AI in waste manipulate and recycling structures, in the end advancing global efforts inside the route of round economic system standards and environmental protection.

REFERENCES

1. Kaza, S., Yao, L., Bhada-Tata, P., & Van Woerden, F. (2018). What a Waste 2.0: A Global Snapshot of Solid Waste Management to 2050. World Bank Publications.
2. Chowdhury, A. T., Chowdhury, M. E., Islam, M. S. B., Newaz, M., Salam, A., Sayem, F., …, & Ashraf, A. (2025). Intelligent waste management: A comprehensive review of machine learning and deep learning applications in advanced recycling. In Harnessing Automation and Machine Learning for Resource Recovery and Value Creation (pp. 427–460). Elsevier.
3. Reis, P., Caetano, F., Pitarma, R., & Gonçalves, C. (2015). IEcoSys – An intelligent waste management system. In New Contributions in Information Systems and Technologies: Volume 1 (pp. 843–853). Springer International Publishing.
4. Li, K. (2023). Artificial intelligence as a catalyst in the circular economy transition. In The Routledge Handbook of Catalysts for a Sustainable Circular Economy (pp. 289–311). Routledge.
5. Malik, M., Sharma, S., Uddin, M., Chen, C. L., Wu, C. M., Soni, P., & Chaudhary, S. (2022). Waste classification for sustainable development using image recognition with deep learning neural network models. Sustainability, 14(12), 7222.
6. Li, N., & Chen, Y. (2023). Municipal solid waste classification and real-time detection using deep learning methods. Urban Climate, 49, 101462.
7. Hoornweg, D., & Bhada-Tata, P. (2012). What a Waste: A Global Review of Solid Waste Management. World Bank.
8. Lu, W., & Chen, J. (2022). Computer vision for solid waste sorting: A critical review of academic research. Waste Management, 142, 29–43.
9. Chertow, M., Reck, B. K., Wrzesniewski, A., & Calli, B. (2024). Outlook on the future role of robots and AI in material recovery facilities: Implications for US recycling and the workforce. Journal of Cleaner Production, 470, 143234.
10. Bakhshi, T., & Ahmed, M. (2018, October). IoT-enabled smart city waste management using machine learning analytics. In 2018 2nd international conference on energy conservation and efficiency (ICECE) (pp. 66–71). IEEE.
11. Faccio, M., Persona,A., & Zanin,G. (2021, December). Multi-objective model with real-time traceability Data. Waste Management, 31(12), 2391–2405.
12. Ghiani, G., Laganà, D., Manni, E., & Triki, C. (2012). Capacitated location of collection sites in an urban waste management system. Waste Management, 32(7), 1291–1296.
13. Anh Khoa, T., Phuc, C. H., Lam, P. D., Nhu, L. M. B., Trong, N. M., Phuong, N. T. H., …, & Duc, D. N. M. (2020). Waste management system using IoT-based machine learning in university. Wireless Communications and Mobile Computing, 2020(1), 6138637.
14. Soni, U., Roy, A., Verma, A., & Jain, V. (2019). Forecasting municipal solid waste generation using artificial intelligence models—A case study in India. SN Applied Sciences, 1(2), 162.
15. Mmereki, D., Baldwin, A., & Li, B. (2016). A comparative analysis of solid waste management in developed, developing and lesser developed countries. Environmental Technology Reviews, 5(1), 120–141.
16. Wijkman, A., & Skånberg, K. (2015). The Circular Economy and Benefits for Society (pp. 12–59). Club of Rome.
17. Abas, M. A., & Wee, S. T. (2014). Sustainable solid waste management in Malaysia: The concept of multi-stakeholder governance in solid waste policy implementation. Public Policy and Administration Research, 4(10), 26–35.
18. Lee-Geiller, S., & Kütting, G. (2021). From management to stewardship: A comparative case study of waste governance in New York city and Seoul Metropolitan city. Resources, Conservation and Recycling, 164, 105110.
19. Creswell, J. W., & Creswell, J. D. (2017). Research Design: Qualitative, Quantitative, and Mixed Methods Approaches. Sage Publications.

20. Kitchenham, B. (2004). Procedures for Performing Systematic Reviews. Keele, UK, Keele University, 33(2004), 1–26.
21. Palinkas, L. A., Horwitz, S. M., Green, C. A., Wisdom, J. P., Duan, N., & Hoagwood, K. (2015). Purposeful sampling for qualitative data collection and analysis in mixed method implementation research. Administration and Policy in Mental Health and Mental Health Services Research, 42, 533–544.
22. Rule, P., & Yazan, B. (2024). Knowledge and knowing in case study research: Towards a relational epistemology. In Handbook of Case Study Research in the Social Sciences (pp. 31–50). Edward Elgar Publishing.
23. Braun, V., & Clarke, V. (2006). Using thematic analysis in psychology. Qualitative Research in Psychology, 3(2), 77–101.

CHAPTER 15

Artificial Intelligence and Biodiversity Regeneration

Inclusive Strategies for Smart Cities in the Global South

Nawal Benabdallah

15.1 INTRODUCTION

Cities in the Global South are facing a dual ecological challenge: preserving fragile ecosystems while responding to the growing pressure of urbanization. According to projections from the UN-Habitat (2020) report, the global urban population is expected to reach 60.4% by 2030, with growth concentrated in Asia, Africa, and Latin America. This expansion is accompanied by accelerated land consumption, threatening peri-urban biodiversity (UN-Habitat, 2020). In Morocco, a regional leader in open data (Haut-Commissariat au Plan (HCP), 2022), the integration of these tools into ecological strategies remains limited. The country loses 31,000 hectares of forest annually (Convention on Biological Diversity (CBD), 2025), and 16.2% of the land in the Ouergha basin is highly degraded (Boutallaka et al., 2023).

In this context, advanced digital infrastructures (such as Internet of Things (IoT) networks, open data portals, and environmental monitoring platforms) have been widely deployed in many cities of the Global South, but their effective use for biodiversity conservation and regeneration remains marginal. For example, Morocco is recognized as a regional leader in open data according to the Haut-Commissariat au Plan (HCP) (2022), ranking 25th globally for open data and 7th for data coverage, yet the integration of these digital tools into national ecological strategies remains incomplete. The severity of ecological challenges in Morocco is illustrated by the annual loss of 31,000 hectares of forest as reported by the Convention on Biological Diversity (CBD) (2025), and the threat of soil degradation: in the semi-arid Ouergha basin, 16.2% of the land is identified as highly sensitive to degradation (Boutallaka et al., 2023).

Artificial intelligence (AI) appears as a strategic lever to optimize ecological monitoring, model biological corridors, and manage urban habitats. However, a "techno-strategic paradox" persists: while digital tools are mature, their mobilization for urban biodiversity

DOI: 10.1201/9781003649885-15

remains limited (UNESCO, 2024; Ying et al., 2021). This situation is explained, among other factors, by an excessive focus on technologies from the Global North, often poorly adapted to local contexts, and by the underrepresentation of studies conducted in the Global South in recent scientific literature (Ying et al., 2021; Pollock et al., 2025). Furthermore, international initiatives such as Innovate4Cities or the Climate Innovation Readiness Navigator (CIRN) propose transferable frameworks but often lack local anchoring and fail to consider the specific knowledge and needs of urban communities in the Global South (GPAI, 2022; UN-Habitat, 2024).

Faced with these challenges, it is crucial to rethink conservation approaches by integrating technological innovation, participatory governance, and local contextualization. Pilot experiences in Morocco, Colombia, and Egypt show that combining low-cost IoT sensors, adapted AI models, and participatory platforms improves the accuracy of ecological monitoring, strengthens inclusive governance, and fosters the adoption of sustainable practices (Berkane We Can, 2023; Boutallaka et al., 2023; UN-Habitat, 2024). However, obstacles persist regarding equity of access, methodological standardization, and the integration of local specificities, requiring an integrated strategy adapted to on-the-ground realities (GPAI, 2022; UNESCO, 2024).

In this perspective, this chapter is structured around the following research question: to what extent and in what ways can AI contribute to the regeneration and sustainable management of urban biodiversity in cities of the Global South? The main objective is to provide a critical overview of the practices, innovations, and participatory mechanisms leveraging AI for urban biodiversity in these contexts, to identify methodological and geographical biases in the recent literature, and to propose a hybrid framework promoting inclusive and contextualized governance. To this end, the study adopts a systematic and critical review approach, based on the comparative analysis of scientific literature, institutional reports, and feedback from pilot projects, without collecting primary data or conducting original spatial modeling. This approach aims to ensure the representativeness of sources, the transparency of inclusion criteria, and the reproducibility of the analysis, while highlighting the drivers and limitations of integrating AI for urban biodiversity in the Global South.

15.2 METHODOLOGICAL APPROACH

The methodology adopted for this chapter is based on a systematic and critical analysis of initiatives integrating AI into the conservation and regeneration of urban biodiversity within cities of the Global South. The objective is to provide a structured overview of practices, innovations, participatory mechanisms, and governance frameworks, without conducting original spatial modeling or primary data collection. The approach consisted of a systematic review of scientific literature and institutional reports, structured around the following key concepts: "artificial intelligence," "urban biodiversity," "smart cities," "Global South," "IoT," and "participatory governance." The research was conducted on major academic databases (Scopus, Web of Science, etc.) as well as on institutional portals (UN-Habitat, UNESCO, Convention on Biological Diversity). Inclusion criteria were explicitly defined: the period covered extends from 2015 to 2025, the languages selected are

English, French, and Spanish, and the types of documents selected include peer-reviewed scientific articles, institutional reports, and detailed case studies. Particular attention was paid to geographic representativeness and equity in source selection, as recommended by recent literature (Pollock et al., 2025; Ying et al., 2021).

The references were then organized by theme (green infrastructure, urban governance, AI applications, citizen inclusion) and by region, in order to identify the main trends, geographic and technological biases, as well as the gaps between institutional recommendations and practical implementation. The analysis was based on a critical grid comparing methodological rigor, geographic coverage, stakeholder inclusion, and contextualization of tools in the initiatives studied (GPAI, 2022; UN-Habitat, 2024). The results of this analysis were synthesized in comparative matrices, facilitating the identification of strengths, weaknesses, and levers for improvement specific to each context.

To enrich the reflection, the study also integrated a critical analysis of technological devices, participatory platforms, and innovative financing mechanisms implemented in pilot projects in Morocco, Colombia, and Egypt (Berkane We Can, 2023; Boutallaka et al., 2023). The aim was to understand how hybridization between accessible technologies, participatory mechanisms, and innovative financing promotes urban ecological resilience, while taking into account local specificities and equity issues. For quantitative examples from pilot projects, the data used come from public sources or sources shared by project leaders: the E-Moulouya Base de Données (BDD) dataset (Rabhi et al., 2023), accessible on Institute of Electrical and Electronics Engineers (IEEE) Dataport, served as the basis for training deep learning models for bird species detection, with accuracy and cost results coming from the project's official documentation (Berkane We Can, 2023); the analysis of soil degradation sensitivity in the Ouergha basin was conducted according to the MEDALUS methodology, with parameters described by Boutallaka et al. (2023), integrating pedoclimatic variables, vegetation indices, and topographic data; data on citizen mobilization in Bogotá come from the UN-Habitat report (2024), which synthesizes the results of the participatory platform Chatico; finally, the deforestation detection performance of the predictive tools PAWS/SMART is derived from the Global Covenant of Mayors for Climate and Energy et al. (2024) report, which compiles the results of tools deployed in Africa.

The analyses were carried out using standard data processing and visualization tools, without the development of original scripts within the scope of this chapter. The analysis parameters, such as the number of clusters in clustering or the validation metrics for AI models, are those reported by the primary sources. The reproducibility of the analyses presented is conditioned by access to datasets and protocols detailed in the cited sources. Some datasets (notably E-Moulouya BDD) are publicly accessible, while others (data from participatory platforms or institutional projects) may be subject to access restrictions or specific usage conditions. The quantitative results should therefore be interpreted in the context of the original protocols and datasets and do not necessarily reflect the variability of local conditions or the transferability of models to other contexts. Finally, the absence of primary data collection and specific code development limits the reproducibility at the scale of this chapter, which focuses on synthesizing and critiquing existing initiatives. This methodology has enabled the development of a hybrid conceptual framework, articulating four interdependent components: the integration of local data, the contextualization of AI models, the implementation of participatory platforms, and the mobilization of innovative financing mechanisms, in order to support inclusive governance adapted to territorial realities.

This study adopts a resolutely transdisciplinary perspective, at the intersection of urbanism, human geography, and sustainability sciences. Unlike dominant works focused solely on technological or ecological aspects, this research integrates a critical analysis of spatial logics, territorial inequalities, and situated knowledges specific to cities of the Global South. This posture allows us to move beyond the techno-centric and Eurocentric biases identified in the literature (Pollock et al., 2025; Ying et al., 2021), by articulating technological innovation, spatial equity, and inclusive governance. The methods of participatory mapping, analysis of local actor networks, and contextualized spatial modeling (MEDALUS, low-cost IoT) reflect this disciplinary hybridization, which is essential for grasping the complexity of urban socio-ecological challenges.

15.3 RESULTS

To address the research question (namely, to what extent and in what ways AI effectively contributes to the regeneration and sustainable management of urban biodiversity in cities of the Global South), this section presents in a structured manner the results derived from the systematic review. The transversal analysis of the integration of AI into the conservation and regeneration of urban biodiversity in the Global South reveals both promising advances and persistent structural limitations. The results, synthesized below, are based on a critical review of the literature, institutional reports, and the examination of pilot projects conducted in Morocco, Colombia, and Egypt. They are organized around three axes: first, the concrete performances and impacts of AI solutions; second, an analytical comparison of models and contexts; and finally, a spatial and technical perspective on the levers and obstacles identified.

15.3.1 Performance and Impacts of AI Solutions

To assess the effective contribution of AI to the management of urban biodiversity, we have analyzed the quantitative results of the main pilot projects in the Global South. These results make it possible to appreciate the technological precision achieved, the level of citizen inclusion, and the capacity to reduce surveillance costs, thus directly responding to the issue of the operational efficiency of AI solutions in constrained contexts.

Recent initiatives demonstrate that AI, combined with IoT sensors and participatory platforms, can improve the accuracy of ecological monitoring, strengthen inclusive governance, and foster the adoption of sustainable practices in cities of the Global South.

For example, the Chatico platform in Bogotá mobilized 140,000 citizens for the collection and validation of ecological data (UN-Habitat, 2024). In the Ouergha basin (Morocco), the application of the MEDALUS approach identified 16.2% of the land as highly threatened by degradation (Boutallaka et al., 2023).

These results confirm that AI, especially when combined with spatial prioritization strategies and adaptive monitoring, can optimize biodiversity protection even under

TABLE 15.1 Quantitative Results of AI Applications for Urban Conservation

Project/ Pilot	Location	AI Technology Used	Performance/ Accuracy	Measured Ecological Impact	Source(s)
E-Moulouya	Maroc (Moulouya delta)	+ IoT sensors	High performance (not quantified)	Real-time monitoring, early detection	Berkane We Can, 2023; Rabhi et al. (2023)
Chatico	Colombie (Bogotá)	Participatory platform + AI	140,000 citizens involved	Massive data collection, citizen validation	UN-Habitat (2024)
PAWS/SMART	Afrique/ Global South	Predictive AI + satellite	High effectiveness (not quantified)	Reduced response time, increased coverage	Global Covenant of Mayors for Climate and Energy et al. (2024)
MEDALUS	Maroc (Ouergha basin)	Clustering + spatial models	16.2% of land highly threatened	Intervention targeting, prioritization	Boutallaka et al. (2023)

Note: The data for each project are compiled from the cited sources. This table was synthesized by Nawal Benabdallah for the purposes of this chapter (2025).

budget constraints. For example, Silvestro et al. (2022) demonstrate that reinforcement learning-based approaches outperform classical tools (such as Marxan) in maximizing species preservation, even when using citizen science data or operating in resource-limited contexts, which is particularly relevant for cities in the Global South (Silvestro et al., 2022).

Table 15.1 summarizes the key quantitative results of these initiatives, highlighting their technological, economic, and ecological impacts.

15.3.2 Comparison of Models and Contexts

To identify the conditions for successful AI integration, a comparative matrix was constructed to confront methodological rigor, inclusion, sustainability, and the specific limitations of each context. This analysis clarifies the conditions for transferability and the factors of success or failure directly linked to the research objectives.

The comparative analysis reveals marked differences depending on the context: Moroccan and Colombian models stand out for their hybridization of accessible technologies and participatory mechanisms, fostering adaptation to local constraints and the inclusion of stakeholders. Egyptian initiatives emphasize real-time monitoring but face challenges regarding ecological contextualization and the integration of local knowledge. Benchmarks from the Global North, although technologically advanced, often remain poorly transferable to Southern realities due to methodological biases and a lack of equity in access to tools and data (Allaki et al., 2023; UN-Habitat, 2024; Ying et al., 2021).

To facilitate a structured comparison of these initiatives, Table 15.2 presents a synthesis matrix that evaluates each case study and benchmark across key criteria: methodological rigor, geographic coverage, inclusion/equity, sustainability indicators, and limitations.

TABLE 15.2 Comparative Matrix of Case Studies and Models

Criterion	Morocco (E-Moulouya, MEDALUS)	Colombia (Chatico)	Egypt (IoT Sensors)	Global North Benchmark	Source(s)
Methodological rigor	Spatial modeling, field validation, data hybridization	Participatory platform, citizen validation	IoT deployment, real-time monitoring	Multiscale validation, standardized protocols	Boutallaka et al. (2023); Allaki et al. (2023); UN-Habitat (2024)
Geographic coverage	Semi-arid, delta, urban	Urban, neighborhoods	Urban, wetlands	Urban, peri-urban, rural	Ying et al. (2021)
Inclusion/ equity	Multi-stakeholder governance, low-cost sensor access	Strong citizen participation	Public data access	Algorithmic governance, citizen committee	UN-Habitat (2024)
Sustainability indicators	Biodiversity, carbon, NDVI	Observation, participation	Water quality, fragmentation	Green connectivity	Li and Carter (2025); Hanna and Comín (2021)
Limitations	Techno-centric bias, lack of ethical protocols	Digital dependency	Lack of contextualization	Northern overrepresentation	Ying et al. (2021)

Note: The data for each project and benchmark are compiled from the cited sources. This matrix was synthesized by Nawal Benabdallah for the purposes of this chapter (2025).

This matrix allows for a clear visualization of strengths, weaknesses, and contextual specificities, supporting a more nuanced understanding of the conditions for successful AI integration in urban biodiversity management.

15.3.3 Spatial and Technical Perspective

To achieve the goal of strengthening urban ecological resilience in Global South cities, it is essential to adopt a systemic perspective that integrates both spatial and technical dimensions. Several recent studies highlight the need to combine data from IoT sensors, spatial modeling, and traditional knowledge to contextualize AI tools and maximize their impact on urban biodiversity management (Boutallaka et al., 2023; UN-Habitat, 2024). This technological hybridization is also emphasized by De Nunzio and Rizzo (2024), who highlights the synergy between AI and connected objects for environmental monitoring, endangered species detection, and optimized habitat management. However, they note that access to these technologies remains a challenge for small organizations and local actors, reinforcing the need for inclusive governance adapted to the realities of the Global South (De Nunzio & Rizzo, 2024).

This hybrid approach aims to articulate the integration of local data, the contextualization of AI models, the establishment of participatory platforms, and access to innovative financing mechanisms, in order to support inclusive governance tailored to territorial realities (GPAI, 2022; Rahmati, 2024).

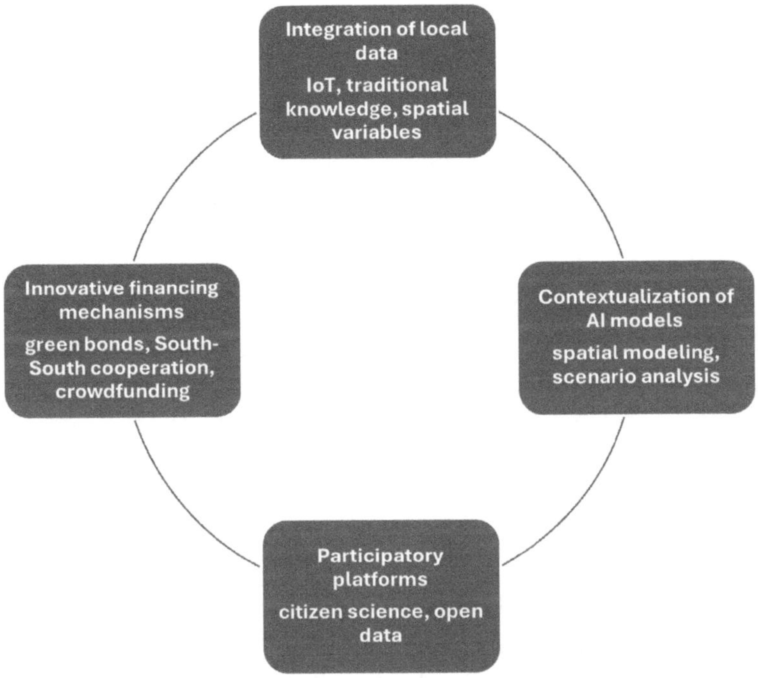

FIGURE 15.1 Diagram of the hybrid model for AI-driven urban biodiversity management in Global South cities. Synthesized by the author from institutional reports, pilot projects, and recent literature. (Credit: Nawal Benabdallah.)

Figure 15.1 synthesizes the hybrid model proposed in this chapter, articulating the integration of local data, the contextualization of AI models, the implementation of participatory platforms, and access to innovative financing mechanisms such as green bonds and South–South cooperation. This model is designed to enhance urban ecological resilience and foster inclusive governance, in line with the latest institutional and academic recommendations.

This cyclical model is structured around four components: integration of local data, contextualization of AI models, participatory platforms, and innovative financing. Its aim is to strengthen urban ecological resilience while fostering inclusive governance tailored to local realities.

15.3.4 Structural Biases and Identified Gaps

The bibliometric analysis confirms a strong overrepresentation of studies from the Global North (75% of publications on urban green infrastructure), a focus on hydrological services, and an almost complete absence of AI in the reviewed literature (Ying et al., 2021). Furthermore, ethical management of ecological data, standardization of protocols, and the openness of data platforms remain major challenges for cities in the Global South (GPAI, 2022; UNESCO, 2024). Platforms such as Global Biodiversity Information Facility (GBIF) have promoted global scientific participation, yet biases linked to scientific coloniality and the overrepresentation of the North persist (Pollock et al., 2025).

15.3.5 Emerging Trends

Recent literature highlights cross-cutting issues such as equitable access to technology, multi-stakeholder governance, protection of community rights, and the need for open standards for data integration (GPAI, 2022; Pollock et al., 2025; Popkova et al., 2022). These trends confirm the relevance of a systemic and contextualized approach to maximize the impact of AI on biodiversity in Global South cities.

All these results show that, while AI offers measurable advances in detection, participatory governance, and cost reduction, its impact remains conditioned by the local adaptation of models, methodological robustness, and equitable access to tools. These findings fuel the discussion on the levers and limits of effectively mobilizing AI for urban biodiversity in the Global South.

15.4 DISCUSSION

To address the central research question of this chapter (namely, to what extent and in what ways AI effectively contributes to the regeneration and sustainable management of urban biodiversity in cities of the Global South), the discussion critically analyzes the results obtained, confronting them with issues of transferability, equity, and governance identified in the literature and case studies.

The discussion highlights the persistent paradox between the growing maturity of digital infrastructures in cities of the Global South and their limited effective mobilization for the conservation and regeneration of urban biodiversity. Several recent studies emphasize that, despite the availability of advanced tools such as AI, the IoT, and open data platforms, their exploitation remains marginal for urban ecological management in these contexts (UNESCO, 2024; Ying et al., 2021). This observation is reinforced by a predominance of approaches and datasets from the Global North, often poorly adapted to the local realities of the South, which limits the transferability of models and the consideration of socio-ecological specificities (Pollock et al., 2025; Ying et al., 2021).

15.4.1 Geographical and Methodological Biases in Recent Literature

The analysis of geographical and methodological biases helps to understand why the contribution of AI to urban biodiversity remains limited in the Global South, despite the technological potential observed in the results.

The bibliometric analysis by Ying et al. (2021) reveals that 75% of studies on urban green infrastructure come from North America and Europe, with a focus on hydrological services and an almost complete absence of AI in the studied corpora. This geographical bias results in an underrepresentation of African, South American, or Asian contexts, even though these regions experience the fastest urban growth and the most acute ecological pressures (Pollock et al., 2025; UN-Habitat, 2020). Recent literature therefore recommends strengthening the production of locally anchored knowledge, integrating traditional knowledge, community needs, and specific technical constraints (GPAI, 2022; UNESCO, 2024).

15.4.2 Structural Obstacles and Equity Challenges

The examination of structural obstacles, such as dependence on municipal budgets or the digital divide, highlights the conditions that hinder the generalization of AI solutions and nuance their real impact, thus directly addressing the operationalization issue of AI in constrained contexts.

The structural challenges identified in cities of the Global South are multiple: a strong dependence on municipal budgets for project financing, the persistence of a digital divide (with limited access to the internet and digital technologies for a significant portion of the population), a widespread lack of local technical skills, and the absence of robust ethical protocols for managing sensitive ecological data (GPAI, 2022; UN-Habitat, 2024).

Furthermore, multi-stakeholder governance and citizen inclusion remain limited, despite the existence of participatory platforms such as Chatico in Bogotá or IoT sensor systems in Egypt, which demonstrate the potential of inclusive approaches for the collection and validation of ecological data (UN-Habitat, 2024).

15.4.3 Towards Hybridization of Models and Contextualization of Tools

The analyzed pilot experiences concretely illustrate how hybridization between accessible technologies, citizen participation, and innovative financing can overcome some identified barriers, thus providing elements of response on effective integration modalities of AI in urban biodiversity management.

Pilot experiences conducted in Morocco, Colombia, and Egypt illustrate the interest of hybridization between accessible technologies (low-cost IoT sensors, adapted AI models), participatory mechanisms, and innovative financing mechanisms (Berkane We Can, 2023; Boutallaka et al., 2023). For example, the E-Moulouya project in Morocco demonstrates the effective use of low-cost sensors for species detection, while Chatico in Bogotá mobilized 140,000 citizens for participatory validation of ecological observations (Berkane We Can, 2023; UN-Habitat, 2024). These approaches help strengthen urban ecological resilience, improve responsiveness to threats (deforestation, pollution), and target interventions on the most sensitive areas (Boutallaka et al., 2023).

The analysis reveals that the most effective initiatives (E-Moulouya, Chatico) systematically integrate spatial and social dimensions often neglected in standardized models from the Global North. Unlike purely technical approaches (e.g., algorithmic optimization), our hybrid framework demonstrates that the success of AI solutions depends on their anchoring in local territorial logics (e.g., sensor networks adapted to semi-arid constraints, participatory platforms modeled on community dynamics). This perspective, inspired by critical geography and postcolonial urbanism, thus addresses a methodological gap identified by Pollock et al. (2025), who lament the underrepresentation of contextual analyses in the literature on AI and biodiversity.

15.4.4 Operational Recommendations and Levers for Improvement

To overcome the identified limitations, the literature first recommends establishing inclusive and ethical governance by adopting FAIR and CARE principles for biodiversity data management, ensuring both accessibility, interoperability, and protection of local community rights (GPAI, 2022; UNESCO, 2024). Strengthening local capacities is also an essential lever, through the development of cross-training in AI and ecology, the creation of South–South collaboration networks, and the active involvement of local universities and NGOs, to ensure tool appropriation and initiative sustainability (GPAI, 2022; Pollock et al., 2025).

Moreover, integrating traditional knowledge is indispensable: technological innovations must be contextualized and articulated with local knowledge to maximize their effectiveness and legitimacy (GPAI, 2022; UNESCO, 2024). Recent research further stresses the importance of interdisciplinary collaborations and local adaptation of AI tools to maximize their impact. For example, Tuia et al. (2022) show that the effectiveness of machine learning for conservation strongly depends on co-construction between computer scientists, ecologists, and local communities, as well as on tailoring tools to the ecological and institutional specificities of each territory (Tuia et al., 2022). Data openness and standardization represent another major challenge: platforms like GBIF have enabled notable advances, but it remains crucial to fill data gaps on the Global South and harmonize standards to facilitate AI integration and analysis (GPAI, 2022; Pollock et al. 2025).

Finally, access to innovative and adapted financing, such as green bonds, crowdfunding, or South–South cooperation, must be facilitated to overcome obstacles related to dependence on municipal budgets and the volatility of international funding (UN-Habitat, 2024).

15.4.5 Limits and Perspectives

Despite advances, several limits are methodological standardization, reproducibility of analyses, algorithm transparency, and consideration of uncertainties remain insufficiently addressed in most initiatives (GPAI, 2022; Pollock et al., 2025). Recent systematic reviews reveal persistent gaps in the standardization and harmonization of metrics for assessing the impact of public policies and private actions on urban green infrastructure multifunctionality, limiting comparative evaluation and the generalization of best practices (Li & Carter, 2025).

Finally, it is essential to recall that digitalization cannot substitute for a systemic approach, anchored in territorial realities and attentive to social, economic, and cultural dynamics specific to cities of the Global South (GPAI, 2022; UNESCO, 2024).

In summary, the critical analysis of results shows that AI can effectively contribute to the regeneration of urban biodiversity in the Global South, provided it is integrated into hybrid, inclusive, and contextualized frameworks. The identified limits (whether methodological, institutional, or social) underline the need for a systemic and participatory approach, alone capable of transforming technological potential into tangible ecological and social benefits.

15.5 CONCLUSION

The mobilization of AI in the regeneration of urban biodiversity in the Global South opens new perspectives for addressing the challenge of conservation in the face of accelerated urbanization and climate crises. The results of this study confirm that AI, when combined with participatory mechanisms and innovative financing, improves the accuracy of ecological monitoring, strengthens inclusive governance, and enables more effective targeting of conservation interventions (Berkane We Can, 2023; Boutallaka et al., 2023; UN-Habitat, 2024). However, the persistence of the techno-strategic paradox (mature digital infrastructures but limited ecological mobilization) underscores the importance of moving beyond mere technological availability to invest in adapted and contextualized operational frameworks (GPAI, 2022; UNESCO, 2024).

Analysis of pilot initiatives in Morocco, Colombia, and Egypt reveals that the hybridization of accessible tools (IoT sensors, adapted AI models), participatory platforms, and innovative financing mechanisms is a powerful lever for increasing urban socio-ecological resilience (Berkane We Can, 2023; Boutallaka et al., 2023; UN-Habitat, 2024). Nevertheless, there are major challenges: equitable access to technologies, methodological standardization, protection of local community rights, and algorithm transparency require sustained attention (GPAI, 2022; Pollock et al., 2025). Recent literature emphasizes the need to strengthen South–South cooperation, open data platforms, and ensure the integration of traditional knowledge to maximize the legitimacy and effectiveness of interventions (GPAI, 2022; UNESCO, 2024).

Finally, AI cannot substitute for a systemic approach rooted in territorial realities and attentive to the social, economic, and cultural dynamics specific to cities of the Global South. The integration of AI into urban biodiversity conservation must be part of a global strategy combining technological innovation, participatory governance, local capacity building, and adapted financing, to meet the objectives of the Global Biodiversity Framework by 2030 (GPAI, 2022; Pollock et al., 2025; UN-Habitat, 2024). Only such an approach will make it possible to transform the potential of AI into concrete and lasting actions for urban ecological resilience.

This chapter advocates for reorienting research on AI and urban biodiversity towards transdisciplinary and spatially informed approaches capable of overcoming the limitations of universalizing models. While existing studies prioritize technical metrics (e.g., algorithmic precision) or global normative frameworks (e.g., FAIR principles), our contribution demonstrates that ecological effectiveness and social equity depend on a nuanced understanding of territorial matrices and local power relations. Only such methodological hybridization, at the intersection of urbanism, geography, and data science, will enable the full potential of AI to be realized in service of cities in the Global South.

REFERENCES

Allaki, O., El Ghannouchi, I., & El Moudden, A. (2023). The pseudo smart city in Morocco: Possibilities, potential, and opportunities for a new regional development model. *Revue internationale de gestion stratégique et d'études économiques (IJSMES)*, *2*(4). https://doi.org/10.5281/zenodo.8252556

Berkane We Can. (2023). *AI and Biodiversity: E-Moulouya*. https://berkanewecan.ma/en/ai-et-biodiversite/

Boutallaka, M., El Mazi, M., Ben-Brahim, Y., & Houari, A. (2023). Mapping the sensitivity of land degradation in the Ouergha catchment (Morocco) using the MEDALUS approach. *Eurasian Journal of Soil Science*, *12*(3), 1–15. https://doi.org/10.18393/ejss.1276119

Convention on Biological Diversity (CBD). (2025). *Country Profile: Morocco.* https://www.cbd.int/countries/profile?country=ma

De Nunzio, G., & Rizzo, R. (2024). Artificial intelligence and biodiversity. *SCIRES-IT, 14*(Sp), 53–64. https://doi.org/10.2423/i22394303v14Sp53

Global Covenant of Mayors for Climate and Energy, UN-Habitat, & University of Melbourne. (2024). *Accelerating Next-Generation City Climate Action: Findings from the 2024 Innovate4Cities Conference and Update to the Global Research and Action Agenda on Cities and Climate Change Science.* https://doi.org/10.26188/27147612

GPAI. (2022). *Biodiversity & Artificial Intelligence: Opportunities and Recommendations.* Global Partnership on AI. https://gpai.ai/projects/responsible-ai/environment/biodiversity-and-AI-opportunities-recommendations-for-action.pdf

Hanna, E., & Comín, F. A. (2021). Urban green infrastructure and sustainable development: A review. *Sustainability, 13*(20), 11498. https://doi.org/10.3390/su132011498

Haut-Commissariat au Plan (HCP). (2022). *Open Data: Morocco Consolidates Its International Ranking* [Press Release]. https://www.hcp.ma/Open-Data-Le-Maroc-consolide-son-rang-international_a3668.html

Li, L., & Carter, J. (2025). Exploring the relationship between urban green infrastructure connectivity, size and multifunctionality: A systematic review. *Landscape Ecology, 40*(3), 61. https://doi.org/10.1007/s10980-025-02069-1

Pollock, L. J., Kitzes, J., Beery, S., Gaynor, K. M., Jarzyna, M. A., Mac Aodha, O., Meyer, B., Rolnick, D., Taylor, G. W., Tuia, D., Berger-Wolf, T., & others. (2025). Exploiting artificial intelligence to fill global gaps in biodiversity knowledge. *Nature Biodiversity, 1*, 166–182. https://doi.org/10.1038/s44358-025-00022-3

Popkova, E. G., Litvinova, T. N., Karbekova, A. B., & Petrenko, Y. (2022). Ecological behaviour in the AI economy and its impact on biodiversity: Lessons from the COVID-19 pandemic and a post-COVID perspective. *Frontiers in Environmental Science, 10*, 975861. https://doi.org/10.3389/fenvs.2022.975861

Rabhi, W., Amara, W., Charouh, Z., El Jaimi, F., Benaboud, H., & Ben Saindou, M. (2023). E-Moulouya BDD: Extended Moulouya Bird Detection Dataset. *IEEE Dataport.* https://doi.org/10.21227/es8y-cr94

Rahmati, Y. (2024). *Artificial Intelligence for Sustainable Urban Biodiversity: A Framework for Monitoring and Conservation.* arXiv:2501.14766 [cs.CY]. https://doi.org/10.48550/arXiv.2501.14766

Silvestro, D., Goria, S., Sterner, T., & Antonelli, A. (2022). Improving biodiversity protection through artificial intelligence. *Nature Sustainability, 5*, 415–424. https://doi.org/10.1038/s41893-022-00851-6

Tuia, D., Kellenberger, B., Beery, S., Blair, R., Costelloe, B. R., Zuffi, S., …, & Berger-Wolf, T. (2022). Perspectives in machine learning for wildlife conservation. *Nature Communications, 13*, 792. https://doi.org/10.1038/s41467-022-27980-y

UNESCO. (2024). *Rapport d'évaluation de l'état de préparation à l'intelligence artificielle: Maroc.* Paris: Organisation des Nations Unies pour l'éducation, la science et la culture. https://unesdoc.unesco.org/ark:/48223/pf0000386790

UN-Habitat. (2020). *World Cities Report 2020: The Value of Sustainable Urbanization.* Nairobi: United Nations Human Settlements Programme (UN-Habitat). https://unhabitat.org/sites/default/files/2020/10/wcr_2020_report.pdf

UN-Habitat. (2024). *World Smart Cities Outlook 2024.* Nairobi: United Nations Human Settlements Programme. https://unhabitat.org/sites/default/files/2024/12/un_smart_city_outlook.pdf

Ying, J., Zhang, X., Zhang, Y., & Bilan, S. (2021). Green infrastructure: Systematic literature review. *Economic Research-Ekonomska Istraživanja, 35*(1), 343–366. https://doi.org/10.1080/1331677X.2021.1893202

## CHAPTER 16

Artificial Intelligence in Sustainable Supply Chain Management

Mohamed Badiy, Charaf Hamidi, Fatima Amounas, Zaynab Khalfi, Salma Gaou, and Abdelaaziz Hessane

16.1 INTRODUCTION

The increasing global demand for sustainable products and services has elevated the importance of supply chain management (SCM) strategies that minimize environmental harm, promote social responsibility, and ensure economic viability (Panigrahi et al., 2019). Today's supply chains are intricate networks facing numerous problems such as resource scarcity, global warming, and shifting consumer values. Increased public pressure alongside government action has, however, shifted industry operations toward a focus on sustainability (Saitone & Sexton, 2017). The impact of the supply chains on the environment comes from expenditures throughout the entire value chain, including but not limited to sourcing, manufacturing, distribution, and logistics. Sustained economic development aggravates the critical problem of greenhouse gas emissions, climate change, and ecosystem decomposition; thus carbon reduction has become an imperative in policy focus. Poorly governed, unsustainable industry practices compound the problem of long-term sustainability of the biosphere. The expansion of carbon markets, coupled with the emergence of new carbon-centric business models, has catalyzed the greater integration of social and environmental responsibility into SCM through the adoption of sustainability-oriented practices (Govindan et al., 2014). This catalyzed the development of sustainable SCM (SSCM), a model that enables companies to balance profit maximization with environmental and social impacts. From acquiring raw materials to product construction, distribution, customer delivery, and even managing the product at the end of its life cycle, SSCM incorporates sustainability principles into every phase of the supply chain (Tundys & Wiśniewski, 2021). Now, artificial intelligence (AI) is gaining attention for its integration into numerous fields and is being thoroughly researched on which indicates that there is an intention for its broader adoption. This also includes SCM. AI comprises machine learning (ML), natural language processing (NLP), computer vision, and robotics which all facilitate

DOI: 10.1201/9781003649885-16

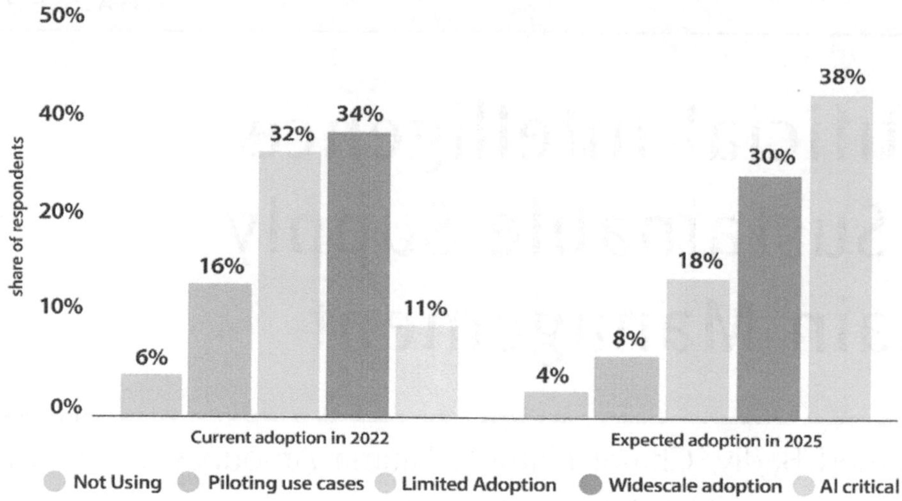

FIGURE 16.1 AI adoption rate in supply chain globally 2022–2025 (Reddy Dolu Surabhi et al., 2024).

a myriad of tasks pertaining to human intelligence, system intricacies, and decision-making. In regard to SSCM, the application of AI has recently become much more useful and popular mainly due to the need for coping with the aggressive demands for sustainability while navigating the intricate systems of international supply chains. Figure 16.1 demonstrates the worldwide rate of adoption of AI technologies in supply chain operations.

This chapter delves deeply into the application of sophisticated AI technologies and their role in promoting sustainability in SCM. We start with an overview of the SSCM system, outlining its defining elements and fundamental building blocks concepts. We analyze relevant AI techniques in SCM and their contributions toward optimally achieving sustainability goals. This is followed by the integration frameworks and system architectures that embed AI and other technologies such as Internet of Things (IoT) and blockchain toward SSCM. The chapter includes case studies and application examples from various industries, providing a comparative analysis of AI-driven sustainability initiatives. We also critically analyze the challenges and ethical considerations including data privacy, algorithmic bias, and implementation barriers that arise when integrating AI into SSCM. Finally, we outline future research directions, highlighting opportunities to integrate AI with technologies like blockchain and IoT, to overcome current limitations and transform supply chains for sustainability. The transformative potential of AI in achieving sustainable supply chains is reflected in the conclusion, emphasizing how, with responsible implementation, AI can be a catalyst for more sustainable, resilient, and ethical supply chain operations.

16.2 THEORETICAL FOUNDATIONS OF SUSTAINABLE SUPPLY CHAIN MANAGEMENT (SSCM)

16.2.1 Defining Sustainable Supply Chain Management: Scope and Objectives

SSCM extends traditional SCM by explicitly incorporating environmental and social objectives alongside economic goals. A widely cited definition by Carter & Rogers (2008) describes SSCM as "*the strategic, transparent integration and achievement of an organization's social, environmental, and economic goals in the systemic coordination of key inter-organizational business processes for improving the long-term economic performance of the individual company and its supply chains*" (Carter & Liane Easton, 2011). This definition highlights that SSCM seeks to achieve triple bottom line outcomes (people, planet, profit) through coordinated efforts both within the firm and across supply chain partners. In action, SSCM includes actions to reduce carbon emissions and waste, ensure fair labor practices and social impact, and maintain cost-effectiveness and efficiency.

16.2.2 The Triple Bottom Line Approach: Principles and Implications

Initially developed the triple bottom line (TBL) framework which serves as the cornerstone of SSCM where value is created in the supply chains on three fronts: deep ecology for the environment, social justice, and business profit (Brandenburg et al., 2019). Figure 16.2 shows the specific types of value allocated to each of the sustainability pillars (Taramsari et al., 2025). Environmental sustainability in the context of supply chains can be defined as "green" procurement, pollution cutback in the factories and transport, energy conservation, and circular economy (recycling and reuse within a closed-loop supply chain). Social sustainability highlights human and ethical dimensions like labor and employment relations health and safety, workplace harassment, community development, and consumer health and safety. Economic sustainability remains essential, it entails long-term viability of the business and supply network, achieved through cost management, innovation, and building resilient operations. SSCM requires balancing potential trade-offs among these dimensions. For example, choosing a supplier with lower carbon

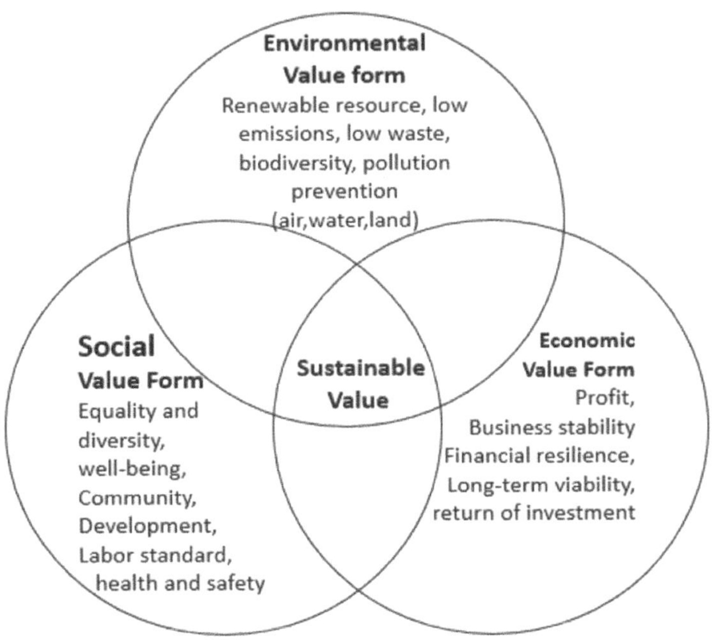

FIGURE 16.2 The triple bottom line (TBL) value forms (Evans et al., 2017).

footprint might involve higher cost, requiring a strategic decision that aligns with both environmental goals and financial constraints (Brandenburg et al., 2019).

16.2.3 Conceptual Models and Frameworks in SSCM

Several conceptual models help in understanding and implementing SSCM. One common framework is the TBL framework, which visualizes sustainability as the intersection of social, environmental, and economic performance (Carter & Liane Easton, 2011). Carter & Rogers (2008) also proposed four supporting facilitators for SSCM: strategy, risk management, organizational culture, and transparency, arguing that a holistic sustainability strategy, proactive risk and contingency planning, a culture of ethics and responsibility, and supply chain transparency are critical enablers of SSCM (Carter & Liane Easton, 2011). Stakeholder theory and institutional theory are often invoked to explain why firms pursue SSCM, noting pressures from stakeholders (customers, governments, NGOs) and normative expectations that drive sustainable practices. Additionally, closed-loop supply chain and circular economy concepts extend SSCM by emphasizing product life cycle management, such as designing supply chains that incorporate product returns, remanufacturing, and recycling to reduce waste (Carter & Rogers, 2008). These foundations set the stage for understanding how AI can be applied AI in SSCM must ultimately serve these sustainability principles and frameworks, enabling better achievement of environmental and social goals without sacrificing efficiency or reliability.

16.2.4 Key Drivers, Barriers, and Challenges in SSCM Implementation

Numerous scholars have emphasized that identifying internal and external drivers and barriers constitutes a critical first step in the implementation of SSCM (Sajjad et al., 2015). At this stage, organizations need to understand the key motivations for opting for SSCM strategies, as well as the challenges they face in the deployment process. As such, the systematic identification of these drivers and barriers plays a pivotal role in ensuring the effective execution of SSCM initiatives.

Supply chains are inherently complex and globalized, which complicates the monitoring and control of sustainability across multiple tiers of suppliers and distributors (Qu & Kim, 2024). Issues such as climate change, resource scarcity, and labor violations in far-reaching supply networks have pressured firms to improve their supply chain sustainability performance (Qu & Kim, 2024). Regulations (e.g., emissions targets, anti-slavery laws) and increasing consumer awareness are major external drivers pushing firms toward SSCM. Internally, companies are discovering that sustainable practices can align with operational efficiency, for instance, reducing energy and material use often lowers costs in the long run, creating a business case for SSCM (Xu & Cong, 2011). However, integrating sustainability can require upfront investment, innovation, and cross-company collaboration. SSCM initiatives may face resistance if they appear to conflict with short-term financial goals (Brandenburg et al., 2019; Carter & Liane Easton, 2011). This is where advanced analytical tools like AI hold promise: by improving the information available and optimizing complex decisions, AI can help identify "win-win" solutions or at least illuminate the trade-offs so that more sustainable choices can be made in a rational, data-driven way.

16.3 ADVANCED AI TECHNIQUES IN SUPPLY CHAIN MANAGEMENT

Advancements in AI have introduced a suite of techniques that are revolutionizing decision-making and information management in supply chains. These AI methods range from data-driven learning algorithms to intelligent physical systems, all of which can be harnessed to address the complexity of SSCM. This part describes key AI tools and their contribution to SCM, employing both technical and academic language to describe their benefits (Figure 16.3).

16.3.1 Machine Learning Applications in Supply Chain Analytics

ML is propelling further developments in SCM by helping organizations derive meaning from data, allowing for insightful decisions to be made. ML augments strategic planning, operational efficiency, and responsiveness throughout the supply chains by discovering concealed patterns and providing precise forecasts (Thejasree et al., 2024). Demand forecasting as one of the most important operational responsibilities within supply chains has been aided through supervised learning techniques like regression models, support vector machines, and neural networks. ML models that incorporate historical sales data, weather data, economic markers, and even social media activity can predict demand far more accurately than conventional methods, decreasing forecast error and eliminating surplus or depleted inventory stocking needed levels buffing efficiency (Yadav et al., 2024). Improved demand forecasts support sustainability by enabling organizations to design more efficient systems and preventing overproduction and waste. Time-series forecasting models (including ARIMA, Prophet, and deep learning models) similarly help optimize inventory control and procurement schedules, ensuring that resources and products flow efficiently with minimal waste. ML is also applied to supply chain risk management classification and anomaly detection algorithms can predict supply disruptions (e.g., identifying if a supplier is likely to fail to comply with sustainability standards or if logistics delays might occur) so that firms can proactively mitigate risks.

16.3.2 Deep Learning and Neural Architectures for Complex Supply Chain Processes

Deep learning, a subset of ML using multi-layer neural networks, has shown superior performance in modeling complex, non-linear relationships in supply chain data. For example, deep neural networks and advanced architectures (CNNs, RNNs, Transformers)

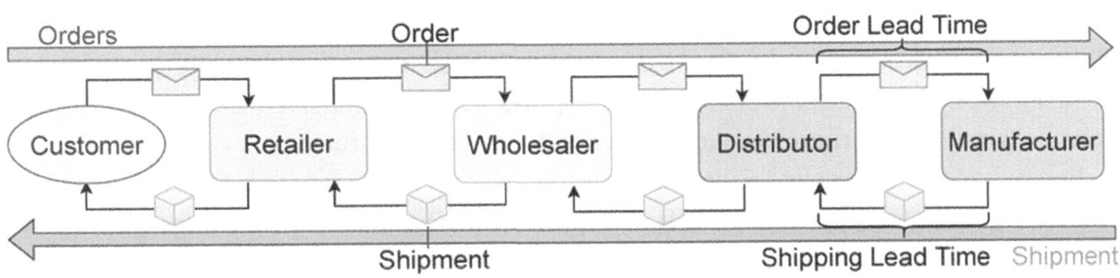

FIGURE 16.3 Multi-position supply chain (Zhang et al., 2024).

can be used to analyze high-dimensional data like images or sensor streams in logistics. Computer vision (an AI capability based on deep learning) can automate quality inspection in manufacturing detecting defects in products on the production line, which reduces waste by catching quality issues early and ensures only good products move forward, supporting both economic and environmental goals. In the context of sustainable manufacturing, researchers have demonstrated how deep learning approaches can optimize processes and improve energy efficiency. Jamwal et al. (2022) proposed a deep learning-based framework that enhances sustainable manufacturing performance by modeling complex process parameters. Deep learning has also been applied to route optimization and transport planning by processing real-time traffic or telematics data to minimize fuel consumption and emissions during distribution.

16.4 ARTIFICIAL INTELLIGENCE-DRIVEN OPTIMIZATION STRATEGIES FOR SUSTAINABLE SUPPLY CHAINS

One of the most impactful uses of AI in SSCM is in optimization, improving supply chain processes and decision-making to achieve better sustainability performance. Traditional supply chain optimization often focuses on cost or service level. AI allows expanding these optimization objectives to include environmental and social metrics and solving much more complex multi-criteria problems. In this section, we discuss how AI-driven optimization is applied to key areas of SCM with sustainability goals in mind.

16.4.1 Optimization of Production and Manufacturing for Environmental Efficiency

In the era of globalized markets and increasing sustainability pressures, manufacturing optimization has become essential not only for cost reduction but also for advancing environmental and social goals. Today's manufacturing enterprises are expected to minimize their ecological footprint while maintaining high levels of efficiency and competitiveness. Although various sustainability dimensions such as energy efficiency, waste reduction, and production agility are widely acknowledged, their effective integration into day-to-day operations remains a challenge that continues to attract attention from both academia and industry (Garbie, 2015).

AI is at the forefront of this transformation, enabling the shift toward smart and sustainable manufacturing. AI-powered scheduling systems can optimize production plans by prioritizing energy-efficient machine states, reducing resource-intensive changeovers, and limiting idle time. For instance, AI can determine the optimal sequencing of orders to minimize energy consumption or material loss, particularly in industries with high variability in batch production. In advanced settings, AI can also synchronize production with the availability of renewable energy, ensuring that energy-intensive tasks align with peak solar or wind output, thereby reducing dependency on fossil fuels.

Beyond operational scheduling, AI enhances process intelligence and decision-making. As highlighted by (Qu & Kim, 2024), AI applications such as intelligent agents and deep learning algorithms contribute to greater supply chain transparency and sustainability tracking. AI also plays a crucial role in quality management, where ML models are used to detect anomalies and identify the root causes of defects. This not only improves product quality but also decreases scrap and rework, leading to substantial savings in materials and energy. Collectively, these AI-driven capabilities are redefining how manufacturers approach sustainability by embedding intelligence into every stage of the production cycle.

16.4.2 AI-Enhanced Supplier Selection and Sustainable Procurement Strategies

Selecting suppliers and sourcing materials are critical strategic decisions in SSCM, and AI can optimize these decisions by processing large amounts of data on supplier performance and risk. Multi-criteria decision models enhanced with AI allow firms to evaluate suppliers not just on cost and quality but also on sustainability criteria (e.g., carbon footprint, labor standards compliance). Techniques like genetic algorithms or simulated annealing can be used to solve complex supplier allocation problems, where the objective function includes minimizing overall supply chain emissions or maximizing a sustainability score. For example, AI can help solve a procurement problem of minimizing total cost and total environmental impact subject to demand constraints – a multi-objective optimization where a Pareto-optimal solution can be identified. Some researchers have proposed decision support systems where AI algorithms rank suppliers by sustainability metrics or even predict a supplier's future compliance issues from past data (Qu & Kim, 2024). This kind of AI-driven optimization ensures that sustainability is quantitatively factored into procurement decisions, rather than being an ad-hoc consideration.

16.4.3 Sustainable Network Design and Strategic Facility Location Optimization

The design of the supply chain network (number, location, and capacity of warehouses, production plants, etc.) has long-term sustainability implications. AI techniques contribute here by evaluating countless design scenarios with respect to cost, service, and environmental criteria. For instance, an AI-enhanced optimization model might consider placing distribution centers closer to customer clusters to reduce transport distance (reducing emissions), while also considering regional renewable energy availability or local community impact. Advanced optimization can integrate life cycle assessment data so that a facility's location decision accounts for things like energy mix of the local grid (affecting manufacturing emissions) or regulations in that region. One study (Qu & Kim, 2024) introduced a simulation-based multi-objective approach for transportation network design under uncertain conditions, again balancing cost and CO_2 output. The inclusion of uncertainty (via scenarios or stochastic optimization) is another strength of AI-driven methods, as they can handle probabilistic inputs and still optimize decisions that are robust against variability (such as demand fluctuations or fuel price changes).

16.5 TECHNOLOGICAL FRAMEWORKS AND INTEGRATION ARCHITECTURES FOR AI-ENABLED SSCM

To harness AI effectively in SSCM, organizations need robust integration frameworks and technical architectures that embed AI into supply chain processes. These frameworks typically involve the convergence of AI with other advanced technologies (such as IoT sensors, big data platforms, and blockchain networks) under unified architectures that support data-driven, real-time, and transparent supply chain operations. This part explains the structure of integration frameworks and the link between various technologies and the construction of a sustainable AI-based supply chain system.

16.5.1 IoT-Based Real-Time Monitoring and Predictive Analytics

Real-time supply chain visibility is a fundamental element of logistics and operations management, ensuring efficiency, flexibility, and strength in today's turbulent global environment. The integration of AI with the IoT has fundamentally changed the SCM landscape, providing the means for comprehensive visibility, predictive analytics, and intelligent automation that enhance proactive decision-making and performance optimization (Taj et al., 2023). AI systems forecast demand surges, potential disruptions, and formulate dynamic routing with inventory optimization through an ever-growing supply of structured and unstructured data streams on data. Concurrently, real-time asset tracking, monitoring of environmental contexts, and all-inclusive anomaly detection are accomplished by IoT technologies such as sensors, RFID tags, and smart equipment. AI and IoT jointly create a cyber-physical interface that supports a digitally rendered supply chain in a physically existing environment, thus enabling adaptive control and real-time situational awareness.

The IoT layer of any integration architecture is responsible for the continuous collection of data like location, temperature, humidity, and fuel consumption, which is sent to the corresponding AI layer for further processing. For instance, sensors on delivery trucks can relay real-time telemetry to AI systems that modify the operational delivery schedule in real time based on the prevailing traffic and fuel consumption efficiency. Similarly, IoT-enabled smart meters within manufacturing facilities can supply granular data to AI systems that manage and optimize energy consumption. Alam (2021) highlights how embedding IoT within supply chains unlocks new levels of operational intelligence, enabling real-time data acquisition and facilitating automated, data-driven decision-making. This real-time visibility allows companies to monitor critical parameters such as temperature and humidity for perishable goods in transit, ensuring product quality and minimizing spoilage. Furthermore, when anomalies are detected such as unexpected delays or environmental deviations, AI-driven systems can autonomously initiate contingency actions like rerouting shipments to prevent loss.

A robust AI+IoT architecture often leverages a combination of edge computing and cloud computing (Figure 16.4). Edge devices (e.g., sensor gateways) can host lightweight AI models that execute local decisions (such as triggering an alert when a threshold is exceeded), while the cloud supports centralized analytics, long-term data storage, and AI model training. This hybrid approach balances real-time responsiveness with strategic insight, allowing supply chains to be both reactive and predictive in nature.

16.5.2 Blockchain-Enabled Transparency and Traceability in Sustainable Supply Chains

Blockchain technology is increasingly considered in SSCM integration to provide transparency and trust in data sharing across supply chain partners. When combined with AI, blockchain can serve as a tamper-evident data layer that ensures the integrity of sustainability-related data (such as proof of origin, emissions records, or labor certifications) that AI

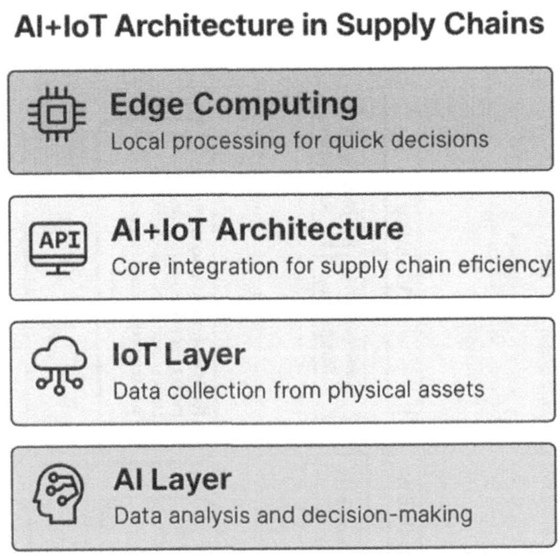

FIGURE 16.4 The associated role of AI and IoT in supply chains.

algorithms analyze. One integration approach is an AI-integrated blockchain network for supply chain traceability (Sherin et al., 2023). For example, Sunmola et al. (2025) propose a conceptual framework where blockchain records of product provenance and sustainability compliance are analyzed by AI to verify and trace sustainability performance at each tier. In such an architecture, AI might automatically flag any anomalies in the blockchain data say a sudden spike in a supplier's reported emissions and could even trigger smart contracts (self-executing agreements on a blockchain) that enforce sustainability clauses (like penalizing a supplier or initiating an audit). The integration frameworks here must handle interoperability: linking enterprise systems and IoT devices to blockchain entries, and linking blockchain data to AI analytics platforms. Future-oriented frameworks envision a "digital twin" of the supply chain on a blockchain, where every product unit's journey is recorded, and AI agents continuously monitor this digital supply chain for compliance and optimization opportunities. Figure 16.5 presents a structured overview of the findings, impacts, and strategic recommendations related to the implementation of blockchain technology in sustainable logistics.

16.5.3 Industry 4.0 Integration for Intelligent and Sustainable Supply Chains

The barriers and enablers associated with the sustainable implementation of Industry 4.0 (I4.0) technologies in supply chains have attracted growing attention from both researchers and practitioners (Belhadi et al., 2022; Sharma et al., 2023). I4.0 defined by the integration of IoT, AI, cloud computing, and cyber-physical systems is fundamentally reshaping supply chains across various sectors, presenting both opportunities and challenges that influence organizational strategies, business models, and operational structures worldwide (Scavarda et al., 2020).

In the context of SSCM, I4.0 offers a powerful digital infrastructure to promote sustainability. As noted by Qu and Kim (2024), I4.0 enables the deployment of AI-integrated big

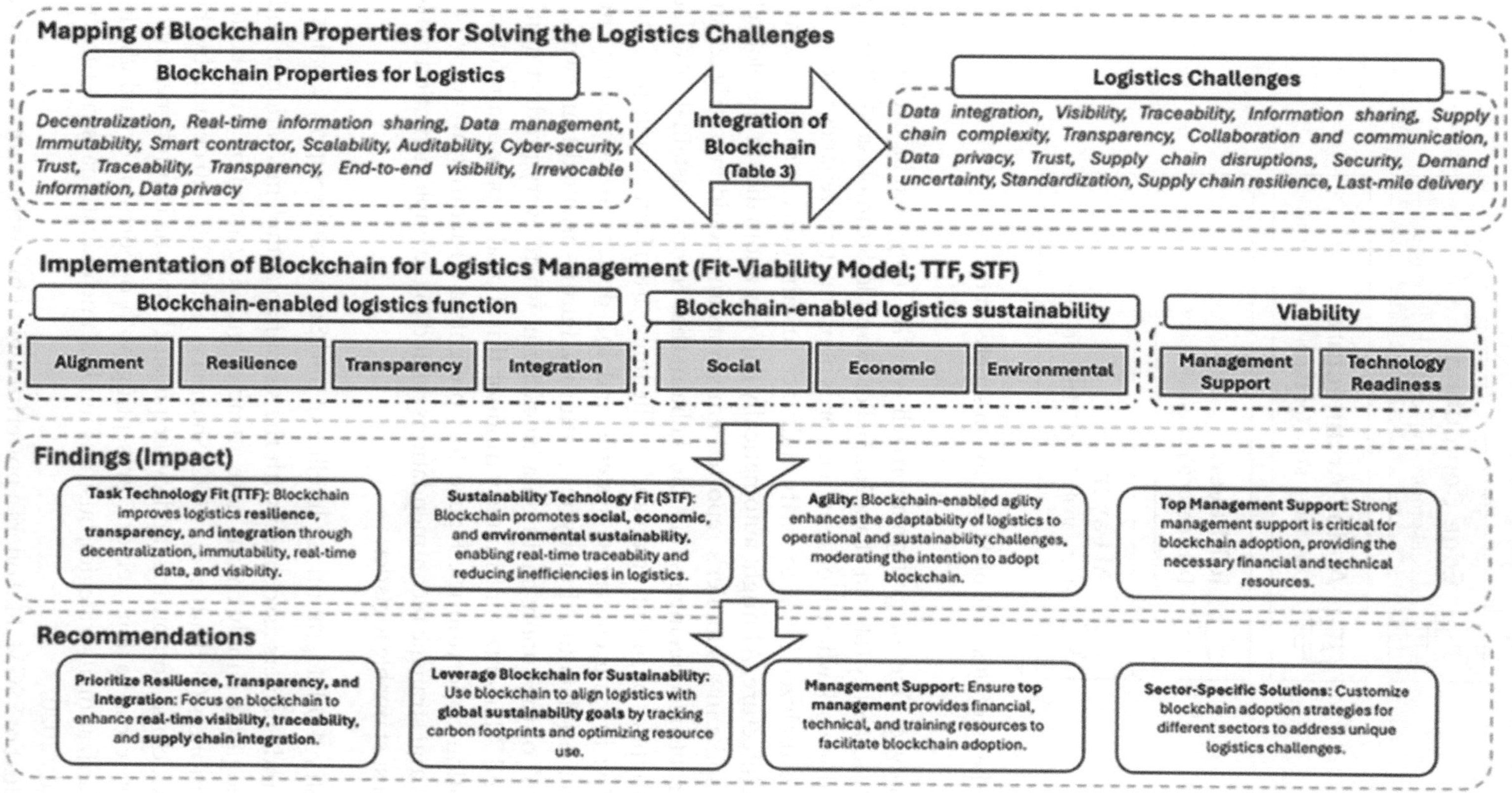

FIGURE 16.5 The findings, impact, and recommendations of the use of blockchain technology in sustainable logistics (Aslam et al., 2024).

data platforms and digital technologies that enhance the responsiveness and environmental performance of supply chains. A practical example includes the use of digital twins virtual representations of physical systems within smart factories to simulate, monitor, and optimize sustainable production scenarios in real time. According to Proposition 7 in Qu and Kim's study, the integration of digital twins with interactive AI agents (such as chatbots) supports agile, resilient, and environmentally sustainable manufacturing. In such architecture, each critical component of the supply chain (e.g., factory floors, warehouses, product batches) is paired with a digital replica that AI algorithms can analyze for scenario testing, waste minimization, and energy optimization. For instance, AI could continuously evaluate alternative production configurations in a digital environment to identify the most eco-efficient setup, which is then deployed in the physical system, thereby enhancing real-time decision-making. Figure 16.6 illustrates how the TBL encompassing environmental, economic, and social dimensions is operationalized within a Sustainable I4.0 framework. This visualization emphasizes the synergy between technological innovation

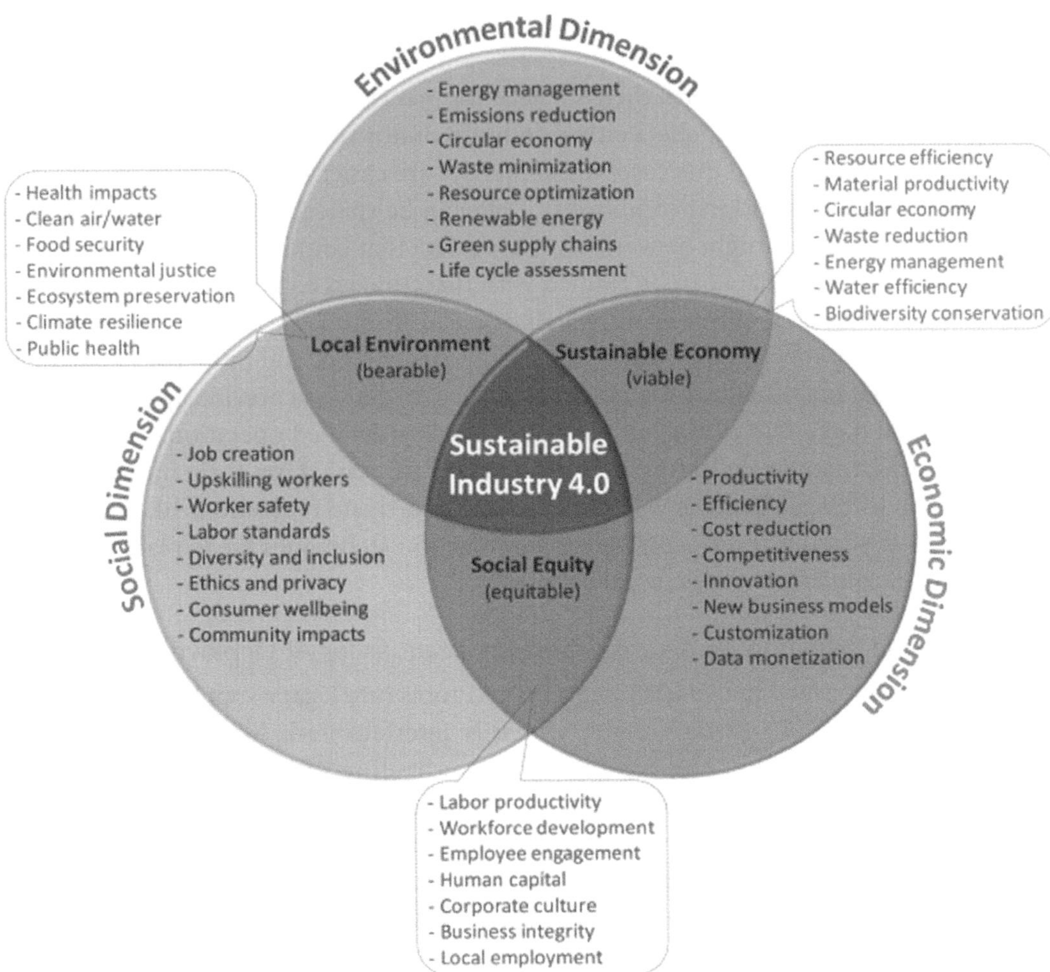

FIGURE 16.6 The triple bottom line on the service of sustainable Industry 4.0 (Khan et al., 2025).

and sustainability imperatives, highlighting how I4.0 can serve not only as a productivity enabler but also as a catalyst for holistic, responsible supply chain transformation.

16.6 AI-ASSISTED SSCM CHALLENGES, LIMITATIONS, AND ETHICAL CONSIDERATIONS

Although AI greatly improves the sustainability of supply chains, it poses new challenges and ethical issues. These challenges cross technical, organizational, and societal boundaries and must all be addressed in order to use AI in SSCM in a responsible and effective manner. In this section, I outline the major challenges and ethical issues such as data privacy, bias, transparency, impediments to implementation, and the absence of focus on social sustainability within the framework of AI (Shahzadi et al., 2024).

16.6.1 Data Governance, Quality, and Accessibility in Sustainable Supply Chains

The performance of an AI system depends on the quality of the data it gets trained on. Through its use in specific sectors like supply chains, data is often stored in isolated pockets within separate organizations and IT systems. One primary issue is acquiring the accurate, high-grade data required for AI algorithms. Take, for instance, the optimization of a supply chain on its carbon emissions; there's a need for precise emission data from every step in the process, but a lot of suppliers either do not monitor their emissions or do not provide that information. As in this case, social metrics are easier to define than to collect or quantify (for instance, working conditions). Data can be sparse, outdated, or deliberately obscured (as companies might be wary of sharing data that could reveal business secrets). Moreover, supply chain data can be riddled with inaccuracies e.g., inventory records that are wrong due to manual entry errors, which can lead AI models astray. Overcoming this requires significant effort in data governance, standardization, and often the deployment of IoT or other data capture technologies in places where data was previously not collected. Even with good data, *data privacy* is a concern: sharing detailed operational data among supply chain partners or collecting data on workers (like tracking their movements for efficiency) can infringe on privacy. Companies must navigate regulations like GDPR and ensure that data sharing agreements and anonymization techniques are in place so that AI use does not violate privacy laws or norms.

16.6.2 System Integration Barriers and Cost Implications for AI Deployment

Many organizations face challenges integrating AI tools into legacy supply chain systems. Existing ERP and SCM software might not easily interface with AI modules, requiring custom integration. The cost and complexity of overhauling systems or even retrofitting AI into them can be high. Small and medium enterprises (SMEs) in a supply chain might lack the resources to adopt AI, creating a gap where only larger players benefit, potentially widening the sustainability performance difference between big and small firms. There is also the question of *scalability* pilot projects might show promise, but scaling AI solutions across a global supply network is non-trivial, requiring infrastructure, training, and support at scale. These practical challenges can slow down AI adoption in SSCM, even if the potential benefits are clear (Singh & Adhikari, 2023). Pankaj Yadav and Naveen (2024) note that significant gaps remain in literature and practice regarding how to optimally implement AI in SCM and that more research is needed to identify and address these barriers. Without careful planning, companies may struggle with pilot purgatory experiments that don't translate into full deployment due to integration challenges.

16.6.3 Algorithmic Bias, Fairness, and Inclusivity Challenges

AI systems can inadvertently encode and perpetuate biases present in training data. In supply chain contexts, this raises ethical issues. For instance, an AI system for supplier selection might be trained on past data that favored suppliers from certain regions or of a certain size, hence the AI might unfairly score small local suppliers lower (because historically they were not used, not necessarily because they are worse) (Li & Li, 2022). This could undermine social sustainability goals, like supporting local economies or minority-owned businesses. Bias can also affect demand forecasting if an AI only learns from historical data, it might not account for shifts such as increased demand for sustainable products (since historically those might have been niche), thus under-stocking them. Ensuring fairness means actively checking AI models for unintended discriminatory patterns and adjusting them, which is technically challenging. There is also concern for labor fairness: algorithms that schedule work or set performance targets could push workers too hard. For example, if an AI optimizes warehouse picking routes and rates purely for efficiency, it might result in workloads that are not humane (a known issue in some e-commerce warehouses). Thus, ethical AI design must include constraints or criteria for worker well-being (e.g., limiting continuous picking time, scheduling adequate breaks) (Figure 16.7).

16.6.4 Addressing the Social Sustainability Gap in AI Applications

As noted earlier, a critical challenge is that current AI applications in supply chain tend to focus on what is more easily measurable (economic and some environmental aspects) and often neglect the social dimension. Research by Qu and Kim (2024) found that while AI in supply chain processes primarily addressed environmental and economic issues, there remains a technological gap in tackling social issues like working conditions and fair

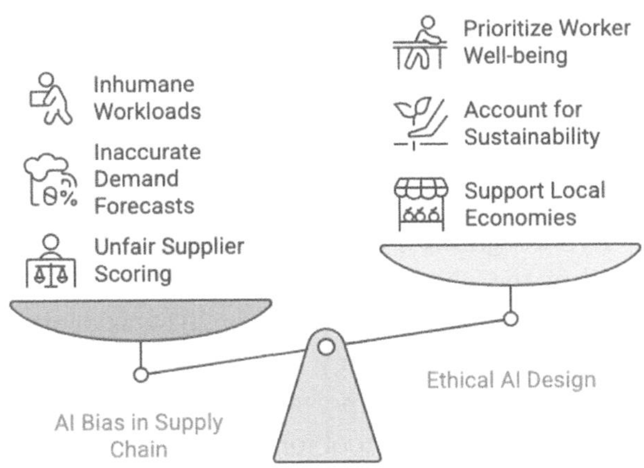

FIGURE 16.7 Balancing AI efficiency with ethical considerations.

dealing (Zejjari & Benhayoun, 2024). This is an ethical concern: if AI is driving decisions purely on efficiency metrics, companies might inadvertently undermine social sustainability (for instance, optimizing for lowest cost could pressure suppliers into labor compromises). Therefore, a major challenge is developing AI methods that can incorporate social metrics (which are qualitative and complex) into optimization and decision-making. It is also an opportunity for future research – creating AI that can detect human rights risks or suggest improvements for worker welfare. Until such solutions mature, companies must be cautious to include human judgment and values to cover social criteria. Dauvergne (2022) provides a critical IPE (international political economy) perspective, questioning whether AI is truly "greening" supply chains or if it is sometimes used to greenwash or shift costs. He argues that if left to the market, AI might be harnessed in ways that improve efficiency (and profits) but still externalize environmental costs, unless there are proper governance mechanisms (Qu & Kim, 2024). This critique reinforces that ethical deployment of AI in SSCM requires aligning AI's objectives with genuine sustainability goals, rather than allowing it to be a tool for superficial improvements or just profit maximization.

16.6.5 Cybersecurity Risks and Vulnerabilities in Digitized Supply Chains

Cyber-attacks pose a critical threat to modern supply chains, as vulnerabilities at any node whether direct or indirect can propagate risk across multiple tiers of the network. The severity of this risk depends heavily on the type of cyber-attack and its ability to cascade through interconnected systems (Dash et al., 2024). A growing body of literature highlights the importance of cybersecurity insurance as a risk mitigation tool; however, few studies have comprehensively considered the combined effects of attack type, two-stage risk propagation, and insurance coverage in optimizing cybersecurity investments across the supply chain. With the ongoing digital transformation of supply chains driven by technologies such as AI, IoT, and cloud computing, the cyber-attack surface has expanded significantly. A malicious breach can result not only in data theft (e.g., proprietary logistics plans, employee records, or customer information) but also in operational disruption. For instance, an attacker feeding false data into an AI system could manipulate its decision-making process, leading to inefficient logistics routes, inaccurate demand forecasts, or even hazardous outcomes. In extreme cases, such compromised systems may cause ethical and environmental consequences, such as the misrouting of chemical shipments or mishandling of temperature-sensitive medical supplies. To address these risks, companies must prioritize robust cybersecurity frameworks that protect AI systems and their underlying data pipelines. Blockchain technologies offer promising solutions for ensuring data integrity and traceability, though they, too, are not immune to attack if misconfigured or poorly managed. According to Cyble threat researchers, 22 of the 25 industry sectors analyzed experienced at least one supply chain attack in 2024, as depicted in Figure 16.8. The most targeted sectors include IT, technology, and aerospace and defense, reflecting their high levels of digitization and strategic importance.

In conclusion, while AI is a powerful tool for SSCM, it must be approached with careful attention to these challenges and ethical issues. A responsible AI framework for supply chains would include data ethics, model ethics (bias mitigation, transparency), and usage

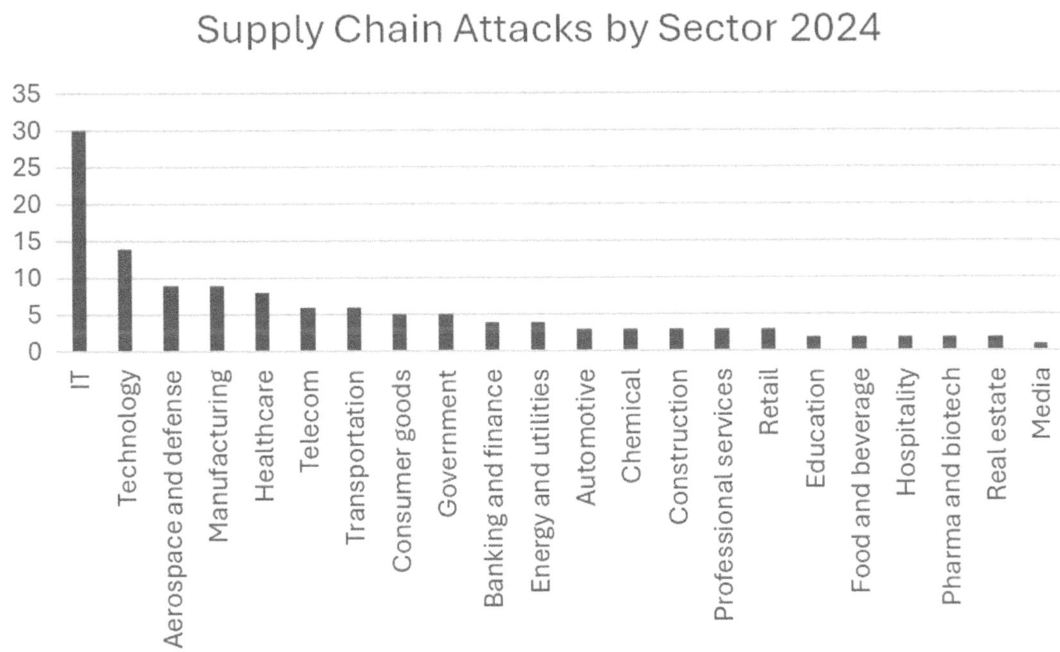

FIGURE 16.8 Number of supply chain cyber-attacks by sector 2024 (Cyble, 2024).

ethics (clear policies on how AI decisions are used and who is accountable). Many of these challenges are active areas of research and industry learning. The next section will consider how future research and developments can address these issues and further enhance the synergy between AI and sustainable supply chains.

16.7 FUTURE RESEARCH DIRECTIONS AND EMERGING OPPORTUNITIES

The intersection of AI and SSCM is a rapidly evolving field, and numerous opportunities exist to deepen its impact in the coming years. Future research and development will likely focus on enhancing integration, addressing current limitations (especially regarding social sustainability and ethics), and leveraging emerging technologies to support SSCM. In this section, we highlight several key future directions and research opportunities.

16.7.1 Integration of Artificial Intelligence, Blockchain, and IoT for Holistic Transparency

One promising avenue is the tighter integration of AI with blockchain and IoT to create fully transparent, traceable, and responsive supply chains. As noted, blockchain can provide a trustworthy ledger of supply chain events (e.g., every time a product changes hands or a sustainability certificate is issued), while IoT provides real-time data on physical flows. Future systems could see AI acting as the intelligence layer on top of these, analyzing blockchain records and IoT sensor feeds to provide insights and automation. For example, AI could automatically verify sustainability compliance by cross-checking IoT data (like temperature logs indicating cold chain integrity) against blockchain records (smart contracts stipulating conditions). Research can explore architectures for autonomous, self-correcting supply chains where AI agents monitor the blockchain for any deviations (like a supplier logging a higher emission output than allowed) and then trigger corrective actions or alerts. Frontiers in research suggest that a joint deployment of blockchain, IoT, and big data analytics (AI) is a key direction (Munir et al., 2022). This could enable what some call "Supply Chain 4.0" – a truly digitized supply chain that is resilient, sustainable, and transparent. There is a need for frameworks that standardize data sharing across these technologies and address scalability (as adding blockchain can sometimes slow transaction speeds, novel consensus mechanisms or off-chain solutions may be needed for high-volume supply chains).

16.7.2 Generative Artificial Intelligence for Innovative Supply Chain Design and Scenario Planning

The recent advances in generative AI (such as GPT models) might find their way into supply chain sustainability as well. For instance, generative AI could be used to generate synthetic data to improve model training where real data is scarce (a possible help for modeling rare disruptions or new products with no history). It could also assist in scenario generation creating plausible future scenarios (like narratives of possible supply chain disruptions or shifts in consumer sentiment) that planners can use for resilience building. Moreover, generative design (an AI technique) could be applied to supply chain network or product design. For example, given sustainability constraints, generative algorithms might propose novel network structures or packaging designs that a human might not think of but that achieve lower environmental impact. This crosses into the innovation domain using AI not just to optimize within given parameters but to create new solutions for sustainable supply chains.

16.7.3 Human–AI Collaboration Models for Enhanced Decision Augmentation in Sustainable Supply Chains

Finally, a more conceptual but crucial research direction is understanding how human decision-makers and AI can best collaborate in SCM. Rather than fully automated supply chains, the foreseeable future is one of augmented intelligence human expertise combined with AI analytics. Studies might examine decision processes to determine which decisions can be automated and which should remain human-supervised, and how to design interactions that complement human strengths (strategic thinking, ethical judgment) with AI strengths (speed, complex computation). This involves research in organizational behavior and cognitive science within the context of AI tools. One practical output could be training programs or interfaces that improve the ability of supply chain professionals to interpret AI outputs and incorporate them into strategy. On the flip side, AI systems might be designed to learn from human feedback continuously, making them better over time in aligning with what human managers consider important but is hard to quantify (like brand reputation or supplier relationship quality).

In summary, the future of AI in sustainable supply chains looks toward deeper integration with technology ecosystems (IoT, blockchain, digital twins), greater scope (covering social aspects and circular economy), improved trust and ethics (explainability, fairness), and innovative uses (generative design, scenario planning). The transformative potential of AI will be realized when these systems not only optimize what is but help imagine and create what could be – enabling supply chains that are not just incrementally greener or fairer but fundamentally redesigned for sustainability. Achieving this will require ongoing interdisciplinary research, pilot programs, and knowledge sharing between academia, industry, and policymakers. The following conclusion will reflect on this transformative potential and the conditions needed to fully harness AI for SSCM.

16.8 CONCLUSION

AI stands as a powerful catalyst for transforming SCM in pursuit of sustainability. This chapter has focused on the application of AI techniques, including ML and deep analytics, intelligent agents, and even robotics, in addressing the environmental, social, and economic challenges of the global supply chains. By employing AI in forecasting, logistics optimization, production scheduling, and supply chain coordination, companies can achieve considerable benefits, including reduced waste and emissions, heightened agility in response to sustainability challenges, and improved transparency and traceability. The conceptual frameworks and case studies that were presented underscore the fact that AI, when thoughtfully applied, is not a panacea, but it does transform what is possible in a world of manual primitive ways of analyzing and optimizing systems into something far more flexible and responsive. A recurring insight AI is amplifying the ability to handle complexity. Sustaining supply chains is complex because they involve balancing several competing goals and managing a vast network of outsourced partners – activities which quickly exceed human cognitive limits. But driven by big data and the IoT, AI systems are capable of cutting through such complexity to deliver clarity and optimal choices. With these capabilities, AI helps devise creative win–win scenarios, including emission reductions while incurring cost savings through optimized route plans or inventory systems. Moreover, AI can illuminate trade-offs explicitly, arming decision-makers with quantitative evidence to make informed choices aligned with corporate sustainability goals and stakeholder expectations.

The chapter also underscores the importance of future research and continuous improvement. AI in SSCM is not a one-time deployment but a journey of learning and adaptation. As more data is gathered and models are refined, the AI's recommendations will improve. Similarly, organizations will learn how best to integrate AI into decision processes. There is considerable room for innovation from new algorithms that better incorporate social and environmental metrics, to new business models that AI might make possible (such as service-based models or sharing platforms that enhance sustainability). Collaboration between academia and industry will be vital to explore these frontiers, as real-world data and feedback loop into research, and research provides cutting-edge methods for trial in practice. In reflecting on the transformative potential of AI, it's clear that technology alone is not enough; what matters is how it is applied. AI can either entrench old ways (making unsustainable practices slightly more efficient) or enable fundamentally new paradigms (like circular supply chains and regenerative business models). The vision that emerges is one of intelligent, sustainable supply networks that are resilient to disruptions, responsive to stakeholder needs, and regenerative for the environment. In such networks, AI acts as an orchestrator coordinating material flows in sync with natural cycles, matching supply with demand in near-perfect alignment, and allocating resources in a way that maximizes shared value for all stakeholders (from raw material producers to end consumers and communities). This represents a shift from the traditional linear, siloed supply chain to a more circular, collaborative, and transparent supply ecosystem.

REFERENCES

Alam, T. (2021). Cloud-based IoT applications and their roles in smart cities. *Smart Cities*, *4*(3), 1196–1219. https://doi.org/10.3390/smartcities4030064

Aslam, J., Lai, K., Kim, Y. B., & Treiblmaier, H. (2024). The implications of blockchain for logistics operations and sustainability. *Journal of Innovation & Knowledge*, *9*(4), 100611. https://doi.org/10.1016/j.jik.2024.100611

Belhadi, A., Kamble, S., Gunasekaran, A., & Mani, V. (2022). Analyzing the mediating role of organizational ambidexterity and digital business transformation on Industry 4.0 capabilities and sustainable supply chain performance. *Supply Chain Management: An International Journal*, *27*(6), 696–711. https://doi.org/10.1108/SCM-04-2021-0152

Brandenburg, M., Gruchmann, T., & Oelze, N. (2019). Sustainable supply chain management—A conceptual framework and future research perspectives. *Sustainability*, *11*(24), 7239. https://doi.org/10.3390/su11247239

Carter, C. R., & Liane Easton, P. (2011). Sustainable supply chain management: Evolution and future directions. *International Journal of Physical Distribution & Logistics Management*, *41*(1), 46–62. https://doi.org/10.1108/09600031111101420

Carter, C. R., & Rogers, D. S. (2008). A framework of sustainable supply chain management: Moving toward new theory. *International Journal of Physical Distribution & Logistics Management*, *38*(5), 360–387. https://doi.org/10.1108/09600030810882816

Cyble. (2024, August). *Supply chain attacks becoming a near-daily occurrence* [Image]. Cyble.

Dash, A., Sarmah, S. P., Tiwari, M. K., Jena, S. K., & Glock, C. H. (2024). Cybersecurity investments in supply chains with two-stage risk propagation. *Computers & Industrial Engineering*, *197*, 110519. https://doi.org/10.1016/j.cie.2024.110519

Dauvergne, P. (2022). Is artificial intelligence greening global supply chains? Exposing the political economy of environmental costs. *Review of International Political Economy*, *29*(3), 696–718. https://doi.org/10.1080/09692290.2020.1814381

Evans, S., Vladimirova, D., Holgado, M., Van Fossen, K., Yang, M., Silva, E. A., & Barlow, C. Y. (2017). Business model innovation for sustainability: Towards a unified perspective for creation of sustainable business models. *Business Strategy and the Environment*, *26*(5), 597–608. https://doi.org/10.1002/bse.1939

Garbie, I. H. (2015). Sustainability optimization in manufacturing enterprises. *Procedia CIRP*, *26*, 504–509. https://doi.org/10.1016/j.procir.2014.07.085

Govindan, K., Azevedo, S. G., Carvalho, H., & Cruz-Machado, V. (2014). Impact of supply chain management practices on sustainability. *Journal of Cleaner Production*, *85*, 212–225. https://doi.org/10.1016/j.jclepro.2014.05.068

Jamwal, A., Agrawal, R., & Sharma, M. (2022). Deep learning for manufacturing sustainability: Models, applications in Industry 4.0 and implications. *International Journal of Information Management Data Insights*, *2*(2), 100107. https://doi.org/10.1016/j.jjimei.2022.100107

Khan, M. I., Yasmeen, T., Khan, M., Hadi, N. U., Asif, M., Farooq, M., & Al-Ghamdi, S. G. (2025). Integrating Industry 4.0 for enhanced sustainability: Pathways and prospects. *Sustainable Production and Consumption*, *54*, 149–189. https://doi.org/10.1016/j.spc.2024.12.012

Li, M., & Li, T. (2022). AI automation and retailer regret in supply chains. *Production and Operations Management*, *31*(1), 83–97. https://doi.org/10.1111/poms.13498

Munir, M. A., Habib, M. S., Hussain, A., Shahbaz, M. A., Qamar, A., Masood, T., Sultan, M., Mujtaba, M. A., Imran, S., Hasan, M., Akhtar, M. S., Uzair Ayub, H. M., & Salman, C. A. (2022). Blockchain adoption for sustainable supply chain management: Economic, environmental, and social perspectives. *Frontiers in Energy Research*, *10*. https://doi.org/10.3389/fenrg.2022.899632

Panigrahi, S. S., Bahinipati, B., & Jain, V. (2019). Sustainable supply chain management. *Management of Environmental Quality: An International Journal*, *30*(5), 1001–1049. https://doi.org/10.1108/MEQ-01-2018-0003

Qu, C., & Kim, E. (2024). Reviewing the roles of AI-integrated technologies in sustainable supply chain management: Research propositions and a framework for future directions. *Sustainability*, *16*(14), 6186. https://doi.org/10.3390/su16146186

Reddy Dolu Surabhi, S. N., Buvvaji, H. V., & Sabbella, V. R. R. (2024). The AI-driven supply chain: Optimizing engine part logistics for maximum efficiency. *Educational Administration Theory and Practices*. https://doi.org/10.53555/kuey.v30i5.4428

Saitone, T. L., & Sexton, R. J. (2017). Agri-food supply chain: Evolution and performance with conflicting consumer and societal demands. *European Review of Agricultural Economics*, *44*(4), 634–657. https://doi.org/10.1093/erae/jbx003

Sajjad, A., Eweje, G., & Tappin, D. (2015). Sustainable supply chain management: Motivators and barriers. *Business Strategy and the Environment*, *24*(7), 643–655. https://doi.org/10.1002/bse.1898

Scavarda, A., Daú, G., Felipe Scavarda, L., Duarte Azevedo, B., & Luis Korzenowski, A. (2020). Social and ecological approaches in urban interfaces: A sharing economy management framework. *Science of The Total Environment*, *713*, 134407. https://doi.org/10.1016/j.scitotenv.2019.134407

Shahzadi, G., Jia, F., Chen, L., & John, A. (2024). AI adoption in supply chain management: A systematic literature review. *Journal of Manufacturing Technology Management*, *35*(6), 1125–1150. https://doi.org/10.1108/JMTM-09-2023-0431

Sharma, M., Luthra, S., Joshi, S., Kumar, A., & Jain, A. (2023). Green logistics driven circular practices adoption in Industry 4.0 Era: A moderating effect of institution pressure and supply chain flexibility. *Journal of Cleaner Production*, *383*, 135284. https://doi.org/10.1016/j.jclepro.2022.135284

Sherin, K., Kaur, N., Joshi, A., B, Ravindar R., Nayak, P., & Srinivas, K. (2023). The Role of AI and Blockchain in Supply Chain Traceability. *2023 3rd International Conference on Advance Computing and Innovative Technologies in Engineering (ICACITE)*, 918–922. https://doi.org/10.1109/ICACITE57410.2023.10183214

Singh, N., & Adhikari, D. (2023). Challenges and solutions in integrating AI with legacy inventory systems. *International Journal for Research in Applied Science and Engineering Technology*, *11*(12), 609–613. https://doi.org/10.22214/ijraset.2023.57376

Sunmola, F., Baryannis, G., Tan, A., Co, K., & Papadakis, E. (2025). Holistic framework for blockchain-based halal compliance in supply chains enabled by artificial intelligence. *Systems*, *13*(1), 21. https://doi.org/10.3390/systems13010021

Taj, S., Imran, A. S., Kastrati, Z., Daudpota, S. M., Memon, R. A., & Ahmed, J. (2023). IoT-based supply chain management: A systematic literature review. *Internet of Things*, *24*, 100982. https://doi.org/10.1016/j.iot.2023.100982

Taramsari, H. B., Hoffenson, S., & Nilchiani, R. (2025). Holistic sustainable design: Incorporating change propagation and triple bottom line sustainability. *Sustainability*, *17*(5), 2274. https://doi.org/10.3390/su17052274

Thejasree, P., Manikandan, N., Vimal, K. E. K., Sivakumar, K., & Krishnamachary, P. C. (2024). *Applications of Machine Learning in Supply Chain Management—A Review* (pp. 73–82). https://doi.org/10.1007/978-981-99-4819-2_6

Tundys, B., & Wiśniewski, T. (2021). Simulation-based analysis of greenhouse gas emissions in sustainable supply chains—Re-design in an approach to supply chain strategy. *Energies*, *14*(12), 3504. https://doi.org/10.3390/en14123504

Xu, K., & Cong, H. (2011). A Framework of Sustainable Supply Chain Management in Beijing Environmental Logistics. *2011 Fourth International Joint Conference on Computational Sciences and Optimization*, 1263–1266. https://doi.org/10.1109/CSO.2011.29

Yadav, A., Garg, R. K., & Sachdeva, A. (2024). Artificial intelligence applications for information management in sustainable supply chain management: A systematic review and future research agenda. *International Journal of Information Management Data Insights*, *4*(2), 100292. https://doi.org/10.1016/j.jjimei.2024.100292

Yadav, P. (2024). Navigating the future: AI's revolutionary drive in the auto industry. *International Journal of Advanced Research in Science, Communication and Technology*, 264–269. https://doi.org/10.48175/IJARSCT-16659

Zejjari, I., & Benhayoun, I. (2024). The use of artificial intelligence to advance sustainable supply chain: Retrospective and future avenues explored through bibliometric analysis. *Discover Sustainability*, *5*(1), 174. https://doi.org/10.1007/s43621-024-00364-6

Zhang, B., Tan, W. J., Cai, W., & Zhang, A. N. (2024). Leveraging multi-agent reinforcement learning for digital transformation in supply chain inventory optimization. *Sustainability*, *16*(22), 9996. https://doi.org/10.3390/su16229996

Index

Note: Locators in *italics* represent figures and **bold** indicate tables in the text.

A

African Development Bank (AfDB), 117
Amazon, 8
ANOVA regression index, **27**
Artificial intelligence (AI), 2, 3, 8, 103, 106–107, 108, 111–112
  AI-powered assessment and feedback systems, 103
  applications in maintenance, 169–170
  applications of, 109
  and biodiversity regeneration in Global South cities, *see* Global South cities, AI and biodiversity regeneration in
  competitiveness, 120
  contribution in teaching–learning of physical sciences, 97, 109
  defined, 94, 106
  generative, for innovative supply chain design and scenario planning, 248
  impacts and challenges on sustainable maintenance, **171**, 171–172
  in industrial sector, 116–117
  integration for holistic transparency, 247–248
  literature review on, **163**, 163–164, 167
  maintenance methods, 162
  modern approach of, 108
  and renewable energy, 147–159
  in service of education, 109
  for sustainability at 2030 World Cup, 19–20
  and sustainable supply chain management, *see* Sustainable supply chain management (SSCM)
  teaching–learning in the era of, 111
  technologies for teaching–learning, 102–103
  transformative capabilities of, 195–196, *196*
  usage in physical education, 102
  varieties of, 106
  and waste management, *see* Waste management, AI-powered
Artificial Intelligence: A Modern Approach (Russell and Norvig), 108
Augmented Reality (AR), 162
  applications in maintenance, 168, *168*
  in education, 111–112
  impacts and challenges on sustainable maintenance, **171**, 171–172
  literature review on, **163**, 163–164
  overview of, 166
Auto-Regressive Distributed Lag (ARDL), 150–151
  cointegration bound test and optimal lag, 154–154, **154–155**
  CUSUM of Squares test, 156–158, **157**
  heteroscedasticity test, 156–158, **157**
  long-run and short-run dynamics, 155–156, **156**
Average variance extracted (AVE), 128
Azure (Microsoft), 7

B

Barcelona smart collection, 204
Bartlett's test of sphericity, 25
Bayesian probabilistic system, 56
Bibliometric analysis, *see* Scientometric analysis
Biodiversity regeneration in Global South cities, AI and, *see* Global South cities, AI and biodiversity regeneration in
"Birds of the Same Feather Tweet Together" (Barberá), 75–76
Blockchain technology, 2–3, 5–6, 141
  integration for holistic transparency, 247–248
  in sustainable supply chain management, 240–241
Brazil 2014 World Cup, 19

C

Cameroonian secondary schools, 53–55
  guidance counselors and student follow-up in, 55

hybrid machine learning model for student performance prediction and guidance, 56–63
Carbon footprint management, 3–4, 6
Circular economy, 166, 201, *202*
Cleaning process, for dataset refinement, 36
Climate Innovation Readiness Navigator (CIRN), 222
Coefficient of determination (R^2), 65
Competitive Sustainable Manufacturing (CSM), 161
Concept of event leveraging, 19
Confusion matrix, 86–88, *87*
Convention on Biological Diversity (CBD), 221
Corporate environmental performance (CEP), 2
bibliometric trends in, 9–10, *9*, **10**
Corporate social responsibility (CSR), 5
Country branding, image effect and, 19
COVID-19 pandemic, 135
X-ray and CT images to detect, *see* X-ray/CT images, for Covid-19 detection
CUSUM of Squares test, 156–158, **157**

D

Data privacy, 244
Data tokenization, 78–79, *79*
Decision tree algorithms, 55–56; *see also* ID3 decision tree algorithm
Deep learning system, 211
for complex supply chain processes, 237–238
Digital disruption, 5
Digital economy, 4
"Digital Economy for Africa" (DE4A) initiative, 1
Digitalisation, 2, 3, 5, 176
Digital simulation, use in physical education, 112
Digital supply chain, 5–6
Digital technologies, 2–3
and group work, 177–179
Digital tools, 5
role in enhancing supply chain sustainability, 5–6
Digital transformation (DT)
bibliometric trends in, 9–10, *9*, **10**
companies achieving improved environmental performance through, **7**
corporate strategy and, 4–5
defined, 1–2
and environmental sustainability, 2–3
sustainable logistics and, 6

E

Echo chambers, on social media, 75–76
Ecological competency, 137
Economic growth (EG), 147–148
Economic spinoff model, 18
Education 4.0, 71
Education 5.0, 75
Electric waste collection vehicles, 201
Ellen MacArthur Foundation, 194
Energy consumption, 3
Environmental Performance Index (EPI) 2024, 2
Environmental sustainability
digital transformation and, 2–3
Environment Risk Analysis (ERA), 166
Epistemic governance, 73–75, *75*
Epistemic Governance in Higher Education (Campbell and Carayannis), 73
Equipment-as-a-service models, 203
Event leveraging, concept of, 19
E-waste, 3

F

Facebook, 72, 77
Feedback
AI-powered assessment and, 103
interpretation, sentiment analysis and, 77–88
social media and, 71–72
traditional, 70
Five-stage maturity framework, *203*, 203–204
Foreign direct investment (FDI), 148
F-statistic, 26–27

G

General government final consumption spending (GGCS), 148
GENIE 2009–2013 program, 94
Global significance test, 26–27
Global South cities, AI and biodiversity regeneration in, 221–230
comparison of models and contexts, 225–226, **226**
methodological approach, 222–224
performance and impacts of AI solutions, 224–225, **225**
spatial and technical perspective, 226–227, *227*
structural biases and identified gaps, 227–228
Global waste crisis, 194
Governance 4.0, 72–73, *73*
summative model for, 91, *91*
Governance, in higher education
epistemic, 73–75, *75*
Governance 4.0, 72–73, *73*
Governance in Higher Education (Hirsch and Weber), 74

Green decision-making, 135–136
  impact of disruptions on, 140
  importance of, 136
  literature review on, 136–137
  methodological approach, 138, *138*
  textual factor analysis with IRaMuTeQ, *139*, 139–140
Greenhouse gas (GHG) emissions, 2, 3
Green logistics, 6
Group work, 176–177
  digital technologies and, 177–179
Guidance counseling, 55

H

Haut-Commissariat au Plan (HCP), 221
Heteroscedasticity test, 156–158, **157**
Higher education, Morocco, *see* Moroccan higher education
Human capital (HC), 147–148
Hybrid machine learning model
  results of, 66, 67
  for student performance prediction and guidance, 56–63
  training and validation, 63–67
Hypothesis test by regression analysis, 25–29

I

ID3 decision tree algorithm, 60–61
  hyperparameters of, **61**
  results of, 66, **66**
IKEA, 8
Image effect and country branding, 19
Industry 4.0 (I4.0), 2, 3, 6, 165
  technologies in SSCM, 241, *243*, 243–244
Industry 5.0, 6, 165
Information and communication technology (ICT), 5, 97, 176
  contribution in teaching–learning of physical sciences, 98–99, 109
  in Morocco banking sector, 176–189
  usage in physical education, 102, 112
Innovate4Cities, 222
Innovation diffusion theory, 120–122
Internet of Things (IoT), 2, 3, 7–8, 141
  integration for holistic transparency, 247–248
  IoT-based real-time monitoring, 240, *241*
Interquartile range (IQR), 84
IRaMuTeQ, 136, 138
  textual factor analysis with, *139*, 139–140

J

Jarque–Bera test, 156

K

K-means, 82–84, *83*
KMO index, 25
K-Nearest Neighbors (KNN), 55–56, 58–59
  hyperparameters of, **61**
  results of, **64**, 64–65
Knowledge transmission, 107–108
Kreativ (IKEA), 8

L

Legacy framework/model, 17
Lemmatization, sentiments stemming and, 80, *80*
Life Cycle Analysis (LCA), 166
Life Cycle Costing (LCC), 166
Linear regression, 59–60
  hyperparameters of, **61**
Logistics, sustainable, 6

M

Machine learning (ML), 3, 8
  algorithms, 9
  applications in supply chain management, 237
  hybrid model, for student performance prediction and guidance, 56–63
  for system optimization, 197
Maintenance
  applications of new technologies in, 167
  artificial intelligence and, 169–170
  augmented reality and, 168, *168*
  defined, 162
  evolution, *164*, 164–165
  impacts and challenges of VR, AR, and AI on, **171**, 171–172
  literature review on, **163**, 163–164, 167
  overview of, *165*, 165–166
  predictive, 162, 169
  as strategic function, 162
  virtual reality and, 169, *169*
Managing Good Governance (Shattock), 71
Mean squared error (MSE), 59
Measurement scales, 25, **26**
Microplastic contamination, 195
Microsoft, 7
MindSphere (Siemens), 7–8
Moroccan higher education
  epistemic governance, 73–75, *75*

Governance 4.0 in, 72–73, *73*
literature review, 71
sentiment analysis and feedback interpretation, 77–88
social media and student feedback platforms, 71–72
Morocco, 97, 109, 221
banking sector, group work and digital technologies in, 176–189
education reform projects in, 94–95
Morocco 2030 World Cup, 19–21
artificial intelligence and emerging technologies for sustainability at, 20–21
empirical framework, 25–31
Moroccan specificities and perspectives of, 21–22
synthesis and formulation of assumptions, 22–25
Multiple linear regression model, 59–60
results of, 65–66
Multiple regression analysis, 28

N

Naive Bayes sentiment analysis model, 82
Natural language processing (NLP), 70, 71–72, 77
Near-infrared (NIR) spectroscopy, 198
Neural architectures for complex supply chain processes, 237–238
New Public Management (NPM), 70, 71
Nigeria
competitive environment in, 116
small business in, 119
Nigerian Stock Exchange, 123–124

O

Open platform communications unified architecture (OPC UA), 202
OUSocial2, 71–72

P

"Path to the Digital Decade" proposal, 1
Pedagogical orientations, physical sciences teaching, 95–96
Physical sciences, teaching of, 94
AI-driven platforms for evaluation and feedback, 112
aims of, 95–96
AI-powered assessment and feedback systems, 103
AI training desired by teachers, 111, *111*
contribution of AI and virtual labs in, 97
contribution of ICT in teaching–learning of, 98–99, 109
existing, 113
experimental activities in, 96
laboratory space, identification of, 99, *99*
place of experience, *100*, 100–101, **101**, *101*
semester system and, 96
targeted teachers, 110, *110*
teacher's level of AI mastery, 110, *110*
3D virtual laboratories and, 112–113
use of ICT and digital simulation in, 112
working conditions, identification of, 99
Practical Work (PW), 96
Predictive analytics, 9, 240
Predictive maintenance (PdM), 162, 169
Preferred Reporting Items for Systematic Reviews and Meta-Analyses (PRISMA) protocol, 138, *138*
Principal component analysis (PCA), 24, 82–84, *83*
Product-Service Systems (PSS) approach, 162

Q

Qatar 2022 World Cup, 19

R

Real laboratory (RL) *vs.* virtual laboratories, 107
Real-time supply chain visibility, 240
Regression analysis
hypothesis test by, 25–29
multiple, 28
results, interpretation of, 29–30, **29**
Renewable energy (RE)
artificial intelligence and, 147–159
Renewable Energy Consumption (REC), 147
RT-PCR (reverse transcription-polymerase chain reaction), 35

S

Scatter plots, 65, *65*
Scientometric analysis
methods of, 36–37
results of, 37–46
Sentiment analysis, and feedback interpretation, 77–88
classification report for, **82**
confusion matrix, 86–88, *87*
data collection, 77–78
results and data analysis, 80–88
sentiment distribution, 84, *85–86*
sentiments stemming and lemmatization, 80, *80*
tokenization, data, 78–79, *79*

Shortwave infrared (SWIR) systems, 198
Showcase effect theory, 16–17
Siemens, 7–8
Singapore, ICT integration into education, 97
Small-and medium-sized firms (SMEs), 117
  empirical literature, 122–123
Small enterprises, in Sub-Saharan Africa, 117–119
  empirical literature, 122–123
  innovation diffusion theory, 120–122
  in Nigeria, 119
  technological innovation, 119–120
Social media, 77; *see also* Facebook
  and student feedback platforms, 71–72
  student homophily on, 75–76
"Social Media" (Azelmad and Harroudi), 72
South Africa 2010 World Cup, 18
Sports events, and tourist attractiveness, 16
  Brazil 2014 World Cup, 19
  concept of event leveraging, 19
  economic spinoff model, 18
  image effect and country branding, 19
  legacy framework, 18
  Qatar 2022 World Cup, 19
  showcase effect theory, 16–17
  South Africa 2010 World Cup, 18
SPSS 25 software, 26–27
SSD (Single Shot Detector), 197
Stationarity test, 152, **152**
Stemming, data, 80, *80*
Structures–Equations Model Partial Least Squares-Structural Equation Modeling (PLS-SEM), 141
Student follow-up in Cameroonian schools, 55
Student homophily on social media, 75–76
Sub-Saharan Africa (SSA), small enterprises in, *see* Small enterprises, in Sub-Saharan Africa
Supply chain management (SCM), 233
  machine learning applications in, 237
Supply chain sustainability, digital tools and, 5–6
Sustainability
  assessment, 161–162
  defined, 161
  literature review on, **163**, 163–164
  overview of, *165*, 165–166
Sustainable decision-making, 136–137; *see also* Green decision-making
Sustainable Development Goals (UN SDG), 1, 116–117, 194
Sustainable logistics, 6
Sustainable sourcing, 5–6
Sustainable supply chain management (SSCM), 233–234
  artificial intelligence-driven optimization strategies for, 238–239
  blockchain technology in, 240–241
  challenges/limitations/ethical considerations, 244–247
  conceptual models and frameworks in, 236
  defined, 234–235
  human–AI collaboration models for enhanced decision augmentation in, 248–249
  implementation, key drivers/barriers/challenges in, 236
  Industry 4.0 (I4.0) technologies in, 241, *243*, 243–244
  IoT-based real-time monitoring, 240, *241*
  system integration barriers and cost implications for AI deployment, 244–245
  technological frameworks and integration architectures for AI-enabled SSCM, 239–244
  theoretical foundations of, 234–236
  triple bottom line framework, *235*, 235–236

T

Take-make-dispose economy, 195
Technological innovation, 116, 119–120
Tesco, 9
Theoretical model test, 26–28
3D virtual laboratories, 112–113
Tokenization, data, 78–79, *79*
TOMRA optical sorting, 204
Tourism sector, sports and, *see* Sports events, and tourist attractiveness
Traditional feedback mechanisms, 70
Trend Topics, 9, *9*, **10**
Triple bottom line (TBL) framework, *235*, 235–236

U

Unified Theory of Acceptance and Use of Technology (UTAUT) model, 182–183, **183**, 186–189
Unilever, 8
United Nations Educational, Scientific and Cultural Organization (UNESCO), 107
Unit root (UR) test, 152, **152**

V

Violin sentiment analysis, 84, *86*
Virtual laboratories (VL)
  contribution in teaching–learning of physical sciences, 97, 109

defined, 107
real laboratory *vs.*, 107
3D, importance of, 112–113
usage in physical education, 102
Virtual Reality (VR), 162
applications in maintenance, 169, *169*
in education, 111–112
impacts and challenges on sustainable maintenance, **171**, 171–172
literature review on, **163**, 163–164
overview of, 166–167
Visualization analysis, 37
VOSviewer software, 37

W

Waste collection logistics, 212, *212*
Waste management, AI-powered, 194–217
AI applications in waste classification and sorting, 211, *211*
AI-driven waste sorting technologies and performance metrics, **213**
applications of AI in waste collection logistics, 212, *212*
computer vision and robotic sorting, 197–199, **198**, *200*
economic and environmental impacts, **213–214**
economic implications, 216
economic viability, 201, **201**
environmental and sustainability outcomes, 217
environmental performance, 200–201
ethical considerations, 215
implementation challenges and enabling factors, 216–217
implementation framework and barriers, 202–204, *203*
machine learning for system optimization, 197
policy and governance frameworks for AI implementation, 212–213
sustainability impact analysis, 200–201
technical performance of AI applications, 216
waste collection logistics, 212, *212*
Word cloud analysis, 88
World Cup 2010 (South Africa), *see* South Africa 2010 World Cup
World Cup 2014 (Brazil), *see* Brazil 2014 World Cup
World Cup 2022 (Qatar), *see* Qatar 2022 World Cup
World Cup 2030 (Morocco), *see* Morocco 2030 World Cup
World Economic Forum (WEF), 117
World Health Organization (WHO), 34

X

X-ray/CT images, for Covid-19 detection, 35
academic institutions contributed to research, 41–42, **42**
authors ranked by research contributions, 44, **44**
cited documents ranked by research contributions, 44–46, **45**
journals quartiles in research publications, 43–44, *43*
leading countries of research contribution, 38–40, *38–40*
publishers ranked by research contributions in, 42–43, **43**
scientometric analysis
methods of, 36–37
results of, 37–46
technique employed for, 41, *41*
trend analysis of publications/citations, 37–38, *37–38*
X-ray transmission, 198

Y

YOLO (You Only Look Once), 197